iacin g/d)[e]	Vitamin B_6 (mg/d)	Folate (μg/d)[f]	Vitamin B_{12} (μg/d)	Pantothenic Acid (mg/d)	Biotin (μg/d)	Choline (mg/d)[g]
2*	0.1*	65*	0.4*	1.7*	5*	125*
4*	0.3*	80*	0.5*	1.8*	6*	150*
6	**0.5**	**150**	**0.9**	2*	8*	200*
8	**0.6**	**200**	**1.2**	3*	12*	250*
12	**1.0**	**300**	**1.8**	4*	20*	375*
16	**1.3**	**400**	**2.4**	5*	25*	550*
16	**1.3**	**400**	**2.4**	5*	30*	550*
16	**1.3**	**400**	**2.4**	5*	30*	550*
16	**1.7**	**400**	**2.4**[h]	5*	30*	550*
16	**1.7**	**400**	**2.4**[h]	5*	30*	550*
12	**1.0**	**300**	**1.8**	4*	20*	375*
14	**1.2**	**400**[i]	**2.4**	5*	25*	400*
14	**1.3**	**400**[i]	**2.4**	5*	30*	425*
14	**1.3**	**400**[i]	**2.4**	5*	30*	425*
14	**1.5**	**400**	**2.4**[h]	5*	30*	425*
14	**1.5**	**400**	**2.4**[h]	5*	30*	425*
18	**1.9**	**600**[j]	**2.6**	6*	30*	450*
18	**1.9**	**600**[j]	**2.6**	6*	30*	450*
18	**1.9**	**600**[j]	**2.6**	6*	30*	450*
17	**2.0**	**500**	**2.8**	7*	35*	550*
17	**2.0**	**500**	**2.8**	7*	35*	550*
17	**2.0**	**500**	**2.8**	7*	35*	550*

[g]Although AIs have been set for choline, there are few data to assess whether a dietary supply of choline is needed at all stages of the life cycle, and it may be that the choline requirement can be met by endogenous synthesis at some of these stages.

[h]Because 10 to 30 percent of older people may malabsorb food-bound B_{12}, it is advisable for those older than 50 years to meet their RDA mainly by consuming foods fortified with B_{12} or a supplement containing B_{12}.

[i]In view of evidence linking folate intake with neural tube defects in the fetus, it is recommended that all women capable of becoming pregnant consume 400 μg from supplements or fortified foods in addition to intake of food folate from a varied diet.

[j]It is assumed that women will continue consuming 400 μg from supplements or fortified food until their pregnancy is confirmed and they enter prenatal care, which ordinarily occurs after the end of the periconceptional period—the critical time for formation of the neural tube.

SOURCES: *Dietary Reference Intakes for Calcium, Phosphorous, Magnesium, Vitamin D, and Fluoride* (1997); *Dietary Reference Intakes for Thiamin, Riboflavin, Niacin, Vitamin B_6, Folate, Vitamin B_{12}, Pantothenic Acid, Biotin, and Choline* (1998); *Dietary Reference Intakes for Vitamin C, Vitamin E, Selenium, and Carotenoids* (2000); *Dietary Reference Intakes for Vitamin A, Vitamin K, Arsenic, Boron, Chromium, Copper, Iodine, Iron, Manganese, Molybdenum, Nickel, Silicon, Vanadium, and Zinc* (2001); *Dietary Reference Intakes for Water, Potassium, Sodium, Chloride, and Sulfate* (2005); and *Dietary Reference Intakes for Calcium and Vitamin D* (2011). These reports may be accessed via www.nap.edu.

Nutrition

FOR A CHANGING WORLD

Nutrition
FOR A CHANGING WORLD

Jamie Pope
Vanderbilt University

Steven Nizielski
Grand Valley State University

Alison McCook

Publisher: Peter Marshall
Publisher, Science: Kate Parker
Senior Development Editor: Elizabeth Marsh
Market Development Manager: Shannon Howard
Executive Marketing Manager: Will Moore
Executive Media Editor: Amanda Nietzel
Marketing Assistant: Cate McCaffery
Associate Editor: Lori Stover
Project Editor: Julio Espin
Director of Design, Content Management: Diana Blume
Cover and Text Designer: Tom Carling, Carling Design, Inc.
Art Manager: Matthew McAdams
Illustrator: Eli Ensor
Photo Editor: Christine Buese
Photo Researchers: Christine Buese, Stephanie Heiman
Production Manager: Paul Rohloff
Composition and Layout: CodeMantra
Printing and Binding: RR Donnelley
Cover Photo: Jason Andrew/Getty Images
Credit is given to the following sources to use the chapter-opener images:
Chapter 1: Natasha Vasilijevic/Gallery Stock; Chapter 2: Wendell and Carolyn/Getty Images; Chapter 3: Science Photo Library/age fotostock; Chapter 4: plainpicture/Rupert Warren; Chapter 5: Fotosearch/age fotostock; Chapter 6: Robert Harding Picture Library Ltd/Alamy; Chapter 7: Kevin Curtis/Science Source; Chapter 8: Monkey Business Images/Shutterstock; Chapter 9: Michael Turek/Gallery Stock; Chapter 10: Kate Williams/Getty Images; Chapter 11: Marco Veringa/Getty Images; Chapter 12: Jasper White/Getty Images; Chapter 13: Anna Omelchenko/Shutterstock; Chapter 14: Time Gainey/Alamy; Chapter 15: Alija/Getty Images; Chapter 16: Chris Keane/Reuters/Corbis; Chapter 17: Gallery Stock; Chapter 18: D. Hurst/Alamy; Chapter 19: Ariel Skelley/Getty Images; Chapter 20: Gabriel B. Tait/KRT/Newscom

Library of Congress Control Number: 2016930064

ISBN-13: 978-1-4641-5288-7
ISBN-10: 1-4641-5288-8

Printed in the United States of America
First Printing

W. H. Freeman
One New York Plaza
New York, NY 10004-1562
www.whfreeman.com

ABOUT THE AUTHORS

Photo Courtesy of Vanderbilt University School of Nursing

Jamie Pope, MS, RD, LDN, is Assistant Professor in Nutritional Sciences in the Vanderbilt University School of Nursing. She has been with Vanderbilt University since 1986 working in the areas of obesity research, weight management, health promotion, heart disease prevention and—since 2000—teaching introductory nutrition courses. Jamie's popular classes bring together undergraduate students from a wide range of majors to learn about the science of nutrition and its application to their personal and professional lives. She was the recipient of the Faculty Innovation in Teaching Award in 2013. Beyond the classroom, Jamie adapted portions of her Vanderbilt courses to produce and offer a Massively Open Online Course (MOOC) on the Coursera learning platform, which attracted more than 175,000 students from all over the world. Outcomes and application of her MOOC to dietetic practice and nutrition education were published in the *Journal of Nutrition and Dietetics* and presented at the Academy of Nutrition and Dietetics Food and Nutrition Exhibition and Conference. She is a long-time member of the Academy of Nutrition and Dietetics and served as media representative for the Tennessee Dietetic Association, during which time she was named as Outstanding Dietitian of the Year for the Nashville District Dietetic Association. She serves on the executive committee for Nutrition Educators of Health Professionals dietetic practice group. Jamie is also a member of the Scientific Advisory Council for the wellness solutions company, Onlife Health. Earlier in her career, Jamie coauthored several books for consumers, such as *The T-Factor Fat Gram Counter,* which spent more than three years on the *New York Times* best-seller list. She has authored or contributed to numerous other scientific and lay press publications. Jamie also served for more than a decade as a corporate nutrition consultant and media spokesperson for food companies such as Chick-fil-A and the makers of Smart Balance products.

Photo Courtesy Grand Valley State University

Steven Nizielski, MS, PhD, earned his Bachelor of Science degree from the University of Minnesota in Wildlife Biology and assisted in research projects involving Siberian tigers and gray wolves before entering graduate school. He earned his Master's and Doctorate degrees at the University of Minnesota in nutrition, with an emphasis in biochemistry. He is currently an Associate Professor in the Department of Biomedical Sciences at Grand Valley State University in Allendale, Michigan, where he teaches introductory nutrition, clinical nutrition, public health, advanced metabolism, life cycle nutrition and sports nutrition courses. His current research seeks to identify cellular adaptations in adipose tissue in response to aging and endurance training. Steve is a fellow of The Obesity Society, and a member of the American Society for Nutrition, and a member of the American Physiological Society. He is an avid competitive cyclist and also enjoys cross-country skiing, hiking, and camping.

Alison McCook has been a science writer and editor for more than 15 years, crafting materials for both general and professional audiences. Her work spans topics from health to molecular mechanisms, and has appeared in well-known publications such as *Reuters, Nature, Discover, Scientific American, Popular Science,* and *The Lancet*. Alison has held staff positions at *Reuters* and Nature and, most recently, she was the Deputy Editor of *The Scientist*, overseeing the entire editorial team. During her tenure, the magazine won two consecutive Magazine of the Year awards, across all categories, from the American Society of Business Publication Editors. Alison resides in Philadelphia, Pennsylvania.

BRIEF CONTENTS

CONTENTS

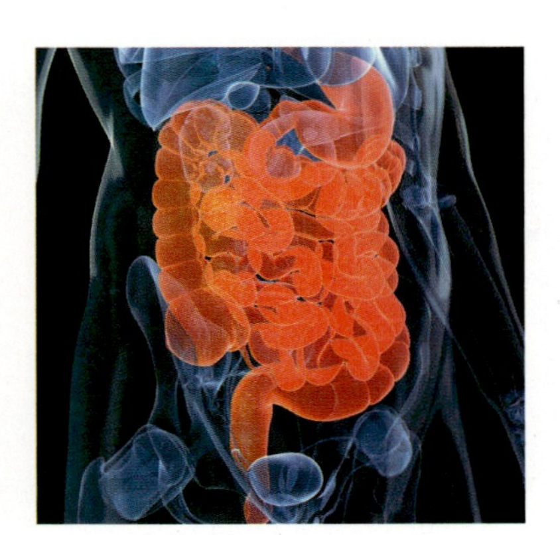

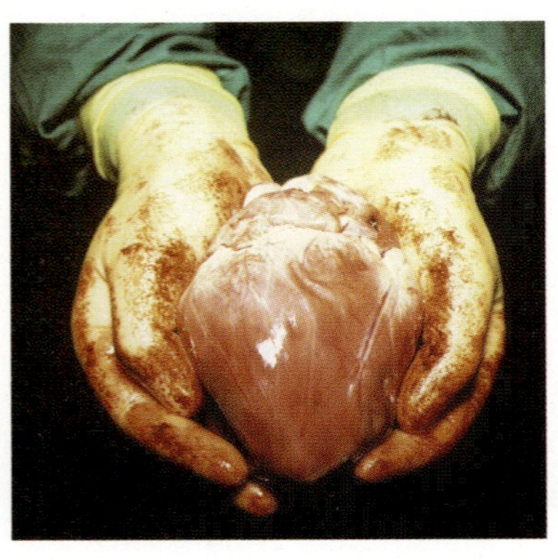

CHAPTER 16

Nutrition and Fitness 365

Eating to Win

Research suggests that when athletes eat may be just as important as what they eat.

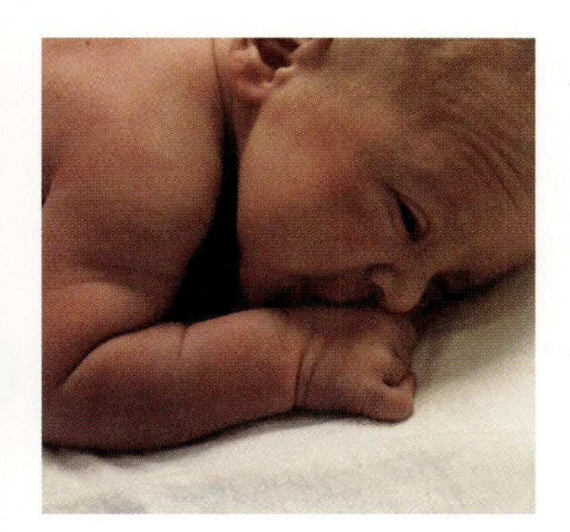

CHAPTER 17

Nutrition for Pregnancy, Breastfeeding, and Infancy 391

Nourishing Mother and Baby

New research suggests that delaying cord clamping after birth preserves iron status.

CHAPTER 18

Childhood Nutrition 415

Food Allergies and Intolerances

As paradoxical as it might seem, could dirt and germs make children healthier for life?

E. COLI
PLEASE
GO AWAY

AUTHORS' LETTER

Dear Reader:

Among the life sciences, nutrition is arguably one of the most dynamic, debated, and discussed. No doubt you've read a headline touting diet as a cure or cause of disease, heard about a new weight loss craze, or have seen a television personality promote a supplement or food plan. So, how do you sort fact from fiction and determine what information or advice is credible and grounded in sound science? And, how do you use the vast amount of information when making personal food choices? We are honored and excited to offer a new and unique textbook that can help students understand the fundamentals of nutrition science, and apply those concepts to everyday life.

Over the past decades, we've taught introductory nutrition to undergraduate students not majoring in nutrition who may or may not have had strong science backgrounds. Our students come from a diverse range of majors with many reasons for taking the class—personal interest, prerequisite for nursing, or other allied health programs, or as a science requirement. Introductory nutrition courses often include athletes, exercise science students, and students for whom the course has been recommended by an academic advisor.

Our hope is that all students will find that this book helps develop and augment their appreciation and understanding of nutrition as a science, increase their understanding of nutrition-related terms and concepts, and advance their comprehension and context of body processes and pathways.

BOOK COVERAGE

We have thought through all of the features of this textbook to make it as useful as possible for our students. To this end, the table of contents includes all of the core topics found in introductory nutrition courses with some innovations. The text opens with an overview of nutrition as a science and discipline, followed by the considerations and recommendations for a healthful diet with subsequent chapters on digestion and absorption, the macronutrients, and the micronutrients.

For carbohydrates and lipids, we offer content in two chapters, one focusing on the classifications, functions, and food sources and another focusing on health and disease implications. For example, blood sugar regulation and diabetes are explored in detail in Chapter 5 after carbohydrates are covered in Chapter 4. We believe this organization will give students a clear understanding of these important macronutrients and showcase the impact they have on health.

The vitamins and minerals can be a challenge for students. To make these critical topics clear and memorable, we offer concise, focused chapters (fat-soluble vitamins, water-soluble vitamins, major minerals and trace minerals) that are enhanced by unique storylines. The text offers exciting—yes, exciting—scientifically applicable chapters on energy balance and obesity, followed by nutrition and fitness. Nutrition during life stages is addressed from pregnancy and beyond. We are particularly excited about offering a chapter devoted to nutrition for the college student. This chapter's emphasis on determinates of eating behavior, and the nutritional needs of young adults, is something we would like to share with not only our students, but also our own children. (We are both parents of young adults during the development and writing of this textbook!)

We have also included unique chapters that, from our experience, students have a lot of questions about and want to know more. These attention-getting chapters build upon core concepts and address relevant topics like plant-based diets, dietary supplements, and functional foods.

BOOK FEATURES AND MEDIA

Knowing that illustrations can often be the best way to grasp a concept, we have put much thought into the artwork. Each chapter contains vibrant, engaging, informative, well-placed, and easy-to-understand Infographics that support and augment concepts. Infographics that address multistep processes—digestion and absorption, and the function of hormones and nutrients—have been designed to facilitate student visualization and learning by accurately breaking these processes down into their component steps. We have also added a question, whenever appropriate, at the end of some Infographics to help the student engage with the material and identify key ideas being presented.

And at the end of each chapter, we've included a variety of classroom-tested assessments and personal application activities that help "bring it home" on a personal level, allowing students to assess their understanding of concepts and terms. The media resources complement the innovations in the book. In the LaunchPad online course space, we've provided robust animations for every chapter, a variety of quizzing options, including adaptive quizzes, and we are very excited to include Analyze My Diet personal diet analysis activities that are **automatically graded** and report to your gradebook. Analyze My Diet exercises are tailored to work with the USDA's SuperTracker program, but can be adapted to use with other food databases and tracking software.

For those instructors who prefer to teach the micronutrients from a functional perspective, Appendix 3 illustrates the roles of vitamins and minerals in energy metabolism, as antioxidants, and in bone and blood health.

ABOUT THE AUTHOR TEAM

Planning, researching, writing, reviewing, and producing this textbook provided the opportunity for two nutrition professionals to collaborate and bring together their unique perspectives and backgrounds. A registered dietitian, with over 15 years of classroom teaching experience at a private university preceded by work in health promotion, research, clinical practice, publishing, and media work collaborating with a nutrition scientist and researcher, who in addition to teaching introductory nutrition and public health courses at large public universities for 20 years, teaches advanced macronutrient metabolism and sports nutrition courses. This teaming up is unique for introductory nutrition textbooks. Science writers and editors have also helped merge the voices of these two experts into an engaging, readable, informative, and effective textbook that meets the goals for instructors of non-major nutrition courses. The chapter stories provide context and content for the science, as well as help provide rationale for "why it matters."

In many ways, this textbook has allowed us to further our own understanding and appreciation of nutrition—evaluating and researching each concept to ensure accuracy, and then translating and presenting that information in a way that students can embrace, understand, apply, and retain has enhanced our own teaching and passion for sharing the ever-evolving and exciting field of nutrition!

Happy and healthy reading!

Jamie Pope and Steve Nizielski

A fresh approach to the introductory nutrition course

Using real stories—about real people and real science—*Nutrition for a Changing World* gives students a relevant and engaging framework for exploring the basic concepts of nutrition.

14 TRACE MINERALS

An abnormal enlargement of the thyroid gland is called a goiter. *The thyroid is a butterfly-shaped gland located at the base of the neck. It secretes several hormones that affect metabolism, growth, and development.*

Biophoto Associates/Science Source

GOITER *enlargement of the thyroid gland, most often caused by lack of iodine in the diet*

TRACE MINERALS *minerals with a daily requirement of less than 100 milligrams (sometimes called microminerals)*

CO-FACTOR *nonprotein inorganic substances that enable enzymes to carry out chemical reactions*

iodine is found in oceans and iodine content in the soil varies by location, which affects the iodine content of crops. In some regions of the world, particularly mountainous areas, or flooded river valleys, iodine-deficient soil is common, increasing the risk of iodine deficiency among people who consume foods primarily from those areas.

Since the days of World War I, many countries have implemented programs to add iodine to table salt and have dramatically reduced the prevalence of iodine deficiency worldwide. However, iodine deficiency is still "one of the most important public health issues globally," according to a 2012 study published by researchers at the Boston Medical Center. In the United States, where much of our table salt is iodized, most people appear to consume sufficient iodine to offset deficiency, although certain groups, such as pregnant women, may be at risk of suboptimal intake and iodine status.

■ ■ ■

INTRODUCING THE TRACE MINERALS

Iodine is categorized as one of the **trace minerals**, because the body requires significantly less of them to meet daily needs than the major minerals. (Trace minerals are also found in smaller amounts in the body.) Minerals with a daily requirement of 100 milligrams or more are considered major (or "macro") minerals, while those with a daily requirement less than 100 milligrams are considered trace (or "micro") minerals. In addition to iodine, the other essential trace minerals are iron, zinc, copper, selenium, molybdenum, manganese, and chromium. Fluoride is another important trace mineral that is required for optimal health, but because it is not required to sustain life it is not considered an essential nutrient. (INFOGRAPHIC 14.1)

Although the body doesn't need large amounts of the trace minerals, they are crucial for health. Trace minerals often act as **co-factors** by binding to enzymes, enabling them to carry out their chemical reactions. Trace minerals are also incorporated into other types of proteins in addition to enzymes; iron, for example, is a component of the oxygen-carrying protein in red blood cells called *hemoglobin*. Trace minerals also serve other roles. Fluoride, for example, has an important structural function in that it hardens the enamel on our teeth, reducing the occurrence of dental caries.

Trace minerals are found in plant-based and animal-based foods, but their bioavailability—the extent to which a mineral can be used by the body—can be influenced by many

312

INFOGRAPHIC 14.1 **Trace Minerals in the Body and Their Functions**

factors, including the form of the food, our nutrition status, our age, and pregnancy status. The safe range of intake for trace minerals is also narrow compared with the major minerals, since they are required in such small amounts. Trace mineral deficiencies can cause vague or varied symptoms that are difficult to diagnose; but, sometimes, inadequacies have severe and even deadly consequences. Pregnant women and children in rapid growth stages are especially affected by inadequate intakes of trace minerals. As with the vitamins and major minerals, it is rare for individuals to overconsume trace minerals through food alone, but intake above the Tolerable Upper Intake Level (UL) is possible through supplementation, resulting in potentially dangerous adverse effects. For example, excess selenium can result in hair loss and brittle nails and in extreme cases, respiratory distress along with kidney and heart failure. In general, by eating a balanced and varied diet, we can get all the trace

313

REAL PEOPLE

REAL STORIES

REAL SCIENCE

Each chapter in *Nutrition for a Changing World* works like a *Scientific American*-style article, with compelling stories providing context for the core science. Beautifully designed Infographics combine memorable images with clear, step-by-step captions to illustrate the chapter's most important ideas. Activities and adaptive quizzes in Macmillan's innovative online course space, LaunchPad macmillan learning, help students apply the science to their personal diets, debunk misconceptions, and prepare for exams.

Real science taught through stories

Current, from-the-headlines stories engage students as they explore the essential topics covered in a nutrition course for non-majors.

Dr. James Levine and his treadmill station. *Levine, a professor of medicine at the Mayo Clinic and Arizona State University, is an expert on the physiology of weight gain and loss.*

AP Photo/Jim Mone

Modern chapter organization, relevant stories, and current science coverage in short chapters

CHAPTER TITLE	STORY TITLE & DESCRIPTION	CHAPTER CONTENT
1 The Science and Scope of Nutrition	**Exploring the Science of Nutrition** *A new field of nutrition research is revealing surprising details about how food affects our genes.*	▪ Introducing the Science of Nutrition ▪ Food Provides Nutrients and Energy ▪ What Is Malnutrition? ▪ Nutrient Intake Recommendations ▪ Energy Recommendations ▪ Nutrition Science in Action ▪ Credible Sources of Nutrition Information ▪ Health Goals for Americans ▪ Assessing the North American Diet
2 Healthy Diets	**From Desert to Oasis** *Are "food deserts" preventing millions of Americans from eating well?*	▪ Healthy Diets Feature Variety, Balance, Adequacy, and Moderation ▪ Understanding the Nutrient Density and Energy Density of Foods ▪ Limit These: Solid Fats and Added Sugars (SoFAS) ▪ 2015 Dietary Guidelines for Americans ▪ Global Nutrition ▪ Understanding the Labeling on Food ▪ Future Food Labels
3 Digestion	**A Gut Feeling** *Is gluten really all that bad?*	▪ Overview of the Digestive Process ▪ Overview of Mechanical and Chemical Digestion ▪ The Path of Digestion from Mouth to Large Intestine ▪ The Role of Bacteria in the Gastrointestinal Tract ▪ Digestive Disorders ▪ Irritable Bowel Syndrome (IBS) and Inflammatory Bowel Disease (IBD) ▪ Awareness of Celiac Disease Is Growing
4 Carbohydrates	**Whole Grain Hype** *Can science help us navigate the perils of the cereal aisle?*	▪ What Are Carbohydrates? ▪ How Grain-Based Foods Measure Up ▪ The Digestion of Carbohydrates ▪ Added Sugars ▪ Dental Caries ▪ Sugar Alternatives ▪ Understanding Fiber ▪ Carbohydrate Intake Recommendations
5 Nutrition and Diabetes	**Medical Miracles** *Nearly 100 years ago, a discovery revolutionized medicine—and saved millions of children as a result.*	▪ What Is Diabetes Mellitus? ▪ Type 2 Diabetes ▪ Diabetes on the Rise ▪ Diabetes Treatment and Prevention
6 The Lipids	**The Skinny on Fat** *How native Greenlanders with a passion for seal blubber showed the world that not all lipids are created equal.*	▪ Overview of the Lipids ▪ Lipid Digestion and Absorption ▪ Lipoprotein Transport ▪ Essential Fatty Acids ▪ Trans Fats and Fat Substitutes ▪ Current Fat Intake and Recommendations ▪ Knowledge of Omega-3 Fatty Acids Is Growing
7 Lipids in Health and Disease	**Death in Bogalusa** *From tragic deaths in a southern town, insights into heart disease.*	▪ Atherosclerosis and Cardiovascular Disease ▪ Risk Factors for Cardiovascular Disease ▪ A Heart-Protective Diet ▪ Fat Intake and Health—Beyond Cardiovascular Disease
8 Protein	**How Much of a Good Thing Do We Need?** *Experts consider how much protein is best for us all.*	▪ Recommendations for Protein Intake ▪ Structure of Protein ▪ Protein Synthesis ▪ Digestion and Absorption of Proteins ▪ Varied Functions of Protein ▪ Protein Turnover and Nitrogen Balance ▪ Seniors May Benefit from Increased Intakes of Protein ▪ Protein Quality ▪ Protein-Deficiency Diseases ▪ High-Protein Diets
9 Plant-Based Diets	**Pass the Plants, Please** *What does a Spanish study say about the benefits of following a Mediterranean-like diet?*	▪ Vegetarian and Semi-Vegetarian Diets ▪ Benefits of a Diet Rich in Plant Foods ▪ How Cancer Develops ▪ Benefits of Phytochemicals ▪ Nutritional Considerations and Concerns Associated with Plant-Based Diets ▪ Plant-Based and Vegetarian Guidelines ▪ How Are Organic Foods Different from Conventional Foods? ▪ The Mediterranean Diet
10 Fat-Soluble Vitamins	**Soak up the Sun?** *An ailing gorilla inspires a lifelong interest in the vitamins.*	▪ Introducing the Vitamins ▪ Bioavailability and Solubility ▪ Properties of Fat-Soluble Vitamins ▪ Vitamin D ▪ Vitamin A ▪ Antioxidants Protect Against Damage from Free Radicals ▪ Vitamin E ▪ Vitamin K

Timely coverage of diabetes and blood glucose regulation

Chapter 6 and 7 work together to explore the science of lipids, and lipids in health and disease.

UNIQUE CHAPTER!

UNIQUE CHAPTER! Includes coverage of cancer and phytochemicals

Fat-soluble and water-soluble vitamins are separate chapters, making the content more manageable for students.

CHAPTER TITLE	STORY TITLE & DESCRIPTION	CHAPTER CONTENT
11 Water-Soluble Vitamins	**It's Not a Germ** *Pioneering research uncovers vitamin deficiency diseases.*	▪ Characteristics of the Water-Soluble Vitamins ▪ The B Vitamins ▪ Choline ▪ Vitamin C
12 Dietary Supplements	**Supplements, Herbs, and Functional Foods** *Surprising studies on the value of vitamin supplements.*	▪ What Are Dietary Supplements? ▪ Regulation of Dietary Supplements ▪ Understanding Supplement Labels ▪ Are Dietary Supplements Harmful? ▪ Pros and Cons of Herbal Supplements ▪ Functional Foods ▪ Let the Buyer Beware
13 Major Minerals and Water	**Potassium Power** *Eating a diet low in sodium and rich in potassium may offer protection from hypertension.*	▪ Overview of the Minerals ▪ Mineral Absorption and Bioavailability ▪ Minerals in Our Food ▪ Calcium, Magnesium, and Phosphorus Have Diverse Structural Roles in the Body ▪ Sulfur Is a Component of Proteins ▪ Sodium, Potassium, and Chloride Maintain Fluid Balance in the Body ▪ Water ▪ Water Intake Recommendations
14 Trace Minerals	**Small Quantities with Big Importance** *How lack of trace minerals can wreak havoc on your health.*	▪ Introducing the Trace Minerals ▪ Iodine: Sources and Functions ▪ Iron ▪ Zinc ▪ Copper ▪ Selenium ▪ Fluoride ▪ Other Trace Minerals: Manganese, Molybdenum, and Chromium ▪ Ultratrace Minerals
15 Energy Balance and Obesity	**The Sitting Disease** *Understanding the causes and consequences of obesity.*	▪ Energy In, Energy Out ▪ The Biology of Hunger ▪ A NEAT Cause of Weight Gain ▪ Lifestyle and Energy Balance ▪ Weight Loss Recommendations ▪ A Moratorium on the Chair
16 Nutrition and Fitness	**Eating to Win** *Research suggests that when athletes eat may be just as important as what they eat.*	Components of Fitness ▪ Fueling the Body ▪ Measures of Exercise Intensity ▪ Dietary Carbohydrates for Endurance Exercise ▪ U.S. Gymnastics Team in Training ▪ Female Athlete Triad ▪ Body Building ▪ Hydrating the Athlete ▪ Sports Supplements ▪ Physical Fitness: Not Just for Athletes
17 Nutrition for Pregnancy, Breastfeeding, and Infancy	**Nourishing Mother and Baby** *New research suggests that delaying cord clamping after birth preserves iron status.*	Changing Nutritional Needs ▪ Energy and Nutrient Needs During Pregnancy ▪ A Healthy Pregnancy ▪ Food Safety ▪ Postnatal Nutrition ▪ Nutrition During Lactation ▪ Formula Feeding ▪ Nutrition for the Growing Child ▪ Iron Deficiency and Cord Clamping
18 Childhood Nutrition	**Food Allergies and Intolerances** *As paradoxical as it might seem, could dirt and germs make children healthier for life?*	▪ Growth and Development in Childhood ▪ What Are Children Eating? ▪ Nutritional Recommendations for Children ▪ Nutrients of Concern in Childhood ▪ Childhood Obesity ▪ Food Allergies
19 The College Years	**Determinants of Eating Behavior, Disordered Eating, and Alcohol** *How food choices are influenced by social norms.*	▪ Growth, Development, and Nutrient Requirements in Late Adolescence ▪ Calcium ▪ Iron ▪ Determinants of Eating Behavior ▪ Eating Challenges on Campus ▪ Eating Disorders ▪ Alcohol ▪ A Successful Transition to College Life
20 Food Safety and Food Security	**Stomach Troubles** *How a series of food crises transformed the U.S. food protection system.*	▪ A Brief History of Food Safety in America ▪ Foodborne Intoxication and Infection ▪ Engineering Food Safety: HACCP and Irradiation ▪ Food Safety and Modernization Act ▪ Food Safety in the Home ▪ Beyond Food Safety: Food Security and Issues Related to Hunger Around the World
21 Nutrition and the Aging Adult	**Live Long and Prosper** *Can you eat your way to a longer, healthier life?*	▪ What Happens When We Age? ▪ Life Expectancy and Lifespan ▪ Blue Zones ▪ Energy Needs and Physical Activity ▪ Special Nutritional Concerns for Older Adults ▪ Nutrient Recommendations for Seniors ▪ Studying the Adventist Lifestyle for Clues to Longevity ▪ Maintaining Physical Strength for a Lifetime ▪ Blue Zone Secrets

UNIQUE CHAPTER!

UNIQUE CHAPTER of great interest to students—debunks misconceptions!

UNIQUE CHAPTER covers the changes and pressures during young adulthood, eating disorders, and alcohol.

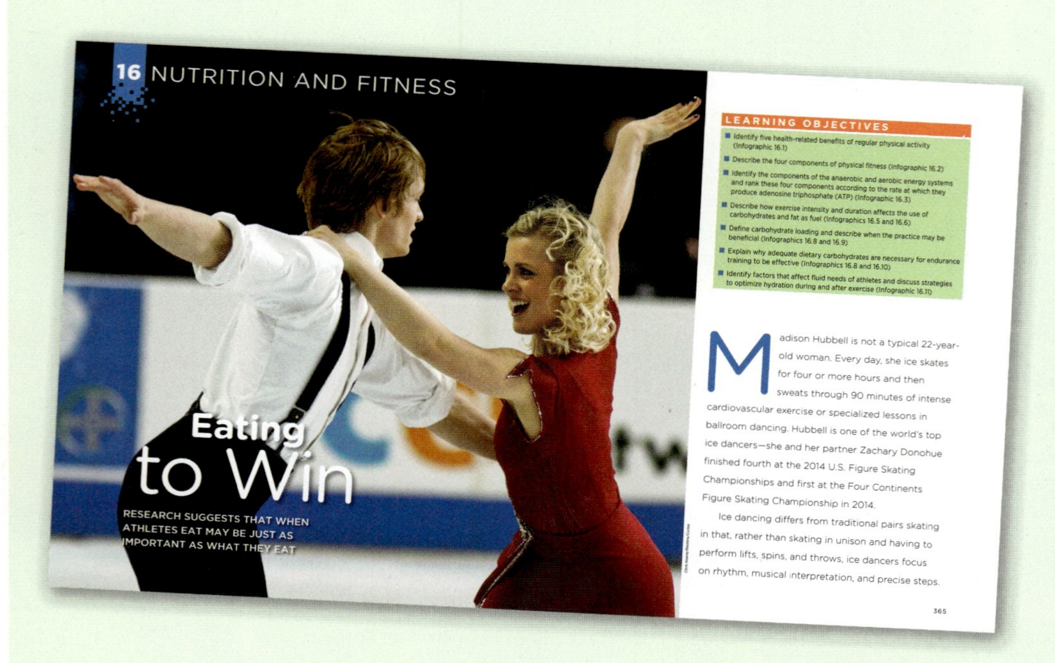

16 NUTRITION AND FITNESS

Eating to Win

RESEARCH SUGGESTS THAT WHEN ATHLETES EAT MAY BE JUST AS IMPORTANT AS WHAT THEY EAT

LEARNING OBJECTIVES

- Identify five health-related benefits of regular physical activity (Infographic 16.1)
- Describe the four components of physical fitness (Infographic 16.2)
- Identify the components of the anaerobic and aerobic energy systems and rank these four components according to the rate at which they produce adenosine triphosphate (ATP) (Infographic 16.3)
- Describe how exercise intensity and duration affects the use of carbohydrates and fat as fuel (Infographics 16.5 and 16.6)
- Define carbohydrate loading and describe when the practice may be beneficial (Infographics 16.8 and 16.9)
- Explain why adequate dietary carbohydrates are necessary for endurance training to be effective (Infographics 16.8 and 16.10)
- Identify factors that affect fluid needs of athletes and discuss strategies to optimize hydration during and after exercise (Infographic 16.11)

Madison Hubbell is not a typical 22-year-old woman. Every day, she ice skates for four or more hours and then sweats through 90 minutes of intense cardiovascular exercise or specialized lessons in ballroom dancing. Hubbell is one of the world's top ice dancers—she and her partner Zachary Donohue finished fourth at the 2014 U.S. Figure Skating Championships and first at the Four Continents Figure Skating Championship in 2014.

Ice dancing differs from traditional pairs skating in that, rather than skating in unison and having to perform lifts, spins, and throws, ice dancers focus on rhythm, musical interpretation, and precise steps.

365

Lively writing and design make content easier to read and remember

Info + graphics = *Infographics*

These "scientific storyboards" combine engaging graphics with step-by-step explanations to illustrate key concepts.

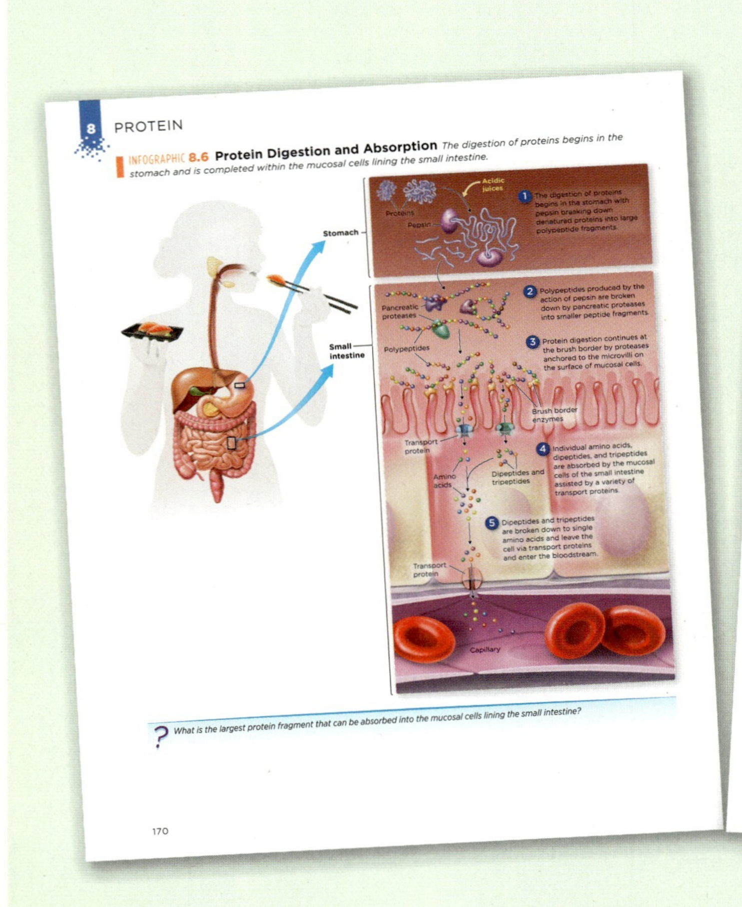

8 PROTEIN

INFOGRAPHIC 8.6 **Protein Digestion and Absorption** *The digestion of proteins begins in the stomach and is completed within the mucosal cells lining the small intestine.*

? *What is the largest protein fragment that can be absorbed into the mucosal cells lining the small intestine?*

170

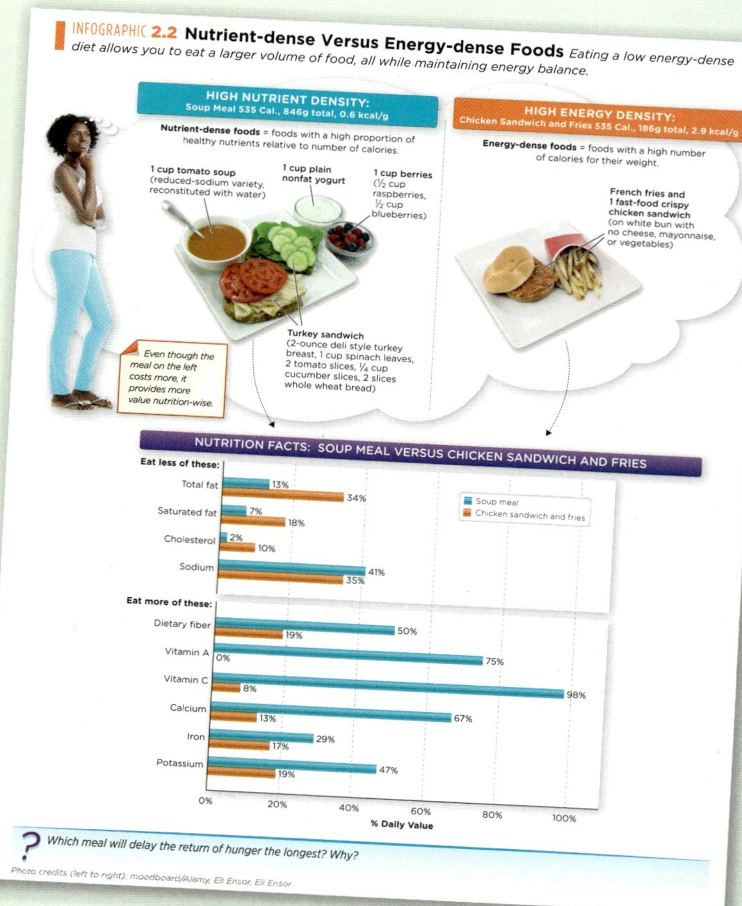

INFOGRAPHIC 2.2 **Nutrient-dense Versus Energy-dense Foods** *Eating a low energy-dense diet allows you to eat a larger volume of food, all while maintaining energy balance.*

? *Which meal will delay the return of hunger the longest? Why?*

Photo credits (left to right): moodboard/Alamy, Eli Ensor, Eli Ensor

New Appendix 3! The unique Infographic has all of the visuals and information you need to present the vitamins and minerals by function in the body.

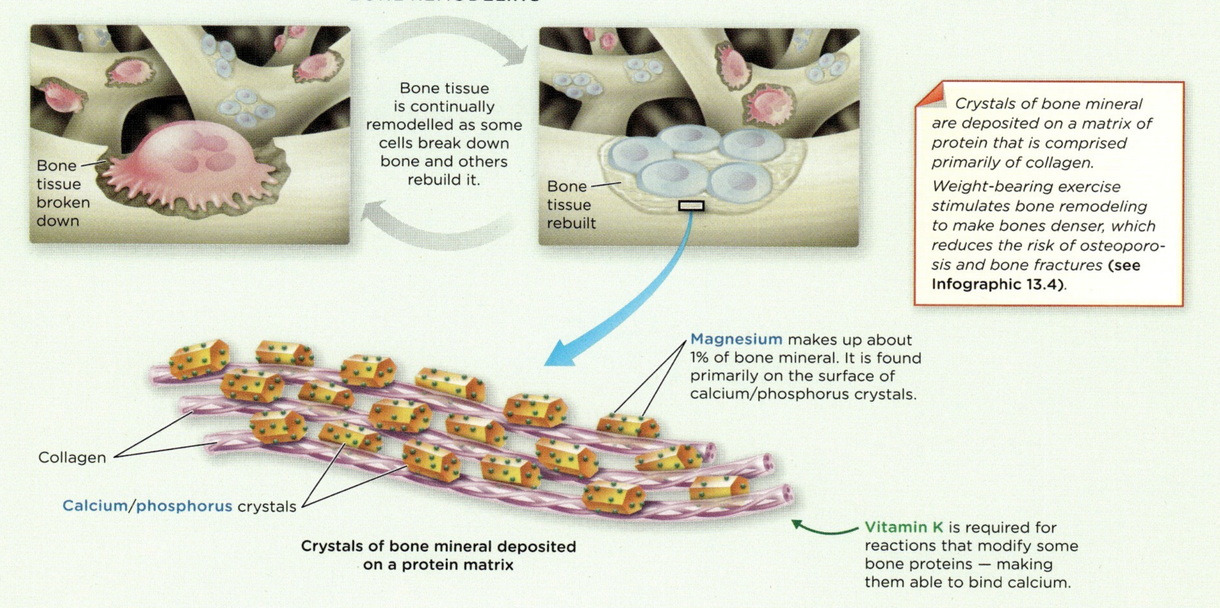

New! Auto-graded Analyze My Diet exercises

Analyze My Diet activities are automatically graded for ease of use in all classrooms and provide instant, accurate feedback to students.

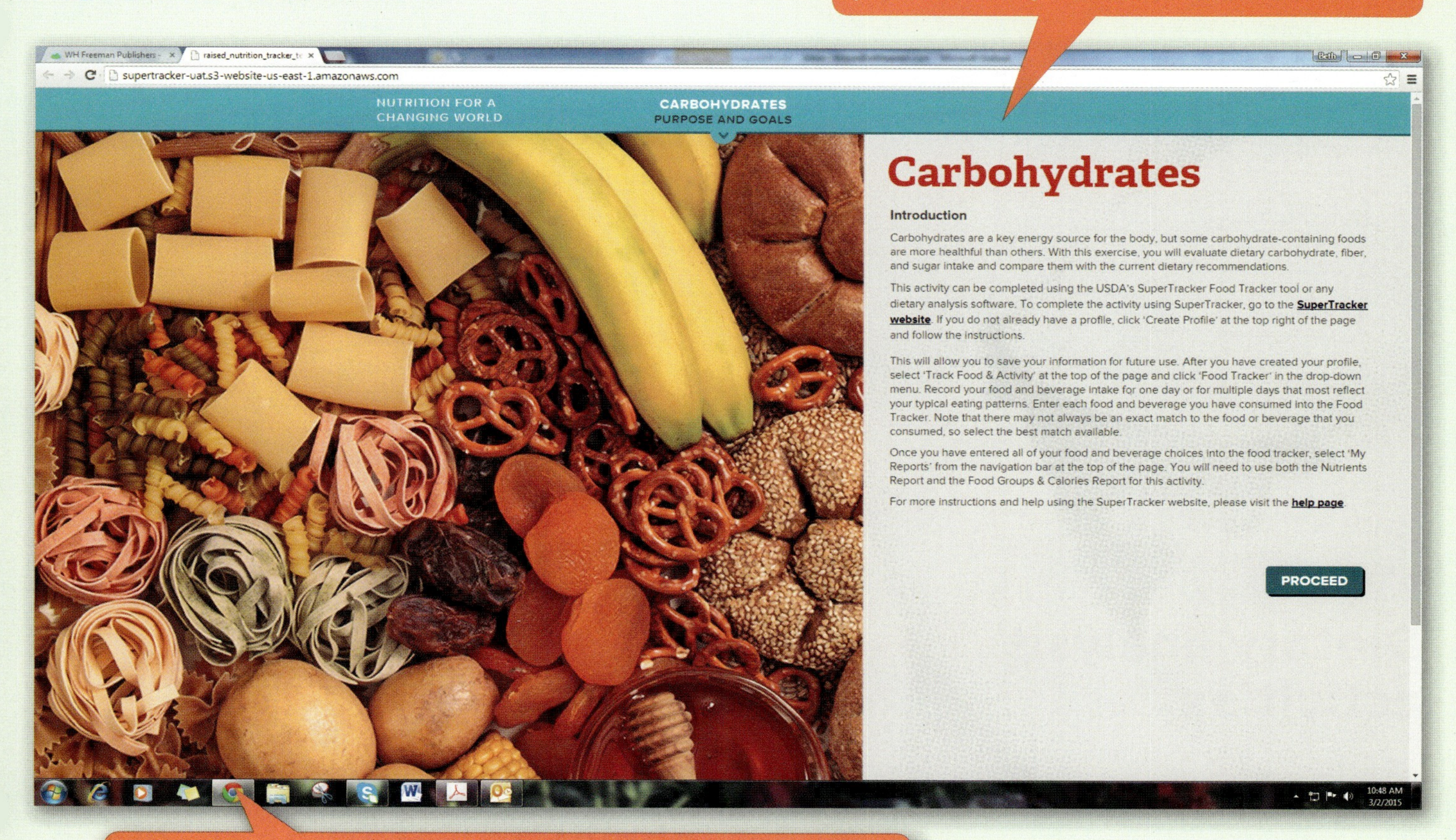

Time-saving and flexible! Select from a range of auto-graded diet analysis exercises and reduce classroom preparation time.

Summaries of KEY IDEAS and NEED TO KNOW help students prepare for assessments and apply what they learned

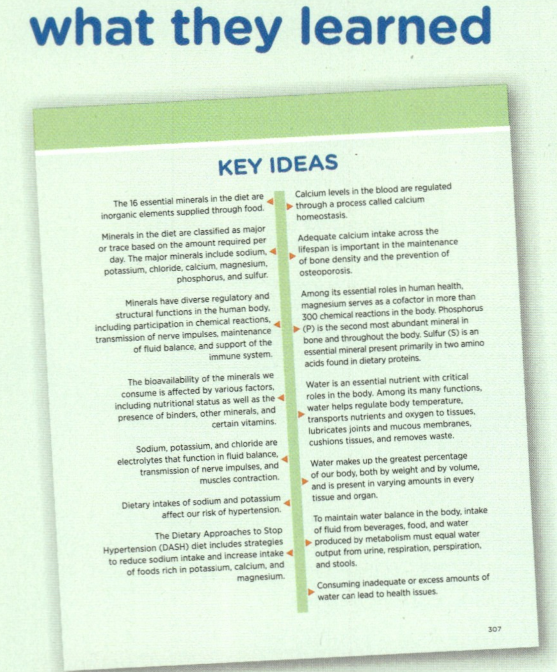

KEY IDEAS

The 16 essential minerals in the diet are inorganic elements supplied through food.

Minerals in the diet are classified as major or trace based on the amount required per day. The major minerals include sodium, potassium, chloride, calcium, magnesium, phosphorus, and sulfur.

Minerals have diverse regulatory and structural functions in the human body, including participation in chemical reactions, transmission of nerve impulses, maintenance of fluid balance, and support of the immune system.

The bioavailability of the minerals we consume is affected by various factors, including nutritional status as well as the presence of binders, other minerals, and certain vitamins.

Sodium, potassium, and chloride are electrolytes that function in fluid balance, transmission of nerve impulses, and muscles contraction.

Dietary intakes of sodium and potassium affect our risk of hypertension.

The Dietary Approaches to Stop Hypertension (DASH) diet includes strategies to reduce sodium intake and increase intake of foods rich in potassium, calcium, and magnesium.

Calcium levels in the blood are regulated through a process called calcium homeostasis.

Adequate calcium intake across the lifespan is important in the maintenance of bone density and the prevention of osteoporosis.

Among its essential roles in human health, magnesium serves as a cofactor in more than 300 chemical reactions in the body. Phosphorus (P) is the second most abundant mineral in bone and throughout the body. Sulfur (S) is an essential mineral present primarily in two amino acids found in dietary proteins.

Water is an essential nutrient with critical roles in the body. Among its many functions, water helps regulate body temperature, transports nutrients and oxygen to tissues, lubricates joints and mucous membranes, cushions tissues, and removes waste.

Water makes up the greatest percentage of our body, both by weight and by volume, and is present in varying amounts in every tissue and organ.

To maintain water balance in the body, intake of fluid from beverages, food, and water produced by metabolism must equal water output from urine, respiration, perspiration, and stools.

Consuming inadequate or excess amounts of water can lead to health issues.

307

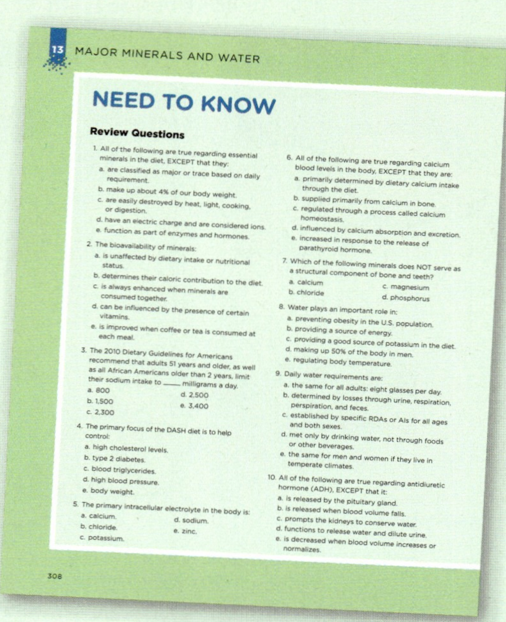

13 MAJOR MINERALS AND WATER

NEED TO KNOW

Review Questions

1. All of the following are true regarding essential minerals in the diet, EXCEPT that they:
   a. are classified as major or trace based on daily requirement.
   b. make up about 4% of our body weight.
   c. are easily destroyed by heat, light, cooking, or digestion.
   d. have an electric charge and are considered ions.
   e. function as part of enzymes and hormones.
2. The bioavailability of minerals:
   a. is unaffected by dietary intake or nutritional status.
   b. determines their caloric contribution to the diet.
   c. is always enhanced when minerals are consumed together.
   d. can be influenced by the presence of certain vitamins.
   e. is improved when coffee or tea is consumed at each meal.
3. The 2010 Dietary Guidelines for Americans recommend that adults 51 years and older, as well as all African Americans older than 2 years, limit their sodium intake to ____ milligrams a day.
   a. 800
   b. 1,500
   c. 2,300
   d. 2,500
   e. 3,400
4. The primary focus of the DASH diet is to help control:
   a. high cholesterol levels.
   b. type 2 diabetes.
   c. blood triglycerides.
   d. high blood pressure.
   e. body weight.
5. The primary intracellular electrolyte in the body is:
   a. calcium.
   b. chloride.
   c. potassium.
   d. sodium.
   e. zinc.
6. All of the following are true regarding calcium blood levels in the body, EXCEPT that they are:
   a. primarily determined by dietary calcium intake through the diet.
   b. supplied primarily from calcium in bone.
   c. regulated through a process called calcium homeostasis.
   d. influenced by calcium absorption and excretion.
   e. increased in response to the release of parathyroid hormone.
7. Which of the following minerals does NOT serve as a structural component of bone and teeth?
   a. calcium
   b. chloride
   c. magnesium
   d. phosphorus
8. Water plays an important role in:
   a. preventing obesity in the U.S. population.
   b. providing a source of energy.
   c. providing a good source of potassium in the diet.
   d. making up 50% of the body in men.
   e. regulating body temperature.
9. Daily water requirements are:
   a. the same for all adults: eight glasses per day.
   b. determined by losses through urine, respiration, perspiration, and feces.
   c. established by specific RDAs or AIs for all ages and both sexes.
   d. met only by drinking water, not through foods or other beverages.
   e. the same for men and women if they live in temperate climates.
10. All of the following are true regarding antidiuretic hormone (ADH), EXCEPT that it:
    a. is released by the pituitary gland.
    b. is released when blood volume falls.
    c. prompts the kidneys to conserve water.
    d. functions to release water and dilute urine.
    e. is decreased when blood volume increases or normalizes.

308

Dietary Analysis Using SuperTracker
Analyzing intake of water-soluble vitamins

The water-soluble vitamins include the B vitamins and vitamin C (and the nutrient choline). Because these vitamins are generally not stored in the body, deficiency conditions of the water-soluble vitamins develop more quickly than for the fat-soluble vitamins. Using SuperTracker, identify the sources of water-soluble vitamins in your diet and compare the amount you get with the current dietary recommendations.

1. Log on to the United States Department of Agriculture (USDA) website at www.supertracker.usda.gov. If you have not done so already, you will need to create a profile to get a personalized diet plan. This profile will allow you to save your information and diet intake for future reference. *Do not use the general plan.*
2. Click the Track Food and Activity option.
3. Record your food and beverage intake for one day that most reflects your typical eating patterns. Enter each food and beverage you consumed into the food tracker. Note that there may not always be an exact match to the food or beverage that you consumed, so select the best match available.
4. Once you have entered all of your food and beverage choices into the food tracker, on the right side of the page under the bar graph, you will see Related Links: View by Meal and Nutrient Intake Report. Print these reports and use them to answer the following questions:
   a. Did you meet the targets for the water-soluble vitamins on the day you selected? If not, which vitamin intakes are below the target numbers?
   b. For each vitamin target that you missed, list two specific foods you could consume to increase your intake of that vitamin.
   c. Was your intake above the targets for any of the water-soluble vitamins? If so, which ones? Are there any issues with toxicity that are associated with those vitamins?
   d. Discuss how the enrichment of grains in the United States has helped eradicate certain water-soluble vitamin deficiencies. What deficiencies are rarely seen in the United States as a result of this policy?
   e. Many foods and beverages are fortified with water-soluble vitamins. Discuss the pros and cons of this practice.

257

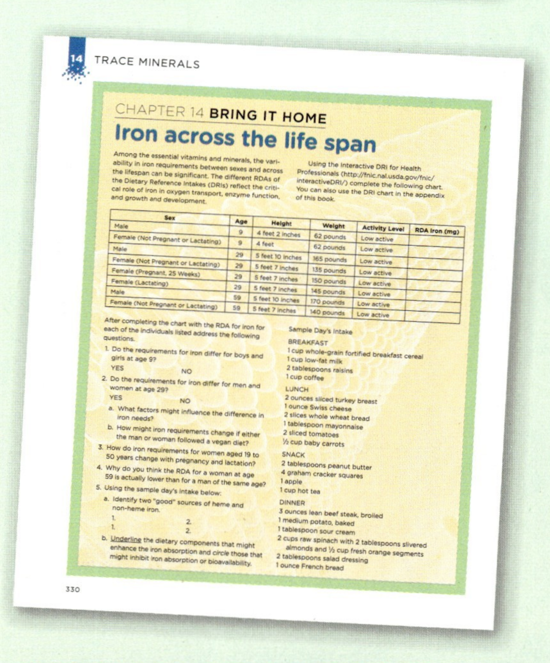

14 TRACE MINERALS

CHAPTER 14 **BRING IT HOME**

Iron across the life span

Among the essential vitamins and minerals, the variability in iron requirements between sexes and across the lifespan can be significant. The different RDAs of the Dietary Reference Intakes (DRIs) reflect the critical role of iron in oxygen transport, enzyme function, and growth and development.

Using the Interactive DRI for Health Professionals (http://fnic.nal.usda.gov/fnic/interactiveDRI/) complete the following chart. You can also use the DRI chart in the appendix of this book.

Sex	Age	Height	Weight	Activity Level	RDA Iron (mg)
Male	9	4 feet 2 inches	62 pounds	Low active	
Female (Not Pregnant or Lactating)	9	4 feet	62 pounds	Low active	
Male	29	5 feet 10 inches	165 pounds	Low active	
Female (Not Pregnant or Lactating)	29	5 feet 7 inches	135 pounds	Low active	
Female (Pregnant, 25 Weeks)	29	5 feet 7 inches	150 pounds	Low active	
Female (Lactating)	29	5 feet 7 inches	145 pounds	Low active	
Male	59	5 feet 10 inches	170 pounds	Low active	
Female (Not Pregnant or Lactating)	59	5 feet 7 inches	140 pounds	Low active	

After completing the chart with the RDA for iron for each of the individuals listed address the following questions.

1. Do the requirements for iron differ for boys and girls at age 9?
   YES NO
2. Do the requirements for iron differ for men and women at age 29?
   YES NO
   a. What factors might influence the difference in iron needs?
   b. How might iron requirements change if either the man or woman followed a vegan diet?
3. How do iron requirements for women aged 19 to 50 years change with pregnancy and lactation?
4. Why do you think the RDA for a woman at age 59 is actually lower than for a man of the same age?
5. Using the sample day's intake below:
   a. Identify two "good" sources of heme and non-heme iron.
      1. 2.
      1. 2.
   b. Underline the dietary components that might enhance the iron absorption and *circle* those that might inhibit iron absorption or bioavailability.

Sample Day's Intake

BREAKFAST
1 cup whole-grain fortified breakfast cereal
1 cup low-fat milk
2 tablespoons raisins
1 cup coffee

LUNCH
2 ounces sliced turkey breast
1 ounce Swiss cheese
2 slices whole wheat bread
1 tablespoon mayonnaise
2 sliced tomatoes
½ cup baby carrots

SNACK
2 tablespoons peanut butter
4 graham cracker squares
1 apple
1 cup hot tea

DINNER
3 ounces lean beef steak, broiled
1 medium potato, baked
1 tablespoon sour cream
2 cups raw spinach with 2 tablespoons slivered almonds and ½ cup fresh orange segments
2 tablespoons salad dressing
1 ounce French bread

330

Dietary analysis activities

In these short, end-of-chapter activities, students use the USDA's SuperTracker to track their food and beverage intake and analyze the results.

BRING IT HOME exercises ask students to think critically about nutrition concepts in the context of their own lives and are easily adapted as in-class activities

Interactive online study tools

Developed with extensive feedback from instructors and students, Macmillan's online course space makes it easy for you to use media in your course. All online resources for the text—including the e-Book—reside in prebuilt, ready-to-assign chapter units. LaunchPad's streamlined interface lets you build an entire course in minutes, and is easily customizable. The Gradebook captures students' scores, while key analytics let you see how your class is doing individually and as a whole.

Analyze My Diet activities built to work with the USDA's SuperTracker open-access database, make grading dietary analysis assignments easier than ever before.

- Student data are auto-calculated and auto-graded, freeing up valuable instructor time.
- A variety of auto-graded activities allow students more opportunities to examine their diets within the context of chapter topics.
- The activities are easily adaptable for use with other dietary analysis databases and software.
- The assignments provide instant and accurate feedback for students.

e-Book

The e-Book for *Nutrition for a Changing World* offers everything found in the printed textbook, plus additional interactivity, such as highlighting and note-taking.

Animations break down key science concepts and processes, and are primarily based on the Infographics in the book.

Consumer activities offer visual, interactive practice in applying nutrition concepts to everyday situations, such as making choices at the grocery store.

In a gamelike format, LearningCurve adaptive quizzing is an engaging and effective assignment that can be used before or after class. LearningCurve offers

- an individualized learning path, with quizzes shaped by each student's correct and incorrect answers.
- a personalized study plan that guides students' preparation for class and for exams.
- feedback for each question with live links to relevant e-Book pages, guiding students to the reading they need to do to improve their areas of weakness.

Instructor Resources

Lecture Presentation Slides with artwork and photos
Test Bank Questions
Chapter Summaries
Clicker Questions
Curated Video Links
In-class activities

ACKNOWLEDGMENTS

Producing a textbook is a large-scale effort and requires the work of many dedicated professionals with differing disciplines and talents. The authors and publishers of this book are indebted to science writer Melinda Wenner Moyer for writing Chapters 8, 9, 11, 12, 13, 14, 17, and 18, and for revising Chapter 3. We also thank science writer Matt Tontonoz for writing Chapters 4, 15, 16, 19, 20, and 21. We thank our delightful publisher Peter Marshall, and our marketing experts Shannon Howard and Will Moore. Beth Marsh, Blythe Robbins, and Andrea Gawrylewski contributed significant editorial expertise to the project—editing, art, text, and media. Karen Misler is a whiz at scheduling. Julio Espin did a terrific job managing the production end of the project. Eli Ensor did a fantastic job of translating our art ideas into reality. We thank Tom Carling for his text and cover design. Lori Stover jumped into the project with enthusiasm and did an amazing job producing My Diet Analysis activities and other media components.

We couldn't produce a book without the guidance of instructors from the nutrition community. For all of you who have read chapters and offered your advice and constructive criticism, thank you! We'd like to acknowledge Annie Wetter, PhD, in particular, who did an excellent job of helping confirm accuracy and providing content advice. Thanks to Lena Goodwin for coordinating all of the chapter reviews.

Two other individuals were very helpful in pulling together essential book components: Hannah Packman compiled the references and glossary for the book, and Aileen McAinsh, PhD, RD, composed the preliminary glossary terms. We thank Anne Cioffi of Hudson Valley Community College for writing the diet analysis activities found at the end of the chapters. In addition to the invaluable contributions of the individuals above, the support and patience of our families and friends amidst long hours and deadlines made this endeavor possible. In particular, Jamie Pope would like to thank her husband, Mark Petty; daughters, Sarah and Glory; and her mother, Nancy Houge. Steve Nizielski would like to thank his wife Myra, and sons Connor and Colin.

We thank the instructors who created course materials in support of this textbook.

Michelle Alexander, *Thomas Nelson Community College, SuperTracker activity for chapter 3*
Jennifer Bess, *Hillsborough Community College, Test Bank*
Wendy Buchan, *California State University–Sacramento, Animations and Lecture Presentation Slides*
Anne Cioffi, *Hudson Valley Community College, SuperTracker Activities*
Michelle Hesse, *James Madison University, LearningCurve*
Sharon Himmelstein, *Central New Mexico Community College, SuperTracker Activities*
Kara Montgomery, *University of South Carolina, Test Bank*
Elizabeth Quintana, *West Virginia University, LearningCurve*
Padmini Shankar, *Georgia Southern University, Test Bank*
Stephen Sowulewski, *J. Sargeant Reynolds Community College, Consumer Activities and Clicker Questions*
Kirsten Straughan, *University of Illinois–Chicago, Test Bank*
Janet Tou, *West Virginia University, Test Bank*
Priya Venkatesan, *Pasadena City College, Curated Videos*

We thank the many reviewers who aided in the development of this text.

Patricía Abraham, *Arkansas State University*
Kwaku Addo, *Prairie View A&M University*
Valerie Adegunleye, *Prince Georges Community College*
Ann Afflerbach, *University of North Texas*
Jeremy Akers, *James Madison University*
Katherine Alaimo, *Michigan State University*
Lisa Alastuey, *University of Houston*
Michelle Alexander, *Thomas Nelson Community College*
Minda Allarde, *Rockland Community College–SUNY*
Amy Allen-Chabot, *Anne Arundel Community College*
Jatin Ambegaonkar, *George Mason University*
Audrey Anastasia, *University of New Hampshire–Manchester*
Susan Anspaugh, *Trine University*

Linda Armstrong, *Normandale Community College*
Vikki Armstrong, *Fayetteville State University*
Malene Arnaud, *Delgado Community College*
Michele Ball, *Louisiana State University*
Jessica Bandy, *Radford University*
Jinan Banna, *University of Hawaii–Manoa*
Carol Barnes, *Mississippi College*
Ana Barreras, *Central New Mexico Community College*
Kristin Bartholomew, *Valencia Community College–West*
Jimikaye Beck, *Metropolitan State College of Denver*
Jeannemarie Beiseigel, *University of St. Thomas*
Tania Beliz, *College of San Mateo*
Marilee Benore, *Univiersity of Michigan Dearborn*
James Bemel, *Utah Valley University*
Lindsey Benucci, *J. Sargeant Reynolds Community College–Downtown*
Brenda Bertrand, *East Carolina University*
Samuel Besong, *Delaware State University*
Jennifer Bess, *Hillsborough Community College*
Gregory Biren, *Rowan University–Glassboro*
Anne Black, *Austin Peay State University*
Laurie E. Black, *Arizona Western College*
Craig James Blalock, *Delgado Community College*
Cynthia Blocksom, *Northern Kentucky University*
Shoshana Bobker, *City University of New York–Bronx Community College*
Linda Boire, *Nassau Community College*
Joye Bond, *Minnesota State University–Mankato*
Tracy Bonoffski, *University of North Carolina–Charlotte*
Adelia Bovell-Benjamin, *Tuskegee University*
Carmen Boyd, *Missouri State University–Springfield*
Mark Brandenburger, *J. Sargeant Reynolds Community College–Downtown*
Karen Brasfield, *Texas State University–San Marcos*
Cindi Brassington, *Quinebaug Valley Community College*
Barbara Brehm-Curtis, *Smith College*
Patricia Brevard, *James Madison University*
Anne Bridges, *University of Alaska–Anchorage*
Tracey Brigman, *University of Georgia*
Molly Brignall, *Highline Community College*
Ardith Brunt, *North Dakota State University*
Wendy Buchan, *California State University–Sacramento*
Nancy Buffum-Herman, *Monroe Community College*
Michael Burg, *California State University–San Marcos*
Sandra Callan, *Community College of Allegheny County*
Jill Campbell, *University of Texas–Tyler*
Maria Carles, *Northern Essex Community College*
Diane Carson, *Chapman University*
Lisa Cassis, *University of Kentucky*
Thomas Castonguay, *University of Maryland–College Park*
Chimene Castor, *Trinity University*
Erin Caudill, *Southeast Community College*
Melissa Chabot, *State University of New York–Buffalo*
Wen-Hsing Cheng, *University of Maryland–College Park*
Susan Chou, *American River College*
Dori Cinque, *Suffolk County Community College–Grant*
Anne Cioffi, *Hudson Valley Community College*
Nicole Clark, *Indiana University of Pennsylvania*
Nica Clark, *Southern Utah University*
Mary Clark, *Monroe Community College*
Donna Clark-Salmonson, *Sierra College*
Beth Collins, *Iowa Central Community College*
Janet Colson, *Middle Tennessee State University*
Bert Connell, *Loma Linda University*
Kevin Cooper, *Sacramento City College*
Jessica Coppola, *Sacramento City College*
Michael Corcoran, *Merrimack College*
Margaret Craig-Schmidt, *Auburn University*
Tracy Crane, *University of Arizona*
Theresa Crossan, *St. Joseph's University*
Cathy Hix Cunningham, *Tennessee Technological University*
Christopher D'Arcy, *Cayuga Community College*
Eileen Daniel, *State University of New York College–Brockport*
Sarah Darrell, *Ivy Tech Community College–Gary*
Patricia Davidson, *West Chester University of Pennsylvania*
Mandi Joy Davidson, *Ivy Tech Community College–Lafayette*
Ann Deblinger, *Passaic County Community College*
Ann Diker, *Metropolitan State College of Denver*
Robert DiSilvestro, *Ohio State University–Columbus*
Betty Dixon, *Armstrong Atlantic State University*
Christine Dobrowolski, *College of the Redwoods*
Heidi Dodson, *North Greenville University*
Rebecca Dority, *Texas Christian University*
Maggi Dorsett, *Butte College*
Courtney Dowell, *Thomas Nelson Community College*
Kamal Dulai, *University of California–Merced*
Rachelle Duncan, *Oklahoma State University*
Stephanie Durbin, *Thomas Nelson Community College*
Nancy Eckert, *Lakes Region Community College*
Michelle Eggers, *Appalachian State University*
Dorelle Engel, *Montgomery College–Rockville*
Barb Erfurt, *Glendale Community College*
Carol Erwin, *Wayne State College*
Alyson Escobar, *Montgomery College*
Claudia Fajardo, *California State University–Northridge*
Olubisi Faoye, *Miami Dade College–Wolfson*
Allen Farrand, *Bellevue College*
David Fell, *Rockland Community College*
Maria Pontes Ferreira, *Wayne State University*
Mark Finke, *Monroe Community College*
Linda Fleming, *Middlesex Community College*
Jerald Foote, *Metropolitan State University of Denver*
Jana Fowler, *Muscatine Community College*
Laura Frank, *Immaculata University*
Jeannie Frazier, *Jacksonville State University*
Jennifer Frere, *Cuesta College*
Siah Fried, *Las Positas College*
Karen Gabrielsen, *Everett Community College*
Jared Garbutt, *American River College*
Mario Garcia-Rios, *Mount Ida College*
Meghan Garrett, *J. Sargeant Reynolds Community College–Downtown*
Jenny Gernhart, *Iowa Central Community College*
Heather Gibbs, *University of Kansas Medical Center*
Susan Gills, *Metropolitan State College of Denver*
Sandria Godwin, *Tennessee State University*
Jill Golden, *Orange Coast College*
Jana Gonsalves, *American River College*
Debra Goodwin, *Jacksonville State University*
Kristin Goss, *Erie Community College–City*
Debalina Goswami, *Texas Tech University*
Artis Grady, *Southern Utah University*
Nancy Graves, *University of Houston*
Margaret Gunther, *Palomar College*
Evette Hackman, *Seattle Pacific University*
Sandra Haggard, *University of Maine–Augusta*
Kristi Haik, *Northern Kentucky University*

Paul Hankamp, *College of San Mateo*
Thorton Hannah, *Texas State University–San Marcos*
Tanya Hargrave-Klein, *Des Moines Area Community College–Boone*
Charlene Harkins, *University of Minnesota–Duluth*
Nancy Harris, *East Carolina University*
Clifton Harris, *West Hills College–Lemoore*
Stephanie Harris, *Case Western Reserve University*
Christy Hawkins, *Thomas Nelson Community College*
Jill Hayes, *Cayuga Community College*
Marianne Heffrin, *Endicott College*
Rachel Hermecz, *Georgia Southern University*
Lisa Herzig, *California State University Fresno*
Jennifer Herzog, *Herkimer County Community College*
Angela Hess, *Bloomsburg University of Pennsylvania*
MIchelle Hesse, *James Madison University*
Elizabeth Hilliard, *North Dakota State University*
Sharon Himmelstein, *Central New Mexico Community College*
Linda Hittleman, *Nassau Community College*
Jessica Hodge, *Folsom Lake College*
Robert Hodge, *Temple University*
David Holben, *Ohio University*
Rachelle Holman, *Carl Albert State College*
Laura Horn, *Cincinnati State Technical and Community College*
Kevin Huggins, *Auburn University*
Robert Humphrey, *Cayuga Community College*
Nancy Hunt, *Lipscomb University*
Sarah Hunter, *Delgado Community College*
Deema Hussein, *Eastfield College*
Dianne Hyson, *California State University–Sacramento*
Dana Jacko, *William Paterson University of New Jersey*
Jean Jackson, *Bluegrass Community and Technical College*
Thunder Jalili, *University of Utah*
Maureen Jamgochian, *College of the Canyons*
Janine Jensen, *Tulsa Community College–Southeast*
Peggy Johnston, *University of Nebraska–Kearney*
Lanae Joubert, *Eastern New Mexico University*
J. Kaufman, *Monroe Community College*
Sanam Kazemi, *California State University–Fullerton*
Shawnee Kelly, *Pennsylvania State University–York*
Shahla Khan, *University of North Florida*
Debbie Kimberlin, *Olivet Nazarene University*
Cheri King, *Community College of Denver*
Vanessa King, *American University*
Darlenen (Dee) Kinney, *University of Cincinnati–Clermont College*
Lynn Parker Klees, *Pennsylvania State University–York*
Allen Knehans, *University of Oklahoma Health Sciences Center*
Michael Kobre, *Cayuga Community College*
Lisa Kobs, *University of Wisconsin La Crosse*
Wanda Koszewski, *University of North Dakota*
Dauna Koval, *Bellevue College*
Gina Kraft, *Oklahoma Baptist University*
Laura Kruskall, *University of Nevada–Las Vegas*
Amy Krystock, *University of New Haven*
Susan Kundrat, *University of Wisconsin–Milwaukee*
Deborah Kupecz, *Community College of Denver*
Colleen Kvaska, *Fullerton College*
Mustapha Lahrach, *Hillsborough Community College*
Kathleen Laquale, *Bridgewater State College*
Merilee Larsen, *Utah Valley University*
Gregory LeBlanc, *Middlesex County College*
Julie Lee, *Western Kentucky University*
Robert Leopard, *Monroe Community College*
Kristine Levy, *Columbus State Community College*
Michele Lewis, *Virginia Tech*
Yanyan Li, *Husson College*
Gina Liberti, *Rockland Community College*
Rosanna Licht, *Pasco-Hernando Community College*
David Lightsey, *Bakersfield College*
Ritamarie Little, *California State University–Northridge*
Geralyne Lopez-de-Victoria, *Midlands Technical College–Airport*
Lynne LoPresto, *Dominican University of California*
Donna Louie, *University of Colorado–Boulder*
Melinda Luis, *Broward Community College–Central*
Mary Lyons, *El Camino College*
Janna Lyons, *Appalachian State University*
Mara Manis, *Hillsborough Community College*
Terri Marcus, *Chattanooga State Technical Community College*
Cindy Marshall, *Saddleback College*
Rose Martin, *Iowa State University*
Theresa Martin, *College of San Mateo*
Virginia Martin, *J. Sargeant Reynolds Community College–Ginter Park*
Mary Martinez, *Central New Mexico Community College*
Pam Massey, *University of Wisconsin–Centers–Marathon*
Suresh Mathews, *Auburn University*
Sarah Mathot, *Santa Ana College*
Deborah Mautone, *Dutchess Community College*
Dorothy Chen Maynard, *California State University–San Bernardino*
Dale McCabe, *Pasadena City College*
Jackie McClelland, *North Carolina State University*
Carrie McConnell, *Community College of Aurora*
Myrtle McCulloch, *Georgetown University*
Richard McIntyre, *Cayuga Community College SUNY System*
Elizabeth Ellen McNeeley, *Ivy Tech Community College–Kokomo*
Mark Meskin, *California State Polytechnic University–Pomona*
Steven Mezik, *Herkimer County Community College*
Allison Miner, *Prince Georges Community College*
Jill Mohr, *Heartland Community College*
Kristin Moline, *Lourdes College*
Lynn Monahan, *West Chester University of Pennsylvania*
Kara Montgomery, *University of South Carolina*
Cherie Moore, *Cuesta College*
Megan Moran, *West Virginia University*
Paul Moore, *Appalachian State University*
Marianne Morano, *County College of Morris*
Sharon Morcos, *Kansas State University*
Holly Morris, *Lehigh Carbon Community College*
Gina Marie Morris, *Frank Phillips College*
Lisa Morse, *Arizona State University*
Elizabeth Morton, *University of South Carolina–Salkehatchie Regional*
Elaine Mostow, *Bergen Community College*
Megan Mullins, *West Virginia Wesleyan College*
Lee Murphy, *The University of Tennessee Institute of Agriculture*
Adelaide Nardone, *Fordham University–Rose Hill*
Eddystone Nebel, *Delgado Community College*

Suzanne Nelson, *The University of Colorado at Boulder*
Carmen Nochera, *Grand Valley State University*
Claire Norton, *University of Massachusetts–Amherst*
Jason O'Briant, *North Carolina Central University*
Valerie O'Brien, *Tulsa Community College–West*
Jill O'Malley, *Erie Community College–City*
Karen Ostenso, *University of Wisconsin–Stout*
Milli Owens, *College of the Sequoias*
Catherine Palmer, *Oklahoma State University–Oklahoma City*
Gloria Payne, *Middle Tennessee State University*
Paula Perez, *North Carolina Central University*
Sara Plaspohl, *Armstrong Atlantic State University*
John Polagruto, *Sacramento City College*
Paola Puig, *Hofstra University*
Virginia Quick, *James Madison University*
Elizabeth Quintana, *West Virginia University Health Science*
Julian Raffoul, *Emory University*
Sudha Raj, *Syracuse University*
Marisa Ramos, *Yuba Community College District*
Linda Rankin, *Idaho State University*
Amanda Reat, *Texas State University–San Marcos*
Mike Reece, *Ozarks Technical Community College*
Tonia Reinhard, *Wayne State University*
Bruce Rengers, *Metropolitan State College of Denver*
Victoria Rethmeier, *Southeast Community College*
Tiffany Ann Ricci, *University of Alaska–Juneau*
Becky Roach, *College of Lake County*
Alexandra Paterson Robins, *Ivy Tech Community College–Columbus*
Jodi Robinson, *Humber College–North*
Linda Rodebaugh, *University of Indianapolis*
Judith Rodriguez, *University of North Florida*
Beverly Roe, *Erie Community College–South*
Elicia Rosen-Fox, *City University of New York–Kingsborough Community College*
Catharine Ross, *Pennsylvania State University-York*
Sara Long Roth, *Southern Illinois University–Carbondale*
Michael Rovito, *University of Central Florida*
Karla Rues, *Ozarks Technical Community College*
Dominique Ruggieri, *Saint Joseph's University*
Thomas Russell, *Delgado Community College*
Kyle Ryan, *Peru State College*
Shauna Salvesen, *Mesa Community College*
Marie Schirmer, *American River College*
Laurie Schlussel, *William* Paterson *University of New Jersey*
Matthew Schmidt, *Southern Utah University*
Karen Schmitz, *Madonna University*
Doris Schomberg, *Alamance Community College*
Julie Schumacher, *Heartland Community College*
Lisa Searing, *Illinois Wesleyan University*
Padmini Shankar, *Georgia Southern University*
Banafshe Sharifian, *California State University–Fullerton*
Jyotsna Sharman, *Radford University*
Karen Shinville, *Heartland Community College*
Christine Sholtey, Waubonsee Community College
Denise Signorelli, *College of Southern Nevada–Las Vegas*
Cheryl Skinner, *Central Carolina Community College*
Carole Sloan, *Henry Ford Community College*
Martha Smallwood, *Abilene Christian University*
Nan Smith, *American University*
Dana Smith, *Blinn College*
Amy Snow, *Greenville Technical College–Barton*
Stephen Sowulewski, *J. Sargeant Reynolds Community College–Downtown*
Leslie Spencer, *Rowan University–Glassboro*
Jennifer Spry-Knutson, *Des Moines Area Community College–Boone*
Diane Stadler, *Oregon Health Sciences University*
Brenda Stagner, *Butte College*
Kathy Stanczyk, *Murray State University*
Jasia Steinmetz, *University of Wisconsin–Stevens Point*
Nicole Stob, *University of Colorado–Boulder*
Noelle Stock, *State University of New York in Potsdam*
Nikolaus Sucher, *Northern Essex Community College*
Larry Sullivan, *Holmes Community College*
Laura Taber, *University of New Mexico*
Jason Talanian, *Fitchburg State College*
Linda Tecklenburg, *Wilmington College*
Jennifer Temple, *State University of New York–Buffalo*
Norman Temple, *Athabasca University*
Kathy Timmons, *Murray State University*
Candice Tinsley, *Chaffey Community College*
Chris Todden, *Baker University*
Janet Tou, *West Virginia University*
Barb Troy, *Marquette University*
Kelly Twitchell, *Mesa Community College*
Ahondju Umadjela, *Langston University*
Wendy Unison-Pace, *J. Sargeant Reynolds Community College–Downtown*
Lindsay Upchurch, *Oklahoma State University–Oklahoma City*
Melinda Valliant, *University of Mississippi*
Benito Velasquez, *Midwestern State University*
Priya Venkatesan, *Pasadena City College*
Andrea Villarreal, *Phoenix College*
Paula Vineyard Most, *John A Logan College*
Eric Vlahov, *University of Tampa*
Ann Volk, *Antelope Valley College*
Susan Wakeman, *Greenville Technical College–Barton*
Laurie Wallace, *Bakersfield College*
Janelle Walter, *Baylor University*
Julie Walton, *Calvin College*
Daryle Wane, *Pasco-Hernando Community College*
Sue Ellen Warren, *El Camino College*
Dana Wassmer, *Cosumnes River College*
Diana Watson-Maile, *East Central University*
Sheldon Oliver Watts, *Temple University*
Kelly Webber, *University of Kentucky*
Jennifer Weddig, *Metropolitan State University of Denver*
Allisha Weeden, *Idaho State University*
Chris Wendtland, *Monroe Community College*
Lisa Werner, *Pima Community College*
Cynthia West, *Marshalltown Community College*
Stan Wilfong, *Baylor University*
Heather Williams, *Truckee Meadows Community College*
Kate Willson, *East Carolina University*
Amelia Wilson, *Purdue University North Central*
Stacie Wing-Gaia, *University of Utah*
Ka Wong, *J. Sargeant Reynolds Community College–Downtown*
Theopholieus Worrell, *Delgado Community College*
Linda Wright, *University of Wisconsin–Milwaukee*
Lauri Wright, *University of South Florida–Tampa*
Crystal Wynn, *J. Sargeant Reynolds Community College–Downtown*
Belinda Zeidler, *Portland State University*
Joseph Zielinski, *SUNY College at Brockport*

1 THE SCIENCE AND SCOPE OF NUTRITION

Exploring the Science of Nutrition

A NEW FIELD IN NUTRITION RESEARCH IS REVEALING SURPRISING DETAILS ABOUT HOW FOOD AFFECTS OUR GENES.

LEARNING OBJECTIVES

- Define the scope and science of nutrition (Infographic 1.1)
- Explain the connection between nutrition and chronic disease (Infographic 1.2)
- Define and identify the major macronutrients and micronutrients (Infographics 1.3 and 1.4)
- Summarize the purpose of the Dietary Reference Intake (DRI) values (Infographic 1.5)
- Distinguish between the different types of DRI values, and what each represents (Infographics 1.6 and 1.7)
- Understand/explain the basis of the scientific method and how it is used in nutrition research (Infographic 1.8)
- Describe three types of experimental design and the primary advantages of each (Infographic 1.9)
- Describe reliable sources of nutrition information (Infographic 1.10)

It was the final months of World War II, and the Dutch people were starving. As a last-ditch effort to hold on to the Netherlands, the Germans had blocked the transport of food from rural areas to western cities. By February, people were eating only a few hundred calories per day, typically a couple of small slices of bread and potatoes. Some people used paper to thicken soup. This period, now known as the Dutch Hunger Winter, lasted from October 1944 until the Netherlands was liberated in May 1945.

Thirty years later, a husband and wife team of scientists at Columbia University decided to investigate whether these extreme nutritional conditions had any long-lasting effects on the Dutch people.

NUTRITION
the study of dietary intake and behavior as well as the nutrients and constituents in food including their use in the body and influence on human health

After examining Dutch army records from 300,000 19-year-old men born around the time of the famine, the researchers made a startling discovery.

They found that men whose mothers endured the famine during the first months of their pregnancy were significantly more likely to be obese. But men whose mothers starved during the last months of their pregnancy, and into the first months of their life, were *less* likely to be obese. The explanation for this pattern wasn't clear—but it appeared as if a woman's diet during pregnancy could potentially have a strong influence on the weight of her future children.

Surviving World War II. *An emaciated Dutch boy, Netherlands, circa 1944.*

Hulton Archive/Getty Images

Other research soon followed. In the late 1980s, scientist David Barker, MD, PhD, found that babies with low birth weight were more likely to develop heart disease later in life—a finding that helped generate the "developmental origins hypothesis," which states that certain diseases originate from conditions during pregnancy and infancy. Specifically, poor nutrition during that crucial time can permanently affect the way the baby's body responds to food throughout his or her life. Subsequent research has also shown that a woman's body weight and diet at the time of conception affects the health of her babies.

Prompted by these findings, nutrition scientists were asking many new questions about how food and nutrients affect the body. Could a mother's nutrition have such long-lasting effects on her unborn child? Are we not, as the common phrase states, *what we eat?*—are we also *what our mothers ate?* And if so, how was this effect carried out in the body?

■ ■ ■

INTRODUCING THE SCIENCE OF NUTRITION

These fascinating questions illustrate the importance and the relevance of the science of **nutrition**, which is an interdisciplinary science that studies factors that affect our food choices, the chemical and physiological processes involved in processing and delivering the chemical components of those foods to cells throughout our body, and ultimately how those chemicals affect our health every day. **(INFOGRAPHIC 1.1)**

A lifetime of good health is a key reason to study the science of nutrition. We eat every day and the substances we consume are broken down to fuel activity and to build and support the body's tissues. How can we fine-tune our food choices to optimize health? How can we maintain strong bones and a high degree of activity well into our later years? How do we avoid heart disease, eat to run a marathon, or support a growing baby during pregnancy? These are questions at the center of the study of nutrition.

INFOGRAPHIC **1.1** **The Science of Nutrition** *The science of nutrition is fundamentally about how foods maintain health and influence the risk of disease.*

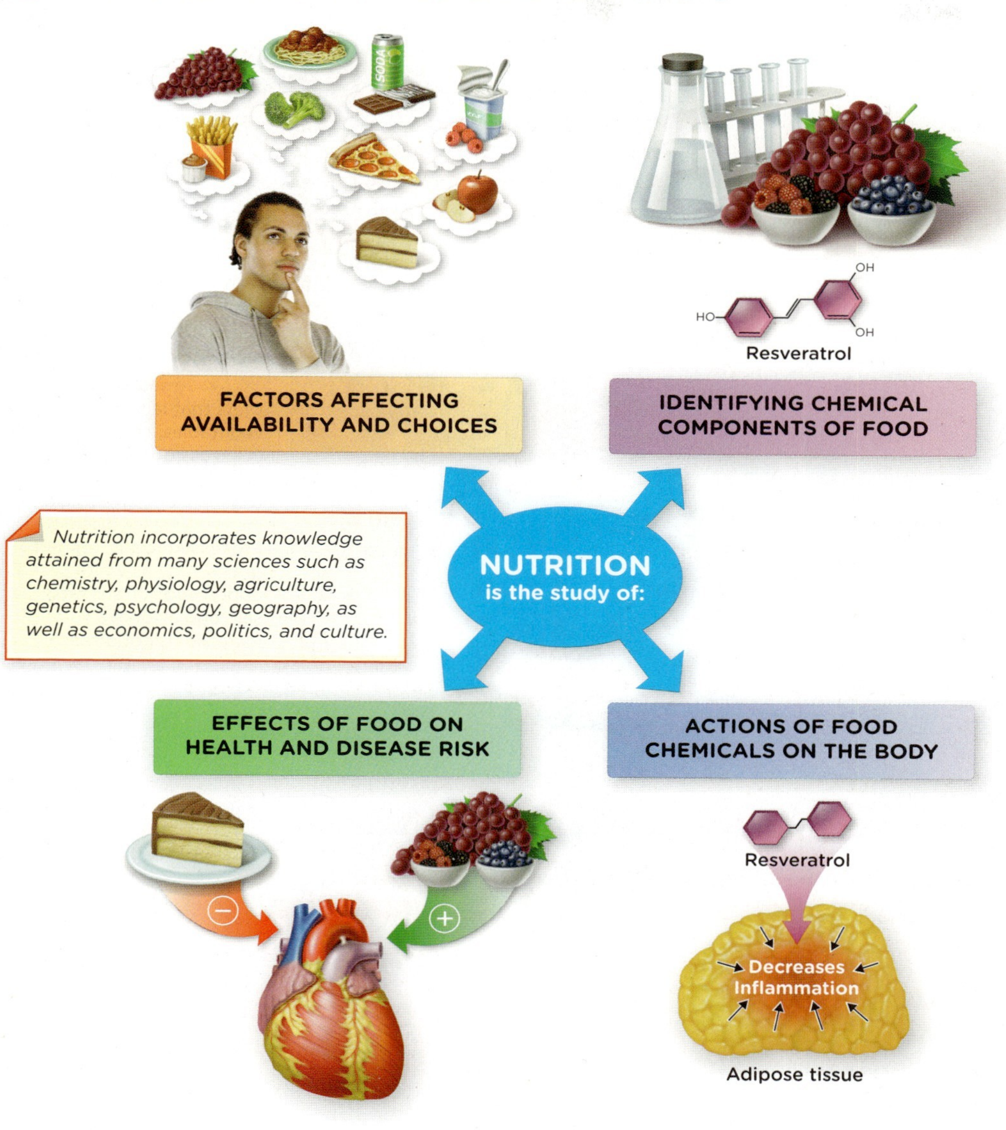

? *Can you think of several factors that affect your daily food choices? (Stay tuned: Chapter 19 provides a full discussion of factors that influence eating behavior.)*

Photo credit (top left): Roger Jegg - Fotodesign-Jegg.de/Shutterstock

FOOD PROVIDES NUTRIENTS AND ENERGY

Nutritionists study food and the components that make up food, including **nutrients**. A nutrient is a chemical substance that is essential for growth and to maintain proper body functioning. The body can actually produce many of its own nutrients, but other **essential nutrients** (such as water) must be supplied through what we eat and drink because the body cannot produce them or enough of them on its own.

There are six classes of nutrients required for the body to function: **carbohydrates**, **proteins**, **fats**, **vitamins**, **minerals**, and **water**. What you eat: the specific quantity and variety of nutrients in foods can promote health or increase your risk of disease. **(INFOGRAPHIC 1.2)**

NUTRIENTS
chemical substances obtained from food that are essential for body function; needed for metabolism, growth, development, reproduction, and tissue maintenance and repair

ESSENTIAL NUTRIENTS
nutrients that must be supplied through food because the body itself cannot produce/synthesize sufficient quantities to meet its needs

CARBOHYDRATES
compounds made up of carbon, hydrogen, and oxygen atoms that are found in foods as either simple sugars or complex carbohydrates (starch and fiber)

PROTEINS
large molecules consisting of carbon, hydrogen, oxygen, and nitrogen assembled in one or more chains of amino acids

FATS
a term for triglycerides, a subclass of lipids, that are the primary form of fat in our food and our bodies

VITAMINS
organic compounds that are required in small quantities for specific functions in the body

MINERALS
inorganic individual chemical elements obtained through foods that are essential in human nutrition

WATER
an essential nutrient that has critical functions in the body

INFOGRAPHIC 1.2 Nutrition and Disease Risk

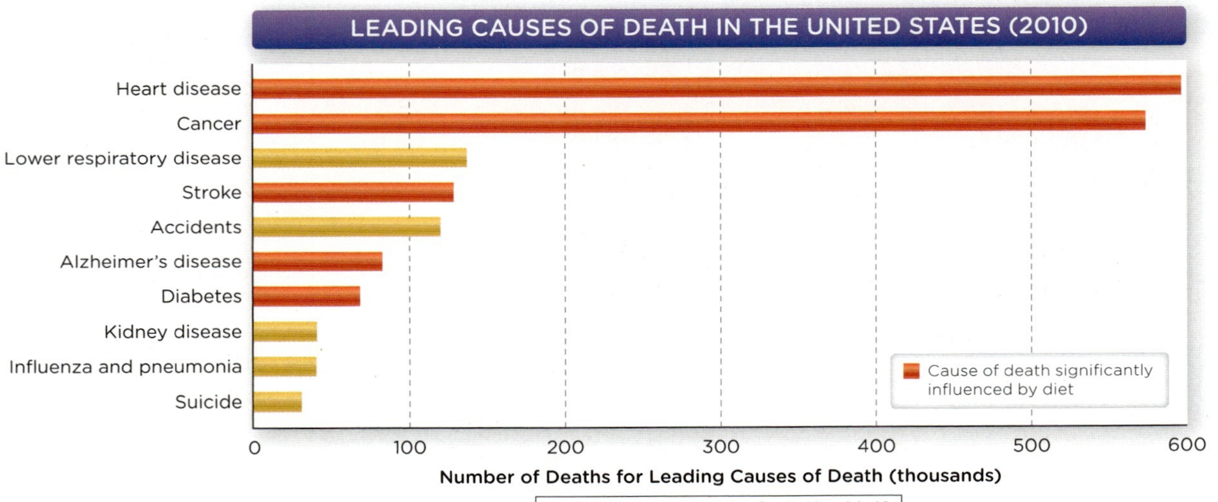

Source: Deaths: Final Data for 2010, table 10

? *What three types of foods increase risk of disease when overconsumed, and what three types of foods increase risk of disease when underconsumed?*

Photo credits (all photos): Eli Ensor

All six classes of nutrients regulate numerous body processes, and most supply the building blocks of key body structures, such as cell membranes, muscles, and bones. They are required for normal growth, development, reproduction, maintenance, repair of cells, and other vital body processes.

Nutrients are divided into two major categories. The first is **macronutrients**, which we need in relatively large quantities to stay healthy—these are carbohydrates, proteins, fats, and water. **(INFOGRAPHIC 1.3)** With the exception of water, they primarily

All six essential nutrients. *This colorful meal contains all 6 essential nutrients: carbohydrates, fats, proteins, vitamins, minerals, and water.*

Tooga/Getty Images

MACRONUTRIENTS *term used to describe nutrients that we require in relatively large daily amounts, e.g., carbohydrates, proteins, water, and fats*

INFOGRAPHIC 1.3 The Macronutrients

Carbohydrates, lipids, protein, and water are required in large amounts. With the exception of water, each of these nutrients provides energy.

The Macronutrients (required in large amounts)

	Carbohydrates	Lipids	Proteins	Water
Energy Content	4 kcal/g	9 kcal/g	4 kcal/g	0
Structural Function	Abundant in cartilage, ligaments, and other joint tissue, sugars are a component of DNA and RNA	Components of the cell membrane, fat deposits shape our body and provide insulation and protection	Major structural component of every cell and tissue in our body	Fills and surrounds every cell
Regulatory Function	Critical source of energy for the brain and red blood cells, helps regulate bowel function	Required for synthesis of hormones and other compounds that regulate many body processes	Regulates fluid balance and facilitates chemical reactions	Controls body temperature and is involved in many chemical reactions

Nutrients are substances that are obtained from food and are required for normal growth, development, reproduction, maintenance, and repair of cells.

Alcohol is not a nutrient—it is a toxin—but does supply energy 7 kcal/g

? *Given that alcohol supplies energy, why is it not considered a nutrient?*

Photo credits (left to right): Maximilian Stock Ltd/age fotostock/Eli Ensor, DUSAN ZIDAR/Shutterstock, Dorling Kindersley/Getty Images, Maria Toutoudaki/Getty Images

INFOGRAPHIC 1.4 Vitamins and Minerals: the Micronutrients

Called **micronutrients** because they are needed in tiny amounts, these substances in animal and plant foods are necessary for proper growth and development; the consequences of a micronutrient deficit can be profound.

The Micronutrients (required in small amounts) — Not a source of energy				
Vitamins			**Minerals**	
Water Soluble		**Fat Soluble**	**Major Minerals**	**Trace Minerals**
• Thiamin • Niacin • Vitamin B_6 • Vitamin C • Riboflavin	• Folate • Pantothenic acid • Vitamin B_{12} • Biotin • Choline	• Vitamin A • Vitamin E • Vitamin D • Vitamin K	• Calcium • Chloride • Magnesium • Phosphorus • Potassium • Sodium	• Chromium • Copper • Fluoride • Iodine • Iron • Manganese • Molybdenum • Selenium • Zinc
Function: • Participate in nearly every chemical reaction in the body • Some function as hormones			**Function:** • Most cooperate with proteins to facilitate chemical reactions • Some participate in nerve impulse transmission and muscle contraction • Some provide body structure	

What is the energy content of vitamins? Of minerals?

ENERGY the capacity to do work; obtained through the breakdown of carbohydrates, proteins, and fats in foods and beverages

KILOCALORIES (KCAL) standard unit to measure energy provided by food

MICRONUTRIENTS term used to describe nutrients essential in our daily diet to maintain good health, but required in only small amounts; e.g., vitamins and minerals

PHYTOCHEMICALS compounds found in plant foods that are physiologically active and beneficial to human health; not considered essential nutrients

supply **energy** and a large portion of the structural components in our body.

When we see calories listed on food labels, these are actually **kilocalories (kcal)** that reference the amount of energy in food. A kilocalorie is equal to 1,000 calories. By convention, when "Calorie" is spelled with a capital "C" it also refers to a kcal. The abbreviation "kcal" will be used throughout this book when discussing specific amounts of energy.

The second category of nutrients is **micronutrients**, which are needed in much smaller amounts—these are vitamins and minerals. They do not supply energy or calories, but are crucial to normal growth and development, even in tiny amounts. (INFOGRAPHIC 1.4)

Another important part of the diet is **phytochemicals** (also known as *phytonutrients*), which are chemicals in plants that are beneficial to human health. Commonly found in vegetables, fruits, and whole grains, these compounds—which number in the thousands—give plants key properties such as color, aroma, or flavor. Lycopene, for example, gives tomatoes and watermelon their red color, and may reduce the risk of cancer in humans. Some phytochemicals promote health because they have either hormonelike actions, or, like the pigment that makes blueberries blue, they repair or prevent damage to cells. Phytochemicals will be discussed in detail in Chapter 9.

These macronutrients, micronutrients, and phytochemicals are extracted from food through the process of digestion (see Chapter 3 for more on digestion), and then absorbed and used by the body. Nutrients play a critical role in maintaining *homeostasis*, the process by which the body maintains a stable, internal environment in the face of external variability. For instance, nutrients in the foods we eat participate in processes that regulate the balance of fluids in the body, our pH, and our body temperature. As long as cells get the nutrients they require, all is well; when cells need more nutrients

Pizza, candy, and chips. *A high-calorie "Westernized" diet is typically low in nutrient-rich fruits, vegetables, whole grains, and fish.*

Dwight Eschliman/Getty Images

than are supplied, however, problems arise.

WHAT IS MALNUTRITION?

Both **undernutrition** and **overnutrition** are forms of **malnutrition**—a state of inadequate or unbalanced nutrition. Sometimes, undernutrition does not stem from a lack of food overall, but from a lack of specific essential nutrients, known as **nutrient deficiency**.

Obesity—a condition characterized by excess body fat and often associated with other health problems—is a classic example of overnutrition. It's a relatively new public health concern—historically, nutrition policy and research has focused on making sure people had enough to eat, so they could meet their nutrient and energy needs. But over the past few decades, policymakers have focused more on the role of diet and nutrition in diseases caused by overconsumption—primarily too many calories, along with too much solid fat (such as animal fats), sugar, and sodium. These are often **chronic diseases** of slow progression, such as heart disease, and diabetes. Chronic diseases are now, by far, the leading causes of mortality worldwide, and diet plays a specific role in the risk, progression, and treatment of nearly every chronic disease. In addition to providing a lot of calories, the highly processed foods common in our "Westernized" diet are generally low in nutrients such as fiber, potassium, calcium, and vitamin D that reduce the risk of chronic diseases.

UNDERNUTRITION
inadequate nourishment caused by insufficient dietary intake of one or more essential nutrients or poor absorption and/or use of nutrients in the body

OVERNUTRITION
excess intake or imbalance of calories and/or essential nutrients relative to need that results in adverse health effects

MALNUTRITION
a state of undernutrition or overnutrition caused by inadequate, excessive, or unbalanced intake of calories and/or essential nutrients

NUTRIENT DEFICIENCY
a condition resulting from insufficient supply of essential nutrients through dietary inadequacy or impaired absorption or use

OBESITY
a condition characterized by accumulation of excess body fat, generally associated with adverse health effects

CHRONIC DISEASE
diseases that are generally slow in progression and of long duration; e.g., heart disease and diabetes

DIETARY REFERENCE INTAKES (DRI) *reference values for vitamins, minerals, macronutrients, and energy that are used to assess and plan the diets of healthy people in the United States and Canada*

ESTIMATED AVERAGE REQUIREMENT (EAR) *the average nutrient intake level estimated to meet the daily requirements of half of the healthy individuals for the different sexes and life-stage groups*

RECOMMENDED DIETARY ALLOWANCE (RDA) *the recommended nutrient intake levels that meet the daily needs and decrease risks of chronic disease in almost all healthy people for different sexes and life-stage groups*

NUTRIENT INTAKE RECOMMENDATIONS

A question just as interesting as the role of nutrition in early life is the one you may frequently ask yourself: *How do I know I'm getting enough nutrients?* To help us choose the foods that create a healthful diet, the Institute of Medicine—an expert committee commissioned by the U.S. government—provides guidance on health and science policy. The Institute of Medicine issues **Dietary Reference Intake (DRI)** values, which are quantitative estimates of nutrient intakes to be used to plan and assess diets for healthy people. These values not only help us avoid nutrient deficiency, they are also intended to help individuals optimize their health, prevent disease, and avoid consuming too much of any one nutrient.

The DRI recommendations for nutrient intakes include four values. **(INFOGRAPHIC 1.5)** The first is the **Estimated Average Requirements (EARs)**, which capture the *average* amount of a nutrient needed by sex and age group. As such, this is an intake that meets the needs of 50% of individuals within that group. But these are not recommendations for each individual, as half of the population would be eating less than they need at this level of intake. Rather, the EARs are used to assess the nutrient adequacy of populations, and are the first step in setting the DRI recommendations.

To establish the second set of values, a safety factor is added to the EAR to create the **Recommended Dietary Allowance (RDA)**, so that the RDA represents the average daily amount of a particular nutrient that meets or exceeds the requirements of nearly all (97% to 98%) healthy individuals.

INFOGRAPHIC 1.5 The Dietary Reference Intakes (DRIs) Have Varied Purposes

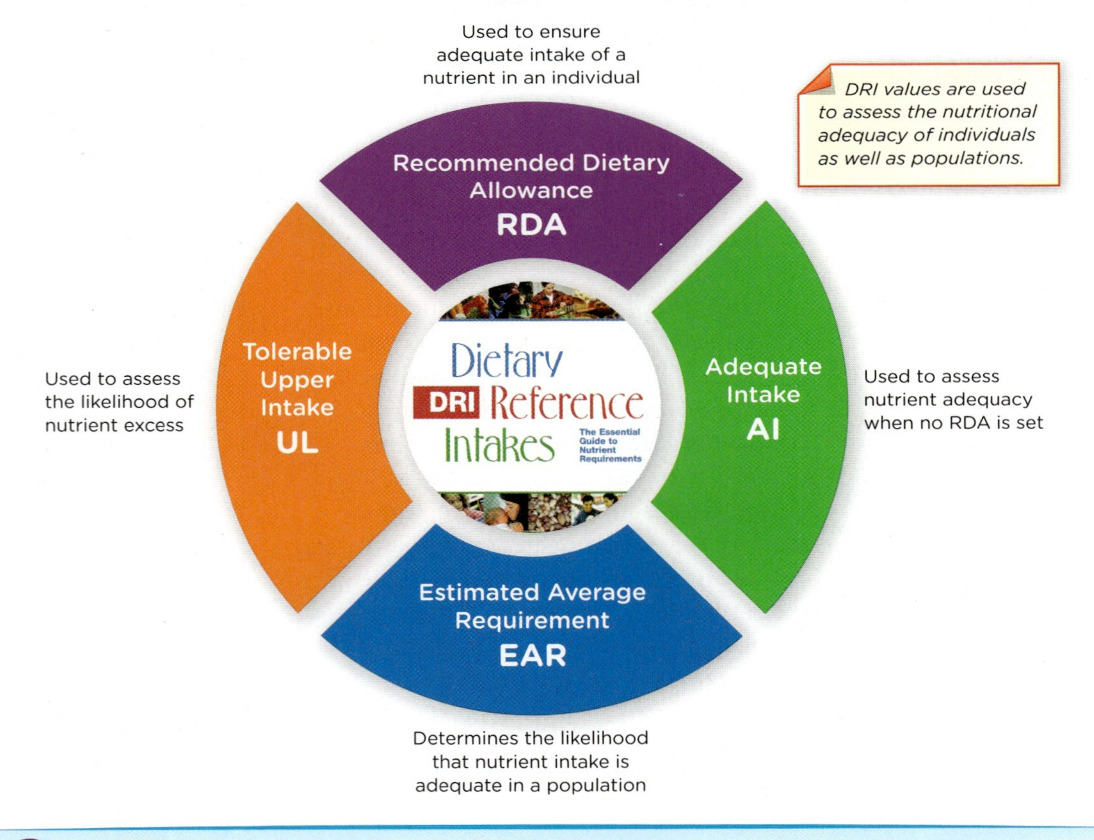

? *What reference value provides information to help us avoid excess consumption of nutrients?*

INFOGRAPHIC 1.6 DRI Values Decoded *The DRIs provide recommendations for nutrient intake needed to meet our body's needs and reduce our risk of chronic disease. They are also meant to help us avoid consuming excess nutrients.*

DIETARY REFERENCE INTAKES (DRI)

DRI Values for Most Nutrients

Reference Value	Description	When Planning Your Diet
Recommended **D**ietary **A**llowance (**RDA**)	The nutrient intake that is sufficient to meet the needs of nearly all healthy individuals in a given age and sex	✓ Goal for **average** daily intake over time
Adequate **I**ntake (**AI**)	Nutrients for which the available data are not sufficient to confidently determine an EAR (and thus an RDA). The AI is the best estimate of the amount that is adequate to meet the needs of the majority of the population based on available data.	✓ Aim for this if an RDA isn't available
Estimated **A**verage **R**equirement (**EAR**)	The nutrient intake that is estimated to meet the needs of 50% of healthy individuals in a given age and sex	✗ Do not use this amount
Upper **L**imit (**UL**)	The highest level of daily nutrient intake that is unlikely to cause adverse effects for nearly all individuals in the population	✗ Do not exceed this amount from all sources, including fortified foods, supplements, and prescription drugs

Additional DRI Values for Energy and Macronutrients

Reference Value	Description	When Planning Your Diet
Acceptable **M**acronutrient **D**istribution **R**ange (**AMDR**)	Intake ranges for energy yielding macronutrients that are consistent with good health, expressed as a percent of total calories	✓ Follow these guidelines for the percent of calories from carbohydrates, fat, and protein
Estimated **E**nergy **R**equirement (**EER**)	The average energy intake predicted to maintain current body weight in a healthy adult of a specific age, sex, weight, height, and activity level. 50% of individuals will have energy needs higher or lower than this value.	✓⚠ Use cautiously as an initial planning goal only

What three reference values provide recommendations for nutrient intake that we should strive to achieve?

When there is insufficient evidence to generate an EAR for a nutrient, the RDA for that nutrient cannot be set and the committee establishes the third set of values: the **Adequate Intake (AI)** value, based on research or observations of the amount of the nutrient individuals typically need. For example, the B vitamin called folate has an established RDA, but the recommended intake for biotin, another B vitamin, is currently provided as an AI.

Because store shelves are now overflowing with food and supplements that supply extra amounts of nutrients, the DRIs also set safe limits of nutrients with its final set of values: **Tolerable Upper Intake Levels (ULs)**, which represent the highest amount of a specific nutrient that most people can consume daily without causing any harm. As research evolves, existing DRIs are modified, and new DRIs for additional nutrients are created. DRIs have been set for macronutrients, micronutrients, electrolytes, and water. (INFOGRAPHIC 1.6)

Putting these all together creates a more complete picture of each nutrient. For example, for adult women and men the EAR for vitamin A is 500 micrograms per day (mcg/day) and 625 mcg/day, respectively; the RDA is 700 mcg/day and 900 mcg/day, respectively; and the UL is 3,000 mcg/day for both sexes. Deficiency or excess of vitamin A

ADEQUATE INTAKE (AI) *estimated value for recommended daily nutrient intake level used when there is insufficient evidence to determine a specific RDA*

TOLERABLE UPPER INTAKE LEVEL (UL) *the maximum amount of nutrient allowed that has been proven to have no risk of side effects*

INFOGRAPHIC 1.7 Adequate and Safe, or Not? *When the intake of a nutrient falls between the RDA and the UL (in green portion of the graph), it is highly likely that intake is both adequate and safe.*

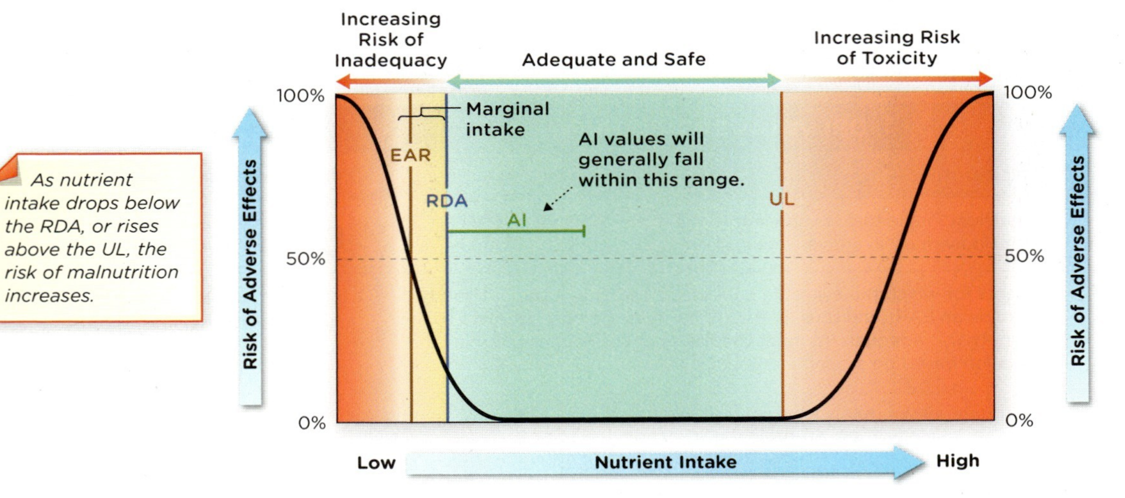

? *What are the two causes of malnutrition shown in this figure?*

can injure the human body. Deficiency impairs immune function, and leads to night blindness, and eventually permanent blindness. At the other end of the spectrum, intakes of vitamin A above the UL cause liver disease and can cause serious birth defects. **(INFOGRAPHIC 1.7)**

ENERGY RECOMMENDATIONS

Two recommendations about energy intake are also included in the DRI. The first is the **Estimated Energy Requirement (EER)**, which is similar to the EAR in that it represents the *average* amount of calories a healthy person of a particular age, sex, weight, height, and level of physical activity needs in order to maintain his or her weight. This value meets the energy requirements of 50% of the population, and actually exceeds the needs of nearly 50% of individuals.

An imbalance in macronutrient intake (particularly of fat and carbohydrates) can increase the risk of several chronic diseases. For this reason the **Acceptable Macronutrient Distribution Ranges (AMDRs)**, have been established to provide a healthy range of intakes for carbohydrates, protein, and total fat (also some specific types of

ESTIMATED ENERGY REQUIREMENT (EER) *estimated number of calories per day required to maintain energy equilibrium in a healthy adult; this value is dependent on age, sex, height, weight, and level of physical activity*

ACCEPTABLE MACRONUTRIENT DISTRIBUTION RANGE (AMDR) *the range of energy intakes that should come from each macronutrient to provide a balanced diet*

Enough pasta! *The AMDRs have been set to provide a healthy range of intakes for carbohydrates, protein, and total fat.*

fat) expressed as a percentage of total calories. Adults can obtain adequate amounts of macronutrients when their carbohydrate intake falls between 45% and 65% of total calories, their protein intake falls between 10% and 35% of total calories, and their fat intake is within 20% and 35% of total calories. Individuals who regularly eat below or above these ranges, put themselves at risk of getting too few essential nutrients or of developing chronic diseases.

But who knows how to use the DRIs to determine what they should eat day to day? Given how difficult it can be to analyze individual diets, most people who use DRIs as a nutritional tool are researchers and registered dietitians. However, some publicly available computer programs and software applications can track meals and provide dietary information, making DRIs a tool that nonexperts can use, too. We will explore some other useful tools in Chapter 2.

■ ■ ■

The DRIs vary depending on the specific nutritional needs of different groups of people—children, adults, men, women, the elderly, and women who are pregnant or nursing. It is this last group that is often most concerned with getting all the nutrition they need.

This is something Robert Waterland experienced first-hand, working in the laboratory of obesity researcher Albert Stunkard as an undergraduate at the University of Pennsylvania in the early 1990s. Waterland was studying physics in college when he read a book that changed everything. It described how diet could boost athletic performance (including marathon running, Waterland's then-pastime). "I started thinking, 'Wow, nutrition is really powerful, and has such an impact on health.' That's how I decided I wanted to study nutrition."

During his time in Stunkard's lab, Waterland helped set up a research study of infants born to mothers who were either of healthy weight or obese. The moms

Robert Waterland, PhD, epigenetics researcher

Courtesy Dr. Robert Waterland

brought their babies into the laboratory, where Waterland and his team spent the day measuring how much food the babies consumed, checking their body composition, and even observing *how* they ate—their sucking style while breast- or bottle-feeding, for instance. The purpose, he says, was to see if any of these variables, measured in infancy, would predict a baby's likelihood of also becoming obese later in life.

Since the moms had to spend all day in the lab, including even the babies' naptimes, they would sometimes sit down with Waterland to chat. Over time, a few of the obese moms asked him if there was anything they could do now, when their babies were very little, to prevent them from becoming obese later in life. "The first time someone asked me, I didn't think much of it," recalls Waterland. "But the second time, I got intrigued."

He went to the library, scanned the scientific literature, and discovered the Dutch famine study. From then on, he was hooked on how early nutrition—even in the womb—could influence a child for life. "For me, that study was the beginning. It just totally captivated me—how is it possible that a person's body can remember an exposure as a fetus or infant, and have it affect their chances of being overweight as an adult?"

Waterland's experimental mice. *One mouse is large and yellow-coated, the other is smaller and dark-coated.*

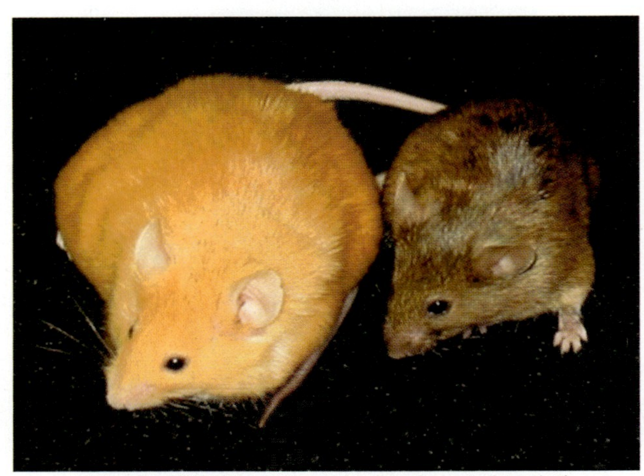

Waterland RA, Epigenetic mechanisms in gastrointestinal development. J Pediatr 2006; 149:S137-S142.

After working in Stunkard's lab, Waterland entered graduate school to continue studying the multigenerational effects of nutrition. During that time, another study was published that deepened his interest even further. This study showed that the diet of pregnant mice appeared to affect the coat colors of their young. Normally, when this strain of mice gives birth, the litter contains babies with coats in a range of colors—some yellow, some brown, and all shades in between. The babies with yellow fur were more likely to eventually become obese and develop diseases, such as cancer, than their brown brothers and sisters. But when scientists fed pregnant mice supplements that contained a mix of folic acid, vitamin B_{12}, and other nutrients, moms gave birth to fewer babies with yellow fur. "That caught my attention," recalls Waterland. To him, this study result was even more compelling evidence that what mothers eat during pregnancy can have lifelong effects on their children.

The field of *genetics* describes how genes encoded in DNA are passed on between generations (from parents to child), but sometimes the DNA in our genes can become modified after it's inherited, which can change traits in the current generation (as well as in subsequent ones). This explains why identical twins—who have the exact same genes—actually have slight differences in their appearance and risk of disease. This fascinating area of study that looks at the cross-generational effects of exposure to nutrients, toxins, and behaviors is called *epigenetics*.

SCIENTIFIC METHOD *a specific series of steps that involves a hypothesis, measurements and data gathering, and interpretation of results*

Waterland was interested in epigenetics; he simply had to understand more about why a mother's diet during pregnancy could affect the future health of her children.

NUTRITION SCIENCE IN ACTION

To begin his investigation, Waterland, like all scientists, followed the **scientific method**. **(INFOGRAPHIC 1.8)** He started with an *observation* and then identified a question or problem to investigate further. In his case, the observation that diet in pregnancy changed mouse babies' coat color and overall health led him to wonder how that occurred. So he talked to his boss at the time, Randy Jirtle at Duke University in North Carolina, and said he wanted to repeat the mouse study, and try to explain this transmission from generation to generation.

Forming a hypothesis

Waterland came up with a testable *hypothesis,* which is a proposed explanation for an observation that can be tested through experimentation. Based on the previous findings, Waterland hypothesized that supplementing the diet in pregnant mice affected mouse babies' coat color and overall health, so his hypothesis maintained that supplementing the diet of pregnant mice with specific nutrients would increase the frequency of the healthier brown-coated offspring. (Waterland and others have found that the high intake of the nutrients used in this study can also have additional affects that are potentially harmful.)

Conducting an experiment

Next, Waterland completed the *experiment* stage of the scientific method. He repeated the mouse experiment, feeding some pregnant mice a supplemented diet, and obtained the same results. Moms who ate the extra nutrients had more babies with brown fur, which previous research had found were at lower risk of obesity and other diseases than those with yellow fur. After performing DNA analyses he found that the supplements were directly responsible for these epigenetic changes. "When I first saw those results that was when I felt just, 'Wow, we really have

INFOGRAPHIC 1.8 The Scientific Method *Scientists follow these steps when pursuing the answers to scientific questions. The best conceived experiments typically generate new questions.*

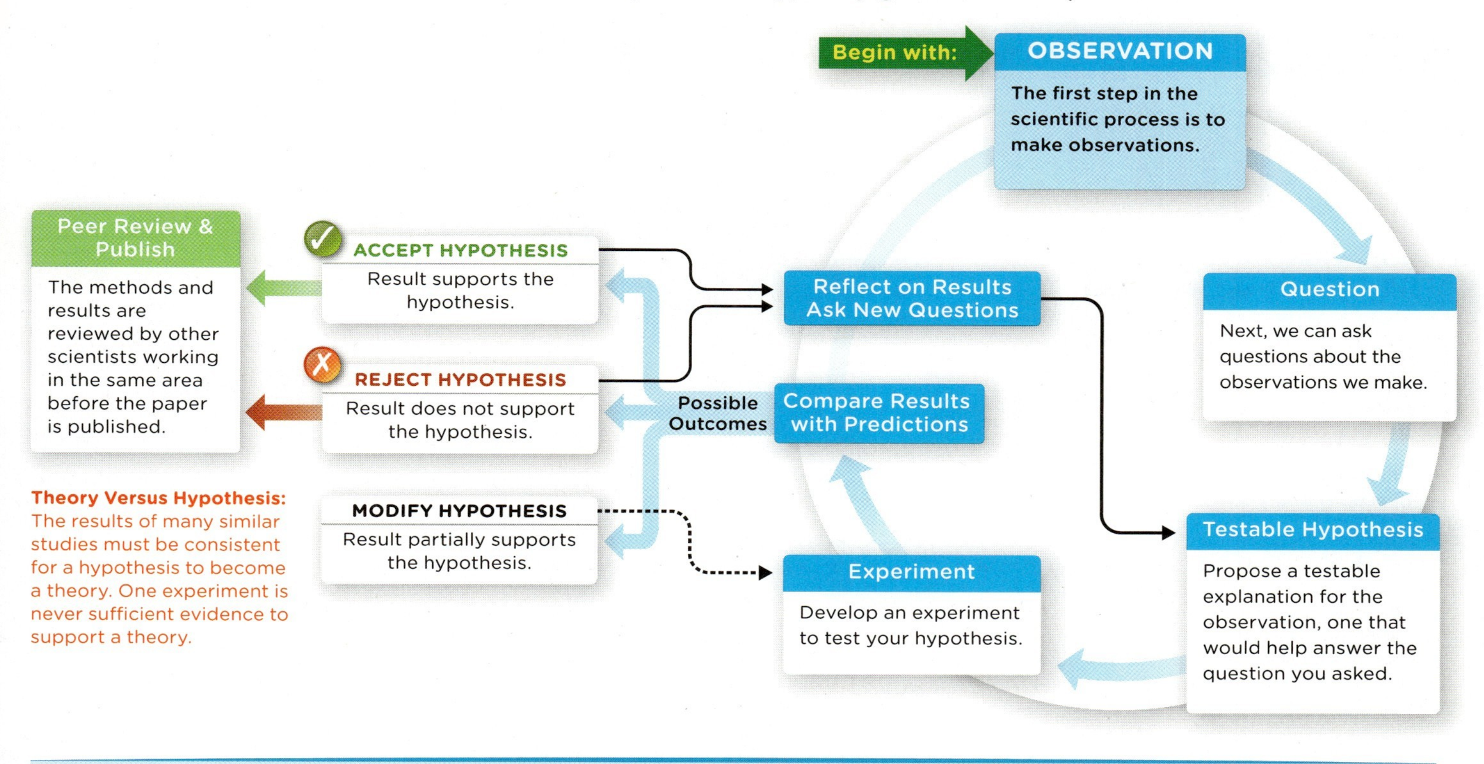

? *Explain why the following hypothesis is not suitable for the design of an experiment: "Dietary factors affect the risk of cancer."*

something here.'" As a final step in the scientific method, Waterland *published* his findings in a scientific journal, so that his findings became another core source of evidence supporting the effect of food on genes—a field known as *nutritional genomics*.

The study contained many of the important elements of a research paper. It had a *control group*, or a group of pregnant mice that received a healthy diet that did not contain any extra supplementation. Waterland then compared the results from the control group with those from mice in the *experimental* or *treatment* group, which received the supplemented diet, to determine that the supplemented diet did produce important epigenetic changes.

In other studies, such as those that look at the effect of medications, people in the control group receive a *placebo* drug, or one that contains no active properties. The purpose of the placebo is to eliminate the **placebo effect**, in which people feel better simply because they take a pill, and therefore have an expectation that they will feel better. To eliminate the placebo effect, it is also critical that people do not know what treatment they are receiving—in other words, they are to the treatment they are receiving. By comparing people receiving a treatment with those taking a pill that has no effect, researchers can determine if the treatment has a true effect, outside of people's expectations. Often, placebos are a key element of *randomized controlled trials* (RCTs), which rigorously compare experimental interventions with controls, and randomly assign people to each category to offset any potential bias. In addition to RCTs, *epidemiological* and animal studies also help us explore and evaluate the role of diet and dietary components in health and disease. (INFOGRAPHIC 1.9)

PLACEBO EFFECT
apparent effect experienced by a patient in response to a "fake" treatment due to the patient's expectation of an effect

INFOGRAPHIC 1.9 Types of Nutrition Studies *Many kinds of research are important in our understanding of the science of nutrition.*

Randomized Controlled Trial

Subjects Randomly assigned

Experimental group receives treatment.

Control group receives placebo.

HOW IT'S DONE
Subjects are recruited and randomly assigned to an experimental group and a control group. The experimental group receives the treatment while the control group receives a placebo. When possible, the subjects are blinded to the treatment they received. At the end of the study, data are collected and analyzed to determine if the treatment altered the outcome compared with those receiving the placebo.

BENEFIT
Results are directly applicable to humans.

Epidemiologic Studies

HOW IT'S DONE
Epidemiological studies observe the association between variables in a population. However, they can never establish cause and effect relationships because variables are neither manipulated nor controlled—there is no intervention.

BENEFIT
Can assess complex interactions amongst genetic, behavioral, and environmental factors.

Experimental Model Systems

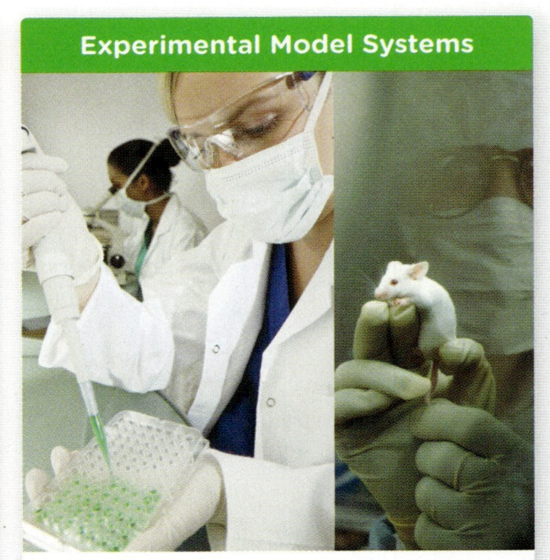

HOW IT'S DONE
Animal experiments, cell culture studies, and biochemical analyses are used as model systems to understand how nutrients and other chemical components of our diet affect physiological processes and their impact on health and disease. Much of what we know about nutrients and the effects of diet on health has been obtained from these types of studies.

BENEFIT
Can be performed when human studies would be too costly, unethical, or require an impractical length of time to complete.

An experimental protocol requires that a nutrient deficiency be induced to assess how this affects memory. What is the most appropriate study design to use to address this question?

Photo credits (left to right): 64/Ocean/Corbis, Darren Baker/Shutterstock, Mark Harmel/Getty Images

Analyzing the results

Because Waterland randomly assigned pregnant mice to receive either the control diet or the supplemented diet, and made sure the mice were the same in every other way, he could determine that the supplements *caused* the change in the patterns of chemical modification. Other studies, however, have to rely on *correlational* evidence. For instance, if Waterland had asked moms who had already given birth to recall if they had taken a prenatal vitamin during their pregnancies, then measured if there were any epigenetic changes in their babies, he might have seen a correlation, but he could never say whether the supplements caused any epigenetic changes. Moms who voluntarily choose to take a vitamin might also exercise more, drink more water, or have a higher socioeconomic status—all of which could also cause epigenetic changes.

Although Waterland is quick to emphasize that the results of his experiment only apply to mice, it is now clear that epigenetic modifications are critical factors affecting the development of many human diseases. The mouse coat color experiment, along with the Dutch famine experiment, have "absolutely affected the way we think about nutrition," says Patrick Stover, PhD, director of the division of nutritional sciences at Cornell University. "Some people seem to eat whatever they want, and never gain weight. This has always been the case in nutrition—people sometimes respond differently to the same foods.

That cliché—we are what we eat—has never really applied," says Stover. "We are just now beginning to understand how genetic differences among individuals are shaped by historical exposures, including food."

CREDIBLE SOURCES OF NUTRITION INFORMATION

As new research emerges that challenges long-held theories, how do scientists know what to trust? Often, they base their assessment of credibility on the strength of the evidence. Major factors that affect this evaluation are the quality of individual studies, the number of studies and the consistency of their results, and the magnitude of the effect. A fundamental principle of the scientific method is that similar results must be demonstrated multiple times before we can be confident that similar cause-and-effect relationships have been identified. The stronger the evidence of a relationship between specific nutrients and disease risk or prevention, the more likely scientists are to recommend that people make changes to their diet as a result of the findings.

Even nonexperts in nutrition today have access to an unprecedented amount of information about the effect of nutrients and food components on the body, genetic or otherwise. Some sources may suggest a particular nutrient is good for cardiovascular health, while other sources argue the exact opposite. So, to whom should you listen?

Experts and educators in nutrition have extensive training, often including clinical or research experience, that equip them to evaluate and translate scientific information into dietary and health advice. One recognized food and nutrition expert is the **registered dietitian**, or RD, (since 2013 RDs can also be called RDNs, which is short for registered dietitian nutritionist) who relies on the research and recommendations of scientists in nutrition, biochemistry, medicine, and the behavioral sciences to provide scientifically valid nutrition information and guidance to individuals and groups. Most RDs work in hospitals and other health-care facilities, perform research, have private practices, or have jobs in public health to help prevent and treat disease.

Quick fix in a drink? *Nutrition quackery abounds. Watch out for claims that sound too good to be true, nonscience-based testimonials, recommendations based upon just one study, or promises of quick "results."*

Paula Thomas/Getty Images

They belong to the Academy of Nutrition and Dietetics, the professional organization of registered dietitians. ("Nutritionists" are not always RDs—always check the background and credentials of people who call themselves nutritionists before seeking or following their advice).

RDs are not the only source of credible nutrition information. Individuals who may not be an RD, but possess an advanced degree in nutrition, nutritional biochemistry, or other related disciplines from accredited universities can also be reliable sources of nutrition information. These individuals may work in universities as faculty (often teaching the courses that future RDs take), government agencies, or research institutions, and are those who carry out the research that informs our understanding of the impact of nutrition on health and disease.

According to the Academy of Nutrition and Dietetics, most people obtain nutrition information from the Internet, television, or magazines. The most trustworthy sources provide evidence that stems from multiple peer-reviewed publications, and has the

REGISTERED DIETITIAN
food and nutrition expert who has met the minimum academic and professional requirements to qualify for the credential

support of a group of experts in the field, described sometimes as "scientific consensus." Be wary of ".com" websites (which might be selling a product related to the nutrition "advice"), and turn mostly to sites that are managed or reviewed by qualified health professionals. **(INFOGRAPHIC 1.10)**

HEALTH GOALS FOR AMERICANS

Health professionals use data provided by government health agencies to understand how effective nutrition and health programs are nationwide. Healthy People 2020, a government-sponsored initiative, identifies measurable health improvement objectives and goals for Americans. The nutrition-related objectives recommend that we should consume a variety of nutrient-dense foods within and across different food groups, emphasizing whole grains, fruits, vegetables, low-fat dairy products, and lean meats and other sources of protein. Americans should eat only as many calories as they need, and should limit their intake of saturated fats, trans fats, cholesterol, added sugars, sodium (salt), and alcohol. Additional information about Healthy People 2020 can be found at www.healthypeople.gov.

INFOGRAPHIC **1.10** **Credible Sources of Nutrition Information** *These agencies provide credible information because their positions rely on the results of peer-reviewed scientific publications and/or the consensus of many credentialed professionals (typically PhDs and MDs).*

Government & Private Agencies/Credentialed Expert Advice

Source	Examples
Nonprofit, Professional Health Organizations	• American Heart Association • American Cancer Society • Academy of Nutrition and Dietetics • American Diabetic Association • American Institute for Cancer Research
Scientific Organizations	• National Academy of Science • American College of Sports Medicine • The Obesity Society • Institute of Medicine (under National Academy of Science)
Government Publications: Nutrition, Diet, and Health Reports	• National Institutes of Health • Surgeon General • Food and Drug Administration • Centers for Disease Control • USDA Food and Nutrition Information Center • USDA Center for Nutrition Policy and Promotion • NIH: National Center for Complementary and Alternative Medicine
Registered Dietitians	• Hospitals • Public Health Departments • Extension Service
Other Nonprofit Organizations	• Sense About Science • Healthwatch-UK

When searching the Internet for reliable information you can limit your search to university and government websites by entering site.edu, or site.gov, respectively.

Publications

Source	Examples
Scientific, Peer-reviewed Journals	• *Obesity* • *American Journal of Physiology: Endocrinology and Metabolism* • *Diabetes Care* • *American Journal of Clinical Nutrition* • *Annual Review of Nutrition* • *Journal of the Academy of Nutrition and Dietetics* • *Journal of Nutrition* • *British Journal of Nutrition* • *Journal of the American College of Nutrition* • *Journal of the American Medical Association* • *European Journal of Nutrition* • *Diabetes* • *Lancet* • *New England Journal of Medicine* • *Journal of the American Medical Association* • *Journal of Clinical Investigation* • *Nature* • *Science* • *Public Health Nutrition* • *International Journal of Sports Nutrition and Exercise Metabolism* • *Medicine & Science in Sports & Exercise*
Other *(Although not peer-reviewed these publications rely on the expertise of the faculty within each of these universities for their content)*	• *Tufts Health and Nutrition Letter* • *Harvard Health Letter* • *Berkeley Wellness Letter*

Why are the three publications that are "letters" considered credible sources of information despite not being peer-reviewed?

***NHANES* Gathering information.** *A researcher conducts a 24-hour recall survey as part of the National Health and Nutrition Examination Survey*

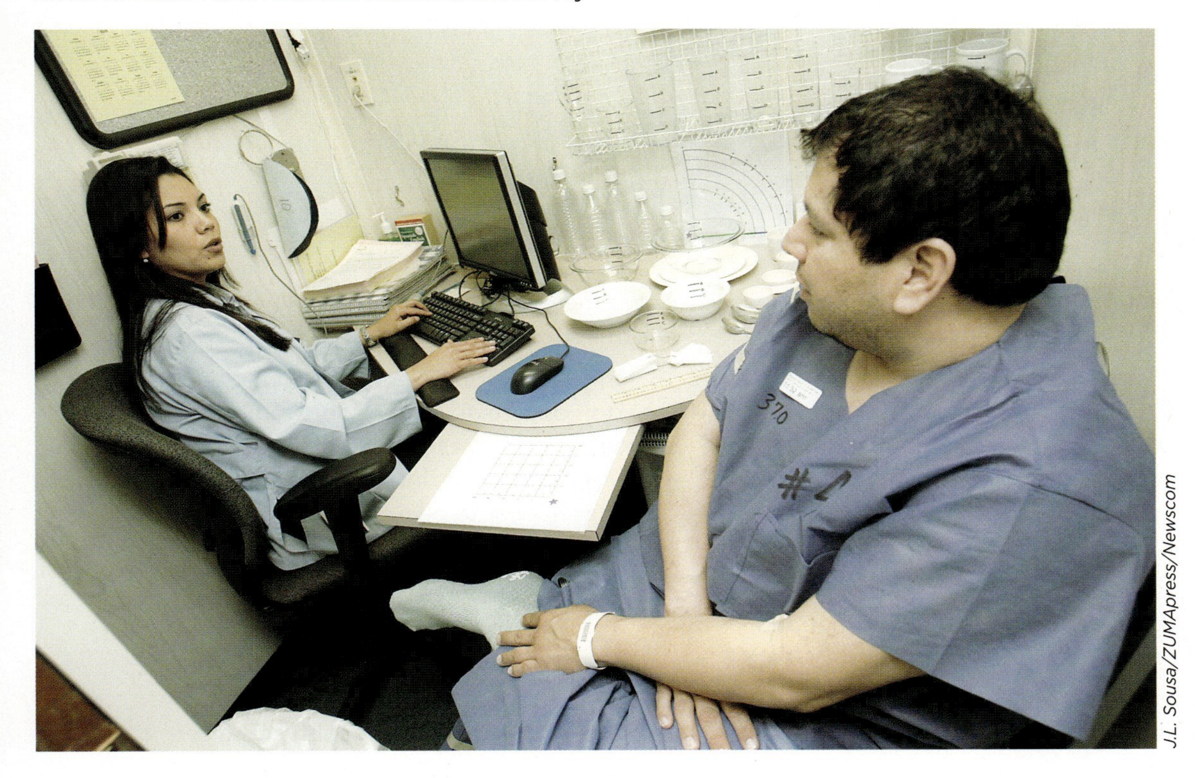

J.L. Sousa/ZUMApress/Newscom

ASSESSING THE NORTH AMERICAN DIET

Unfortunately, the dietary goals set forth in Healthy People 2020 don't align with what many Americans eat. Over the years, the U.S. government has monitored the health of Americans through the **National Health and Nutrition Examination Survey (NHANES)**, a survey that collects data about food and nutrient consumption. And the picture isn't always pretty.

Although the U.S. food supply is abundant, we are experiencing major shortfalls in certain essential nutrients and excesses in others. Approximately half of Americans don't get enough of the mineral magnesium, another 40% have low vitamin A intake, while another 30% are not getting enough vitamin C. We consume less than half of the recommended amount (at least seven cups per week) of red, orange, and deep-green vegetables. Snacks provide one-third of all daily calories from empty calories (calories with few nutrients)—namely, solid fats (butter, shortening, fat in meat), added sugars, (syrups and other caloric sweeteners) and refined starches. For adults and children older than two years, nutrition surveys reveal that most Americans significantly exceed recommended limits for these empty calorie foods. Grain-based desserts, such as cake or cookies, account for a greater proportion of daily calories than any other food group.

These nationwide data are very concerning to nutrition scientists, including Waterland, now an associate professor at the USDA/ARS Children's Nutrition Research Center at Baylor College of Medicine.

Although Waterland studies the effects of nutrients on future generations, he urges us to eat in a way that promotes our own health and well-being. Waterland explains: "The more variety of things you eat, the better, because there's still so much that we don't understand about nutrition," says Waterland. "The safest bet is to eat a wide variety of foods, so you're exposing your body to all the potential benefits. Not just the epigenetic benefits, but others that we know about, and still others we haven't discovered yet."

NATIONAL HEALTH AND NUTRITION EXAMINATION SURVEY (NHANES)
a group of studies designed to assess the health and nutritional status of adults and children in the United States

CHAPTER 1 **BRING IT HOME**

Evaluating sources of nutrition information

The Internet is full of nutrition information—however, not all of it is valid, useful, or accurate. Before you follow any nutrition advice, carefully evaluate the source.

1. On an Internet search engine, look up a nutrition question that interests you. For example, "Do some foods make you smarter?" "What are the best food sources of calcium?" What article did you select?
2. On what kind of Web site does the information appear? Is it a .com, .edu, or .gov?
3. Is the author of the article listed? If not, can you determine who might have written it?
4. Is there any evidence that the author of the information has some authority in the field of nutrition or health? What are the author's qualifications, credentials, and connections to the subject?
5. Who is the intended audience for this information? Is the audience a scholarly one? If so, does the article include a clear bibliography that you can consult for further sources?
6. Are there clues that the author is biased? For example, is the author selling or promoting a product? Is the author taking a personal stand on a social or political issue? Describe how the the article is biased, or provide evidence that it is unbiased.
7. How timely is the information? When was the information published?
8. To verify the credibility of the source, search for additional texts on the topic to find related sources, and sources that cite this source to better understand its intellectual value. Overall, does this appear to be a credible article? Explain.

Take It Further

The Academy of Nutrition and Dietetics publishes Position and Practice Papers on a variety of nutrition topics as related to health and disease. Visit www.eatright.org to learn more about the organization. Under the "eatrightPro" tab on the home page you will find a link to the Position and Practice Papers. After accessing this link, click on "What Is an Academy Position Paper?" and browse the list of position papers. Choose one and read the abstract and skim the paper.

1. What Position Paper was of interest to you and why?
2. Was any other professional organization involved in the development and publication of this Position Paper? If so, what organization(s) partnered with the Academy of Nutrition and Dietetics?
3. List two to three pertinent facts or findings from the Position Paper abstract or article.

KEY IDEAS

Nutrition science is an interdisciplinary field that studies dietary intake and behavior, as well as the nutrients and constituents in food, their use by the body, and their influence on human health.

Nutrients are chemical substances obtained from food that are required for normal growth, development, reproduction, and tissue maintenance and repair. They include the macronutrients—carbohydrates, proteins, fats, and water—and the micronutrients, vitamins and minerals.

Phytochemicals are a dietary constituent found in plant foods that are beneficial to human health.

Unbalanced nutrition, or malnutrition, can be the consequence of undernutrition, which may result in nutrient deficiencies, or overnutrition, which may result in accumulation of excess body fat or obesity.

Diet plays a significant role in the risk, progression, and treatment of most chronic diseases. Excess consumption of calories, animal fats, sugar, and sodium is linked with increased risk of the leading causes of mortality.

The Dietary Reference Intake values (DRIs) are issued by the Institute of Medicine and provide quantitative estimates of recommended nutrient intake for individuals on the basis of sex, age, and life stage.

The DRIs are intended to help prevent nutrient deficiency, promote overall health, prevent disease, and avoid excess intake. To support these goals and provide more comprehensive nutrient recommendations, the DRIs encompass Estimated Average Requirements (EARs), Recommended Dietary Allowances (RDAs), Adequate Intakes (AIs), and Tolerable Upper Intake Levels (ULs).

The DRIs establish Acceptable Macronutrient Distribution Ranges (AMDRs) for carbohydrates, proteins, and fats expressed as a percent of total calories.

Estimated Energy Requirements (EERs) are also established within the DRIs as average energy intakes needed to maintain weight in a healthy person of a particular age, sex, weight, height, and physical activity.

Nutrition research employs the scientific method. Research design and consistency across studies are important determinants of the quality, reliability, and applicability of study findings.

Registered dietitian (RD) (also called RDNs—registered dietitian nutritionist) is the standard legal definition for someone having extensive knowledge and training in foods, nutrition, and dietetics.

Healthy People 2020 is a government-sponsored initiative that provides science-based, 10-year national objectives for improving the health of all Americans.

The health and diets of Americans are monitored through government surveys like the National Health and Examination Survey (NHANES) that collects data about food and nutrient intake.

NEED TO KNOW

Review Questions

1. Each of the six classes of nutrients:
   a. provide energy for the body.
   b. are essential for growth, development, and reproduction.
   c. are required in similar quantities by the body.
   d. can be produced by the body if dietary intake is inadequate.

2. All of the following are true regarding phytochemicals, EXCEPT:
   a. they are classified as an essential micronutrient.
   b. there are thousands of these compounds found in foods.
   c. they are linked with reduced risk of cancer.
   d. they are found primarily in plant foods.

3. Homeostasis can be defined as:
   a. the process of digestion, absorption, and excretion of nutrients.
   b. the interaction of nutrients resulting in reduced absorption and utilization.
   c. the maintenance of a stable, internal environment in the face of external variability.
   d. the balance and distribution of carbohydrate, protein, and fat intake in the overall diet.

4. Current nutrition research and policy focuses primarily on:
   a. diseases of overconsumption.
   b. nutrient deficiency diseases.
   c. prevention of infectious diseases.
   d. dietary supplements and herbs.

5. The focus of the RDA reference value:
   a. is to only prevent deficiency diseases due to the prevalence of undernutrition.
   b. is on nutritional supplements rather than food intake.
   c. is to provide the amounts of nutrients required to restore health after major illness.
   d. is to prevent deficiency diseases as well as reduce the risk of chronic diet-related diseases.

6. All of the following are true regarding the Recommended Dietary Allowance (RDA), EXCEPT:
   a. it represents the average amount of calories allowed for maintenance of a healthy body weight.
   b. it represents the average daily amount of a nutrient that meets the needs of nearly all healthy individuals.
   c. the nutrient requirement for most people is less than the RDA for individual nutrients.
   d. it cannot be established if there is insufficient evidence to generate an EAR for that nutrient.

7. The Tolerable Upper Intake Level (UL) is defined as:
   a. the optimal level of nutrient intake a person should consume each day.
   b. the estimate of a safe daily level of nutrient intake that should not be exceeded.
   c. the amount of excessive intake consumed from supplements and fortified foods.
   d. the level at which calorie intake exceeds expenditure.

8. Randomized control trials encompass all of the following characteristics, EXCEPT:
   a. they are retrospective in nature and classified as epidemiological studies.
   b. they compare the effects of an intervention between experimental and control groups.
   c. they randomly assign people to receive or not receive an intervention or treatment.
   d. they may incorporate a placebo to avoid the expectation of results, or placebo effect.

9. The name of the professional organization of registered dietitians in the United States is the:
   a. American Association of Dietitians and Nutritionists.
   b. Academy of Nutrition and Dietetics.
   c. American College of Dietetic Professionals.
   d. National Food and Nutrition Association.

10. Healthy People 2020 is:
    a. an ongoing government survey that collects data on the intake and eating patterns of Americans.
    b. a public school-based program that translates the DRIs into food and exercise interventions.
    c. a government-sponsored initiative that sets goals and guidelines for the health of Americans.
    d. a corporate and consumer partnership aimed at improving the healthfulness of commercial food products.

Take It Further

What are three specific ways you can evaluate the credibility and scientific validity of nutrition information or claims in the media or on the Internet?

Dietary Analysis Using SuperTracker

Get Ready to Track Your Diet!

In this first SuperTracker exercise you will get familiar with the MyPlate component of the SuperTracker tool.

1. Log onto the United States Department of Agriculture (USDA) website at www.choosemyplate.gov. On the top of the home page, you will see the heading "Interactive Tools." Click this link.
2. You will need to create a profile for yourself to get a personalized diet plan. This profile will allow you to save your information and diet intake for future reference. You will be using the SuperTracker tool for the assignments in each of the chapters so choose a user name and password that will be easy for you to remember.
3. Return to the home page and click on the MyPlate link.
4. Click on the fruits category.
5. Click on the vegetables group.
6. Click on the grains group.
7. Click on the protein group.
8. Click on the dairy group.
9. Click on the oils group.
   a. What are two health benefits of eating fruit?
   b. What are two health benefits of eating vegetables?
   c. What are two examples of whole grain foods and two examples of refined grain foods?
   d. List two examples of animal sources of protein and two examples of plant sources of protein.
   e. What are two nutrients commonly found in dairy foods?
   f. According to MyPlate, what is the difference between oils and fats? List two examples of oils. List two foods that are naturally high in oils.

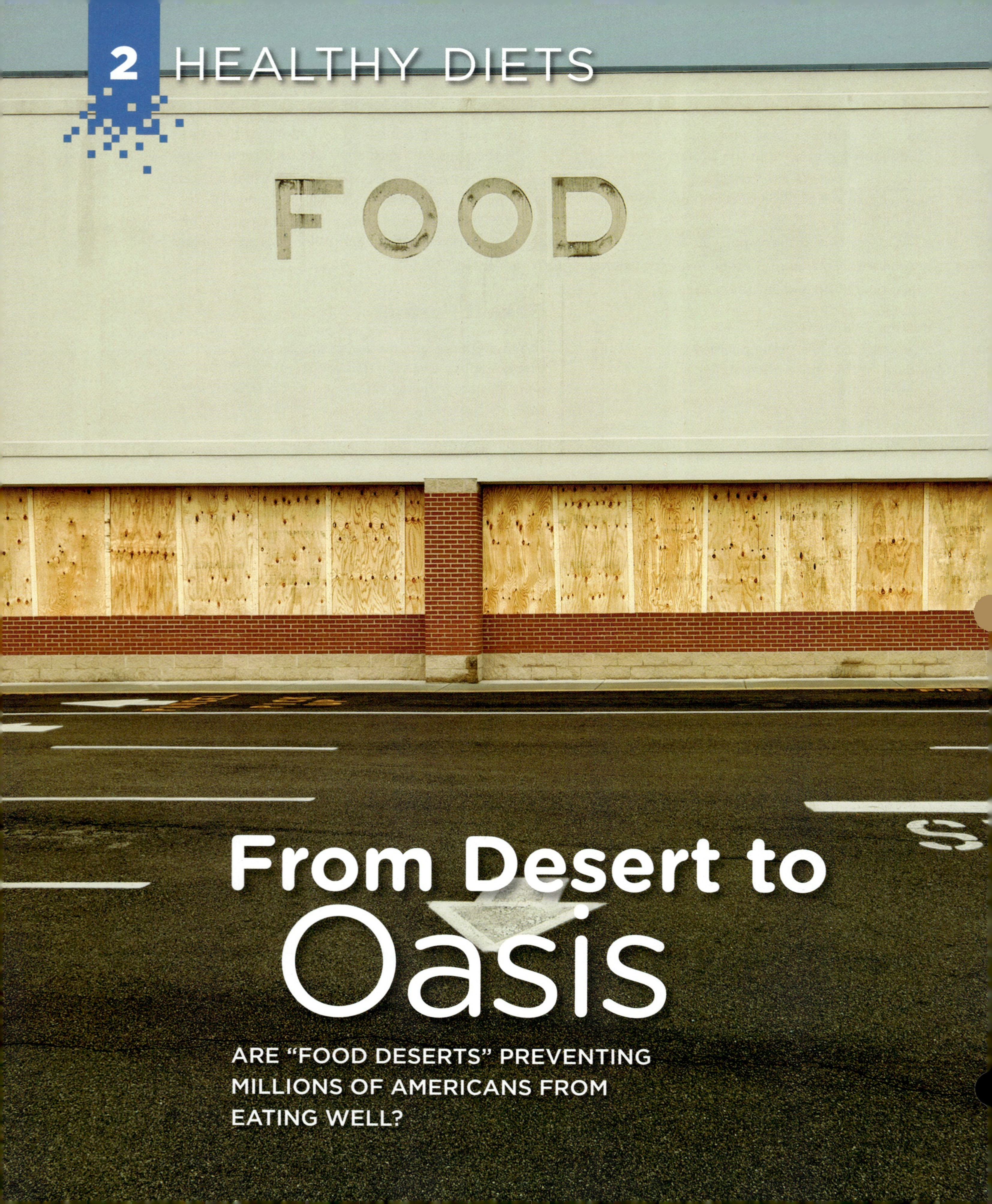

From Desert to Oasis

ARE "FOOD DESERTS" PREVENTING MILLIONS OF AMERICANS FROM EATING WELL?

In some communities, full-service grocery stores are hard to find.

Wendell and Carolyn/Getty Images

LEARNING OBJECTIVES

- Identify the primary characteristics of a healthy diet (Infographic 2.1)
- Define nutrient and energy density and describe why it is necessary to consider these factors when making food selections (Infographics 2.2 and 2.3)
- Identify the key excesses and inadequacies of the current average American diet (Infographic 2.4)
- List the core recommendations of the Dietary Guidelines for Americans and discuss the significant changes that have occurred in dietary recommendations to Americans over time (Infographics 2.5, Infographic 2.6 and 2.7)
- Describe how the USDA's MyPlate can be used to design a healthy diet (Infographic 2.8)
- Explain what characteristics of a healthy diet are common throughout the world (Infographic 2.10)
- Identify the information that is required on food labels and describe how this information can be used to select healthier foods (Infographic 2.11)
- Identify the types of claims that can be made on food labels and discuss how their use is regulated by the FDA (Infographic 2.12)

A few years ago, Mari Gallagher met a group of 9-year-old boys living in Alabama who had never seen a strawberry. As a researcher who specializes in urban health, Gallagher was speaking at a local community center about the importance of eating well. She knew many people in the area were struggling to provide their families with healthy foods. To get a better idea of what kids were eating, Gallagher asked a group of young boys what their favorite foods were. Naturally, they listed things that many kids enjoy, like potato chips, cookies, pizza,

Mari Gallagher, food researcher

Eric Shropshire Photography

and other treats. Then she asked them if they enjoyed foods like strawberries or grapes. "And they just looked at me like they didn't know what those were," she recalls. "Then they said they had never had them."

The young boys Gallagher met are among the millions of Americans living in a *food desert*—a neighborhood or community with little access to a variety of affordable, healthy foods, such as bananas, oranges, and other fresh vegetables and fruits. In urban areas, there may be corner stores with packaged snack foods and sweets. But getting to a grocery store that carries a wide variety of brands and products, along with abundant produce, dairy, and meat selections often requires a long bus ride, which some families don't have time for or the money to afford. So they rely mostly on chips, fast food, and **processed food** (often found in boxes or cans with many unfamiliar ingredients) to meet their energy (caloric) needs.

PROCESSED FOOD *any food that is altered from its raw form through processing such as canning, cooking, freezing, or milling; processing often involves adding ingredients such as sodium-containing additives and preservatives*

HEALTHY DIET *an eating pattern characterized by variety, balance, adequacy, and moderation that promotes health and reduces risk of chronic disease*

If these boys had never seen common fruits before and had limited access to other nutritious foods, chances are they were not eating a **healthy diet**. A healthy diet consists of a variety of foods, chosen from all the major food groups—vegetables, fruits, grains, protein, and dairy—and provides the energy and essential nutrients we need. All food groups are rich in certain nutrients but not others, and play an important role in dietary adequacy. Any time a food group is eliminated because of access problems (or by personal choice), it is important to understand the role that group plays nutritionally and for health promotion, and to devise a plan to meet those nutritional needs.

The catch is, even those with ready access to a wide variety of foods don't necessarily make choices that yield appropriate amounts of energy and essential nutrients that they need. Availability is only part of the equation. People make food choices based on many other factors like convenience, taste, price, emotions, and cultural and social influences to name a few.

HEALTHY DIETS FEATURE VARIETY, BALANCE, ADEQUACY, AND MODERATION

All types of healthy diets have a few qualities in common: a *variety* of foods, *balanced* across food groups and macronutrients (carbohydrates, protein, and fats), *adequate* amounts that provide the calories and essential nutrients necessary to maintain and promote optimal health. Considering the high rates of obesity in the United States, another key component to remember is *moderation*—not overindulging in any one type of food or in potentially harmful foods, such as those that contain unhealthy levels of fat, sugar, and salt. (INFOGRAPHIC 2.1)

First, let's consider variety: Quite simply, this means choosing different foods, even within the individual food groups. For example, the more varied your consumption of fruits and vegetables, the more likely you are to get a broad range of essential nutrients and health benefits from compounds, such as phytochemicals found in those foods (see Chapter 1). One way to ensure variety is to follow the advice to "eat a rainbow" by choosing foods of different colors, such as red peppers, green spinach, and orange carrots. Here's a clue that food deserts don't provide the basics of a healthy diet: Most of the food at corner convenience stores and fast-food restaurants is beige and brown—bread, soda, chips, and French fries, for instance.

INFOGRAPHIC 2.1 Components of a Healthy Diet

What core characteristic of a healthy diet is best promoted by eating a colorful diet?

Photo credit: Eli Ensor

Another important component of a healthy diet is balance: the right proportion of foods from each of the food groups, and the appropriate amounts of calories, macronutrients, vitamins, and minerals. When we eat a variety of foods from each food group, it is easiest to achieve a balanced diet. Again, living in a food desert can make it difficult to eat a balanced diet, since most of the food available has disproportionately large amounts of calories, unhealthy fats, and refined carbohydrates, such as foods with added sugar and white bread.

Indeed, food deserts often provide people with more-than-adequate calories—there are plenty of sodas, "junk" foods, and other calorie-rich options to choose from. But adequacy doesn't just mean getting enough calories; it also includes getting enough essential nutrients such as vitamins, minerals, and proteins, as well as other beneficial components of a healthy diet, such as fiber and phytochemicals. Recall from Chapter 1 that the Dietary Reference Intakes (DRIs) help us determine adequate and appropriate intake for these dietary components based on our age, sex, and life stage.

UNDERSTANDING THE NUTRIENT DENSITY AND ENERGY DENSITY OF FOODS

Healthy diets are full of foods with high **nutrient density**; these foods contain many beneficial nutrients relative to calorie content. Nutrient-dense foods are a "good deal" nutritionally in that they provide many nutrients at a low calorie "cost." Healthy diets also take into consideration the **energy density** of foods. (INFOGRAPHIC 2.2) Foods that are rich in calories relative to weight are considered energy dense. Nutrient-rich nuts fit this bill, but many foods are dense in calories and low in nutrients, such as cookies and chips. Unfortunately, food deserts are generally well-supplied with energy-dense, convenience-store foods, whereas nutrient-dense fruits and vegetables are harder to find.

The water, fiber, and fat content of foods is the primary factor that determines energy density. As the water and fiber content of

NUTRIENT DENSITY
the amount of nutrients supplied by a food in relation to the number of calories in that food; black beans, for example, provide much protein, fiber, vitamins and minerals relative to their calories

ENERGY DENSITY
the amount of energy or calories in a given weight of food; generally presented as the number of calories in a gram (kcal/g)

INFOGRAPHIC 2.2 Nutrient-dense Versus Energy-dense Foods *Eating a low energy-dense diet allows you to eat a larger volume of food, all while maintaining energy balance.*

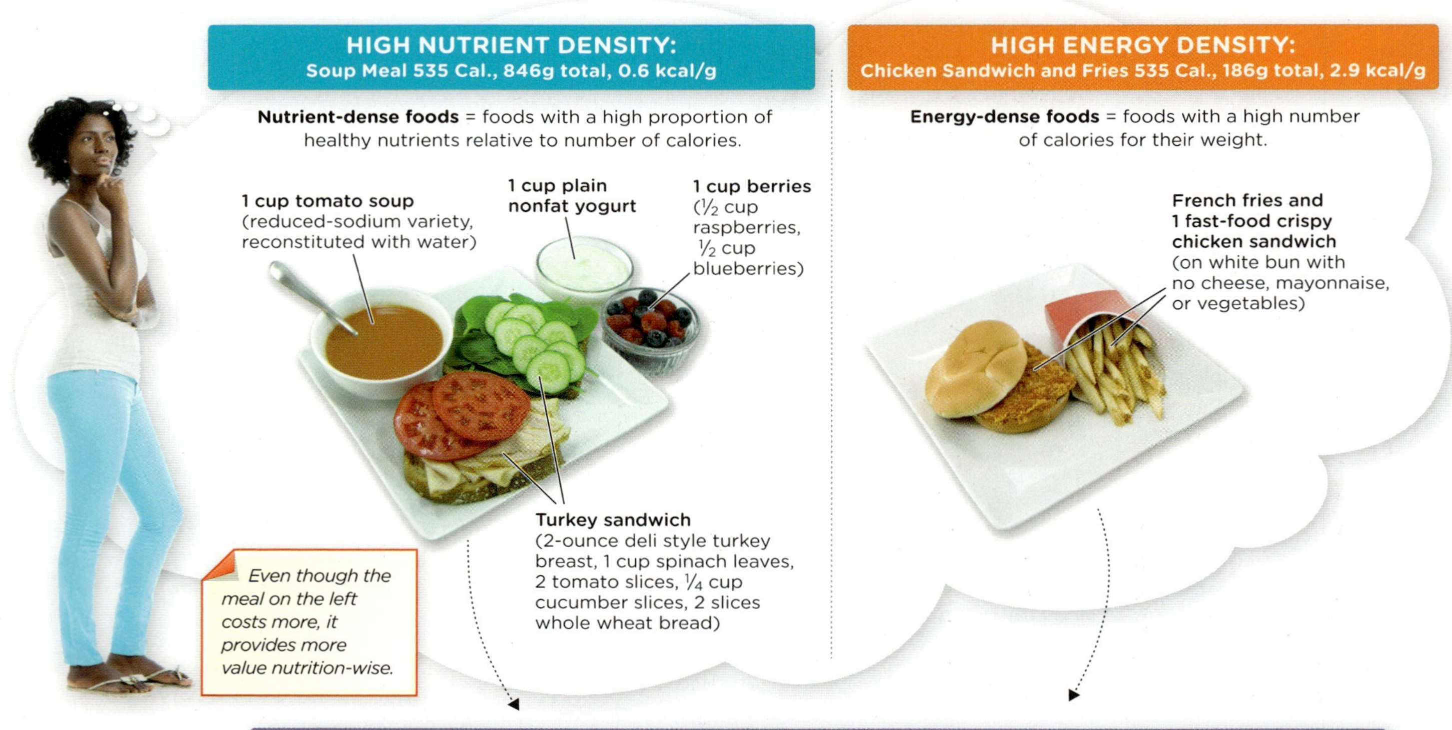

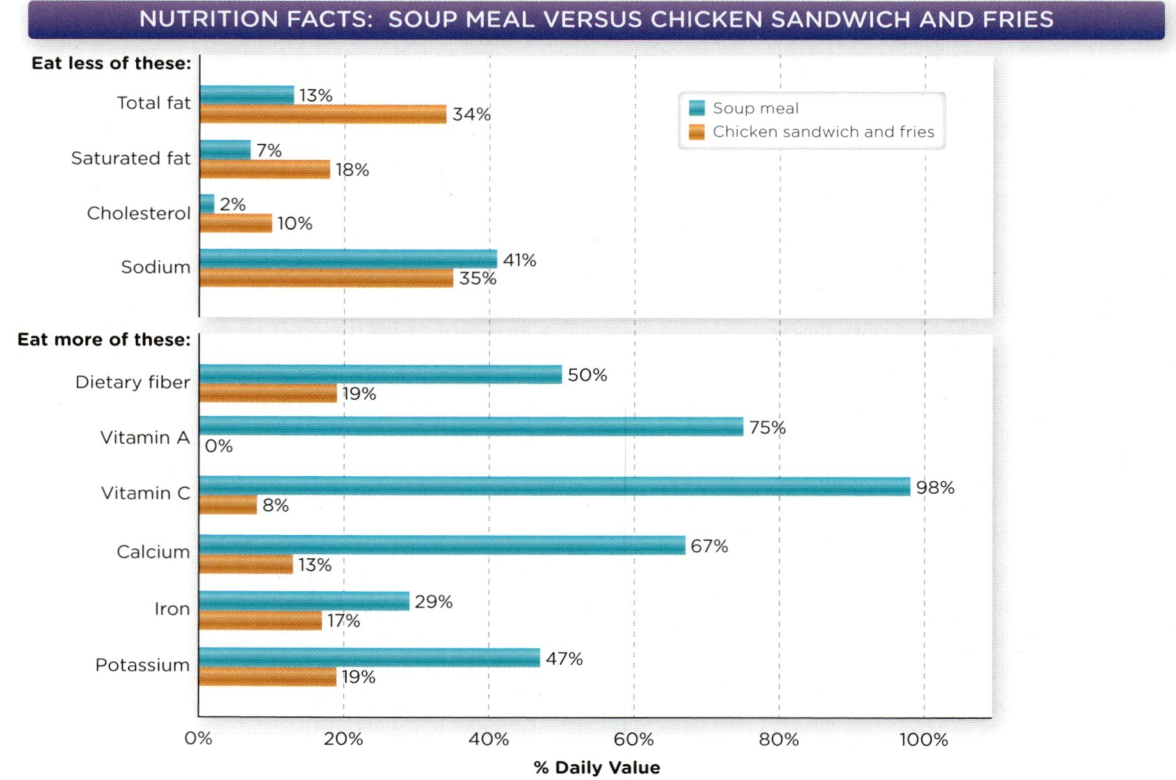

Which meal will delay the return of hunger the longest? Why?

Photo credits (left to right): moodboard/Alamy, Eli Ensor, Eli Ensor

food increases, they generally decrease the energy density of food by adding weight and volume but no (or very few) calories. Fat content has the opposite effect—the more fat is added to food, the more energy dense it becomes, since fat has more than twice as many calories per gram as either protein or carbohydrates. In general, as the energy density of foods increases, the nutrient density decreases. Most vegetables are nutrient dense as they provide lots of essential nutrients relative to their calorie content. **(INFOGRAPHIC 2.3)**

■ ■ ■

Although historically there have always been people with little access to a variety of nutrient-dense foods, the problem of

INFOGRAPHIC **2.3** **Food Composition and Energy Density** *The energy density of foods is influenced strongly by fat and water content. Increasing amounts of fat in foods increases energy density, while increasing water content in foods decreases energy density.*

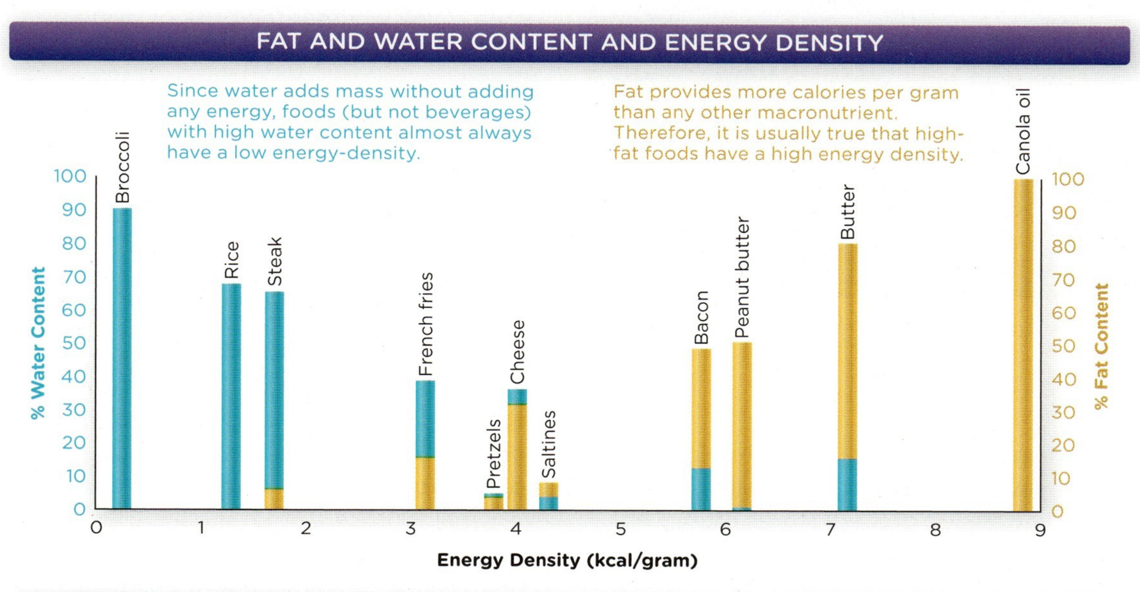

? *Why does butter have slightly fewer calories per gram than canola oil?*

Convenient but not nutritious. *Convenience stores are a source of food in neighborhoods without a full-service grocery store, but the goods they stock are often calorie-dense and nutrient poor.*

Aurora Photos/Robert Harding

food deserts in many cities started in the 1950s, says Gallagher, when wealthy residents moved away and grocery stores often went with them. Even some rural areas have become at risk of becoming food deserts, since local farms that used to provide a variety of produce have since been consolidated and converted into farms that grow mostly corn or soybeans.

One city that Gallagher studied is Chicago, Illinois. Chicago in the 1990s could be a desolate place. Beautiful historic buildings had become decrepit gathering places for drinking and street fights. Gallagher—who has a Master's degree in urban planning and community development—would pass block after block with no major grocery store. It's not prejudice that causes grocery stores to shy away from poor communities, she says, it's likely just a business decision. "If a developer goes around and sees there are no grocery stores, they assume people don't want or need one; that there is no market for grocery stores."

ADDED SUGARS *are those sugars that are added to foods during processing, food preparation, or at the table; not those that occur naturally in foods.*

But over time, Gallagher participated in projects such as a successful urban community garden, which suggested that people living in neglected areas would eat healthy foods if they became available. She began analyzing block-by-block data on the location of grocery stores in Chicago, trying to identify patterns.

In 2006, over lunch with a representative of a bank headquartered in Chicago, she casually mentioned the project she was working on. Perhaps she could analyze data to determine whether their risk of dying of various diseases is at all related to the types of nearby stores. Intrigued, "he funded it right away," recalls Gallagher.

Little did either of them know what she would find—and the firestorm it would spark.

■ ■ ■

LIMIT THESE: SOLID FATS AND ADDED SUGARS

Regardless of where you live, making food choices requires some planning and thought. Today, eating is all about options—every major grocery store is filled with innumerable types of bread, breakfast cereal, tomato sauce, and any other item imaginable. Most people can identify nutrient-dense foods like vegetables, fruits and whole grains. Realistically, though, there are many reasons why people don't always opt for nutrient-dense foods. When you're hungry, it's often easier and cheaper to grab whatever is handy, like a cheeseburger or a candy bar. Plus, after a stressful day, the salty or sweet taste can feel like a reward. Such foods get most of their calories from saturated fats and **added sugars**. Added sugars are found in sugar-sweetened drinks like sodas, energy drinks, and sports drinks, and desserts. Average Americans typically get approximately one-third of their calories from solid fats and added sugars. **(INFOGRAPHIC 2.4)**

INFOGRAPHIC 2.4 How Do Americans Stack Up Against Food and Nutrient Recommendations?

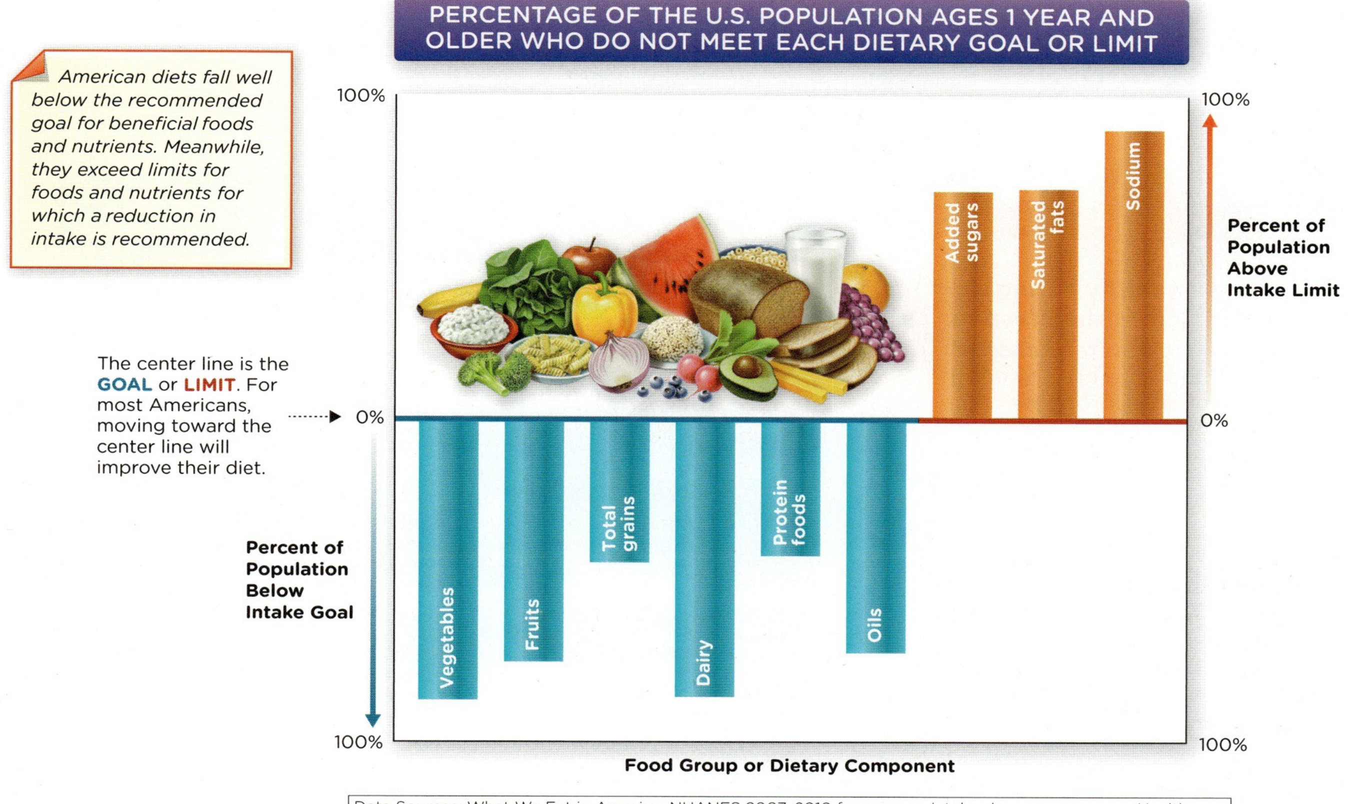

Data Sources: What We Eat in America, NHANES 2007–2010 for average intakes by age-sex group. Healthy U.S.-Style Food Patterns, which vary based on age, sex, and activity level, for recommended intakes and limits.

What foods in your diet could you decrease the consumption of to reduce your intake of added sugars and saturated fats?

Solid fats include butter, beef fat, chicken fat, pork fat, stick margarine, and shortening. The fat in milk is also considered to be solid fat. Solid fats are foods, such as butter, but also food ingredients, such as shortening in cookies or cakes, or the hydrogenated oils used to fry foods. A good way to reduce solid fat intake is to eat lean meats, and significantly reduce the intake of cheese, pizza, desserts, and pastries.

Solid fats are high in saturated fats, which are strongly linked to an increased risk of heart disease, but contribute few essential nutrients and no fiber. Solid fats contribute approximately 19% of total calories in the American diet, and reducing their consumption is an important way to reduce the intake of saturated fats, excess calories, and the risk of heart disease.

Although it's fine to eat some energy-dense foods in moderation, for many Americans, consuming excessive amounts of energy-dense foods and minimal amounts of nutrient-dense foods is a daily norm—contributing not only to unbalanced nutrition, but also to obesity and other chronic diseases.

SOLID FATS *contain high amounts of saturated fats, which make them solid at room temperature. Reducing dietary solid fats is an important way to reduce saturated fat intake and excess calories.*

INFOGRAPHIC 2.5 The 2015-2020 Dietary Guidelines for Americans *The main purpose of the Dietary Guidelines is to inform the development of Federal food, nutrition, and health policies and programs. The primary audiences are policymakers, as well as nutrition and health professionals, not the general public. MyPlate translates the Dietary Guidelines into easily implemented recommendations for the general public.*

1 Follow a healthy eating pattern across the lifespan. All food and beverage choices matter. Choose a healthy eating pattern at an appropriate calorie level to help achieve and maintain a healthy body weight, support nutrient adequacy, and reduce the risk of chronic disease.

2 Focus on variety, nutrient density, and amount. To meet nutrient needs within calorie limits, choose a variety of nutrient-dense foods across and within all food groups in recommended amounts.

3 Limit calories from added sugars and saturated fats and reduce sodium intake. Consume an eating pattern low in added sugars, saturated fats, and sodium. Cut back on foods and beverages higher in these components to amounts that fit within healthy eating patterns.

4 Shift to healthier food and beverage choices. Choose nutrient-dense foods and beverages across and within all food groups in place of less healthy choices. Consider cultural and personal preferences to make these shifts easier to accomplish and maintain.

5 Support healthy eating patterns for all. Everyone has a role in helping to create and support healthy eating patterns in multiple settings nationwide, from home to school to work to communities.

Follow a healthy eating pattern over time to help support a healthy body weight and reduce the risk of chronic disease.

A HEALTHY EATING PATTERN ***INCLUDES:***

MyPlate messages for consumers:

FRUITS
Focus on whole fruits with little or no added sugar. Enjoy fruit as a snack or dessert.

VEGETABLES
Consume a variety of vegetables from all of the subgroups-dark green, red & orange, legumes, starchy, and other. Limit the use of salt, butter, or creamy sauces.

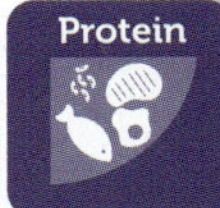

PROTEIN
Vary your protein routine. Include a variety of protein foods, including seafood, lean meats and poultry, eggs, legumes, and nuts, seeds, and soy products.

DAIRY
Substitute fat-free or low-fat milk and yogurt for cheese and sour cream.

GRAINS
Make half your grains whole grains. Limit grain desserts and snacks that contribute to intakes of added sugars and saturated fat.

OILS
A healthy eating pattern includes oils. Use oils like canola, olive, & others instead of solid fats (like butter, and stick margarine, shortening, lard, and coconut oil).

A HEALTHY EATING PATTERN ***LIMITS:***

SATURATED FATS and TRANS FATS
Limit the intake of saturated fat to 10% of total calories.

ADDED SUGARS
Limit the intake of added sugars to 10% of total calories. Drink water instead of sugary drinks.

SODIUM
Limit the intake of sodium to 2300 mg per day, 1500 mg per day for those with prehypertension or hypertension.

Give an example of a specific grain product that will often contain a significant amount of saturated fat.

Credits: U.S. Department of Health and Human Services and U.S. Department of Agriculture. 2015 – 2020 Dietary Guidelines for Americans. 8th Edition. December 2015. http://health.gov/dietaryguidelines/2015/guidelines/.

DIETARY GUIDELINES FOR AMERICANS

Sometimes, the best option isn't immediately obvious, so government agencies issue guidelines that make it easier to identify what's most nutritious. Those guidelines change with the times—during widespread economic crisis and food rationing, the focus was on adequate nutrient intake. But, today, cheap calories are more abundant than ever, so the current guidelines focus on how to get the nutrients we need to lead a healthy lifestyle and reduce the risk of chronic disease.

To help people make nutritious choices, experts created tools such as the **Dietary Guidelines for Americans (DGAs)**, which provide essential advice for how to eat healthfully and reduce the risk of chronic diseases. Since their creation in 1980, the DGAs have been updated every five years by the United States Department of Agriculture (USDA) and the Department of Health and Human Services (HHS), according to the newest science-based information about nutrition and health, and they provide the core of federal food and nutrition education programs. **(INFOGRAPHIC 2.5)**

The 2015 *Dietary Guidelines for Americans* (DGAs) focus on healthy eating patterns—the combinations of all the foods and drinks that we consume over time, and not on specific foods or nutrients. Three healthy eating patterns are provided as examples, U.S.-style, Mediterranean-style, and Vegetarian, which can be adapted to meet personal preferences. It is emphasized that all foods consumed as part of a healthy eating pattern fit together like the pieces of a puzzle to meet our nutrient needs without exceeding limits for sodium, saturated fat, added sugars, and total calories. All the pieces of this food puzzle are necessary to promote good health and prevent disease.

The current Dietary Guidelines also focus on improving the diet of Americans by encouraging small *shifts* in eating habits to align our diet with these healthy eating patterns. Emphasis is placed on the need to substitute nutrient-dense foods and beverages for less healthy choices (those that are high in saturated fat, added sugars and/or sodium, and calories). **(INFOGRAPHIC 2.6)** Specifically, the DGAs recommend that Americans consume a variety of fruits, vegetables, and protein foods (including fish, poultry, legumes, nuts, and lean meats). They also recommend that we replace refined grains with whole grains, and solid fats with oils; and limit the consumption of saturated fat (to 10% of total calories) and sodium. The limit for sodium is set at 2,300 milligrams per day for adults and children 14 years and older because excess sodium intake can increase the risk of high blood pressure (hypertension) and heart disease. For adults with prehypertension or hypertension it is recommended that they further reduce sodium intake to 1,500 mg per day to allow for greater reductions in blood pressure. For the first time, the 2015 DGAs also specifically recommend that added sugars be limited to 10% of total calories. Because added sugars contribute calories to the diet but no essential nutrients, their consumption above recommended levels can make it difficult to meet nutrient requirements without consuming excess calories. Lastly, because the environments we live, learn, and work in dramatically influence our behaviors, the Dietary Guidelines remind us that everyone has a role in supporting individuals as they shift their food and activity choices to improve their health.

To put all of this food advice into action, the USDA, the agency that regulates farming and food production, updates tools to make it easier for people to consolidate dietary advice when choosing meals. During the 1940s, for instance, the government promoted the Basic 7, in which people were encouraged to eat something from each of seven categories of foods (meat, milk, green/yellow vegetables, etc.). This system was designed to

DIETARY GUIDELINES FOR AMERICANS (DGAs)
national health guidelines that provide information and advice, based on scientific evidence, on how to choose a healthy eating plan

INFOGRAPHIC 2.6 Health Benefits Result From Small Shifts in Eating Habits *The 2015 edition of the Dietary Guidelines focuses on shifts to emphasize the need to make substitutions — that is, choosing nutrient-dense foods and beverages in place of less healthy choices — rather than increasing intake overall.*

Photo credits: Eli Ensor

provide a foundation to help Americans meet nutrient needs when food was scarce. The Basic 7 eventually became the Basic 4 in the 1960s and 1970s, which painted a simpler picture of a nutritious diet but lacked specific guidance about fats, sugar, and calories. In the 1990s, a new *Food Pyramid* ranked the five food groups by how much of each to consume per day. The iconic pyramid illustrated graphically the key concepts of variety, moderation, and proportion. After 20 years of promoting variations on the Food Pyramid, in 2011 the USDA released **MyPlate**—a new tool to help consumers make better food choices. **(INFOGRAPHIC 2.7)**

MyPlate is an illustration of a healthy meal on a "plate" divided into vegetables and fruits (slightly more vegetables than fruits), just under one-quarter lean protein sources, and just over one-quarter from grains (at least half of which should be whole grains). One serving of dairy is indicated in a serving off to the side—as a cup of milk or carton of yogurt. The website www.ChooseMyPlate.gov provides tips and tools for tracking and moderating the intake of calories, solid fats and added sugars, and more. This website is where you can access a personalized diet plan (My Daily Food Plan), with more detailed recommendations about specific foods to eat each day and each week that essentially mirror those recommendations found in the DGA. **(INFOGRAPHIC 2.8)**

But none of this advice about healthy diets means anything to families who have little or no access to nutritious foods.

■ ■ ■

In 2006, after spending months reviewing data from across Chicago, Gallagher found that more than 600,000 residents were in that very situation. They were living in food deserts—large, mostly poor areas with little or no access to grocery stores that sell healthy foods, including dairy, meat, and produce. And these residents, mostly African-Americans, were more likely to be obese and die prematurely from diabetes. The same pattern of disease was true even when Gallagher used statistical tools to remove the influence of race, income, education, and other factors that could affect health—independent of access to nutritious foods.

Soon after she released the report on her consulting company's website and issued a press release, Gallagher was inundated by calls from the media; she talked about her research on CNN. "It was the starkness of the findings that caught everyone's attention."

She realized the problem wasn't just in Chicago. In Birmingham, Alabama, she found that nearly 90,000 people live in food deserts, including more than 20,000 children **(INFOGRAPHIC 2.9)**—and those who had equal or better access to stores that sell only unhealthy foods were more likely to lose years of their lives to diseases such as cancer and others linked to dietary causes. In Washington, D.C., she found that pregnant women who lived at least one half-mile from the nearest full-service grocery store were 10% more likely to give birth to an overweight baby; when they lived at least one mile from a grocery store that risk rose to 20%.

In Pennsylvania, the entire city of Chester—with more than 34,000 residents—has not had a single full-service grocery store since 2001. "We have heard stories of a mom who had to get two taxicabs for her and her four kids to bring home all their bags from the grocery store," says Marlo DelSordo of Philabundance, a local food bank and hunger relief organization. "She could only afford to go once or twice per month."

The residents of food deserts were finding it impossible to eat nutrient-dense diets. "If you have to take a couple of buses to get a head of lettuce, it's not feasible for you to eat a healthy diet," says Gallagher.

MyPLATE
visual presentation of foods from five food groups "on a dinner plate" to represent the ideal balance that will provide a spectrum of nutrients

INFOGRAPHIC 2.7 Evolution of U.S. Food Guides and MyPlate *MyPlate is the most recent graphic representation of healthy eating; it models food choices in step with current nutrition and health concerns in the*

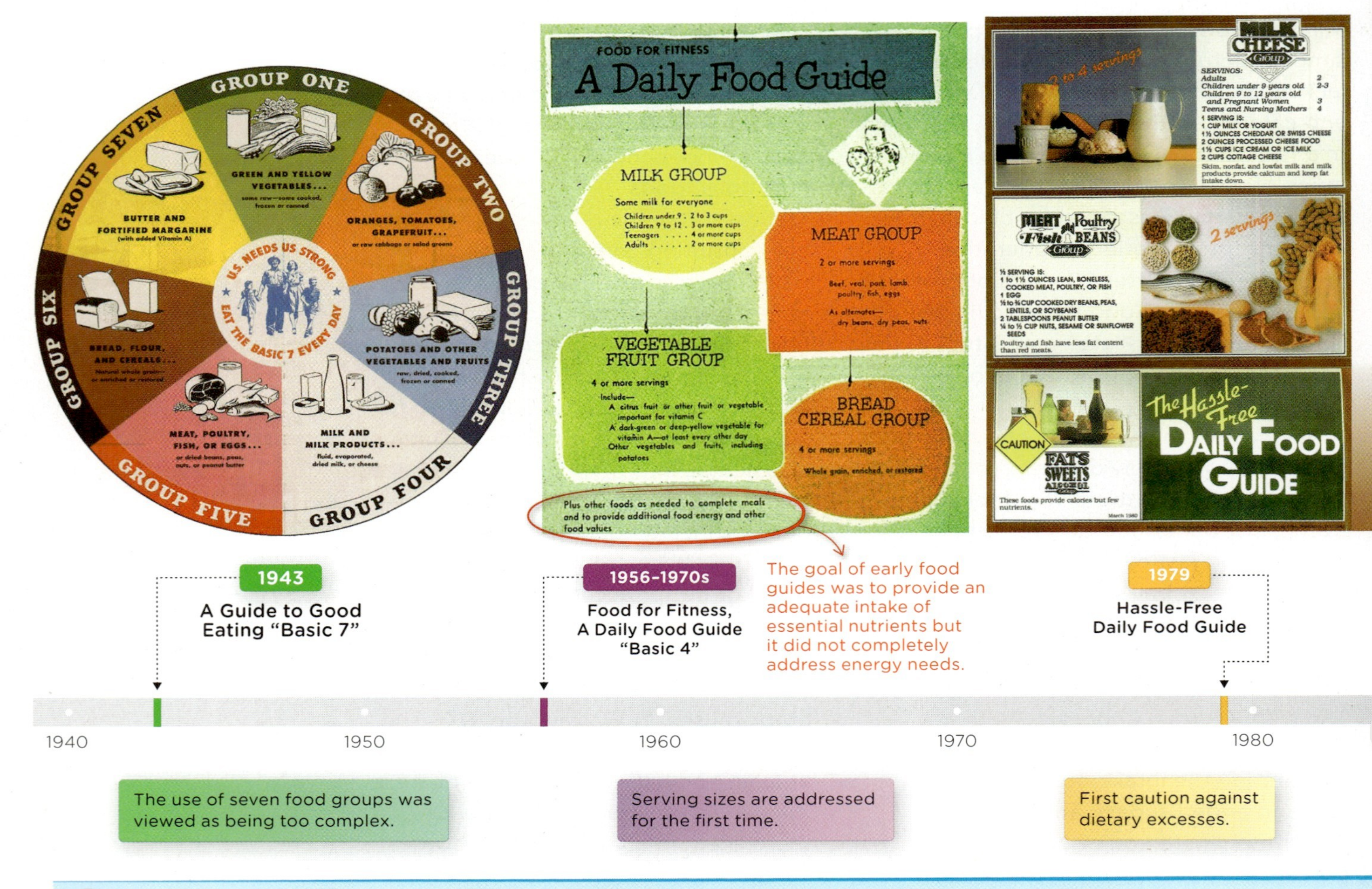

? *What are the food groups currently represented in the MyPlate graphic and in what proportion are they shown?*

Photo credits (left to right): National Archives/Records of the Office of Government Reports, USDA/National Agricultural Library, USDA/National Agricultural Library, USDA/Center for Nutrition Policy and Promotion, USDA/Center for Nutrition Policy and Promotion

GLOBAL NUTRITION

It wasn't just the United States that put food deserts on the map. In 2002, researchers described what happened to residents in a deprived area in Leeds, United Kingdom, after the opening of a major grocery store. Nearly half of people they surveyed switched to the new store as their main food source, and their diets improved.

The researchers estimated that the best measure of a healthy diet was the amount of vegetables and fruits consumed, a measure that also works when evaluating diets from around the world. Geographic location and culture have an enormous impact on what people eat—for example, Americans rarely eat rice for breakfast, although it is common in many Asian countries—but all share the same essential features.

Specifically, most healthy diet plans focus on eating more plant foods, including vegetables, fruits, whole grains, and beans. Meals should include lean proteins from a variety of sources, and healthy fats, which

United States. Dietary recommendations to Americans have evolved over time in response to societal conditions, how effective they are, and advancements in science that identify new dietary challenges that affect health.

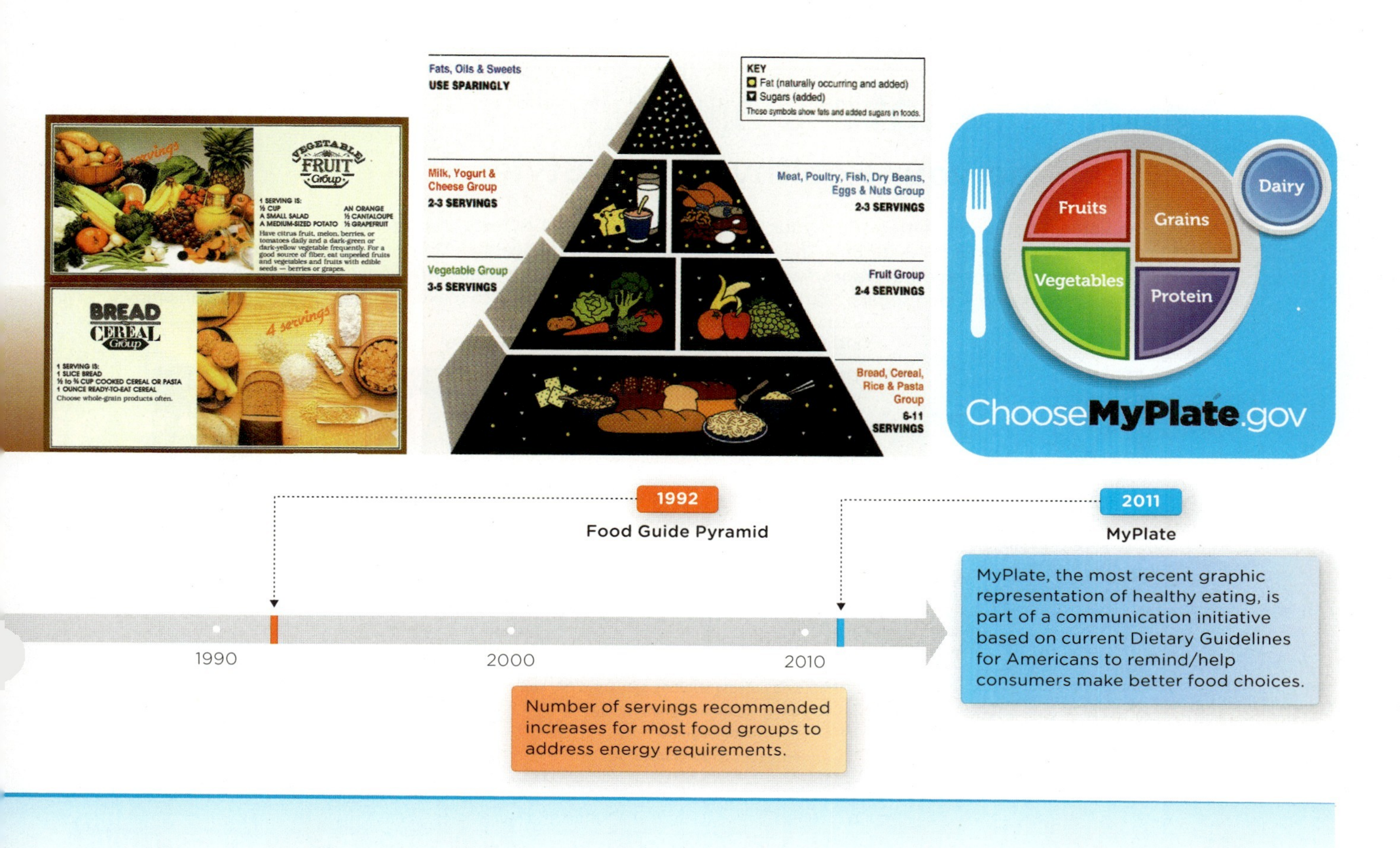

are present in foods such as avocados, olive oil, nuts, and some types of fish. Healthy eaters should minimize their intake of unhealthy fats (lard, butter, fatty meats), sugar, and salt; eat appropriate portion sizes; and stay physically active.

Obesity has become such a global problem that even the World Health Organization (WHO), the agency of the United Nations concerned with global health, has issued recommendations to countries about preparing national policies and dietary guidelines. Along with the features common to all dietary guidelines, the WHO suggests that food policies should encourage people to achieve and maintain a healthy weight, and reduce the intake of saturated fats and trans-fatty acids (also known as trans fats; see Chapter 6).

Countries don't just rely on pyramids or plates to provide advice—in France, dietary guidelines are illustrated as stairs. Foods you can eat frequently are found at the top of the stairs, those that should only be consumed in small quantities are

INFOGRAPHIC 2.8 MyPlate Daily Checklist and Portion Sizes *MyPlate Daily Checklist is an online tool provided at ChooseMyPlate.gov. The plan below is for someone requiring 2,200 kcal/day. For a food plan specific to your energy needs, use "SuperTracker" where you can enter your personal information (https://www.supertracker.usda.gov/createprofile.aspx). Not all foods within a food group count equally toward the daily intake goals because some foods are more airy or more concentrated. Use the visual cues to determine portion sizes.*

MyPlate Daily Checklist

Find your Healthy Eating Style

Everything you eat and drink matters. Find your healthy eating style that reflects your preferences, culture, traditions, and budget—and maintain it for a lifetime! The right mix can help you be healthier now and into the future. The key is choosing a variety of foods and beverages from each food group—*and making sure that each choice is limited in saturated fat, sodium, and added sugars.* Start with small changes—**"MyWins"**—to make healthier choices you can enjoy.

Food Group Amounts for 2,200 Calories a Day

Fruits 	Vegetables 	Grains 	Protein 	Dairy 
2 cups	**3 cups**	**7 ounces**	**6 ounces**	**3 cups**
Focus on whole fruits	Vary your veggies	Make half your grains whole grains	Vary your protein routine	Move to low-fat or fat-free milk or yogurt
Focus on whole fruits that are fresh, frozen, canned, or dried.	Choose a variety of colorful fresh, frozen, and canned vegetables—make sure to include dark green, red, and orange choices.	Find whole-grain foods by reading the Nutrition Facts label and ingredients list.	Mix up your protein foods to include seafood, beans and peas, unsalted nuts and seeds, soy products, eggs, and lean meats and poultry.	Choose fat-free milk, yogurt, and soy beverages (soy milk) to cut back on your saturated fat.

It is important to consume all the vegetable subgroups because each subgroup contributes a different combination of nutrients.

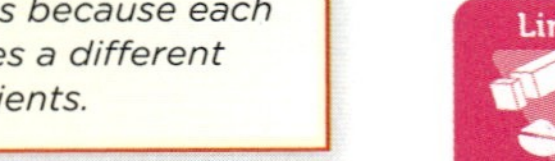

Limit

Drink and eat less sodium, saturated fat, and added sugars. Limit:

- Sodium to **2,200 milligrams** a day.
- Saturated fat to **24 grams** a day.
- Added sugars to **55 grams** a day.

VISUAL CUES FOR ESTIMATING PORTION SIZES

Golf ball = ¼ cup or 2 ounces.

Baseball = 1 cup

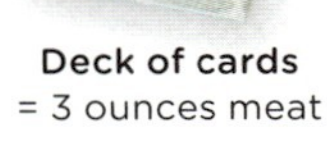

Deck of cards = 3 ounces meat

Palmful of nuts = 1 ounce

6 dice = 1 ounce cheese

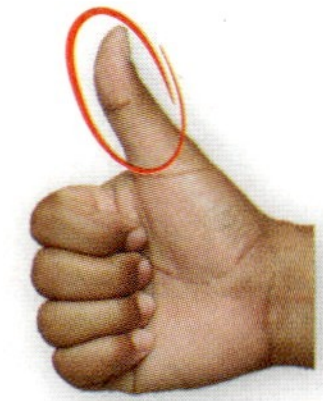

Thumb = 1 tbsp. or ½ ounce

PORTION SIZE EQUIVALENTS IDENTIFY THE AMOUNTS OF FOODS FROM EACH FOOD GROUP WITH SIMILAR NUTRITION CONTENT

Grains (what counts as an ounce?)
- 1 slice of bread
- ½ cup of cooked pasta or rice
- 8 grams of whole grains per ounce is approximately "half whole grain"
- 1 cup of ready to eat cereal
- 1 mini bagel (large bagel = 4 ounces)
- 3 cups popped corn

Vegetables (what counts as a cup?)
- 1 cup raw or cooked vegetables
- 1 cup of vegetable juice
- 2 cups of raw, leafy greens

Fruit (what counts as a cup?)
- 1 cup of fruit
- 1 cup of 100% juice
- ½ cup of dried fruit

Dairy (what counts as a cup?)
- 1 cup of milk or yogurt
- 1 cup calcium-fortified soymilk
- 1½ ounces hard cheese (cheddar, mozzarella, parmesan)
- 2 ounces processed cheese (American)
- 2 cups cottage cheese
- 1½ cups ice cream
- 1 cup frozen yogurt

Protein (What counts as an ounce?)
- 1 ounce of meat, poultry, fish
- 1 large egg
- 1 tbsp. peanut butter
- ½ ounce of nuts or seeds
- ¼ cup cooked beans or peas
- 2 tbsp. hummus

How much peanut butter counts as an ounce of protein? How would you estimate that quantity?

Photo credits (top—all): Center for Nutrition Policy and Promotion/USDA; (bottom—baseball): Pavel Hlystov/Shutterstock; (bottom—thumb): foto76/Shutterstock; (bottom—all others): Eli Ensor

INFOGRAPHIC 2.9 Food Desert Map *The two maps on the left show the proximity of full-service groceries to two groups for whom healthy food is often difficult to procure: low-income households and those without access to a vehicle. Scientists are still exploring the links between food deserts and health by investigating how the nonavailability of fresh food may spur obesity, diabetes, and other diet-related conditions (shown in the maps on the right).*

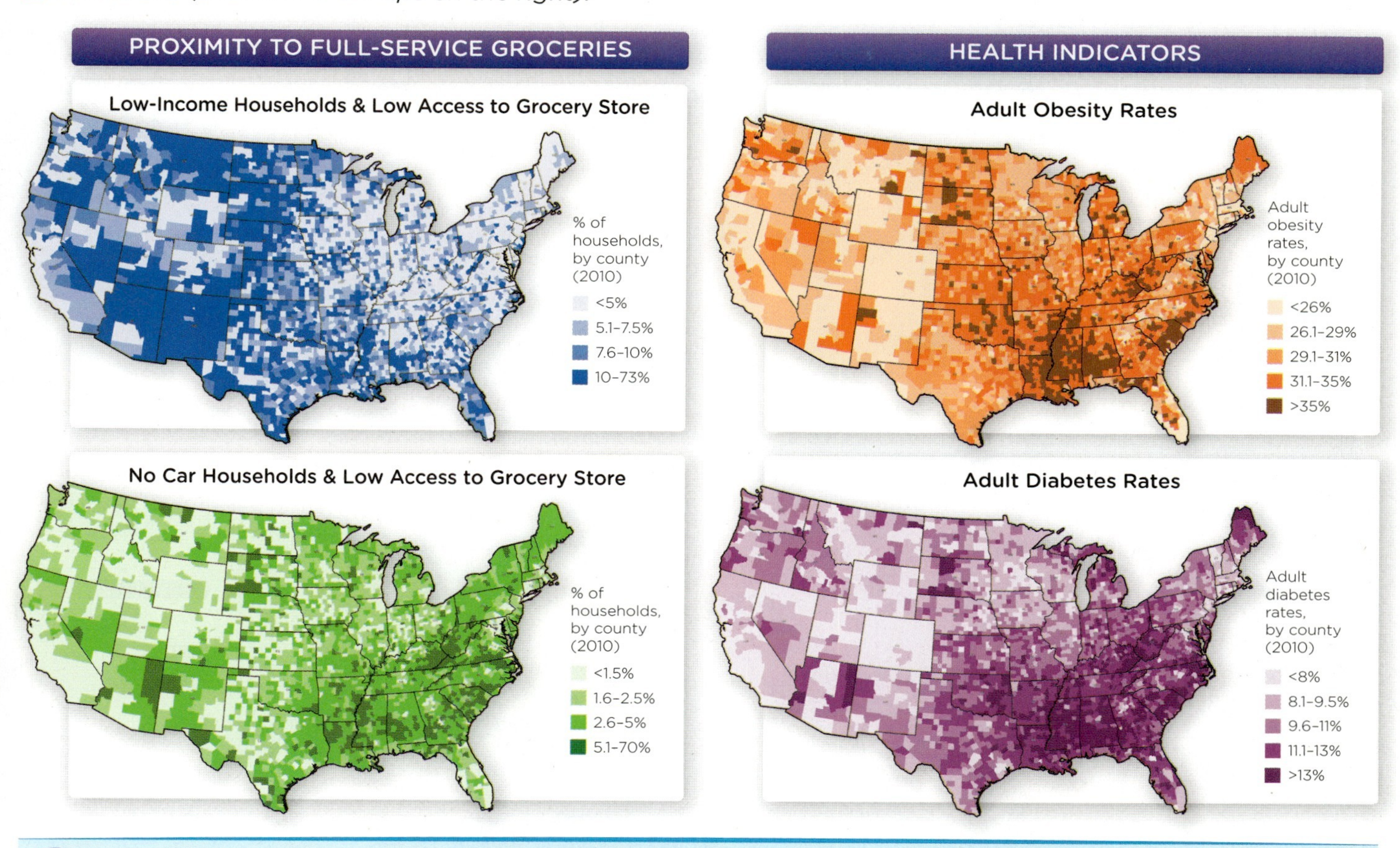

? *What areas of the country have a high levels of diabetes and also obesity?*

Source: USDA/ERS

found at the bottom, and those to avoid are hidden in a magnifying glass. **(INFOGRAPHIC 2.10)**

■ ■ ■

Back in the United States, Mari Gallagher's identification of food deserts helped put this issue on the map, but not everyone is convinced that food deserts are a significant cause of poor nutrition and obesity. To collect nationwide data, Helen Lee, PhD, a researcher at MDRC, a nonprofit social policy research organization, launched a study that compared the location of food stores with residents' income and the health of their children. In 2012, she found that poor neighborhoods did, in fact, have more fast-food and convenience stores per square mile than wealthier neighborhoods, but also more grocery stores than wealthier neighborhoods, including those that sold meat and fresh produce. Some of these might be ethnic stores that cater to certain groups, but they carried a variety of healthy options. And a sample of kids from a range of different neighborhoods followed from kindergarten to fifth grade showed that those with better access to nutritious foods in their neighborhoods were no less likely to become obese.

INFOGRAPHIC 2.10 Food Guides Around the World *Despite cultural and agricultural differences, international recommendations have common messages for their populations in choosing and consuming a healthy diet.*

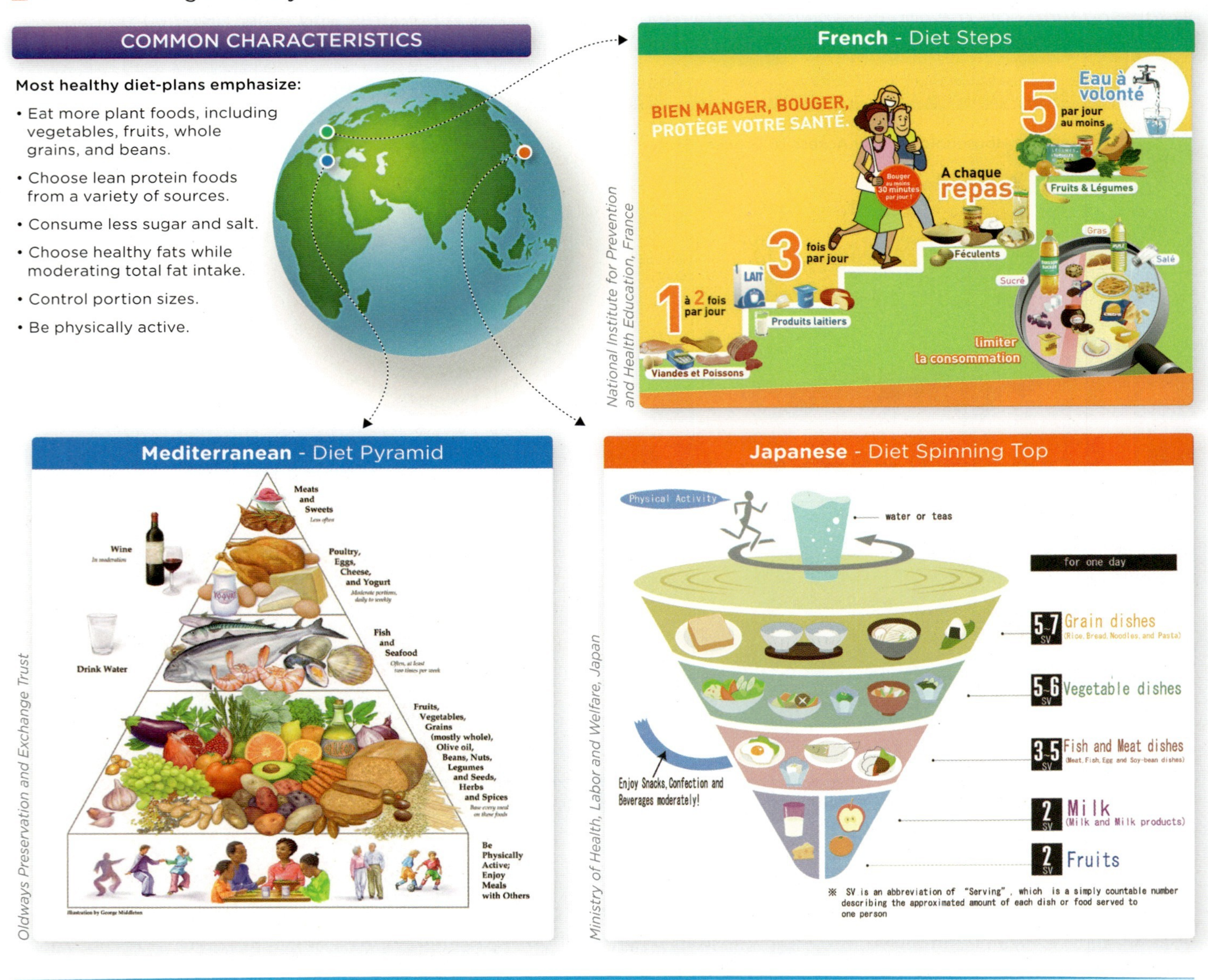

Name 3 ways each of the international diet and lifestyle recommendations differ from those provided by MyPlate.

Other research has produced mixed evidence regarding the relationship between what kids actually eat and the types of foods available to them.

The trouble with much of food desert research, explains Lee, is that it simply *correlates* food deserts to income and obesity, but doesn't show that lack of access to nutritious foods is the *cause* of residents' unhealthy eating habits. "Grocery stores have a lot of healthy foods, but they have a lot of bad ones, too. You can spend your whole budget in the chips and processed food aisles, easily." People choose foods based on many factors other than access, such as budget.

Gallagher also acknowledges that simply giving people access to fresh food doesn't mean they'll eat it—but it's a start. "We have stressed throughout the course of our work that simply plopping down a grocery store does not mean that these problems are instantly solved," says Gallagher. "But if we have access, we also have the power to choose an apple over a candy bar, at least a little more often."

UNDERSTANDING THE LABELING ON FOOD

To assist people in making nutritious purchases at the grocery store, consumer, public health, and medical organizations led a campaign in the 1980s for legislation that required **food labeling** using a standardized format.

In 1990, after several years of effort, the U.S. Congress passed the **Nutrition Labeling and Education Act (NLEA)** giving the **Food and Drug Administration (FDA)** the authority to require products sold in the United States to provide detailed nutrition information, as well as requiring nutrition and health claims for foods to comply with government standards.

Food labels must include a *Nutrition Facts Panel* that provides specific information about the calorie content and nutritional values for specific components. On every food label, you will see serving size, number of servings, and number of calories per serving, as well as information on the amount of dietary fat, cholesterol, dietary fiber, dietary sodium, carbohydrates, proteins, and at least two vitamins and minerals in each serving. (INFOGRAPHIC 2.11)

Percent Daily Value (%DV)

The **percent Daily Value (%DV)** serves as a guide to the level of key nutrients in one serving of food. Developed specifically for use on food labels, DVs can help guide consumers toward better selections. Expressed as a percentage of recommended intakes within a 2,000 kilocalories per day diet, consumers can see how foods might add up in helping them meet—or, in some cases, not exceed—dietary goals. Of course, people's daily caloric needs vary, but the %DV can still serve as a frame of reference.

For the nutrients that we sometimes eat too much of (total fat, saturated fat, cholesterol, and sodium) the DV represents the *maximum* amount to eat per day. In contrast, the DVs for carbohydrates and the nutrients listed below it on the food label (such as fiber, calcium, and iron) represent the *minimum* amount that we should consume each day, and our goal is to consume at least 100% of the DV for these nutrients. For other nutrients, like trans fats and sugars, there is not enough research to establish a specific reference DV.

Ingredients list

The NLEA also requires an **ingredients list**, which is another useful tool when making purchases, as it provides more specific information about what the food product contains. The ingredients must be listed in descending order of amount, measured by weight. Many foods include added colors; without them, colas wouldn't be brown, and mint-flavored ice cream wouldn't be green. Any pigment or coloring derived from natural materials can be listed simply as "artificial colors," but any coloring derived from synthetic chemicals must be listed by name. Other label requirements are that all juices must indicate the percentage of fruit juice content, and all foods must list any ingredient that could cause an allergic reaction, such as milk, peanuts, sulfites, and eggs.

Before 1990, manufacturers had not been required to provide details of any nutritional content, and any labeling that occurred was not consistent. Since the NLEA went into effect, not much has changed in the appearance and content of the food label; in 2006, however, the FDA began requiring manufacturers to list the amount of trans fat present in their foods. The requirement for nutrition labeling also serves as an incentive for food manufacturers to produce or modify their food products to improve the nutritional profile.

Claims on food labels

The NLEA also regulates the claims that can appear on food and dietary supplement labels that inform consumers of the health-related attributes of these products. For instance, manufacturers will make **nutrient claims** that

FOOD LABELING
the declaration on a food package that describes the nutrient content and serving size of a food

NUTRITION LABELING AND EDUCATION ACT (NLEA)
an act that allows the FDA to require nutrition labeling of most prepared foods and of dietary supplements

FOOD AND DRUG ADMINISTRATION (FDA)
government agency responsible for the supply of safe food, regulation of additives, and labeling

PERCENT DAILY VALUE (%DV)
an estimation of the amount of a specific nutrient contained in one serving, expressed as a percentage of the Daily Value, based on a daily intake of 2000 kcal; DVs were developed specifically for nutrition labels

INGREDIENTS LIST
a list of ingredients on a food package presented in descending order of amount, measured by weight, according to the guidelines set out in NLEA

NUTRIENT CLAIMS
declarations on food packages to indicate a possibly beneficial level of nutrient (e.g., "high fiber," "low fat," etc.), federally regulated to be consistent with labeling laws

INFOGRAPHIC 2.11 Navigating the Nutrition Facts Panel *The Nutrition Facts Panel provides key nutrition information, and is consistently presented on food products.*

SERVING SIZE: All the information on the label is based on a single serving. Compare the quantity you usually eat to the size of the serving on the label.

Nutrition Facts
Serving Size 4 oz cup (112g / about 1/6 Box)
(Makes about 1 cup)
Servings Per Container about 6

CALORIES (kcal): This is the amount of the total energy in one serving of the food, as well as the energy from fat. Remember that the upper limit of the AMDR for fat is 35% of calories.

Amount Per Serving
Calories 360 Calories From Fat 110

To determine the % of calories from fat: divide fat calories by total calories and then multiply by 100.

LIMIT THESE NUTRIENTS: The goal is to stay below 100% of these nutrients per day. Most Americans eat enough or too much of these nutrients. Excess consumption of these nutrients can increase the risk of several chronic diseases.

	% Daily Value*
Total Fat 12g	**18%**
Saturated Fat 0g	**18%**
Trans Fat 0.5g	
Cholesterol 20mg	**7%**
Sodium 870mg	**36%**

% DAILY VALUE: The %DVs are based on a 2,000-calorie (kcal) diet and indicate how much one serving contributes to the total daily diet (Daily Value) of these nutrients. The %DVs make it easy to compare similar products to choose healthier options. Just be certain that the serving sizes and calorie contents are similar.

Daily Value Quick Guide
- 5% or less is low
- 20% or more is high

GET ENOUGH OF THESE NUTRIENTS: Getting enough of these nutrients can improve overall health and may reduce the risk of several chronic diseases.

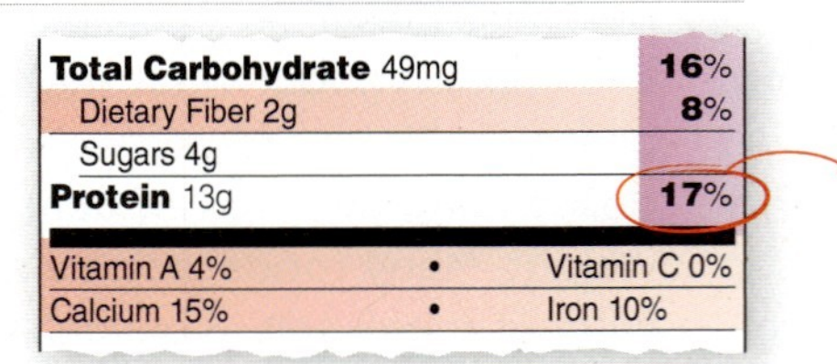

Total Carbohydrate 49mg	**16%**
Dietary Fiber 2g	**8%**
Sugars 4g	
Protein 13g	**17%**
Vitamin A 4% •	Vitamin C 0%
Calcium 15% •	Iron 10%

The %DV for protein is not required unless the product makes a protein claim, or the product is used for children or infants less than 4 years of age.

Eli Ensor

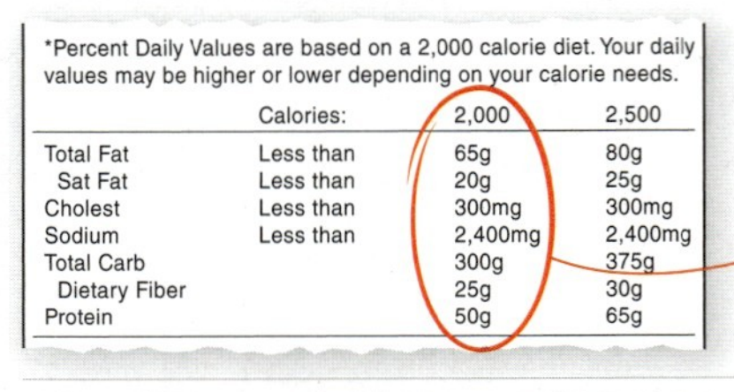

*Percent Daily Values are based on a 2,000 calorie diet. Your daily values may be higher or lower depending on your calorie needs.

	Calories:	2,000	2,500
Total Fat	Less than	65g	80g
Sat Fat	Less than	20g	25g
Cholest	Less than	300mg	300mg
Sodium	Less than	2,400mg	2,400mg
Total Carb		300g	375g
Dietary Fiber		25g	30g
Protein		50g	65g

FOOTNOTE: The footnote provides the Daily Value (DV) of several important nutrients for both a 2,000-calorie (kcal) and a 2,500-calorie (kcal) diet. Note that only some of the DVs change with different calorie diets. The portion of the footnote indicating that the "%DVs are based on a 2,000-calorie (kcal) diet" must be on all food labels. The rest of the information may not be included if the label is too small.

*Calories per gram:
Fat 9 • Carbohydrate 4 • Protein 4

CALORIES PER GRAM: Indicates the number of calories (kcal) per gram of fat, carbohydrate, and protein.

INGREDIENTS: ENRICHED MACARONI PRODUCT (WHEAT FLOUR, NIACIN, FERROUS SULFATE [IRON], THIAMIN MONONITRATE [VITAMIN B1], RIBOFLAVIN [VITAMIN B2], FOLIC ACID); CHEESE SAUCE (MILK, WHEY, WATER, WHEY PROTEIN CONCENTRATE, CANOLA OIL, MILK PROTEIN CONCENTRATE, SODIUM PHOSPHATE, SALT, CONTAINS LESS THAN 2% OF LACTIC ACID, SODIUM ALGINATE, SODIUM ACID AS A PRESERVATIVE, OLEORESIN PAPRIKA [COLOR], ANNATTO [COLOR], NATURAL FLAVOR, CHEESE CULTURE, ENZYMES)

INGREDIENTS LIST: Ingredients are listed in descending order of weight.

Calculate the percent of calories from fat and from carbohydrates for this product.

HEALTH CLAIMS
a statement on a packaged food or dietary supplement that indicates a link between a food, food component, or dietary supplement and a reduction in the risk of a disease; all health claims must be approved by the FDA

describe the level of a nutrient in a food using terms such as "low," "high," "excellent source," "light," or "reduced." The FDA regulates the use of these terms to ensure that they are used consistently across all types of food products.

Health claims describe the link between a food, food component, or dietary supplement substance, and a reduction in the risk of a disease. All health claims that appear on food products and dietary supplement labels must be approved by the FDA, which ensures they are supported by a significant amount of science before appearing on a label. Recently, the agency began allowing claims based on new research, but these must include a statement indicating that the evidence supporting the claim is limited, or even weak.

Finally, **structure/function claims** describe the role of a nutrient or dietary substance in maintaining normal structures, function, and health, but may not have any relationship to disease. "Heart healthy," or "helps support immunity" are examples of structure/function claims. Such claims are not evaluated or regulated by the FDA. (INFOGRAPHIC 2.12) Chapter 12 discusses structure/function claims in more detail.

STRUCTURE/FUNCTION CLAIMS *a statement on the label of a packaged food or dietary supplement about how that product might affect the human body's structure ("calcium builds strong bones") or function ("antioxidants maintain cell integrity")*

INFOGRAPHIC 2.12 Food Label Regulations

The Food and Drug Administration (FDA) regulates the claims that can be made on food and dietary supplement labels.

LABELING REGULATIONS FOR SPECIFIC TYPES OF CLAIMS

Nutrient Content Claims

- Most nutrient content claims apply only to nutrients with a DV.
- Exceptions are relative claims that compare content of other foods (i.e., "reduced," or "light").
- The word "healthy" may only be used for foods that provide healthy levels of fat, cholesterol, and sodium.

Health Claims

- Must be approved by the FDA.
- Must always use "may" or "might" to describe the ability of the product to reduce the risk of the disease.
- Not allowed on products containing high amounts of sodium, total or saturated fat, or cholesterol.
- Not allowed on foods with little natural nutritional value (the "jelly bean rule").

Structure/Function Claims

- Must not make any link (directly or implied) to a disease or health-related condition.
- They are supposed to be truthful and not misleading.
- Not reviewed or authorized by the FDA.
- When on a dietary supplement label:
  - the manufacturer must have evidence that the claim is truthful.
  - a disclaimer must state that the FDA has not evaluated the claim, and that the product is not intended to diagnose, treat, cure, or prevent any disease.

Would the claim "prevents diabetes" be allowed on a food label without FDA approval? Why or why not?

Food labels don't always make it easier to eat healthfully. A recent survey of more than 25,000 people worldwide found that more than half have difficulty understanding food labels. And just because labels are there doesn't mean people will use them; other research has shown that less than 25% of teenagers choose foods based on nutritional labels, and men who say they read food labels are only slightly leaner than men who don't.

FUTURE FOOD LABELS

To address ongoing concerns about food labels, experts are debating ways to revamp labels and make them more useful to the average American. Serving sizes sometimes seem unrealistically small, and people who eat 2 to 3 servings in one sitting may not realize the nutrient values should be multiplied. Some experts have proposed putting calorie and serving size information in larger type at the top of the label, so that it's immediately clear. In 2014, the FDA issued proposed changes to the Nutrition Facts Panel. **(INFOGRAPHIC 2.13)**

In addition to government planning, companies and advocacy groups are experimenting with different ways to present nutrition information and recommendations that they hope will be easier to follow, such as symbols that appear on the front of packages. For instance, the United Kingdom has established a "traffic light" system, where the amount of calories, sugar, fat, saturated fat, and salt are presented on the front of the package in green, yellow, or red according to the healthfulness of the choice. However, manufacturers are not obligated to use the system.

In the United States, some grocery stores are displaying the NuVal Nutrition

INFOGRAPHIC 2.13 Proposed Food Label Changes

FDA PROPOSED LABEL CHANGES

- Require information about the amount of "added sugars" in a food product.
- Update serving size requirements to reflect the amounts people currently consume. Present calorie and nutrition information for the whole package of certain food products that could be consumed in one sitting.
- Present "dual column" labels to indicate both "per serving" and "per package" calorie and nutrition information for larger packages that could be consumed in one sitting or multiple sittings.
- Require the inclusion of potassium and vitamin D, nutrients that some in the U.S. population are not getting enough of, which puts them at higher risk for chronic disease. Vitamins A and C would no longer be required on the label, though manufacturers could declare them voluntarily.
- Revise the Daily Values for a variety of nutrients such as sodium, dietary fiber, and vitamin D. Daily Values are used to calculate the Percent Daily Value on the label, which helps consumers understand the nutrition information in the context of a total daily diet.
- While continuing to require "Total Fat," "Saturated Fat," and "Trans Fat" on the label, "Calories from Fat" would be removed because research shows the type of fat is more important than the amount.
- Refresh the format to emphasize certain elements, such as calories, serving sizes, and Percent Daily Value, which are important in addressing current public health problems like obesity and heart disease.

PROPOSED FORMAT

Nutrition Facts

8 servings per container

Serving size 2/3 cup (55g)

Amount per 2/3 cup

Calories 230

% DV*	
12%	**Total Fat** 8g
5%	Saturated Fat 1g
	Trans Fat 1g
0%	**Cholesterol** 0mg
7%	**Sodium** 160mg
12%	**Total Carbs** 37g
14%	Dietary Fiber 4g
	Sugars 1g
	Added Sugars 0g
	Protein 3g
10%	Vitamin D 2mcg
20%	Calcium 260mg
45%	Iron 8mg
5%	Potassium 235mg

* Footnote on Daily Values (DV) and calories reference to be inserted here.

Which nutrients have been added to the food label? Which have been removed?

Scoring System, in which a food's price tag is accompanied by a rating of 1 to 100 based on more than 30 factors, including fiber, phytochemicals, calories, and healthy and unhealthy fats. In other supermarkets, consumers might see Guiding Stars that rate the nutritional quality of food with one to three stars using information from the Nutrition Facts Panel and the ingredients list. As of December of 2015, the government also requires large chain restaurants to list total calorie content for all standard menu items on the menu or menu board. In addition, beginning in December of 2016, the calorie content of foods sold in some vending machines will have to be available to the consumer prior to purchasing the food.

But to Gallagher, none of these tools will help people make better food choices if they don't have any access to more nutritious options. Since 2006, when results of her first study of Chicago's food deserts were released, she has helped the city identify six sites that would benefit significantly from a large grocery store. At one of those sites, her data suggest that adding a grocery store would take 24,000 people out of a food desert; people who could potentially collectively gain 13 years of life by avoiding liver disease, 15 years by forestalling diabetes, 58 years by averting diet-related cancer, and more than 100 years by eluding cardiovascular disease. But improving access doesn't have to be as major a project as building an entirely new store, she says; simply upgrading corner stores by adding fresh foods or establishing mobile grocery stores that travel to neighborhoods in need could make a big difference.

In 2010, Walmart hired Gallagher to develop a plan that would entirely eliminate all food deserts from Chicago. That same year, drug-store chain Walgreens announced it would start to sell fruits and vegetables in 10 areas identified as food deserts. Thanks to the increase in access, since 2006 food deserts are affecting approximately 250,000 fewer people. "In the next 10 or 15 years, I see the problem of food deserts being solved. But we'll still be dealing with the other issues that drive unhealthy habits."

That's because access is not the "silver bullet" to get people to make healthy choices and lose weight, she says; just building grocery stores where there are none will not solve all the country's nutrition woes. So she and her colleagues are conducting experiments to find ways to encourage shoppers to make healthy choices, such as providing selection information for food products, or changing the placement of nutritious food in grocery stores.

If policymakers don't address the problem of food deserts, they can't expect people to follow their dietary advice, says Gallagher. "It is disingenuous to preach 'follow the food pyramid' or 'eat your fruits and vegetables' if you don't provide access to those foods."

CHAPTER 2 BRING IT HOME

What's in a claim?

With your grocery list in hand, you reach for crackers and notice a reduced-fat variety of your favorite brand. Using the two product labels below compare the Nutrition Facts Panel and ingredients listing to address these questions:

1. Complete this table using the Nutrition Facts Panels for the original and reduced-fat cracker varieties shown below.

	Calories	Total Fat (g)	Total Fat %DV	Saturated Fat (g)	Saturated Fat %DV	Sodium (mg)	Sodium %DV	Carbo-hydrate (g)	Dietary Fiber (g)	Sugars (g)
Crackers—original										
Crackers—reduced fat										
Differences (+/−)										

2. What is the difference between the daily value (DV) for total and saturated fat? How might calorie content make a difference when comparing DVs of two similar products?
3. Besides differences in total and saturated fat content, are there any other differences that you observe?
4. Review the ingredient list for each of these products. What are the differences in ingredients or the order of ingredients between the two products?
5. Which of these two products would you consider purchasing and why?
6. Would you consider these crackers a food product that should be a regular choice or an occasional part of one's diet? Explain.

BUTTERY CRACKERS

Nutrition Facts
Serving Size 5 crackers (15g)

Amount Per Serving	
Calories 80	Calories From Fat 40
	% Daily Value*
Total Fat 4.5g	**7%**
Saturated Fat 1g	**5%**
Cholesterol 0mg	**0%**
Sodium 105mg	**4%**
Total Carbohydrate 10g	**3%**
Dietary Fiber 0g	**0%**
Sugars 1g	
Protein 0g	
Vitamin A 0% •	Vitamin C 0%
Calcium 2% •	Iron 2%

INGREDIENTS: UNBLEACHED ENRICHED, FLOUR (WHEAT FLOUR, NIACIN, REDUCED IRON, THIAMINE MONONITRATE {VITAMIN B1}, RIBOFLAVIN {VITAMIN B2}, FOLIC ACID), SOYBEAN OIL, SUGAR, PARTIALLY HYDROGENATED COTTONSEED OIL, SALT, LEAVENING (CALCIUM PHOSPHATE AND/OR BAKING SODA), HIGH FRUCTOSE CORN SYRUP, SOY LECITHIN, MALTED BARLEY FLOUR, NATURAL FLAVOR.

***REDUCED FAT* BUTTERY CRACKERS**

Nutrition Facts
Serving Size 5 crackers (15g)

Amount Per Serving	
Calories 70	Calories From Fat 18
	% Daily Value*
Total Fat 2g	**3%**
Saturated Fat 0g	**0%**
Cholesterol 0mg	**0%**
Sodium 150mg	**6%**
Total Carbohydrate 11g	**4%**
Dietary Fiber 0g	**0%**
Sugars 2g	
Protein 1g	
Vitamin A 0% •	Vitamin C 0%
Calcium 2% •	Iron 4%

INGREDIENTS: UNBLEACHED ENRICHED FLOUR (WHEAT FLOUR, NIACIN, REDUCED IRON, THAIMINE MONONITRATE [B1], RIBOFLAVIN [B2], FOLIC ACID), SOYBEAN OIL, SUGAR, PARTIALLY HYDROGENATED COTTONSEED OIL, LEAVENING (CALCIUM PHOSOPATE AND/OR BAKING SODA), SALT, HIGH FRUCTOSE CORN SYRUP, SOY LECITHIN, NATURAL FLAVOR.

Svetlana Foote/Shutterstock

KEY IDEAS

A healthy diet is a pattern of eating characterized by variety, balance, adequacy, and moderation that promotes health and reduces risk of chronic disease.

A healthy diet emphasizes nutrient-dense foods that provide a higher proportion of nutrients relative to calories.

A healthy diet balances energy-dense and nutrient-dense foods. Energy density is a measure of the calorie content of a food relative to a given weight.

The Dietary Guidelines for Americans (DGAs) provide evidence-based recommendations for healthy eating patterns that focus on variety and nutrient-density and limit calories from added sugars and saturated fats and reduce sodium intake tp reduce risk of chronic disease and promote overall health. Revised every five years, the DGAs are the cornerstone of federal nutrition policy and nutrition education initiatives.

MyPlate (at www.ChooseMyPlate.gov) is based on the DGAs and provides a visual representation of the ideal balance of food groups.

International food guides share common recommendations of consuming more plant-based foods, choosing lean protein sources, limiting intake of sugar and sodium, minimizing intake of unhealthy fats, controlling portion sizes, and increasing physical activity.

The Nutrition Labeling and Education Act (NLEA) of 1990 gives the Food and Drug Administration (FDA) the authority to oversee the labeling of food products and standardizes the presentation of nutrition information on food labels.

The Nutrition Facts Panel provides specific information about the serving size, calorie content, and other mandatory nutrition information.

The percent Daily Value (or %DV) on the Nutrition Facts Panel is the percent of a nutrient provided by a standard serving of a food in relation to the approximate goal for that nutrient. Percent Daily Values are based on a 2,000-calorie diet.

The NLEA established guidelines for the three types of claims that can be used on food and dietary supplement labels: nutrient content claims, health claims, and structure/function claims.

Nutrient, health, and structure/function claims on a food or dietary supplement label are statements regulated by the FDA that refer to nutrient content, potential health benefits, or the specific structure or function effects of a food or food component in the body. Health claims are evaluated and approved by the FDA; structure/function claims are not.

NEED TO KNOW

Review Questions

1. Which of the following is true with regard to a healthy diet?
   a. A healthy diet emphasizes nutrient-dense over energy-dense foods.
   b. A healthy diet provides calories and nutrients in amounts necessary to promote good health.
   c. A healthy diet is characterized by adequacy, balance, variety, and moderation.
   d. All of the above.

2. Emphasizing nutrient-dense foods and reducing intake of energy-dense foods while meeting overall energy needs typically results in:
   a. less dietary fiber and reduced overall volume of food intake.
   b. increased intake of dietary fat and risk of overweight and obesity.
   c. greater likelihood of achieving recommended intake of essential nutrients.
   d. dietary inadequacies that may contribute to nutrient deficiencies.

3. The Dietary Guidelines for Americans (DGAs) are characterized by all of the following, EXCEPT:
   a. they are updated every 10 years.
   b. they are science-based guidelines to promote health and reduce risk of chronic disease.
   c. they stress consumption of nutrient-dense foods.
   d. they encourage limiting added sugar intake to 10% of total calories.

4. The DGAs recommend that sodium intake should:
   a. be limited to 3,000 milligrams/day for all Americans older than 2 years.
   b. be restricted to 1,500 milligrams/day for individuals with high blood pressure.
   c. be balanced with overall calorie intake to prevent fluid retention.
   d. be limited only if older than 30 years.

5. All of the following are true regarding the USDA MyPlate food guide, EXCEPT:
   a. it replaced the Food Guide Pyramid in 2011.
   b. it has an online web site to help individualize recommendations.
   c. it is designed to depict food choices across food groups at meals.
   d. it is designed specifically for use by children rather than adults.

6. The World Health Organization's dietary guidelines:
   a. reinforce recommendations from other countries around the world.
   b. focus exclusively on malnutrition in the form of nutrient deficiency disease.
   c. contradict the Dietary Guidelines for Americans.
   d. recommend complete avoidance of sugar and salt.

7. The Nutrition Labeling and Education Act of 1990 requires all of the following, EXCEPT:
   a. health claims must be approved by FDA before use on food labels.
   b. listing of ingredients in a food product on the label.
   c. using a standardized Nutrition Facts Panel on food products.
   d. warning if a food product contains excessive amounts of sugar or sodium.

8. Percent daily value (%DV) on processed food packages is:
   a. developed specifically for use on food labels.
   b. based on an average 1,800-calorie intake.
   c. the level of nutrients that should not be exceeded.
   d. established by food manufacturers.

9. On a nutrition label, the list of ingredients:
   a. is in alphabetical order.
   b. begins with the ingredient that comprises the highest proportion of the product's weight.

c. begins with the ingredient with the highest caloric density.

d. begins with any potential ingredient that might cause an allergic reaction.

10. Currently, health claims that can appear on food labels:

a. can guarantee that consumption of a food will reduce risk of specific diseases.

b. have been approved by the FDA.

c. can appear on any processed food even if high in saturated fat, cholesterol, or sodium.

d. are based exclusively on research and evidence provided by the food manufacturer.

Take It Further

More than 60% of U.S. consumers report referring to the Nutrition Facts Panel when selecting food products. Even if you do not typically look to food labels to guide your food selections and purchases, consider the information included on a standardized label. Discuss what you typically (or might) look at first on a food label and why. How does the information on the food label help you determine whether to purchase a particular product?

Dietary Analysis Using SuperTracker

Diet planning and diet analysis using food groups

In this exercise you will evaluate a day's worth of meals for distribution and abundance of foods from each of the food groups.

1. Log onto the United States Department of Agriculture (USDA) website at www.supertracker.usda.gov. If you have not done so already, you will need to create a profile to get a personalized diet plan. This profile will allow you to save your information and diet intake for future reference. Do not use the general plan.
2. Click the Track Food and Activity option.
3. Record your food and beverage intake for one day that most reflects your typical eating patterns. Enter each food and beverage you consumed into the food tracker. Note that there may not always be an exact match to the food or beverage that you consumed, so select the best match available.
4. Once you have entered all of your food and beverage choices into the food tracker, on the right side of the page under the bar graph, you will see Related Links: View by Meal. Print this report and use it to answer the following questions:

a. Did you meet the minimum recommended servings for each food group (including oils) on the day you selected?

b. What food groups, if any, are you lacking? List three specific diet changes you can make now to meet the recommended servings for the food groups you are missing or lacking.

c. Discuss how the following factors influence your food choices: taste preferences, food trends and advertising, time available for meals, eating habits, and the cost of food. Which factor is the most important and why?

d. According to the food tracker, did you go over the allowance for empty calorie foods for this day? Is this typical of your normal eating habits?

e. List two pieces of information that you look for on food labels. Why are these important to you?

f. What have you learned about your eating behaviors from doing this assignment that you weren't aware of before?

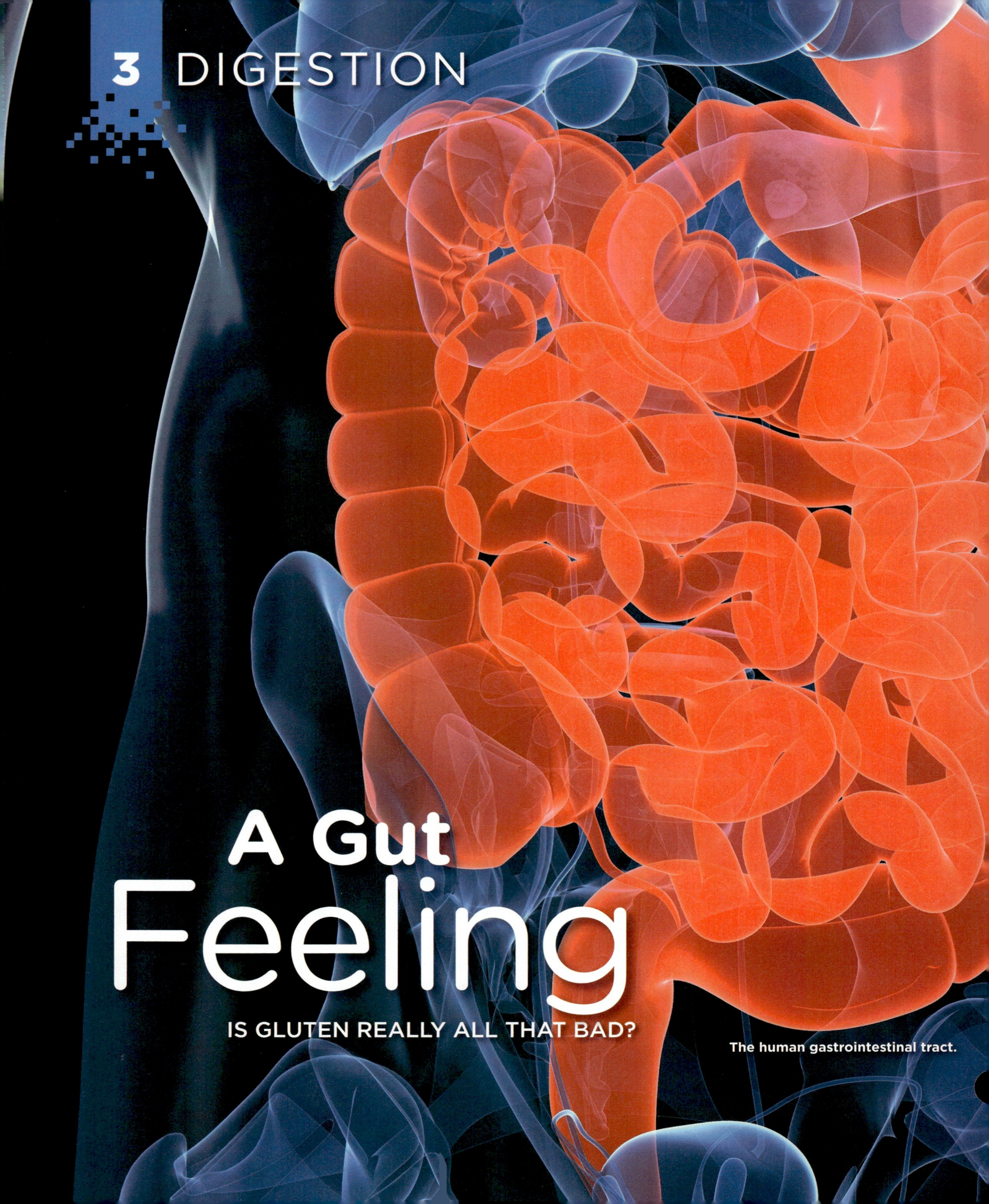

A Gut Feeling

IS GLUTEN REALLY ALL THAT BAD?

The human gastrointestinal tract.

LEARNING OBJECTIVES

- Summarize the body's organization from cells to organ systems (Infographic 3.1)
- Explain the four main functions of the gastrointestinal (GI) tract (Infographic 3.2)
- Identify and describe the function of the organs and accessory organs of the GI tract (Infographic 3.3)
- Describe mechanisms by which food is propelled through the GI tract (Infographic 3.4)
- Explain the importance of enzymes in digestion (Infographic 3.5)
- Describe how the structure of the small intestine facilitates its role in digestion and absorption (Infographic 3.7)
- Describe three ways nutrients can be transported across a cell membrane (Infographic 3.8)
- Describe the purpose of the circulatory system (Infographic 3.9)
- Explain the absorptive and secretory functions of the large intestine (Infographic 3.10)
- Identify three nutrition-related GI disorders and summarize potential nutritional (or dietary) implications and interventions (Infographic 3.11)

Soon after gastroenterologist Joseph Murray, MD, moved to the United States from Ireland in 1988, he began to notice something strange about his patients. Many of them came to him complaining about stomach problems. Some had lost dramatic amounts of weight. After running tests, Murray often discovered that the patients' blood showed the tell-tale signs of celiac disease, in which the body mounts an immune response against gluten, a protein found in wheat and several other grains.

Joseph Murray, MD, Professor of Medicine, Department of Gastroenterology, Mayo Clinic, Rochester, Minnesota

Murray was surprised. Although celiac disease was common in Ireland, it was considered rare in North America. Whenever he had the opportunity, Murray tested blood and tissue samples from patients to look for signs of celiac disease. What he found shocked him: From 1991 to 2001, the number of new diagnoses increased 10-fold in the geographic area around where he now works at the Mayo Clinic in Rochester, Minnesota. Since then, the number has risen further.

At first, Murray assumed that the rise in diagnoses was a result of the fact that he and other doctors were simply looking for it more—screening more people and using better tests.

INFOGRAPHIC **3.1** **Cells, Tissues, Organs, and Organ Systems** *Numerous cells representing every tissue type are organized to form organs and organ systems. Cells within the various organs of the digestive system produce substances that are needed to process the nutrients in the foods we eat into units that can be taken up (absorbed) by the cells lining the gastrointestinal (GI) tract. Those absorptive cells then export nutrients into the circulatory system for distribution to the rest of the body.*

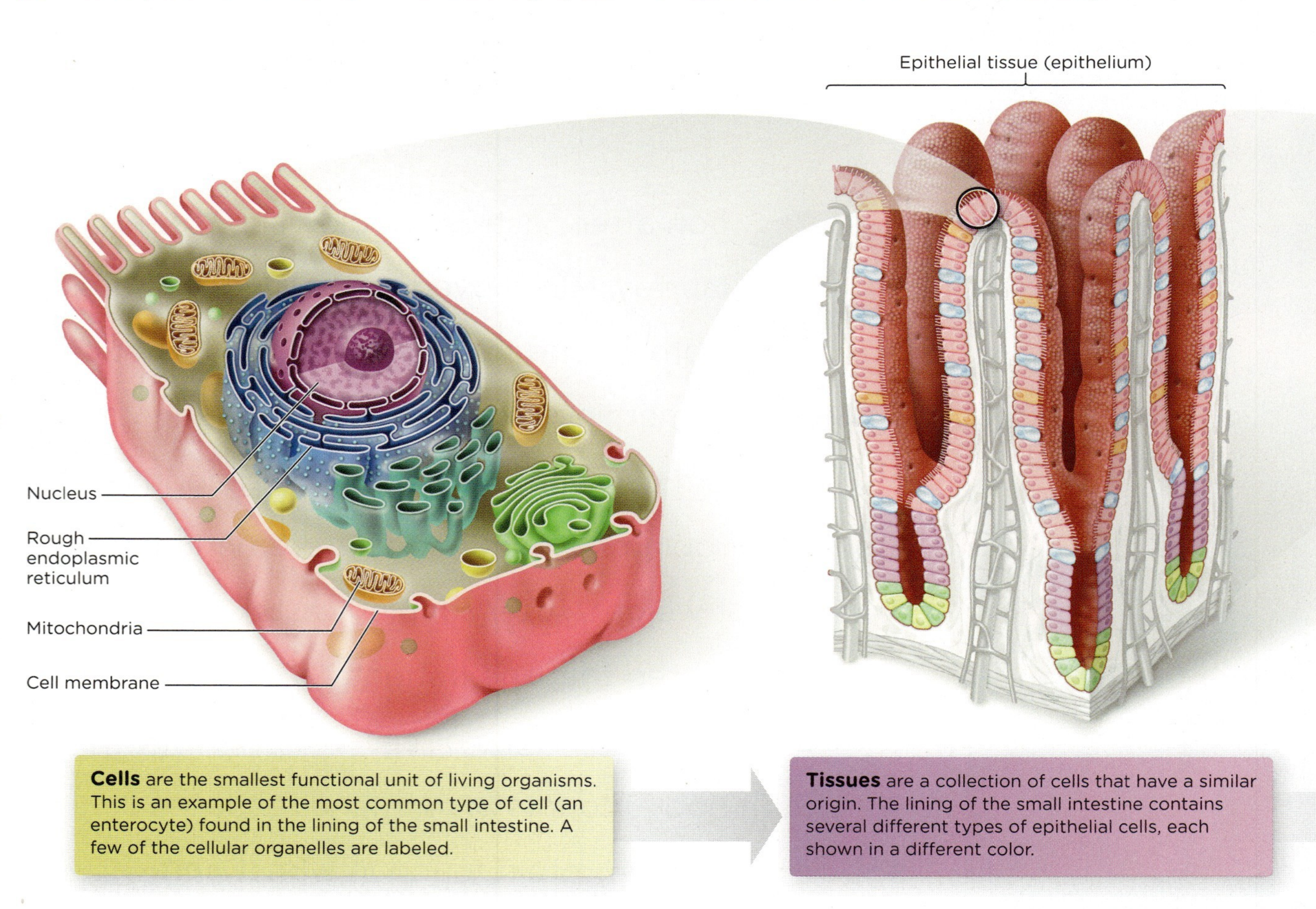

What is the hierarchical organization of cells in the human body and what benefit does this organization provide?

But now it seems clear to Murray and other doctors that, for as yet unknown reasons, celiac disease has become much more common in the United States in recent decades. And this rise has worrying implications.

The protein gluten is difficult to avoid: It's found not only in wheat, rye, and barley, but also in many medicines, vitamins, and even lip balms. And when those with celiac disease consume gluten, the consequences can go far beyond an annoying stomachache. The presence of the protein in the digestive tract of those with celiac disease causes their immune system to attack the lining of the small intestine, thereby impairing nutrient absorption. Ultimately, people with celiac disease who eat gluten can become malnourished no matter how much food they eat, and they can suffer from stunted growth, anemia, and bone loss.

The devastating health impact of untreated celiac disease illustrates the importance of **digestion**, one of the key processes of the human body. If a person's digestive system isn't working properly, he can eat a nutrient-rich diet but still end up malnourished. Digestion extracts vitamins, minerals, and other nutrients from the foods we consume to provide the body with the nutrients it needs for growth, maintenance, reproduction, repair, and continuous renewal of cells, tissues, and organs. (INFOGRAPHIC 3.1) The health of an individual's digestive system, then, is just as important as a nourishing diet.

DIGESTION
the process of breaking food down to its smallest units in order for the nutrients to be absorbed

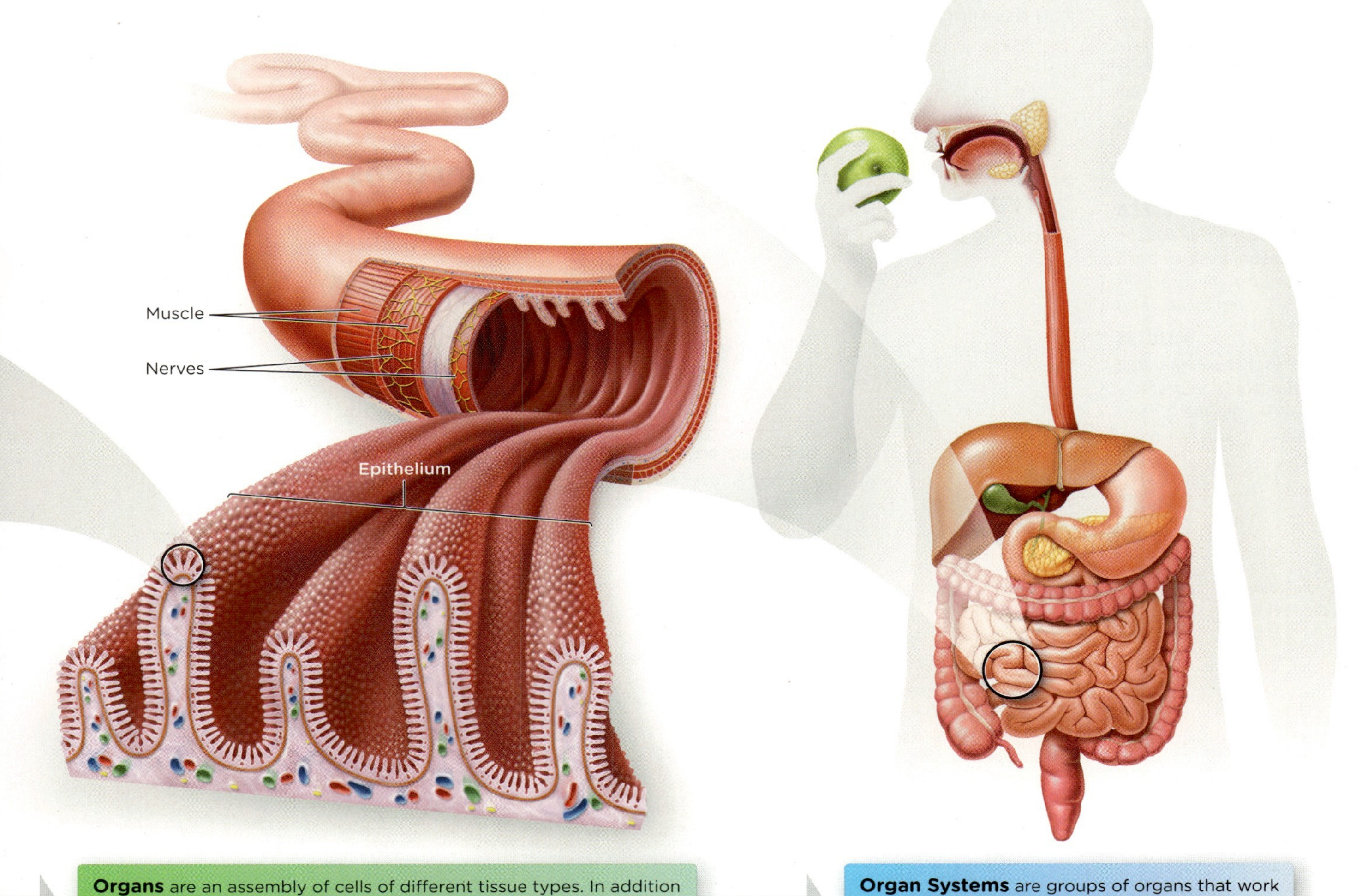

Organs are an assembly of cells of different tissue types. In addition to epithelial cells, the small intestine is composed of cells from each of the three remaining types of tissues: *connective*, *nervous*, and *muscle* tissues. Connective tissue cells are not shown here.

Organ Systems are groups of organs that work together to carry out one or more functions. The small intestine is just one of many organs that make up the gastrointestinal system.

INGESTION
the process of taking food or other substances into the gastrointestinal tract via the mouth

GASTROINTESTINAL TRACT
tubular channel extending from the mouth to the anus where digestion and absorption occur; also called the alimentary tract

ABSORPTION
the process by which nutrients and other substances are removed from the lumen of the gastrointestinal tract to enter the blood stream or lymphatic system

EXCRETION
elimination of waste from the body; digestive waste includes any food stuff not absorbed by the intestine

LUMEN
inner space of the GI tract lined with mucosal cells

MUCOSA
tissues that line the gastrointestinal tract, made up of mucosal cells (epithelial cells)

MUCUS
a viscous solution that lubricates and protects the GI tract

OVERVIEW OF THE DIGESTIVE PROCESS

There are four basic stages to the digestive process: First, food is **ingested** through the mouth and enters the tubular channel called the alimentary canal, also called the **gastrointestinal tract**. The gastrointestinal (GI) tract extends from the mouth to the anus, and includes the mouth, esophagus, stomach, small intestine, and large intestine. Second, as food passes down the GI tract it is *digested*, or broken down into smaller units. Third, these smaller units of nutrients are then **absorbed**, meaning that they pass into the bloodstream or lymphatic system, which transports them throughout the body. Lastly, anything left over is **excreted** through the opening (the anus) at the other end. **(INFOGRAPHIC 3.2)**

The primary function of the digestive system is to digest, or break down, the nutrients in our foods into compounds that are small enough to be taken up into our body. The inside space of the GI tract is called the **lumen**, which is lined with a layer of mucosal cells called the **mucosa**, so named because it is protected by a layer of thick fluid called **mucus**. *Mucosal cells* allow our body to absorb nutrients, so that they can be transported by the blood or lymph to where they are needed.

■ ■ ■

Over time, the digestive problems caused by celiac disease can predispose individuals to an early death, as Murray discovered in his research. Between 1948 and 1952, a group of researchers collected blood samples from young Air Force recruits in Wyoming—roughly

INFOGRAPHIC 3.2 An Overview of the Functions of the Digestive Tract *The digestive tract has four main functions that allow us to use nutrients in our foods.*

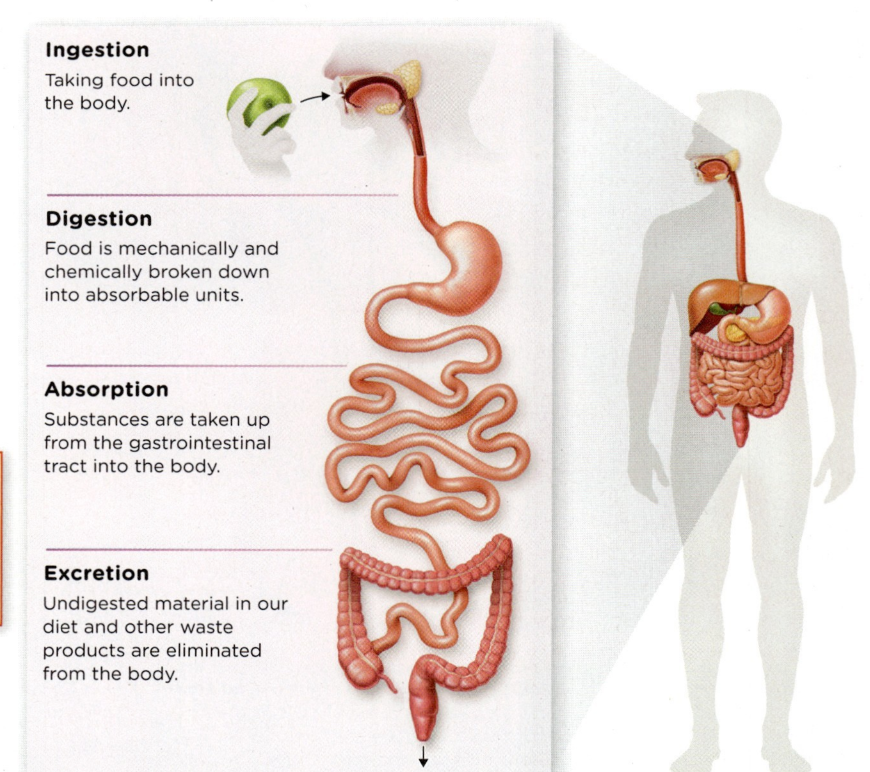

The time it takes for food to travel through the gastrointestinal tract varies widely, but generally takes between 24 and 72 hours.

Why is the section of the GI tract that is responsible for absorption so elongated?

Physicians working in the lab at Francis E. Warren Air Force Base, near Cheyenne, Wyoming, approximately 1949.

U.S. Army Medical Department, Office of Medical History

9,000 in total. Afterward, one of the researchers stuck the thousands of samples in the back of a freezer in Ohio, where drops of water from a condenser created a frozen, hermetic seal over the container, preserving them perfectly. Decades later, they were passed on to Murray, who tested them for levels of certain types of antibodies that indicate the body is reacting to gluten. Among other things, Murray discovered that men who carried the antibodies as young recruits were several times more likely to have died over the subsequent decades than men who didn't.

What goes wrong, digestively, in patients with celiac disease and how does this threaten their long-term health? Let's walk through the digestive process in greater detail to find out.

Gastrointestinal Tract and Accessory Organs

As mentioned, the GI tract extends from the mouth to the anus and includes the mouth, esophagus, stomach, small intestine, and large intestine. Because many of these organs fold over onto themselves within the body, the entire GI tract is much longer than it appears—in fact, it extends for approximately 8 to 9 meters (26 to 30 feet).

The **digestive system** includes the organs of the GI tract and *accessory organs*: the salivary glands, liver, gallbladder, and pancreas, all of which secrete fluids containing a variety of agents, such as enzymes and acids, that aid in digestion. **(INFOGRAPHIC 3.3)**

OVERVIEW OF MECHANICAL AND CHEMICAL DIGESTION

Digestion is comprised of two processes: mechanical and chemical digestion. Working together, these two processes break down the nutrients in food into manageable pieces so that they can be processed by our bodies, but each has its own unique spin on the process. **Mechanical digestion** is the physical fragmentation of foods into small particles, while **chemical digestion** breaks chemical bonds to cleave large molecules into smaller ones.

Mechanical digestion begins in the mouth, where teeth crush and tear food into small bits. It continues in the stomach, as forceful contractions vigorously churn food. This churning action disperses the small food fragments and exposes large molecules to the digestive fluids that will dismantle them. In many cases, mechanical digestion is all that is required to release many vitamins, minerals, and phytochemicals from foods that are then taken up by mucosal cells lining the small intestine.

Gastrointestinal Motility

Motility is a term used to describe the contractions of the GI tract's smooth muscles

DIGESTIVE SYSTEM
system responsible for digestion, made up of digestive tract (mouth, esophagus, stomach, small intestine and large intestine), and accessory organs (salivary glands, liver, gallbladder, and pancreas)

MECHANICAL DIGESTION
physical breakdown of food by mastication (chewing) and mixing with digestive fluids

CHEMICAL DIGESTION
digestion that involves enzymes and other chemical substances released from salivary glands, stomach, pancreas, and gallbladder

INFOGRAPHIC 3.3 The Digestive System *The gastrointestinal tract (mouth, esophagus, stomach, small intestine, large intestine) is responsible for the processes of ingestion, digestion, absorption, and excretion. Accessory organs that support digestion include the salivary glands, liver, gallbladder, and pancreas.*

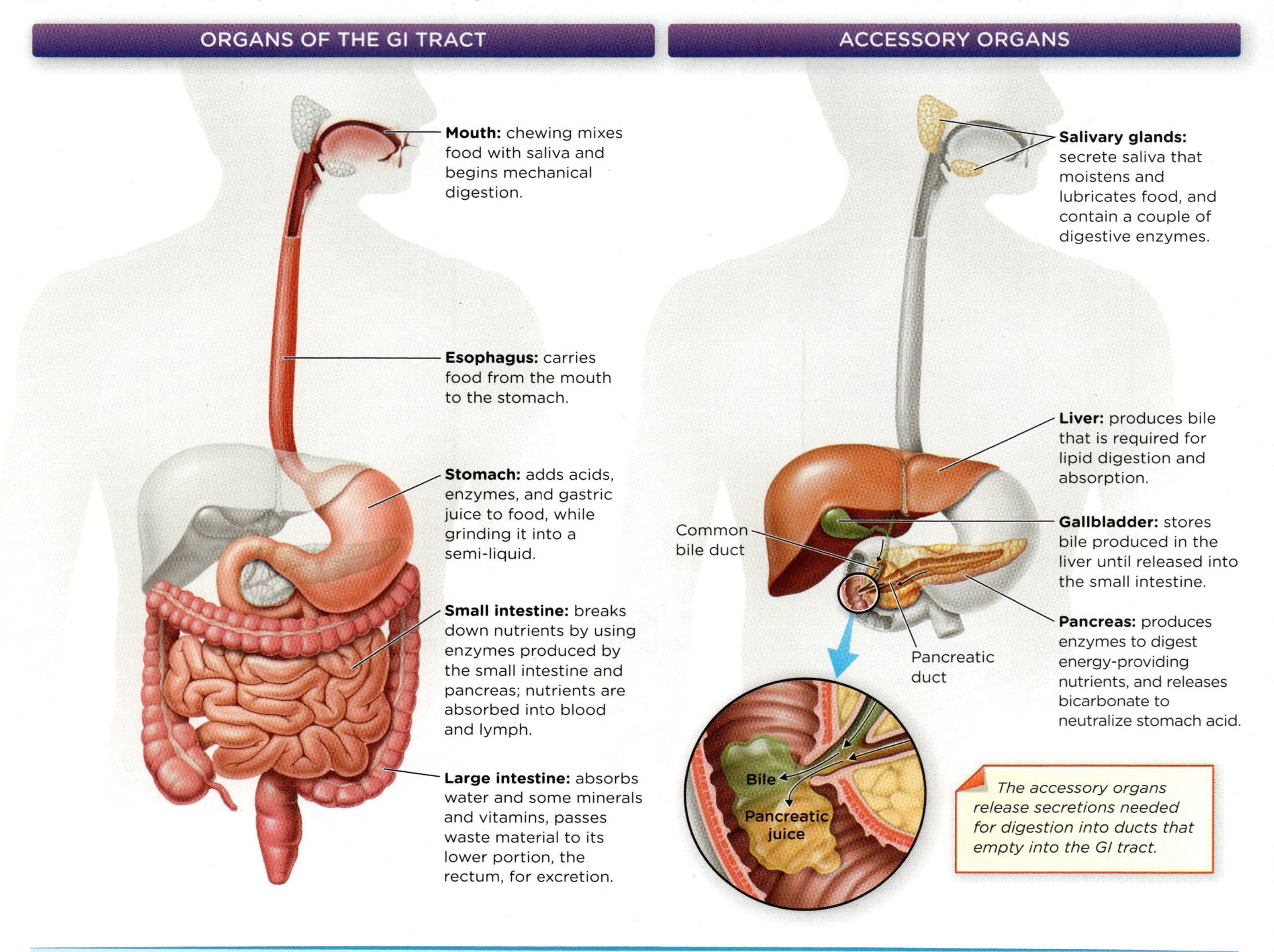

What accessory organ produces a compound that is required for lipid digestion and absorption, but is not an enzyme?

PERISTALSIS *rhythmic, wavelike contractions of the smooth muscle of the GI tract*

SEGMENTATION *circular muscles in the small intestine contract to mix intestinal contents with digestive fluids and bring nutrients in the intestinal fluid in contact with the intestine's absorptive surface*

that mix food with digestive fluids and propel food along the length of the tract. There are two fundamental patterns of these muscle contractions: **peristalsis** and **segmentation**. Peristalsis creates propulsive muscle contractions to move food forward through the complete length of the GI tract, from the esophagus to the anus. Segmentation, in contrast, occurs when circular muscles in the small intestine contract in an uncoordinated fashion so that fluid contents gently slosh back and forth between the segments. These contractions serve to mix intestinal contents with digestive fluids and bring nutrients in the intestinal fluid in contact with the intestine's absorptive surface. Similar segmentation contractions also occur in the colon. (INFOGRAPHIC 3.4)

INFOGRAPHIC 3.4 Peristalsis and Segmentation Propel and Mix the Contents of the Gastrointestinal (GI) Tract

PERISTALSIS: FORWARD PROPULSION

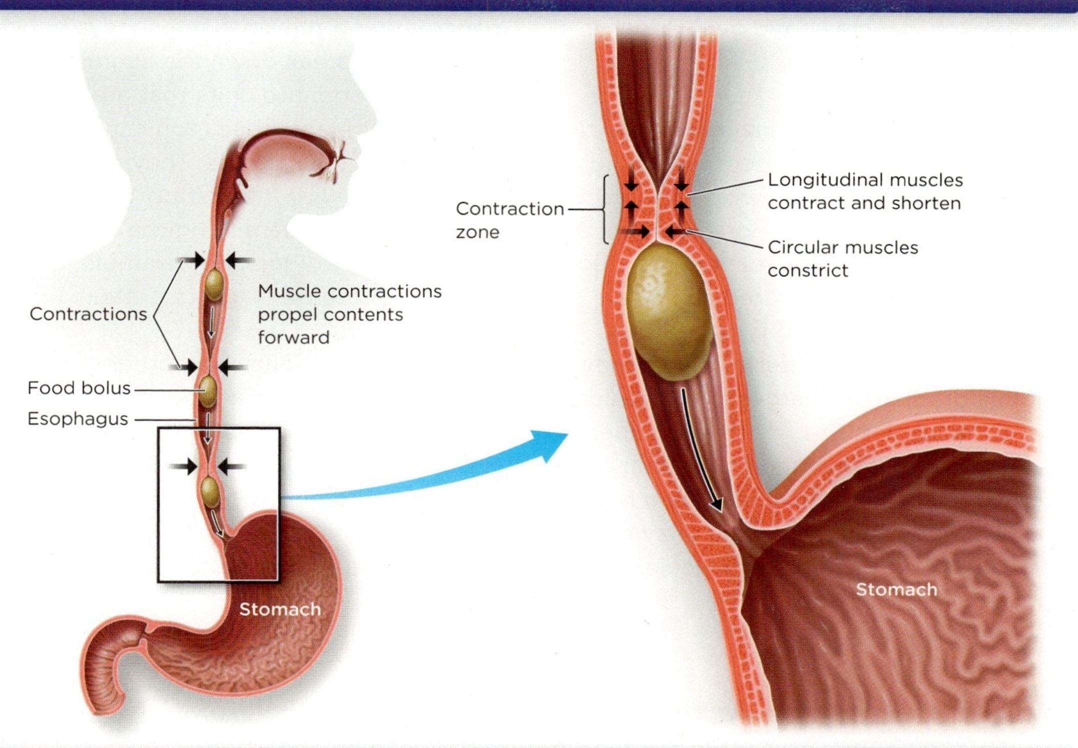

SEGMENTATION: MIXING

Circular muscle contractions

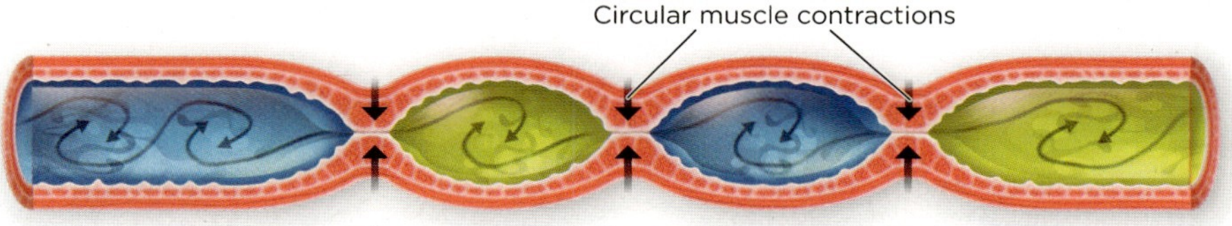

Rings of circular muscles in the small intestine contract at irregular intervals, while adjacent muscles relax.

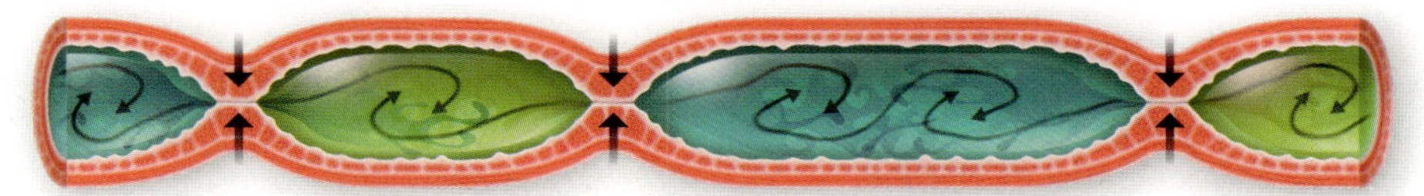

These contractions cause intestinal fluid to gently slosh back and forth between segments, mixing the contents of the fluid with digestive juices.

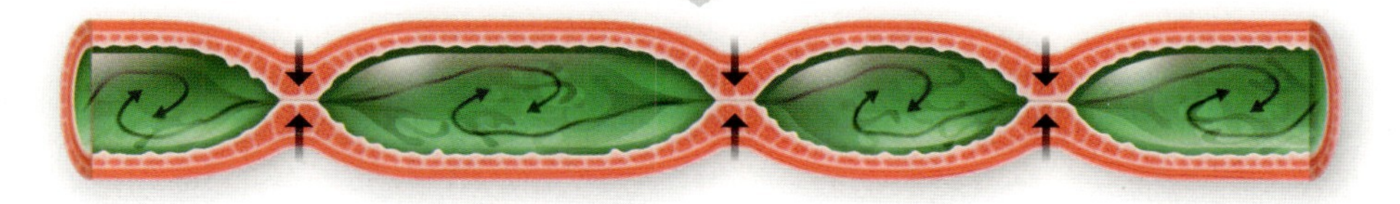

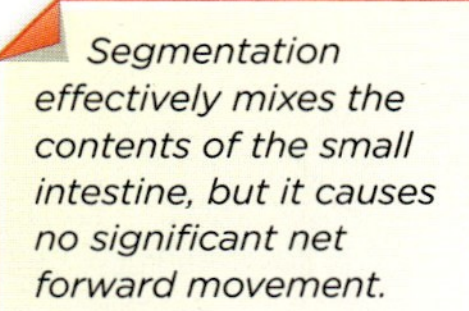

Segmentation effectively mixes the contents of the small intestine, but it causes no significant net forward movement.

? *Identify which type of contraction (segmentation or peristalsis) is responsible for the elimination of feces.*

ENZYMES
protein molecules that catalyze, or speed up, the rate at which a chemical reaction produces new compounds with altered chemical structures; enzyme names end in the suffix "-ase"

HORMONES
chemical substances that serve as messengers in the control and regulation of body processes

Enzymes and Hormones in Digestion
Chemical digestion is the form of digestion that involves enzymes and other substances released from salivary glands, as well as from the stomach, pancreas, and gallbladder. It takes place in the mouth, small intestine, and stomach.

Without the work of **enzymes**, digestion (and many other body processes) could not occur. An enzyme is a protein molecule that functions to catalyze, or speed up, the rate at which a chemical reaction produces new compounds with altered chemical structures. These reactions may subtly alter the chemical structure, or they may produce dramatically larger or dramatically smaller molecules. Enzymes have a specific shape that will fit only molecules that have a coordinating shape, like matching pieces of a puzzle. As facilitators, the enzymes bind to their coordinating molecule, initiate a chemical reaction, and move on. Thus enzymes can participate in these chemical reactions many times without being altered themselves. The salivary glands, stomach, pancreas, and the small intestine produce enzymes to break down the chemical compounds in food into units that are small enough to be absorbed. You can recognize enzymes by their names: They end in the suffix "-ase," with the first portion of the enzyme's name identifying the type of molecule they digest. For example, the enzyme "protease" digests protein. **(INFOGRAPHIC 3.5)**

In addition to enzymes, other chemicals, such as hormones, are produced by the various organs of the digestive system. **Hormones** are your body's chemical messengers. Together with the nervous system, these hormones regulate motility, appetite, and the release of secretions into the GI tract.

INFOGRAPHIC **3.5** **A Model of Enzyme Action** *Enzymes are required for the chemical digestion (breakdown) of many of the compounds present in our food before they can be absorbed.*

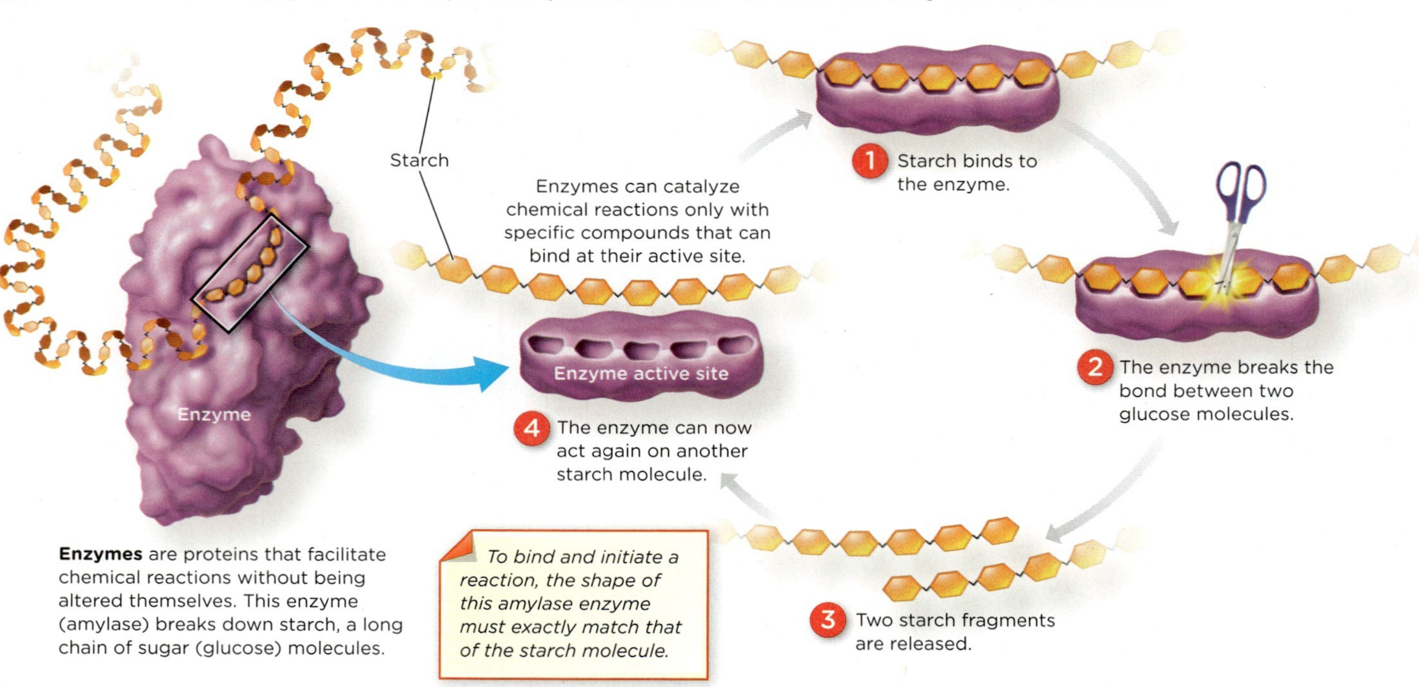

Why won't this enzyme break down lipids or proteins?

THE PATH OF DIGESTION FROM MOUTH TO LARGE INTESTINE

The object of digestion is to break down large molecules into smaller units, and the first stop is the mouth. Picture a hamburger poised to enter your *mouth*, or oral cavity. The first bite, or even the mouth-watering anticipation of the first bite, starts the mechanical and chemical processes of digestion.

As you chew your mouthful of hamburger, the teeth tear and crush the bread and meat thus beginning mechanical digestion, which increases the surface area available for the enzymes to begin their work of chemical digestion. While we chew, the salivary glands near the jaw and under the tongue release **saliva**, which both lubricates the mouth and esophagus and contains salivary **amylase**, an enzyme that starts digesting the carbohydrates in the bun, as well as a **lipase**, which begins digesting fats in the meat. The *tongue* mixes saliva with the foods in the mouth and pushes food to the back of the mouth to initiate swallowing.

The tongue does more than help mix and swallow the food, though—it's also part of the reason we enjoy that savory, juicy burger. The tongue contains **taste buds** that identify or sense foods on the basis of specific flavors or tastes. Food is a combination of five primary tastes: sweet, sour, salty, bitter, and *umami*, a more recently recognized savory flavor that's harder to identify but may be an element of your burger, as it's found in meats, seafood, cheese, and some vegetables. The number of taste buds we have declines with age, which explains why children—whose tongues are coated in taste-sensing

The first bite. *The mouth-watering anticipation of eating starts the processes of digestion.*

Pavel L Photo and Video/Shutterstock

SALIVA
fluid secreted from salivary glands in the mouth to moisten food and provide lubrication

AMYLASES
enzymes that break down starch into smaller polysaccharides and disaccharides

LIPASES
enzymes that break down fats (triglycerides) by releasing one or more fatty acids

TASTE BUDS
taste receptor cells found on the tongue within the papillae that are involved in sensing foods on the basis of specific flavors (tastes), such as sweet, sour, salty, bitter, and umami

Corbis

How many tastes in this meal? *The tongue's taste buds identify five primary tastes: sweet, sour, salty, bitter, and umami.*

BOLUS
a masticated, round lump of food, lubricated in the mouth by mixing with saliva

SPHINCTER
a ring-like muscle that relaxes or contracts to open or close a bodily passageway

PROTEASES
enzymes that break down proteins

CHYME
semi-liquefied, partially digested contents that leave the stomach a few teaspoons at a time to enter the small intestine

VILLI
fingerlike projections that protrude from the absorptive mucosal cells of the small intestine into the lumen of the GI tract; responsible for increasing the available surface area for absorption

MICROVILLI
very small projections that protrude from the absorptive mucosal cells of the villi in the small intestine; responsible for increasing the surface area for absorption twenty-fold

cells—are often averse to strong flavors and may prefer to eat their burgers plain (or with a little ketchup) than with spicy mayonnaise or hot sauce.

These taste buds don't just tell us if something is too bitter or salty; they also create signals that tell the rest of the GI tract to prepare for the next steps of digestion.

Food Passes from the Esophagus to the Stomach

Once the hamburger has been chewed and coated in saliva, it becomes a soft, moist lump of food known as a **bolus**, which is swallowed and passed through the throat. It then enters the esophagus, a roughly 10-inch-long muscular tube that transports the bolus of burger from the mouth to the stomach, relying on gravity and peristalsis. Positioned at the junction of the esophagus and the stomach is a circular muscle (**sphincter**) that normally functions as a one-way valve (as do the other sphincters in the body). As food nears the *lower-esophageal sphincter,* it relaxes to let food pass into the stomach; otherwise, it stays tightly closed to prevent foods and secretions from the stomach from moving backwards into the esophagus.

The bolus then enters the *stomach*, a muscular, J-shaped sack that can accommodate up to four cups of food and generally takes about two to four hours to empty. The stomach secretes *gastric juices* that contain *hydrochloric acid* (HCL), which help to unfold the proteins in the hamburger through chemical digestion. The gastric juices also contain enzymes such as **proteases**, which digest proteins, and lipase, which continues to digest fat. Importantly, the gastric juices also contain mucus, which lubricates the food and protects the stomach lining from acidity. The particularly forceful peristaltic contractions of the stomach vigorously churn the hamburger bolus, fragmenting it into fine particles that are dispersed throughout the gastric fluid. This essentially grinds the food into a semi-liquid mass called **chyme**. A few teaspoons at a time, chyme is passed along to the small intestine, moving through the partially relaxed *pyloric sphincter* that functions as a sieve, allowing only small food particles (generally smaller than 1 mm) to pass through. **(INFOGRAPHIC 3.6)**

How quickly the stomach empties into the small intestine depends on the composition and quantity of the foods and fluids you consume. A hamburger typically spends 24 to 72 hours going from the mouth to the anus, but this transit time can change because of illness, medication, how active you are, and even your emotional state. Food with more fiber, for instance, slows emptying from the stomach, helping you feel full. But as fiber passes into the large intestine, it can also stimulate propulsive contractions, which speeds up the transit of the intestinal contents through the rest of the digestive system. These combined effects of fiber are important and help explain why nutrition professionals recommend healthy amounts of vegetables, fruits and other high-fiber foods in the diet.

The Structural Features of the Small Intestine Facilitate Absorption of Nutrients

The *small intestine* is the primary site for the digestion of food and the absorption of nutrients and as we shall see, it is where digestion goes awry in those with celiac disease. The small intestine has three sections: the *duodenum*, the first portion of the small intestine after the stomach; the *jejenum*, the middle portion; and the *ileum*, the last and longest portion. The small intestine isn't actually "small" at all; it is a coiled hollow tube approximately 20 feet long and one-and-a-half inches wide, and several structural features give it a surface area approximately equal to the size of a tennis court. For example, the internal *circular folds* of the small intestine carry fingerlike projections called **villi** that increase its surface area for absorption. These villi are densely covered with fine, hairlike projections called **microvilli** that further increase surface area and the efficiency of absorption. Because these

INFOGRAPHIC 3.6 **The Structure and Function of the Stomach** *The stomach possesses the strongest muscles in the gastrointestinal (GI) tract and it is the major site for mechanical digestion.*

THE MUSCULAR STOMACH

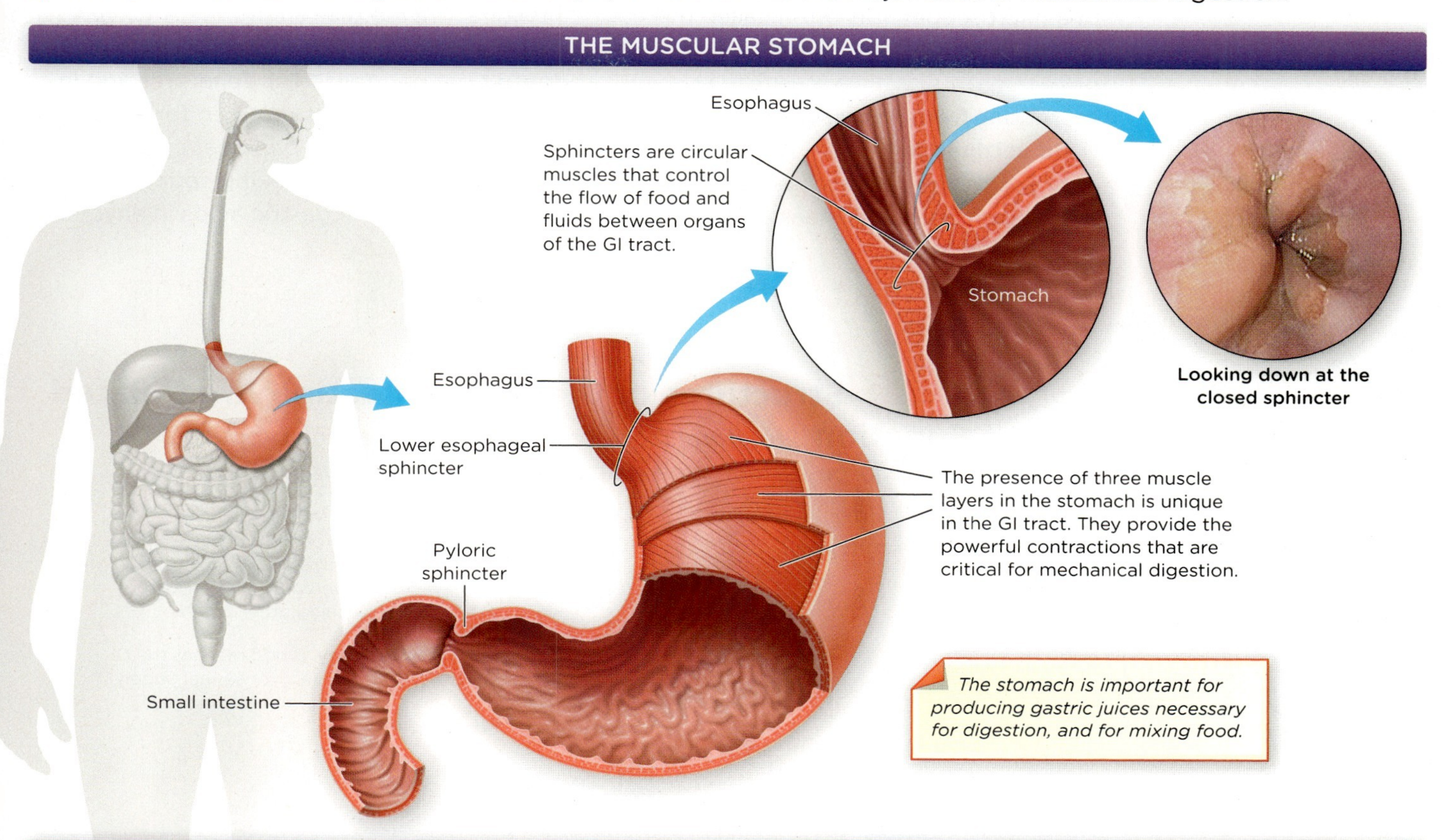

The stomach is important for producing gastric juices necessary for digestion, and for mixing food.

DIGESTION IN THE STOMACH

Peristaltic contractions churn and breakdown food.

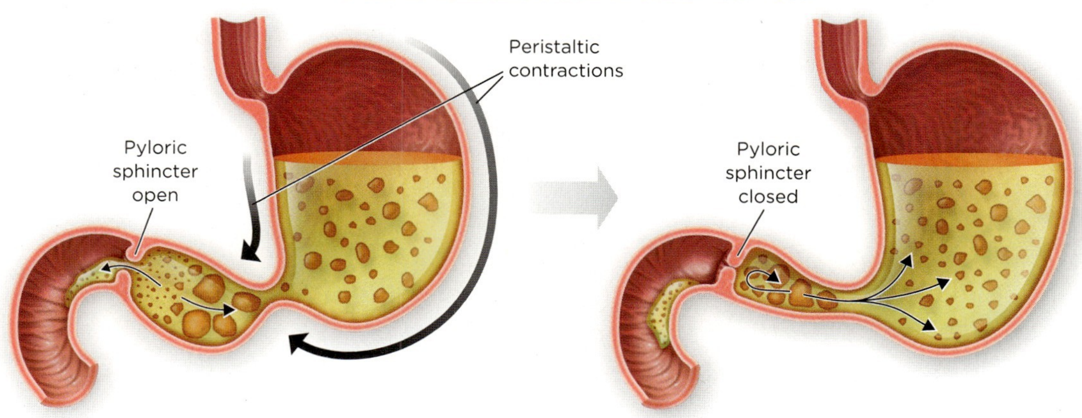

A small amount of chyme is pushed through the partially open sphincter.

As the sphincter closes, the jetlike backward flow of chyme grinds large food particles into smaller fragments.

? *What structural feature is found in the stomach that is responsible for the strength of its muscle contractions?*

Photo credits (sphincter): Gastrolab/Science Source

structures give the lining of the small intestine a brushlike appearance (when viewed with a microscope), it is often called the **brush border**. (INFOGRAPHIC 3.7)

Secretions from Accessory Organs Aid in Digestion

Secretions from accessory organs such as the pancreas and gallbladder play an important role in the digestion of the hamburger within the lumen of the small intestine. Chyme that enters the small intestine from the stomach is very acidic, and if it doesn't get neutralized, it denatures and inactivates the enzymes required for digestion. The pancreas releases *pancreatic juice* that contains bicarbonate (baking soda), a base that neutralizes the gastric acids in chyme. Refer to INFOGRAPHIC 3.3.

INFOGRAPHIC 3.7 Structures of the Small Intestine Are Related to Its Function *Structural features of the small intestine produce an extraordinarily large surface area that enhances digestion and absorption.*

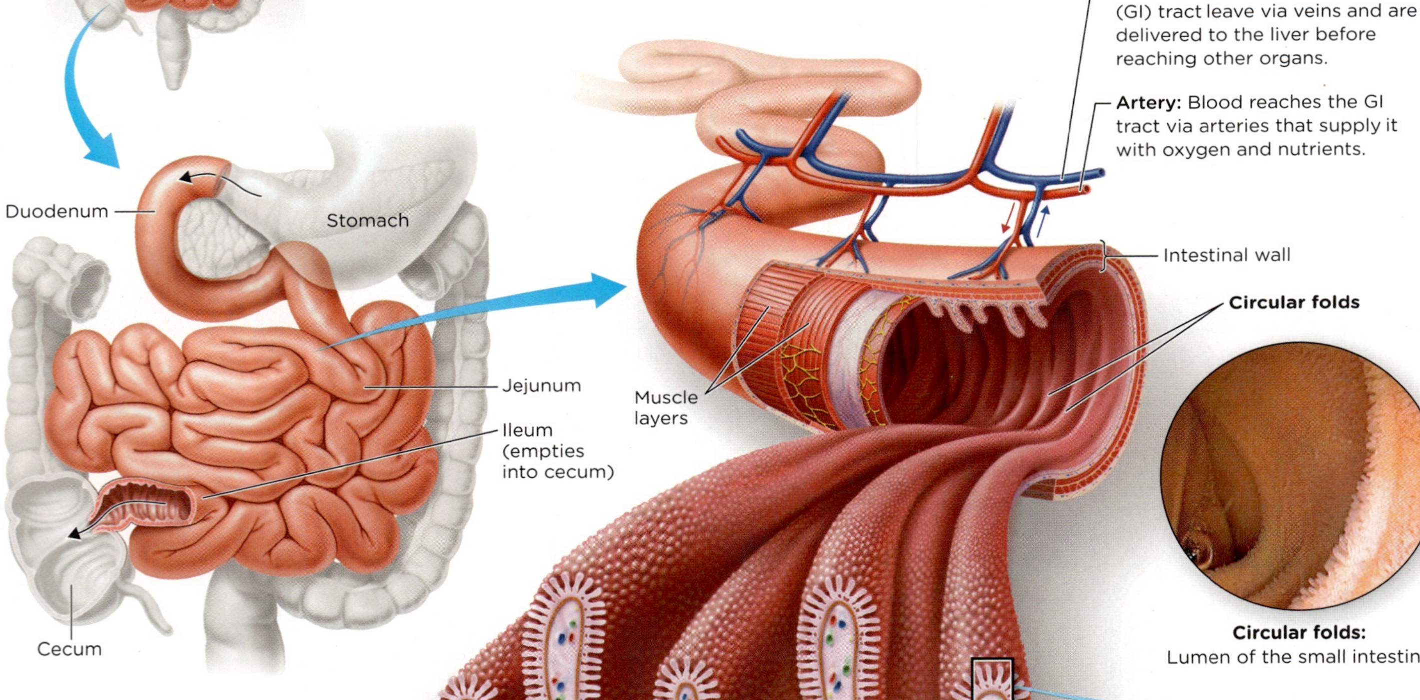

List the three structural features of the small intestine that cause it to have a much larger surface area than a simple pipe of the same length would have.

Photo credits (circular folds): David M. Martin, M.D./Science Source; (microvilli): Science Photo Library

The liver and the gallbladder also help to digest the many lipids found in the hamburger. The liver produces **bile**—which is stored in the gallbladder, a small, pear-shaped sac below the liver in the right upper abdomen—that contains substances critical for effective lipid digestion. The hormone *cholecystokinin* (CCK) is released from the small intestine in response to the hamburger's fats and protein in the small intestine; it stimulates the gallbladder to release bile and the pancreas to secrete juice into the lumen of the small intestine. In addition to bicarbonate, pancreatic juice contains enzymes—lipases, proteases, and amylase—that break the hamburger's large fat, protein, and carbohydrate molecules into smaller ones. Enzymes located in the brush border complete the digestion of the burger's carbohydrates and proteins.

Transport of Nutrients across the Cell Membrane

To enter the mucosal cells lining the GI tract, water and small amounts of a few other nutrients can pass directly through the cell membrane by **simple diffusion**. The cell membrane serves as the boundary that holds the content of body's cells in place and keeps their internal structures safe, so that cells function properly. The membrane

BRUSH BORDER
name for the microvilli-covered surface of the small intestine that functions in the absorption of nutrients

BILE
a fluid produced in the liver, concentrated and stored in the gallbladder, and secreted into the small intestine in response to food present in stomach; bile promotes the digestion of fat by emulsifying it, which allows lipase easier access

SIMPLE DIFFUSION
movement of a substance across a cell membrane, down a concentration gradient

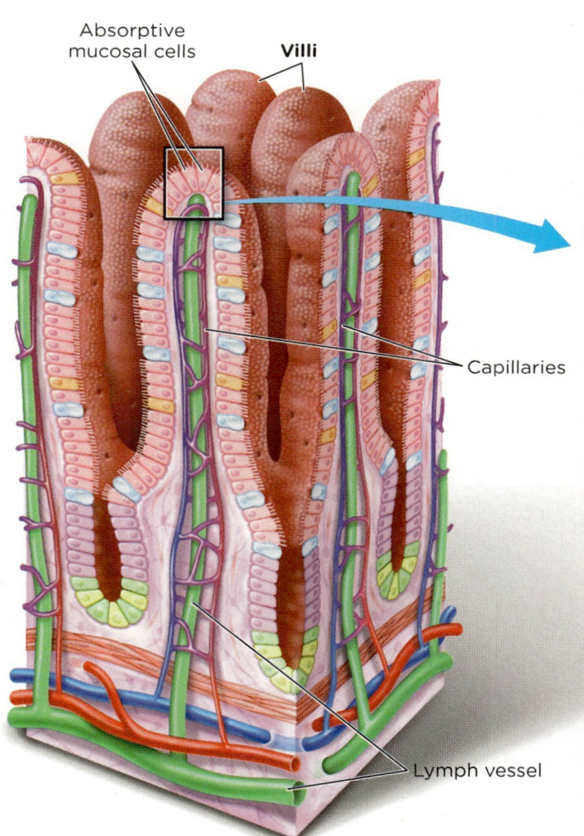

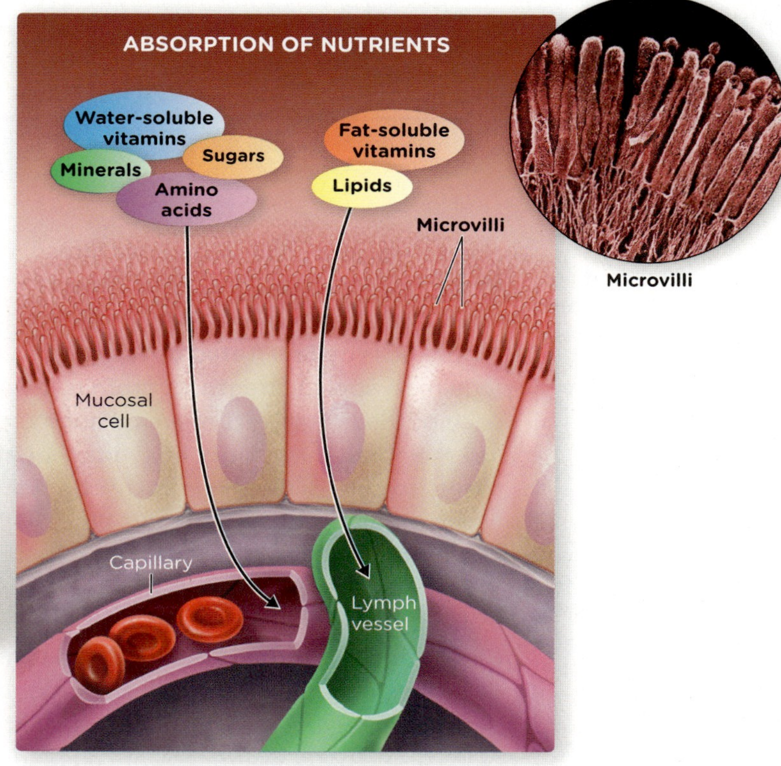

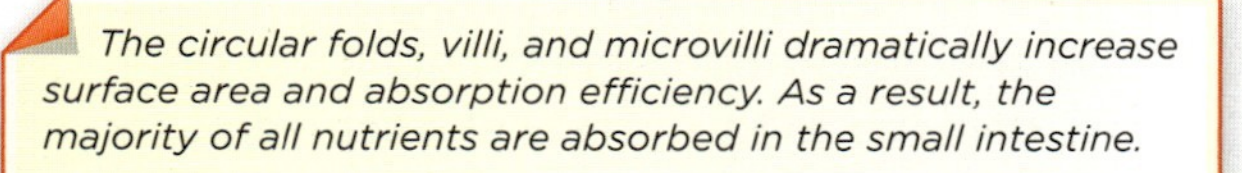
The circular folds, villi, and microvilli dramatically increase surface area and absorption efficiency. As a result, the majority of all nutrients are absorbed in the small intestine.

INFOGRAPHIC 3.8 Transport Mechanisms Involved in Nutrient Absorption *Several different mechanisms are used to transport nutrients into cells lining the lumen of the gastrointestinal tract.*

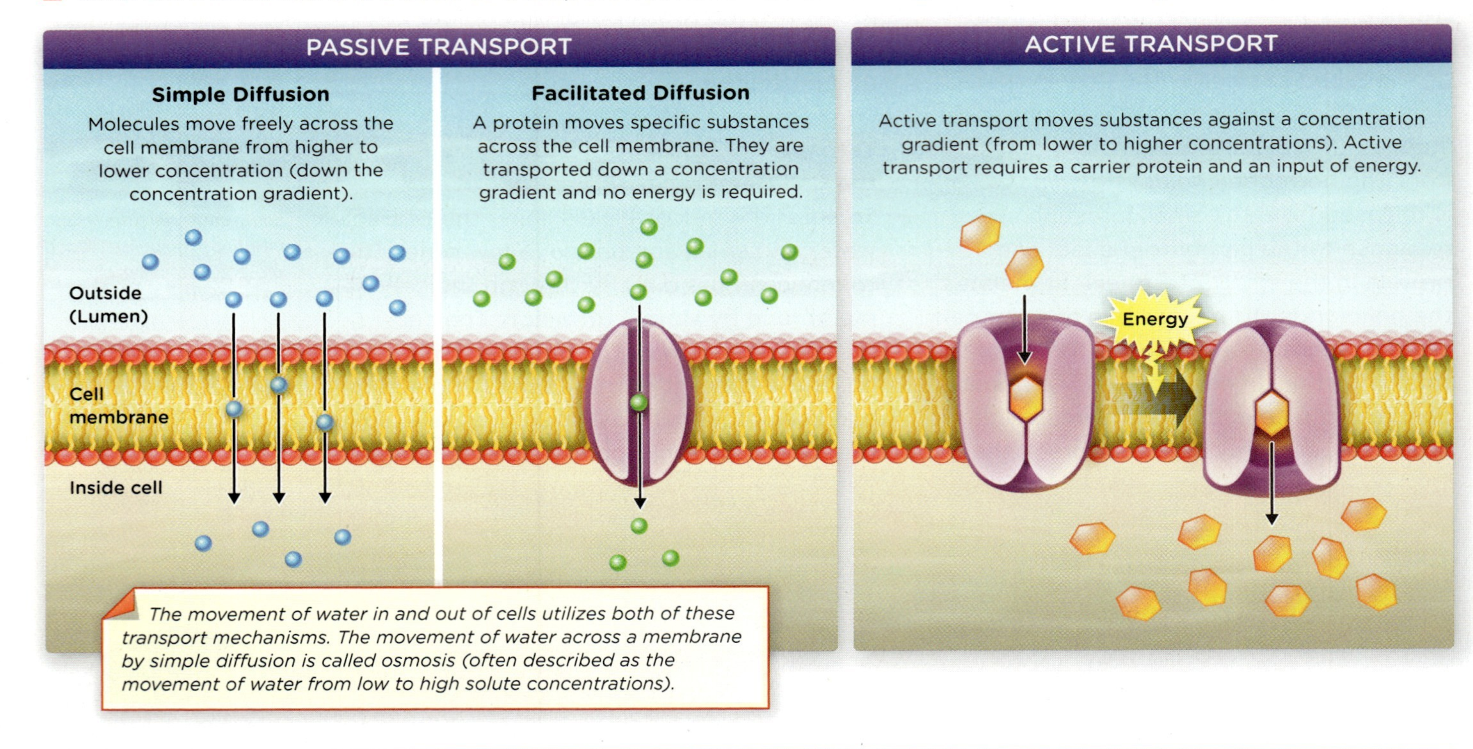

What is the only transport mechanism that can transport substances against a concentration gradient?

FACILITATED DIFFUSION *movement of a substance across a cell membrane, down a concentration gradient, with the assistance of a specific transport protein*

ACTIVE TRANSPORT *the energy-requiring, carrier-mediated process of transporting a substance across a cell membrane against a concentration gradient*

CIRCULATORY SYSTEM *a system made up of veins, arteries, capillaries, heart and lymphatic vessels; responsible for movement of blood and lymph throughout the body*

LYMPHATIC SYSTEM *a system of vessels in which the products of fat digestion, among other things, are transported from the GI tract to the blood*

also serves as a semi-permeable filter through which nutrients can enter and wastes can be excreted. Many nutrients enter the cells by **facilitated diffusion**, which requires a specific transport protein to help each of these nutrients move through the cell membrane. The entry of even other nutrients to cells depends on **active transport**, which requires both a transport protein and energy to bring them across a membrane. (INFOGRAPHIC 3.8)

Circulation of Nutrients in Blood and Lymph

Once inside the cells of the brush border of the small intestine, nutrients must reach the areas of the body where they are needed. The **circulatory system** which includes both the blood and the **lymphatic system** distribute vital nutrients to the tissues and organs. For example, carbohydrates, amino acids, minerals, and water-soluble vitamins enter directly into blood vessels, where they are transported to the liver before reaching other organs. Most fats (and some vitamins) first enter lymphatic vessels before they find their way into the blood. As a result, they reach the liver only after circulating throughout the rest of the body. (INFOGRAPHIC 3.9)

■ ■ ■

For people with celiac disease, however, these last few steps do not progress properly, and the delicate and essential process of absorption becomes disrupted. The gluten found in the hamburger bun sparks an immune reaction that triggers the person's immune cells to attack other body cells. These so-called autoimmune cells begin destroying the villi of the small intestine. This leaves a flattened surface on the small intestine that compromises digestion and absorption of nutrients. Over time, this damage to the intestinal tract increases the risk for certain types of intestinal cancers, including intestinal lymphomas, which could help explain the higher mortality rate among people suffering from celiac disease.

INFOGRAPHIC 3.9 **The Circulatory System** *is the organ system that allows blood and lymph to circulate—delivering oxygen, nutrients, wastes, and immune cells to their appropriate destinations throughout the body.*

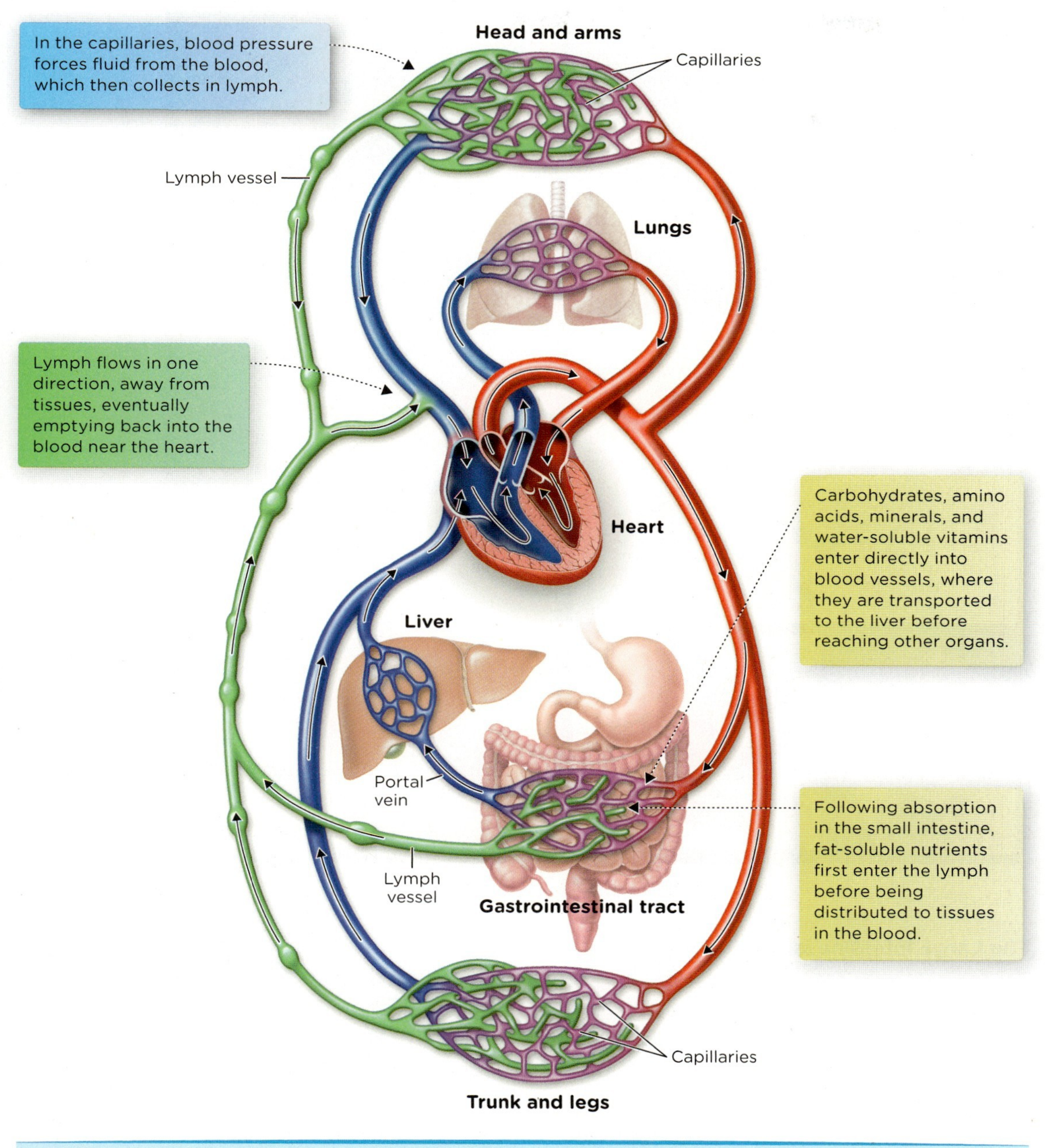

? *How do lipids that leave the small intestine in lymph get to tissues throughout the body?*

Grains with gluten. *The protein gluten is found in wheat, rye, and barley. Gluten causes inflammation in the small intestines of people with celiac disease, which may affect up to 1% of the population in the United States.*

Valentyn Volkov/Alamy

The Large Intestine is the End of the Gastrointestinal Tract

From the small intestine, any undigested nutrients that remain in the chyme are passed into the *large intestine*, which consists of the *cecum*, *colon*, and the *rectum*. Here, little digestion or absorption take place; there are no villi. Secreted mucus protects and lubricates the lining, making it easier for everything that's leftover to be excreted as *feces*. **(INFOGRAPHIC 3.10)**

But before that happens, the large intestine will extract electrolytes (sodium, chloride, and potassium), some fatty acids, vitamins (K, biotin, and folate), and water. In addition, the large intestine contains more than 1,000 species of bacteria that feed on undigested fiber and starch; since humans have no enzymes that digest dietary fiber, these bacteria perform some of that function, producing gas and short-chain fatty acids in the process. Some bacteria play a role in preventing disease, reducing the activity of other bacteria that may cause it.

PROBIOTICS *live, beneficial bacteria found in fermented foods that can restore or maintain a healthy balance of "friendly" bacteria in GI tract*

THE ROLE OF BACTERIA IN THE GASTROINTESTINAL TRACT

Some of the most helpful bacteria are known as **probiotics**—a word you may have seen on food or beverage labels. Food manufacturers highlight the probiotics in their products because these bacteria help restore or maintain a healthy balance of "friendly" bacteria in the GI tract. Although some manufacturers add probiotics to their

INFOGRAPHIC 3.10 The Large Intestine

The primary role of the large intestine is to form and store feces, although water, electrolytes, and a limited number of fatty acids are absorbed from the large intestine.

THE LARGE INTESTINE

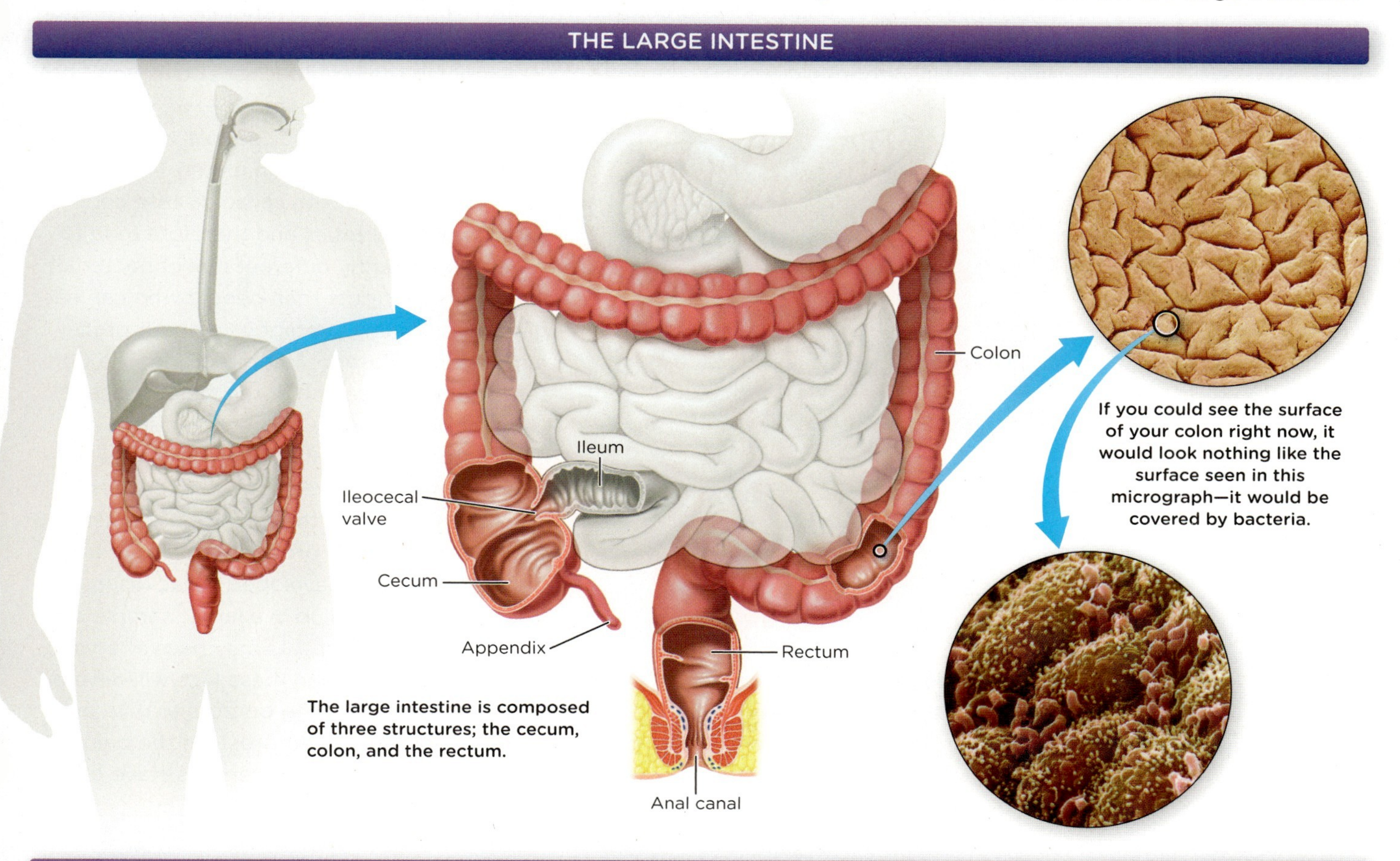

ABSORPTIVE AND SECRETORY FUNCTIONS OF THE COLON

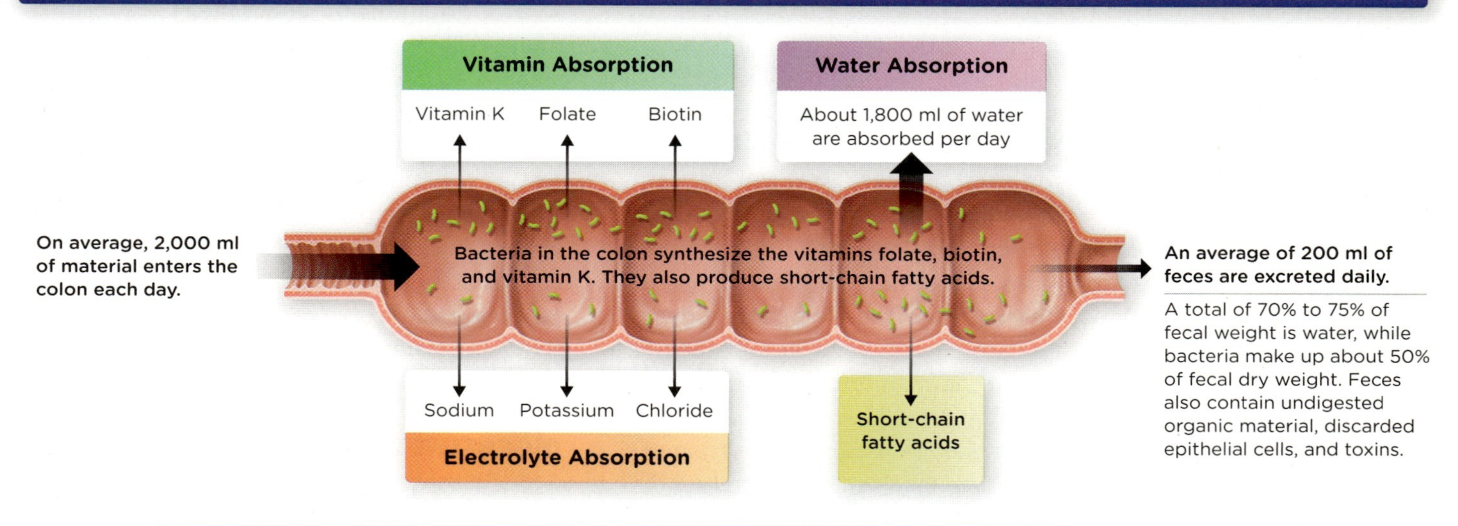

? *What is the name of the structure that prevents the movement of fecal material into the small intestine?*

Photo credits (top to bottom): Susumu Nishinaga/Science Source, Biomedical Imaging Unit, Southampton General Hospital/Science Photo Library

Live and active cultures. *The bacteria listed on the label of this yogurt convert pasteurized milk to yogurt during fermentation, and as probiotics they help maintain the natural balance of organisms in the intestines.*

The Photo Works/Alamy

products, probiotics are found naturally in fermented foods, such as dairy products like yogurts, buttermilk, and cottage cheese. Other fermented sources of probiotics include soy, tempeh, miso, and sauerkraut.

A healthy diet should also contain plenty of **prebiotics**, certain types of nondigestible carbohydrates that healthy bacteria use to boost their growth in the large intestine. (You can think of prebiotics as substances that "feed" or nourish good bacteria.) Eating prebiotics may prevent and treat diarrhea and colon cancer, boost the absorption of minerals, reduce levels of fat in the blood, and help control blood glucose. Some of these benefits may result from the short-chain fatty acids that are produced as bacteria use the prebiotics. These short-chain fatty acids, which are only 3 or 4 carbons in length, are then absorbed and provide us with a little energy, as well. Sources of prebiotics include chicory, Jerusalem artichokes, whole-grain rye, oats, wheat, and barley, leeks, garlic, and onion. Chapter 12 contains more detailed information about both probiotics and prebiotics.

PREBIOTICS *nondigestible carbohydrates broken down by colon bacteria that foster the growth of good bacteria*

VOMITING *forcible ejection of contents of stomach through the mouth; may be self-induced or due to sickness or food-borne illness*

Scanning the prebiotic list above, you'll notice that some of those foods—rye, wheat, and barley—are damaging for people with celiac disease, since they contain gluten. Not surprisingly, then, people with celiac disease consume fewer prebiotics and therefore tend to have a significantly different mix of bacterial species in their guts than people who can eat gluten-containing prebiotics without problems. Alessio Fasano, MD, director of the Center for Celiac Research at the Massachusetts General Hospital for Children in Boston, recently studied the guts of a small number of infants genetically predisposed to celiac disease and found that the balance of bacteria and their metabolic products shifted just before the infants developed celiac disease. If these results hold up in a larger sample of infants, he says, scientists may have a "crystal ball that predicts when these kids will develop autoimmunity," allowing doctors to give probiotics or prebiotics that restore the balance of their gut bacteria and perhaps stave off the disorder.

Even though celiac disease does its damage in the small intestine, its effects are felt throughout the body. That's because the immune cells that start attacking the villi of the small intestine eventually go somewhere else. "Celiac starts in the small intestine, but any organ in the body can be affected," Fasano says. He has seen patients with celiac disease and no stomach problems, for instance, but who have anemia or damage to the liver. The pancreas can also be damaged by these immune cells, rendering it less able to produce hormones that regulate metabolism, such as insulin.

■ ■ ■

DIGESTIVE DISORDERS

Celiac disease isn't the only condition that can arise within the GI tract, of course. Most of us are familiar with the most common problems. **Vomiting**—the forceful movement of stomach contents from the stomach through the mouth and out of the body—can be caused by many things including emotional stress, motion sickness, viruses, or foodborne bacteria.

Chapter 20 contains more information about foodborne illnesses.

Diarrhea, Constipation, and Hemorrhoids

Many digestive disorders include the symptom of **diarrhea**, which is defined as frequent loose and watery bowel movements that occur more than three times a day. **Constipation**, having a bowel movement fewer than three times per week, is another symptom common in digestive disorders. In otherwise healthy people, a low-fiber diet, medication, or dehydration may be the culprit. Consuming sufficient fiber will soften stools since it holds water and so decreases the risk of both constipation and painful inflamed veins called **hemorrhoids**.

Problems Associated with Weakened or Relaxed Lower Esophogeal Sphincters

Acid reflux is a condition in which the sphincter muscles separating the stomach and the esophagus relax, so that food and stomach acid are occasionally regurgitated back into the esophagus. The stomach acid can cause a burning feeling called *heartburn*. **Gastroesophageal reflux disease (GERD)** is a recurrent and more serious form of acid reflux that is accompanied by inflammation and/or erosion of the esophageal lining along with pain and gastrointestinal issues. Individuals with untreated GERD can be at risk of cancer or other complications. Fortunately, over-the-counter and prescription drugs can treat GERD, and sufferers can control symptoms by avoiding foods that cause discomfort, including alcohol, onions, chocolate, citrus fruits, and large or fatty meals.

Gallstones

Painful *gallstones* can develop when substances in bile crystallize into small pebbles within the gallbladder. The gallbladder may not be able to empty properly causing pain in the upper abdomen. Recurrent gallstones generally require the surgical removal of the gallbladder.

Diverticular Disease

Diverticular disease is a condition characterized by pouches or pockets in the wall or lining of any portion of the digestive tract, which can develop when the inner layer of the digestive tract pushes through weak spots in the outer layer. (INFOGRAPHIC 3.11)

IRRITABLE BOWEL SYNDROME (IBS) AND INFLAMMATORY BOWEL DISEASE (IBD)

Irritable bowel syndrome (IBS) is a group of symptoms including abdominal pain, bloating, diarrhea, and other discomforts caused by changes in how the gastrointestinal tract works. This disorder affects the muscle contraction of the colon (large intestine). Although it can be very uncomfortable and inconvenient, IBS does not lead to serious disease, such as cancer, and it does not permanently harm the large intestine. IBS is one of the most common disorders diagnosed by doctors, up to 20% of adults in the United States have symptoms of IBS. Certain foods may trigger IBS symptoms, so individuals with the condition must take note of the foods or beverages, such as caffeine, alcohol, fatty foods, and some nonnutritive sweeteners, that may cause difficulty. When diarrhea occurs, water consumption must be increased to avoid dehydration. An increase in dietary fiber can help deal with the symptom of constipation. Some medications are prescribed for individuals with IBS, and probiotics may be helpful for some people.

Inflammatory bowel disease (IBD) is a broad term that describes serious, chronic conditions that are caused by an abnormal response by the body's immune system, which causes inflammation of the GI tract. The most common IBDs are ulcerative colitis and Crohn's disease. In ulcerative colitis, the mucosa of the intestine becomes irritated and swollen and open wounds (**ulcers**) develop. Ulcerative colitis is often most severe in the lower colon (the rectum), which can cause diarrhea. Crohn's disease most often affects the lower portions of the small intestine and parts of the large intestine. However, it can attack any part of the digestive tract.

Individuals with IBDs are at risk for serious or even life-threatening complications, such as bowel obstruction (caused by the narrowing of the intestinal wall); ulcers in the digestive track, including the mouth and anus; fistulas (an abnormal connection between the intestine and skin or other organs); malnutrition because of poor absorption; colon cancer; and problems in parts of the body outside of the GI tract, such

DIARRHEA
loose, watery stools on more than three occasions in a 24-hour period

CONSTIPATION
difficulty, or reduced frequency, of stool passage through intestines

HEMORRHOIDS
swollen or inflamed veins in anus or lower rectum

ACID REFLUX
the regurgitation of acid content from the stomach into the esophagus; characterized by a burning feeling in the chest called heartburn

GASTROESOPHAGEAL REFLUX DISEASE (GERD)
a recurrent and more serious form of acid reflux, accompanied by inflammation and/or erosion of esophageal lining

DIVERTICULAR DISEASE
condition in which there are small pouches or pockets in the wall or lining of the colon; a single pouch is called a diverticulum

IRRITABLE BOWEL SYNDROME (IBS)
a group of symptoms that occur together: abdominal pain or discomfort, along with diarrhea and/or constipation

INFLAMMATORY BOWEL DISEASE (IBD)
general name for diseases that cause inflammation and irritation of the gastrointestinal tract; examples include Crohn's disease and ulcerative colitis

ULCER
irritation or perforation of stomach (gastric) or small intestinal (duodenal) mucosal wall, caused by Helicobacter pylori *infection, decreased mucus production, or impaired removal of stomach acid*

INFOGRAPHIC 3.11 Common Digestive Disorders *Digestive disorders affect millions of people each year and decrease quality of life and overall health. See http://www.nutrition.gov/nutrition-and-health-issues/digestive-disorders.*

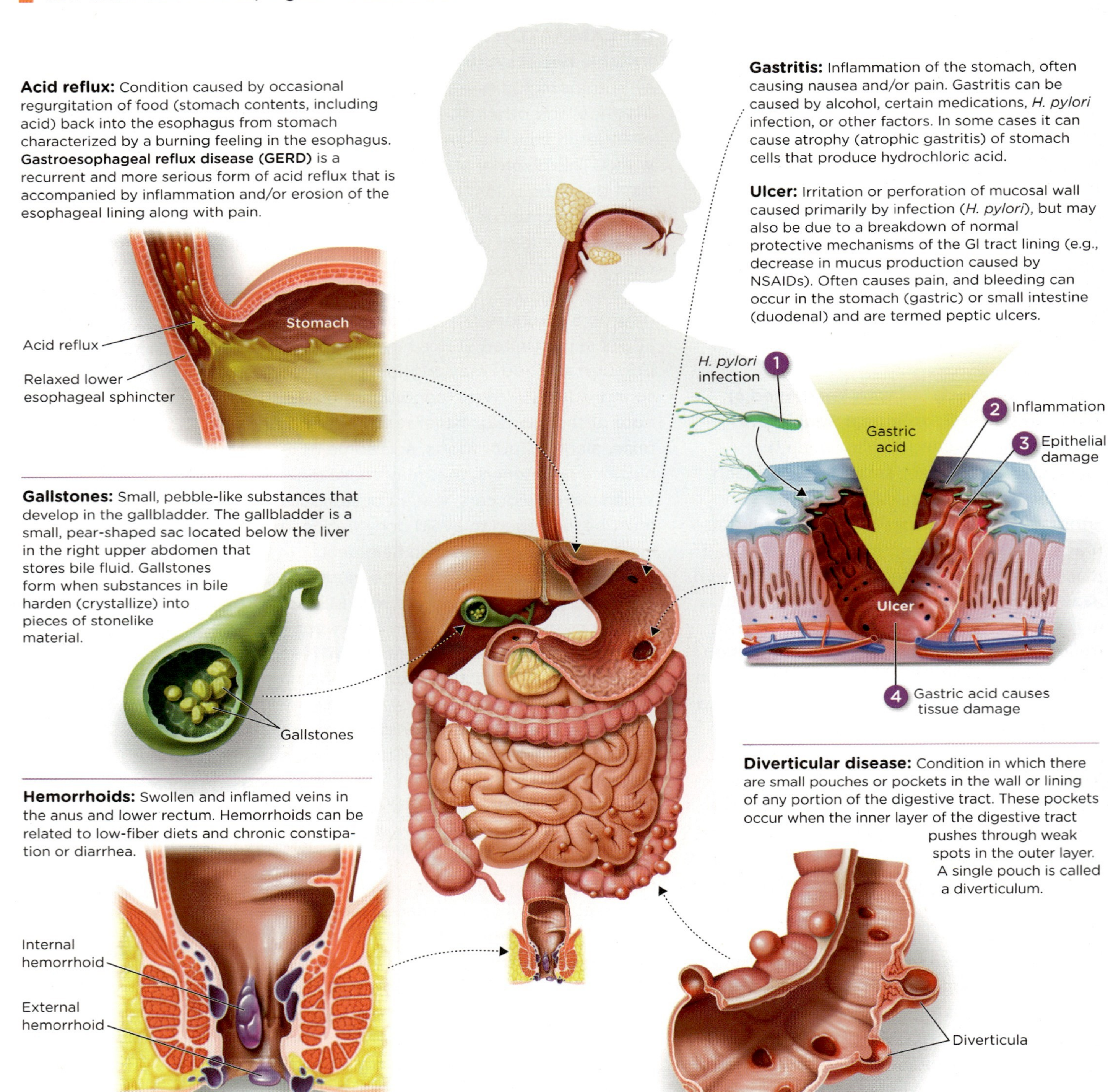

as kidney stones or arthritis. A person with IBD needs a nourishing diet to deal with potential nutrient deficiencies brought about by absorption issues, as well as assistance when significant weight loss occurs. People suffering with IBDs are treated with anti-inflammatory drugs, and often require surgical procedures to deal with medical complications.

AWARENESS OF CELIAC DISEASE IS GROWING

Chances are, however, that the digestive condition you have heard about most frequently is celiac disease, in part because going "gluten-free" has become a popular diet fad among people both with and without the condition. Celiac disease is also attracting attention because, as researchers have recently discovered, some people with the condition don't exhibit the typical symptoms of diarrhea or weight loss, which makes it even harder to identify and diagnose. One other major symptom of celiac disease, for instance, is a specific type of skin rash; when these patients cut gluten out of their diets, the rash disappears. Many people with celiac disease also develop neurological problems, making it more difficult for them to walk. Some of these people will experience minimal or no digestive issues.

■ ■ ■

Once Murray began to spread the news that celiac disease could manifest itself in mysterious ways, and that untreated celiac disease could increase the risks of all sorts of health problems, people started paying a lot more attention to gluten. Slowly, gluten became transformed from a protein most people didn't know existed to one of the most talked about food ingredients in the country. The hype surrounding gluten doesn't really apply to the vast majority of people—studies show that celiac disease may affect only approximately 1% of Americans. Other individuals may have GI-related or other symptoms when they consume gluten and experience what has been termed *non-celiac gluten sensitivity*: a real but poorly understood condition that at present is primarily self-diagnosed. The actual incidence rate is unclear, but a recent study confirmed the condition in fewer than 10% of individuals who reported gluten sensitivity. Thus, many people appear to be avoiding gluten unnecessarily with any perceived benefits coming from heightened awareness of their overall food choices and a lower intake of refined and processed foods. Many of the gluten-free products are processed foods (snacks and baking mixes) that are more expensive and less nutrient-dense than whole foods.

Gone gluten-free. *Grocery store shelves are stocked with gluten-free alternative food products.*

Daniel Acker/Bloomberg via Getty Images

Unfortunately, no one yet knows exactly why the incidence of this immune condition is rising, says Murray, although there are many possibilities. Celiac disease tends to appear more commonly in people with other immune disorders, which suggests that something in the environment may be disrupting people's immunity. In other words, the rising incidence of celiac disease may be a symptom of a much larger and more pervasive environmental health problem. But the good news is that celiac disease is largely treatable. By avoiding one little protein, people with celiac disease can not only avoid unpleasant stomachaches but also reduce risk of other autoimmune issues and certain cancers. And for the few people with celiac disease, not eating gluten will allow them to better absorb nutrients from their diet and improve their health.

CHAPTER 3 **BRING IT HOME**

Focus on digestive disorders

According to the National Digestive Diseases Information Clearinghouse (NDDIC) of the United States Department of Health and Human Services (http://digestive.niddk.nih.gov/statistics/statistics.aspx), digestive disorders affect more than 60 million people and account for 50 million visits to primary care providers and 20 million hospitalizations per year. The annual economic impact in the United States is estimated at more than $140 billion. Diet and food choice can contribute to the cause, prevention, and treatment of the various digestive disorders.

More than likely, you or a family member has experienced one or more of the digestive disorders listed in the table below. Using information provided on page 68, in Infographic 3.11, and through the NDDIC website at http://digestive.niddk.nih.gov/ddiseases/a-z.aspx complete the table for at least three of the digestive disorders with a brief description of the disorder, possible causes (primarily as related to diet and health habits), and a summary of potential prevention and/or treatment of dietary interventions. The table has been completed for constipation as an example.

DIGESTIVE DISORDERS—DESCRIPTION	POSSIBLE CAUSES	PREVENTION AND/OR TREATMENT—DIETARY INTERVENTIONS
Constipation—*Difficulty, or reduced frequency, of stool passage through intestines*	*Lack of dietary fiber and lack of physical activity*	*Consume a diet with adequate fiber (20–35 grams recommended) along with sufficient fluids and physical activity*
Diarrhea		
Diverticular disease		
Gallstones		
Gastritis		
Gastroesophageal reflux disease (GERD)		
Heartburn		
Hemorrhoids		
Inflammatory bowel disease (IBD)		
Irritable bowel syndrome (IBS)		
Ulcer		
Vomiting		

KEY IDEAS

The process of digestion makes nutrients available for absorption and use in the body.

The primary organs of the gastrointestinal (GI) tract include the mouth, esophagus, stomach, small intestine, and large intestine. The lumen, or inner space of the GI tract, is lined by mucosal cells that allow our body to absorb nutrients, so that they can be transported by the blood or lymph to where they are needed.

The digestive system includes accessory organs that secrete fluids containing a variety of agents that aid in digestion and include the salivary glands, liver, gallbladder, and pancreas.

Mechanical digestion, which begins in the mouth, involves the physical fragmentation of foods into small particles. Food is mixed with digestive fluids, propelled along the GI tract through two patterns of contractions of the smooth muscles called peristalsis and segmentation.

Chemical digestion involves specific enzymes and other secretions that break down the chemical compounds in food.

Enzymes are protein molecules that catalyze, or speed up, the rate at which a chemical reaction produces new compounds with altered chemical structures.

Hormones are chemical messengers that participate in the control and regulation of body processes.

The stomach secretes gastric juices that contain hydrochloric acid and nutrient-specific enzymes that, together with peristaltic contractions, break down food into a semi-liquid called chyme.

The small intestine is the primary site where the chemical digestion of food and the absorption of nutrients are accomplished. The small intestine has structural features—villi and microvilli (often called the brush border)—that increase its absorptive area.

Nutrients enter mucosal cells along the GI tract through simple diffusion and facilitated diffusion, as well as through active transport.

From the brush border, nutrients either enter the blood immediately, or first enter the lymphatic system and then are distributed to tissues and organs via the blood.

Probiotics are live, beneficial bacteria found in fermented foods that can restore or maintain a healthy balance of "friendly" bacteria in the GI tract. Prebiotics are nondigestible carbohydrates broken down by bacteria in the large intestine that foster the growth of beneficial bacteria.

Digestive disorders affect millions of people each year and diminish quality of life and overall health. Diet may play an important role in the prevention and treatment of these conditions.

NEED TO KNOW

Review Questions

1. The primary organ for digestion and absorption in the body is the:
   a. stomach.
   b. liver.
   c. colon.
   d. small intestine.

2. All of the following are functions of the mouth in digestion, EXCEPT:
   a. the mechanical breakup of food.
   b. the moistening and mixing of food.
   c. the secretion of protease to begin protein digestion.
   d. the secretion of amylase to begin starch digestion.

3. Chemical digestion of food includes:
   a. chewing of food in the mouth.
   b. gastric acids in the stomach.
   c. peristalsis and segmentation.
   d. the incorporation of fatty acids into chylomicrons.

4. All of the following are true regarding enzymes that participate in the process of digestion, EXCEPT:
   a. they are secreted by the salivary glands, stomach, pancreas, and small intestine.
   b. they speed up chemical reactions that break down food.
   c. they are changed by the chemical reactions they facilitate.
   d. they can participate in chemical reactions many times.

5. Chemical substances that act as messengers between organs to cause the release of secretions needed for digestion and other body processes are termed:
   a. chylomicrons.
   b. lipoproteins.
   c. enzymes.
   d. hormones.

6. During digestion, the stomach gradually ejects small amounts of semi-liquefied content called _____ into the _____.
   a. chyme; small intestine
   b. bolus; esophagus
   c. chylomicrons; colon
   d. bile; jejunum

7. Structures that increase the surface area of the small intestine are called:
   a. chyme.
   b. villi.
   c. mucosa cells.
   d. sphincters.

8. Absorption is the process by which:
   a. excess calories are incorporated into adipose tissue.
   b. nutrients are converted into ATP.
   c. food moves from the esophagus into the stomach.
   d. nutrients pass through the intestinal mucosa into the blood or lymph.

9. Both facilitated diffusion and active transport of nutrients into cells require:
   a. specific transport proteins.
   b. energy.
   c. transport fatty acids.
   d. enzymes.

10. From inside the cells of the brush border most fats are absorbed into:
    a. blood vessels.
    b. lymph vessels.
    c. triglycerides.
    d. arterial cholesterol deposits.

11. Food sources of prebiotics that foster growth of good bacteria in the large intestine include all of the following, EXCEPT:
    a. oats.
    b. fermented products.
    c. barley.
    d. wheat.

12. A digestive disorder that affects the muscle contractions of the colon and is characterized by a group of symptoms that include abdominal pain, bloating, diarrhea, and other possible GI symptoms is termed:
    a. celiac disease.
    b. gastrointestinal reflux disease.
    c. irritable bowel syndrome.
    d. diverticular disease.

Take It Further

Describe the differences in the muscular movements of peristalsis and segmentation, and the way they assist in digestion. After digestion, the next step is absorption by the mucosal cells of the small intestine. Describe the features of the small intestine that make it possible to maximize the absorption of nutrients.

Dietary Analysis Using SuperTracker

Take charge of your digestive health

Now that you have learned about the structures and functions associated with the digestive process, you can begin to think critically about how the foods you eat may or may not support a healthy digestive environment.

To begin, keep a journal of everything you eat for three full days. It is important to include all food and drink items you consume during this three-day period, taking note of the amount of each item you consume.

After completing your three-day food journal, you will enter/analyze your food using the SuperTracker.

To do this, follow the steps below:

1. Log onto the United States Department of Agriculture (USDA) website at www.supertracker.usda.gov. If you have not done so already, you will need to create a profile to get a personalized diet plan. This profile will allow you to save your information and diet intake for future reference. *Do not use the general plan.*
2. Click the Track Food and Activity option.
3. After you have entered your food for each day, you can then use the "My Reports" feature to analyze your intake. The "Food Groups and Calories" report shows your average intake for each food group. It also shows average calorie intake. The "Nutrients Report" shows an average intake of specific nutrients (such as sodium, calcium, and vitamin D), as well as the foods you consumed that provide the highest or lowest amount of each nutrient.
4. Once you have had a chance to generate and review these two reports, reflect on the results of your analysis by answering the following questions:
    a. In which food groups are you meeting your daily targets? Based on what you have learned in this chapter, what, if any, are the digestive benefits of consuming your target amount in these groups?
    b. In which food groups are you falling short of your daily targets? What are the possible digestive implications of not meeting your daily targets in these groups?
    c. Does your analysis show that you are eating foods that offer additional digestive benefits, such as probiotics and prebiotics? Which foods in your intake fall in this category?
    d. Based on your analysis, write two goals that relate to making dietary choices that promote proper digestion.

Whole Grain Hype

CAN SCIENCE HELP US NAVIGATE THE PERILS OF THE CEREAL AISLE?

plainpicture/Rupert Warren

LEARNING OBJECTIVES

- Identify the primary functions of carbohydrates in food and in the body (Infographic 4.1)
- Describe the classifications of dietary carbohydrates and their chemical composition (Infographic 4.2)
- Define "whole grain" and explain what occurs when grain is refined (Infographic 4.5)
- Explain how carbohydrates are synthesized in plants and animals (Infographics 4.2 and 4.4)
- Outline the steps in carbohydrate digestion (Infographic 4.6)
- Identify sources of added sugar in the U.S. diet (Infographic 4.7)
- Describe the differences between nutritive and non-nutritive alternative sweeteners and identify examples of each category (Infographic 4.8)
- Identify the types and sources of fiber and describe its health benefits (Infographics 4.9 and 4.10)
- Identify sources of carbohydrates in foods and describe the dietary recommendations for carbohydrate intake (Infographics 4.3 and 4.11)

When Rebecca Mozaffarian and her colleagues at the Harvard School of Public Health were asked to come up with guidelines for healthful snacks to serve in schools, they quickly ran into a problem: it was easy to make certain food recommendations—avoid sugar-sweetened beverages, serve plenty of vegetables and fruits—but much harder to come up with clear recommendations for foods made from grains. Grains include things like wheat, barley, rice, corn, and

WHOLE GRAIN *cereal grains, or foods made from cereal grains, that contain all the essential parts (starchy endosperm, germ, and bran) of the entire grain seed in its original proportions*

REFINED GRAIN *cereal grains that have been milled, a process that removes the bran and germ; white flour is an example*

ENRICHED GRAIN *cereal grains that lost nutrients during processing but have vitamins and minerals added back in; refined grains are often enriched*

Whole grain wheat kernels and ears of wheat. *The wheat kernels can be ground to make flour.*

Elena Elisseeva/Alamy

other cereal grasses that humans use to make everything from granola bars, breads, and breakfast cereals, to pasta, crackers, and cupcakes.

Scientists generally agree that eating "whole" grains is better than eating "refined" or "enriched" grains. A **whole grain** is one where all the edible parts of the original plant seed (grains are seeds) are found in the food product, including the energy-rich endosperm, the oil-rich germ, and the fiber-rich bran coating. **Refined grains** are stripped of their bran and the germ, leaving only the endosperm. Since many of the beneficial nutrients and phytochemicals are found in the bran and germ, refined grains are much less healthful. **Enriched grains** have some vitamins and minerals added back in, but they still pale nutritionally in comparison with whole grains.

Consuming whole grains has been linked to a host of health benefits, including a reduced risk of heart disease, obesity, diabetes—even cancer. The Dietary Guidelines for Americans advise that Americans should consume at least half their grains as whole grains.

Given their importance to health, Mozaffarian and her colleagues knew they wanted to say something about whole grains in their recommendations but soon realized that there was no easy

Foods that boast whole grain content vary in overall nutritional quality and healthfulness.

Kristoffer Tripplaar/Alamy

INFOGRAPHIC 4.1 Overview of the Carbohydrates

COMPOSITION

- Carbon, hydrogen, and oxygen

- Composed of one or more sugar (saccharide) units

- Contains 4 kcal per gram

FUNCTIONS IN FOOD

- Source of fiber
- Adds sweetness and flavor

FUNCTIONS IN THE BODY

- Source of energy for all cells in the body
- Indispensable source of energy for the brain, red blood cells, and muscles during intense exercise
- Important for intestinal health
- Reduces the use of protein for energy

forsterforest/ age fotostock

Fruits, vegetables, grains, and milk and milk products are sources of carbohydrates.

The Institute of Medicine recommends adults consume 45% to 65% of their daily calories from carbohydrates. Since carbohydrates contain 4 kcal in each gram, about how many carbohydrate kcals should you consume each day? (Base your calculations on a 2,000-kcal daily intake.)

way to identify healthful whole grain foods. Although many food products tout their "whole grain" ingredients, and some even include a "stamp of approval," some whole grain food choices are much better than others. When it comes to whole grain foods, separating science from marketing—wheat from chaff—can be tricky.

"We were sitting around thinking 'People are probably struggling with this issue,'" Mozaffarian says. That's when she got the idea for a scientific study: compare different ways of identifying whole grain foods to see which ones were most useful in identifying healthful food choices. She was particularly interested in distinguishing the types of carbohydrates found in the products.

To appreciate why Mozaffarian and her colleagues were interested in carbohydrates, it helps to understand the different forms that carbohydrates can take and their different properties.

WHAT ARE CARBOHYDRATES?

Carbohydrates are molecules made up of carbon, hydrogen, and oxygen arranged as one or more sugar units. They are abundant in grains and other plant foods, as well as in milk and some milk products, and are key sources of fuel for the body, providing four kcal of energy per gram. Carbohydrates are one source of energy for exercising muscles; they are the exclusive source of energy for red blood cells and provide a significant portion of the energy needed by the brain. (INFOGRAPHIC 4.1)

Structure and function of carbohydrates

Depending on their size, carbohydrates can be classified as either *simple* or *complex.* Simple carbohydrates, also known as sugars, are short carbohydrates made up of one or two sugar units. Sugars that are made up of one sugar unit are called **monosaccharides**; sugars made up of two sugar units are called **disaccharides**.

MONOSACCHARIDE
a carbohydrate that consists of only one sugar molecule; for example, glucose, fructose, and galactose

DISACCHARIDE
a carbohydrate that consists of two sugar molecules; for example, maltose, sucrose, and lactose

INFOGRAPHIC 4.2 Simple and Complex Carbohydrates *Carbohydrates can be classified as simple carbohydrates (sugars), or complex carbohydrates (starch and fiber in plants, and glycogen).*

SIMPLE CARBOHYDRATES

Monosaccharides: made up of one sugar unit	Disaccharides: made up of two sugar units
Glucose circulates in the blood stream. It is found in fruits, vegetables, and honey.	**Maltose** is formed in large amounts as a product of starch digestion; however, very little is found in the foods we eat.
Fructose is found in fruits, vegetables, honey.	**Sucrose** is otherwise known as "table sugar." It is also found in fruits and vegetables.
Galactose is one of the monosaccharides that make up milk sugar.	**Lactose** is often called "milk sugar" as it is found only in milk, yogurt, and other dairy products.

COMPLEX CARBOHYDRATES (POLYSACCHARIDES)

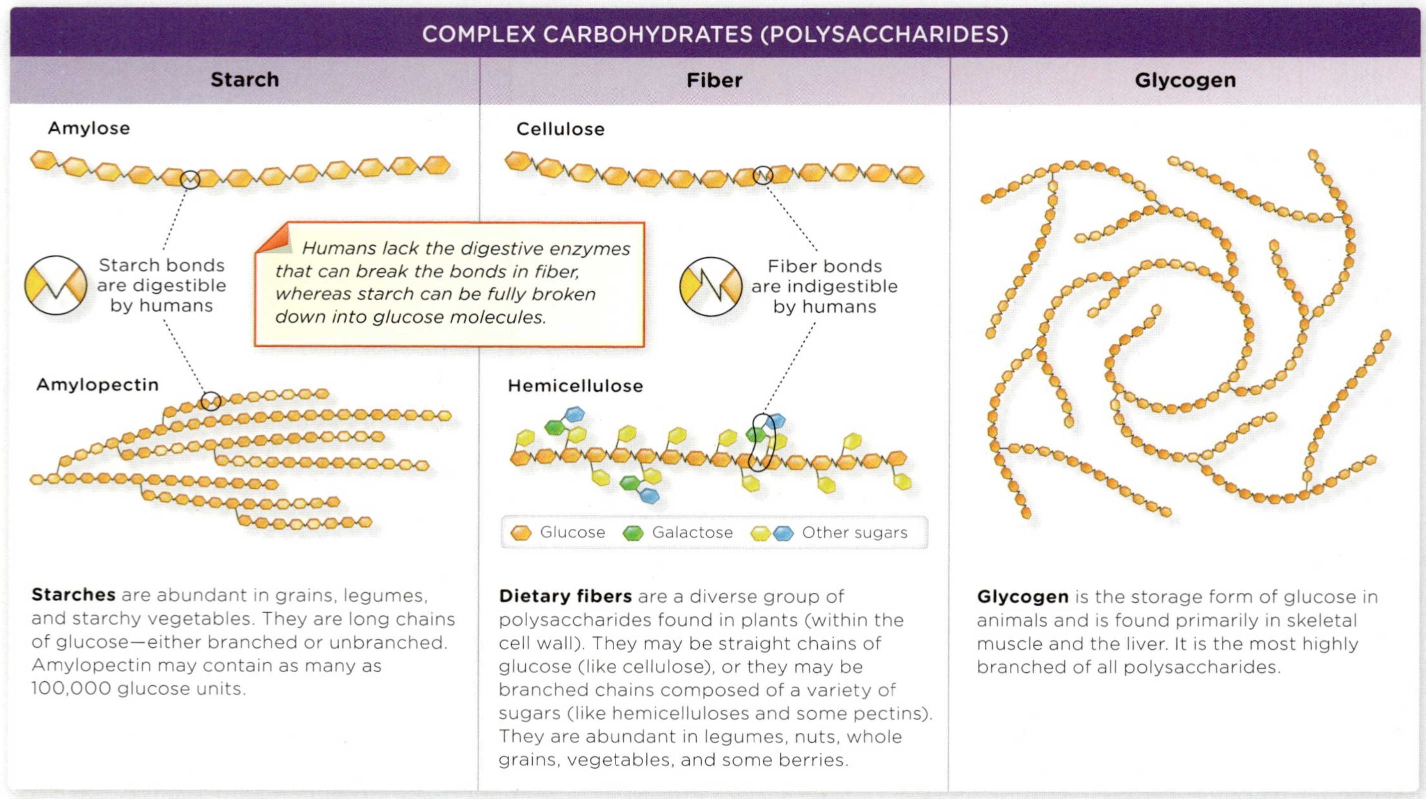

In what form are carbohydrates stored in our body?

The three most abundant monosaccharides are *fructose, glucose,* and *galactose*. Fructose is the sweetest of the sugars and is found in fruits and vegetables, and in honey. Fructose is also used to sweeten foods and beverages by food manufacturers who process corn to create high-fructose corn syrup. Glucose is the most abundant sugar in our diet because it is part of every disaccharide, starches, and most fibers. Galactose is one of the two monosaccharides (the other is glucose) that make up the sugar found in milk.

The disaccharides are made up of pairs of monosaccharides and are called *lactose, maltose*, and *sucrose*. The mammary glands of humans and other mammals synthesize lactose (galactose and glucose), which is incorporated into milk to feed offspring. *Maltose* is composed of two glucose units and is produced when starch is broken down in digestion. *Sucrose*, or "table sugar," is a disaccharide composed of glucose and fructose. We must digest dissacharides into their component monosaccharides before they can be absorbed by cells of the intestines.

Complex carbohydrates are composed of three or more monosaccharides linked together. They often take the form of long or branched chains. Humans and other animals break down some complex carbohydrates into individual monosaccharides during the process of digestion. These units are absorbed by the cells of the intestine and dumped into the blood for all body cells to pick up and use for energy. **Oligosaccharides** contain 3 to 10 linked monosaccharides, while **polysaccharides** are chains (or polymers) of more than 10 monosaccharides.

Starch, fiber, and glycogen

Most complex carbohydrates in our diet come from plants. Depending on how the sugars are bonded together, polysaccharides in plants may function as a source of stored energy (starch) or as structural material (fiber). **Starches** include amylose and amylopectin, found in grains, legumes, and starchy vegetables such as peas, potatoes, and corn. Starches are a form of stored energy for the plant. **Fibers** include cellulose and hemicellulose, which make up plant cell walls and impart structure to the plant. Fiber-rich foods are numerous and include whole grains and vegetables such as broccoli and green beans. Humans lack the digestive enzymes to break down plant fiber, so this material passes undigested through the digestive tract.

Humans and other animals also store glucose for later use. The storage form of carbohydrates in animals is called **glycogen**. Like starch, it is made up of linked glucose sugars, but is the most heavily branched of the polysaccharides. Glycogen is made and stored primarily in liver and muscle. When glucose is needed to maintain proper blood sugar levels, the liver breaks down the glycogen into glucose and releases it into the blood. Glycogen in skeletal muscles is broken down to supply fuel for contracting muscles during intense exercise. (INFOGRAPHIC 4.2)

Sources of carbohydrates

Carbohydrate-rich foods are a staple for most humans on earth. That's because carbohydrate-rich plant foods are typically a plentiful and cheap source of energy. Plant foods supply most of the carbohydrates we consume. Milk and milk products and honey are the only significant animal source of carbohydrates in the American diet. (INFOGRAPHIC 4.3)

The energy found within carbohydrates comes, ultimately, from the sun. Photosynthesis enables plants to capture the energy of sunlight and convert it into chemical energy in the form of glucose

OLIGOSACCHARIDE
a short-chain carbohydrate that consists of 3 to 10 monosaccharide units joined together

POLYSACCHARIDE
a long-chain carbohydrate that consists of more than 10 monosaccharides joined together; tend not to have a sweet taste (unlike monosaccharides and disaccharides) and can be found in foods such as whole grain breads, dried beans, and starchy vegetables

STARCH
polysaccharide made up of many glucose units joined together by digestible bonds; amylose and amylopectin are examples

FIBER
includes cellulose and hemicellulose, which make up plant cell walls and impart structure to the plant; humans lack the digestive enzymes to break down plant fiber, so it passes undigested through the digestive tract

GLYCOGEN
a polysaccharide consisting of many glucose molecules; glycogen acts as the storage form of glucose in animal tissues (liver and muscle)

INFOGRAPHIC 4.3 Carbohydrate Content of Commonly Eaten Foods

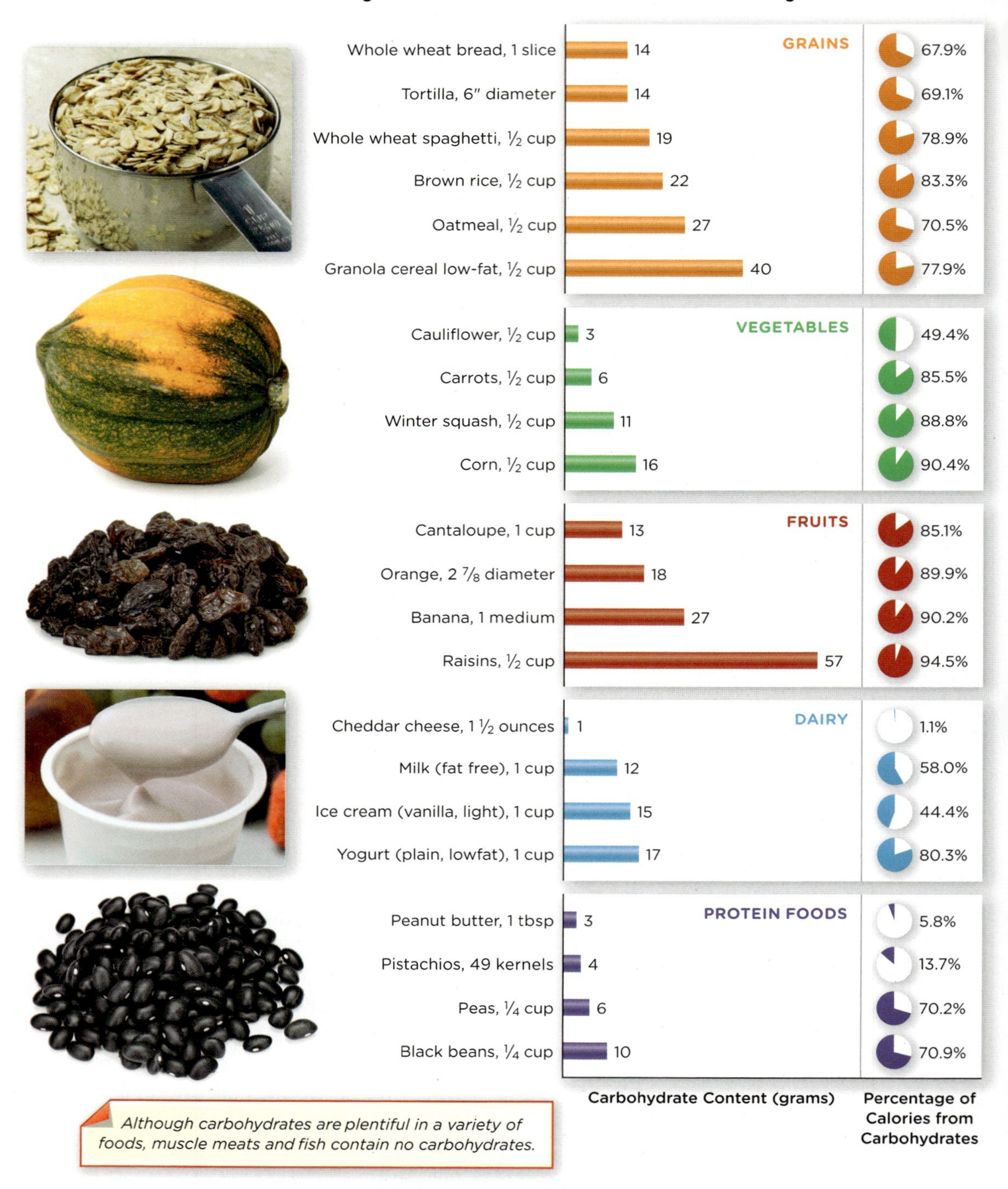

? *What food has the highest carbohydrate and protein content?*

Photo credits (top to bottom): Julie Woodhouse/Alamy, Eli Ensor, Eli Ensor, Karan Kapoor/Getty Images, 4kodiak/Getty Images

INFOGRAPHIC 4.4 Photosynthesis: Nature's Carbohydrate Production Process

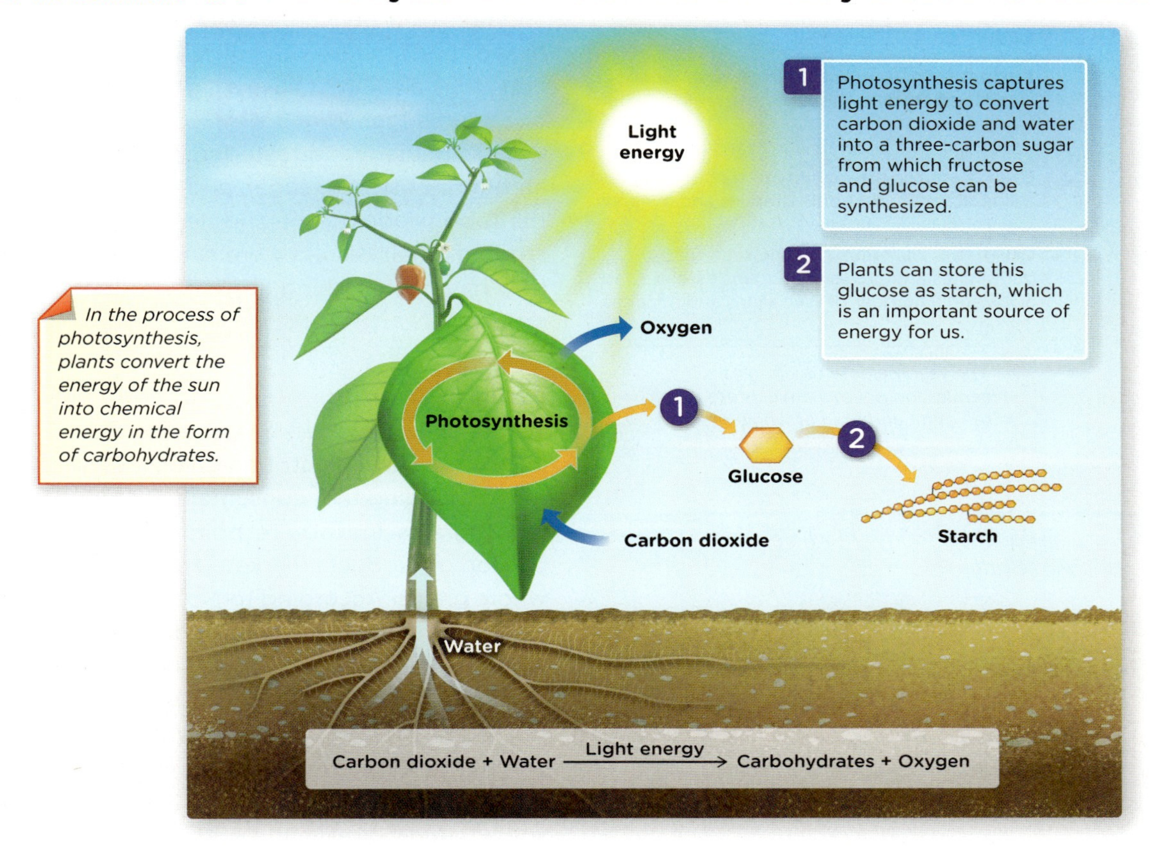

? *A green plant can carry out photosynthesis so long as it has what three things?*

and fructose. Plants then link glucose molecules together to form starch and some dietary fibers. **(INFOGRAPHIC 4.4)**

Refining of grains

In grains, most of the starch is contained within the endosperm, which, if the seed was planted, would provide energy to the growing plant embryo, or germ. This energy is not readily accessible to humans who wish to consume the grain, however, since it is surrounded by the fibrous bran coating. That's why, for millennia, humans have crushed, cracked, popped, pulverized, milled, or otherwise modified whole grains to expose the inner endosperm. Without such modifications, many grains would be either unpleasant to eat or would pass largely undigested through the digestive tract.

Beginning in the nineteenth century, "refining" grains became a common practice. Sophisticated milling machines allowed humans to separate the starchy endosperm from the tough bran and oily germ. The purified endosperm could then be ground into white flour. This practice helped keep flour from spoiling, since oil

INFOGRAPHIC 4.5 Anatomy of a Whole Grain

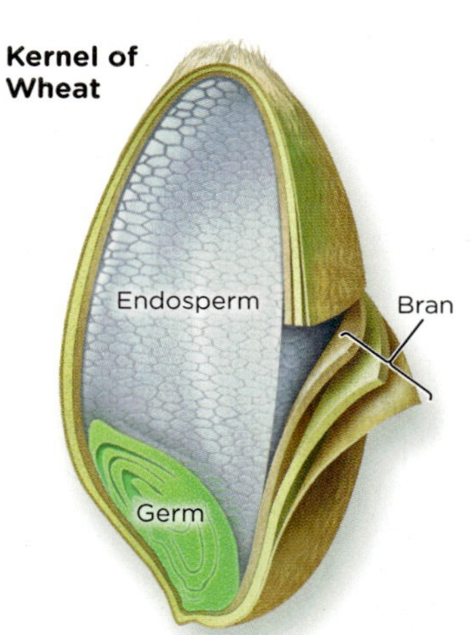

Endosperm: Contains the highest amount of starch and protein. This is all that remains when a grain is refined.

Bran: Contains the majority of dietary fiber and a significant amount of B vitamins and minerals.

Germ: It is the embryo of the seed that germinates and grows. It contains essential fatty acids and a number of B vitamins and minerals.

The vast majority of vitamins, minerals, and phytochemicals are found in the germ and bran of whole grains.

? *How many different ways can you tell if foods are whole grain (for example, if a bread is whole grain)?*

quickly turns rancid when exposed to air, and resulted in white bread that was soft and smooth in texture. **(INFOGRAPHIC 4.5)**

The downside of refining grains is that the bran and germ contain the majority of healthful vitamins, minerals, phytochemicals, and fiber. Not only are many modern grain-based foods made with nutrient-poor refined flour, they also often have a significant amount of added sugar. That's why Mozaffarian was particularly interested in the relative amounts of starch, sugar, and fiber in the food products she and her colleagues studied. Their results were surprising, and may have you looking at grain products differently.

Photocuisine/Alamy

The ingredients list shows all the ingredients in a packaged food. To identify whole grains, look for these words: brown rice, bulgur, millet, teff, oatmeal, rolled oats, whole grain barley, whole grain corn, whole grain sorghum, whole grain triticale, whole oats, whole rye, whole wheat, and wild rice.

Grains of Truth

Whole grain hype can be traced to the mid-2000s, when nutrition guidelines started recommending that consumers eat more whole grains. Industry responded with an explosion of products marketed as containing whole grains. According to Mozaffarian, in 2010, the number of new products advertising their "whole grain" status was nearly 20 times higher than the number in 2000. But many of these products contain a mixture of whole grains, refined grains, and sugars. "That's when you started seeing stuff like whole grain Goldfish crackers, whole grain Froot Loops, and so on," Mozaffarian notes.

To help consumers make more informed choices, the USDA's MyPlate program recommends that people look for whole grains listed first on the ingredient list (ingredients are listed in decreasing order of quantity). But that means being able to recognize any one of 29 ingredients classified as whole grains in the USDA's MyPlate Servings Database, including oats, bulgur, whole wheat flour, brown rice flour, whole grain corn, and wheat bran.

If the product bears the Basic Stamp, it contains a minimum of 8g of whole grains per serving. Even if a product contains large amounts of whole grain (20g, 37g, etc.), it will use the Basic Stamp if it also contains extra bran, germ, or refined flour.

® Whole Grain Stamps are a trademark of Oldways Preservation Trust and the Whole Grains Council, www.wholegrainscouncil.org

If a product bears the ***100% Stamp,*** *then all its grain ingredients are whole grains. There is a minimum requirement of 16g—a full serving—of whole grain per labeled serving, for products using the 100% Stamp.*

MyPlate also recommends that consumers choose products with fewer added sugars, but to do that consumers must be able to recognize as many as 21 different ingredients as a kind of sugar, including sugar, brown sugar, corn syrup, high-fructose corn syrup, malt syrup, maple syrup, fruit juice concentrate, honey, molasses, and dextrose.

To help address the confusion, the Whole Grains Council, a nongovernmental organization funded by food companies, created the Whole Grain Stamp. Used in 36 countries, the Whole Grain Stamp is a front-of-package icon that tells consumers that a product contains at least 8 grams of whole grains per serving.

A different approach, advocated by the American Heart Association, is to look for products with a total carbohydrate-to-fiber ratio of less than 10:1—the approximate ratio of carbs and fiber in whole wheat flour. Most white all-purpose flour, for example, has a ratio of over 20:1. This measure is intended to help consumers identify healthy whole grain foods without having to memorize long lists of specific ingredients. But with all these different recommendations out there, how is a consumer supposed to know which to use and trust?

HOW GRAIN-BASED FOODS MEASURE UP

To conduct her study, Mozaffarian identified 545 grain products from two grocery store chains in the Boston area (Walmart and Stop & Shop). The products fell into eight categories of frequently consumed grains, including bread, bagels, English muffins, cereals, crackers, cereal bars, granola bars, and chips. Looking at the list of ingredients and nutrition information, she then asked which of five common ways of defining whole grain foods fared best in terms of identifying the most healthful products.

When all the results were in, Mozaffarian made some surprising discoveries. She found, for example, that the Whole Grain Stamp was indeed useful for identifying foods with whole grains. However, she found that many foods with the stamp were also very high in added sugar, making them less healthful overall. This is because added sugars contribute energy (kcal) to foods but not micronutrients or phytochemicals; thus added sugars reduce the nutrient density of foods, even those that are made with whole grains.

Likewise, looking for whole grains (for example, whole corn meal, whole

LACTOSE INTOLERANCE
a condition characterized by diminished levels of the enzyme lactase and subsequent reduced ability to digest the disaccharide lactase

wheat flour, bulgar) as the first ingredient on food labels identified foods with ample amounts of whole grains, but also tended to identify foods with added sugar. Looking for the word "whole" before grain ingredients further down the list was even less helpful.

It may seem confusing to condemn carbohydrate-rich foods for their added sugar content. After all, haven't we just learned that all complex carbohydrates are, in essence, chains of linked sugar molecules? The problem is that sugar—whether naturally found in a product or added to it—is a source of energy. And when it comes to energy, it is indeed possible to have too much of a good thing. When foods rich in refined carbohydrates and added sugars are plentiful in our diet, our ability to meet our nutrient needs within our energy budget is reduced. Over time, ingesting more calories than we expend in activity will cause weight gain.

As we will see, different carbohydrate-rich foods also vary in the amount of time it takes to digest them into simple sugars, which affects how many calories we are prone to consume.

Got lactase? *To digest the disaccharide lactose in milk, we need the enzyme lactase, which breaks lactose into glucose and galactose in the small intestine. In most people, however, lactase levels decline with age.*

Gallery Stock

THE DIGESTION OF CARBOHYDRATES

Before complex carbohydrates can be used for energy, they need to be broken down into their component sugar molecules. Only monosaccharides, such as glucose, can be absorbed by the cells of the small intestine.

Enzymes in carbohydrate digestion

The body uses specific enzymes to break down carbohydrates. Different enzymes are specific for different carbohydrates. In the mouth, the enzyme salivary amylase breaks down starch molecules into shorter polysaccharides. In the small intestine, the enzyme pancreatic amylase digests starch into oligosaccharides and maltose; *maltase* is the enzyme that breaks maltose down into two glucose units; *sucrase* digests sucrose into fructose and glucose; and *lactase* breaks down lactose into glucose and galactose. These monosaccharides can be absorbed by the cells lining the small intestine, and from there travel into the blood. **(INFOGRAPHIC 4.6)**

Some people experience gastrointestinal discomfort after drinking milk or consuming some dairy foods, a condition called **lactose intolerance**, which is the result of producing low levels of lactase in the intestines. In most people, lactase levels decline with age, but different populations of humans make differing amounts of lactase as adults, and so some people are more prone to lactose intolerance than others. Luckily, the treatment for lactose intolerance is fairly straightforward: Cut back on dairy products. There are also many lactose-free dairy alternatives on the market, as well as milk products like "Lactaid" that incorporate the enzyme lactase to improve digestibility.

Glucose in the body

Of the several monosaccharides released from carbohydrates through the process of digestion, glucose is the one that all of our cells use as fuel. Fructose and galactose are taken up primarily by the liver where they can be converted into glucose, which then may be exported back into the blood for distribution to cells throughout the body.

INFOGRAPHIC 4.6 Carbohydrate Digestion and Absorption

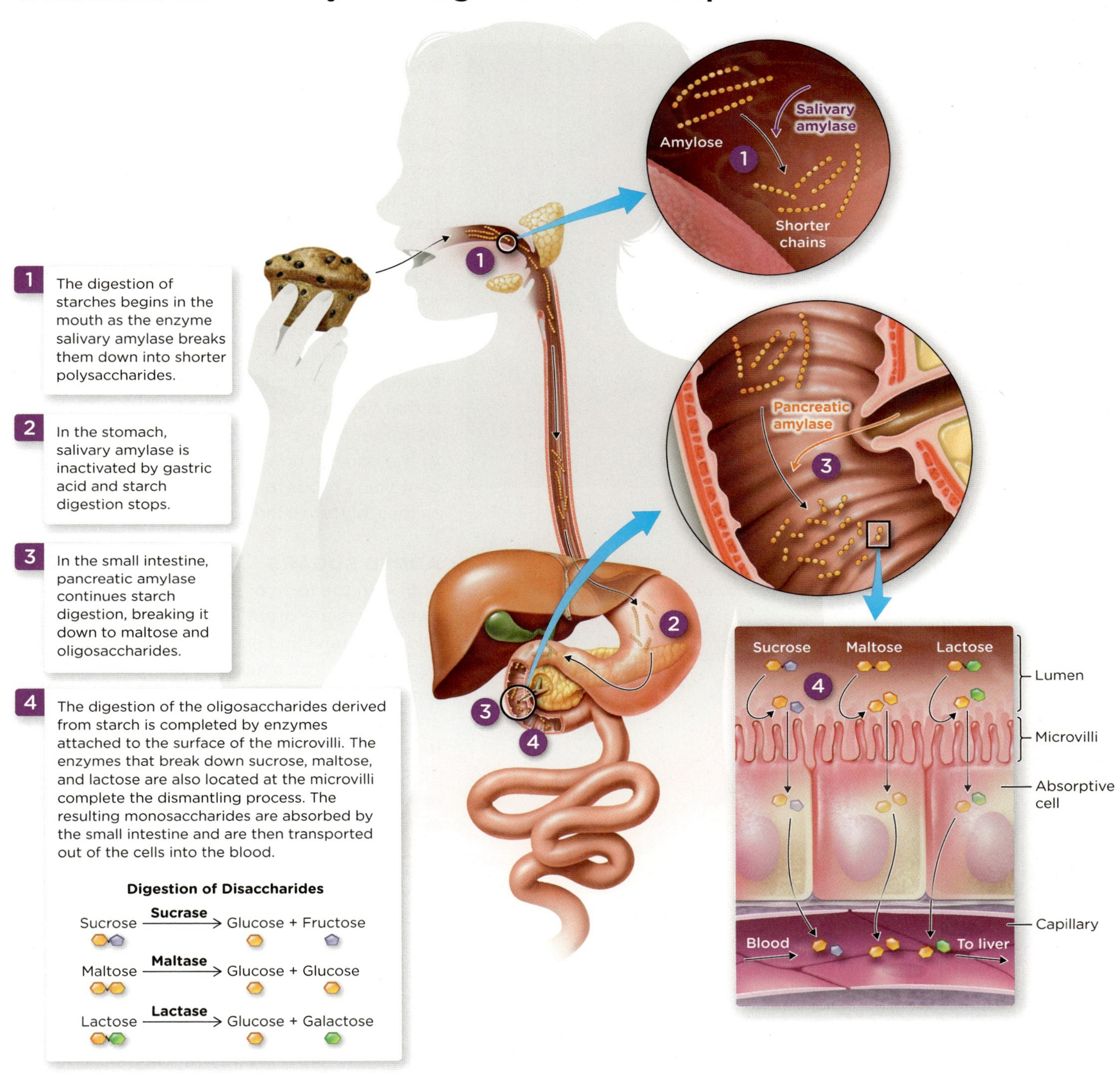

To be absorbed, carbohydrates must be in what form?

Any glucose the body doesn't use to meet its immediate energy needs is stored for later use. The liver and skeletal muscles use this excess glucose to synthesize glycogen. The liver will break down this glycogen into glucose when needed to maintain normal blood sugar levels, and skeletal muscles will use it to fuel muscle contractions during intense exercise.

Fueling demanding exercise. *Skeletal muscles relies heavily on glycogen to fuel muscle contractions during intense exercise.*

Image Source/Getty Images

How much added sugar? *Each of these foods has many calories from added sugar.*

(a) *12-ounce soda: 132 calories of added sugar*

Jesus Ayala/Getty Images

(b) *1 cup canned peaches in heavy syrup: 115 calories of added sugar*

supermimicry/Getty Images

(c) *1 donut: 74 calories of added sugar*

Jens Mortensen/Gallery Stock

(d) *½ cup vanilla ice cream: 48 calories of added sugar*

Andrew Unangst/Getty Images

Different carbohydrate-rich foods are broken down into glucose more quickly than others, depending on how resistant their starches are to digestive enzymes. When we eat foods with resistant starches, such as whole grain oatmeal, pinto beans, or cashews, for example, the sugars within them are released slowly over the course of hours, helping to keep our hunger satisfied. By contrast, when we eat foods with less-resistant starches, such as a piece of most breads or a baked potato, the sugar is released quickly—almost as quickly as if you ate pure glucose. What this means is that blood sugar levels rise quickly, insulin levels spike, and hunger returns sooner than you might expect. The end result is that we end up eating more calories than we might have otherwise. (We will return to the issue of how much carbohydrate-containing foods raise blood glucose, known as their glycemic index, in Chapter 5.)

ADDED SUGARS

Sugars in the form of syrups and other caloric sweeteners are often added to food products add sweetness and enhance palatability. These added sugars also have functional roles in preserving foods, browning of baked items, and improving texture and appearance. On a food label, they are listed by a wide variety of names such as "brown sugar," "corn sweetener," "dextrose," "fructose," "fruit juice concentrate," "glucose," "honey," "invert sugar," "lactose," "maltose," "molasses," "raw sugar," "sucrose," and "table sugar". Another common and controversial added sugar is high fructose corn syrup (HFCS) which differs slightly from sucrose or table sugar in that it contains 55% versus 50% fructose in relation to glucose. Added sugars do not include naturally occurring sugars like those found in fruit or milk. **(INFOGRAPHIC 4.7)**

Added sugars aren't just found in desserts and sweet snacks; they also lurk unexpectedly in processed foods such as ketchup, sandwich bread, and pizza. And, as Mozaffarian discovered, they are also found in many foods advertised as containing whole grains. Sugars are added to whole grain foods to improve consumer acceptance. To help consumers better assess

INFOGRAPHIC **4.7** **Sources of Added Sugar in the Diet** *Beverages (not milk or 100% fruit juice) account for the majority of the added sugar in foods consumed in the U.S. by people ages 2 years and older.*

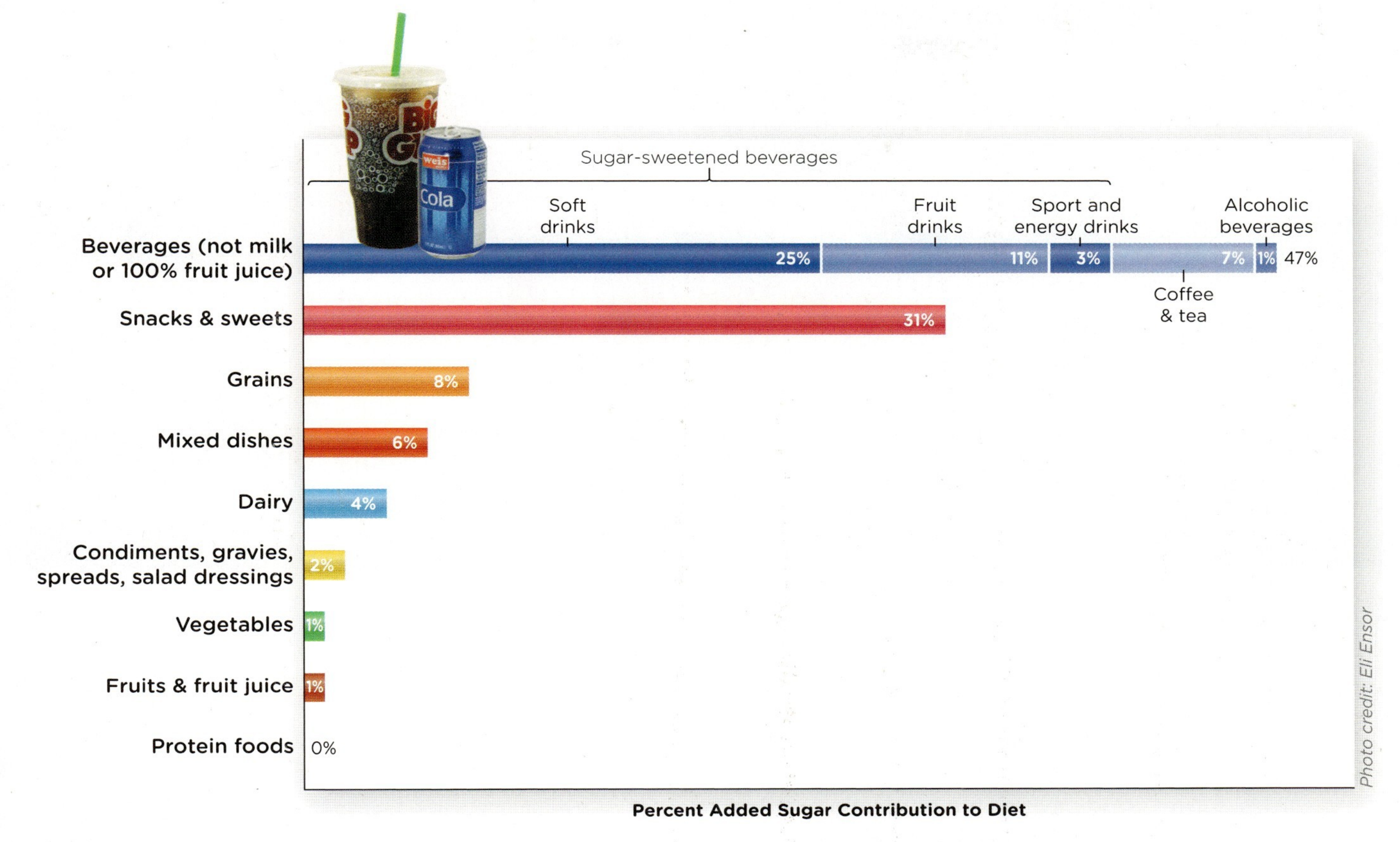

Source: What We Eat in America (WWEIA) Food Category analyses for the 2015 Dietary Guidelines Advisory Committee. Estimates based on day 1 dietary recalls from WWEIA, NHANES 2009-2010.

What food category does not contribute added sugar? Which type of beverage contributes the most added sugar?

added sugars in foods, proposed food label changes by the FDA (see chapter 2) would require the inclusion of "added sugars" on all packaged foods.

Most Americans consume too much added sugar, eating on average about 270 calories per day or 13% of total calories. There are two main problems associated with this excessive consumption of foods with added sugars. The addition of sugars to foods adds substantial calories without adding significant bulk; this increases the energy density of foods and is likely one contributing factor to the increased prevalence of obesity around the world. Added sugars also provide no additional vitamins or minerals, and are therefore empty calories that decrease the nutrient density of foods. To help people move towards healthy eating patterns and not exceed calorie limits, the 2015 Dietary Guidelines for Americans recommend consuming less than 10% of total calories from added sugar (see chapter 2.)

DENTAL CARIES

And, of course, eating too much added sugar promotes cavities, also known as **dental caries**. Individuals are especially at risk if they frequently expose their teeth to sugar by eating sweet snacks and "sticky" foods, sucking hard candy, or by slowly sipping sugary drinks. The more our teeth are exposed

DENTAL CARIES *also called cavities or tooth decay; the progressive destruction of tooth enamel and ultimately the tooth itself, through the action of bacteria on carbohydrates in the mouth*

Checking the teeth for cavities. *Children (and adults) who have the habit of eating sticky, sugar-laden treats are at risk of developing dental cavities. Good dental care and avoidance of sugary treats can prevent dental caries (right).*

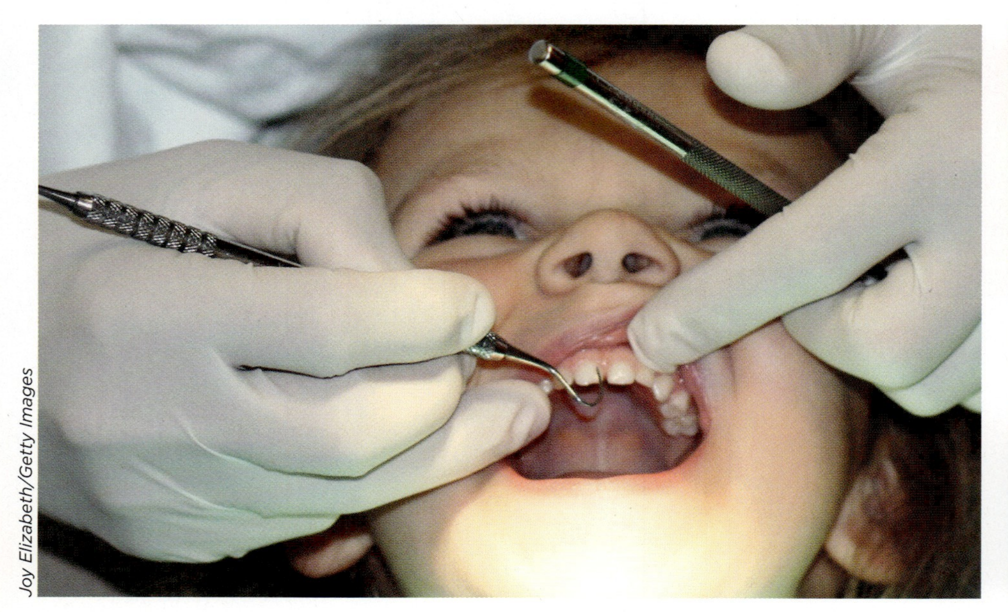

Joy Elizabeth/Getty Images

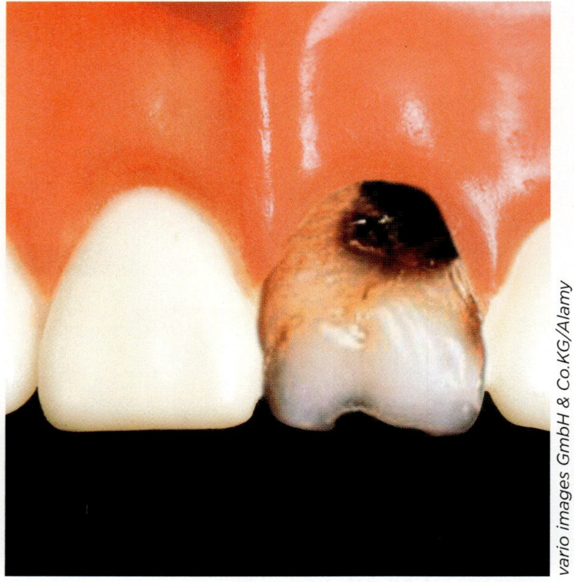

vario images GmbH & Co.KG/Alamy

to sugar, the more the bacteria on our teeth metabolize the sugars and produce acids that dissolve tooth enamel and decay teeth. And cavities are not just a problem in childhood—85% of people 18 years and older have had cavities at sometime in their life.

SUGAR ALTERNATIVES

Because excess sugar consumption is also associated with chronic diseases such as obesity and cardiovascular disease, the food industry has created a handful of sugar alternatives that are regulated by the United States Food and Drug Administration (FDA), which has deemed them Generally Recognized As Safe (GRAS) for use in foods. There are two main types: those that are a source of calories, the nutritive sweeteners; and those that are not, called the non-nutritive sweeteners. Nutritive sweeteners include the *polyols*, which are alcohol forms of sugars. Sorbitol (the alcohol form of glucose) and mannitol (the alcohol form of mannose), are made from naturally occurring sugars in plants, and because they are poorly absorbed by the body, they provide fewer than 4 kcal per gram. Because of their poor absorption, excessive consumption of sugar alcohols often causes diarrhea.

Non-nutritive sweeteners include the widely used aspartame products (Nutrasweet and Equal). Aspartame is made by linking two amino acids, and like protein, it provides 4 kcal per gram. However, it is so intensely sweet (about 200 times sweeter than sucrose) that so little is used to achieve the same level of sweetness as sucrose that it provides no significant amount of calories. Another popular non-nutritive sweetener is stevia extract (Truvia), made from the sweet leaves of subtropical stevia plants. Examples of other non-nutritive sweeteners found in foods and in the "sugar" bowl at your restaurant table include sucralose (Splenda) and saccharin (Sweet'N Low). (INFOGRAPHIC **4.8**)

INFOGRAPHIC 4.8 Alternative Sweeteners

Food manufacturers have turned to alternative sweeteners as a way to replace sugar.

Nutritive Sweeteners				
Sweetener	**Trade Name**	**Kcal/g**	**Sweetness Relative to Sucrose**	**Uses & Highlights**
Tagatose	Naturlose	1.5	0.75–0.92	A monosaccharide almost identical to fructose. Provides fewer calories because it is poorly absorbed. It also occurs naturally in foods (dairy) in small amounts. Is used much like sucrose to provide both bulk and sweetness in foods such as ice cream, cakes, and candies.
Sugar Alcohol Sweeteners (Polyols) *Provide reduced calories because they are poorly absorbed. Also occur naturally in foods.*				
Sorbitol	Sorbitol	2.6	0.5–0.7	Used in sugarless gums, chocolate candies, and ice cream. It is not metabolized by bacteria in the mouth and therefore does not promote tooth decay. Prunes naturally contain high amounts. Likely to cause a laxative effect when consumed at ≥ 50 g/day.
Mannitol	Mannitol	1.6	0.5–0.7	Used primarily in making chewable tablets. Likely to cause a laxative effect when consumed at ≥ 20 g/day.
Xylitol	Xylitol	2.4	1	Used in mouthwash, sugarless gums, and candies. Like sorbitol it does not promote tooth decay. Causes a cooling sensation in the mouth when used in chewing gums and hard candies (as do sorbitol and mannitol). The laxative effect appears to be slightly less than that of sorbitol.

Non-nutritive Sweeteners				
Sweetener	**Trade Name**	**Kcal/g**	**Sweetness Relative to Sucrose**	**Uses & Highlights**
Acesulfame K	Sunnet, Sweet One	0	200X	Long-lasting and heat stable. It is used in a wide variety of products, particularly in sugar-free beverages and desserts.
Aspartame	Equal, NutraSweet	0	160–220X	Widely used in sugar-free soft drinks. Composed of two amino acids (aspartate, phenylalanine). Can withstand elevated temperatures for only a brief period but is destroyed at baking temperatures. When in solution it is not as stable as other sweeteners.
Neotame	*Used infrequently*	0	7,000–13,000X	Very similar in structure to aspartame. Much greater stability in solution and can withstand high temperatures encountered during baking.
Saccharin	Sweet'N Low Sugar Twin	0	300X	Discovered in 1878. Widely used in sugar-free soft drinks and as a tabletop sweetener. Can be used in baking without losing its sweetness. Once listed as a possible carcinogen, it has since been shown to not cause cancer in humans.
Stevia	Pure Via Truvia	0	250X	Rebaudioside A (rebiana) is the active compound that is isolated from the leaves of the South American plant stevia. Approved for use in the United States in 2008. Used primarily in beverages, as a tabletop sweetener, and in yogurt. It is heat stable.
Sucralose	Splenda	0	600X	It is made from sucrose by replacing 3 $^{-}$OH groups with chlorine. It is used as a tabletop sweetener, and it is widely used in beverages where it is remarkably stable over long periods. It is also used as a tabletop sweetener.

Sources: 1) Fitch C, Keim KS; Academy of Nutrition and Dietetics. Position of the Academy of Nutrition and Dietetics: use of nutritive and nonnutritive sweeteners. J Acad Nutr Diet. 112: 739-58, 2012. 2) Alternative Sweeteners, 4th ed. Nabors, Lyn O'Brien editor. CRC Press. 2011.

You talk with someone who complains of constant upset "stomach" and diarrhea. You learn that they chew sugar-free gum and eat sugar-free breath mints all day long, as well as eat sugar-free ice cream. What is likely causing her discomfort?

UNDERSTANDING FIBER

The best measure to identify healthful whole grain foods, Mozaffarian found, was a total carbohydrates to fiber ratio of 10:1 (or less). This is approximately the ratio of total carbohydrate to fiber in whole wheat flour. Foods with a ratio higher than this are likely to contain added sugars and refined grains.

For example, a product with 40 grams of carbohydrates and 5 grams of fiber would have a ratio of 8:1, which is less than 10:1. This is a favorable ratio. This measure tends to identify foods that have significant whole grains, low sodium, low trans fats, and low added sugar—in short, the healthiest products.

Fiber recommendations

In addition to reflecting the presence of whole grains versus refined grains, fiber is important in its own right. In fact, many of the benefits of eating whole grain foods likely come from the fiber within them. The Institute of Medicine, the health arm of the U.S. National Academy of Sciences, recommends that men younger than 50 years should consume 38 grams of fiber per day; women younger than 50 years should consume 25 grams of fiber a day. Only about 5% of Americans meet these goals consuming an average of 17 grams of fiber a day.

SOLUBLE FIBER *a type of fiber that dissolves in water and often forms a viscous gel that acts to slow digestion and lower blood cholesterol and the risk of heart disease; they are also often readily fermented by bacteria in the colon*

INSOLUBLE FIBER *a type of fiber that does not dissolve in water, increases transit time through the GI tract, and contributes "bulk" to stool, fostering regular bowel movements*

DIETARY FIBER *naturally occurring carbohydrates and lignin from plants that either cannot be digested in the intestinal tract or for which digestion is delayed*

FUNCTIONAL FIBER *nondigestible carbohydrates isolated from plants and animals and added to foods, which have a beneficial effect on health, such as psyllium and pectin*

Fiber includes cellulose and other nonstarch polysaccharides and oligosaccharides that cannot be broken down by human digestive enzymes. Fiber is typically found in and around plant cell walls. It is most abundant in legumes; nuts and seeds; berries; many vegetables; fruit coverings, such as apple peels; and the bran surrounding grains like wheat, oats, rice, and rye.

Types of fiber

There are several ways to categorize fiber. One way is by the solubility of the fiber in water. **Soluble fiber** dissolves in water, often forming viscous gels and it is

Calculate total carbohydrate to fiber ratio. *The American Heart Association recommendation is to look for products with a total carbohydrate-to-fiber ratio of less than 10:1.*

Step 1. *Find the total number of grams of carbohydrate on the label (23g), as well as the grams of fiber (5g).*

Step 2. *Write the expression as 23:5, then divide the first number by the second to get the ratio. In this case the ratio is 4.6 to 1.*

Nutrition Facts
Serving Size 3/4 Cup (27g)

Amount Per Serving	Cereal	With 1/2 Cup Skim Milk
Calories	90	130
Calories from Fat	10	10
	% Daily Value	
Total Fat 1g*	**2%**	**2%**
Saturated Fat 0g	**0%**	**0%**
Trans Fat 0g	**0%**	**0%**
Cholesterol 0mg	**0%**	**0%**
Sodium 190mg	**8%**	**11%**
Potassium 85mg	**2%**	**8%**
Total Carbohydrate 23g	**8%**	**10%**
Dietary Fiber 5g	**20%**	**20%**
Sugars 5g		
Protein 2g		
Vitamin A	0%	4%
Vitamin C	10%	15%
Calcium	0%	15%
Iron	2%	2%

Editorial Image, LLC/Alamy

typically readily fermented (broken down) by bacteria in the colon. In contrast, **insoluble fiber** does not dissolve in water and is typically poorly fermented. It passes through the gastrointestinal tract relatively intact.

The source of fiber is another way that fiber is classified. **Dietary fiber** is undigestible carbohydrates that are present naturally in intact plant foods. **Functional fiber** is nondigestible carbohydrates that have been added to a food product and have health benefits that are similar to those of dietary fiber. When you add up a food's dietary fiber and its functional fiber, you get its total fiber. **(INFOGRAPHIC 4.9)**

Benefits of fiber

Fiber (particularly insoluble fiber) is healthful in part because it softens

■ INFOGRAPHIC 4.9 Understanding the Fibers in Foods

CATEGORIES OF FIBER

Fiber is often categorized by its solubility in water (*soluble* versus *insoluble*).

Insoluble fiber is always the dominant type of fiber that is naturally present in our foods. Therefore, all high-fiber foods are good sources of insoluble fiber.

Best sources of ***soluble fiber*** are:

- Whole grain oats, barley, and rye
- Citrus fruits
- Legumes
- Mangos
- Avocados
- Pears
- Apples
- Psyllium

The Institute of Medicine developed the following definitions to classify fiber in the food supply:

Dietary Fiber: Nondigestible carbohydrates and lignin that occur naturally and are intact in plants

Functional Fiber: Nondigestible carbohydrates that have beneficial physiological effects in humans but are extracted from plants or animals, or synthesized and then added to foods

Total Fiber: The sum of dietary and functional fibers

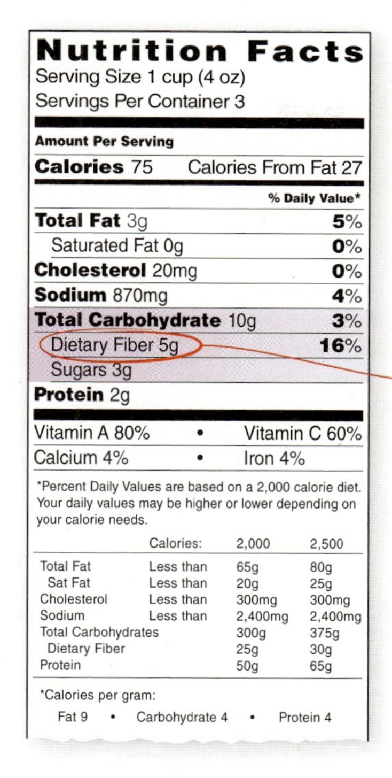

Nutrition Facts
Serving Size 1 cup (4 oz)
Servings Per Container 3

Amount Per Serving
Calories 75 Calories From Fat 27

	% Daily Value*
Total Fat 3g	**5%**
Saturated Fat 0g	**0%**
Cholesterol 20mg	**0%**
Sodium 870mg	**4%**
Total Carbohydrate 10g	**3%**
Dietary Fiber 5g	**16%**
Sugars 3g	
Protein 2g	

Vitamin A 80% • Vitamin C 60%
Calcium 4% • Iron 4%

*Percent Daily Values are based on a 2,000 calorie diet. Your daily values may be higher or lower depending on your calorie needs.

	Calories:	2,000	2,500
Total Fat	Less than	65g	80g
Sat Fat	Less than	20g	25g
Cholesterol	Less than	300mg	300mg
Sodium	Less than	2,400mg	2,400mg
Total Carbohydrates		300g	375g
Dietary Fiber		25g	30g
Protein		50g	65g

*Calories per gram:
Fat 9 • Carbohydrate 4 • Protein 4

A Nutrition Facts Panel does not identify whether the fiber is insoluble or soluble, but contains valuable information about sugar and total fiber in a food.

TOTAL CARBOHYDRATES on the food label includes the total amount of starch, sugars, and dietary fiber in one serving.

Subtract fiber from total carbohydrates to determine the amount of carbohydrates that can be digested and absorbed.

What type of soluble fiber have you consumed most recently?

stools and thereby helps maintain regular bowel movements and reduces the risk of hemorrhoids and diverticular disease. Insoluble fiber has also been shown to decrease the risk of diabetes.

A higher intake of soluble fiber has been shown to reduce the risk of coronary heart disease. Diets high in soluble fiber also slow the emptying of food from the stomach into the small intestine, which may extend the sensation of fullness following a meal. Many soluble fibers also slow digestion and absorption, and reduce the rise in blood glucose following a carbohydrate-containing meal, which may improve blood glucose control in those with diabetes. In addition, nutrition surveys demonstrate that high fiber diets are more likely to be nutrient-dense and lower in saturated fats, sodium and added sugar. (INFOGRAPHIC 4.10)

CARBOHYDRATE INTAKE RECOMMENDATIONS

Given the importance of carbohydrates to health, all nutrition guidelines provide recommendations for carbohydrate intake. In addition to encouraging consumption of carbohydrate-rich plant foods, the 2015 Dietary Guidelines for Americans recommends that people consume at least half of all grains as whole grains, and to increase whole grain intake by replacing refined grains with whole grains. The Institute of Medicine recommends that people consume carbohydrates within a certain range, the Acceptable Macronutrient Distribution Range (AMDR) of 45% to 65% of total calories, which is associated with a reduced risk of chronic disease while providing adequate amounts of essential nutrients. There is also a recommended dietary allowance (RDA)

INFOGRAPHIC 4.10 Increasing Fiber Intake: What's Your Strategy?

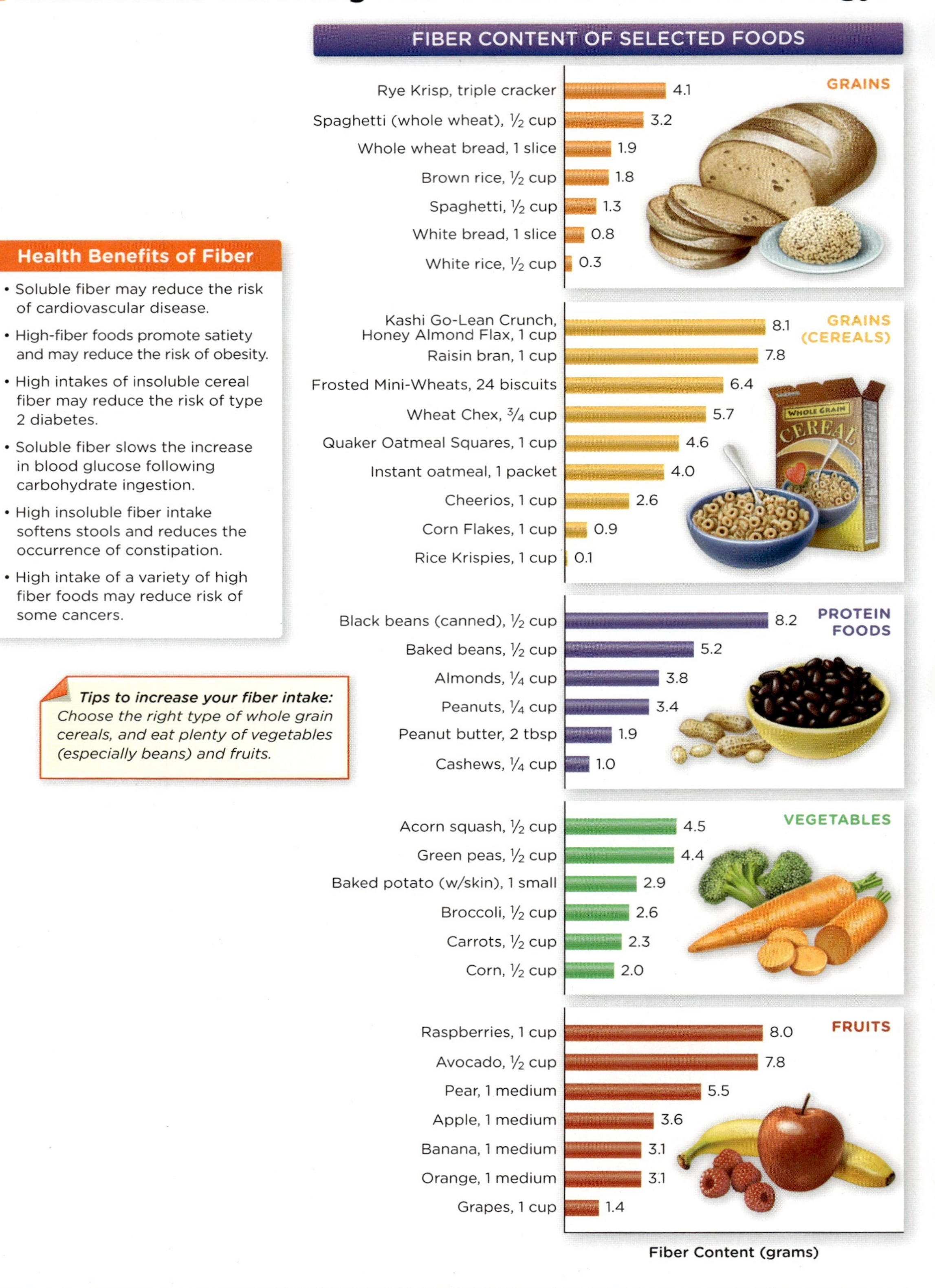

What two or three high-fiber foods could you use to replace lower fiber foods that you eat every day? How many additional grams of fiber would you consume if you made these substitutions?

INFOGRAPHIC 4.11 Dietary Recommendations for Carbohydrate Intake

Source	Total Carbohydrates	Fiber	Added Sugars
2015 Dietary Guidelines for Americans	Consume at least half of all grains as whole grains. Increase whole grain intake by replacing refined grains with whole grains.		Consume less than 10% of total calories from added sugars.
Institute of Medicine Recommendations (DRIs)	**RDA:** 130 g/day **AMDR:** 45%–65% of total calories • 2,000 kcal/day diet: 225–325 g/day • 2,500 kcal/day diet: 281–406 g/day	**AI:** 14 g per 1,000 kcal/day Women / Men Age 19–50: 25 g/day / 38 g/day Age ≤51: 21 g/day / 30 g/day	≤ 25% of total calories consumed
American Heart Association (2009)			Women ≤ 100 kcal/day (25 g); Men ≤ 150 kcal/day (38 g)
World Health Organization (2003)			≤ 10% of energy intake

You can help meet dietary recommendations for carbohydrates by limiting added sugars like sweets and soda, eating more fruits and vegetables, and by checking the ingredient list on foods for the words "whole" or "whole grain."

Why do the RDA and AMDR for total dietary carbohydrates conflict with each other?

for carbohydrates, which is based on the amount of carbohydrates that is needed for the brain to function properly. (INFOGRAPHIC 4.11)

■ ■ ■

When Mozaffarian published her study, she got some push-back from industry. The Whole Grains Council, which controls the Whole Grain Stamp, said it was unfair to criticize the stamp for not doing something it was never designed to do. The stamp, they said, is only supposed to identify products that contain a meaningful amount of whole grains, not to rate the overall healthfulness of the product. "Their point was, 'Well, if you're going to have a Pop Tart, it may as well be a whole grain Pop Tart,'" says Mozaffarian. Furthermore, an approach that requires consumers to do math might also be less useful than the easy-to-see stamp, a point that Mozaffarian readily concedes.

But Mozaffarian stands by the study. Though she gives credit to the Whole Grains Council for helping to raise awareness of whole grain foods, and for getting more whole grains into the food supply, she also thinks that the stamp has limitations as well. "They're focused on just one thing, in this case whole grains, and ignoring other things such as sodium and total sugars and all of these other issues. It could be a little bit short-sighted."

CHAPTER 4 **BRING IT HOME**

Focus on fiber

The following sample menu provides less than half of the recommended daily intake of dietary fiber for a 20-year-old college student. Create new meals and snacks to boost the student's fiber intake (and improve overall nutrient density and healthfulness). For each suggestion, record the total fiber. You can use the USDA nutrient database http://ndb.nal.usda.gov/ to determine the fiber content of the foods you add, as well as other foods in the meals and snacks. After completing your alternate menu, answer the questions below.

BREAKFAST
1-½ cups sugar-frosted cornflakes
1 cup 1% milk
1 slice white toast
2 teaspoons margarine
1 tablespoon jelly
1 cup coffee
2 teaspoons sugar
1 tablespoon half and half

LUNCH
Fast-food hamburger with cheese
1 medium order French fries
12 ounces cola

SNACK
1 ounce cheese snack crackers
1 cup apple juice

DINNER
2 slices pepperoni pizza (¼ large pizza)
2 cups tossed salad (iceberg lettuce)
2 tablespoons ranch dressing
1 2-inch-square brownie
½ cup vanilla ice cream

SNACK
2 ounces M&Ms

1. According to the Dietary Reference Intake values (DRIs), what is the Adequate Intake (AI) for total fiber for a 20-year-old *man*? With your suggested changes and additions, did the fiber content for this day's intake meet fiber recommendations?
2. According to the DRIs, what is the AI for total fiber for a 20-year-old *woman*? With your suggested changes and additions, did the fiber content for this day's intake meet fiber recommendations?
3. Which of the foods you suggested contained the highest amount of *soluble* fiber? List two benefits of diets rich in soluble fiber.
4. Which of the foods you suggested contained the highest amount of *insoluble* fiber? List two benefits of diets rich in insoluble fiber.
5. Identify at least five ways to boost fiber intake in your overall diet.

Take It Further

The original menu provides more than 150 grams of added sugar, which is more than 20% of the total calories. The Dietary Guidelines recommend that we all reduce our intake of solid fats and added sugars (SoFAS).

1. What are some of the potential nutritional and health effects of consuming excessive amounts of added sugar?
2. In your opinion, what are the five foods from the menu provided that contribute the most added sugar?
3. Identify at least five ways to reduce added sugar in the original menu provided. Remember to suggest lower sugar alternatives, not just eliminate foods.

KEY IDEAS

Carbohydrates, which are composed of carbon, hydrogen, and oxygen, contain 4 kcal per gram and are found in plant foods and milk products.

In addition to serving as a primary energy source, carbohydrates are also a source of fiber, add sweetness and flavor to food, and reduce the use of protein for energy.

The Institute of Medicine recommends that carbohydrates make up 45% to 65% of total daily calorie intake.

Carbohydrates can be classified as simple carbohydrates (sugars) or complex carbohydrates (starch and fiber).

Simple carbohydrates include single sugar molecules called monosaccharides (glucose, fructose, and galactose) or two sugar units called disaccharides (sucrose, lactose, and maltose). The monosaccharide glucose is the most abundant sugar circulating in the blood and serves as an essential energy source for the cells of the body. The disaccharide sucrose, or "table sugar," is composed of glucose and fructose.

Lactose is the primary form of sugar found in milk. Some individuals lack or have diminished levels of lactase, the enzyme needed to break down lactose, so they experience lactose intolerance with gastrointestinal symptoms when they consume milk products.

Complex carbohydrates include polysaccharides that function as a stored form of energy (starch) or indigestible material (fiber), depending on whether humans can break the bonds linking the sugars.

Carbohydrate digestion is accomplished through the action of specific enzymes, which start their work in the mouth and complete it in the small intestine.

Fiber is indigestible by humans and has multiple health benefits. It is often categorized by its solubility in water. Insoluble fiber aids digestive health and soluble fiber may lower the risk of cardiovascular disease.

Fiber may also be categorized by its source in food. Dietary fiber is the fiber naturally found in intact plant foods; functional fiber is extracted or synthesized and added to foods during the manufacturing process. Together, dietary fiber and functional fiber make up total fiber. The DRIs recommend a daily total fiber intake of 25 grams for women and 38 grams for men.

Glycogen, a highly branched polysaccharide, is the storage form of glucose in humans and other animals. Glycogen is stored primarily in the liver and muscles: Liver glycogen can help maintain blood glucose and muscle glycogen can supply fuel for contracting muscles during intense exercise.

The Dietary Guidelines for Americans (DGA) recommend that we reduce intake of added sugars to less than 10% of total calories. Excess intake of sugars increases empty calories and the risk of dental caries.

Alternative sweeteners are regulated and considered "generally recognized as safe" (GRAS) by the FDA. Alternative sweeteners include nutritive sweeteners (sugar alcohols) and several non-nutritive sweeteners that are calorie-free and hundreds of times sweeter than sucrose.

NEED TO KNOW

Review Questions

1. The primary function of carbohydrates in the body is:
   a. growth and development.
   b. as a source of energy.
   c. as a source of linoleic acid.
   d. to enhance satiety.

2. Carbohydrates are found primarily in plant sources. The only other food source for carbohydrates includes:
   a. meats.
   b. eggs.
   c. dairy foods.
   d. olive oil.

3. According to the DRIs, approximately _____ of total calories in the diet should come from carbohydrates, particularly complex carbohydrates.
   a. 25%–45%
   b. 35%–55%
   c. 45%–65%
   d. 55%–75%

4. The disaccharide sucrose is formed from:
   a. glucose and fructose.
   b. glucose and glucose.
   c. glucose and galactose.
   d. fructose and maltose.

5. Which of the following carbohydrates are classified as polysaccharides?
   a. sucrose and lactose
   b. xylitol and mannitol
   c. starch and dietary fiber
   d. high-fructose corn syrup

6. Polysaccharides in plant foods may function as a stored form of energy (starch) or indigestible material (fiber) depending on:
   a. the presence of water.
   b. how many monosaccharides are present.
   c. whether the food is raw or cooked.
   d. the type of bonds between the monosaccharides.

7. Scientists and health authorities categorize fiber according to its source in foods, whether naturally occurring or added during manufacturing, as well as by:
   a. the number of calories it provides.
   b. its solubility in water.
   c. the number of monosaccharide molecules.
   d. its digestibility.

8. Glycogen is the storage form of glucose in humans and other animals found primarily in:
   a. the liver and skeletal muscles.
   b. adipose and muscle tissue.
   c. insoluble fiber and lignin.
   d. the blood and brain.

9. Which of the following foods would be most likely to increase risk of dental caries?
   a. alcoholic beverages
   b. white bread
   c. marshmallows
   d. diet soda

10. The digestion of starch:
    a. is dependent on the enzyme sucrase.
    b. begins in the mouth with the enzyme amylase.
    c. cannot occur in humans due to lack of specific enzymes.
    d. is unnecessary because starch is absorbed directly into the blood.

11. The cause of lactose intolerance is:
    a. a food allergy during childhood.
    b. excessive consumption of high-fructose corn syrup.
    c. insufficient levels of the enzyme lactase.
    d. overproduction of the enzyme lactase.

12. All of the following are true with regard to non-nutritive sweeteners, EXCEPT that they:
    a. are regulated by the FDA.
    b. are essentially calorie-free.
    c. include sorbitol and mannitol.
    d. are hundreds of times sweeter than sucrose.

Take It Further

Many Americans consume excessive amounts of added sugar. Identify three foods or beverages that you believe are the main sources of added sugar in the typical U.S. diet. What are three possible consequences of excess sugar consumption?

Dietary Analysis Using SuperTracker

Whole grain, fiber-rich, or not? Know the carbohydrates in your diet.

Carbohydrates are a key energy source for the body, but some carbohydrate-containing foods are more healthful than others. With this exercise, you will evaluate dietary carbohydrate, fiber, and sugar intake and compare them with the current dietary recommendations.

1. Log onto the USDA website at www.supertracker.usda.gov
2. You will need to create a profile for yourself to get a personalized diet plan. This profile will allow you to save your information and diet intake for future reference. *Do not use the general plan.*
3. Click the Track Food and Activity option.
4. Record your food and beverage intake for one day that most reflects your typical eating patterns. Enter each food and beverage you consumed into the food tracker. Note that there may not always be an exact match to the food or beverage that you consumed, so select the best match available.
5. Once you have entered all of your food and beverage choices into the food tracker, you will see (on the right side of the page under the bar graph): Related Links: View by Meal and Nutrient Intake Report. Print these two reports and use them to answer the following questions:
    a. How many grams of carbohydrate did you consume? Did you meet the minimum of 130 grams?
    b. Was your carbohydrate intake within the recommended 45% to 65% of total calories?
    c. How many grams of fiber did you consume? Is this below your target number? If so, what foods could you include in your diet to get this number up to the recommended amounts?
    d. According to the View by Meal report, how many calories of added sugars did you consume? Is this above the daily limit?
    e. What information could you look for on food labels to be sure that you are purchasing whole grain products?

A granola bar contains 100 calories, and eight grams of sugar. How many calories in the bar are from the sugar? Is this more than one-third of the calories?

Medical Miracles

NEARLY 100 YEARS AGO, A DISCOVERY REVOLUTIONIZED MEDICINE—AND SAVED MILLIONS OF CHILDREN AS A RESULT.

Fotosearch/age fotostock

LEARNING OBJECTIVES

- Describe how blood glucose is regulated (Infographic 5.1)
- Explain what happens when sensitivity to insulin is impaired (Infographic 5.2)
- Explain the difference between diabetes and prediabetes (Infographic 5.3)
- List health issues that are common in people with poorly controlled diabetes (Infographic 5.4)
- Describe how gestational diabetes differs from type 1 or type 2 diabetes (Infographic 5.5)
- Describe factors that increase risk for type 2 and gestational diabetes and those that protect against it and understand how these factors relate to the Dietary Guidelines for Americans (Infographic 5.6)

For many years after the Joslin Diabetes Center in Boston opened in 1898, the waiting room was deathly quiet. Children and their parents sat in silence, waiting for their names to be called, praying that the doctor would give them good news. They rarely got it.

The children had one of the most terrifying diseases of the time—diabetes. These children would slowly and painfully waste away, often lapsing in and out of consciousness.

And doctors could do nothing about it. Children with diabetes "were living under a sentence of death," says Michael Bliss, PhD, distinguished professor emeritus at the University of Toronto. "The only real question was how long it would be until you died."

In 1922, with nothing left to lose, one little boy's parents agreed to let the doctors inject their son with an experimental treatment. Within minutes the boy regained consciousness and looked alert.

WHAT IS DIABETES MELLITUS?

Diabetes mellitus, more commonly known simply as diabetes, is a group of diseases that affect how the body uses blood glucose, and left untreated, can result in life-threatening complications.

The underlying cause of diabetes was unknown until a crucial clue came at the end of the nineteenth century, when German scientists discovered that removing the pancreas from a dog caused it to develop diabetes and die. This suggested that the pancreas isn't just an organ that supplies digestive enzymes—it also helps regulate the body's levels of blood glucose (commonly called blood sugar).

■ ■ ■

Most of us know someone with diabetes, and we may know that diabetes affects young people and old people, and has something to do with blood sugar, and that people living with the disease pay close attention to diet and may take medications. But what are the different types of diabetes, and what can you do to prevent diabetes? Understanding what happens when blood glucose is regulated normally, and what happens when it goes awry, is a good place to start.

INSULIN
a hormone produced in the pancreas that removes glucose from the bloodstream for use by the cells

GLUCAGON
a hormone produced in the pancreas that increases glucose availability in the blood in response to low blood glucose

TYPE 1 DIABETES MELLITUS
an autoimmune disease characterized by elevated blood glucose levels, caused by destruction of the cells in the pancreas that normally produce insulin

KETONES (KETONE BODIES)
compounds synthesized from fatty acids by the liver when insulin levels are low

KETOACIDOSIS
the formation of excess ketone bodies from fatty acids when there is a relative absence of insulin

HYPERGLYCEMIA
higher than normal blood glucose levels (fasting plasma glucose ≥ 100 mg/dl).

Blood glucose is tightly regulated

After we eat, levels of glucose in our blood rise. This stimulates the pancreas to secrete the crucial hormone **insulin**, which directly stimulates the cells in skeletal muscle, cardiac muscle, and adipose tissue (but not other tissues) to take up glucose from the blood. Because more glucose may be available than is needed, the body is able to store the excess energy for later use. Insulin is a potent *anabolic* hormone in that it stimulates the synthesis of large molecules. Insulin promotes the storage of excess glucose (recall glucose is a single sugar molecule) as glycogen (a molecule made up of many glucose molecules bonded together) in the liver and in skeletal muscle. Because there is a limit to how much glycogen can be stored, insulin also promotes the conversion of glucose to fat in the liver and adipose tissue.

Hours after you've eaten and levels of glucose in the blood have fallen, the pancreas releases another hormone, **glucagon**, which signals liver cells to release glucose into the blood to maintain blood glucose levels. The liver produces and exports this glucose by breaking down glycogen, as well as by synthesizing new glucose molecules (primarily from amino acids). (INFOGRAPHIC 5.1)

Blood glucose regulation is disrupted in diabetes

We've noted the importance of insulin in blood glucose regulation. In diabetes, the hormone insulin is either absent or the cells of the body are resistant to its action. In **type 1 diabetes**, the pancreatic cells that secrete insulin are destroyed by the body's immune system, leading to a major deficiency of insulin that typically progresses to its complete absence. Type 1 diabetes is an *autoimmune disease,* like celiac disease (covered in Chapter 3). In an autoimmune disease, the body's immune system attacks or destroys its own cells. Type 1 diabetes is not preventable, and between 5% and 10% of individuals with diabetes have type 1.

In type 1 diabetes, the extremely low levels of insulin limit the body's ability to use glucose. To minimize the breakdown of muscle protein for energy (and prevent muscle wasting), fatty acids are released from adipose tissue. However, the brain cannot use fatty acids effectively for energy, so the liver converts some of these fatty acids into a source of energy the brain can use called **ketone bodies**. If the diabetes is left untreated, over time these ketones build up in the bloodstream, increasing the acidity of blood and causing a condition called **ketoacidosis**, which can lead to coma and death.

When the pancreas becomes unable to produce insulin, muscle and adipose tissues cannot appropriately take up glucose, and blood glucose levels skyrocket—a condition known as **hyperglycemia**. When blood

INFOGRAPHIC 5.1 Regulation of Blood Glucose *For the body to function properly, blood glucose levels are maintained in a narrow range, primarily through the actions of the pancreatic hormones insulin and glucagon.*

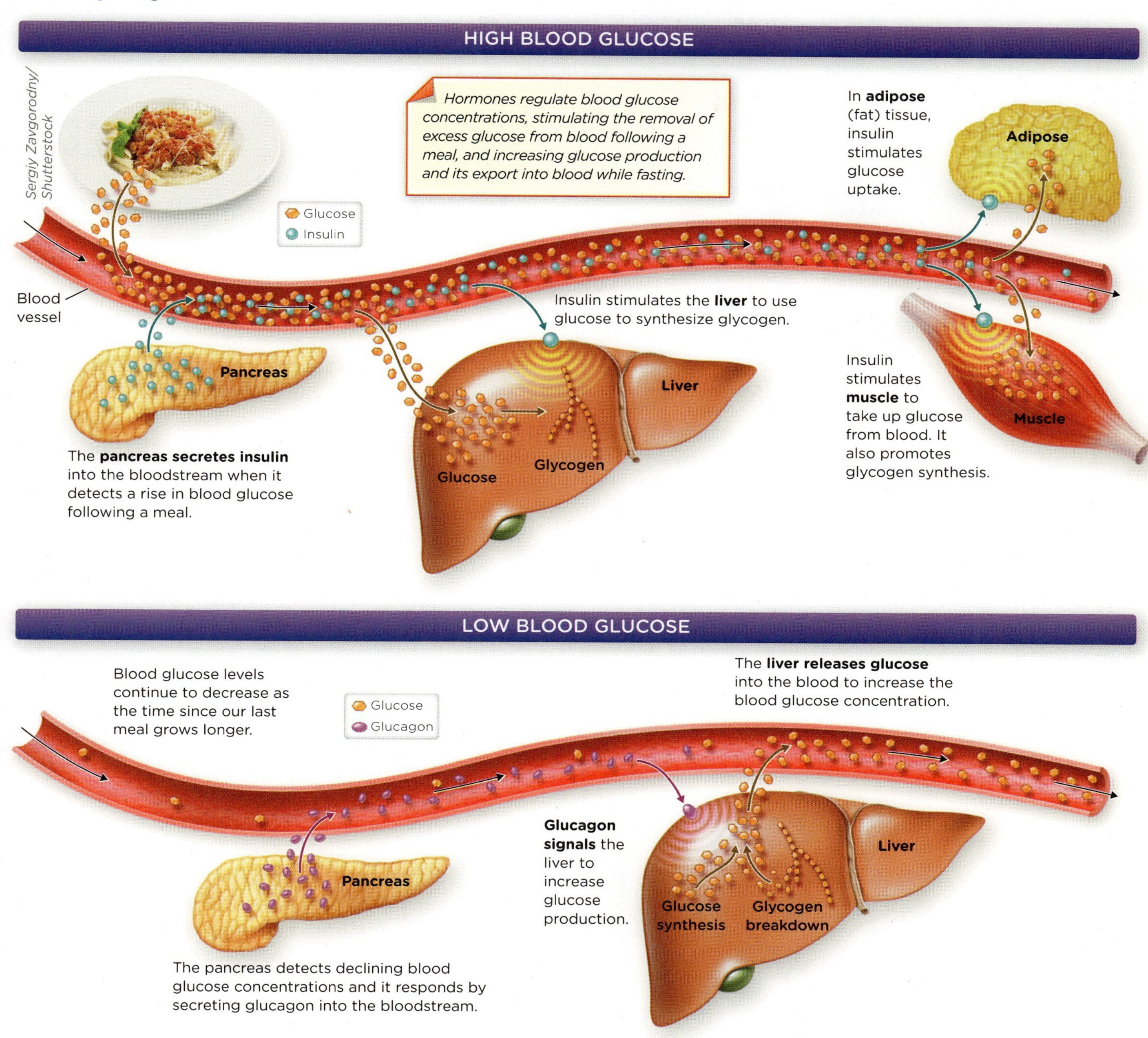

What tissues or organs are stimulated by insulin to take up glucose from blood into their cells?

glucose levels rise too high the glucose enters urine faster than the kidneys can transport it back into the blood and the excess glucose then "spills over" into urine and is excreted. This loss of glucose in urine deprives the body of a significant portion of the calories consumed throughout the day. The presence of all this glucose in urine also draws additional water into the urine causing increased urine volume and rapid rates of water loss. In fact, the word *diabetes* comes from a Greek word meaning to siphon—to pass through,

JL before and after insulin treatment.

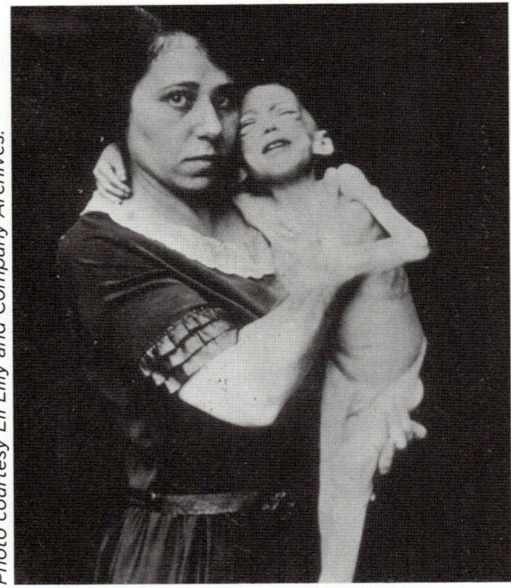

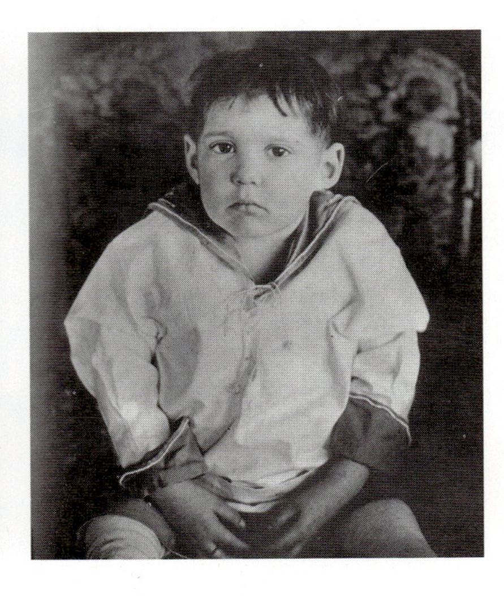

because of the excessive urination and fluid intake; while the word *mellitus* is derived from a Latin word meaning sweet or honeyed, because of the high sugar content of the urine.

■ ■ ■

Finding a treatment for type 1 diabetes. *Frederick Banting and Charles Best isolated a hormone from the pancreas that would later be called insulin. They are shown at the University of Toronto with one of the dogs used in their experiments.*

Science Source

Turn-of the-century scientists and doctors were desperate to find a cure for the dreadful disease. In 1920, a Canadian doctor named Frederick Banting read an article in a medical journal that sparked an idea for a radical experiment. He brought the idea to the head of physiology at the University of Toronto, John Macleod, who was instantly skeptical. But Banting wore him down, and in 1921, against all odds, he received permission to proceed. He got a small laboratory, 10 dogs, and an assistant named Charles Best.

In a race against time, Banting and Best began their experiments. They knew that the pancreas was producing insulin, so they wanted to find a way to isolate the pancreatic cells that produced insulin. They theorized that if they could purify extracts of these cells to produce insulin itself, they could give them to dying children to make up for what their bodies lacked.

The researchers disrupted the digestive function of the pancreas of one of the dogs, to prevent digestive enzymes from destroying the insulin that they were seeking isolate. They removed the pancreas of one of the dogs, ground it up into a soupy mixture, filtered it, and injected that final product into another dog that had diabetes. Immediately, the dog's blood glucose levels fell and it appeared to be healthier. With a few daily injections, the dog remained symptom-free.

If this treatment were to work, Banting and Best reasoned, they needed more pancreas than a handful of dogs could supply. (They had resorted to catching stray dogs on the street.) So they turned to slaughterhouses, procuring fresh pancreas from cows and pigs. That appeared to work just as well, without having to disable the organ's digestive function first. A biochemist was brought on board to help purify the extracts. Now, with a larger supply of insulin, they could keep several dogs with diabetes alive.

Then, of course, the researchers wanted to find out if this treatment worked in people. For that, they needed a brave volunteer.

Around that time, a 14-year-old boy named Leonard Thompson was admitted to Toronto General Hospital, which was associated with the University of Toronto. He weighed only 65 pounds, was weak and lethargic, and was likely weeks away from dying from uncontrolled diabetes. He had no other options, so his parents consented to the experimental treatment. In January 1922, the researchers injected him with their purified insulin and held their breath.

Amazingly, Thompson immediately got stronger. Encouraged, other families brought in their children with diabetes, and those children also reacted well to the medicine. A new era of medicine was born; the next year, Banting and Macleod won the Nobel Prize for Physiology or Medicine.

The researchers soon realized that in order to supply all the insulin that children needed worldwide they had to find a way to mass-produce it. They didn't even know if they could make enough to keep alive the handful of patients who'd already started treatment. Thankfully, the research director of the pharmaceutical company Eli Lilly offered to help. By 1923, there was enough insulin to treat every person with diabetes in North America.

Risky behavior. *Although our race and family history is out of our control, we can decrease our risk of type 2 diabetes through physical activity, and by making food choices that support a healthy body weight.*

picturelibrary/Alamy

TYPE 2 DIABETES

Interestingly, in the past several decades, the number of people with diabetes has doubled. Some of that increase is due to an aging population, as people with type 1 diabetes now live longer. But most of the dramatic rise in diabetes is due to an increase in a second form of the disease—**type 2 diabetes**. Its increase is primarily attributed to the increased prevalence of obesity and decreased physical activity of our population—and not just in adults, incidence in those under 20 years of age is increasing as well. Today, 9 of 10 people with diabetes have type 2 diabetes.

Although type 1 diabetes is typically diagnosed in children and young adults, type 2 is most frequently seen in adults. Unlike type 1, in which insulin-producing pancreatic cells are destroyed and there is little or no insulin made, type 2 diabetes occurs when cells are less sensitive to the effects of insulin, even if the pancreas is able to produce normal amounts. *Obesity* is by far the most significant risk factor for the disease: 80% to 90% of people with type 2 diabetes are obese. In addition to having excess body fat, physical inactivity is another factor that significantly affects the risk of developing diabetes. However, genetics and lifestyle factors also potently affect an individual's risk of disease. If one identical twin has diabetes, the other twin will also have the disease 75% of the time. In addition, the risk of diabetes is higher among African-Americans, Hispanics and Latinos, American-Indians, Asian-Americans, and Pacific Islanders than among white people.

Type 2 diabetes begins with the development of **insulin resistance**. Most often this occurs because excess adipose tissue produces hormone-like substances that circulate throughout the body and interfere with the signal that insulin sends into cells to stimulate glucose uptake. As a result, muscle and fat cells do not respond properly to insulin and poorly take up glucose from the blood. The body then needs higher levels of insulin to help glucose enter cells. The cells of the pancreas produce more of the hormone to try to meet the demand. In some insulin-resistant individuals the pancreas can no longer keep up with the increased demand

TYPE 2 DIABETES MELLITUS
a condition characterized by elevated blood sugar levels due to insulin sensitivity (or resistance) and some impairment of insulin secretion from the pancreas

INSULIN RESISTANCE
a condition in which cells have a decreased sensitivity to insulin, resulting in impaired glucose uptake from blood, increased blood glucose levels, and further insulin release from the pancreas

INFOGRAPHIC 5.2 Insulin Sensitivity and Insulin Resistance *When insulin resistance develops, the muscle and fat cells do not effectively take up glucose.*

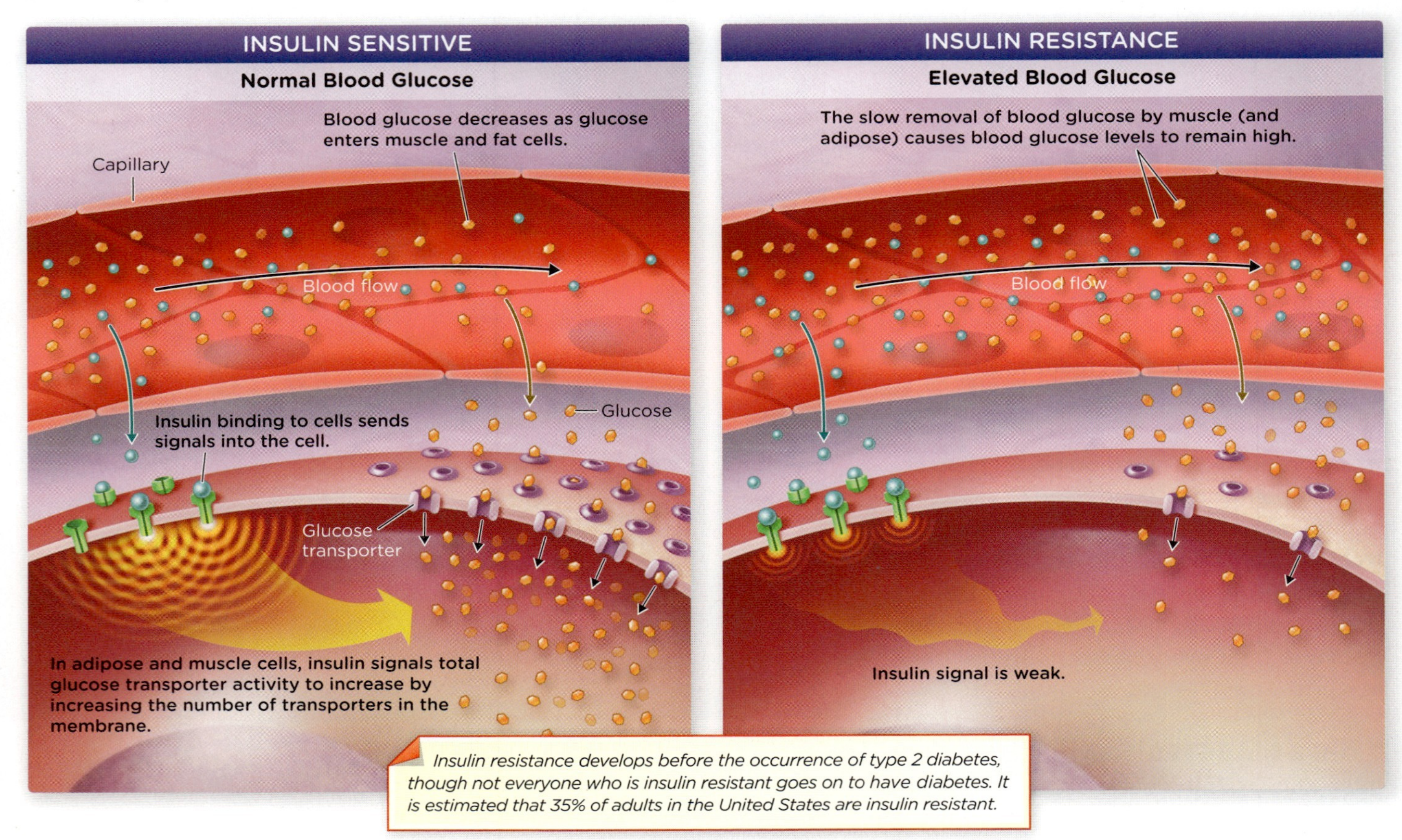

Insulin resistance develops before the occurrence of type 2 diabetes, though not everyone who is insulin resistant goes on to have diabetes. It is estimated that 35% of adults in the United States are insulin resistant.

How does insulin resistance in the liver contribute to elevated blood glucose?

PREDIABETES
a condition of higher-than-normal blood glucose levels, but not high enough to be diagnosed as diabetes; characterized by fasting plasma glucose level of 100 mg to 125 mg per dl of blood

ORAL GLUCOSE TOLERANCE TEST (OGTT)
a test used to diagnose prediabetes and diabetes; it measures the body's response to glucose in the bloodstream

for insulin, and blood glucose concentrations increase. Insulin resistance also blocks insulin's ability to adequately suppress liver glucose production, contributing further to the increase in blood glucose levels. **(INFOGRAPHIC 5.2)**

Without intervention and lifestyle changes, many people who have insulin resistance go on to develop type 2 diabetes. However, they may first develop **prediabetes**, a condition in which blood glucose levels are moderately elevated above levels that are considered normal or desirable. For example, after fasting for eight hours, the blood glucose levels of someone with prediabetes would measure 100 mg to 125 mg per 100 ml blood, which is elevated but still less than the fasting level of at least 126 mg per ml that is required for a diagnosis of diabetes. After ingesting glucose as part of an **oral glucose tolerance test (OGTT)**, the same person would have elevated blood glucose levels in the 140 mg to 199 mg per 100 ml range—just under the 200 mg/100 ml diagnostic level for diabetes. **(INFOGRAPHIC 5.3)** According to the American Diabetes Association, more than 33% of American adults older than 20 have prediabetes and fewer than 10% have been told they have it. Prediabetes not only increases the risk of type 2 diabetes, but also heart disease, stroke, and other conditions associated with elevated blood glucose levels. However, as we will see, there are lifestyle strategies, including weight control, physical activity, and dietary modifications that can normalize blood glucose levels and prevent or delay onset or impact of type 2 diabetes.

INFOGRAPHIC 5.3 Diagnosing Diabetes *Two key tests are used to measure problems related to blood sugar regulation: the fasting blood glucose and oral glucose tolerance tests.*

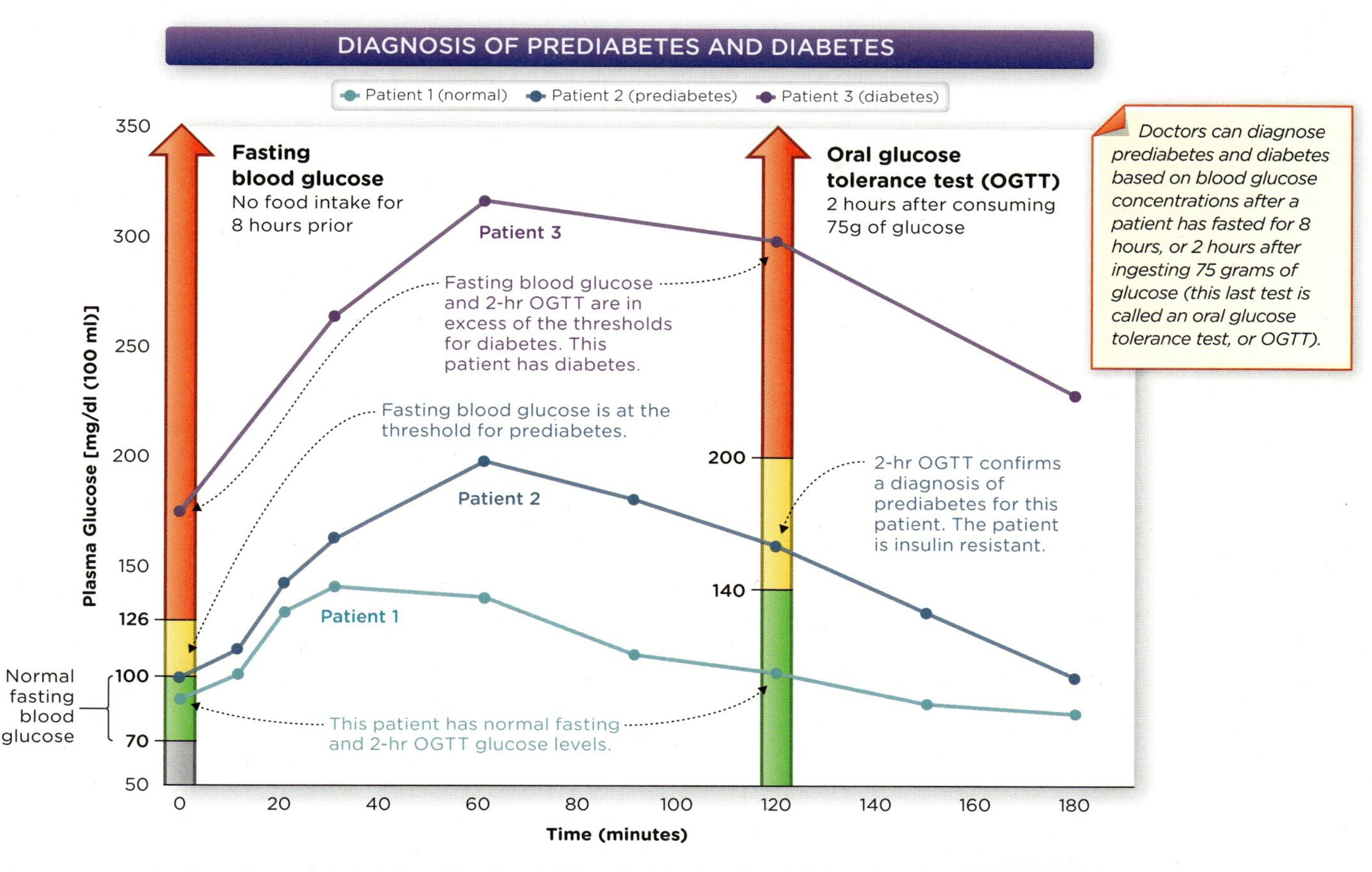

What condition causes blood glucose to remain at higher-than-normal concentrations in the prediabetic patient?

Because all forms of diabetes decrease the body's ability to remove glucose from the blood, each can cause blood glucose concentrations to increase to dangerous levels, leading to similar symptoms—feeling constantly thirsty (as the kidneys excrete the extra glucose through urine), blurred vision (as extra glucose enters the eye), and weight loss (due to low insulin levels and muscle cells relying on burning fat for energy). Over time, high levels of blood glucose can damage cells and organs, creating long-term problems in the heart, blood vessels, kidneys, eyes, and nerves. The excess glucose causes blood vessels to thicken and lose some of their elasticity, making these people more prone to heart attacks and stroke. Even today, with all the vast improvements in treatment, diabetes is the number one cause of adult blindness, and is responsible for more than 60% of lower limb amputations that don't result from trauma. **(INFOGRAPHIC 5.4)**

Gestational diabetes

Pregnant women can develop a form of diabetes known as **gestational diabetes**. Approximately 18% of expectant mothers will experience high blood glucose levels for the first time during pregnancy, and this can affect their pregnancy—increasing the rate of complications or causing the baby to grow too large. Obese women are at higher risk, so are often screened early in their pregnancies. Extra exercise can reduce the risk, while high-fiber and a low-*glycemic index* diet can help keep blood glucose at healthy levels. For many women, gestational diabetes goes away once they give birth, but they will

GESTATIONAL DIABETES
a condition of elevated blood glucose levels arising in approximately 18% of all pregnant women, most of whom revert to normal blood glucose levels after delivery

INFOGRAPHIC **5.4** **Complications of Diabetes**

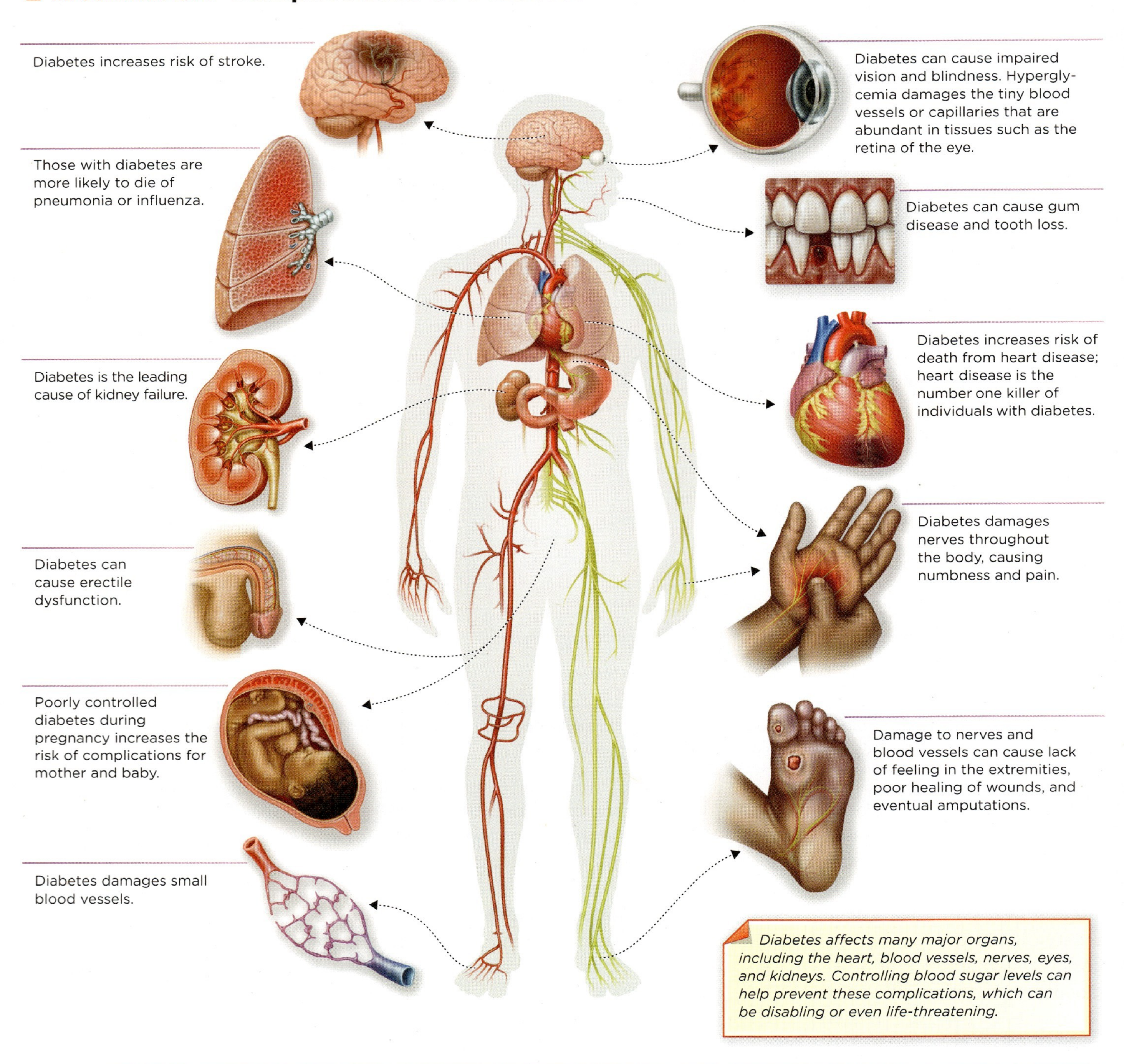

What is the leading cause of death among those with diabetes?

remain at a higher risk of developing type 2 diabetes later in life. (INFOGRAPHIC 5.5)

DIABETES ON THE RISE

For decades, doctors mostly distinguished type 1 and type 2 diabetes by the age of its onset—type 1 was something diagnosed in children, type 2 in adults. That distinction was fairly reliable; early in her career, Lori Laffel, MD, Chief of the Pediatric, Adolescent, and Young Adult Section at the Joslin Diabetes Center, only saw the "occasional" child with

INFOGRAPHIC 5.5 Gestational Diabetes *Like other types of diabetes, gestational diabetes affects how the cells of the body use glucose. High blood sugar can affect the health of the mother and infant.*

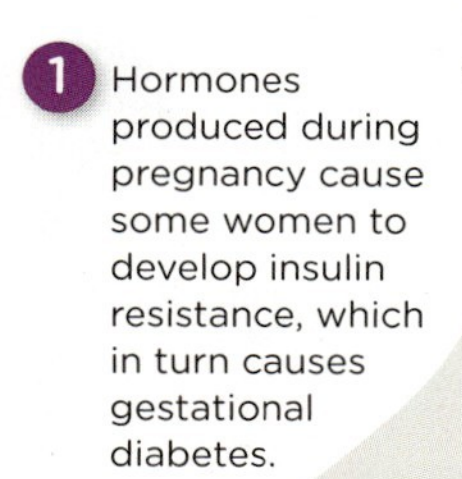

1 Hormones produced during pregnancy cause some women to develop insulin resistance, which in turn causes gestational diabetes.

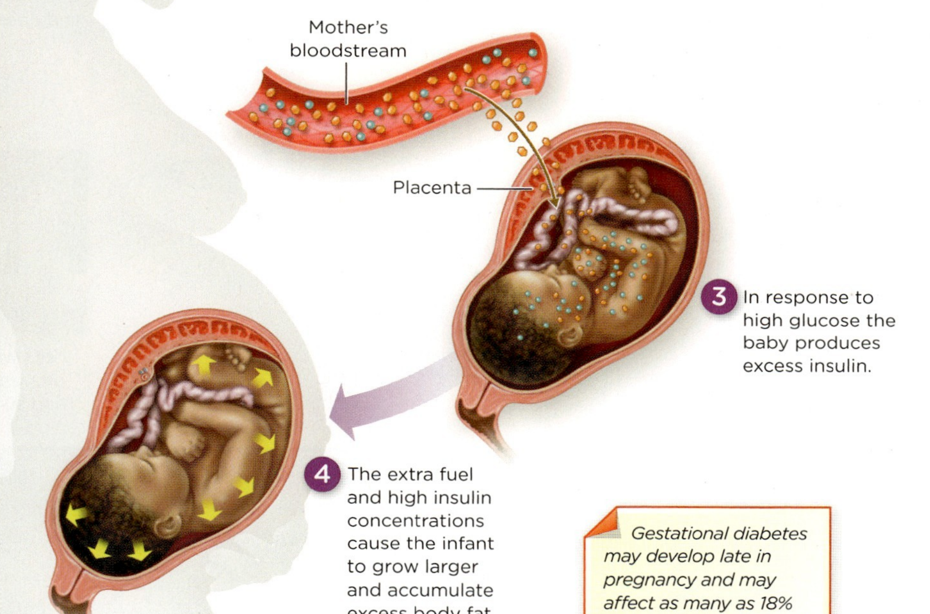

2 Gestational diabetes causes glucose in the mother's blood to increase and cross the placenta. The developing infant is then exposed to the excess glucose.

3 In response to high glucose the baby produces excess insulin.

4 The extra fuel and high insulin concentrations cause the infant to grow larger and accumulate excess body fat.

Gestational diabetes may develop late in pregnancy and may affect as many as 18% of pregnancies.

Risk Factors	Consequences
• Obese prior to pregnancy • Older than 25 years • Prediabetic • Family history of type 2 diabetes • Non-White race	**For the mother:** • Higher incidence of cesarean section • High blood pressure during pregnancy • Increased risk of developing diabetes in 10–20 years **For the infant:** • Risk of shoulder injury during birth • Higher risk of breathing problems • Increased risk of becoming an obese child • Increased risk of developing diabetes as an adult

What risk factors for gestational diabetes are similar to risk factors for type 2 diabetes?

INFOGRAPHIC 5.6 Tip the Balance in Favor of Diabetes Prevention *The Dietary Guidelines for Americans recommends consuming fewer calories, making informed food choices, and being physically active. Following the guidelines can help people attain and maintain a healthy weight, and reduce their risk of chronic diseases, such as diabetes.*

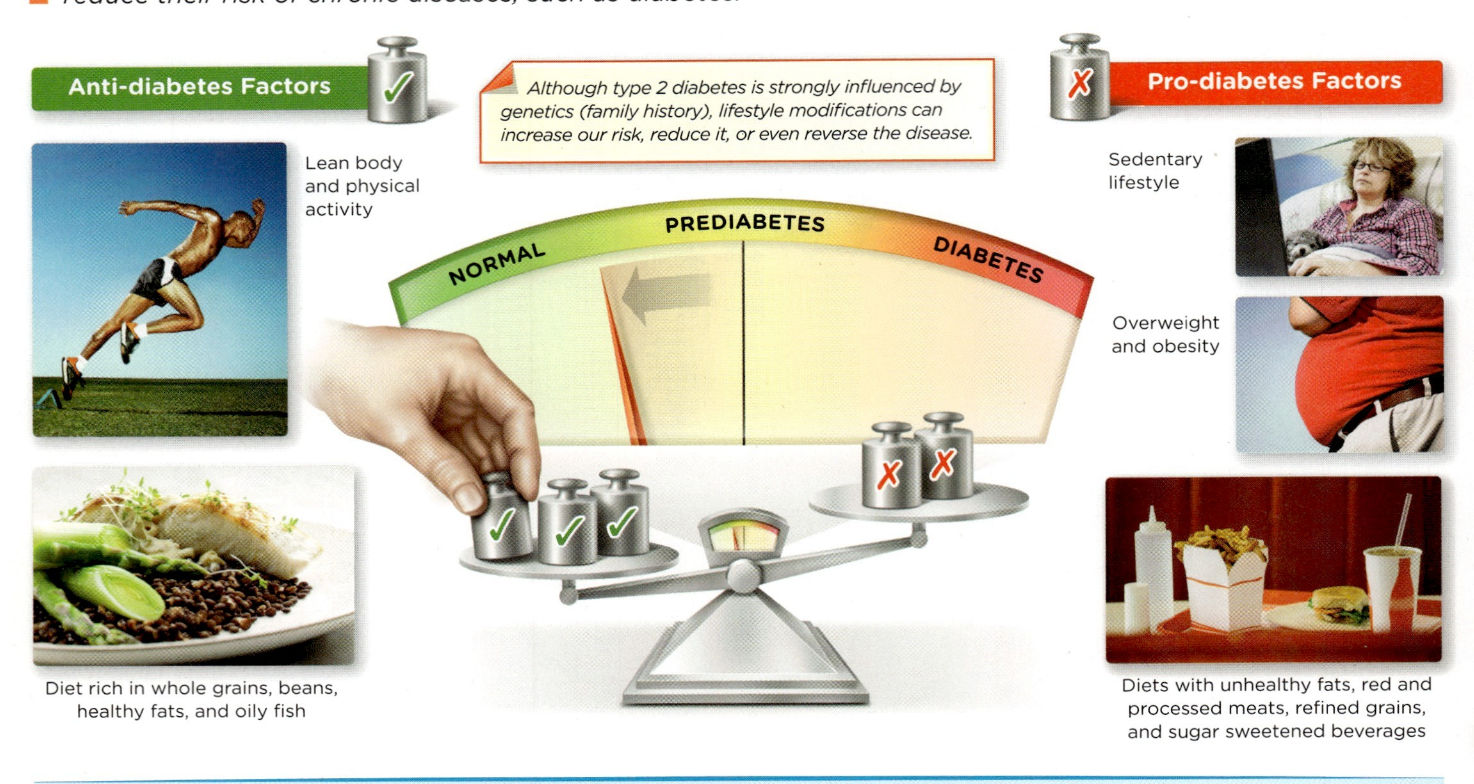

? *What do you think is the most difficult habit to adopt in an "anti-diabetes" lifestyle?*

Photo credits (left side; top to bottom): Karan Kapoor/Getty Images, Richard Semik/Shutterstock; (right side; top to bottom): Paul Vasarhelyi/Shutterstock, Suzanne Tucker/Shutterstock, Corey Hendrickson/Gallery Stock

type 2 diabetes. But in the last 20 years, that's all changed.

"Where we used to see literally a handful of children with type 2 diabetes each year, we now see ten-fold that number," says Laffel. "And that's what makes it an epidemic." At the same time, over the last 20 to 30 years, "we have also seen an epidemic rise in the number of children developing type 1 diabetes."

And they're developing it at younger and younger ages, she adds. And it's not entirely clear why. For type 2, lifestyle and body weight play major roles, so when kids spend more time in front of a screen and less time playing, while consuming an energy-dense, often nutrient-poor diet, their risk increases. Dietary factors associated with increased risk of type 2 diabetes are excess calories, a low intake of whole grains, and high intakes of refined carbohydrates, sugar-sweetened beverages, and trans fat and saturated fat. "Without a doubt, the overeating, decreased physical activity, and increase in sedentary behaviors contributes to the rise in both obesity and type 2 diabetes among children," says Laffel. (INFOGRAPHIC 5.6)

Worldwide, approximately 371 million people have diabetes, primarily type 2; by 2030, that number is expected to increase to 550 million. In the United States alone, approximately 26 million people have diabetes, and an estimated 3% of American adults have diabetes that is undiagnosed.

DIABETES TREATMENT AND PREVENTION

People with type 1 diabetes must receive injections of insulin to manage their blood glucose levels. An insulin pill isn't yet on the horizon—as a protein, insulin would be

Managing diabetes. *Individuals with type 1 diabetes require regular doses of insulin to be delivered into the blood stream through injection or through a pump. A glucometer is a device that individuals with type 1 and type 2 use to track the amount of glucose present in the blood.*

Bruno Boissonnet/Science Source

Anne-Marie Palmer/Alamy

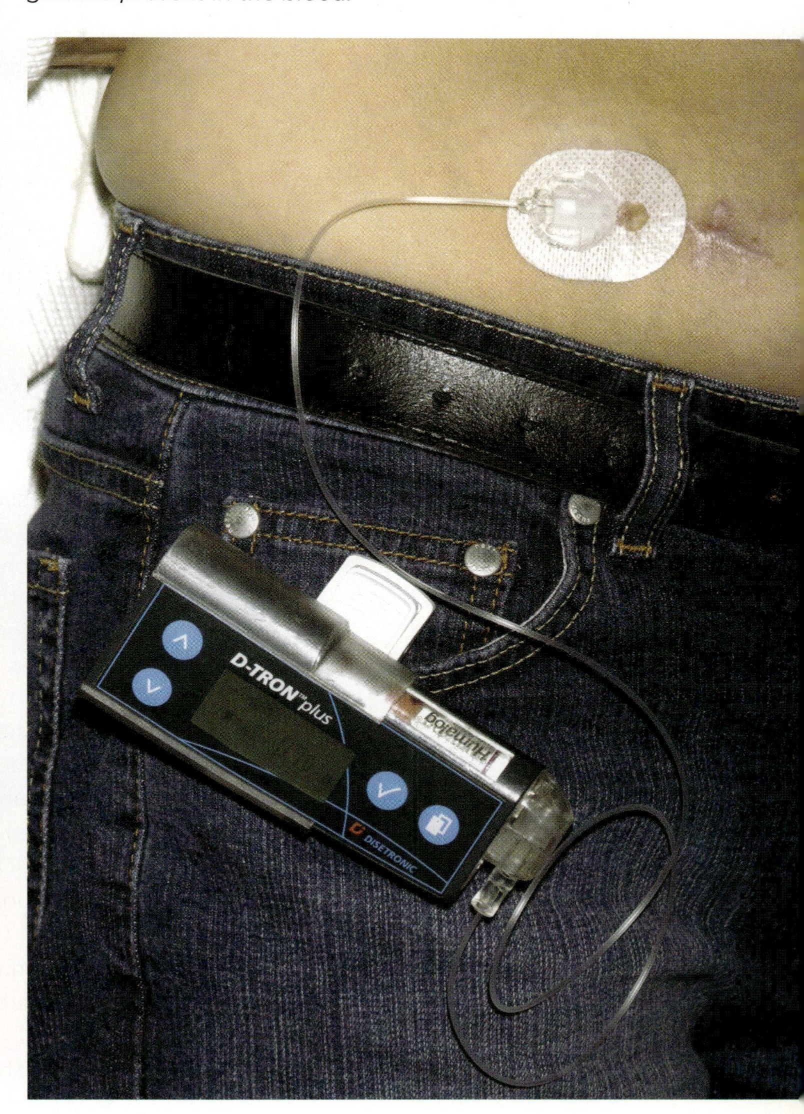

Dr. P. Marazzi/Science Source

digested if it passed through the GI tract. People with type 1 diabetes either give themselves injections several times per day, or have a medical device known as an *insulin pump* that delivers insulin as needed, particularly following a meal or snack. Regular monitoring of blood glucose levels with a glucose meter gives people with diabetes a way to track the impact of their food choices and activity level, so that they can adjust insulin injections accordingly to maintain desirable blood glucose levels. Type 2 diabetes, however, generally does not require insulin injections and can often be managed, and even prevented or reversed, by lifestyle modifications (including diet and physical activity) that promote insulin sensitivity and a healthy body weight.

Different foods have different effects on blood glucose, even when they contain the same amount of starch and sugars. Specifically, how the body processes particular carbohydrates dictates how much and at what

Simple steps to prevent diabetes. *Exercise, achieving a healthy weight, smoking cessation, and eating a balanced diet can help prevent diabetes.*

rate glucose will be released into the bloodstream and therefore how much insulin is released. For example, starches in foods can be partially "trapped" within the physical structure of grains or in the internal portion of larger food particles. These physical structures can make it difficult for digestive enzymes to gain access to the starch, slowing digestion and the subsequent appearance of glucose in blood. Such starches, that remain undigested and enter the large intestine, are called **resistant starches**. Diets high in resistant starch may improve insulin sensitivity and make people feel full for longer periods. Examples of foods high in resistant starches include beans, under-ripe bananas, whole grain kernels, and pasta.

At the other extreme, starches in some foods—such as boiled potatoes and many breakfast cereals—are digested so quickly that blood glucose rises nearly as rapidly as what is seen after consuming an equal amount of pure glucose.

Surprisingly, sucrose, lactose, and fructose and foods that contain these sugars (such as candies, dairy foods, or fruit) often produce a less dramatic rise in blood glucose than starchy foods. This is because fructose and galactose (a monosaccharide in lactose) have no immediate impact on blood glucose levels, until they are converted to glucose in the liver. Diets high in fiber, especially soluble fiber, also seem to help control blood glucose levels. One study found that people who consumed 50 grams of fiber per day with at least half as soluble fiber were able to control their blood glucose better than those who ate less fiber.

Since food sources of carbohydrates have different effects on blood glucose levels, Laffel tells children and their families to pay attention to a measure known as **glycemic index (GI)** which ranks foods by how quickly and to what degree they raise blood glucose levels. The amount of available carbohydrate (starch and sugars) present in the food is compared with the equivalent amount of glucose, which has a GI of 100. Foods with a high GI raise blood glucose quickly and cause considerable spikes in blood glucose levels, while foods with a low GI cause a smaller and more gradual increase. Low GI diets help individuals with diabetes better control blood glucose levels and are associated with reduced risk of developing type 2 diabetes.

Although gylcemic index can be a helpful measure to consider, it has some limitations. For example, we rarely eat carbohydrate containing foods in consistently the same amounts (smaller or larger amounts would alter the GI) or eat these foods by themselves. And even carbs with high GIs have less effect on blood glucose when consumed with protein, fat, or fiber, which all delay gastric emptying. Even factors such as the ripeness of a fruit or how

RESISTANT STARCH
a starch that remains intact after cooking, is not broken down by human digestive enzymes, and is not absorbed from the intestines

GLYCEMIC INDEX
a number used to rank carbohydrate foods by their ability to raise blood glucose levels compared with a reference standard

a carbohydrate is processed can alter its GI. (INFOGRAPHIC 5.7)

The GI of a food, however, often does not reflect the amount of carbohydrates contained in the servings of food that we typically consume. Foods with a low to moderate GI can still cause a dramatic rise in blood glucose levels if consumed in large amounts. So consumers can also consult the **glycemic load (GL)**, which calculates the effect of the actual serving size of food on blood glucose (by multiplying the GI by a food's grams of carbohydrates, divided by 100). The GL is a more useful assessment of a food's effect on blood glucose.

Some research suggests that diets emphasizing low-GI and low-GL foods in the overall diet may help manage diabetes, control body weight, and improve other health conditions. In addition, low-GI foods are often rich in fiber and other nutrients, which have clear benefits for health.

Registered dietitians and other health care providers may also teach their patients with diabetes about **carbohydrate counting**, or "carb counting," a meal-planning technique that can help manage blood glucose levels. If part of the person's diabetes management is to keep blood glucose levels in a desirable

GLYCEMIC LOAD
the extent of increase in blood glucose levels, calculated by multiplying glycemic index by the carbohydrate content of a food

CARBOHYDRATE COUNTING
a method to track carbohydrates consumed so that those with diabetes can appropriately balance physical activity and medication to manage blood glucose levels

INFOGRAPHIC 5.7 Glycemic Index of Commonly Consumed Foods *The glycemic index (GI) ranks foods by their blood glucose response following ingestion.*

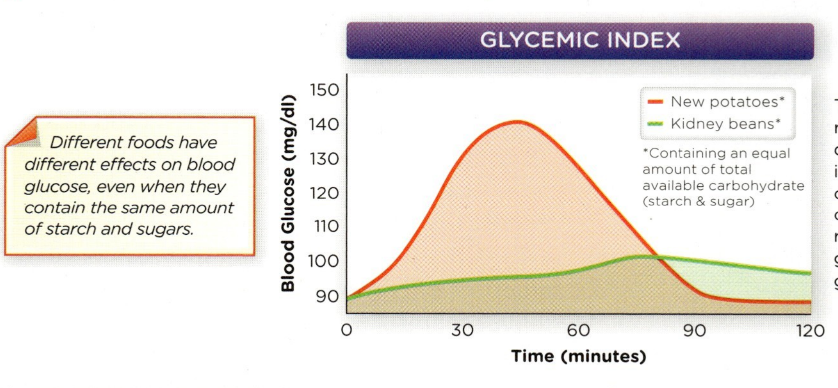

The glycemic index ranks foods by how dramatically they increase blood glucose concentrations compared with that of a reference food (typically glucose, which has a glycemic index of 100).

LOW GI (55 or less)		MEDIUM GI (56 to 69)		HIGH GI (70 and above)	
Oatmeal	55	Shredded wheat	69	Potato (baked)	96
Sweet corn	52	Brown rice	68	Corn flakes	92
Banana	51	Pancake	66	Gatorade	89
Spaghetti (white)	49	Popcorn	65	Rice Krispies	88
Spaghetti (whole grain)	48	Raisins	64	Pretzels	84
Nectarine	43	Sucrose	63	Rye bread	78
Milk chocolate bar	43	Cherries	63	Watermelon	76
Chocolate cake, frosted	39	Potato, french fries	63	Waffle (frozen)	76
Carrots (boiled)	39	Cola, soft drink	63	Doughnut, cake	76
Pear	38	Cranberries (dried, sweetened)	62	Bread (white)	74
Ice cream (Low fat)	37	Raisin bran	61	Bread (whole wheat)	74
Apples	37	Honey	61	White rice	72
Skim milk	35	Bran muffin	60	Jam and peanut butter toast	72
Pinto beans	33	Rye bread (60% whole kernels)	57	Sweet potato	70

Foods with high sugar content often have a low GI because the need to convert fructose to glucose in the liver causes a delay in its appearance in the blood as glucose.

In some foods starches are trapped within the matrix of the food and digestion is slowed, lowering the GI.

Many foods with high starch content have a high GI because starches can often be very rapidly digested to yield glucose.

? *What are two additional foods that are high in starch and have a high GI, and two additional foods that are high in sugar and have a low GI?*

Photo credits (left to right): Gregor Schuster/Getty Images, Glow Cuisine/Getty Images, hh5800/Getty Images, Daniel Grill/Tetra Images/Corbis, Gentl and Hyers/Getty Images, Dean Turner/Getty Images, Joff Lee/Getty Images, ifong/Shutterstock

Benefits of fiber. *Diets high in fiber, especially soluble fiber, seem to help control blood glucose.*

foodfolio/Alamy

range, they can determine the total amount of carbohydrates (in grams) to include each day according to their energy needs and activity level. By portioning out that total among the day's meals and snacks, they are better able to align their diet and medication. But those with diabetes have to do more than just watch their carbohydrate intake. Individuals with diabetes have a much higher risk of death from heart disease than those without diabetes; therefore, they also have to keep track of their fat intake, limiting saturated fat to less than 7% of their total calories, and eating as little trans fat as possible. Some fats are heart healthy, though, and doctors will often tell people with diabetes to eat two or more weekly servings of fish, which provide omega-3 polyunsaturated fatty acids.

Physical activity also helps improve insulin sensitivity, lower blood glucose levels, and keep weight off over the long term. Exercise alone may not be enough—but that, plus weight control and a varied and balanced diet, provide the foundation for managing and preventing type 2 diabetes.

When those lifestyle changes aren't enough, oral medications are available to lower blood glucose levels of individuals with type 2 diabetes, and some people may require insulin injections to get their glucose under control.

Hypoglycemia

It is possible to overmedicate patients with diabetes; they can develop **hypoglycemia**, in which their blood glucose levels drop too low (diagnosed when < 70mg/100 ml blood), causing symptoms ranging from irritability, headache, hunger, weakness or fatigue, sweating, and rapid heartbeat.

There are different types of hypoglycemia—*fasting hypoglycemia*, for instance, occurs when people have not eaten, have drunk too much alcohol, or have underlying hormonal conditions or tumors, such as pancreatic tumors. People with *reactive hypoglycemia*, in contrast, experience symptoms when they eat large amounts of carbohydrates, causing a huge release of insulin and rapid drop in blood glucose. (In response, the adrenal glands release **epinephrine**—also known as adrenaline—which, like glucagon, functions to raise blood glucose by stimulating glucose synthesis in the liver.) The best way to manage or prevent hypoglycemia is to not eat too many carb-heavy meals or snacks, include some protein along with carbohydrates, and emphasize foods that contain fiber, particularly soluble fiber.

■ ■ ■

Today, millions of people take insulin and, along with appropriate diet and physical activity, are able to manage their diabetes and lead productive, healthy, and happy lives.

Although the work of Banting and his colleagues to develop insulin revolutionized diabetes therapy, scientists are continuing to investigate ways to treat diabetes without relying on cumbersome and unpleasant injections—and hopefully prevent and cure it altogether. "Insulin therapy is like the

HYPOGLYCEMIA
abnormally low blood glucose levels, resulting in symptoms of anxiety, hunger, sweating, and heart palpitations (fasting plasma glucose < 70 mg/dl)

EPINEPHRINE (ADRENALINE)
a hormone released from the adrenal glands to help the body prepare for a fight-or-flight response by increasing glucose availability in the blood

difference between automobiles between 1922 and 2013—we've got vastly better automobiles," says Bliss. "But is there something better than the automobile? Is there something better than insulin that can effectively cure diabetes, and reverse the damage done? Scientists have kept trying and trying." And in the past several decades, the development and widespread use of effective and safe synthetic human insulin has made the dependence on cow and pig pancreatic cells obsolete.

Research toward a cure for type 1 diabetes is ongoing and varied in approach. Efforts have been made to understand the immune system attack that causes type 1 diabetes, and to intervene in that process. Another important effort is to regenerate the cells that produce insulin through the use of embryonic or adult stem cells.

While the development of insulin therapy revolutionized diabetes treatment, today we face an increasing epidemic of type 2 diabetes and concern about the long-term consequences of living with diabetes. Scientists are therefore continuing to investigate better ways to treat all different types of diabetes with the hope of one day preventing and curing it altogether.

CHAPTER 5 **BRING IT HOME**

Focus on type 2 diabetes

Type 2 diabetes affects at least one in 10 U.S. adults 20 years or older—and at least one in four older than 65 years. To better understand your risk of developing type 2 diabetes, as well as strategies for prevention and treatment, visit the website for the American Diabetes Association at www.diabetes.org.

1. Take the Type 2 Diabetes Risk Test found at www.diabetes.org/are-you-at-risk/ to address these questions:
   a. What was your score (on their scale from 1 to 10, with 10 representing those at highest risk)?
   b. List any risk factors identified under Your Risk Factors, as well as the points earned for each.
   c. Click on Next Steps to learn more about your risk factors and ways to reduce risk. List at least one way each of these contributes to risk and at least one strategy to help prevent type 2 diabetes.
2. Go to www.diabetes.org/diabetes-basics/ to address these questions and learn more about the symptoms of and myths about type 2 diabetes.
   a. Click on Symptoms (www.diabetes.org/diabetes-basics/symptoms/) and list five common symptoms of diabetes.
   b. Click on Diabetes Myths (www.diabetes.org/diabetes-basics/myths/). Read over the diabetes myths, choose one, and briefly describe how the information provided clarifies the common misconception.
3. Go to Diabetes Meal Plans and a Healthy Diet at http://www.diabetes.org/food-and-fitness/food/planning-meals/diabetes-meal-plans-and-a-healthy-diet.html and use the information provided to answer these questions:
   a. What is a diabetes meal plan?
   b. Briefly describe the "Create Your Plate" and "Carbohydrate Counting" meal planning tools.

Take It Further

1. Share the Type 2 Diabetes Risk Test with a family member or friend.
2. Take the Type 2 Diabetes Risk Test for an individual with the following profile: 44-year-old male with a family history of diabetes, high blood pressure, 5′10″ tall, and 210 pounds. What is his score? Identify three lifestyle strategies to help lower this individual's risk of developing type 2 diabetes.

KEY IDEAS

Diabetes mellitus, more commonly known as simply diabetes, is a disease that disrupts the body's ability to adequately regulate glucose metabolism.

Insulin and glucagon are hormones secreted by the pancreas that regulate glucose metabolism and blood glucose levels.

Insulin enables cells in the skeletal muscle, cardiac muscle, and adipose tissue to take up glucose from the blood. Insulin also promotes the conversion of excess glucose to glycogen in the liver and muscle. Additionally, as glycogen storage is limited, insulin stimulates the conversion of glucose to fat in the liver and adipose tissue.

Glucagon is responsible for increasing glucose availability in the blood when blood glucose levels fall by signaling the liver to break down glycogen or synthesize new glucose molecules.

Type 1 diabetes is an autoimmune disease characterized by the destruction of insulin-producing cells in the pancreas and increased levels of blood glucose (hyperglycemia).

Type 2 diabetes is the most common form of diabetes and is characterized by insulin resistance and hyperglycemia.

Insulin resistance is a common condition in which cells lose their sensitivity to insulin and often precedes development of type 2 diabetes.

A diagnosis of diabetes is based on blood glucose concentrations of at least 126 mg per 100 ml of blood after an 8-hour fast, and an oral glucose tolerance test result of at least 200 mg per 100 ml 2 hours after the ingestion of 75 grams of glucose.

Individuals with prediabetes or glucose levels higher than normal (typically 100–125 mg/dl), but not high enough to be diagnosed with type 2 diabetes are more likely to develop some form of diabetes and are at higher risk of heart disease and stroke.

Diabetes affects many major organs, including the heart, blood vessels, nerves, eyes, and kidneys. Controlling blood glucose levels can help prevent these complications, which can be disabling or even life-threatening.

Gestational diabetes is a common form of diabetes that afflicts women during pregnancy, particularly women who are overweight or obese. Most women revert to normal blood glucose levels following delivery, but may be at increased risk of developing type 2 diabetes later in life.

Although type 2 diabetes is influenced by genetics, lifestyle modifications including a diet rich in whole grains, beans, and healthy fats, along with regular physical activity and maintenance of a healthy body weight, can reduce the risk of or potentially reverse the disease.

Diets that emphasize foods with resistant starch (foods that are resistant to digestion because of their physical structure) may improve insulin sensitivity, satiety, and blood glucose control.

Food sources of carbohydrates can have different effects on blood glucose levels. The glycemic index (GI) is a ranking of how food affects blood glucose relative to the effect of an equivalent amount of carbohydrate. Although the clinical relevance of using GI is unclear, a more applicable measure may be the glycemic load (GL), which indicates the effect of typical portions on blood glucose.

Hypoglycemia, or low blood glucose levels, may result from overmedication in individuals with diabetes, in response to prolonged periods of not eating, to consuming excessive amounts of alcohol, or in rare cases, from abnormalities in the way the body produces and responds to insulin.

NEED TO KNOW

Review Questions

1. All of the following are true for the hormone insulin, EXCEPT that it:
   a. functions to lower blood glucose levels.
   b. enhances storage of excess glucose to fat in adipose tissue.
   c. enhances conversion of excess glucose into glycogen in liver and muscles.
   d. is required for release of glucose from glycogen stores.
   e. is secreted by cells in the pancreas.

2. All of the following are true for the hormone glucagon, EXCEPT that it:
   a. triggers the synthesis of glycogen in the liver.
   b. signals the liver to release glucose into the blood.
   c. is secreted from the pancreas.
   d. works with insulin in regulating blood sugar levels.
   e. signals glucose production from amino acids.

3. In type 1 diabetes, cells in the pancreas:
   a. are less responsive to effects of circulating insulin.
   b. secrete excessive amounts of insulin.
   c. secrete excessive amounts of glucagon.
   d. are destroyed by the body's immune system.
   e. multiply to provide sufficient amounts of insulin.

4. Type 2 diabetes:
   a. is characterized by an absence of insulin production by the pancreas.
   b. is characterized by insulin resistance and obesity.
   c. is caused by excessive intake of sugar or sucrose.
   d. occurs only during pregnancy and disappears after delivery.
   e. is an autoimmune disease.

5. Ketone bodies in the blood:
   a. are synthesized from the breakdown of amino acids.
   b. stimulate the action of insulin in glucose uptake by the body's cells.
   c. can be used as an energy source when glucose use is impaired.
   d. can increase the alkalinity of the blood to dangerous levels.
   e. are correlated with low blood insulin levels.

6. All of the following are true with regard to insulin resistance, EXCEPT that it:
   a. develops in the latter stages of type 2 diabetes.
   b. can occur even when the pancreas produces normal amounts of insulin.
   c. is associated with excess adipose tissue and obesity.
   d. blocks insulin's ability to adequately suppress liver glucose production.
   e. results in impaired removal of excess glucose from blood.

7. The elevated blood glucose level that is needed for a diagnosis of diabetes following an oral glucose tolerance test is:
   a. 100 mg/100 ml blood.
   b. 125 mg/100 ml blood.
   c. 150 mg/100 ml blood.
   d. 175 mg/100 ml blood.
   e. 200 mg/110 ml blood.

8. All of the following are true with regard to complications from diabetes, EXCEPT that they:
   a. can be minimized or avoided with proper blood glucose management.
   b. are much more prevalent in type 1 than in type 2 diabetes.
   c. can result in the primary cause of adult blindness.
   d. increased risk of heart attack and stroke due to the effect on blood vessels.
   e. can cause weight loss and increased burning of fat for energy.

9. Expectant mothers with gestational diabetes:
   a. usually give birth to a full-term baby that is significantly smaller than normal birth weight.
   b. have increased risk of complications during pregnancy.
   c. are at the same risk for developing the condition whether they are underweight, normal, or overweight.
   d. are almost always still diabetic following delivery.
   e. represent only about 2% of mothers.

10. All of the following are dietary factors associated with increased risk of developing type 2 diabetes, EXCEPT:
   a. sugar-sweetened beverage consumption.
   b. high intake of refined carbohydrates.
   c. excess calorie intake.
   d. unsaturated fat intake.
   e. low intake of whole grains.

11. Diets that emphasize foods with a low glycemic index (GI):
   a. may be low in dietary fiber.
   b. increase the risk of hyperglycemia.
   c. may help in blood glucose control.
   d. are generally high in sugar.
   e. increase risk of obesity.

12. The type of hypoglycemia that may occur in a few individuals following a meal is termed:
   a. reactive hypoglycemia.
   b. insulin resistant hypoglycemia.
   c. autoimmune hypoglycemia.
   d. fasting hypoglycemia.
   e. metabolic syndrome.

Take It Further

Describe the roles of insulin and glucagon in maintaining a desirable blood glucose level.

Dietary Analysis Using SuperTracker

Analyzing specific dietary carbohydrates

Complete this diet analysis to identify specific dietary carbohydrates that are associated with metabolic disorders.

1. Log onto the United States Department of Agriculture (USDA) website at www.supertracker.usda.gov. If you have not done so already, you will need to create a profile to get a personalized diet plan. This profile will allow you to save your information and diet intake for future reference. *Do not use the general plan.*
2. Click the Track Food and Activity option.
3. For one day, record food and beverage intake that most reflects your typical eating patterns. Enter each food and beverage you consume into the food tracker. Note that there may not always be an exact match to the food or beverage that you consume, so select the best match available.
4. Once you have entered all of your food and beverage choices into the food tracker, on the right side of the page under the bar graph, you will see Related Links: View by Meal and Nutrient Intake Report. Print these reports and use them to answer the following questions:
   a. How many grams of carbohydrate did you consume? Did you reach your target level?
   b. Was your carbohydrate intake within the 45% to 65% of calories that is recommended?
   c. How many grams of added sugars did you consume? How would this amount need to be modified for a person with diabetes?
   d. Identify two dietary changes a person with diabetes would be advised to make to help keep their blood sugar under control.
   e. Next, analyze your lactose intake. How many foods did you consume that contain lactose? How would this need to be modified for a person with lactose intolerance?

6 THE LIPIDS

The Skinny on Fat

HOW NATIVE GREENLANDERS WITH A PASSION FOR SEAL BLUBBER SHOWED THE WORLD THAT NOT ALL LIPIDS ARE CREATED EQUAL.

Robert Harding Picture Library Ltd/Alamy

LEARNING OBJECTIVES

- Identify the four major categories of dietary lipids (Infographic 6.1)
- Describe the structural differences between saturated, monounsaturated, and polyunsaturated fats (Infographic 6.2)
- Identify the types of foods that are rich in monounsaturated fat, polyunsaturated fat, and saturated fat (Infographic 6.3)
- Name two roles of phospholipids in the body (Infographic 6.4)
- Describe the process of lipid digestion and explain how emulsification assists in the process (Infographic 6.5)
- List the four major lipoproteins and describe their functions in the transport of lipids (Infographics 6.6 and 6.7)
- Identify the two essential fatty acids, their primary structural difference, and food sources of each in the diet (Infographic 6.8)
- Describe the Acceptable Macronutrient Distribution Range for lipids and how to use a Nutrition Facts Panel to evaluate a food's fat content (Infographic 6.9)
- Describe sources of saturated fat in the U.S. diet (Infographic 6.10)

Even in the summer, Greenland is cold. The first time Jørn Dyerberg visited, as a young doctor in 1970, the temperature hovered around 40 degrees Fahrenheit in mid-August; on other trips, such as in April, the thermometer would drop below zero.

But Dyerberg was not deterred by the weather—he had traveled to one of the coldest places on Earth for a mission. He was there to understand why native Greenlanders—the Inuit—were much less likely to die of heart disease than similarly aged people living in

DIETARY FAT *compound found in plant and animal foods that serves as an important energy source, and, among other functions, is necessary for absorption and transport of fat-soluble vitamins*

LIPIDS *structurally diverse group of naturally occurring molecules that are generally insoluble in water, but are soluble in organic solvents; examples include fatty acids, triglycerides, sterols, and phospholipids*

Denmark (of which Greenland was a colony at the time), despite eating a high-fat diet rich in seal blubber and fatty fish.

It was a mystery. At the time, research evidence showed that high-fat diets were associated with an increase in blood cholesterol levels, leading to numerous health problems—in particular, heart disease. This made fat public enemy number one, as heart disease was the largest killer in Denmark, the United States, and many other countries. The popularity of low-fat diets, which emphasize carbohydrates and allow fats only sparingly, was on the rise.

But the relationship between **dietary fat** intake and health was not always consistent. Some populations—such as the Greeks, who followed a Mediterranean diet (to learn more about the Mediterranean diet and its benefits see Chapter 9)—ate relatively large amounts of fat, but had low rates of heart disease. In 1969, the head of Dyerberg's hospital, Hans Olaf Bang, read an article about the Inuit, who ate regular lunches of raw, slippery seal organs that were chock full of fat and cholesterol, but typically died of infections such as tuberculosis, rarely of heart disease.

At the time, Dyerberg was 32 years old and working on his PhD, investigating a new way to screen blood for different types of fats. When Bang asked him to come to Greenland to collect blood samples from the Inuit to see what was going on, he jumped at the chance.

After collecting blood samples from 130 adults, Dyerberg and Bang ran the blood through a machine that analyzed its *lipid* content. They saw that, despite their high-fat diet, Inuit's levels of all blood lipids were lower than those found in Danish adults. But there was one group of lipids that was found in much higher amounts in Inuit adults. This finding contradicted everything experts believed at the time—that the more lipids you had in your blood, the higher your risk of heart disease.

The Inuit blood profile couldn't be the result of genetics, the researchers knew—whenever Inuit adults moved to Denmark and began eating a diet typical of that nation, their blood lipids looked like those of the average Dane.

The researchers were puzzled. "We knew high intake of dietary fat increases cholesterol," says Dyerberg. "So why did the Inuit have such a healthy profile?"

Clearly, the role of dietary fat in the body was more complicated than many believed.

■ ■ ■

Lipids researcher, Jørn Dyerberg, Professor Emeritus, Department of Human Nutrition, University of Copenhagen

Press Photo, University of Copenhagen

OVERVIEW OF THE LIPIDS

Scientists have long known that what we commonly call "fats" are a subclass of **lipids**, a group of compounds made up of carbon, hydrogen, and a small amount of oxygen that generally can't mix or dissolve in water (they are water insoluble).

Lipids play important roles in the body, such as acting as a major component of cell membranes, giving them flexibility and integrity. They also facilitate the transport of nutrients, including fat-soluble vitamins (A, D, E, and K) and phytochemicals, as well as enhancing their absorption. Adipose tissue also cushions, protects, and insulates the kidney, heart, and other organs, shielding them from temperature extremes and acting as a natural shock absorber.

Fats are a concentrated source of energy, providing 9 kcal per gram, typically accounting for one-third or more of our total calorie intake. Fat is significantly more energy dense than carbohydrates and protein, which provide only 4 kcal per gram.

Lipids are diverse in structure and function. We will discuss the four most common lipid classes: fatty acids, triglycerides, sterols (such as chole*sterol*), and phospholipids (such as lecithin). Although some use the term "lipid" interchangeably with "fat," the word "fat," more precisely, refers to triglycerides, which make up 95% of all lipids in our foods and 99% of the stored fat in our bodies. (INFOGRAPHIC 6.1)

Dietary fats come from plant and animal sources. *Dietary fats contain 9 kcal per gram and contribute to the taste and texture of foods.*

alexpro9500/Shutterstock

It's no wonder the Inuit often enjoyed blubbery seal meat and other fatty foods—fats increase the palatability (taste and

INFOGRAPHIC 6.1 The Four Major Categories of Dietary Lipids

Lipids are a structurally diverse group of naturally occurring molecules that are generally, but not always, insoluble (or poorly soluble) in water.

Fatty Acids

A major energy source.

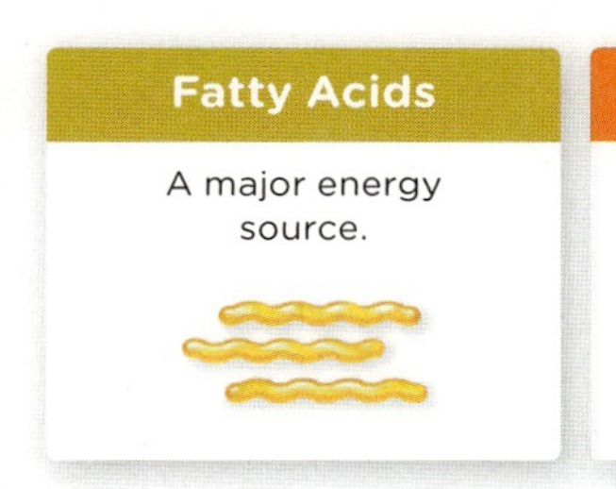

Triglycerides

The most abundant lipid in our diet and storage form of fat in our bodies.

Sterols

Cholesterol is the primary dietary sterol.

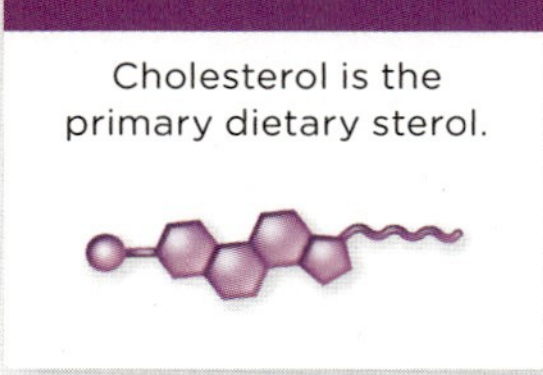

Phospholipids

The primary lipid in cell membranes.

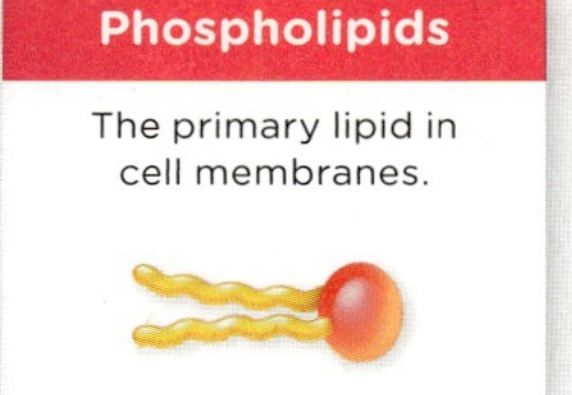

Not only are fatty acids a class of lipid, they are also components of both triglycerides and phospholipids.

Which of the lipid classes has a distinctly different structure from the others?

FATTY ACID
a chain of carbon atoms with hydrogen atoms attached, includes a carboxyl group on one end and a methyl group on the other

CARBOXYL GROUP
the acid group attached to one end of the fatty acid chain

METHYL GROUP
a group of three hydrogen atoms bonded to a carbon atom found at one end (the "omega" end) of the fatty acid chain

SATURATED FATTY ACID
a fatty acid that contains no double bonds between carbons in the carbon chain and carries the maximum number of hydrogen atoms

UNSATURATED FATTY ACIDS
a fatty acid that has at least one double bond between carbons in the carbon chain and has fewer than the maximum possible number of hydrogen atoms

MONOUNSATURATED FATTY ACID
a fatty acid with only one double bond between carbons in the carbon chain

POLYUNSATURATED FATTY ACID
a fatty acid with two or more double bonds between carbons in the carbon chain

TRIGLYCERIDES
storage form of fat, made up of three fatty acid chains attached to the three carbons on a glycerol molecule

GLYCEROL
a three-carbon compound that makes up the backbone of a triglyceride molecule

flavor) of food and contribute to texture and aroma. Eating fat also contributes to the sensation of feeling full (satiety) in part by slowing gastric emptying, keeping food in the stomach a bit longer. Relative to high-carb, low-fat foods, carbohydrate foods with higher amounts of fat are digested and absorbed over longer periods, resulting in a lower glycemic response (see Chapter 5).

Dietary fat confers some health benefits, but because many tasty foods are also fatty foods, it can be easy to eat too much fat. Any excess fat in the diet is efficiently deposited as fat in the body (adipose tissue), making high-fat, energy-dense diets one reason why people tend to gain weight.

When Dyerberg and Bang made their historic trip to Greenland, they were eager to learn about the fish-oriented dietary pattern of the Inuit, and how it affected their health. Chapter 7 discusses what we know today about the effects of dietary lipids on health and disease.

Fatty Acids

Fatty acids are a type of lipid, but they are also the primary components of both triglycerides and phospholipids. Fatty acids consist of a chain of carbon atoms with hydrogen atoms attached to each carbon atom (a hydrocarbon chain): a **carboxyl group** attached to one end of the fatty acid chain, and a **methyl group** at the other end (three hydrogens bonded to a carbon atom—also called the "omega" end of the fatty acid).

Fatty acids differ in length, and so they are categorized by the length of their hydrocarbon chains, as well as in their degree of *saturation* (how many hydrogen atoms fill the available bonds with carbon). Short-chain fatty acids have fewer than 6 carbons; medium-chain fatty acids have 6 to 12 carbons, and long-chain fatty acids have more than 12 carbons. Both properties, degree of saturation, and chain length determine their function in the body and role in health and disease.

Some fatty acids are **saturated**, with hydrogen atoms filling every possible bond with carbon atoms. Saturated fatty acids are relatively solid at room temperature, and most commonly found in animal products (such as meats and dairy), as well as some vegetable products (such as coconut oil, palm kernel oil, palm oil, and cocoa butter). The rest are **unsaturated**, with less hydrogen and one or more double bond (or point of unsaturation) between carbon atoms, and are generally liquid at room temperature. Unsaturated fats are found most abundantly in plant foods, such as seeds, nuts, grains, and most vegetable oils. Fatty acids with one point of unsaturation are called **monounsaturated fatty acids** (abundant in olive and canola oils and nuts); those with more than one point of unsaturation are called **polyunsaturated fatty acids** (abundant in corn, safflower, sunflower, sesame, and soybean oils). The arrangement of the hydrogen atoms on either side of the double bonds can be in either a "cis" or a "trans" orientation, which has important health implications, as will be discussed later in the chapter. **(INFOGRAPHIC 6.2)**

Triglycerides

Triglycerides are lipids made up of three fatty acid chains bound to one **glycerol**, a small three-carbon molecule that makes up the glycerol backbone of each triglyceride. The fatty acid chains form the "tail" of the triglyceride. All triglycerides are composed of a mix of short-chain, medium-chain, and long-chain fatty acids. They seldom contain exclusively one type of fatty acid.

Triglycerides in foods supply energy and may also carry certain fat-soluble

INFOGRAPHIC 6.2 The Structure of Triglycerides and Fatty Acids *Food fats, or triglycerides, are a mixture of many different fatty acids. The balance of the different fatty acids determines the physical property of the food fat at room temperature.*

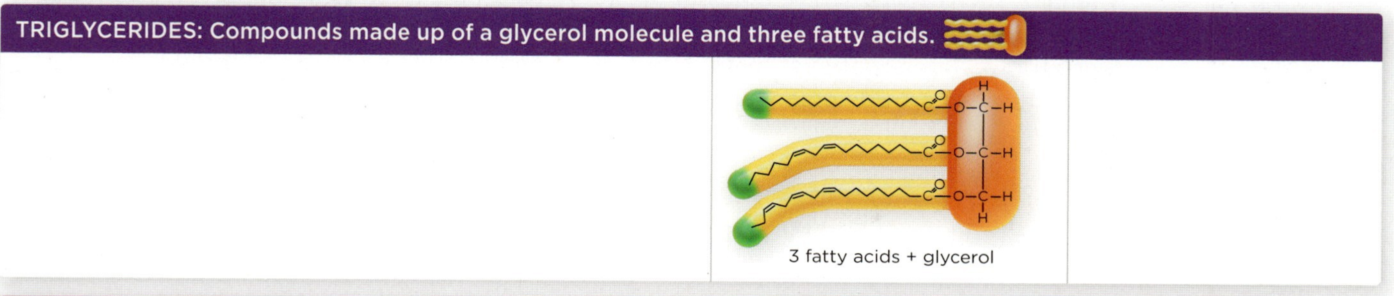

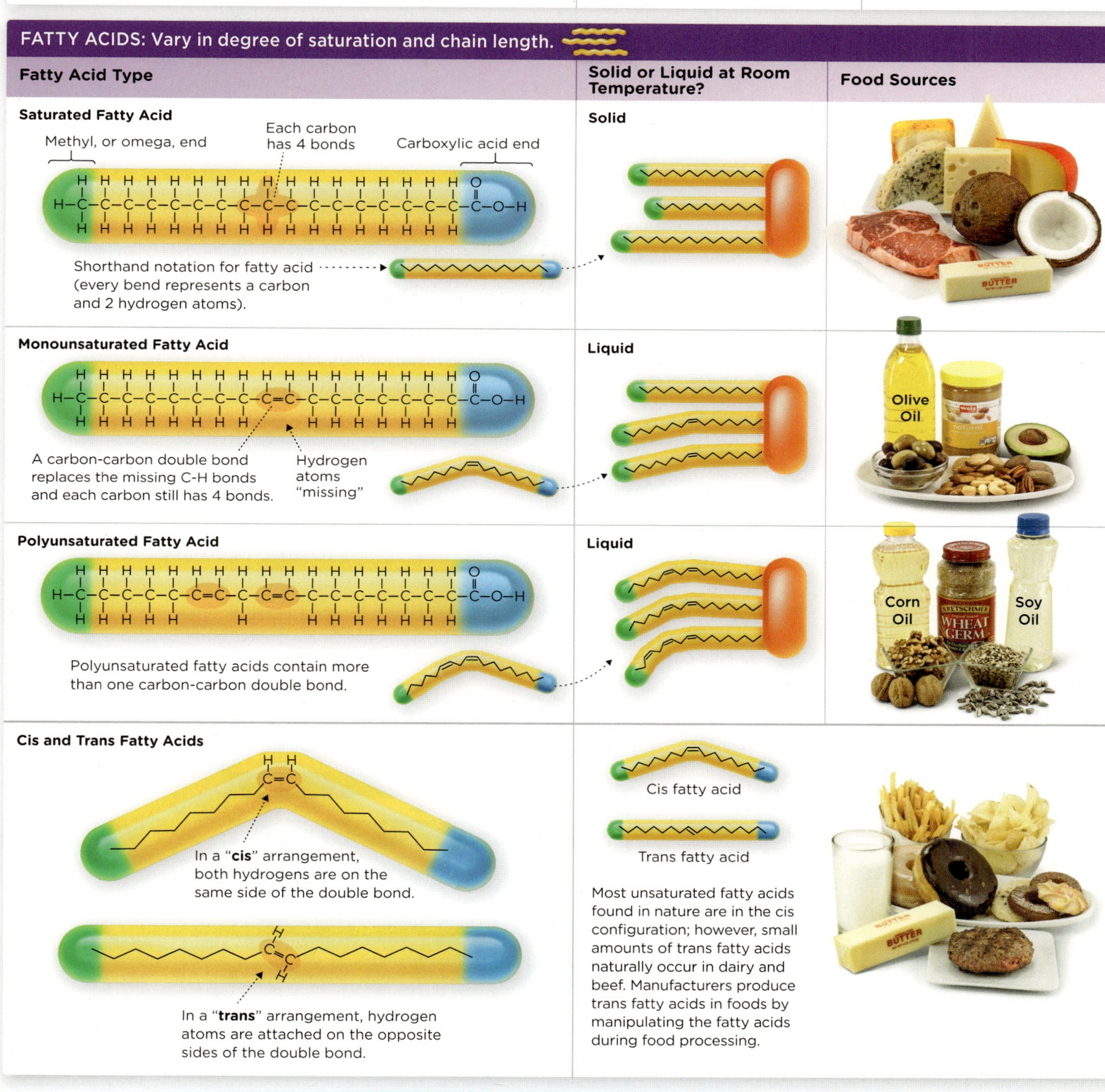

How are trans fatty acids both similar to and different from saturated and unsaturated fatty acids?

Photo credit (all photos): Eli Ensor

vitamins, such as vitamins A, D, E, and K (covered in more detail in Chapter 10). Since triglycerides consist of three fatty acids, they can provide essential fatty acids, the fatty acids our bodies need but cannot synthesize in sufficient amounts, and so must be obtained through our diets. **(INFOGRAPHIC 6.3)**

INFOGRAPHIC 6.3 Fatty Acid Composition of Common Sources of Dietary Fats

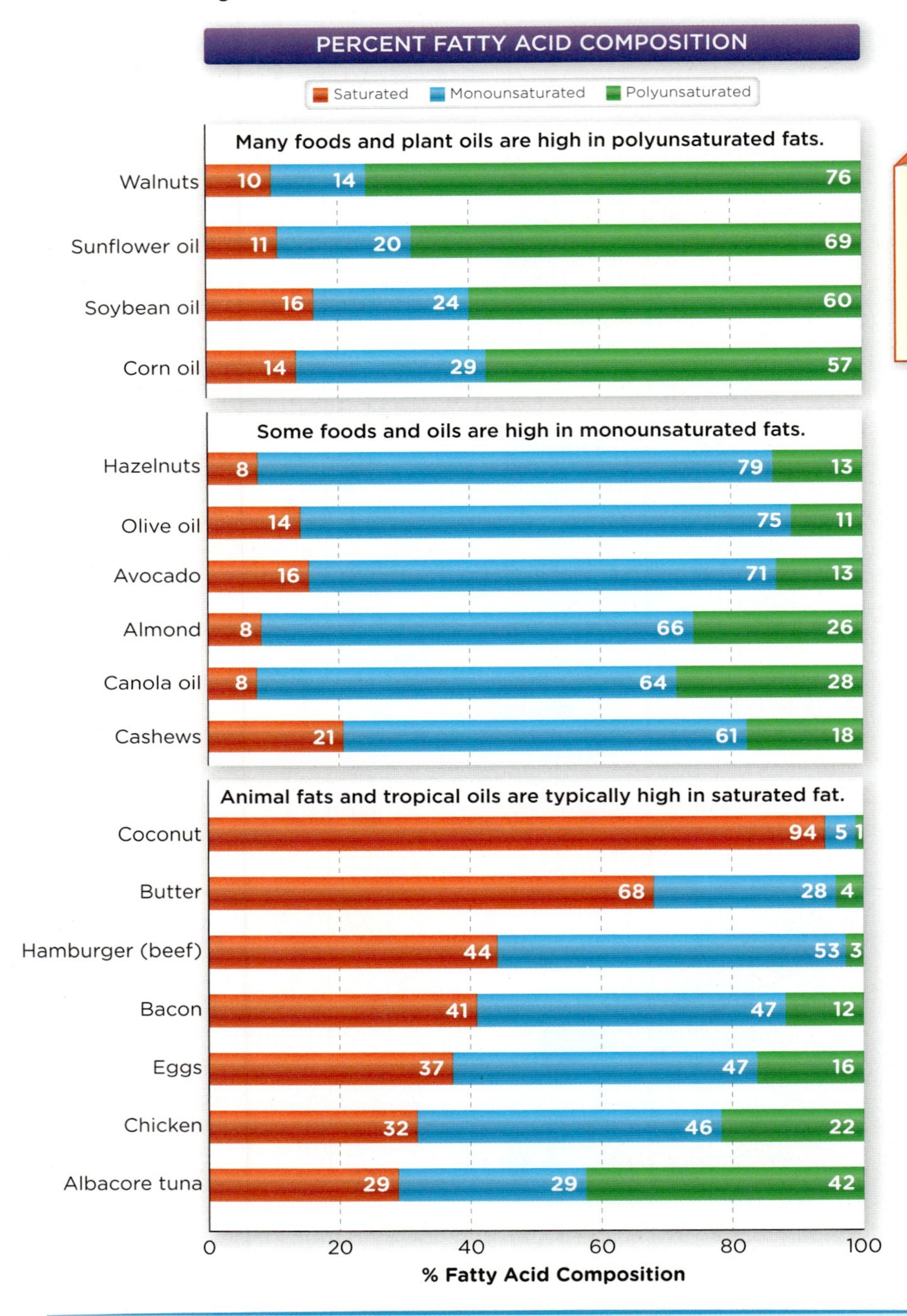

? *Which category of foods tends to be rich in monounsatured fats as well as low in saturated fats?*

Sterols

Chemically, **sterols** are complex lipids with four interconnected carbon rings with a hydrocarbon side chain.

The most discussed sterol is **cholesterol**. A molecule with varied functions, cholesterol is a critical component of our cell membranes and is also needed as a precursor for the synthesis of bile acids, vitamin D, and steroid hormones such as estrogen and testosterone, but it does not provide any energy.

Cholesterol is synthesized in nearly every tissue in the body, but in particularly large quantities by the liver. Approximately 75% of the cholesterol in blood is made in our body, which provides all the cholesterol needed for body functions. Although a dietary source of cholesterol is not required, we consume cholesterol in animal foods, such as meats and dairy products. Indeed, the presence of cholesterol in cell membranes is one distinguishing characteristic between plant and animal cells, thus dietary cholesterol is found only in foods of animal origin. Plants synthesize other types of sterols (and the closely related stanols), but these types are poorly absorbed by the body and can actually interfere with and lower cholesterol absorption. For this reason, some spreads and other food products are fortified with plant sterols or stanols, to help lower cholesterol levels in the body. Chapter 7 further explores the relationship of cholesterol and other dietary fats on blood cholesterol.

joesayhello/Shutterstock

A cholesterol-rich egg bacon cheeseburger. *Dietary sources of cholesterol are not needed, and only come from foods of animal origin.*

Phospholipids

Phospholipids, the fourth most common category of lipids (after fatty acids, triglycerides, and sterols), are also the primary component of cell membranes and the structures that transport lipids in the blood. Because the body can produce phospholipids, they are not considered an essential nutrient. Phospholipids are similar to triglycerides in structure: they have a glycerol backbone, but with two rather than three attached fatty acids. Attached at the third position is a phosphate group and one of several water-soluble "head-groups." Together the glycerol backbone, the phosphate, and the head group create a water-soluble head on the phospholipid, while the fatty acid "tails" at the other end are soluble in lipids but not in water. **(INFOGRAPHIC 6.4)**

This two-sided structural arrangement of phospholipids—one end water-soluble and the other end fat-soluble—allows phospholipids to suspend fat in water. The fat-soluble tails of phospholipids will surround small fat droplets, leaving the water-soluble head of the phospholipid facing outward toward the water. Placing this water-friendly coating around the lipid droplets allows them to remain stably dispersed (mixed) in the watery environment. An example of a phospholipid is **lecithin**. In addition to being the most abundant phospholipid in our body, lecithin is also frequently added to food products

STEROLS
complex lipids with interconnected carbon rings with an oxygen, and a side chain consisting of carbons and hydrogens; a precursor for synthesis of steroid hormones

CHOLESTEROL
a sterol that is produced by the body and required for steroid production and cell membrane function

PHOSPHOLIPID
a molecule that is both hydrophobic (water fearing) and hydrophilic (water loving) and is required to form cell membranes; lecithin, which can be found in egg yolks, liver, and some plant products, is a phospholipid

LECITHIN
the most abundant phospholipid in the body; frequently added to food products like salad dressings as an emulsifier

INFOGRAPHIC 6.4 **Phosphopholipids** *are a critical component of cell membranes. They also play a central role in transporting lipids throughout the body. Despite their importance, they are not essential nutrients, as we are able to synthesize all that we need.*

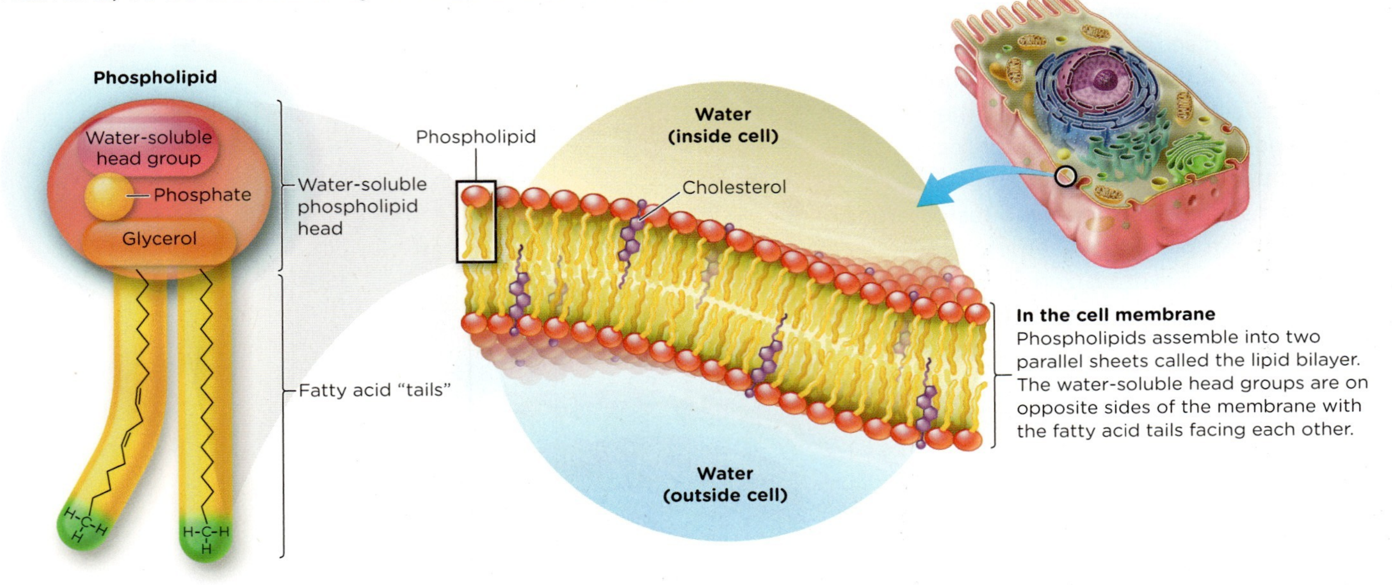

? *Why do the fatty acid "tails" face each other within the cell membrane?*

like salad dressings as an *emulsifier* because of its ability to keep water and lipids from separating.

■ ■ ■

Dyerberg and Bang suspected that there might be something about the mix of fatty acids in the high-fat Inuit diet that protected them from heart disease. Back in their lab in Denmark, the scientists examined blood samples more closely, using an instrument that separated fatty acids by their boiling point. Short, saturated fatty acids appeared through the machine first, followed by longer fatty acids, and, finally, long, unsaturated fatty acids. Immediately, the researchers noticed remarkable differences between the pattern of fats in samples from the Inuit and from people living in Denmark. In particular, the readout from Inuit adults contained a huge spike toward the end, showing their blood contained large amounts of an unknown fatty acid that was relatively long and unsaturated.

The researchers scratched their heads. What was this mysterious fatty acid, and how was it protecting the Inuit from the number one cause of death in other parts of the world?

Even before Dyerberg and Bang began collecting blood samples from native Greenlanders, they suspected the then-popular belief that all fats were equally bad for health could not be entirely true. They were beginning to understand the important roles that different fats play in the body.

LIPID DIGESTION AND ABSORPTION

To carry out their functions in the body, most dietary lipids must first be digested. A small amount of fat digestion takes place

Large oil droplets form in water. *Most lipids are not soluble in water and tend to clump together.*

in the mouth and stomach (via enzymes called lipases), but most occurs in the small intestine. However, a problem must be overcome for digestion to proceed. Fats cannot mix with water because lipids are not water soluble; the fat tends to clump together in the intestine's watery environment. The body's solution to that problem is **emulsification**.

Emulsification aids digestion by breaking up large fat globules into smaller droplets so that fat-digesting enzymes can operate efficiently. Bile acids produced in the liver make emulsification possible. Bile acids have a water-soluble and fat-soluble "face." These two-faced molecules attach to lipids, so that the lipids remain suspended in water instead of clumping together. **Lipases** produced in the pancreas can then digest triglycerides into monoglycerides and free fatty acids. These lipids, along with bile acids, form structures called *micelles* that deliver dietary lipids to the mucosal cells of the small intestine so that they can be absorbed.

Once inside mucosal cells, fatty acids and monoglycerides (which, as their name implies, contain only one fatty acid chain) are reassembled into triglycerides. (INFOGRAPHIC 6.5)

LIPOPROTEIN TRANSPORT

In the mucosal cells, triglycerides and other dietary lipids are incorporated into a lipid transport particle called a **chylomicron**, which is one type of **lipoprotein**. Lipoproteins are protein-containing spherical particles that act as the primary carriers of lipids in blood. Lipids are not typically soluble in water and require carriers for transport, and so cannot travel freely in the bloodstream. As a result, clinical laboratory tests that monitor blood lipid levels typically measure amounts and types of lipids present in lipoproteins in blood. Unlike other lipoproteins, however, chylomicrons are so large that they cannot enter blood immediately after their formation, so they first enter the lymphatic system, which then delivers them into the bloodstream.

Lipoproteins are classified by their density, which is determined by the relative ratios of triglycerides, cholesterol, and proteins present. This classification scheme also separates them by their functions. Chylomicrons (the largest and least-dense particles) and **very low-density lipoproteins (VLDLs)** are similar in that they both primarily transport triglycerides to adipose tissue, and cardiac and skeletal muscles. They differ, however, in their site of origin; VLDLs originate from the liver, while chylomicrons

EMULSIFICATION
a process that allows lipids—fats—to mix with water

LIPASE
enzyme that removes fatty acids from the glycerol backbone of triglycerides

CHYLOMICRON
a very large lipoprotein that transports triglycerides and other dietary lipids away from the small intestine, first in the lymph and then in the blood

LIPOPROTEINS
particles formed by the assembly of proteins and phospholipids that transport lipids in lymph and in blood

VERY LOW-DENSITY LIPOPROTEIN (VLDL)
a lipoprotein responsible for transporting primarily triglycerides to adipose tissue, cardiac muscle, and skeletal muscles

INFOGRAPHIC 6.5 Lipid Digestion and Absorption *The majority of the lipids in our diet are in the form of triglycerides, which must be digested before they can be absorbed. Most triglyceride digestion occurs in the small intestine by pancreatic lipase, with the assistance of bile acids produced by the liver.*

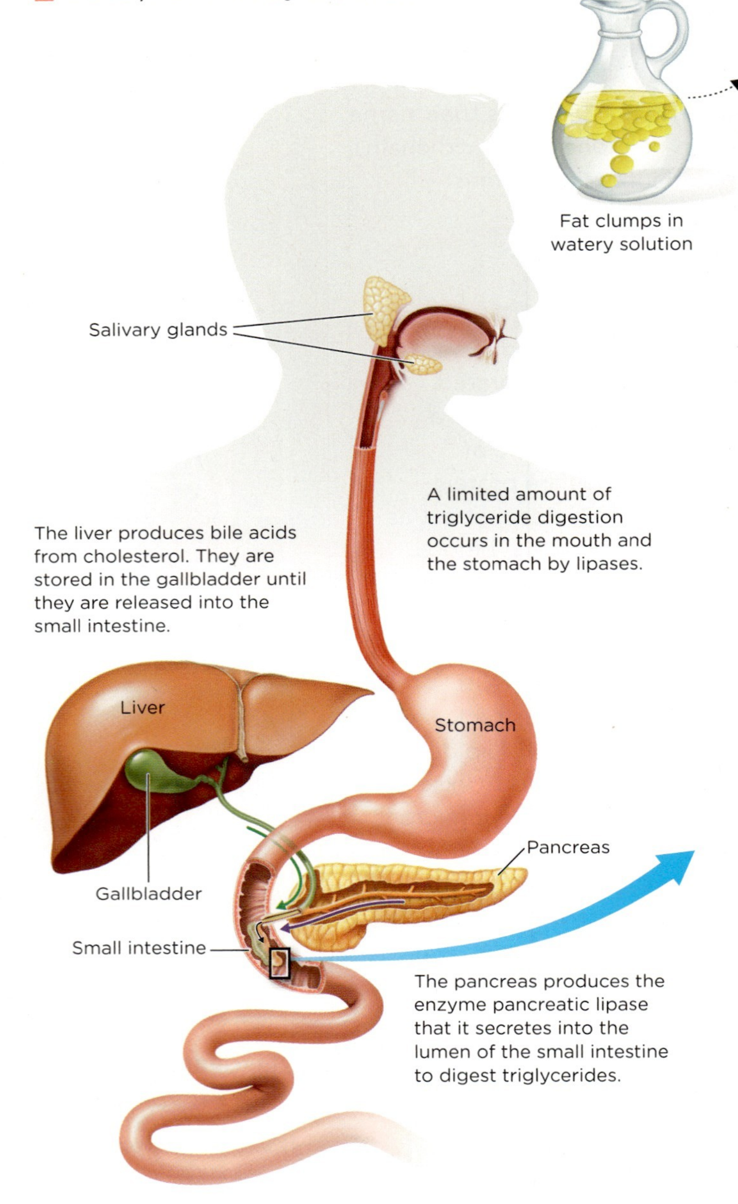

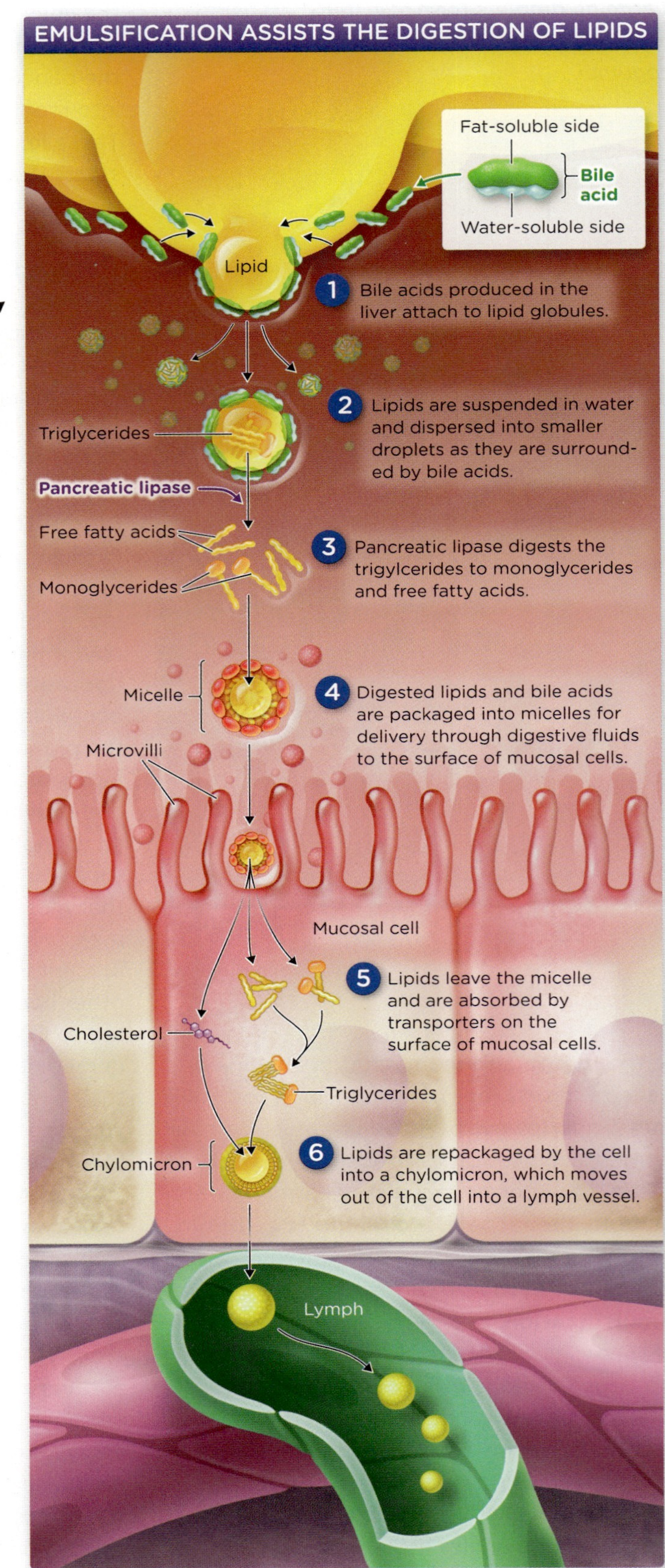

Why might someone who has had their gallbladder surgically removed need to be careful to avoid consuming excess fat?

INFOGRAPHIC **6.6** **Lipoprotein Structure and Function** *Lipoproteins consist of a lipid core of triglycerides and cholesterol surrounded by a single layer of phospholipids containing both cholesterol and proteins.*

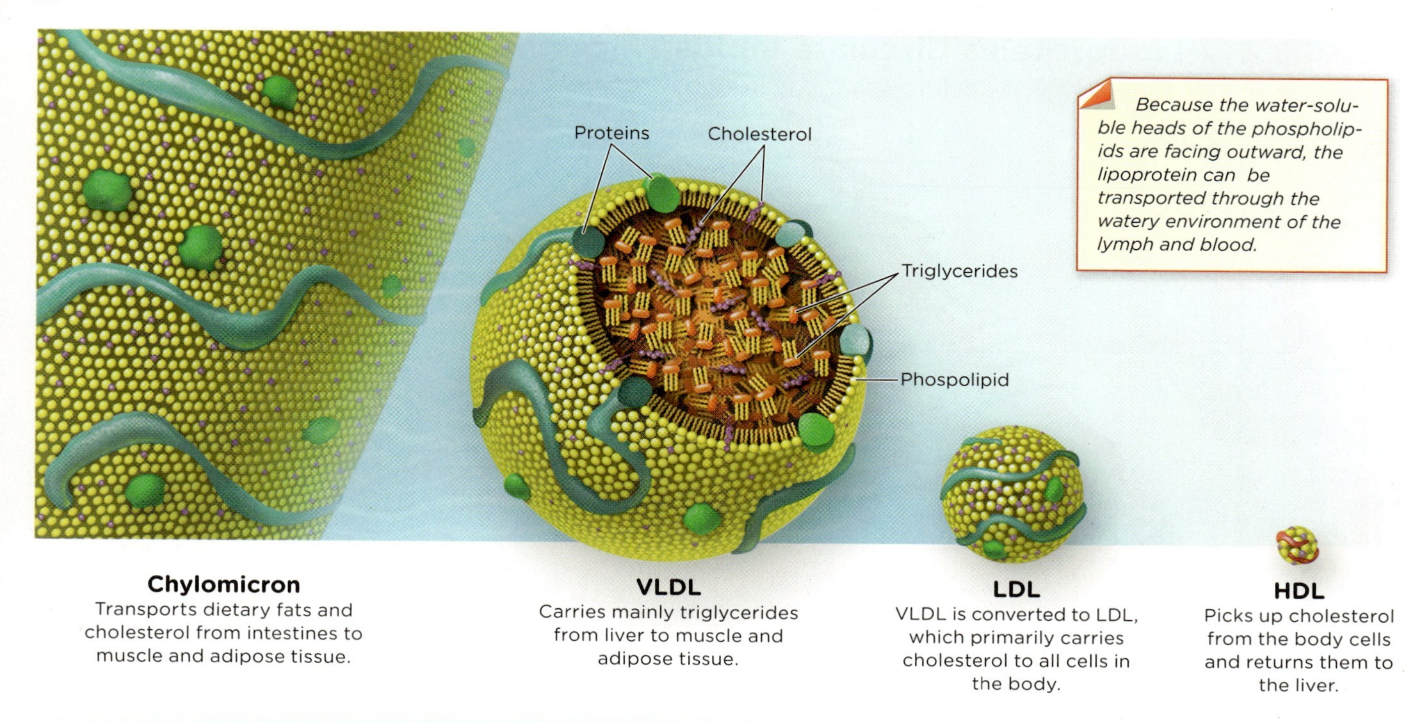

Which lipoprotein transports fat and cholesterol from dietary sources?

originate in the small intestine, and carry essentially all lipids present in our diet. **Low-density lipoproteins (LDLs)** transport cholesterol from the liver to all cells and are often referred to as "bad" cholesterol, as elevated levels are associated with an increase in the risk of heart disease. (INFOGRAPHIC **6.6**)

In contrast, the Inuit had high levels of the so-called "good" cholesterol, or **high-density lipoproteins (HDLs)**. HDL contains a lot of protein, so rather than delivering cholesterol around the body, HDL picks up as much excess cholesterol as it can from the cells and tissues and takes it back to the liver, which then either uses the cholesterol to make bile acids, excretes it directly into bile, or recycles it. This action is thought to explain why high levels of HDL are associated with low risk for heart disease. The commonly held view of cholesterol, as being either "good" or "bad," however, is potentially misleading. Cholesterol is necessary for numerous essential functions throughout the body, and since most cells in the body receive cholesterol transported by LDLs, even LDLs are necessary. However, when the LDL level rises too high and HDL level drops too low, it is this ratio of lipoproteins (LDL to HDL) that is "bad" because high LDL increases the risk of cardiovascular disease. Chapter 7 will further explore lipoprotein levels in the blood and their relationship to the risk of heart disease. (INFOGRAPHIC **6.7**)

■ ■ ■

LOW-DENSITY LIPOPROTEIN (LDL) *a lipoprotein responsible for transporting primarily cholesterol from the liver through the bloodstream to the tissues*

HIGH-DENSITY LIPOPROTEIN (HDL) *a lipoprotein responsible for transporting cholesterol from the bloodstream and tissues back to the liver*

INFOGRAPHIC **6.7** **Lipoproteins Circulate Lipids Throughout the Body** *Each lipoprotein serves as a vehicle that carries lipids through the watery environment of the bloodstream.*

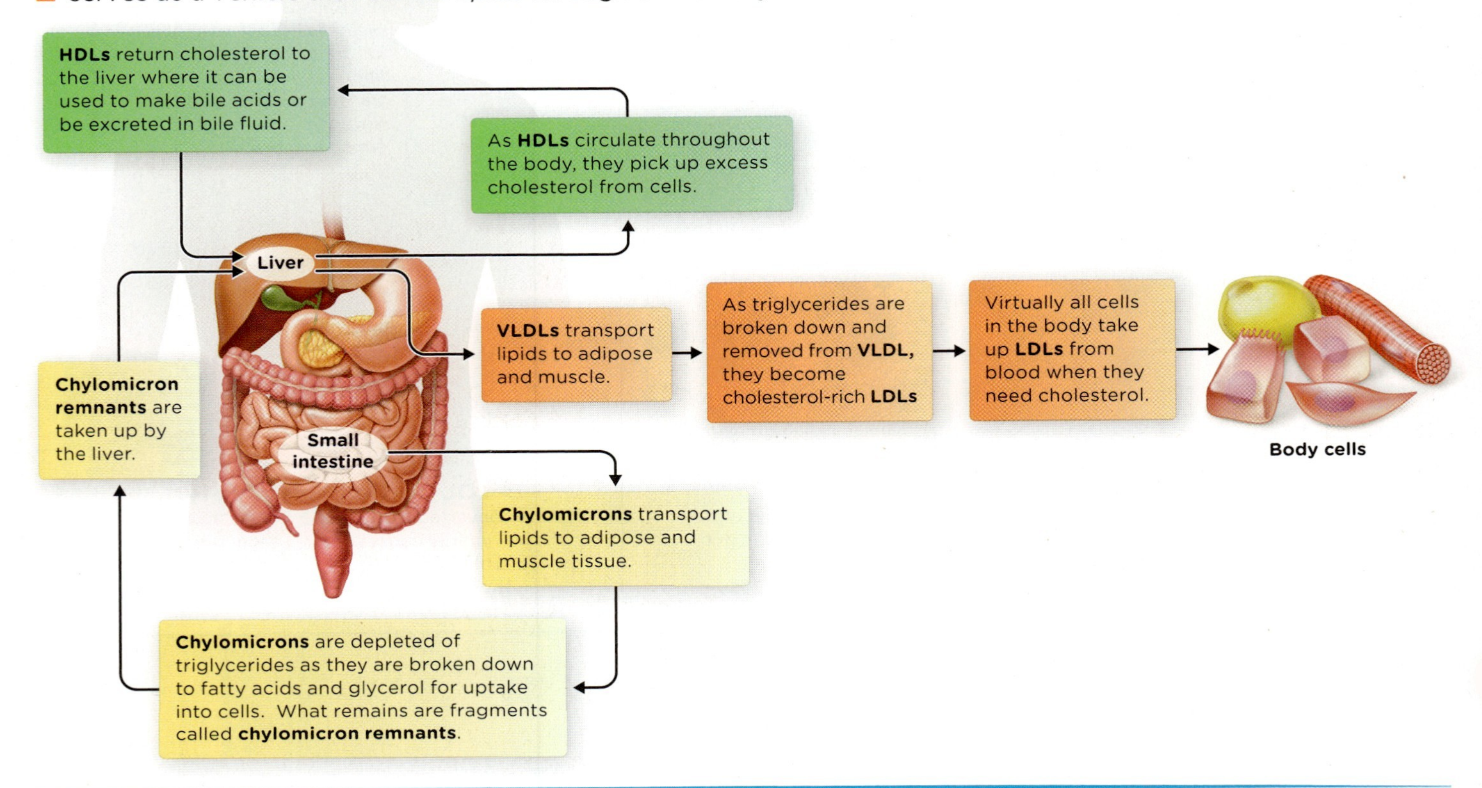

Which lipoprotein picks up excess cholesterol from cells and delivers it to the liver for use or excretion?

EICOSAPENTAENOIC ACID (EPA) *a 20-carbon omega-3 fatty acid that can be produced in the body by the metabolism of the essential fatty acid alpha-linolenic acid, or provided in the diet by oily fish*

DOCOSAHEXAENOIC ACID (DHA) *a 22-carbon omega-3 fatty acid that can be produced in the body from the essential fatty acid alpha-linopenic acid, also found in oily fish*

Dyerberg and Bang knew they needed extra help to identify the mysterious, long-chain fatty acid in the Greenlanders' blood samples. So Dyerberg bought a plane ticket for Minnesota to visit the lab of Ralph Holman, PhD, an international expert on lipids.

As soon as he arrived, Dyerberg held the results up to the chemist. Immediately, Holman identified the mystery fatty acid: That's **eicosapentaenoic acid**, or **EPA**, he told the young scientist. And the other spike at the end of the graph, representing another long-chain fatty acid present in higher levels in the Inuit than in Danish adults is **docosahexaenoic acid**, or **DHA**. You get both from eating fatty fish.

As Dyerberg left the lab, his head was swimming. He kept repeating the words over and over: "Ei-co. . . ei-co-sa. . . eicosapentaenoic acid. Docosahexaenoic acid." He'd never heard them before.

ESSENTIAL FATTY ACIDS

The human body needs fatty acids; most are easily supplied in sufficient amounts through our diet, and if not we can make them from excess carbohydrates and proteins. There are two fatty acids that are

Seek these types of fats. *Omega-3 fatty acids are found in fatty fish, such as salmon, and in some oils, such as canola oil.*

considered **essential fatty acids**, because they cannot be synthesized by humans, and must be supplied by diet. These are **linoleic acid** and **alpha-linolenic acid** (commonly called linolenic acid, for short). Both linoleic and linolenic acid are long-chain polyunsaturated fatty acids with 18 carbon molecules. Linoleic acid is an example of an **omega-6 fatty acid**, so called because the first double bond of the carbon chain is placed in the sixth position, counting from the omega end of the fatty acid. It's needed for normal growth and for synthesis of important hormonelike compounds called *eicosanoids.*

By far, linoleic acid is the most abundant polyunsaturated fatty acid in our diet. The primary sources are cooking oils, salad dressings, nuts, and seeds.

Linolenic acid is an **omega-3 fatty acid**—with the first double bond of the carbon chain located in the third position. Like phospholipids and sterols, it's a structural component of cell membranes. Linolenic acid is found in walnuts, flax seeds, soy, canola oil, and chia seeds. In relation to linoleic acid, intake of linolenic is low in the diets of most Americans, who consume only about one-tenth of our essential fatty acids in this omega-3 form.

One of the most important functions of essential fatty acids is to provide parent compounds to produce the hormonelike eicosanoids, compounds that regulate blood pressure, inflammation, and even pain. However, the effect of an eicosanoid on body functions depends on the fatty acid from which it is made. Two polyunsaturated fatty acids that are directly used for eicosanoid synthesis are arachidonic acid and EPA, which are made from linoleic and linolenic acids, respectively.

Blood clotting and inflammation is promoted when there is excess production of eicosanoids from arachidonic acid (an omega-6 fatty acid), compared with those produced from EPA (an omega-3 fatty acid). Blood clotting and inflammation are decreased when the production of eicosanoids from EPA increases. Because of these opposing effects, it is desirable to have a proper balance of omega-6 and omega-3 fatty acids in the diet. Although neither EPA nor DHA are essential fatty acids—in fact DHA can be made from EPA—we do not effectively convert linolenic acid to EPA or DHA. We can more efficiently boost EPA and DHA concentrations in our body by regularly eating fish, particularly fatty fish like salmon, albacore tuna, trout, and sardines, which are excellent sources of these fatty acids. **(INFOGRAPHIC 6.8)**

■ ■ ■

ESSENTIAL FATTY ACIDS
linoleic acid (omega-6, for example) and linolenic acid (omega-3); required in the diet because they cannot be synthesized by the human body

LINOLEIC ACID
an 18-carbon omega-6 polyunsaturated essential fatty acid

LINOLENIC ACID (ALPHA-LINOLENIC ACID)
an 18-carbon omega-3 polyunsaturated essential fatty acid; modified in the body to produce EPA and DHA

OMEGA-6 FATTY ACID
a polyunsaturated fatty acid that has the first double bond at the sixth carbon molecule from the methyl end of the carbon chain; needed for normal growth

OMEGA-3 FATTY ACID
a polyunsaturated fatty acid that has the first double bond at the third carbon molecule from the methyl end of the chain; associated with a decreased risk of cardiovascular disease and improved brain function

INFOGRAPHIC 6.8 Essential Fatty Acids and Their Food Sources

ESSENTIAL FATTY ACIDS

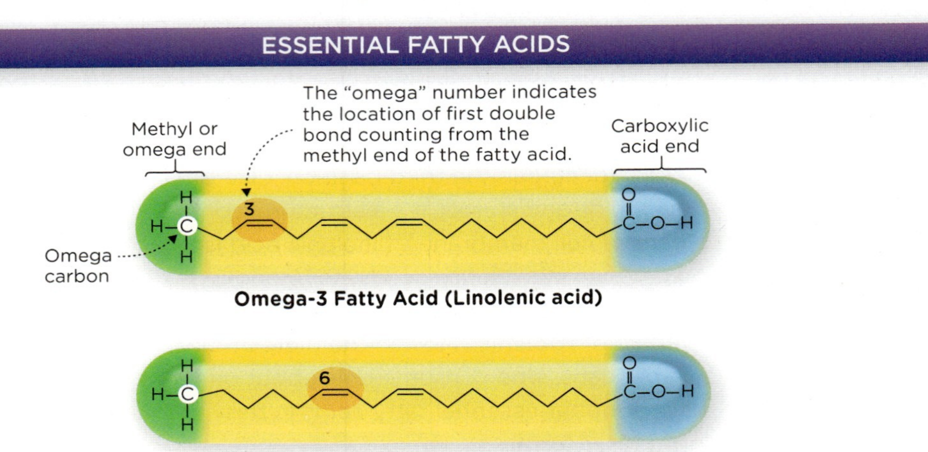

FOOD SOURCES OF ESSENTIAL FATTY ACIDS

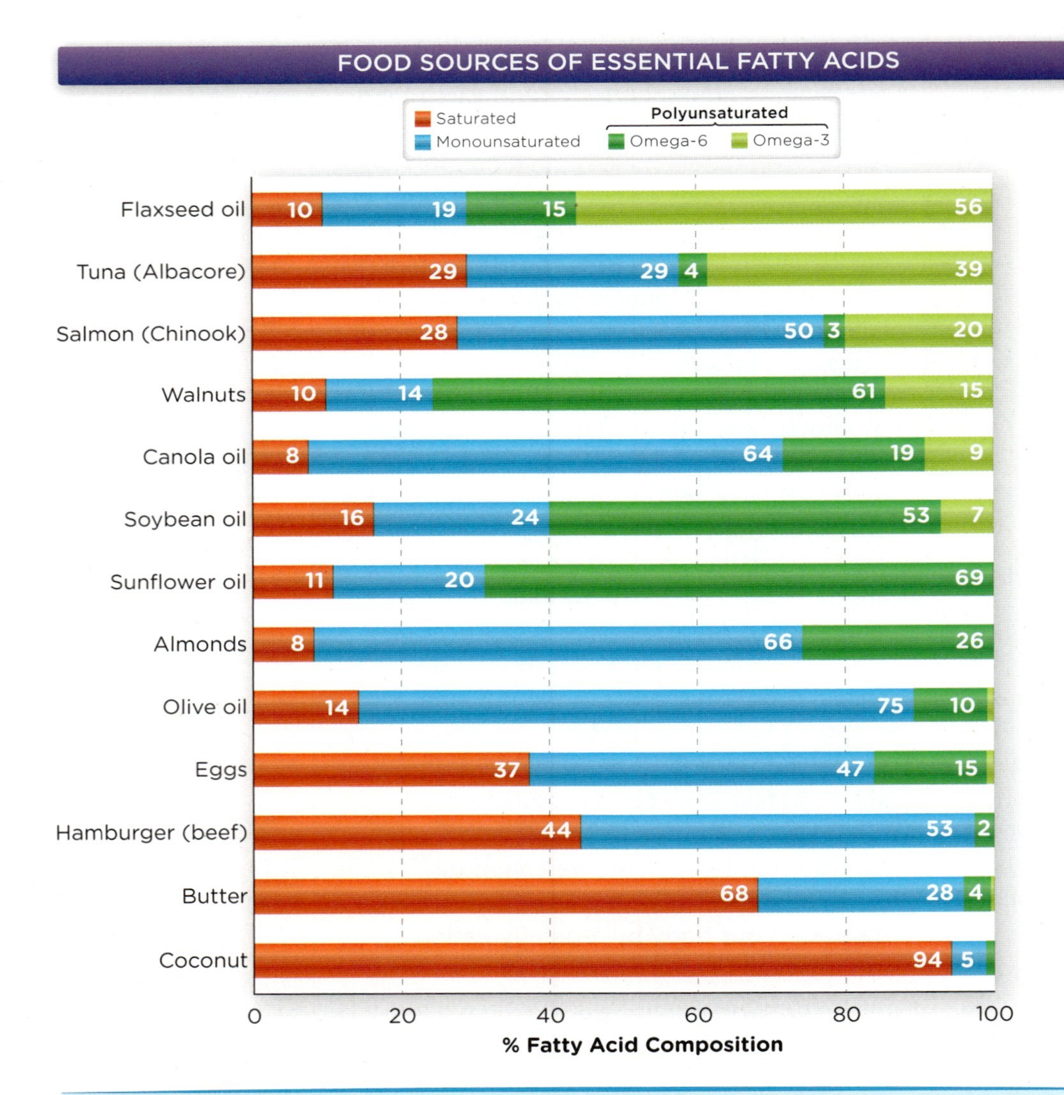

? *What is a good source of omega-3 fatty acids other than fish?*

EPA and DHA in the bodies of the Inuit stemmed from their high intakes of fish, and marine mammals that eat lots of fish (such as seals). To better understand this diet, the scientists asked the Inuit to surrender their lunches so they could analyze the food for fatty acid content. Some would hand over plastic bags dripping with seal intestines and kidneys; one fisherman submitted raw liver and an eye. "The eye was a delicacy," says Dyerberg. "We got one, and he ate the other one." Even raw liver and intestine—if turned inside out and cleaned—is tasty, he recalls. "I've done that with the seal intestine many times. I can tell you it's good."

Sure enough, these lunch samples were packed full of EPA and DHA. But what about them was protecting the Inuit's heart and blood vessels from cardiovascular disease? Were EPA and DHA more protective than other polyunsaturated fatty acids?

TRANS FATS AND FAT SUBSTITUTES

Most people who eat a Western diet—such as Americans and many Europeans—eat plenty of fat, just not necessarily the right kinds. Much of the fat we consume has undergone a process of **hydrogenation**, which makes unsaturated fats more solid and stable by chemically adding hydrogen to double bonds between carbon atoms on the fatty acid chain and thus makes them more saturated. This renders them harder at room temperature and more resistant to spoiling. This process can also improve taste, texture, and increase their culinary applications.

One outcome of *partial* hydrogenation, however, is the structure of some of the unsaturated fatty acids can be converted from its natural "cis" configuration to another known as "trans." These so-called **trans fatty acids** raise levels of "bad" cholesterol, and lower levels of "good" cholesterol in blood, and increase the risk of heart disease more than any other type of fat. As little as 1% of total calories from trans fats can increase the risk of heart disease, stroke, or sudden death from these and other causes. Although a small amount of trans fat is naturally present in full-fat dairy products and beef, most of the trans fat we consume comes from the partially hydrogenated oils used to make processed foods that need a longer shelf life. The majority of trans fats in the U.S. diet are found in commercial baked goods (cakes, cookies, pies, biscuits, and doughnuts), packaged snacks such as crackers and popcorn, margarine, frozen pizzas, and fried potatoes. Given the risks of trans fats, as of 2006, all Nutrition Facts Panels must include the trans fat content of food products and more recently, the U.S. Food and Drug Administration has proposed a ban on their addition to any food products.

When purchasing processed foods, it is important to look beyond the grams of trans fat listed on the nutrition facts panel because foods with less than half a gram per serving can claim 0 grams trans fat. To minimize your intake of trans fats, avoid purchasing foods that list partially hydrogenated oils in the list of ingredients. Increasing awareness of the dangers of trans fats has resulted in more frequent use of fully hydrogenated oils. When oils are fully hydrogenated all double bonds are eliminated and no trans fats are produced. **(INFOGRAPHIC 6.9)**

CURRENT FAT INTAKE AND RECOMMENDATIONS

Our understanding of the role of fat in the diet is constantly changing. The dietary recommendations for fat intake have changed over the years as new scientific evidence and population studies are released and research continues to evolve.

This makes it difficult for the average consumer to know what to think about dietary fat, says Susan Allport, author of *The Queen*

HYDROGENATION
chemical process by which hydrogen molecules are added to unsaturated fatty acids

TRANS FATTY ACIDS
fatty acids created by adding hydrogen to liquid vegetable oils (partial hydrogenation) to make them more solid; associated with an increased risk of chronic diseases, such as heart disease and cancer

INFOGRAPHIC 6.9 The AMDR for total fat is 20% to 35% of our total calories and most of those calories should come from unsaturated fats.

What's Your Strategy for Choosing Healthier Fats?

In an attempt to eat healthier, people too often purchase low- or reduced-fat foods that are often high in sugar or unhealthy fats. Try a different strategy—replace harmful fats with those that are beneficial. As you work to reduce your consumption of harmful fats, ***food labels can help guide your choices.***

Baked Goods, Savory Snacks, Margarine, Fried Foods

TRANS FATS, HYDROGENATED OR PARTIALLY HYDROGENATED OILS

HARMFUL

Red Meat, Butter, Cheese, Ice Cream

SATURATED FATS

Photo credit (all photos): Eli Ensor

Golden Flaky Biscuits

Nutrition Facts

Serving Size 59g
Servings Per Container 20

Amount Per Serving	
Calories 170	Calories From Fat 70
	% Daily Value*
Total Fat 8g	**12%**
Saturated Fat 4g	**20%**
Trans Fat 0g	
Cholesterol 20mg	**0%**
Sodium 550mg	**23%**
Total Carbohydrate 22mg	**7%**
Dietary Fiber <1g	**2%**
Sugars 2g	
Protein 4g	

Ingredients: Enriched Flour Bleached, Water, Soybean and Palm Oil, Sugar, Hydrogenated Palm Oil, Baking Powder (sodium acid pyrophosphate, baking soda, sodium aluminum phosphate). Contains 2% or less of: Partially Hydrogenated Soybean Oil, Salt, Dextrose.

% OF CALORIES FROM FAT: Look beyond the grams of total fat, determine the percentage of total calories that are provided by fat. To do this divide fat calories by total calories and multiply by 100. **(70 ÷ 170) × 100 = 41% of calories from fat.** That's above the upper end of the AMDR range of 35% for fat.

UNHEALTHY SATURATED AND TRANS FATS: Even when total fat is low, unhealthy fats may be high. Half of the fat in this product is from saturated fats. Note that the %DV for saturated fats is much higher than for total fat.

CHECK FOR SOURCES OF TRANS FAT: Labels can claim "0 g" of trans fat as long as the food contains <0.5 g per serving. Check the list of ingredients for partially hydrogenated oils that can add up when one consumes several servings of processed foods.

CHECK FOR SOURCES OF SATURATED FAT: Hydrogenated and tropical oils (coconut, palm, and palm kernel oils) are sources of saturated fat.

? *What parts of Nutrition Facts panel can help you to reduce your intake of saturated fat and trans fat? How can it help you identify unsaturated fat content? How can a food label help you understand whether a food product is within the AMDR range for fat?*

of Fats, about the discovery of the benefits of omega-3s. Over the last few decades, people have heard nothing but conflicting messages about fat, she says—get rid of it entirely, eat a little bit, or eat only some of certain types. "Which story are they supposed to believe? Is it the trans fat, is it the animal fat [that's harmful]? I think people have gotten a bit fed up with the changes, and so they're throwing them all out."

Since the early 1970s, Americans have eaten more low-fat foods, which have a smaller percentage of calories from fat (and saturated fat). But the total amount of fat consumed has not decreased, because we simply eat more food (and calories) than we did before. Today, approximately one-third of the calories American men and women consume come from fat. Surveys suggest that we are eating less fat from some sources, such as meat and dairy (opting for low-fat milk, for instance), but getting more fat in our diets from other sources, such as high-fat snack foods and baked goods.

To promote health and reduce the risk of chronic disease, the Acceptable Macronutrient Distribution Range (AMDR) for total fat has been set at 20% to 35% of our total calories. Most of that should come from unsaturated fats; we should eat as little saturated fat and trans fat as possible. **(INFOGRAPHIC 6.10)** Chapter 7 further describes current dietary recommendations and strategies to promote health and reduce the risk of heart disease.

"We can overconsume fat, but it's not the evil thing we once thought it was, as long as we make sure to emphasize essential fatty acids, particularly omega-3s," says Allport. Americans generally get enough omega-6s in their diets, which are found in many common foods, such as vegetable oils, cereals, eggs, and poultry.

Some people who want to reduce their fat intake turn to **fat substitutes**, which imitate the taste, texture, and cooking properties of fats but provide fewer calories. Use of these products in place of their higher-fat counterparts, may help reduce dietary fat intake, and lower blood cholesterol, but not without potential side effects.

Fat substitutes from carbohydrates, protein, and vegetable oils are considered safe; those from other substances must be tested for safety. One common substitute is *Olestra*, which has chemical components similar to those of triglycerides in a different configuration, causing them to not be digested or absorbed by the body. But there are consequences: The U.S. Food and Drug Administration requires any products that contain Olestra to read: "Olestra may cause abdominal cramping and loose stools. Olestra inhibits the

FAT SUBSTITUTE
additive that replaces fat in foods; contains fewer or no calories, but has a similar texture and produces similar sensation in mouth

INFOGRAPHIC 6.10 Sources of Saturated Fat in the U.S. Population *Burgers and sandwiches account for 19% of the saturated fats in the diets of people 2 and older.*

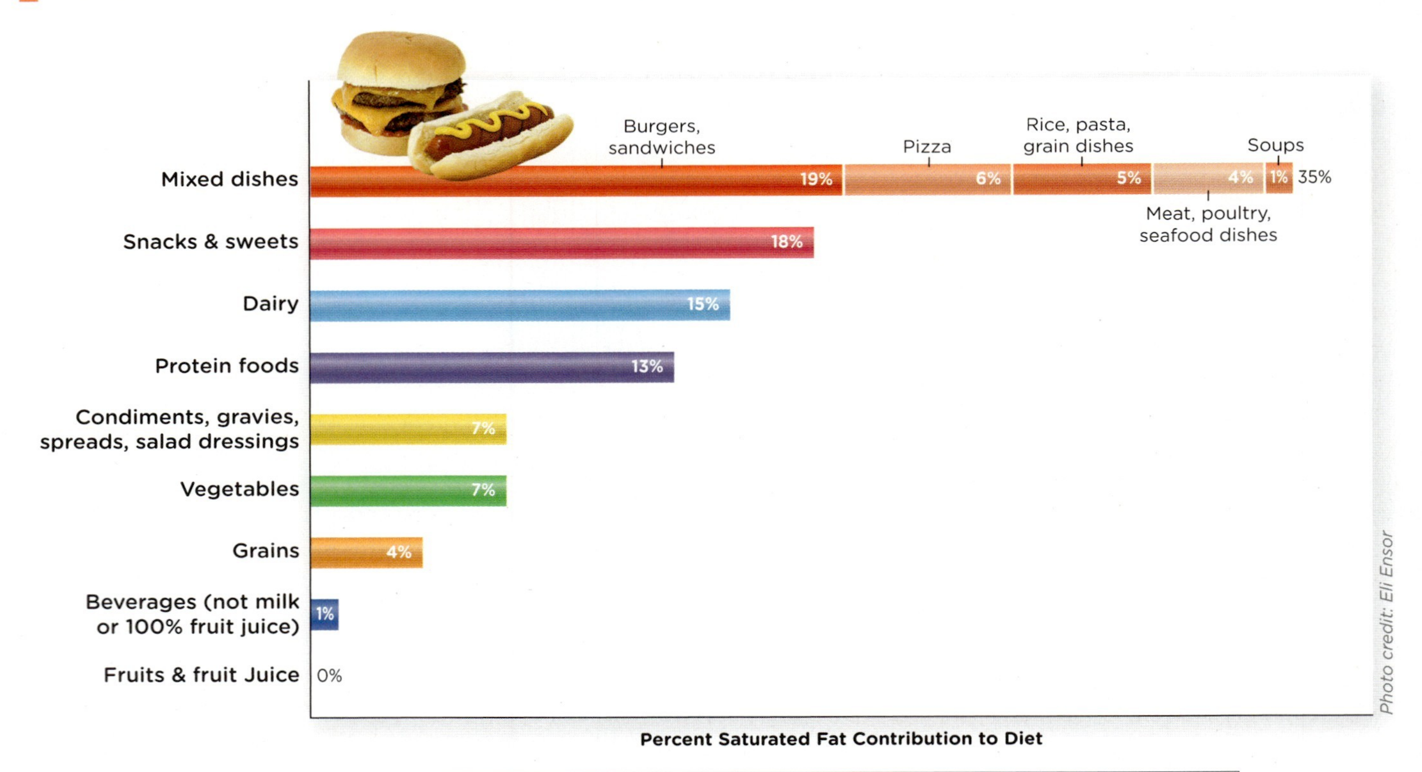

Source: What We Eat in America (WWEIA) Food Category analyses for the 2015 Dietary Guidelines Advisory Committee. Estimates based on day 1 dietary recalls from WWEIA, NHANES 2009-2010.

absorption of some vitamins and other nutrients."

■ ■ ■

In general, most people are aware that there is a link between dietary fat and disease, primarily obesity and heart disease. However, the majority are confused about, or lack an awareness of, the types of fat—their differences, their sources, and their impact on health.

This was even truer in the 1970s, when Dyerberg and Bang were struggling to find out whether DHA and EPA offered specific protections against heart disease and other conditions that aren't offered by other fats, even other polyunsaturated fats. When their technician, Aase Brøndum Nielsen, collected blood samples from the Inuit, she found it took the Inuit twice as long as the Danish adults to stop bleeding after the needle was removed. The Inuit also had a reputation for frequent nosebleeds. Something about their blood was different.

Around that time, a group of scientists discovered compounds derived from long-chain fatty acids called *prostaglandins*, which are released during injury or stress. (This discovery eventually won them the Nobel Prize.) Aspirin and similar pain medicines work by blocking the synthesis of prostaglandins, which are one class of eicosanoids. This reduction in prostaglandin synthesis has the effect of not only reducing pain, but also reducing blood clotting; a build-up of blood clots can cause cardiovascular problems. In the lab one day, Bang and Dyerberg came up with an idea: What if EPA and DHA somehow interfere with these potentially deadly blood clots? Sure enough, when they added EPA to blood samples, they were less likely to form clumps.

KNOWLEDGE OF OMEGA-3 FATTY ACIDS IS GROWING

Since then, research on the benefits of omega-3 fatty acids has exploded. According to Dyerberg, in 1971 he and Bang were the first to publish any research on the topic; by 2012, there were nearly 25,000 research papers about omega-3 fatty acids. He says, "I've had the privilege of opening up a new avenue of human nutrition research." Research has identified other ways in which EPA and DHA exert health effects: by reducing blood pressure, by lowering the levels of other blood fats, and by reducing inflammation.

Initially, scientists were primarily concerned with cardiovascular disease—but Dyerberg says that these omega-3 fatty acids may also have beneficial effects on our brains, as well, by improving eyesight, cognitive function, and even possibly IQ.

In general, most people living in America, Europe, and other Western countries do not get enough omega-3 fatty acids in their diets from fatty fish and plant sources of linolenic acid. This isn't just happening in America. Only a few years after Dyerberg and Bang's groundbreaking work in Greenland, everything began changing for the Inuit. When the scientists returned in 1982, some Inuit had obtained phones and generators that supplied electricity for cooking. Every time he went back, he saw more snowmobiles and fewer dog sleds; fishing poles began to be replaced by large shops that stocked typical American or Danish food. Rates of smoking and alcohol use began to skyrocket. "Today, only really remote areas have the Inuit eating in a traditional way," he says. "Most Inuit try to eat much as we do." Not surprisingly, their rates of heart disease have risen accordingly, and now match those seen in other Western countries.

As is often true for information regarding the impact of diet on health, this story continues to evolve. A recent analysis indicates that the number of deaths attributed to coronary artery disease (CAD) in the Inuit was not likely very accurate. There is also a general lack of good information examining the affect of fish oils on CAD risk in Caucasians compared to native peoples inhabiting regions of the Arctic. While the result of many recent studies disputes the ability of fish oils to reduce the risk of heart disease, there does appear to be enough evidence to support the idea that an adequate intake of fish oils may indeed reduce the risk of some forms of heart disease in some specific subgroups within our population. Despite these recent questions, the work of Dyerberg and Bang had a critically important role in developing our current understanding of the significance that EPA and DHA play in our body.

Their story does tell us that, although eating too much fat is never a good thing, ingesting too little of specific types of fats can also be detrimental, says Allport. "Having a sufficient amount of fat on your body protects you," she says. "Those are our reserves—we absolutely need them. We don't need as much as many people have, but it's not a good guy–bad guy kind of thing. We have to have a healthy appreciation for a healthy amount and type of fat."

CHAPTER 6 **BRING IT HOME**

Choosing added or cooking fats

Do you spread butter or margarine on your toast in the morning? At a restaurant, do you dip your bread in olive oil or spread butter on it? Do you add butter or margarine to your baked potato? When you prepare a meal, what type of fat do you use to sauté vegetables? Using the food labels provided here, answer the questions about these commonly used fats.

Consider

1. To moderate saturated fat intake, which of the fats would you use sparingly?
2. Which of these fats contain trans fatty acids? What word(s) might you find in the ingredient listing that would indicate the presence of trans fat?
3. Of these fats, why is butter the only source of dietary cholesterol?
4. At a restaurant, you have the option of spreading butter on your bread or dipping your bread in a seasoned olive oil. Which would you choose *and why*?
5. You observe a friend prepare a pasta dish using a large amount of olive oil. Your friend says, "This stuff is great for you!" How might you respond? (Hint: Note the percent Daily Value for total fat on the label for olive oil.)
6. Despite its high concentration of saturated fat, coconut oil has been advertised as a healthy fat and promoted on television for multiple benefits. Explore some of the science behind the claims for coconut oil by searching "coconut oil" in www.pubmed.gov. After reading 2 or 3 abstracts about the role of coconut oil in health, do you feel the evidence is sufficient to recommend the use of coconut oil as a cooking fat on a regular basis? What might you tell a friend who asks, "Is coconut oil healthy?"

BUTTER

Nutrition Facts	
Serving Size 1 Tbsp	
Amount Per Serving	
Calories 100	Calories From Fat 100
	% Daily Value*
Total Fat 11g	**17%**
Saturated Fat 7g	**35%**
Monounsaturated Fat 3g	
Polyunsaturated Fat 0g	
Trans Fat 0g	
Cholesterol 30mg	**10%**
Sodium 100mg	**5%**
Total Carbohydrate 0g	**0%**
Dietary Fiber 0g	**0%**
Sugars 0g	
Protein 0g	

CANOLA OIL

Nutrition Facts	
Serving Size 1 Tbsp	
Amount Per Serving	
Calories 125	Calories From Fat 125
	% Daily Value*
Total Fat 14g	**22%**
Saturated Fat 1g	**5%**
Monounsaturated Fat 9g	
Polyunsaturated Fat 4g	
Trans Fat 0g	
Cholesterol 0mg	**0%**
Sodium 0mg	**0%**
Total Carbohydrate 0g	**0%**
Dietary Fiber 0g	**0%**
Sugars 0g	
Protein 0g	

COCONUT OIL

Nutrition Facts	
Serving Size 1 Tbsp	
Amount Per Serving	
Calories 117	Calories From Fat 117
	% Daily Value*
Total Fat 13g	**20%**
Saturated Fat 12g	**60%**
Monounsaturated Fat 1g	
Polyunsaturated Fat 0g	
Trans Fat 0g	
Cholesterol 0mg	**0%**
Sodium 0mg	**0%**
Total Carbohydrate 0g	**0%**
Dietary Fiber 0g	**0%**
Sugars 0g	
Protein 0g	

MARGARINE (SPREAD)

Nutrition Facts	
Serving Size 1 Tbsp	
Amount Per Serving	
Calories 75	Calories From Fat 75
	% Daily Value*
Total Fat 8g	**14%**
Saturated Fat 2g	**10%**
Monounsaturated Fat 2g	
Polyunsaturated Fat 4g	
Trans fats 1g	
Cholesterol 0mg	**0%**
Sodium 90mg	**4%**
Total Carbohydrate 0g	**0%**
Dietary Fiber 0g	**0%**
Sugars 0g	
Protein 0g	

MARGARINE (STICK)

Nutrition Facts	
Serving Size 1 Tbsp	
Amount Per Serving	
Calories 100	Calories From Fat 100
	% Daily Value*
Total Fat 11g	**17%**
Saturated Fat 2g	**5%**
Monounsaturated Fat 5g	
Polyunsaturated Fat 3g	
Trans fats 2g	
Cholesterol 0mg	**0%**
Sodium 90mg	**4%**
Total Carbohydrate 0g	**0%**
Dietary Fiber 0g	**0%**
Sugars 0g	
Protein 0g	

OLIVE OIL

Nutrition Facts	
Serving Size 1 Tbsp	
Amount Per Serving	
Calories 120	Calories From Fat 120
	% Daily Value*
Total Fat 14g	**22%**
Saturated Fat 2g	**10%**
Monounsaturated Fat 10g	
Polyunsaturated Fat 1g	
Trans Fat 0g	
Cholesterol 0mg	**0%**
Sodium 0mg	**0%**
Total Carbohydrate 0g	**0%**
Dietary Fiber 0g	**0%**
Sugars 0g	
Protein 0g	

KEY IDEAS

Lipids comprise a group of structurally diverse compounds composed of carbon, hydrogen, and a small amount of oxygen that are insoluble in water.

Among their varied functions in the body, lipids serve as a component of cell membranes and in nutrient transport, as protection and insulation of internal organs, and as precursors for hormones.

Four classes of lipids include fatty acids, triglycerides, sterols, and phospholipids.

Fatty acids are the primary components of triglycerides and phospholipids. Dependent upon the number of hydrogen atoms filling their carbon bonds, fatty acids are either saturated or unsaturated with one or more double bonds determining their function in the body, as well as their role in health and disease.

Triglycerides, commonly referred to as "fat," are the primary storage form of lipid in our bodies and the primary source of lipid in our diet, providing a concentrated energy source, essential fatty acids, and carriers of fat-soluble vitamins.

The essential fatty acids include linoleic acid (an omega-6 fatty acid) and alpha-linolenic acid (an omega-3 fatty acid), which are required for the synthesis of longer-chain fatty acids and, in turn, the production of eicosanoids.

Cholesterol is a sterol synthesized primarily by the liver in amounts sufficient to meet our needs, but is also consumed through foods of animal origin.

Most lipid digestion occurs in the small intestine facilitated by bile acids, the process of emulsification, and the action of lipases.

Protein-rich lipoproteins transport lipids in the bloodstream and include chylomicrons, low-density lipoproteins (LDLs), high-density lipoproteins (HDLs), and very low-density lipoproteins (VLDLs). Lipoproteins are classified according to their density and by their function; they differ in which lipids they carry to which body parts.

Depending on the type and amount in the diet, fat can have positive and negative health effects.

The current recommended fat intake is 20% to 35% of total calories.

Although some trans fatty acids are naturally occurring, most are produced through partial hydrogenation of polyunsaturated fats, and increase the risk of heart disease more than any type of fat.

NEED TO KNOW

Review Questions

1. Characteristics of lipids include all of the following, EXCEPT that they are:
   a. structurally similar compounds with specific functions.
   b. composed of carbon, hydrogen, and oxygen.
   c. include triglycerides, phospholipids, and sterols.
   d. are generally insoluble in water.
2. Dietary fat contributes to satiety by:
   a. increasing the glycemic response of a meal or snack.
   b. stimulating motility in the small intestine.
   c. prolonging the time food stays in the stomach.
   d. producing bulk to fill the stomach.
3. Structurally, fatty acids:
   a. consist of carbon chains of similar lengths.
   b. differ in their degree of saturation.
   c. are composed of carbon rings with a hydrocarbon side chain.
   d. have a hydrogen and an oxygen attached to each carbon.
4. A serving of cookies has seven grams of fat. How many calories from fat would these cookies provide?
   a. 7 calories
   b. 28 calories
   c. 49 calories
   d. 63 calories
   e. 70 calories
5. Cholesterol is a sterol that:
   a. must be consumed in the diet to meet needs.
   b. provides nine calories per gram.
   c. is found in plant foods that are high in saturated fats.
   d. functions as a precursor for synthesis of steroid hormones.
6. Emulsification of fats in the small intestine requires the presence of:
   a. bile acids.
   b. hydrochloric acid.
   c. lipases.
   d. cholesterol.
   e. insulin.
7. Low-density lipoproteins (LDLs):
   a. transport cholesterol from the body's tissues back to the liver.
   b. transport cholesterol to essentially all cells in the body.
   c. lower the risk of heart disease as their levels increase.
   d. aid in the digestion of lipids in the small intestine.
   e. are derived from chylomicrons as triglycerides are removed.
8. Linoleic acid and linolenic acid:
   a. can be synthesized from phospholipids in the body.
   b. are both classified as omega-3 fatty acids.
   c. must be consumed through the diet to meet the body's needs.
   d. are both saturated fatty acids.
   e. are both trans fatty acids.
9. All of the following are true statements regarding trans fatty acids, EXCEPT:
   a. trans fatty acids are produced by complete hydrogenation of unsaturated fatty acids.
   b. trans fatty acids are present in small amounts in meats and full-fat dairy foods.
   c. trans fatty acids can increase LDL cholesterol levels.
   d. trans fatty acids can make packaged foods more resistant to spoilage, increasing shelf-life.
   e. trans fatty acids are listed on the Nutrition Facts Panel.
10. What is the AMDR for total calories from dietary fat for adults?
    a. 5% to 20%
    b. 10% to 25%
    c. 15% to 30%
    d. 20% to 35%
    e. 25% to 40%

Take It Further

Trans fatty acids can be found in partially hydrogenated foods like stick margarine, commercially baked pastries and cakes, and many packaged cookies and crackers. Recently, the FDA proposed a possible ban on trans fat in the American food supply. Currently, foods that contain less than 0.5 grams of trans fat can list content on the Nutrition Facts Panel as 0. Discuss the rationale for the proposal to ban trans fat and include your perspective on whether a ban should be imposed by the FDA or whether concerned consumers should refer to the Nutrition Facts Panel.

Dietary Analysis Using SuperTracker

Getting to know the lipids in your lunch

Lipids, or fats, are a key component of any healthy diet, but it's important to know if we're getting too little or too much. In this exercise, you will evaluate your lipid intake, break down the various types of fats you consume on an average day, and compare them with the current dietary recommendations.

1. Log onto the United States Department of Agriculture (USDA) website at www.supertracker.usda.gov. If you have not done so already, you will need to create a profile to get a personalized diet plan. This profile will allow you to save your information and diet intake for future reference. *Do not use the general plan.*
2. Click the Track Food and Activity option.
3. Record your food and beverage intake for one day that most reflects your typical eating patterns. Enter each food and beverage you consumed into the food tracker. Note that there may not always be an exact match to the food or beverage that you consumed, so select the best match available.
4. Once you have entered all of your food and beverage choices into the food tracker, on the right side of the page under the bar graph, you will see Related Links: View by Meal and Nutrient Intake Report. Print these reports and use them to answer the following questions:

   a. What percent of your total calories came from dietary fat? Is this within the recommended 20% to 35%? What foods contributed most to your overall intake of dietary fat?

   b. What percent of your total calories came from saturated fat? Is it more than 10%?

   c. What percent of total calories came from monounsaturated fat? Is this greater than your intake of saturated fat?

   d. What percent of total calories came from polyunsaturated fat? Is this greater than your intake of saturated fat?

   e. How many grams of linoleic acid did you consume? Is this within the dietary recommendations?

   f. How many grams of linolenic acid did you consume? Is this within the dietary recommendations?

   g. How many milligrams of cholesterol did you consume? Is this above the recommended 300 mg daily limit? What foods contributed the most cholesterol to your diet on this day?

   h. Considering what you've learned about trans fatty acids, what foods did you consume that might be sources of this type of fatty acid?

   i. What *specific* dietary changes could you make to improve upon the results of this day's analysis? List at least three food substitutions that would not only help keep total fat intake within recommended range, but boost unsaturated fat intake over saturated. For example, choosing fish or poultry rather than ground beef.

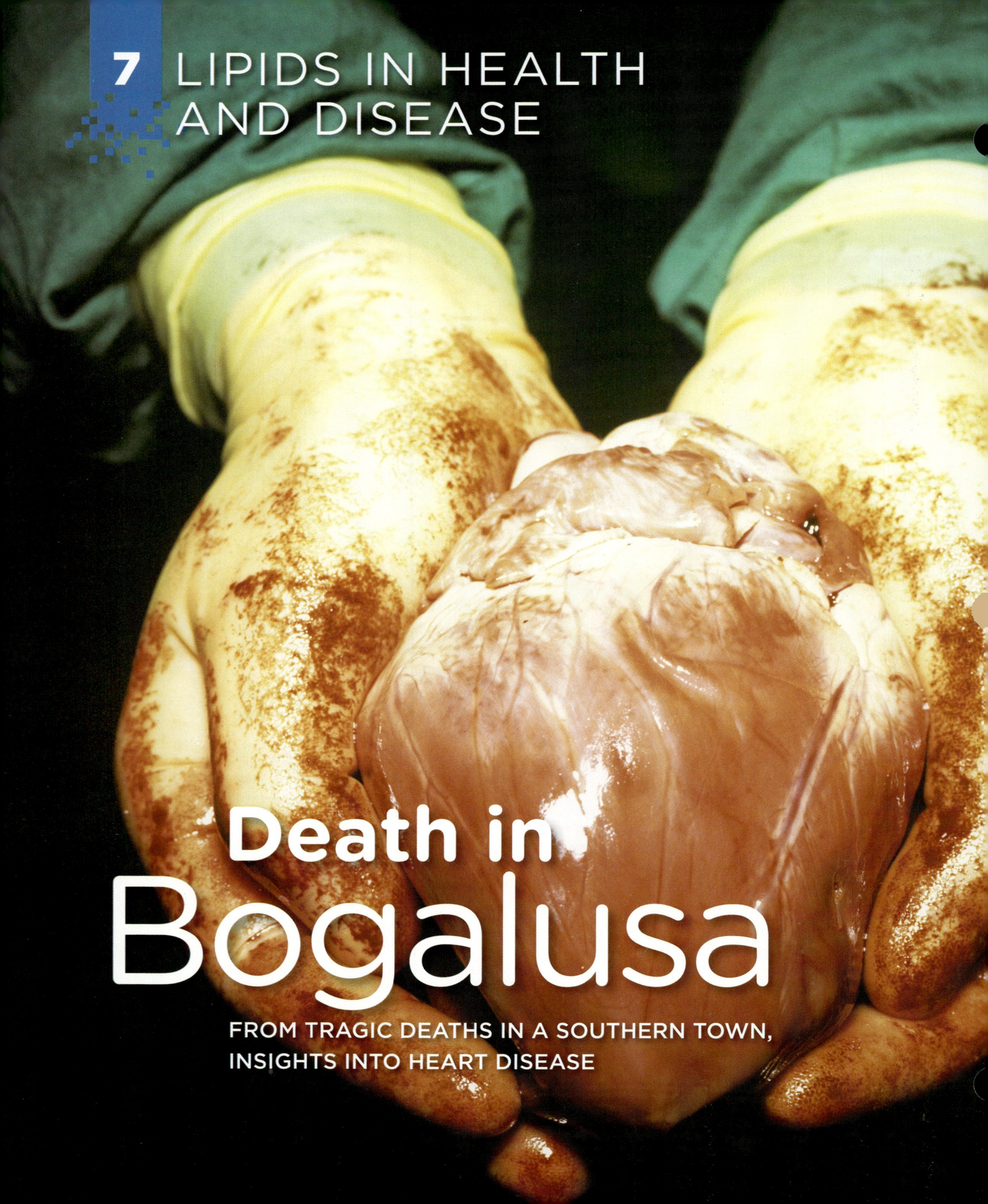

7 LIPIDS IN HEALTH AND DISEASE

Death in Bogalusa

FROM TRAGIC DEATHS IN A SOUTHERN TOWN, INSIGHTS INTO HEART DISEASE

Kevin Curtis/Science Source

LEARNING OBJECTIVES

- Summarize the events that lead to the development of atherosclerosis (Infographic 7.2)
- Identify at least five risk factors that affect the initiation or progression of cardiovascular disease (Infographic 7.4)
- Describe how total cholesterol, low-density lipoproteins, and high-density lipoproteins interact to affect the risk of cardiovascular disease (Infographic 7.5)
- List the cluster of risk factors associated with metabolic syndrome (Infographic 7.6)
- Discuss the implications of and recommendations for intake of saturated fatty acids, trans fatty acids, and unsaturated fatty acids in relation to cardiovascular disease (Infographics 7.7 and 7.8)
- Summarize the dietary strategies that reduce the risk of heart disease (Infographic 7.9)

Some scientific discoveries begin in a laboratory. This one began in a funeral home in 1978.

Under glaring lights, in the back room of the Cook-Richmond Funeral Home in Bogalusa, Louisiana, two pathologists hovered over the lifeless body of a young African-American male. There was no morgue in Bogalusa, so an autopsy was being conducted here, with newspapers spread beneath the body. The pathologists sliced through skin and muscle with scalpels, looking for the usual suspects—internal bleeding, broken bones—then wrote up their report for the coroner. The pathologists weren't quite done, however.

Gerald Berenson, MD, *Research Professor, Department of Epidemiology, Tulane University School of Public Health and Tropical Medicine*

Before closing up, they carefully removed the victim's heart, packaged it in saline, and prepared it for a trip to a medical school in New Orleans. It was an unusual step, not standard for an autopsy. But the pathologists were heeding the instructions of a prominent local physician, who had the blessing of the family.

Since 1972, Gerald Berenson, a cardiologist with the Louisiana State University School of Medicine, had spearheaded what was then a novel investigation: an epidemiological study of heart disease in Bogalusa, a rural town of about 16,000 people, located 60 miles north of New Orleans.

The study began with a pretty simple idea: Follow a large group of children over time and correlate their physical and lifestyle attributes with their risk for developing heart disease in adulthood. The biggest hurdle was a logistical one—how to enlist the thousands of children necessary to produce a robust data set, and keep them coming back for evaluation year after year.

And Berenson wanted to do more than make statistical correlations: He also wanted to document the progression of heart disease directly. For that he needed a different type of evidence.

The heart that Berenson obtained in 1978 was one of more than 200 such organs collected from young people in Bogalusa over the next 20 years. Some died in car accidents. Others drowned, or were stricken with pneumonia. Still others were slain by bullets. The victims were as young as 2 and as old as 39.

In fact, nearly every young person who died, of whatever cause, was autopsied and their heart was removed. Together, they transformed our understanding of heart disease.

Bogalusa is typical of many rural Southern towns, in which poverty and poor education have exacerbated problems of public health. More than 40% of individuals in Bogalusa live below the poverty line. Sadly, poverty is linked to a number of health problems, most notably *heart disease*, known more formally

INFOGRAPHIC 7.1 Fast Facts About Cardiovascular Disease (CVD) and Selected Risk Factors in the United States *CVD includes heart diseases and vascular disease affecting the brain.*

OF ADULTS IN THE UNITED STATES HAVE HIGH BLOOD PRESSURE

OF ALL DEATHS IN THE UNITED STATES ARE CAUSED BY CVD

OF ADULTS IN THE UNITED STATES HAVE CVD

OF ADULTS IN THE UNITED STATES HAVE HIGH BLOOD CHOLESTEROL

Heart Disease and Stroke Statistics—2013 Update. A Report from the American Heart Association

as **cardiovascular disease (CVD)**. CVD refers to conditions that impair the heart, arteries, veins, and capillaries, which move blood throughout the human body. For more than 100 years, CVD has been the number one cause of death in the United States; today an estimated 1 in 3 Americans have some form of the disease. In 2009, more than 2,000 Americans died of CVD every day, equating to an average of 1 death due to CVD every 40 seconds. It's not just a disease of old age, either—more than 150,000 people who died of CVD in 2009 were younger than 65 years. It's those early roots of heart disease—that initially take hold in the youngest victims—that Berenson and his team were after. (INFOGRAPHIC 7.1)

Atherosclerosis and blood clot. *The extensive fatty deposit (brown) on the artery wall has narrowed the lumen (internal space), aiding the formation of the blood clot (red). If the coronary artery becomes completely blocked, this can lead to a fatal heart attack.*

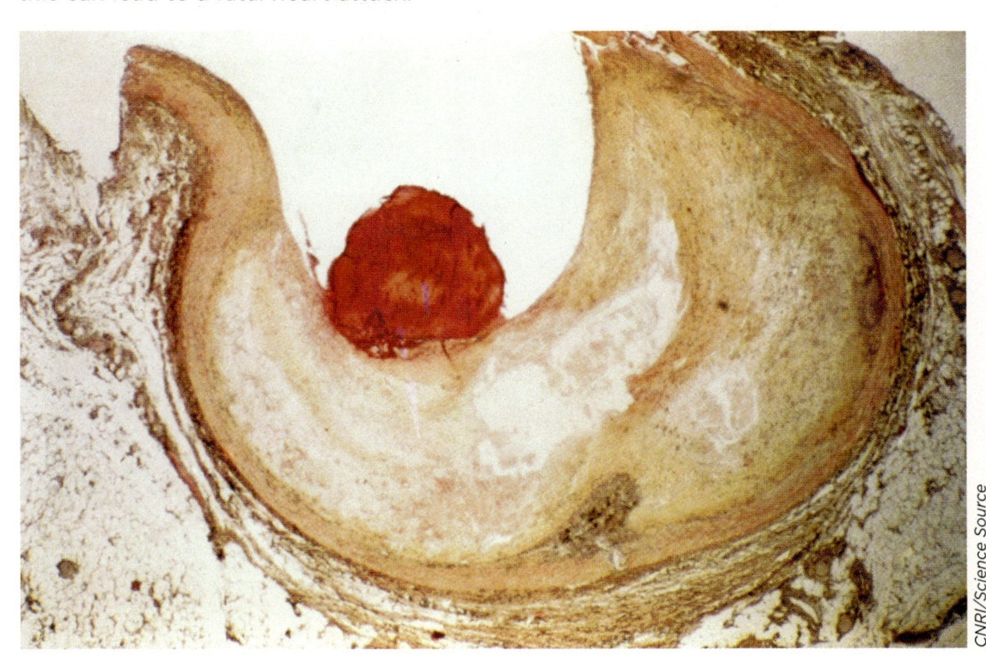

CNRI/Science Source

ATHEROSCLEROSIS AND CARDIOVASCULAR DISEASE

Although often called "heart disease," CVD strikes the blood vessels—*veins* and *arteries*—as well. But even before people experience major events such as heart attack and stroke, the long-term narrowing and loss of elasticity in blood vessels are caused by **atherosclerosis**. Atherosclerosis is an inflammatory disease characterized by the accumulation of fatty plaque in the walls of arteries and blood vessels that generally develops over the course of several decades. This process is typically caused by the presence of elevated levels of cholesterol-rich *low-density lipoproteins* (LDLs) in the blood. As blood levels of LDL rise, they infiltrate the artery wall, where the LDLs are likely to become *oxidized* (by reacting with unstable oxygen-containing molecules). Oxidized LDLs cause injury to cells that line the vessel wall, and this initiates an inflammatory process that attracts white blood cells called *macrophages* inside the arterial lining. Other factors, such as smoking or hypertension, can also cause injury to the arterial wall and trigger inflammation. Inside the vessel wall, macrophages take up the oxidized LDLs in a rapid and uncontrolled fashion. These fat-laden macrophages (now called *foam cells*) die and deposit their accumulated lipids within the wall of the artery, promoting further inflammation.

As a result, the lining of the blood vessel becomes more prone to develop a waxy accumulation of cholesterol and triglycerides, known as a **plaque**. Over time, plaque development, loss of elasticity, and thickening in the blood vessel walls may make it difficult for blood to flow through the vessel. This "traffic jam" increases the chances of forming blood clots that either block flow at that location or break off and travel through the bloodstream, blocking blood flow elsewhere, causing tissue damage and tissue death. When blood flow is blocked in the *coronary artery*, which supplies blood to the heart, people experience a *heart attack*, or **myocardial infarction**. Most cases of **stroke** result when a clot impairs the supply of blood to the brain. (INFOGRAPHIC 7.2)

■ ■ ■

For the most part, young people do not have heart attacks (and those that do usually suffer from rare genetic conditions). But just because most teenagers do not die from heart disease does not mean they can ignore heart health. That's because heart disease can be insidiously gaining a foothold long before we have any obvious symptoms.

CARDIOVASCULAR DISEASE (CVD)
a group of conditions that impair the heart and blood vessels

ATHEROSCLEROSIS
a type of cardiovascular disease characterized by the narrowing and loss of elasticity of blood vessel walls, caused by accumulation of plaque and inflammation of tissue

PLAQUE
deposits of cholesterol, triglycerides, and cell materials that accumulate within the arterial wall

MYOCARDIAL INFARCTION (HEART ATTACK)
damage to heart tissue caused by decreased blood flow to the coronary arteries

STROKE
cerebral event that occurs when blood vessels supplying the brain are damaged or blocked

INFOGRAPHIC 7.2 Development of Atherosclerosis and Its Consequences *The development of atherosclerosis often begins with an injury to an arterial lining, and excess LDL cholesterol in circulation promotes its progression. Atherosclerosis is a major risk factor for heart attack and stroke.*

CLOT FORMATION

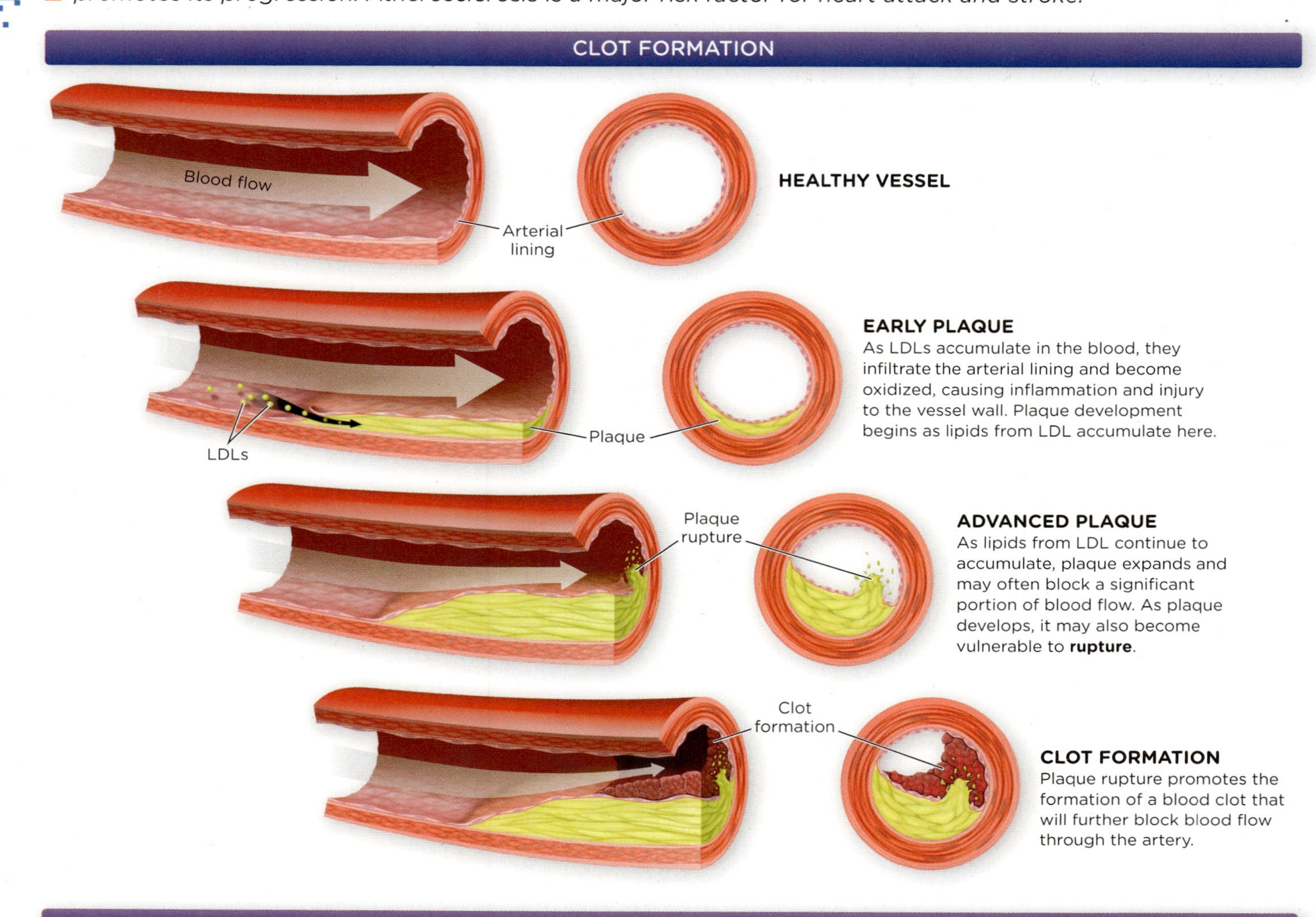

ATHEROSCLEROSIS CAN LEAD TO HEART ATTACK OR STROKE

HEART ATTACK

If the clot completely blocks blood flow in an artery supplying blood to the heart, this will cause a heart attack (myocardial infarction). Cells that are deprived of blood will be damaged and die, permanently weakening the heart.

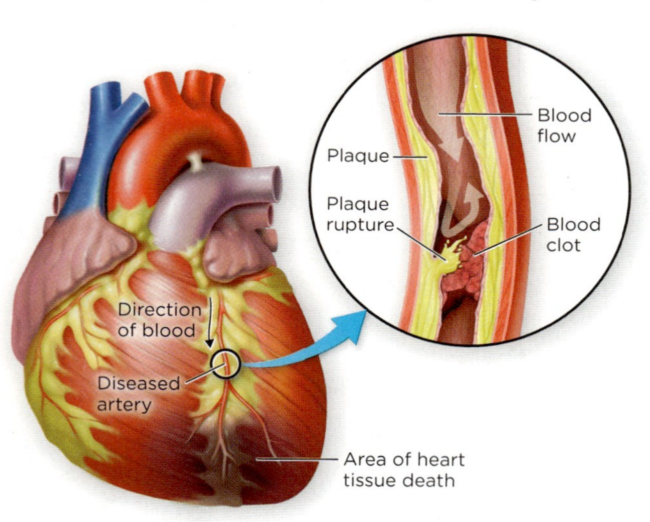

STROKE

An interruption of blood flow to the brain causes the death of brain tissue.

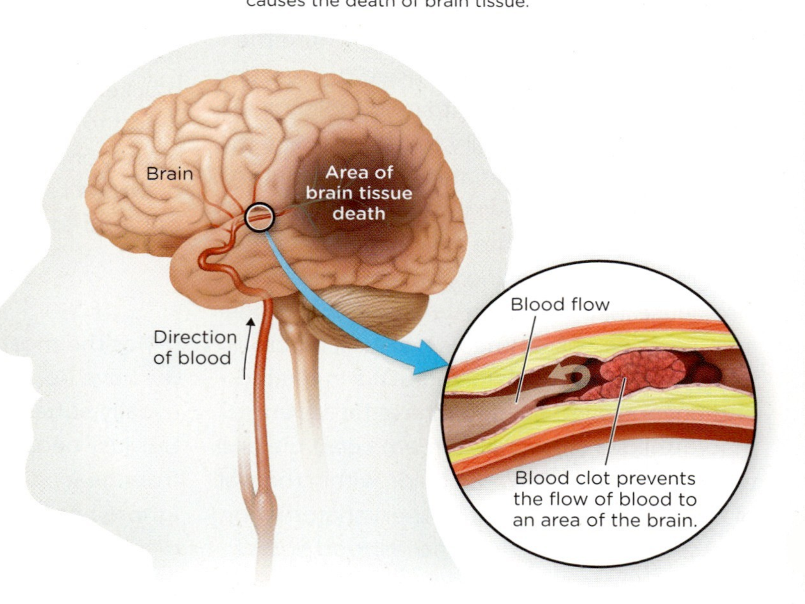

? *Where in the blood vessel do LDLs have to be located to promote the development of atherosclerosis?*

INFOGRAPHIC 7.3 Atherosclerosis in Cerebral Arteries Is a Contributing Factor to Alzheimer Disease

These images show the significant narrowing of cerebral arteries due to atherosclerosis.

Healthy Artery

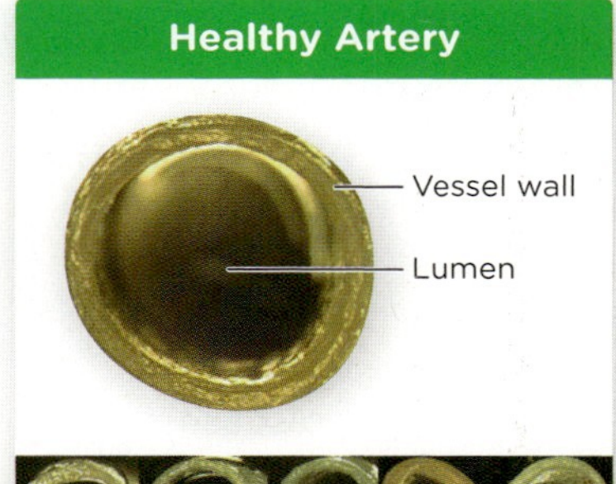

Artery Narrowed by Plaque

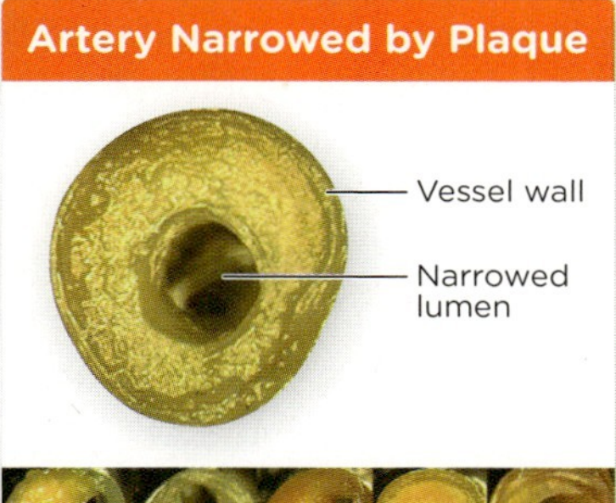

Individuals with Alzheimer disease experience greater narrowing of the arteries in the brain than do healthy individuals.

Why might atherosclerosis in cerebral arteries contribute to dementia?

When Gerald Berenson started his study in 1972, this was far from accepted wisdom. Cardiologists—influenced by the prevailing practice of the day—focused more on the treatment of CVD than on its prevention. But Berenson had been trained in pediatrics as well as cardiology, and he knew that it was important to understand the beginnings of heart disease, as well as its endings.

He decided to follow children and young people ages 5 to 17 years in the local school system, monitoring them twice yearly. Nurses measured their height, weight, smoking history, blood pressure, and blood lipid levels, among other variables. After age 17, they returned for evaluation periodically, up to age 45. Later, it was decided that any of the study participants who died would be autopsied.

From small beginnings, the study grew to enroll some 16,000 individuals. It is now one of the longest-running CVD studies in the world, the only one with a biracial (black-white) study population. Known as the Bogalusa Heart Study, it was the first study to demonstrate that heart disease begins in childhood.

Although the exact cause of CVD is often unknown, the Bogalusa study has shown there are certain **risk factors** that increase young people's chances of developing problems—and the more they have, the higher their risk.

RISK FACTORS FOR CARDIOVASCULAR DISEASE

You can't do anything about some risk factors for CVD. For example, individuals with a family history of heart disease at an early age are more likely to develop CVD, as are African Americans, and both men and women are at increased risk as they age. But there are many risk factors people of all ages can modify, most significantly through diet and exercise.

RISK FACTOR
any characteristic, condition, or behavior that increases the likelihood of developing a particular disease

Modifiable risk factors include smoking; a diet high in saturated and trans fats, cholesterol, sodium, and added sugar; a sedentary lifestyle; obesity; excessive alcohol consumption (see chapter 19 to learn more about health implications of alcohol consumption); and high blood pressure. Having high blood pressure (hypertension)—a condition in which blood pushes with excessive force against artery walls—is correlated closely with the extent of atherosclerosis found in the arteries. (Hypertension is discussed in more detail in Chapter 13.) **(INFOGRAPHIC 7.4)**

Though many risk factors for CVD are modifiable, more than 90% of Americans eat a "poor" diet for heart health, according to the American Heart Association. For example, the concentrations of cholesterol and triglycerides in blood are a major risk factor for CVD. Blood cholesterol levels measuring less than 200 milligrams of cholesterol per deciliter of blood (mg/dl)—the standard units of measurement for cholesterol—are considered healthy. Yet the average cholesterol level for a U.S. adult is 200 mg/dl—meaning the average American has a borderline high level. People with high levels of total blood cholesterol have approximately twice the risk of heart disease as those who do not. Normal triglyceride levels vary by age and sex, but anything above 200 mg/dl is considered high. High triglyceride levels are common in people with heart disease or diabetes.

INFOGRAPHIC 7.4 Modifiable Risk Factors for Cardiovascular Disease

Appropriate diet and lifestyle choices may reduce our risk of CVD by about 80%.

Poor Diets: Diets high in trans fats, saturated fats, and cholesterol, and low in polyunsaturated fats, vegetables, fruits, and whole grains increase the risk of CVD. Excess intake of sodium can lead to hypertension, which increases the risk of CVD.

Smoking: Smokers are two to four times more likely to develop heart disease or experience a stroke than nonsmokers. Exposure to secondhand smoke at home or work increases the risk of heart disease.

Physical Inactivity: It is estimated that 150 minutes of moderate- to vigorous-intensity exercise per week will reduce the risk of CVD mortality by 30% to 35% compared with those who are physically inactive.

Excessive Alcohol Consumption: Excessive alcohol consumption causes hypertension that dramatically increases the risk of stroke. Drinking large amounts can also cause the heart to enlarge and heart muscles to thin and weaken. Heavy or at-risk alcohol use is defined as more than 3 drinks a day or 7 per week for women, and 4 drinks a day or 14 per week for men.

Obesity: Particularly central or abdominal obesity is a major independent risk factor for CVD. It also increases the occurrence of other CVD risk factors (hypertension, diabetes, high blood cholesterol, and high triglycerides).

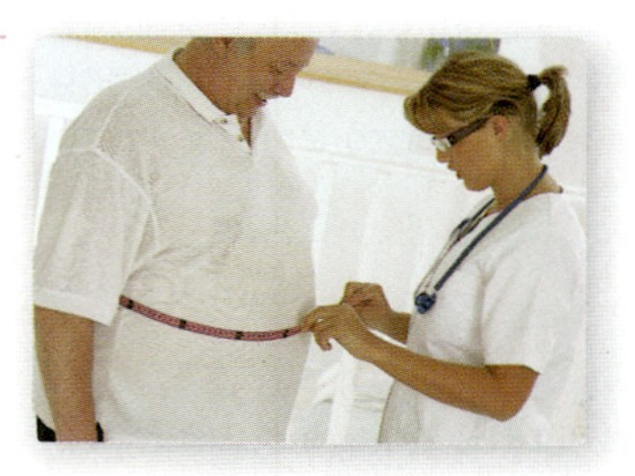

Heart-Related Conditions: The risk of CVD is increased by high blood pressure, blood glucose, LDL cholesterol, and triglycerides, and by low HDL cholesterol. Improving one's diet and exercising regularly will help manage these conditions.

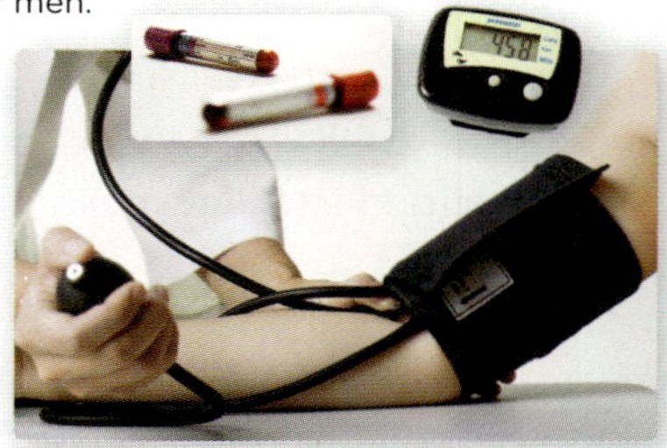

Unmodifiable risk factors for CVD are age, race, and family history of CVD (genetics).

If you exercised five days per week, how many minutes would you have to exercise each day to meet the recommendations to reduce your risk of cardiovascular disease?

Photo credits (left side — top to bottom): mipstudio/Shutterstock, maxriesgo/Shutterstock, JPC-PROD/Shutterstock; (right side — top to bottom): Datacraft Co Ltd/Getty Images, Alexey Lysenko/Shutterstock, Stephen Smith/Getty Images, Denis Dryashkin/Shutterstock, Michael Krinke/Getty Images

INFOGRAPHIC **7.5** **Classification of Cholesterol and Triglyceride Concentration in Blood** *The concentrations of total cholesterol, HDL cholesterol, LDL cholesterol and triglycerides in blood significantly affect risk of cardiovascular disease.*

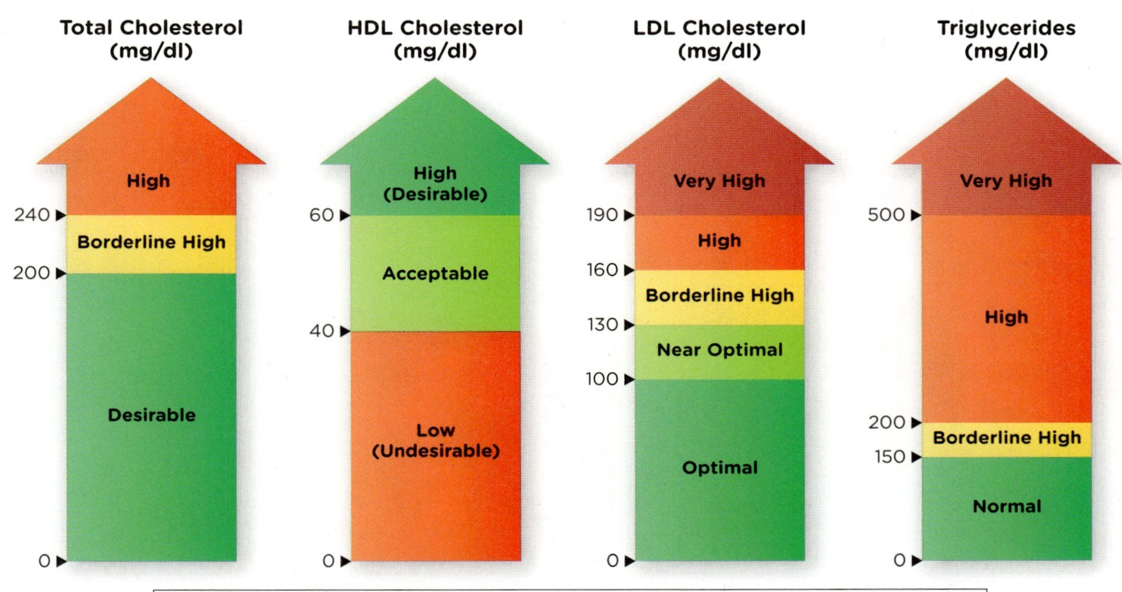

The figures in this table are provided by the National Cholesterol Education Program (NCEP) of the NIH.

For which of these values does the risk of heart disease decrease as the value increases?

One of the most important tools Berenson and his Bogalusa study team used to evaluate healthy diets was children's intake of fat and cholesterol. Besides asking kids about their diets, the Bogalusa researchers also directly measured levels of lipids in their blood, by looking at the *lipoproteins* that transport lipids around the body. Likewise, physicians use a diagnostic test called a lipid panel to provide information about total cholesterol, *high-density lipoproteins* (HDLs), LDLs, and triglycerides in the blood.

Recall from Chapter 6 that lipoproteins contain different proportions of proteins, cholesterol, triglycerides, and other components that affect their density and function—as well as their impact on the risk of CVD. The LDLs ferry cholesterol to all of the cells of the body. This cholesterol forms the root of the plaques that slowly constrict blood flow through the vessels that feed the heart and brain; the lower your LDL cholesterol level, the lower your risk of heart attack and stroke. If you have other risk factors for CVD, it's best to keep your LDL level below 100 mg/dl (some experts recommend levels closer to 70 mg/dl). In some cases, diet and exercise may not suffice, in which case people also take medications that lower their total cholesterol and LDL levels.

HDLs, in contrast, bring cholesterol from tissues back to the liver, where it is processed and eliminated; not surprisingly, higher levels of HDL have been associated with a lower risk of heart disease. In fact, low levels of HDLs—less than 40 mg/dl in men and 50 mg/dl in women—can even increase the risk of CVD. **(INFOGRAPHIC 7.5)**

In general, the risk for CVD increases with the number of risk factors you have. People who are carrying around a cluster of some of the most dangerous risk factors for CVD are said to have **metabolic syndrome**. For example, having an elevated blood glucose level is a serious condition, but when coupled with excessive amounts of abdominal fat and high blood pressure, a person has a greater

METABOLIC SYNDROME
a cluster of risk factors associated with the development of cardiovascular disease and type 2 diabetes

INFOGRAPHIC 7.6 Factors Associated with Metabolic Syndrome *Metabolic syndrome is a cluster of risk factors associated with the development of cardiovascular disease and type 2 diabetes.*

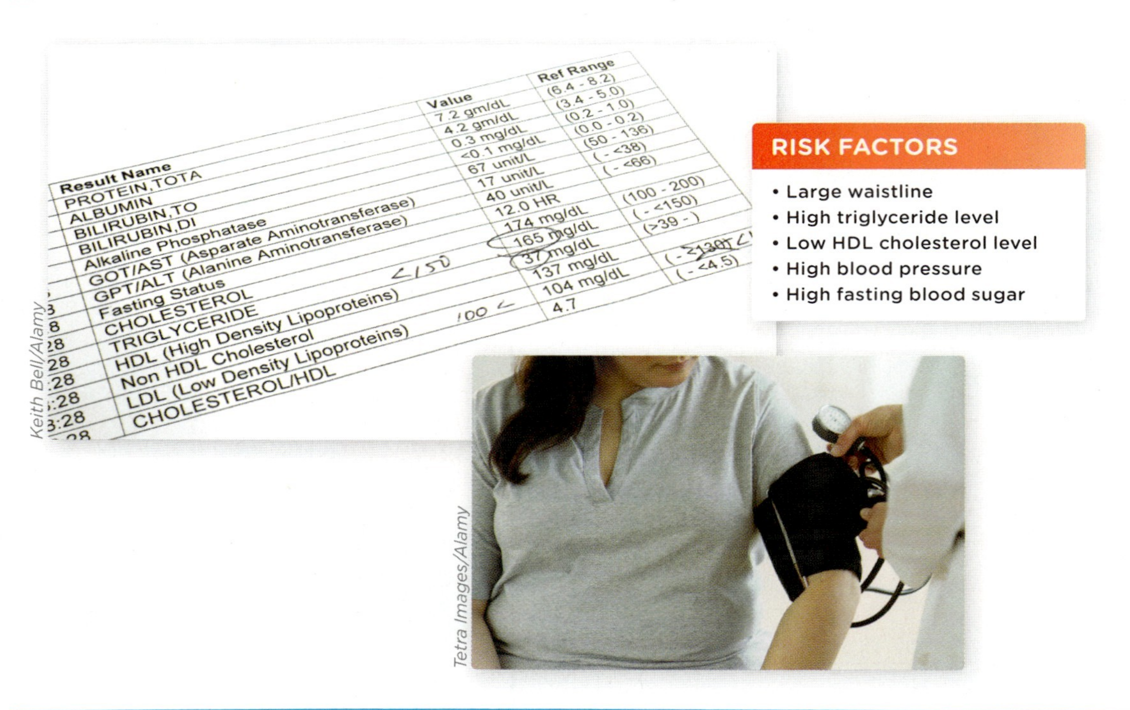

What is considered a normal triglyceride level?

chance of cardiovascular problems because of the combination of risk factors. Not surprisingly, this makes them much more likely to have a heart attack or stroke, and five times more likely to develop type 2 diabetes. To be given a diagnosis of metabolic syndrome individuals must have at least three of the following problems: excessive abdominal fat, high blood pressure, elevated levels of triglycerides in the blood, low levels of HDL, and elevated blood glucose levels (either prediabetes or type 2 diabetes—see Chapter 5). Furthermore, the prevalence of metabolic syndrome is on the rise. Today, it affects more than 50 million adults in the United States, and approximately 25% of adults worldwide. **(INFOGRAPHIC 7.6)**

■ ■ ■

To learn more about the development of CVD, pathologist William Newman and a colleague would drive the 66 miles up from Louisiana State University School of Medicine in New Orleans to Bogalusa to perform the autopsy whenever a young person in Bogalusa died. What Newman and his colleagues saw when they opened the hearts of the young people shocked them.

Although most people think of heart disease as something that affects adults, virtually all of the young people of Bogalusa already had *fatty streaks*—the precursors to plaques—lining the blood vessels of their hearts. As soon as Newman and his colleagues cut open a vessel, they knew what they were seeing—the streaks "look a bit yellow," Newman said. By adding a specific dye, he got a better sense of the shapes and sizes of the streaks to confirm his belief. There was no doubt—these children had early forms of atherosclerosis.

To quantify the extent of atherosclerosis, each pathologist would look at every autopsy specimen and assign the stained fatty streaks a score, and the scores were then averaged together. The study was blinded, so the pathologists didn't know the source of the tissues beforehand.

The researchers then compared the results of those autopsies with data collected from

INFOGRAPHIC 7.7 Sources of Saturated Fat in the U.S. Population, 2005–2006

The percent contribution of specific foods to total saturated fat intake for individuals 2 years and older.

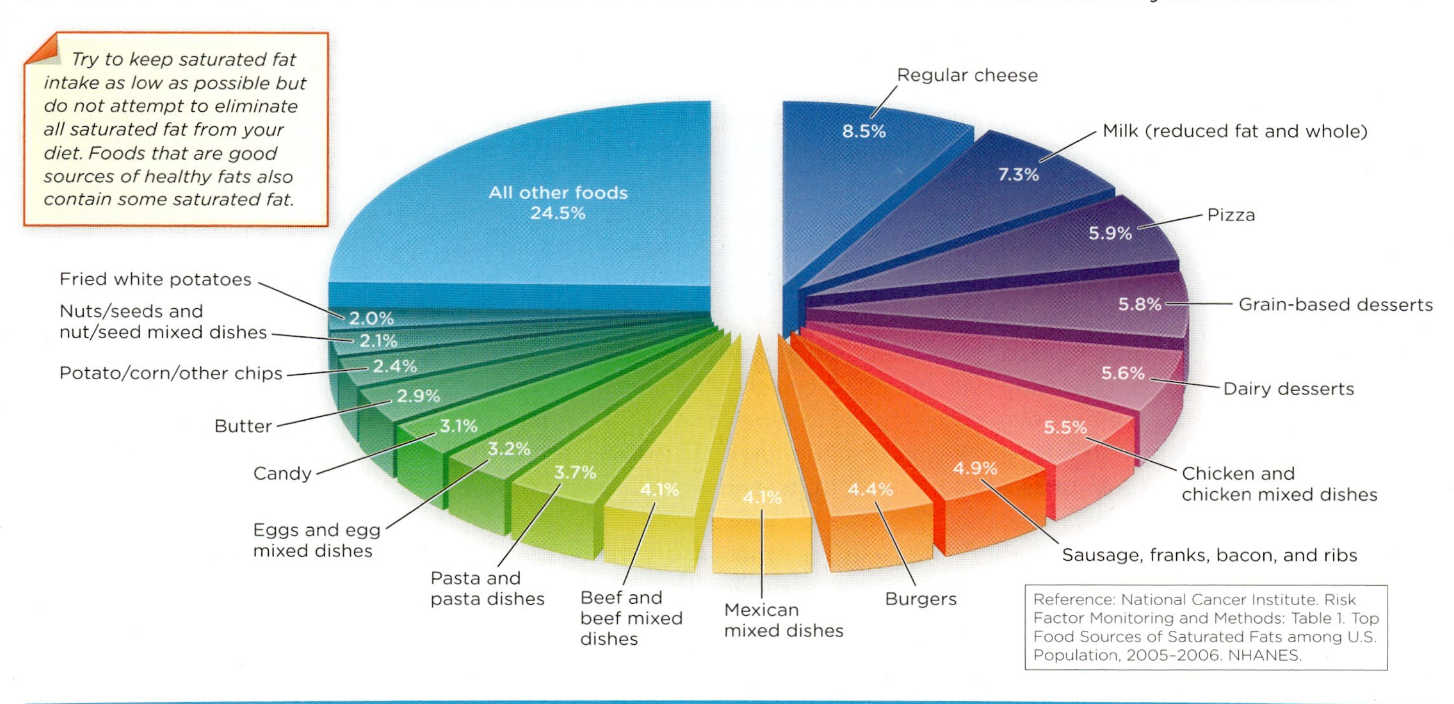

Reference: National Cancer Institute. Risk Factor Monitoring and Methods: Table 1. Top Food Sources of Saturated Fats among U.S. Population, 2005–2006. NHANES.

? *What percent of total saturated fat intake is from the consumption of red meats?*

young people during the study, and identified a number of risk factors associated with the risk of early atherosclerosis: smoking, obesity, hypertension, and high levels of cholesterol and triglycerides in the blood.

When all the known risk factors were considered, a clear pattern emerged: "We found that those individuals who had higher levels of known risk factors on the average had more fatty streaks in their coronary arteries and the aortas than individuals who had lower levels," says Newman. In other words, though many of them were not even old enough to vote, these young people already had telltale signs of heart disease. Some even had developed full-blown plaques that were beginning to obstruct blood flow. Had these young people lived, those with the higher levels of atherosclerosis would have been at risk for a heart attack or other complications of heart disease.

"The autopsy studies were really landmark studies," says Peter Katzmarzyk, an epidemiologist at the Pennington Biomedical Research Center in Baton Rouge. Before them, he says, scientists didn't really understand that things like heart attack and stroke have their genesis in childhood. "The Bogalusa Heart Study really put that on the map," he says. And it made a strong case that averting severe consequences of heart disease later in life would mean addressing the lifestyle choices we make when we are young.

The science behind the influence of dietary fat and fatty acids on CVD is complex and continually evolving. As early as the 1950s, scientists and the public were talking about the dangers of high-fat diets, particularly those that contain saturated fatty acids—but, recently, we've begun to realize the picture isn't quite so straightforward. **(INFOGRAPHIC 7.7)**

A HEART-PROTECTIVE DIET

Although saturated fatty acids do, indeed, have the potential to raise total and LDL cholesterol levels, an important factor to consider is what people who cut back on saturated fatty acids eat *instead*. For instance, if people replace high-fat foods with lower-fat options that are high in carbohydrates, particularly refined starches, added sugar, and potatoes (which happens frequently), then they aren't likely going to see health benefits. Low-fat diets high in carbohydrates, particularly when low in fiber and high in sugar, actually increase triglyceride levels in the blood. And some recent analyses even suggest that some sources of saturated fatty acids, such as those in dairy products and coconut oil, don't necessarily negatively affect blood lipids levels and increase the risk of CVD. However, based on current evidence, the best overall strategy is to replace saturated fatty acids with unsaturated ones, and to emphasize vegetables, fruits, and whole grains in the diet. **(INFOGRAPHIC 7.8)**

The Dietary Guidelines for Americans recommend limiting overall saturated fat intake to 10% or less of total daily calories. In contrast to earlier editions, the 2015 Guidelines do not establish specific limits for dietary cholesterol, but qualify that eating as little as possible is advisable within a healthy eating pattern (see Chapter 2). For those with increased risk for heart disease and high blood cholesterol levels, recommendations from the National Cholesterol Education Program (NCEP) of the National Institutes of Health are to keep calories from saturated fat under 7%. NCEP created the *Therapeutic Lifestyle Changes* (TLC) program, which incorporates diet, physical activity, and weight management to lower blood cholesterol levels and reduce the risk of heart disease. Also endorsed by the American Heart Association, the TLC dietary component focuses on lowering saturated fat and dietary cholesterol along with other heart-healthy strategies, such as eating foods that have added plant sterols and stanols and contain sufficient soluble fiber.

When the Bogalusa Heart Study began in 1972, Berenson and his colleagues were interested in identifying risk factors that contribute to heart disease in childhood, in part because other studies had identified risk factors in adulthood. The Nurses' Health Study, for instance, has followed more than 120,000 women since 1976, looking for links between their

INFOGRAPHIC 7.8 Dietary Fat Replacement and Heart Disease Risk

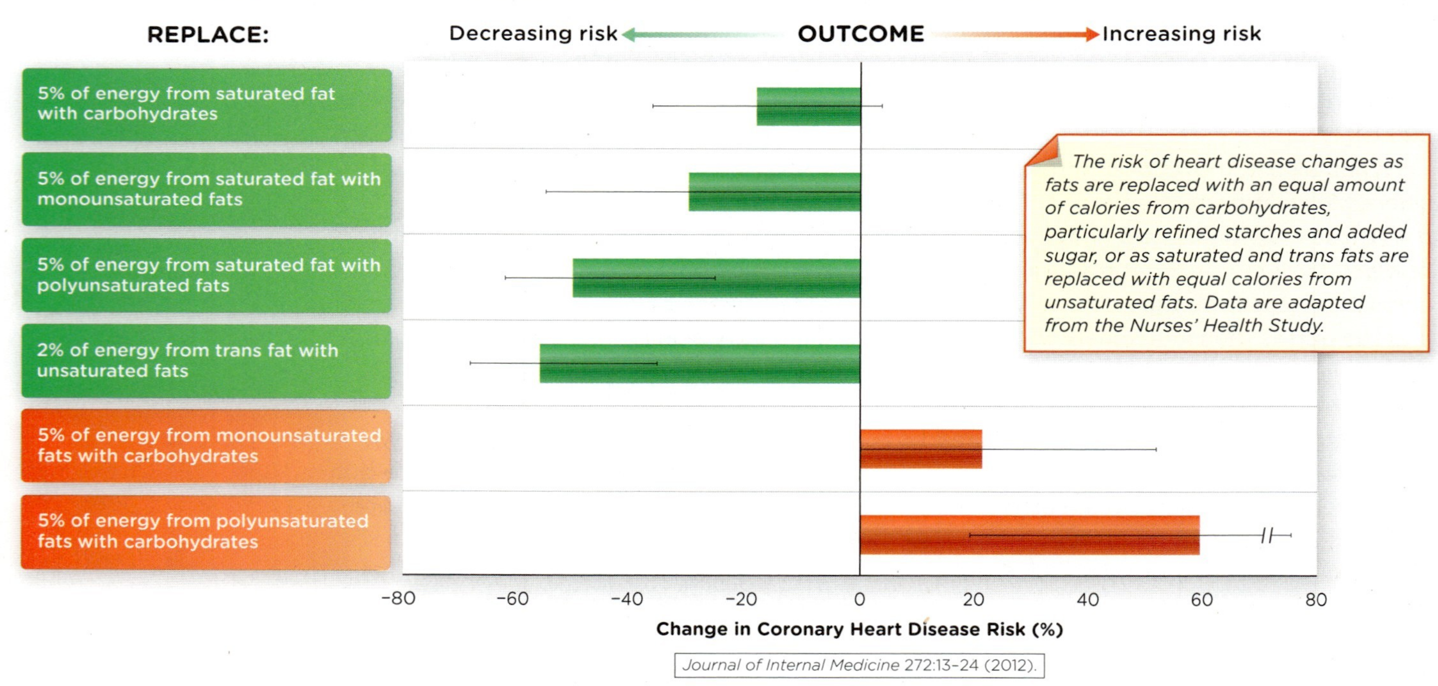

? *What is the most effective dietary substitution that we can make to reduce the risk of coronary heart disease?*

diets and lifestyles and their risks of various diseases. When the researchers examined women's fat intake, they found that women who ate the most *polyunsaturated fatty acids* were the least likely to develop heart disease.

Research has also found that *monounsaturated fatty acids* can improve blood lipid profiles, including raising HDL cholesterol and lowering overall triglyceride levels, which may reduce the risk of heart disease. But you can get too much of a good thing—there is no consistent rationale or recommendation for the specific amount of monounsaturated fatty acids we should obtain from our diets, but the American Heart Association suggests people get no more than 25% to 35% of their total calories from fat, most of which should come from unsaturated fatty acids, and that saturated fat not exceed 7% of total calories.

Furthermore, different unsaturated fats have different effects on the body, as well. *Omega-6 polyunsaturated fatty acids*, for instance, appear to improve blood lipid profiles, especially when they take the place of saturated fatty acids, reducing the risk of heart disease. The American Heart Association suggests this type of fat should make up no less than 5% to 10% of our total calories, given that they appear to reduce inflammation, a culprit in the initiation of plaque development.

Some research suggests *omega-3 fatty acids* may hold particular health benefits, but the picture remains somewhat unclear—low levels of the omega-3 docosahexaenoic acid (DHA) in the diet may increase the risk of Alzheimer disease and other forms of dementia, but it's not known whether DHA supplements hold any benefits. However, increasing the intake of DHA and EPA (eicosapentaenoic acid, another omega-3 fatty acid) will generally lower the level of circulating triglycerides.

Everyone's diet contains a *ratio* of omega-6 and omega-3 fatty acids—meaning the amount of each, relative to each other. The lower the ratio, the better (such as 5:1 or 4:1), but that's difficult to achieve, given that omega-6-rich foods are much more common in the U.S. diet. Most Americans, and others who follow a Western diet, eat much more omega-6 fatty acids than omega-3 fatty acids, with a typical ratio of 10:1.

But for some dietary fats, the effect on health is clear—trans fatty acids, for instance, help lengthen a food's shelf life and are therefore often found in many commercial cakes, cookies, pies, and pastries. But these fats also raise LDL *and* lower HDL, along with raising total cholesterol levels. And the more trans fatty acids you eat, the higher your risk—even just a 2% increase in calories from trans fats can boost your risk of cardiovascular problems by 23%. In the Nurses' Health Study, women who ate diets high in trans fats had a significantly higher risk of heart disease.

Since trans fats are nearly impossible to avoid completely (they are naturally present in dairy and meat products), the American Heart Association recommends limiting them to less than 1% of your total calorie intake. But that's a tough goal to meet, since many hydrogenated or partially hydrogenated oils and food products that include them contain small amounts of trans fat.

One way to monitor your intake is by reading food labels, which are now required to list trans fat content. Still, products with less than 0.5 gram of trans fats are allowed to claim they contain 0 grams trans fat—and eating several of those throughout the day could cause people to unknowingly exceed the maximum recommendation set by the American Heart Association. Still, the amount of trans fats Americans eat has dropped significantly since these labeling laws took effect, likely due to a combination of consumers' increased awareness and manufacturers trying to lower the amount of trans fats present in foods, since it must now be declared. The FDA has also issued a preliminary determination that partially hydrogenated oils are no longer "generally recognized as safe." Recall from Chapter 6 that the manufacturing practice of partial hydrogenation can chemically modify a fat so that it takes on a trans formation.

But people don't make a meal of trans fats, saturated fats, or omega-3 and omega-6 fatty acids; instead, they think in terms of *food*. Some foods are naturally rich in heart-healthy fats, such as nuts. Unsalted peanuts and some tree nuts, specifically walnuts, almonds, pecans, and pistachios, are particularly beneficial, and eating at least five ounces per week (roughly 900 calories) is consistently associated with a lower risk of heart disease. As long as you don't simply add nuts to the foods you're already eating—thereby increasing your fat and calorie intake—they are a good component of a healthy diet, contributing unsaturated fatty acids, fiber, protein, folate, minerals, antioxidants, and phytochemicals. (INFOGRAPHIC 7.9)

Another great food source of healthy fats is fish, especially oily, cold-water fish such as anchovies, sardines, trout, albacore tuna, and salmon. The American Heart Association recommends at least two weekly servings of these fish, which provide omega-3 fatty acids, protein, and other key nutrients. Research shows fish oil—rich in these healthy fats—lowers triglyceride levels, blood pressure, and heart rate, so it's not surprising that eating fish regularly is associated with a lower risk of CVD. Many companies now sell fish oil as a supplement, but it's not yet clear whether these supplements yield the same health benefits as whole fish. Since some types of fish have high levels of mercury and other environmental pollutants, it's important to limit consumption of some types, such as shark, swordfish, king mackerel, and tilefish, which contain high levels of mercury.

FAT INTAKE AND HEALTH—BEYOND CARDIOVASCULAR DISEASE

Since it began, the Bogalusa Heart Study has generated thousands of research studies, and helped train hundreds of students in medicine and public health. Although the study focused on heart disease in childhood, including the effects of diet, eating unhealthy foods is not just associated with heart problems. Many studies have shown that people who eat high-fat diets, particularly diets high in animal fats, are more likely to develop *cancer*. Diet is one of the biggest risk factors for many diseases—including cancer—that we have some control over. (Chapter 9 will provide more information about cancer's links to diet.) Indeed, more than 30% of cancers in adults could be delayed or even prevented by healthy diet, regular exercise, and maintaining a healthy weight. Not surprisingly, a heart-healthy diet is also one that protects against cancer: Specifically, a diet rich in plant-based foods (such as fruits and vegetables), whole grains, and fish appears to be protective, while few fruits and vegetables, extra portions of processed meat, sodium, alcohol, refined carbohydrates, and high amounts of total fat have the opposite effect. Naturally, watching your calorie intake and having a healthy body weight also help.

High-fat diets may also increase the risk of obesity, although that idea remains somewhat controversial. Fat is calorie dense, and any extra is more likely to be stored as body fat than excess protein or carbohydrate. Obesity is a particular problem in Louisiana, which is one of a series of states (those located between Texas and Florida) that epidemiologists refer to as the "diabetes belt," "stroke belt," or "obesity belt." For some reason—no one knows exactly why—CVD seems to cluster in this region.

This is a phenomenon that's now become all too familiar to Gerald Berenson, Principal Investigator of the Bogalusa Heart Study. The many house calls he's made to study participants in Louisiana over the decades have helped him retain an air of the old-fashioned family doctor. "If there was any kind of medical problems I went and examined them myself and took care of them," Berenson recalled in an interview with Tulane's *Global Health News* in 2012. Initially, he had to examine more than 4,000 patients.

Because he grew up there, Berenson knew Bogalusa like the back of his hand. And that firsthand knowledge of this town less than two hours north of New Orleans proved crucial in solidifying support for the study. In fact, nearly everyone in the town has participated in the study in some way: Teachers and nurses at local schools serve as study liaisons; the pathologists who conducted the autopsies had Berenson as an instructor in medical school. Even Berenson and the coroner were old friends, and it was because of

INFOGRAPHIC 7.9 Diet Strategies to Reduce the Risk of Heart Disease *Select foods that can reduce your risk of heart disease.*

Consumption of **plant sterols or stanols** reduces cholesterol absorption and thereby reduces blood cholesterol. They are added to some food products, which are allowed to display an FDA-approved health claim stating that they may reduce the risk of heart disease.

Plant-based diets that provide abundant deeply colored fruit and vegetables supply thousands of phytochemicals (flavonoids) that reduce LDL oxidation and blood clotting.

Oily, cold-water fish such as salmon, trout, and albacore tuna contain **omega-3 fatty acids,** which can lower blood triglycerides and delay clotting.

Moderate alcohol consumption raises HDL cholesterol and reduces the risk of CVD. Moderate drinking is one drink for a woman and no more than two for a man.

Nut consumption has consistently been found to reduce the risk of heart disease. A recent study observed that those who ate nuts nearly every day had a 20% lower risk of dying of heart disease and cancer than those who did not eat nuts.

Whole grain oats, barley, and rye, as well as legumes (beans) are good sources of **soluble fiber,** which can lower blood cholesterol concentrations and reduce the risk of heart disease.

Substituting unsaturated oils for saturated fats improves blood lipids and reduces the risk of CVD.

What two types of foods have been shown to reduce blood clotting? The frequent consumption of what type of food has recently been shown to reduce the risk of death from both heart disease and cancer?

Photo credits (top): Photo Researchers; (left side — top to bottom): Cristian Baitg/Getty Images, svariophoto/Shutterstock, Douglas Allen/Getty Images; (right side — top to bottom): C Squared Studios/Getty Images, Janine Lamontagne/Getty Images, mama_mia/Shutterstock

that relationship that Berenson was able to work out the arrangement that enabled the heart autopsies to be performed. "Eighty percent of the known deaths in the area we were able to autopsy," says Berenson proudly. "Nobody gets that kind of rate."

But to truly help the children in his neighborhood, he knows he needs to address more than just their cardiovascular risk factors. In one very depressing statistic, Berenson found that many kids in Bogalusa start smoking as early as the third grade. That's why Berenson is on a new mission: to get heart disease prevention taught in elementary schools, alongside standard subjects like reading and math. He and his colleagues have developed a curriculum called Health Ahead/Heart Smart that builds on the lessons of the Bogalusa Heart Study and attempts to apply them in practical ways to help *prevent* heart disease.

It's this personalized approach to medicine, combined with Berenson's brand of southern tenacity, that has ensured success of the study over the years. "I'm often asked 'Why Bogalusa?'" The answer is simple, he says: "It's where I'm from."

CHAPTER 7 BRING IT HOME

Fast-food facts

The term "fast food" brings to mind high-calorie, high-fat, and low-nutrition menu items. However, for many, particularly young to middle-aged Americans, eating out is not only considered necessary, but it provides a significant source of their overall nutrition. In a 2013 Gallup poll, 8 in 10 Americans reported eating at fast-food restaurants at least monthly; almost half said they eat fast food at least weekly. Only 4% said they never eat at fast-food restaurants.

Is it possible to find healthful menu items at chain restaurants? Although greater variety and control of ingredients and preparation methods can usually be achieved by preparing food at home, with a bit of planning and awareness, you can still make positive choices when eating out.

Explore and consider

Visit the corporate website for a major fast-food chain (e.g., www.mcdonalds.com; www.wendys.com; www.subway.com; www.in-n-out.com; or many others) and find the nutrition information or facts. Using the chart provided, compare a typical meal you or a friend might order (for example, cheeseburger, large fries, and large cola) with another meal that has a better overall nutrient profile and a more positive—or, at least, less negative—potential impact on blood lipid profile (total, LDL, and HDL cholesterol levels). Nutrition information and suggestions can also be found at http://www.fastfoodnutrition.org/.

Typical restaurant meal at

Food Item or Beverage	Serving Size	Calories	Fat (g)	Percent Calories from Fat* (for meal only)	Saturated Fat (g)	Trans Fat (g), If Available	Fiber (g)	Sodium (mg)
	TOTALS							

Healthier alternative at same restaurant

Food Item or Beverage	Serving Size	Calories	Fat (g)	Percent Calories from Fat* (for meal only)	Saturated Fat (g)	Trans Fat (g), If Available	Fiber (g)	Sodium (mg)
	TOTALS							
	DIFFERENCES (+/−)							

*To calculate percentage calories from fat, multiply grams of fat by 9 and then divide by the total calories times 100.

Consider

1. As you reviewed the nutrition facts information for the restaurant you chose, did calorie or fat content of menu items surprise you? Why or why not?
2. Briefly comment on the differences between the typical and healthier meals you chose.
3. With regard to total fat, saturated fat, and trans fat, discuss the potential impact of the meals you chose on blood lipid levels and heart disease risk.
4. Do you feel it is possible to find a reasonably healthy meal at a fast-food restaurant? Why or why not?
5. Do you feel the posting of calorie information on menu boards will make a difference in what customers order? Do you think requiring calorie information has the potential of lowering risk of heart disease and obesity in the United States? Why or why not?
6. What might you propose on a food industry, policy, or consumer level to encourage not only healthier options, but improved individual food choice behavior at fast-food restaurants?

KEY IDEAS

Atherosclerosis, a thickening and hardening of the arteries along with plaque development along blood vessel walls, is a major cause of heart attack and stroke.

Atherosclerosis is a form of cardiovascular disease (CVD) that often begins when an injury to a vessel wall triggers inflammation and low-density-lipoprotein (LDL) cholesterol infiltration, which results in plaque accumulation.

CVD risk factors increase the risk of vessel injury, as well as increase the rate and extent of the progression of plaque buildup and arterial blockage.

The presence of risk factors increases the risk of developing and promoting CVD. Some risk factors cannot be modified, like race, age, and a family history of heart disease, but certain lifestyle choices can significantly affect the risk of CVD.

Modifiable risk factors for CVD include smoking; a diet high in saturated and trans fats, cholesterol, sodium, and added sugar; a sedentary lifestyle; obesity; diabetes; elevated lipid levels in the blood (cholesterol, LDL, and triglycerides); excessive alcohol consumption; and high blood pressure.

The concentrations of various lipids in the blood correlate with the risk of atherosclerosis and CVD. A high level of total blood cholesterol is a major risk factor, particularly when accompanied by high levels of LDL cholesterol and low levels of high-density lipoprotein (HDL) cholesterol.

Metabolic syndrome is a cluster of risk factors associated with the development of CVD and type 2 diabetes. Diagnosis involves the presence of at least three of the following abnormalities: excessive abdominal fat, high blood pressure, elevated levels of triglycerides in the blood, low levels of HDL, and elevated blood glucose levels.

Diet, including the intake of fatty acids and cholesterol, plays a critical role in the development or prevention of CVD.

Available evidence suggests that replacing saturated fats with monounsaturated and polyunsaturated fatty acids may be beneficial.

The Therapeutic Lifestyle Changes (TLC) program was created by the National Institutes of Health's National Cholesterol Education Program and is also endorsed by the American Heart Association as a heart-healthy regimen that can lower blood cholesterol and reduce the risk of CVD.

The dietary component of the TLC program focuses on lowering saturated fat and dietary cholesterol levels along with other heart-healthy strategies, such as eating foods that contain plant sterols and stanols and sufficient soluble fiber.

Additional dietary strategies to improve blood lipid profiles and reduce the risk of CVD include minimizing intake of trans fatty acids and lowering the ratio of omega-6 fatty acids to omega-3 fatty acids.

The science behind the biological and health effects of dietary fat and fatty acids is complex and continually evolving.

NEED TO KNOW

Review Questions

1. Atherosclerosis is theorized to begin with:
   a. a heart attack.
   b. injury to the lining of the artery.
   c. excessive sugar intake over time.
   d. blood clot formation.
   e. a stroke.

2. All of the following are true with regard to inflammation of blood vessel walls, EXCEPT that it:
   a. can be triggered by damage or injury to the blood vessel wall.
   b. occurs when white blood cells move to vessel walls in response to LDLs in the blood.
   c. causes further damage to blood vessel walls.
   d. makes blood vessels more prone to accumulation of plaque.
   e. helps keep vessel walls flexible and protects against heart attack.

3. Modifiable risk factors do NOT include:
   a. diets that are high in trans fat, sodium, and added sugar.
   b. smoking, sedentary lifestyle, obesity, and high blood pressure.
   c. consumption of excessive amounts of alcohol.
   d. risk factors like family history, increasing age, and race.

4. For adults, a healthy blood cholesterol level is considered to be less than:
   a. 100 mg/dl.
   b. 120 mg/dl.
   c. 160 mg/dl.
   d. 200 mg/dl.
   e. 240 mg/dl.

5. A higher risk of CVD is associated with higher levels of:
   a. linolenic acid.
   b. chylomicrons.
   c. red blood cells.
   d. LDLs.
   e. HDLs.

6. A primary function of HDLs is to:
   a. aid in the digestion of lipids in the small intestine.
   b. transport cholesterol to all cells of the body.
   c. transport cholesterol from tissues to the liver.
   d. transport triglycerides to adipose tissue.
   e. link amino acids to form proteins.

7. All of the following are among the cluster of conditions that characterize metabolic syndrome, EXCEPT:
   a. excessive abdominal fat.
   b. high blood pressure.
   c. prediabetes or type 2 diabetes.
   d. low levels of LDL cholesterol in the blood.
   e. elevated levels of triglycerides in the blood.

8. Current evidence indicates that reducing saturated fats in the diet is most effective in reducing the risk of CVD when:
   a. saturated fats are replaced by unsaturated fats.
   b. all dietary cholesterol is eliminated.
   c. saturated fats are replaced by carbohydrate-rich foods.
   d. saturated fats are replaced by low-fat breads and cereals.
   e. total fat intake is kept under 20% of calories from fat.

9. The Therapeutic Lifestyle Changes (TLC) dietary component for individuals with increased risk of CVD focuses on all of the following, EXCEPT:
   a. foods that contain plant sterols and stanols.
   b. sufficient soluble fiber intake.
   c. lowering intake of monounsaturated fats.
   d. lowering intake of dietary cholesterol.
   e. lowering saturated fat to less than 7% of total calories.

10. Trans fatty acids:
    a. are recommended to make up at least 5% of daily calories from fat.
    b. shorten the shelf life of food products.
    c. are currently not required to be listed on the Nutrition Facts panel of food labels.
    d. are found only in processed foods, not found naturally in any other foods.
    e. can raise LDLs and lower HDLs.

Take It Further

Contrast two 45-year-old men, one with a low risk of CVD and one with an elevated risk of CVD. List five ways in which diet, exercise, and other lifestyle choices might have increased the risk of CVD for one of these men more than the other.

Dietary Analysis Using SuperTracker

Focus on fats for heart health

The lipid content of our diet is a key component of any healthy diet, but it's important to know if you're getting too little or too much. In this exercise, you will evaluate your lipid intake, break down the various types of fats you consume on an average day, and compare them with heart-healthy dietary recommendations. You will also be evaluating your diet for CVD risk and will identify appropriate changes to reduce these risks.

1. Log onto the United States Department of Agriculture (USDA) website at www.supertracker.usda.gov. Click this link. If you have not done so already, you will need to create a profile to get a personalized diet plan. This profile will allow you to save your information and diet intake for future reference. *Do not use the general plan.*
2. Click the Track Food and Activity option.
3. Record your food and beverage intake for one day that most reflects your typical eating patterns. Enter each food and beverage you consume into the food tracker. Note that there may not always be an exact match to the food or beverage that you consumed, so select the best match available.
4. Once you have entered all of your food and beverage choices into the food tracker, on the right side of the page under the bar graph you will see Related Links: View by Meal and Nutrient Intake Report. Print these reports and use them to answer the following questions.

You may use the analysis and reports you generated for the Supertracker activity in Chapter 6 to help you complete this activity.

a. What percent of your total calories came from dietary fat? How does this compare with the 25% to 35% recommended by the American Heart Association (AHA)?
b. What percent of your total calorie intake came from saturated fat? Does this exceed the AHA's limit of 7% of total calories? If so, what foods contributed the most saturated fat to your diet?
c. What percent of your total calorie intake came from monounsaturated fat? Is this greater than your intake of saturated fat?
d. What percent of your total calorie intake came from polyunsaturated fat? Is this greater than your intake of saturated fat?
e. Did you meet the recommended target for omega-3 fatty acids? Identify at least three food sources of omega-3 fatty acids.
f. How much dietary cholesterol did you consume on this day? Does this exceed the limit of 300 mg? If so, what foods contributed the most cholesterol to your diet?
g. Health experts recommend that we minimize our intake of trans fats. What information can you look for on food labels to indicate whether a food product contains trans fat?
h. Did you meet the target for dietary fiber for this day? Sufficient fiber intake, particularly soluble fiber, is an important part of a heart-healthy diet. Identify at least three foods that provide an excellent source of soluble fiber.

8 PROTEIN

How Much of a Good Thing Do We Need?

EXPERTS CONSIDER HOW MUCH PROTEIN IS BEST FOR US ALL.

Monkey Business Images/Shutterstock

LEARNING OBJECTIVES

- Discuss at least four functions of protein in the body (Infographic 8.1)
- Distinguish between essential and nonessential amino acids (Infographic 8.2)
- Describe the primary steps in protein synthesis and what determines the shape of a protein (Infographics 8.3 and 8.4)
- Explain denaturation and how it may alter protein function (Infographic 8.5)
- Summarize protein digestion and absorption (Infographic 8.6)
- Identify the Recommended Dietary Allowances and Acceptable Macronutrient Distribution Ranges for protein for adults (page 164)
- Explain protein turnover and how amino acids may be used for energy (Infographics 8.7 and 8.8)
- Describe protein deficiency diseases and identify regions in the world where this condition is prevalent (Infographic 8.12)

Sometimes it's nice to start the day with a fried-egg sandwich, topped with cheese, lettuce, tomatoes, and fried onions. But imagine starting every day with *three* of those sandwiches. Then add a five-egg omelet, a bowl of grits, three slices of French toast, three chocolate chip pancakes, and two cups of coffee.

This was the regular breakfast of Michael Phelps—the American swimmer who has won more Olympic medals than any other athlete in history—as he was training for the 2008 games in Beijing. In all, it was a 4,000-calorie meal, replete with fat, carbohydrates, and other nutrients to fuel his hours-long training sessions. But many athletes believe that the one key element is protein.

Without protein, we wouldn't be able to breathe, contract our muscles, or complete numerous basic functions. It carries out the biological instructions written in our genes. Given protein's vital importance, it's crucial that people get enough in their diet—but, unlike our intake of some of the other nutrients (such as certain vitamins and minerals)—getting enough protein typically isn't a problem in the United States. Most of us eat plenty of protein to meet our needs—and in some cases, much more.

This seems particularly true for competitive athletes, like Michael Phelps. The amount of protein he ate from eggs alone, just during breakfast, nearly matches what most 150-pound adults need for an entire day. And that's not counting the cheese, any milk he adds to his coffee, and other extras he squeezes in. Plus, there's lunch and dinner.

Many athletes who don't have the time and Olympic staff to help them plan appropriate diets choose to boost their protein intake by mixing powders, guzzling shakes, and noshing on high-protein bars. If more is better, they reason, why not take in as much as possible?

But some experts disagree over whether that extra protein is serving any useful purpose in the body. If not, athletes—sometimes on strict budgets—may be wasting their money on supplements and extra servings of protein-rich foods. Plus, different organs and systems use protein in different ways, and it's not clear whether such mega-doses, over the long haul, disrupt vital processes in the body. Could a protein-rich diet hurt more than just your budget?

■ ■ ■

__What's your number?__ Adults with a healthy body weight can meet the RDA for protein by consuming 0.8 grams of protein per kilogram of body weight. The calculation is simple. Divide your current body weight in pounds by 2.2 to convert to kilograms, and then multiply that number by 0.8.

Do the math

If weight in pounds is 170: 170/2.2 = 77.3 kg

77.3 kg × 0.8 = 61.84 grams of protein per day

See Appendix 1 for a listing of Recommended Dietary Allowances for protein for all age groups.

Monkey Business Images/Shutterstock

Whether you're an athlete, or maintain a lightly active lifestyle, adequate protein is critically important to your body for many reasons. Proteins are responsible for the majority of dynamic and adaptive processes that keep us alive and functioning. Much of the structural material in the body is provided by protein; its constituents are found in muscle, bone, hair, skin, and fingernails. They also carry out critical functions, such as facilitating chemical reactions through enzymes, regulating most body and cellular functions by hormones, maintaining fluid balance, fighting off bacteria, and blood clotting. (INFOGRAPHIC 8.1)

RECOMMENDATIONS FOR PROTEIN INTAKE

Protein needs are relatively high during growth and development periods, such as infancy or during pregnancy. But for adults with a healthy body weight, the current recommended dietary allowance (RDA) equates to 0.8 grams of protein per kilogram of body weight per day (g/kg/d). For example, the RDA for protein for a male or female weighing 70 kilograms (154 lbs; 1 kg = 2.2 lbs) would be 56 grams, approximately the amount of protein in an 8-ounce steak or a 4-egg omelet with ham and cheese.

The Acceptable Macronutrient Distribution Range (AMDR) for protein is 10% to 35% of

INFOGRAPHIC 8.1 Roles of Protein in the Body *The thousands of proteins in the body serve to carry out and regulate many critical processes and provide structural material for cells, and on a larger scale, allow the body to move.*

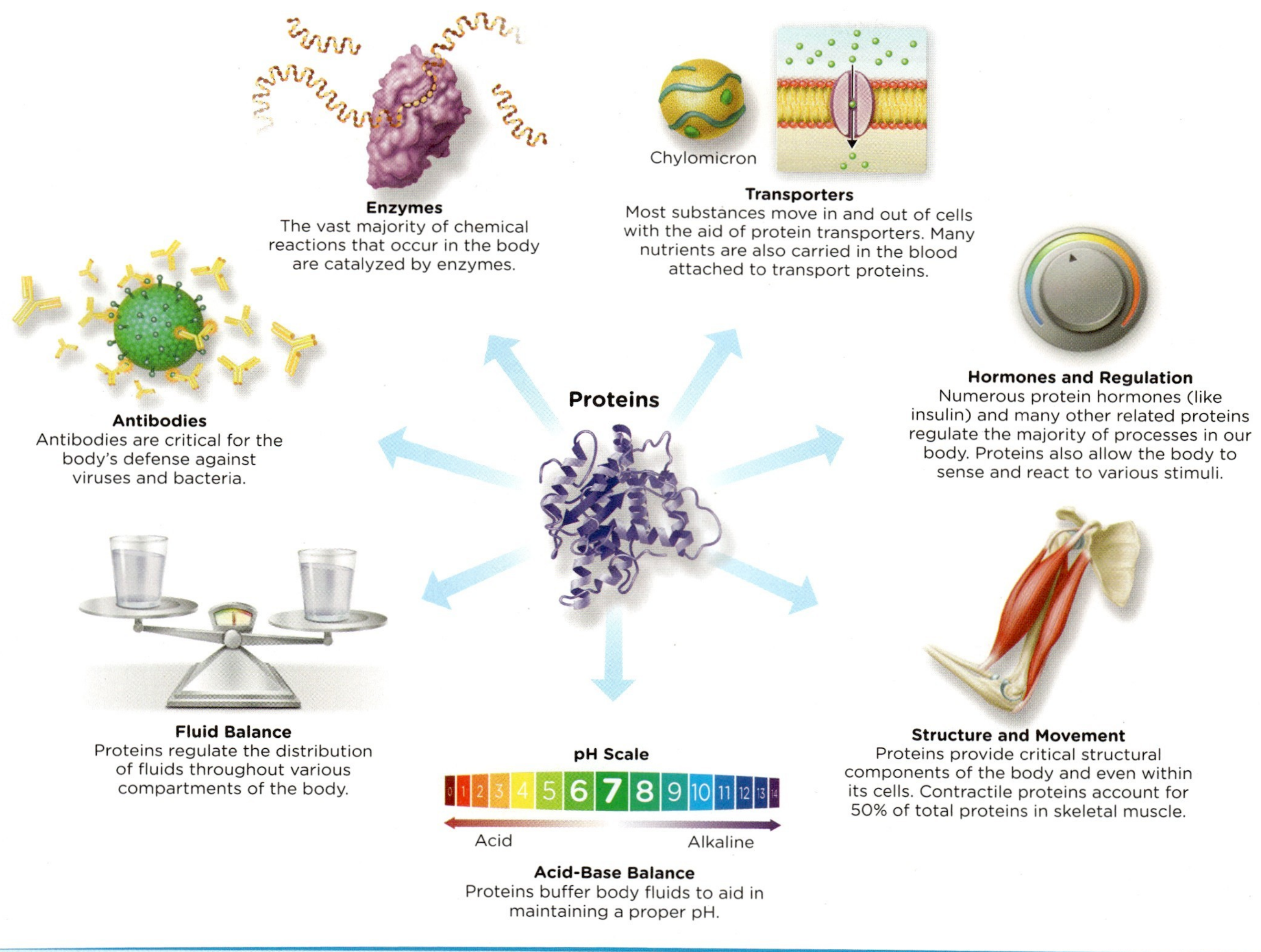

What do proteins do in the plasma membrane of the cell?

total kcal. In the United States, recent nutrition surveys report an average protein intake close to 16% of total calories, with men consuming about 100 grams and women consuming close to 70 grams. These protein intake levels are within the AMDR, but well above the RDA for most people. The overconsumption of protein by teen boys and adult men in particular has prompted the 2015 Dietary Guidelines for Americans to recommend that they specifically reduce consumption of protein foods by decreasing intake of meats, poultry and eggs.

STRUCTURE OF PROTEIN

Like carbohydrates and fats, **protein** is composed of carbon, hydrogen, and oxygen. But there's another key element that distinguishes protein from other macronutrients: nitrogen. The nitrogen is supplied by **amino acids**, which are the building blocks of protein.

Every amino acid consists of a central atom of carbon (C), an amino group, which contains nitrogen (written as the chemical formula NH_2); an acid group (COOH), and a variable *side chain*. Side chains may be as simple as a single atom, or a group of as many as 19 atoms. It's the side chains that distinguish one amino acid from another;

PROTEIN
a large polymer made up of a chain of amino acids; consists of carbon, hydrogen, oxygen, and nitrogen

AMINO ACIDS
molecules of carbon, hydrogen, oxygen, and nitrogen that join together to form a protein

INFOGRAPHIC 8.2 Amino Acids Are the Building Blocks of Protein

Proteins are macronutrients made up of amino acids. They contain carbon, hydrogen, oxygen, and nitrogen.

AMINO ACID STRUCTURE

All amino acids contain a central carbon atom, an amino group that contains an atom of nitrogen, an acid group, a hydrogen atom, and a side chain.

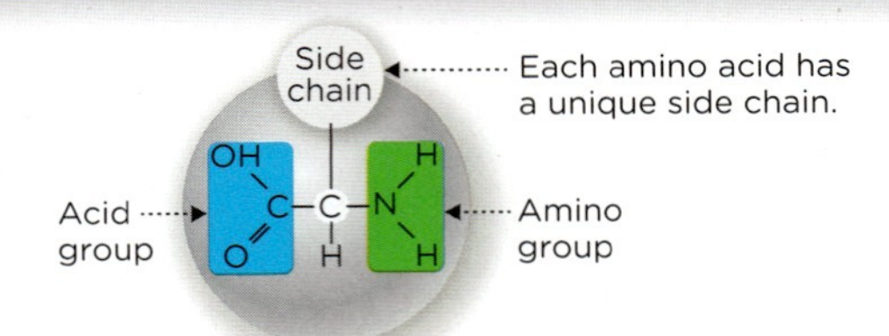

General structure of an amino acid

UNIQUE AMINO ACID SIDE CHAINS

Only the side chain differs for each of the 20 amino acids, giving each its unique properties.

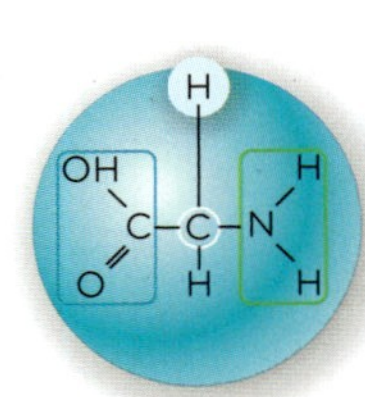

Glycine (Gly)

Serine (Ser)

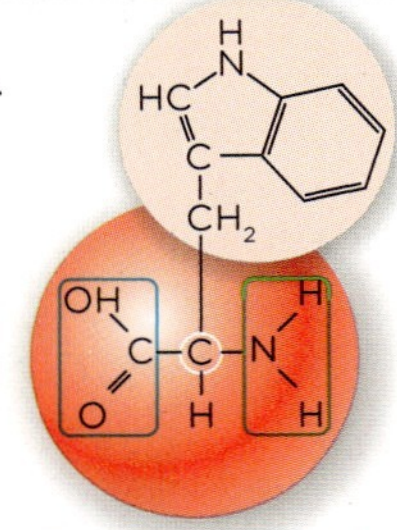

Tryptophan (Trp)

PEPTIDE BONDS

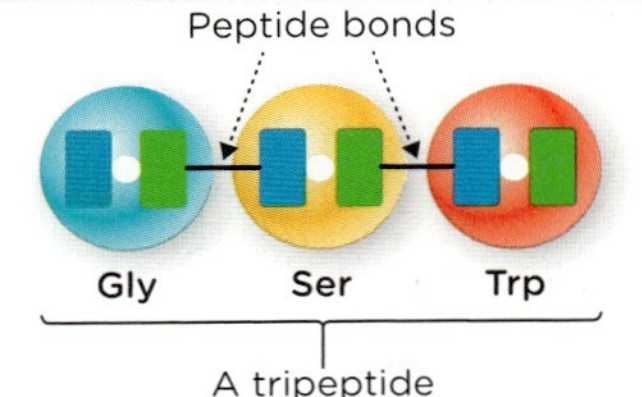

A peptide bond forms when the acid group (COOH) of one amino acid is joined with the amino group (NH_2) of another amino acid.

AMINO ACIDS OF THE HUMAN BODY

Nine amino acids must be obtained from food, and are called essential amino acids.

Essential Amino Acids		Nonessential Amino Acids	
• Histidine	• Phenylalanine	• Alanine	• Glutamine
• Isoleucine	• Threonine	• Arganine	• Glycine
• Leucine	• Tryptophan	• Asparagine	• Proline
• Lysine	• Valine	• Aspartic acid	• Serine
• Methionine		• Cysteine	• Tyrosine
		• Glutamic acid	

? *What is a chain of amino acids called?*

PEPTIDE BOND
the bond that forms between the acid group of one amino acid and the amino group of another amino acid

PEPTIDE
short chain of amino acids attached together

ESSENTIAL AMINO ACIDS
amino acids that cannot be produced by the human body and therefore must be obtained from food

NONESSENTIAL AMINO ACIDS
amino acids that the body can make and therefore need not be obtained through diet

PROTEIN SYNTHESIS
the process of building peptide chains and proteins from amino acids using information provided by genes in a two-step process of transcription and translation

TRANSCRIPTION
process by which information encoded in genes (DNA) is used to make messenger RNA

DEOXYRIBONUCLEIC ACID (DNA)
nucleic acid that stores the body's genetic information; made of a double strand of nucleotide subunits

for example, tryptophan from serine. (INFOGRAPHIC 8.2)

Approximately 22,000 different proteins carry out the structural and functional roles of the body. Proteins are synthesized by linking up to 20 different amino acids by **peptide bonds** into chains of varying lengths. Short chains of amino acids are called **peptides**: a dipeptide has two amino acids, a tripeptide has three amino acids, and a polypeptide has many amino acids. All proteins are polypeptides. The sequence of the amino acids and the types of amino acids distinguish one protein from another. This can be compared with how the sequence and distribution of letters in the alphabet distinguish one word from another.

We need all of those 20 amino acids to make the necessary proteins, but we can get them in different ways. Nine of the 20 are considered **essential amino acids**—we must get them from the foods we eat because they cannot be produced by the human body. The rest are **nonessential amino acids**—sometimes called "dispensable" amino acids—because they can be manufactured by the body.

PROTEIN SYNTHESIS

When we eat protein in foods, the body breaks them down into amino acids and then uses the amino acids to produce the particular protein needed. The nearly 22,000 proteins produced in the human body are generated through a two-step process of **protein synthesis**. In the first step of protein synthesis, known as **transcription**, segments of **deoxyribonucleic acid (DNA)** called genes provide the "instructions" for the assembly of amino acids into particular proteins. These instructions are transcribed into **messenger ribonucleic acid (mRNA)**. In the second step of protein synthesis, called **translation**, ribosomes translate the instructions into proteins. **(INFOGRAPHIC 8.3)**

MESSENGER RIBONUCLEIC ACID (mRNA)
the type of RNA that carries the genetic code for a specific protein from the nucleus to the cytoplasm where proteins are made

TRANSLATION
process by which mRNA is decoded by ribosomes to synthesize proteins in the correct amino acid sequence

INFOGRAPHIC **8.3** **Protein Synthesis** *In cells, sequences of DNA called genes provide the instructions for the synthesis of every protein in our body. This is a two-step process that begins in the nucleus with gene transcription and is completed in the cytoplasm with translation.*

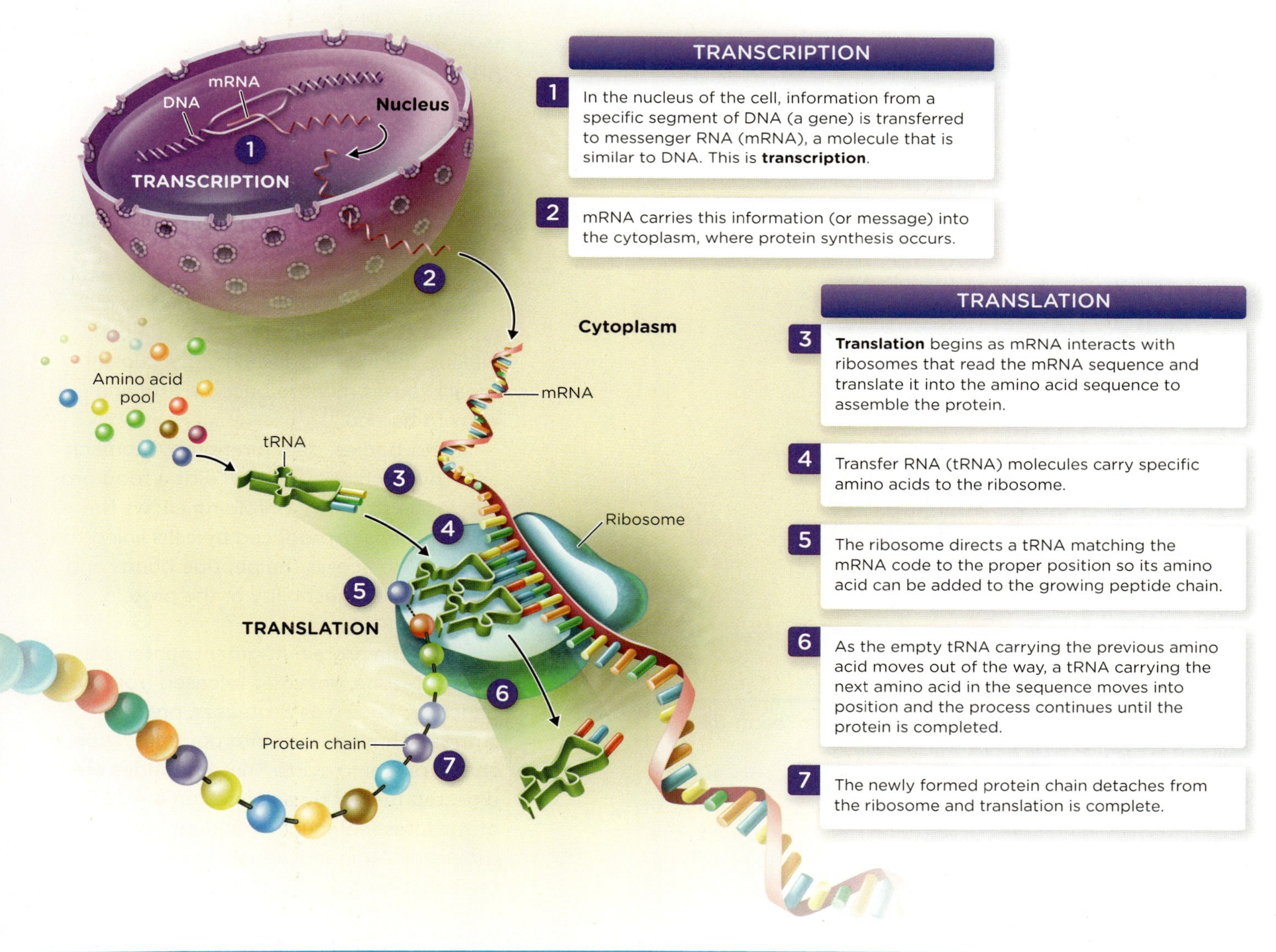

? *What information does mRNA provide during protein translation?*

INFOGRAPHIC 8.4 Protein Folding *The overall shape of a protein molecule determines its function, and how it interacts with other molecules. For proteins to function properly, they must retain their three-dimensional shape.*

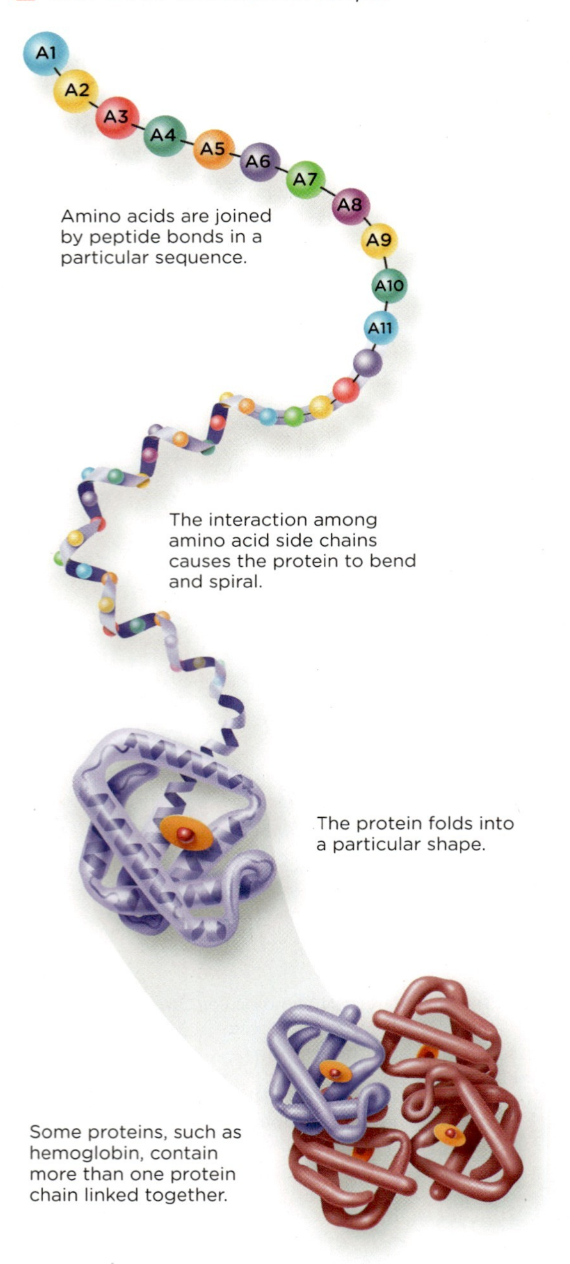

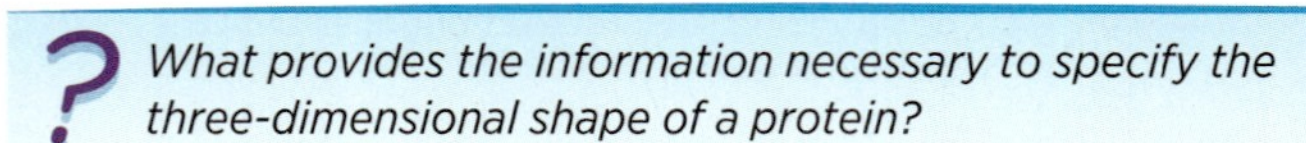

Once the synthesis of the protein is completed, it has not necessarily taken its final form. The unique nature of each amino acid in the sequence prevents the protein from remaining in a straight line. Interactions between these amino acids in the sequence cause the protein to fold into a three-dimensional shape. The shapes determine the function of the protein. (INFOGRAPHIC 8.4)

DIGESTION AND ABSORPTION OF PROTEINS

The original shape of the protein is not necessarily permanent. Heat, light, change in pH, alcohol, or motion—such as beating egg whites—will change the three-dimensional structure, a process known as **denaturation**. Very high fevers or extreme changes in the pH of blood may be sufficient to denature proteins within our body and cause them not to function properly. For example, heat or a change in pH could destroy the function of enzymes or damage the protein hemoglobin, which carries oxygen in the blood.

It's easy to visualize protein denaturation in the kitchen—adding lemon juice to milk renders it more acidic (changes the pH), making the milk curdle. Similarly, cooking (heat) causes meat to become firm and eggs to harden.

In our bodies, denaturation by stomach acid (hydrochloric acid) unfolds a food protein in the stomach. Unfolding allows the enzyme *pepsin,* produced by cells lining the stomach, to access the peptide bonds and increase the digestibility of the protein. (INFOGRAPHIC 8.5)

Once the protein fragments enter the small intestine, enzymes released from the pancreas, known as *proteases*, break the strings of amino acids into peptides, short chains of amino acids. These peptides are digested further by enzymes on the surface of the intestinal mucosa, and dipeptides, tripeptides, and individual amino acids are then absorbed by mucosal cells of the small

Sunnyside up. *You can watch denaturation in action by observing an egg change from a gelatinous liquid to a rubbery solid when heated.*

Valentina Proskurina/Shutterstock

DENATURATION
disruption of the three-dimensional shape of a protein, usually by heat, light, acid, or chemical reaction

INFOGRAPHIC **8.5 Protein Denaturation** *Denaturation by stomach acid allows the enzyme pepsin to cut protein into shorter strands.*

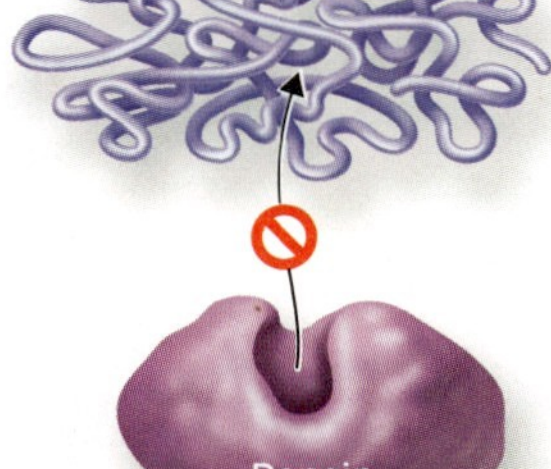

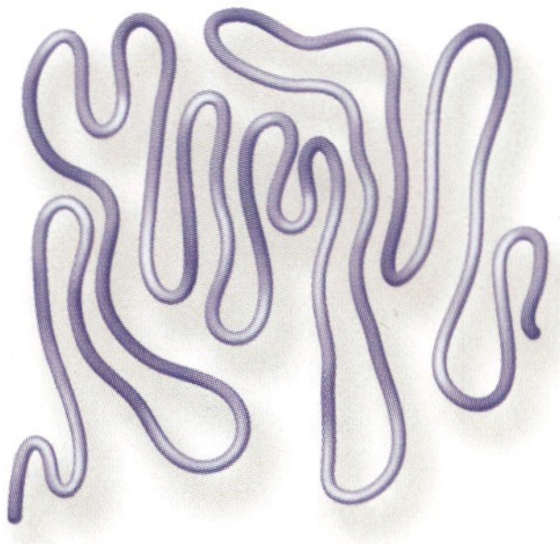

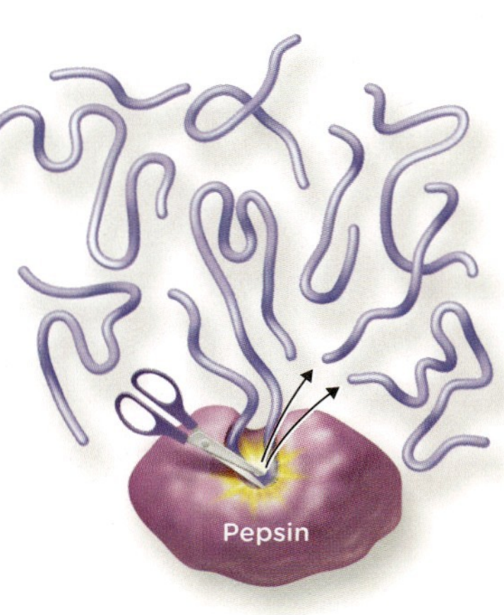

Atrophic gastritis is a condition that results in insufficient hydrochloric acid production by the stomach. Describe why this condition would decrease the efficiency of protein digestion.

intestine with the assistance of various transport proteins. From the mucosal cells the amino acids are transported into the blood by additional transport proteins. **(INFOGRAPHIC 8.6)**

VARIED FUNCTIONS OF PROTEIN

Given the essential role of protein in body function and structure, it's no wonder that athletes have often assumed that more is always better. This became crystal clear to

INFOGRAPHIC 8.6 Protein Digestion and Absorption *The digestion of proteins begins in the stomach and is completed within the mucosal cells lining the small intestine.*

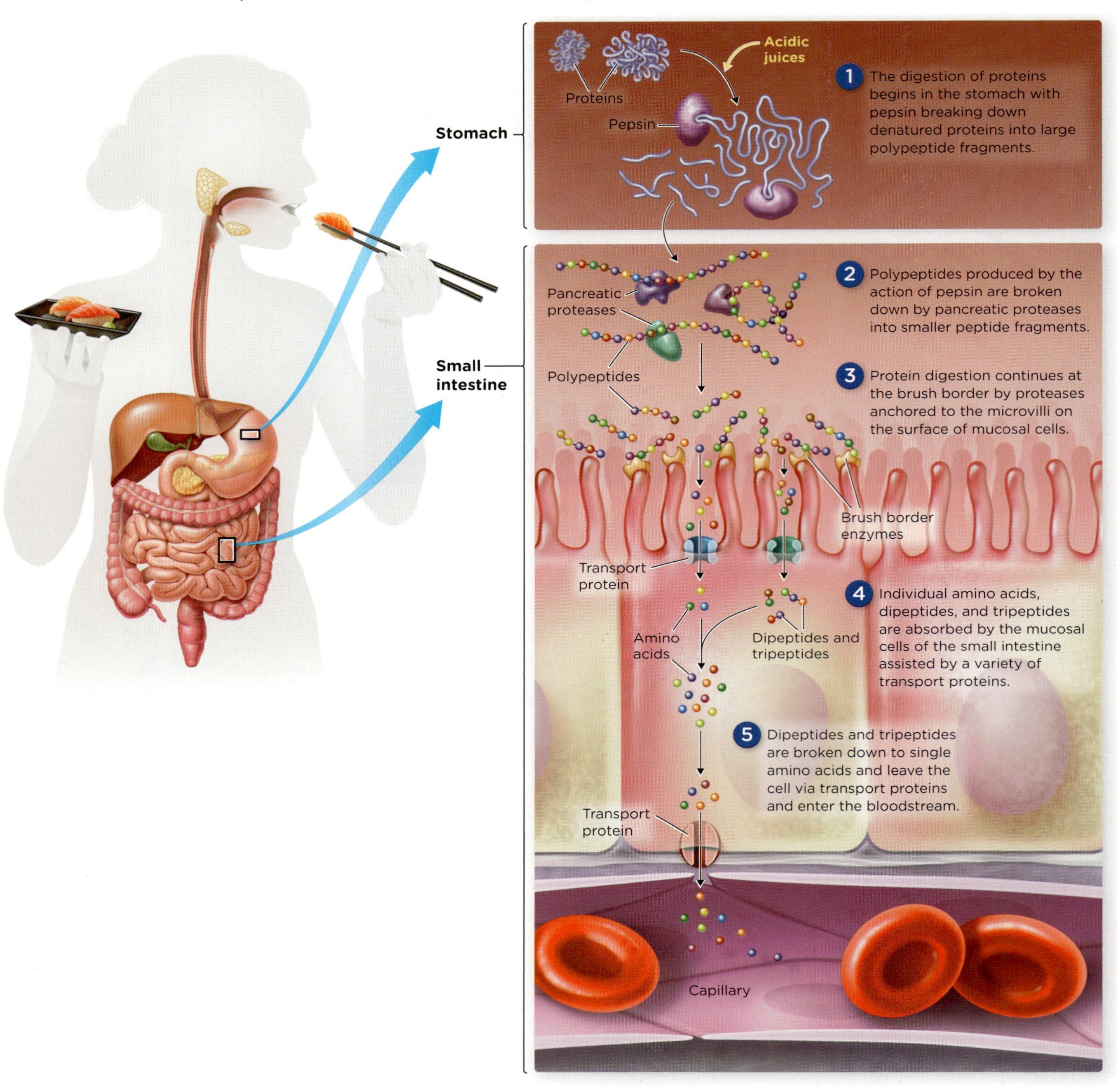

? *What is the largest protein fragment that can be absorbed into the mucosal cells lining the small intestine?*

Robert Wolfe, PhD, who holds the Jane and Edward Warmack Chair in Nutritional Longevity at the University of Arkansas for Medical Sciences. After talking to bodybuilders at a conference, he calculated that they were getting more than five times the amount of protein recommended for a normally active person.

These bodybuilders ate huge amounts of protein because they thought they needed it to build and maintain their muscles. Indeed, protein serves as a structural part of skeletal muscle and bone (along with many tissues, organs, blood cells, hair, and nails). In addition, two proteins, actin and myosin, compose about 50% of total muscle proteins and they are the primary proteins that enable muscles to contract.

Although protein intake at five times the recommended intake level is certainly excessive and unnecessary, there are "some studies showing that if you're doing weight lifting or other power sports, a higher protein intake helps you gain muscle faster," says Wolfe.

But proteins have functions that extend beyond their contractile role in muscles: enzymes catalyze chemical reactions in the body, hormones regulate nearly every bodily function, hemoglobin transports oxygen, and albumin transports a variety of nutrients. Proteins also have central roles in immunity and blood clotting. And proteins help maintain a proper balance of fluid around the body, both inside and outside cells, which wards off swelling (such as in the ankles) and supports a normal blood pressure.

Uncomfortable fluid retention. *Edema is swelling caused by excess fluid trapped in body tissues. When a protein in blood, called albumin, gets too low, fluid is retained in extremities and edema can occur.*

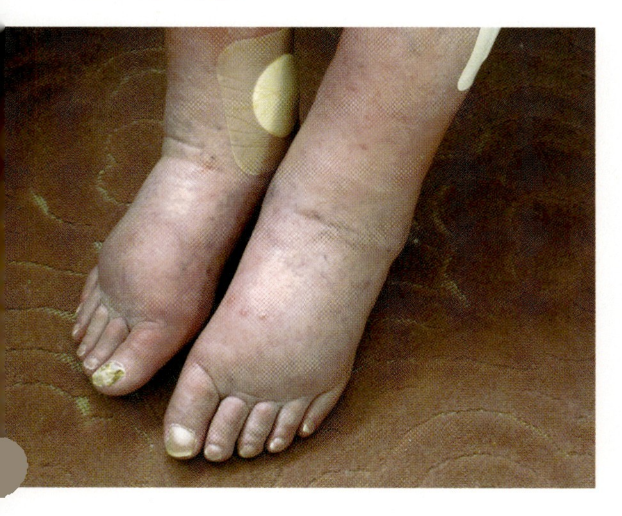

PROTEIN TURNOVER AND NITROGEN BALANCE

The body needs amino acids from the diet to replace proteins that are lost when cells from our skin and those lining our gastrointestinal tract are shed. Dietary proteins are also needed to allow for the accumulation of additional body protein mass that occurs with growth, pregnancy, increasing muscle mass, and to support the growth of hair and nails, as well as for wound healing.

It is important to emphasize that proteins are synthesized as needed to support necessary body functions, so that consuming protein in excess of need will not increase the amount of proteins made. In other words, excess amino acids are not stored in our body as proteins. Excess amino acids are used as an energy source or stored as fat.

Proteins in the body are constantly being broken down and reassembled in a process called **protein turnover**. In fact, many of the amino acids used to make proteins don't come from the food we eat each day, but are drawn from a pool of amino acids obtained from the breakdown of the body's own proteins. Although we consume about 70 grams to 100 grams of protein daily, approximately 300 grams of proteins in cells and fluids throughout the body are broken down and resynthesized each day. Most of the amino acids released by the breakdown of body proteins are reused in the production of new proteins, but some amino acids are metabolized (chemically altered), which prevents them from being used for protein synthesis. These modified amino acids must be replaced

PROTEIN TURNOVER *the continuous breakdown and resynthesis of proteins in the body*

INFOGRAPHIC 8.7 Protein Turnover *Proteins throughout the body are continually broken down into amino acids and resynthesized.*

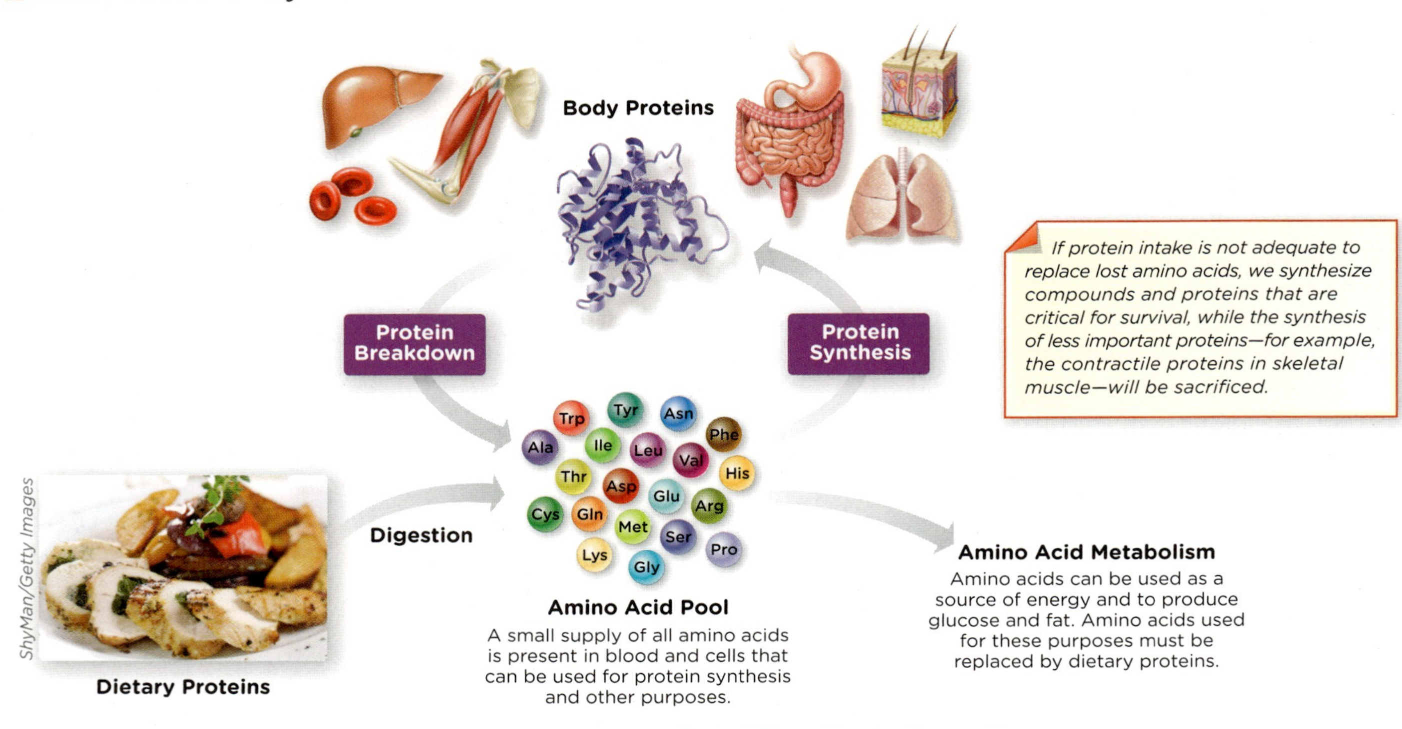

? *If amino acids generated from protein breakdown are reused for protein synthesis, why is it necessary for a nongrowing adult to eat any protein?*

with dietary proteins to provide sufficient amino acids to remake all the body proteins that were broken down. **(INFOGRAPHIC 8.7)**

In many cases, the metabolism of amino acids requires that they first be stripped of their amino group, which leaves a *carbon skeleton.* This carbon skeleton is used primarily to synthesize glucose to maintain blood glucose levels when carbohydrate intake is low, or to a lesser degree, as a direct source of energy. In times of energy abundance, the body may also convert amino acids to fatty acids that are then stored in adipose tissue as triglycerides. **(INFOGRAPHIC 8.8)**

When amino acids are used for energy, or to synthesize glucose or fatty acids, the amino group that was stripped off must be disposed of; otherwise it would accumulate in the body as ammonia, which is toxic. To prevent this, the liver converts ammonia to a less toxic substance called urea. Urea is then released into the blood, filtered by the kidneys, and excreted in urine.

Scientists can measure urea in urine to study protein turnover in what is called a *nitrogen balance study*. As the name implies, **nitrogen balance** reflects if a body is gaining, losing, or maintaining protein. Protein intake is measured, as well as nitrogen output in urine and feces. Nitrogen losses from less significant sources (sweat, skin, hair, nails, breath, saliva, mucus, and other secretions) are estimated. As nongrowing, weight-stable adults, we have approximately the same amount of total protein in our body from day to day, so we are in nitrogen balance, such that the amount of nitrogen we

NITROGEN BALANCE *a measure of nitrogen intake (primarily from protein) minus nitrogen excretion.*

INFOGRAPHIC 8.8 Fate of an Amino Acid *Amino acids are metabolized (chemically altered) to produce many important compounds. When used as a source of energy, or to synthesize glucose or fat, the first step in their metabolism is to remove the amino group.*

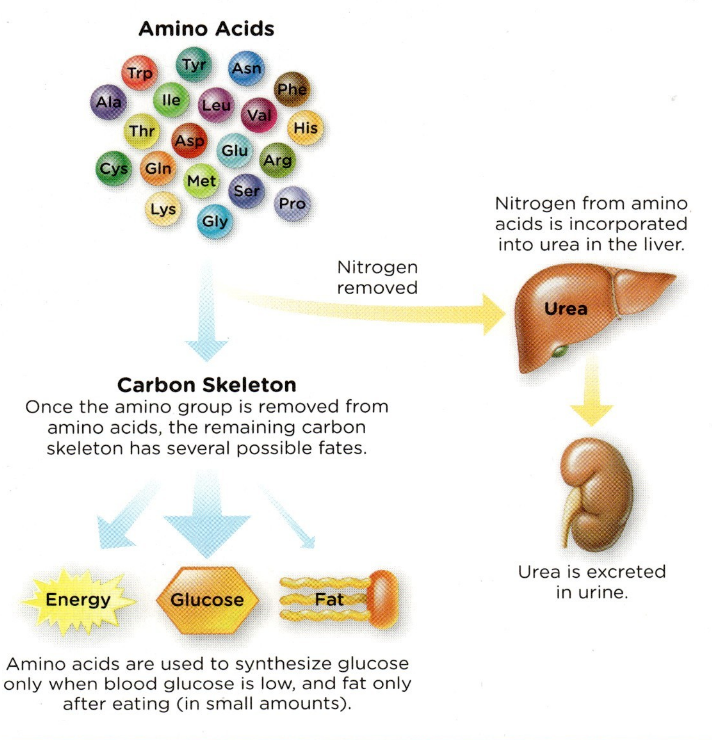

? *What is a toxic waste product of amino acid breakdown that must be excreted from the body?*

consume (N_{in}) is equal to the nitrogen we excrete (N_{out}), or $N_{in} = N_{out}$. In contrast, a growing child, a pregnant woman, or someone who is just starting a resistance-training (weight-lifting) program will be increasing their mass of total body proteins, so they must excrete less nitrogen than they consume, or N_{in} is greater than N_{out}. **(INFOGRAPHIC 8.9)**

■ ■ ■

Instead of lab-based nitrogen balance tests, researchers have typically measured protein requirements for competitive athletes in their normal environment, says Linda Lamont, PhD, professor of kinesiology at the University of Rhode Island in Kingston. They may give athletes some packaged meals, or ask them to write down everything they are eating, and give them a container to collect their urine. Based on these experiments, some researchers have concluded that highly active athletes might need more than the 0.8 g/kg/d RDA for protein. Instead, they might recommend that, for performance advantage, athletes get between 1.2 and 1.7 g/kg/d to help with endurance and strength, respectively. For the average 150-pound adult (approximately 68 kg), that's a shift from about 50 to more than 100 g/kg/d. However, most athletes do not need to make significant alterations in their diet to meet this level of protein intake as the average intake of many American's falls within this range. Furthermore, the increased calorie intake of athletes naturally leads to an increase in protein intake as well.

INFOGRAPHIC 8.9 **Nitrogen Balance** *Three factors are required to retain body proteins (lean body mass): adequate diet, hormones and growth factors, plus muscle contractions.*

POSITIVE NITROGEN BALANCE

For tissue growth to occur, less nitrogen must be excreted than what is taken in. This means that protein synthesis must exceed protein breakdown. (Nitrogen in > Nitrogen out)

NITROGEN EQUILIBRIUM

In an active, healthy adult, body weight and lean body mass are not changing and protein synthesis and breakdown must be equal. For body weight and lean body mass to remain constant, the amount of nitrogen excreted must be equal to what is consumed. (Nitrogen in = Nitrogen out)

NEGATIVE NITROGEN BALANCE

The decrease in activity that often accompanies aging results in the loss of skeletal muscle mass. Any illness or dietary inadequacies that may exist will further accelerate this loss as well as lead to a loss of proteins from many other tissues and organs. Since body mass is decreasing, particularly lean body mass, protein breakdown exceeds protein synthesis. (Nitrogen in < Nitrogen out)

What will be the nitrogen balance status of someone who is on a low-calorie diet to lose excess body fat?

This methodology has Lamont concerned. For one, when researchers repeat the experiments in the laboratory on the same athletes, they often get different results. That's not a surprise, she notes—a field experiment performed outside a controlled laboratory setting can't possibly measure all of the protein athletes consume and excrete. Furthermore, protein metabolism is affected by sex, age, total calorie intake, and factors related to exercise, such as how long and hard you work out, she adds. There is even evidence that athletes may need less of a protein-rich diet than the average person, since regular training renders the body more efficient at using amino acids.

However, many researchers and nutrition professionals believe extra amounts of protein can help highly competitive athletes who need to get maximum performance out of their bodies. How much protein athletes should eat is really two questions, says Stuart Phillips, PhD, professor of kinesiology at McMaster University in Ontario—how much they have to eat to replace what they lose during exercise and normal metabolism throughout the day, and how much will boost their performance. "Athletes don't *need* any more protein than the average person. What an athlete can benefit from," he says, "is a higher protein intake than the average person—which I call an *optimal protein intake*."

SENIORS MAY BENEFIT FROM HIGHER INTAKES OF PROTEIN

Athletes aren't the only individuals who may benefit from protein intakes that exceed the RDA. A higher protein intake for adults older than 65 or 70 years appears to be of benefit in maintaining a healthy body weight and protecting against frailty. Because of decreases in physical activity and a number of other factors, most elderly individuals experience a loss of lean body mass, primarily from skeletal muscle. This decline is of particular concern in the elderly as the loss of lean body mass is associated with a high risk of disability and death. A number of studies demonstrate that slightly higher protein intakes (approximately 1.2 g/kg/d) in adults older than 65 years can effectively reduce this loss of lean body mass, improve functionality, and reduce the risk of disability and death, particularly when combined with a resistance training program. However, recent research indicates that for middle-aged adults (ages 50 to 65) higher protein intakes do not offer added benefit. In this age group, consuming protein in line with the RDA (0.7–0.8 g/kg/d) is actually beneficial for the prevention of cancer, overall mortality, and possibly diabetes. This level of protein intake is significantly less than most U.S. adults currently consume. Health benefits of both a higher protein intake in older adults and lower protein intake for middle-aged adults are maximized when plant foods make up the majority of food intake.

Adults over 65 have special protein needs. *Protein intakes of approximately 1.2 g/kg/d in adults older than 65 years can effectively reduce loss of lean body mass, improve functionality, and reduce the risk of disability and death, especially when combined with a resistance training program.*

Anne Clark/Getty Images

PROTEIN QUALITY

More than 60% of the protein Americans consume is from animal products, such as meat and dairy foods. Combined with the protein found in plant foods, most of us exceed the

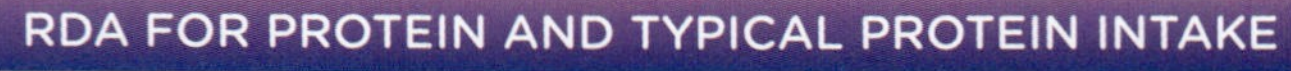

INFOGRAPHIC 8.10 Protein: What's Your Strategy? *Most Americans exceed their RDA for protein.*

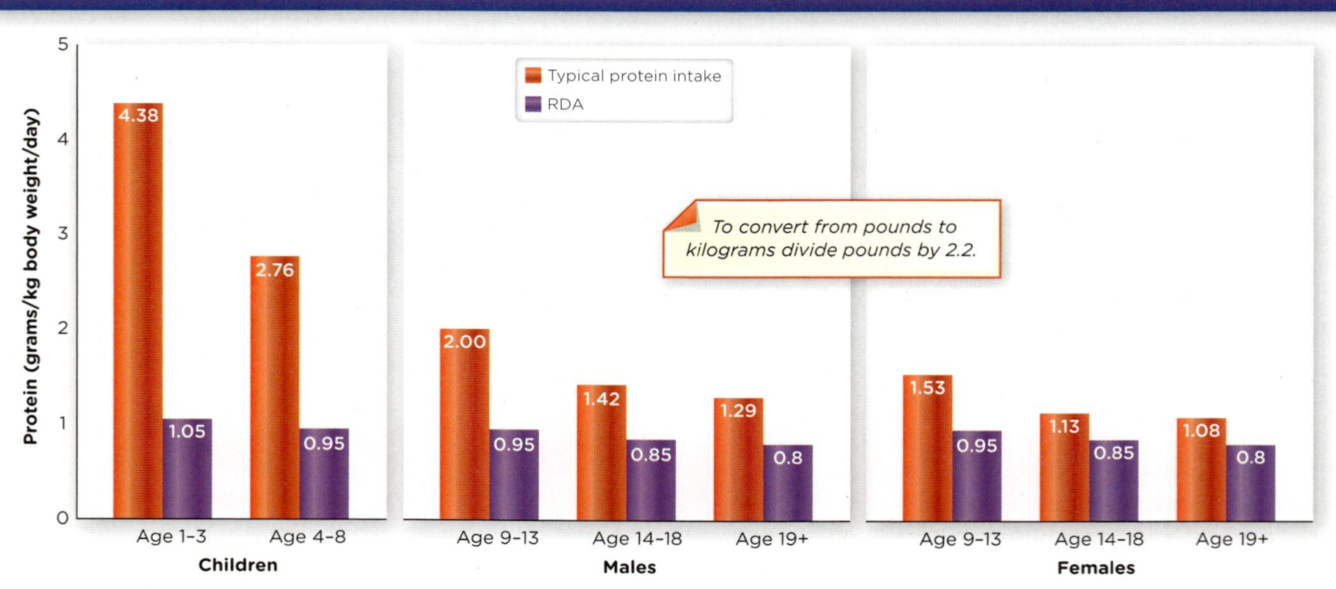

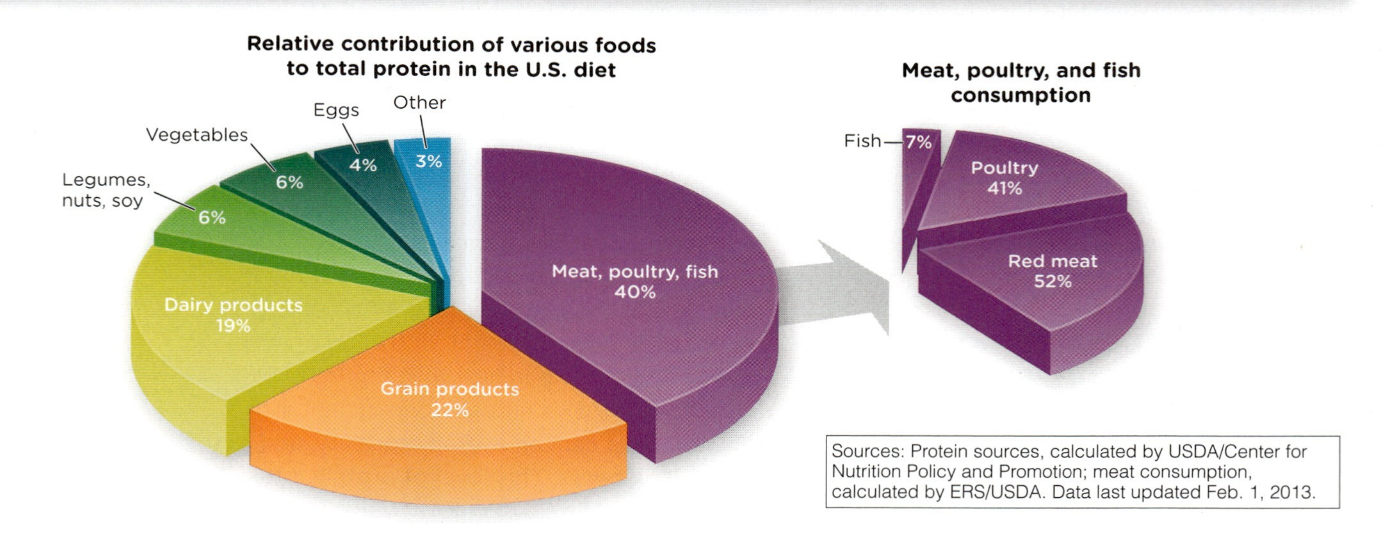

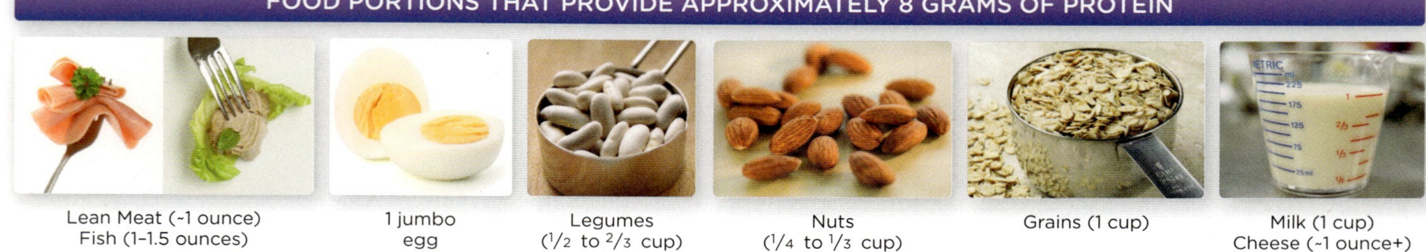

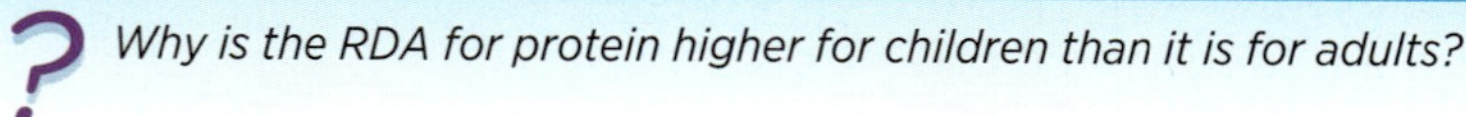

? *Why is the RDA for protein higher for children than it is for adults?*

Photo credits (left to right): Viktor1/Shutterstock, DJ Srki/Shutterstock, Nattika/Shutterstock, Tim Hill/Alamy, Daniel Grill/Getty Images, Julie Woodhouse/Alamy, makuromi/Shutterstock

Beans and rice. *This simple complementary protein dish is easy to prepare and delicious.*

Juanmonino/Getty Images

RDA for protein. **(INFOGRAPHIC 8.10)** Adopting a more plant-based diet may help align our protein intake with the RDA and provide other health advantages. (See Chapter 9 for more information.)

When bodybuilders and Olympic athletes try to eat more protein, it's no coincidence that many of them eat eggs. Eggs are a food with high **protein quality**; that is, eggs contain all of the nine essential amino acids in amounts and proportions that best support protein synthesis in the body. Animal foods (meat, fish, dairy, and eggs) provide high-quality protein, as do some plant foods like soy, quinoa, and amaranth. These foods are considered *complete* proteins in that they contain all nine essential amino acids in the appropriate proportions. In contrast, *incomplete* proteins are foods with lower protein quality; they lack or supply low amounts of one or more essential amino acids. Without an adequate supply of all nine essential amino acids, protein synthesis is disrupted or limited—the amino acid in the shortest supply relative to its requirement is referred to as the *limiting amino acid.*

Most plant foods are incomplete proteins and they vary in the types and proportions of amino acids they contain. However, all the essential amino acids needed for protein synthesis can be supplied by consuming two or more "complementary" protein plant foods. Chapter 9 will further explore plant-based diets and protein complementation.

An example of protein complementation is rice and beans, which have different limiting amino acids. The amino acids methionine and cysteine are low in beans, but adequately supplied in rice. And as long as complementary protein foods are consumed within the context of an overall varied and balanced diet, they don't even have to be eaten at the same meal. This helps explain why strict vegetarians (vegans) can still meet their protein requirements. **(INFOGRAPHIC 8.11)**

PROTEIN-DEFICIENCY DISEASES

Even though most people get more protein than the RDA recommends, it's possible to get too little. This prevents us from being able to synthesize the proteins we need to meet physiological demands, and this results

PROTEIN QUALITY *a measure of how well a protein meets our needs for protein synthesis; based on the proportion of essential amino acids present*

INFOGRAPHIC 8.11 Choose Your Proteins Wisely *Making good choices about the foods that provide our dietary proteins is about more than just limiting saturated fat. Animal products provide no fiber, and only low levels of polyunsaturated fatty acids (PUFAs).*

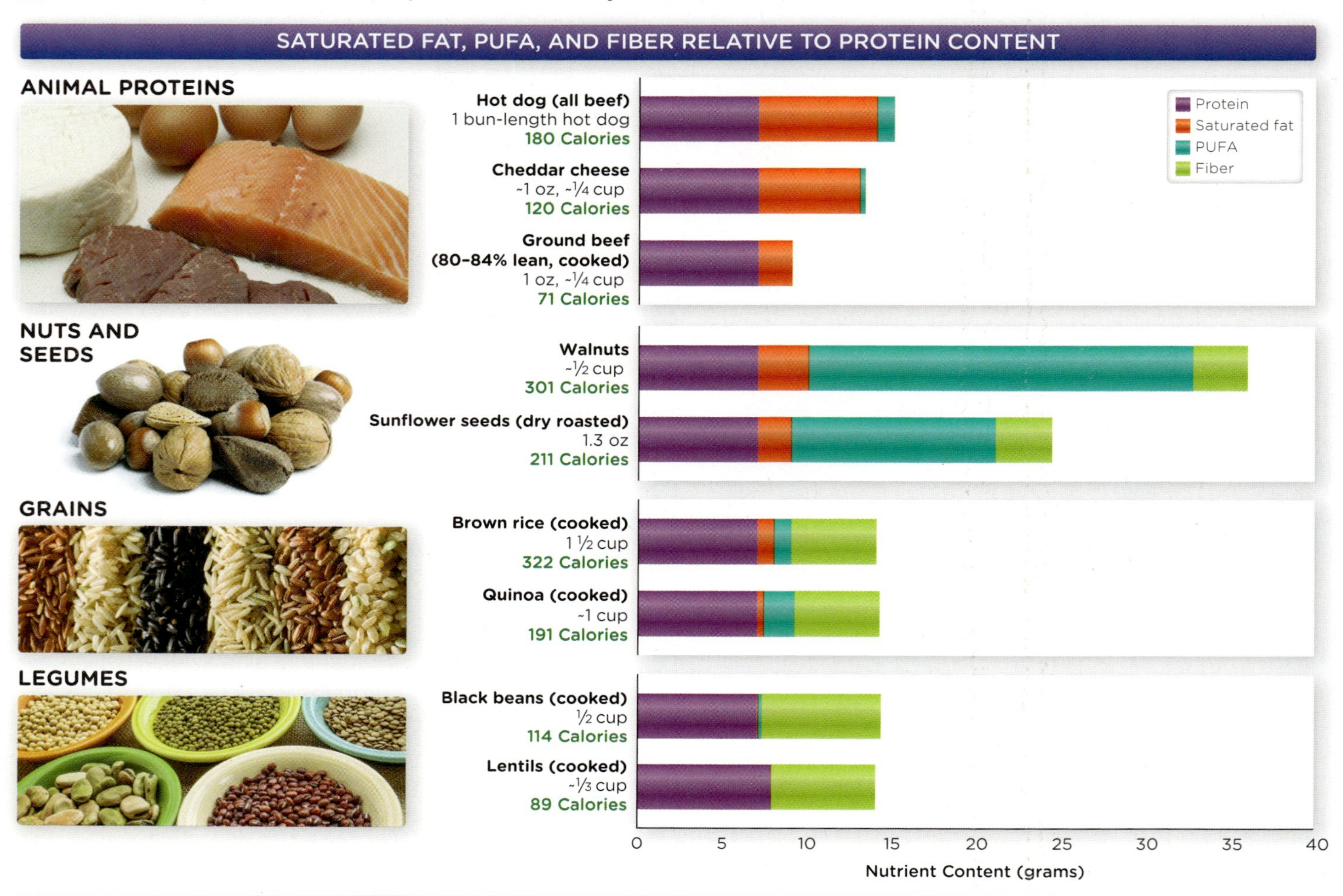

Which group contains the most fiber? The most PUFA?

Photo credits (top to bottom): Celso Pupo/Shutterstock, YinYang/Getty Images, Francesco Tonelli/Alamy, marekuliasz/Shutterstock

KWASHIORKOR *a condition caused by an inadequate protein intake with reasonable caloric (energy) intake characterized by a swollen belly*

MARASMUS *a condition caused by inadequate intake of protein, calories, and overall nutrients characterized by emaciation ("skin and bones" appearance)*

in detrimental effects in body structure and function. In one type of protein deficiency, **kwashiorkor**, people don't eat enough protein but they meet their energy needs, usually through carbohydrates. This is seen in the swollen bellies (from edema and a fatty liver) of children in developing countries who eat mostly grains, and who are often shorter, more prone to infections, and experience changes in hair color and flaky skin. The other type of deficiency, **marasmus**, occurs when people are deficient in most nutrients, including protein. They will waste to "skin and bones" and have little body fat. In malnutrition hotspots, the causes of protein malnutrition are a complex mixture of environmental, economic, social, and political factors. Poor weather such as drought, flooding, as well as hazards such as earthquakes and tsunamis can disrupt the food supply. Wars, political unrest, poverty, and the needs of refugees seeking safety can set off food emergencies.

In developed countries such as the United States, most protein deficiencies occur in alcoholics who swap food for alcohol, and in people with long-term illnesses, malabsorption, or injuries that prevent them from eating well. We will further explore protein and energy deficiency in Chapter 20. **(INFOGRAPHIC 8.12)**

INFOGRAPHIC 8.12 Kwashiorkor and Marasmus

KWASHIORKOR

Mauro Fermariello/Science Photo Library

Kwashiorkor often results when a child is abruptly weaned, and the high-quality protein of breast milk is replaced with a starchy gruel that is low in both protein quality and quantity. Because calorie intake may not be drastically low, these children have generally not lost a lot of subcutaneous fat. They often develop edema and a fatty liver (that may obscure a loss of adipose tissue) because they are unable to adequately synthesize proteins needed to maintain fluid balance or transport fat from the liver.

MARASMUS

Peter Menzel/Science Photo Library

Marasmus results from a severe and prolonged deficiency of both calorie and protein intake. It is characterized by a depletion of fat stores and serious muscle wasting. When severe protein-energy malnutrition occurs in childhood, during periods of rapid brain growth, some permanent brain damage is typically observed.

MALNUTRITION HOTSPOTS

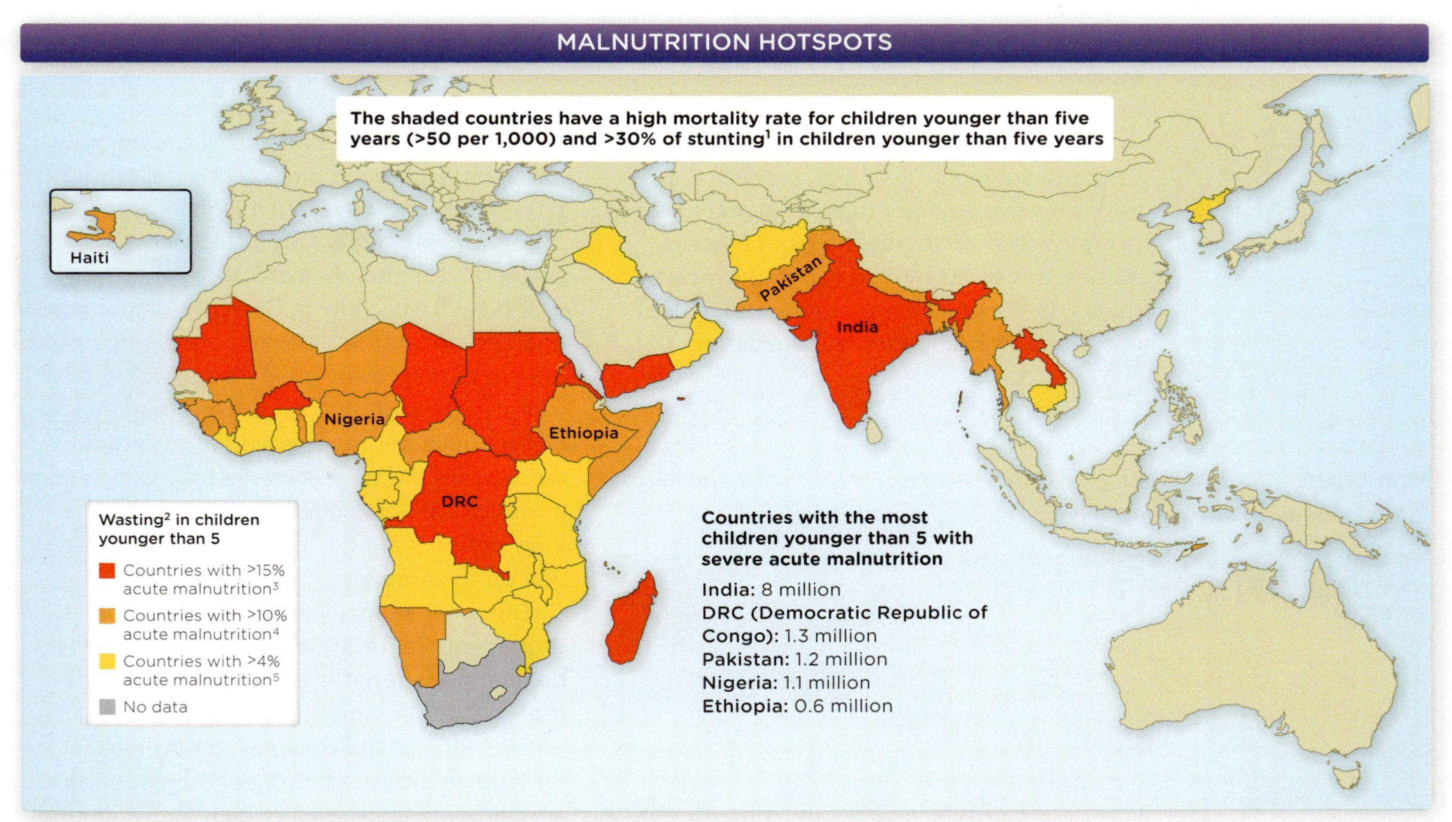

[1] Stunting — Growth retardation, indicated by low height for age.
[2] Wasting — Emaciation or thinness as measured by low weight for one's height.
[3] Burkina Faso, Chad, Democratic Republic of Congo, Eritrea, India, Lao People's Democratic Republic, Madagascar, Mauritania, Sudan, Yemen.
[4] Bangladesh, Central Africa Republic, Comoros, Ethiopia, Guinea, Guinea Bissau, Haiti, Mali, Myanmar, Namibia, Nepal, Niger, Nigeria, Pakistan, Sierra Leone, Somalia, Timor-Leste, Togo.
[5] Afghanistan, Angola, Benin, Burundi, Cambodia, Cameroon, Republic of Congo, Côte d'Ivoire, Equatorial Guinea, Ghana, Iraq, Kenya, Democratic People's Republic of Korea, Liberia, Malawi, Mozambique, Rwanda, Tanzania, Uganda, Zambia, Zimbabwe.

Sources for map: Population reference Bureau 2007 World Population Data. WHO Analyses of national nutritional surveys done 2001-2006. UNICEF — The State of the World's Children 2008.

HIGH-PROTEIN DIETS

Extra protein isn't just attractive to athletes, who want to maintain or increase muscle mass—people trying to lose weight often turn to high-protein, low-carbohydrate diets, which promote getting 30% to 50% of total calories from protein. The potent ability of dietary proteins to reduce hunger and extend the time until we feel hungry again (satiety) is likely the main reason why these diets often produce rapid weight loss. However, long-term studies of weight loss suggest high-protein diets may help take weight off at first, but over the long haul, studies show little difference between diets higher and lower in protein in maintaining weight loss.

Recall that the AMDR for protein is 10% to 35% of total calories; in healthy people, intakes towards the upper end of the range are not associated with any negative effects. What is likely more important is the source of the protein. For instance, some studies have connected diets that are high in animal proteins with an increased risk of kidney stones, and diets high in red and processed meats, in particular, are associated with an increased risk of diabetes, cancer, and heart disease. However, it is unlikely that the increased risk is due to the increased intake of protein per se, but rather the constituents of these meat products. Processed meats are high in sodium and other preservatives that may be the underlying cause of the increased risk of chronic disease, while red meats in general are high in unhealthy fats. In contrast to the apparent ill effects of consuming a large amount of red and processed meats, diets with higher intakes of fish are generally associated with a lower risk of these diseases.

Replacing red and processed meats with fish and poultry is linked to reductions in the risk of diabetes, cancer, and heart disease, and even greater risk reductions are seen when the meat is replaced with beans, legumes, and nuts. Substituting with fish and poultry will reduce the intake of unhealthy fats, while substituting with nuts and beans will have the added benefits of providing excellent sources of dietary fiber, essential fatty acids, and phytochemicals.

High-protein diets are often rich in saturated fat and cholesterol (refer to Infographic 8.11), but low in fiber, phytochemicals, and certain vitamins and minerals, since people are eating fewer grains, vegetables, and fruits. As a result, the American Heart Association does not recommend high-protein diets, since the typical intake may put people at risk of cardiovascular problems. "Over a long period, eating a lot of protein can set you up for problems," says Dr. Lamont.

But what seems like a big increase in protein intake for athletes may not actually be so high when it is evaluated in light of their overall calorie intake, says Robert Wolfe, PhD, of the University of Arkansas for Medical Sciences. Someone competing at sports typically eats a much larger amount of calories overall, not just from protein—so doubling their protein intake may simply keep pace with other components of their overall diet, he says. "That seems like a big increase in protein, but you have to realize that total energy expenditure has also at least doubled," he says. And even though some protein sources are "beefed up" with

High-protein diets. *Some high-protein diets don't provide the variety of foods needed to adequately meet all nutritional needs.*

MSPhotographic/Shutterstock

A fortunate inheritance. *Olympian Michael Phelps requires not only a healthy diet, and an extraordinary amount of hard work, but exceptional genes to excel in swimming.*

Simon Bruty/Sports Illustrated/Getty Images

saturated fat and other unhealthy components, not all are. If athletes need extra calories for their workouts, there are worse sources than protein, adds Wolfe. "You have to consider protein not just as an isolated nutrient, but as one of three macronutrients in the body," says Wolfe. "If you don't eat more protein when you exercise and burn more calories, then you're eating more carbohydrates and possibly more fat." But some athletes who want to boost protein intake to levels far above the RDA often rely on protein-packed powders, rather than protein-containing foods, which they mix into shakes or smoothies, or sprinkle over meals.

When Stuart Phillips, PhD, professor of kinesiology at McMaster University in Ontario, lectures to varsity-level athletes, he asks them: What factors do you believe most contribute to your performance? Sometimes, he'll even draw a pyramid, and ask them to fill it in with the most important factors at the bottom, representing the largest influence. Most place training, nutrition, and supplements at the bottom of the pyramid. "You're wrong," he says. Then he draws another pyramid. In the largest portion at the bottom he writes "genetics"—your biology is a major determinant of your athletic ability. But that's not the only one; the next level of the pyramid is training. Finally, in the portion that's at the top, he writes "nutrition." Maybe he'll add a little dot to represent supplements. Learn more about dietary supplements in Chapter 12.

■ ■ ■

Even Michael Phelps, who ingested almost unimaginable amounts of food—including protein—to prepare for the Beijing Olympics, moderated his diet for the next Olympic Games in London. There, he typically started the day with a bowl of oatmeal, a large omelet with ham and cheese, and fruit. Although he didn't win as many medals that time around, his diet likely had little to do with it, says Wolfe, who spent part of the 1990s studying female members of the U.S. Olympic swimming team. He agrees that diet is only one part of what makes an athlete successful. Phelps "could probably change his diet pretty drastically and still been a darn good swimmer," says Wolfe. "That wasn't why he was an Olympic champion—it was just one component. And from my experience with Olympic athletes, probably not a large component."

CHAPTER 8 **BRING IT HOME**

Protein power?

Imagine you are a registered dietitian and sports nutritionist for a university-level athletic program and you educate student athletes on how to maintain a healthy diet that both supplies necessary nutrients and enhances sports performance. Jake, a soccer player, requests a meeting to address some questions about his diet. Jake is 20 years old; 5 feet, 11 inches tall; and weighs 175 pounds. His practice and game schedule demands well over an hour of vigorous activity each day and he also lifts weights 3 or 4 days each week. You interview Jake and obtain a 24-hour recall of a typical day's intake as follows.

Jake's 24-hour recall

BREAKFAST
2 slices whole grain bread
2 tablespoons peanut butter
1 orange
1 cup chocolate milk

LUNCH
Grilled chicken sandwich
(3-ounce chicken breast)
Sandwich bun
1 tablespoon mayonnaise
2 slices tomato, lettuce
1 cup French fries
12 ounces soda

SNACK
2 ounces pretzels
1 apple
1 ounce cheese stick

DINNER
2 cups pasta
1 cup marinara sauce
3 one-ounce meatballs
2 cups tossed vegetable salad
2 tablespoons Ranch dressing

TOTAL: 3,226 calories, 158 grams protein

Consider

1. What is Jake's RDA for protein? (Hint: To determine weight in kilograms, divide weight in pounds by 2.2.)
2. What is the AMDR for protein? Explain how the RDA for protein fits within the acceptable range for protein intake.
3. Do you feel the RDA meets Jake's protein needs? From what you've learned in this chapter, discuss the rationale behind your answer.
4. Visit www.supertracker.usda.gov and set up a profile for Jake using his age, sex, height, weight, and activity level. Don't register—just submit after setting up profile.
5. Click on "View your plan" under *Welcome, Jake*.
6. How many calories does Jake need each day?
7. Given his calorie requirements, what is the protein AMDR for Jake expressed in calories from protein and grams of protein? (Note: This will be a range, not an absolute number.)
8. After analyzing Jake's intake using Supertracker, you determine that Jake's typical intake is 3,226 calories with 158 grams of protein. Does this fall within his AMDR? (To calculate percent of calories from protein, multiply grams of protein by 4 and divide by total calories.)
9. Jake tells you that several of his teammates add protein powders to beverages to supplement their protein intake. He asks if he also needs to purchase protein supplements to make sure he is getting sufficient protein. What do you tell Jake?
10. If Jake's intake remains the same, thus meeting his calorie needs, but he chooses to add 25 additional grams of protein through a protein supplement beverage, explain what will likely happen to the additional protein in his body.

KEY IDEAS

- Protein has many critical roles in the body's structure and processes, including catalyzing chemical reactions (enzymes), regulating body functions (hormones), and transporting substances in the blood. It also has central roles in immunity, fluid balance, and blood clotting.
- Proteins are composed of carbon, hydrogen, and oxygen, but also contain nitrogen, which is supplied through amino acids, the building blocks of protein.
- Protein needs are determined by sex, life stage, and other factors. The RDA for adults is 0.8 g/kg/d. The AMDR for protein is 10% to 35% of total calories.
- Proteins are complex structures synthesized by linking up to 20 amino acids into chains of varying sequence and length. Nine of these amino acids are considered essential because they must be supplied through the diet. The remaining 11 nonessential amino acids can be manufactured by the body.
- Sequences of DNA called genes provide the instructions for the synthesis of every protein in the body. This is a two-step process that begins in the nucleus with gene transcription and is completed in the cytoplasm with translation.
- The overall shape of a protein molecule determines its function, and how it interacts with other molecules.
- The process of denaturation alters the shape and function of proteins.
- The digestion of protein begins in the stomach, where proteins are denatured and fragmented, continues in the lumen of the small intestine, and is completed within the mucosal cells of the small intestine.
- Proteins in the body are constantly being broken down into amino acids and reassembled in a process called protein turnover.
- Nitrogen balance is a reflection of protein intake versus protein breakdown and is influenced by diet, hormones and growth factors, and muscle contractions.
- Depending on the proportion of each of the essential amino acids present (protein quality), foods can be classified as complete or incomplete proteins.
- Inadequate protein intake can have health and metabolic consequences.

NEED TO KNOW

Review Questions

1. Like carbohydrates and fat, protein is composed of carbon, hydrogen, and oxygen. Protein differs in that it also contains:
   a. calcium.
   b. chloride.
   c. nitrogen.
   d. phosphorus.
   e. sodium.

2. Functions of protein do NOT include:
   a. serving as an energy source.
   b. serving as a structural component of hair and fingernails.
   c. serving as a store of excess amino acids.
   d. serving as hormones.
   e. catalyzing chemical reactions in the body.

3. There are 20 amino acids present in proteins; of these _____ are considered essential and _____ are considered nonessential amino acids.
   a. 7; 13
   b. 13; 7
   c. 15; 5
   d. 9; 11
   e. 11; 9

4. Nonessential amino acids:
   a. are incomplete proteins.
   b. are extra essential amino acids.
   c. can be manufactured in the body.
   d. enhance muscle development.
   e. are found only in plant foods.

5. The Institute of Medicine has established that a safe range of protein intake is:
   a. 10% to 25% of total calories.
   b. 10% to 35% of total calories.
   c. 15% to 30% of total calories.
   d. 15% to 35% of total calories.
   e. 20% to 40% of total calories.

6. The second step of protein synthesis in which the order of amino acids added to the growing protein chain is read from (or dictated by) the information in the mRNA is termed:
   a. phosphorylation.
   b. gluconeogenesis.
   c. transcription.
   d. translation.
   e. deamination.

7. The shape of a protein is a determinant of its _____ in the body.
   a. storage form
   b. energy potential
   c. function
   d. heat production
   e. permanence

8. Protein denaturation refers to:
   a. the lack of one or more essential amino acids in a food.
   b. the process of adding an amino acid to a food that is not naturally present.
   c. a change in shape of protein structure due to heat, light, motion, or change in pH.
   d. breakdown of muscle mass during caloric restriction.
   e. the conversion of protein to glucose or fatty acids.

9. Nick lifts weights and takes a daily protein supplement *in addition* to his 30% protein weight-maintenance diet. Taking the protein supplement:
   a. is an effective practice to add muscle weight without exercise.
   b. is not necessary because his intake is already in excess of the RDA.
   c. provides additional calories that can contribute to fat stores.
   d. will increase his athletic performance.
   e. Both b and c are true.

10. Each of the following food choices is a source of complete proteins, EXCEPT:
    a. beef.
    b. black beans.
    c. eggs.
    d. quinoa.
    e. yogurt.

11. Marasmus:
    a. is characterized by deficiency in most nutrients, including protein.
    b. is characterized by sufficient total caloric intake, but deficient intake in protein.
    c. is associated with a swollen belly appearance.
    d. only occurs in children.
    e. may occur when consuming excess calories from carbohydrates.

Take It Further

Discuss the possible implications of consuming protein at levels near or above the upper end of the AMDR range. Include a discussion of how the dietary sources of protein might affect potential risk.

Dietary Analysis Using SuperTracker

Protein: Intake and Implications

Using Supertracker, track your intake of dietary protein and compare it to the current dietary recommendations.

1. Log onto the United States Department of Agriculture (USDA) website at www.supertracker.usda.gov. If you have not done so already, you will need to create a profile to get a personalized diet plan. This profile will allow you to save your information and diet intake for future reference. *Do not use the general plan.*
2. Click the Track Food and Activity option. We will just focus on tracking food for this activity.
3. Record your food and beverage intake for one day that most reflects your typical eating patterns. Enter each food and beverage you consumed into the food tracker. Note that there may not always be an exact match to the food or beverage that you consumed, so select the best match available.
4. Once you have entered all of your food and beverage choices into the food tracker, on the right side of the page under the bar graph, you will see Related Links: View by Meal and Nutrient Intake Report. Print these reports and use them to answer the following questions:
    a. Did your dietary intake of protein meet your target amount? If not, were you over the target or under the target?
    b. If the target was not met, list two dietary changes you could make to meet the target amount.
    c. Did you meet your target for the percent of total calories for protein? If not, were you over the target or under the target?
    d. Next, look at your dietary sources of protein. Did the majority of them come from plant sources or animal sources?
    e. Identify two potential health concerns associated with a diet high in animal protein.

Pass the Plants, Please

WHAT DOES A SPANISH STUDY SAY ABOUT THE BENEFITS OF FOLLOWING A MEDITERRANEAN-LIKE DIET?

Michael Turek/Gallery Stock

LEARNING OBJECTIVES

- Compare and contrast the types of vegetarian and semi-vegetarian diets (Infographic 9.1)
- Describe the steps that occur during the development of cancer (Infographic 9.2)
- List the recommendations that reduce the risk of cancer (Infographic 9.3)
- Provide examples of two specific phytochemicals and their significant dietary sources (Infographics 9.4 and 9.5)
- Identify challenges that vegetarians may experience in meeting their requirements for specific nutrients and provide strategies to alleviate these concerns (Infographics 9.6 and 9.7)
- Compare and contrast the dietary recommendations and potential benefits of a vegetarian and a Mediterranean diet (Infographic 9.9)

In April 2013, a team of researchers published the results of a rigorous randomized controlled clinical trial that *Forbes* magazine proclaimed would "undoubtedly have a major effect in the field of nutrition." The scientists, based in Spain, had spent years investigating the impact of a Mediterranean-like diet on the risk of heart attack, stroke, and death from heart disease compared with the effect of a traditional low-fat diet, the latter of which has long been recommended for its heart-healthy benefits in the United States and elsewhere.

The Spanish researchers assigned nearly 7,500 people at high risk of heart disease to three

Two groups of participants in the Spanish PREDIMED study consumed daily servings of either nuts or olive oil.

Kristin Duvall/Getty Images

different groups. Two of the groups were told to eat a Mediterranean-like diet—rich in fruit, vegetables, nuts, olive oil, and whole grains and low in processed and red meats, dairy products, and sweets—while people in the third group, the control group, were told to reduce their intake of all types of fat, and consume lean meats, low-fat dairy products, cereals, potatoes, pasta, rice, fruits, and vegetables. Individuals in one of the Mediterranean diet groups were given weekly gifts of high-quality olive oil to encourage their consumption of this rich source of monounsaturated fat, while participants in the other Mediterranean diet group were given weekly gifts of mixed nuts to foster an increase of their nut intake.

MEDITERRANEAN DIET *a dietary pattern traditionally followed in Mediterranean countries that has been proven to have health benefits*

PLANT-BASED DIET *a diet that emphasizes whole plant foods, limits processed foods, and may or may not include foods of animal origin*

Although the researchers had planned to observe the individuals and collect data for five years, the trial was so successful that they had to stop it short. The Mediterranean diet seemed to work *so well* to prevent heart attacks, strokes, and heart disease–related deaths that it would have been considered unethical to allow the people in the control group to continue their assigned low-fat regimen, which was clearly much less effective. Indeed, the Mediterranean dieters who were given olive oil had a 30% reduced risk of heart problems over the five-year period compared with the low-fat dieters, while those who had been given mixed nuts had a 28% reduced risk of heart problems or heart-related death compared with the low-fat dieters.

Lead researcher Emilio Ros says he was quite surprised by the diet's efficacy, considering that the people eating the Mediterranean-like diet weren't directed to do other heart-healthy things, like limit their caloric intake or exercise. The trial showed "that a healthy dietary pattern such as the Mediterranean diet is as potent as modern drugs to reduce cardiovascular risk."

■ ■ ■

A **Mediterranean diet** is similar to a **plant-based diet**. Both eating plans emphasize the consumption of fresh fruits and vegetables, and the avoidance of processed or refined foods, but plant-based diets limit the consumption of animal foods or omit them entirely. Mediterranean diets, however, include fish and occasional consumption of lean protein sources like poultry. Specifics aside, most nutritionists agree that Americans should be eating far more plant foods than they are—and the general public seems to know this, too. According to recent Gallup surveys, Americans say that "eating more fruits and vegetables" is the most important change they can make to their diet, and that eating fewer animal foods, particularly

INFOGRAPHIC 9.1 Types of Vegetarian Diets

Vegetarians do not consume gelatin (a primary component of Jell-O) because it is a protein that has been isolated from the skin and bones of animals.

	VEGETARIANS Consume no meat, poultry, fish, shellfish, animal broths, gelatin, or lard			**SEMI-VEGETARIANS** Consume eggs and dairy, plus:		
Vegetarian Diet Type	**Vegan**	**Lacto-vegetarian**	**Lacto-ovo-vegetarian**	**Pescatarian**	**Quasi-vegetarian**	**Flexitarian**
Animal Foods Eaten	None	Dairy products	Eggs & dairy products	Fish	Poultry & fish	Limited amounts of meat, fish, & poultry

What do you think may be the most common difference in the reasons given for why people chose to be vegetarians versus semi-vegetarians?

Photo credits (left to right): Carlos Gawronski/Getty Images, John Scott/Getty Images, ajafoto/Getty Images, Evlakhov Valeriy/Shutterstock, Evlakhov Valeriy/Shutterstock

red meats and processed meats that are smoked or cured or salted, would likely be advantageous, too.

VEGETARIAN AND SEMI-VEGETARIAN DIETS

All true **vegetarian** diets completely eliminate meat, poultry, fish, and shellfish. The *lacto-vegetarian* diet consists of plant foods plus dairy (lacto) products. **Lacto-ovo vegetarians** consume plant foods plus dairy products and eggs (ovo). A **vegan** diet, the most limited type, excludes all foods of animal origin, including dairy, eggs, and honey. Individuals who restrict their consumption of some meats may refer to themselves as vegetarians, but they are more appropriately considered *semi-vegetarians*. This includes those who exclude only red meat but eat all other animal products (*quasi-vegetarians*), those who's diet excludes red-meat and poultry, but includes fish and shell fish (**pescatarians**), and those who eat a mostly plant-based diet but who occasionally eat meat too (*flexitarians*). **(INFOGRAPHIC 9.1)**

Not only can vegetarians meet the recommended intakes for various nutrients with a bit of dietary planning, but they can also benefit from a reduced incidence of chronic disease and lower mortality rates compared with people eating a typical Western diet of high meat intake, and refined grains. However, the decision to "go vegetarian" does not necessarily *guarantee* better health. Some vegetarians eat primarily grain-based foods or legumes without much attention to variety or overall nutrition, which is not necessarily good for them. The quality of the diet depends significantly on the distribution of the nutrients being consumed and the types of food choices being made.

VEGETARIAN
a diet consisting of plant-based foods, which excludes all meats, fish, and shellfish, but may include dairy products and eggs

LACTO-OVO VEGETARIAN
a vegetarian diet consisting of plant foods plus dairy (lacto) and egg (ovo) products

VEGAN
a vegetarian diet that eliminates all foods of animal origin

PESCATARIAN
a semi-vegetarian diet that excludes meats and poultry, but includes plant foods, dairy foods, eggs, fish, and shellfish

Taking the meat out of a meal doesn't necessarily improve its nutrient profile. What substitutions could make this meatless meal more nutrient dense?

Michael C. Gray/Shutterstock

BENEFITS OF A DIET RICH IN PLANT FOODS

Although people worldwide follow plant-based diets for various reasons—cultural, ethical, environmental, and religious—there are also plenty of health reasons to do so. Studies show that vegetarians have lower total blood cholesterol, low-density lipoprotein levels, and blood pressure, all of which reduce their cardiovascular risks; indeed, research suggests that vegetarians have a lower risk of obesity, heart disease, hypertension, cancer, type 2 diabetes, and mortality than those consuming a typical western diet. Vegetarians also tend to have a higher intake of nutrients including vitamins C and E, magnesium, potassium, folate, antioxidants, and phytochemicals.

According to the 2010 Dietary Guidelines for Americans, "In prospective studies of adults, compared to non-vegetarian eating patterns, vegetarian-style eating patterns have been associated with improved health outcomes—lower levels of obesity, a reduced risk of cardiovascular disease, and lower total mortality." Affirming the nutritional and health benefits of plant-rich diets, the 2015 Dietary Guidelines for Americans include the Healthy Vegetarian Eating Pattern that provides recommendations for those who follow a vegetarian pattern.

The Vegetarian Pattern is similar to the Healthy U.S.-Style Pattern (see Chapter 2), but is higher in calcium and fiber and lower in vitamin D. Also, the Vegetarian Pattern increases consumption of soy products, legumes, nuts and seeds, and whole grains and eliminates meats, poultry, and seafood.

To reflect the habits of most U.S. vegetarians, dairy foods and eggs are included, but a vegan variation is provided. Echoing earlier Guidelines, the 2015 edition cites reduced risk of cardiovascular disease, lower rates of obesity and lower total mortality associated with vegetarian-style eating patterns.

How and why plant-based diets are beneficial is an area of active research. A plant-based diet may benefit mortality because of general patterns of lower body weight, decreased consumption of processed (smoked, salted, cured) meat, and an increased consumption of plant foods. For example, a diet rich in meat may be at the expense of fruits and vegetables and the fiber and other biologically active nutrients they contain. Some processed meats contain carcinogenic (cancer-causing) compounds formed during cooking or produced during processing to preserve color or flavor.

INFOGRAPHIC 9.2 Cancer Development Is a Multistep Process *The stages of cancer include initiation, promotion, and progression.*

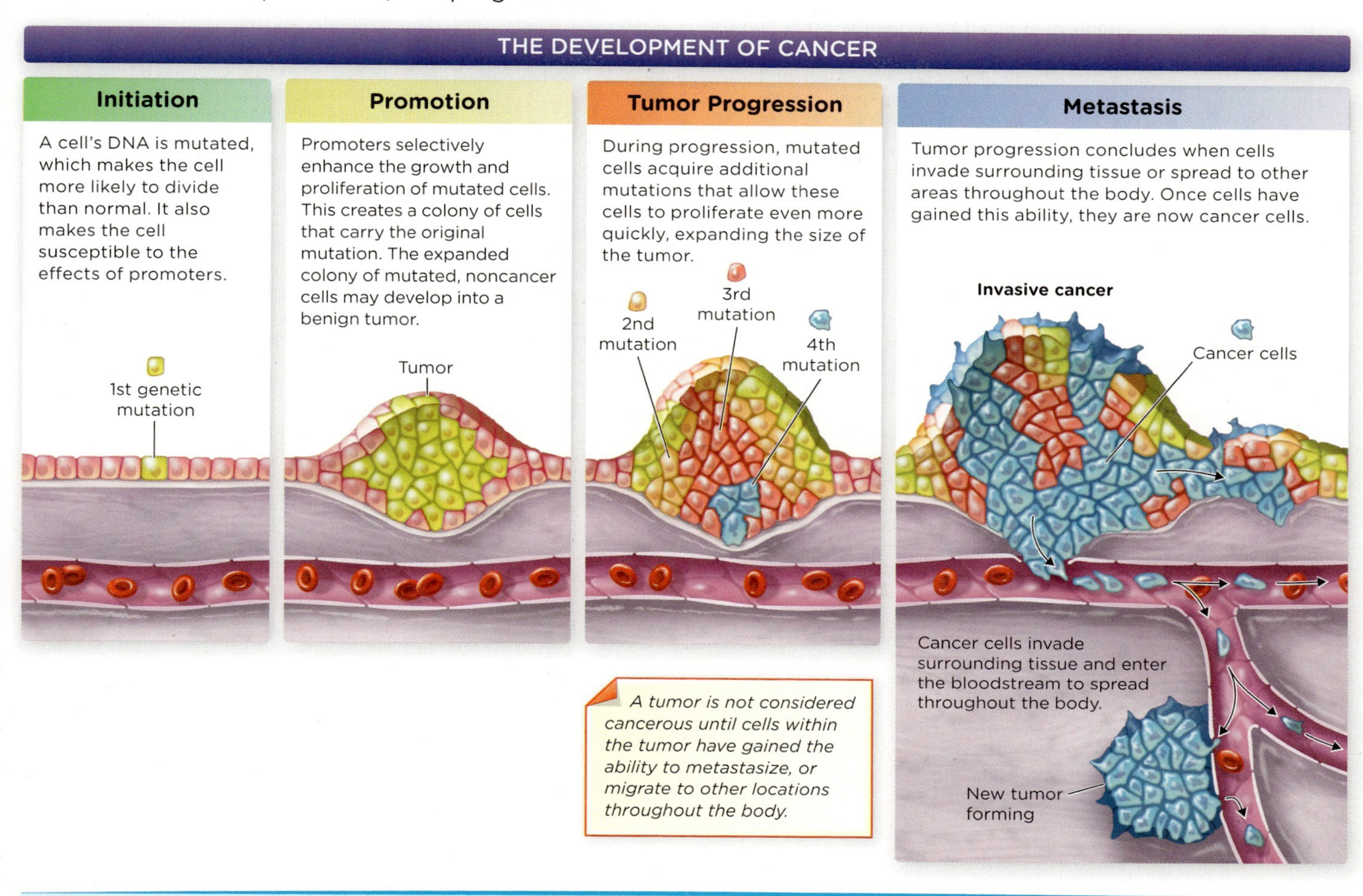

A tumor is not considered cancerous until cells within the tumor have gained the ability to metastasize, or migrate to other locations throughout the body.

The development of all cancers requires what initiating event?

HOW CANCER DEVELOPS

Studies find that plant-based diets reduce the risk of **cancer**, the second leading cause of death in the United States, ranking just below cardiovascular disease. (See Chapter 7.)

Cancer is a group of conditions that result from the uncontrolled growth or division of abnormal cells that invade a part or parts of the body. Although scientists don't fully understand how cancer develops, most agree that it is a multistep process. **(INFOGRAPHIC 9.2)**

First, during the process of *initiation*, DNA inside a cell undergoes a **mutation** that alters the DNA sequence permanently. Any cells produced from the division of the mutated cell will also carry the mutation.

Once a cell has been mutated, it is susceptible to the effects of promoters. During *promotion*, various factors (like inflammation or some chemicals) selectively promote these mutated cells to divide and proliferate more quickly than they should. Finally, in the process of *progression*, the mutated cells acquire additional mutations that allow the cancer cells to migrate to, and invade, other tissues (**metastasis**) where they can form new tumors that disrupt the function of body organs and tissues.

Although many cancers are influenced by genetic factors, cancer risk is also affected by environmental factors, including lifestyle choices—so in that sense, dietary

CANCER
the uncontrolled growth or division of abnormal cells that invade a part or parts of the body

MUTATION
a permanent change in the DNA sequence of a gene

METASTASIS
during the process of cancer progression, mutated cells acquire additional mutations that allow the cancer cells to migrate to, and invade, other tissues

INFOGRAPHIC 9.3 Recommendations to Reduce the Risk of Cancer

Achieve or Maintain a Healthy Body Weight

Increased body fat is strongly linked to an increased risk of cancers of the colon and rectum, esophagus, endometrium, pancreas, kidney, breast, and gallbladder.

Limit Consumption of Energy-Dense Foods and Avoid Sugary Drinks

Overconsumption of these foods likely contributes to weight gain and therefore cancers associated with increased body fat.

Limit Consumption of Red Meat and Avoid Processed Meat

Overconsumption of red and processed meats is strongly linked to an increased risk of colorectal cancer. Limit red meat intake to no more than 18 ounces a week.

Limit Salt Intake

Salt and salt-preserved foods are likely a cause of stomach cancer.

Encourage Infant Breastfeeding

Having been breastfed as an infant reduces the risk of children becoming overweight and obese. Mothers who breastfed their infants have a lower risk of breast cancer.

Be Physically Active

Physical activity protects against colon, endometrial, and postmenopausal breast cancer. Because physical activity also protects against weight gain, it also indirectly protects against those cancers whose risk is increased by obesity.

Eat Mostly Foods of Plant Origin

Most diets that protect against cancer are composed primarily of foods of plant origins. Nonstarchy vegetables and/or fruit probably protect against cancers of the mouth, esophagus, stomach, and lung. Foods containing fiber protect against colorectal cancer.

Limit Alcoholic Drinks

There is no amount of alcohol consumption that does not increase the risk of cancer. Alcohol consumption increases the risk of mouth, pharynx, larynx, esophagus, and colorectal cancers.

Aim to Meet Nutritional Needs Through Diet Alone

High-dose supplements may increase the risk of some cancers. In others cases, intake of nutrients from foods is found to be protective while nutrient supplements are not.

Avoid Consuming Moldy Grains, Legumes, and Other Foods

Some molds produce aflatoxins that are potent cancer-causing compounds.

Reference: "The Second Expert Report, Food, Nutrition, Physical Activity, and the Prevention of Cancer: a Global Perspective", from the World Cancer Research Fund, and the American Institute for Cancer Research. http://www.dietandcancerreport.org/cancer_prevention_recommendations/index.php

In what area do you see the greatest need for improvement to reduce your risk of cancer?

Photo credits (left side — top to bottom): Michael Phillips/Getty Images, Sage Corson/Getty Images, Lew Robertson/Getty Images, Danny Smythe/Alamy, szefei/Shutterstock; (right side — top to bottom): ajkkafe/Shutterstock, Adisa/Shutterstock, Shaiith/Shutterstock, vgstudio/Shutterstock, Tom Hoenig/Getty Images

PHYTOCHEMICALS *compounds found in plant foods that are physiologically active and beneficial to human health; not considered essential nutrients*

choices can both increase and decrease risk. For example, eating a diet rich in vegetables, fruits, and legumes has been linked with a reduced risk of mouth, esophageal, stomach, and colon cancer. In contrast, alcohol consumption is associated with an increased risk of several types of cancer. **(INFOGRAPHIC 9.3)**

BENEFITS OF PHYTOCHEMICALS

It is widely known that a diet rich in vegetables, fruits, legumes, and whole grains reduces the risk of cancer, heart disease, and other illnesses. Plants are rich in **phytochemicals**, which are chemicals that can have antioxidant or hormone-like actions and are associated

Eat a rainbow. *Plant pigments are a rich source of phytochemicals.*

Kunal Mehta/Shutterstock

with many health benefits. Fruits, vegetables, and whole grains can contain thousands of these compounds, which give them their color, aroma, and flavor. Because phytochemicals are often color-specific, similarly colored foods, such as carrots and sweet potatoes, often contain similar types of phytochemicals—hence the recommendation to "eat a rainbow" of foods to ensure the consumption of a variety of phytochemicals.

It is not yet possible to pinpoint a single phytochemical-powered bullet against cancer or cardiovascular disease, however. The sheer number of phytochemicals in plants, the complexity of the chemical processes in which they are involved, the way the chemicals interact, and the way they are processed by the body all make it difficult to find out which phytochemicals in foods may fight cancer and other diseases, which may have no effect, and which may even be harmful.

Polyphenols

Although there are many types of phytochemicals, *polyphenols* are the most abundant and diverse phytochemicals in our diet. Found in a wide variety of foods, polyphenols are particularly rich in berries, coffee, tea, red wine, cocoa powder, nuts, and spices; numerous fruits and vegetables are also good sources.

Many polyphenols have anti-inflammatory effects, and diets rich in polyphenols are associated with a reduced risk of chronic diseases such as cardiovascular disease, diabetes, osteoporosis, and neurological-related disorders. And polyphenols may reverse, suppress, or prevent the development of cancer. There are thought to be many mechanisms of action—for example, polyphenols may be able to interrupt or reverse cancer development by interrupting cellular communication systems, thus stopping the initiation or promotion of cancer. Some

INFOGRAPHIC **9.4** **Classification of the Major Dietary Polyphenols** *Polyphenols are the most abundant phytochemicals in the diet. The polyphenols called flavonoids may explain some of the health benefits associated with fruit- and vegetable-rich diets.*

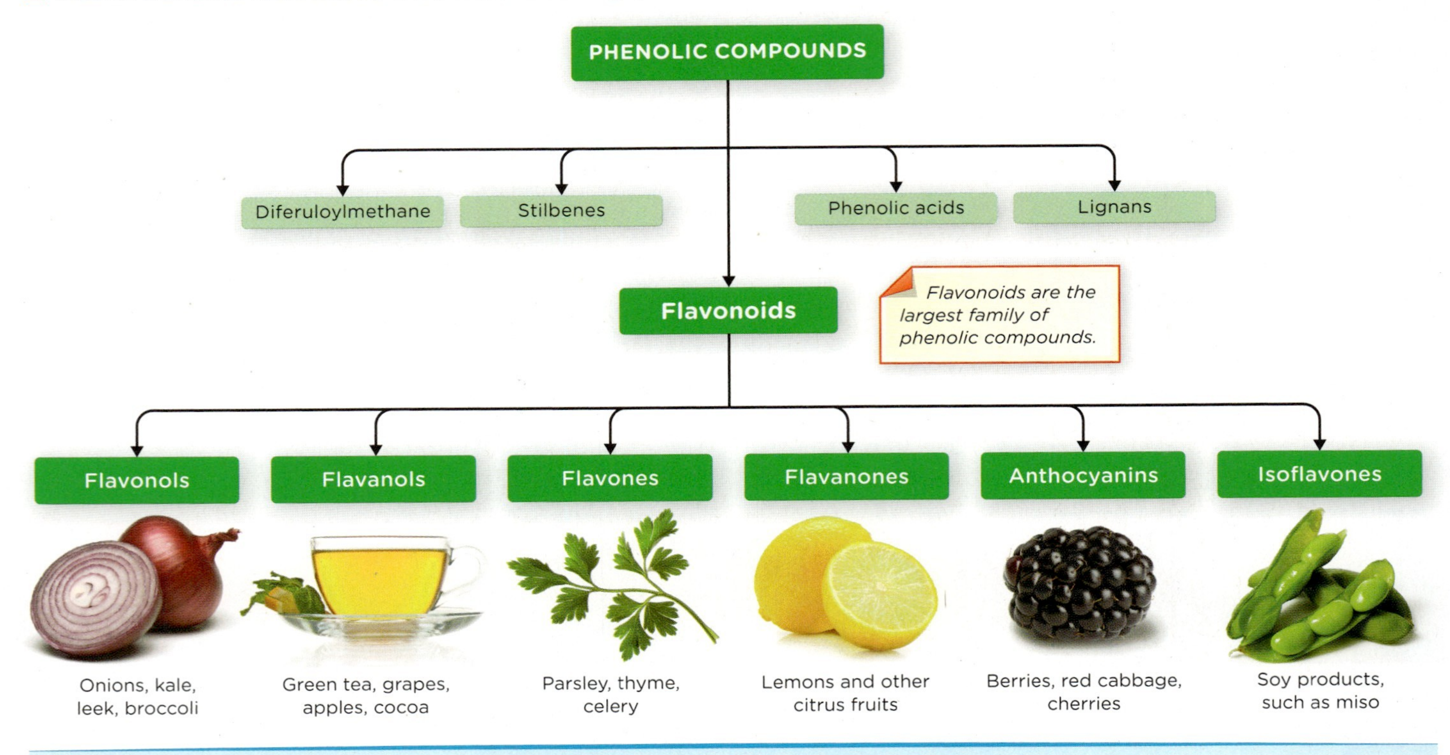

? *What are the two ways you could incorporate more flavonoids into your diet?*

Photo credits (left to right): Kaan Ates/Getty Images, dem10/Getty Images, Valeria Titova/Getty Images, Ben Hung/Getty Images, julichka/Getty Images, kyoshino/Getty Images

polyphenols can also trigger cell death (apoptosis) in cancer cells. **(INFOGRAPHIC 9.4)**

Resveratrol is a polyphenol found in grapes and red wine. Resveratrol has received attention because animal studies have suggested the compound lengthens lifespan. Some people surmise that it might be responsible for the "French paradox," the fact that the French, who consume many rich and high-fat foods but drink a lot of red wine, tend to die less often of coronary heart disease than individuals from other Western countries. More research is necessary as wine and other dietary sources of resveratrol may not provide the health benefits suggested, or the benefits may stem from other chemical constituents.

Polyphenols are divided into several chemical classes, with flavonoids being the most abundant. Flavonoids are further divided into subclasses, including anthocyanins (abundant in berries), isoflavones (abundant in soy products and licorice), and flavanols (abundant in dark chocolate and cocoa). Some research suggests that regularly consuming moderate amounts (1 to 2 ounces a day) of dark chocolate may have beneficial effects on blood pressure, insulin resistance, and the risk of cardiovascular disease.

In addition to having antioxidant and anti-inflammatory effects, phytochemicals can also have hormone-like effects. In particular, two groups of polyphenols, isoflavones and lignans (found in flaxseed), are believed to mimic the actions of the female hormone estrogen.

Carotenoids

Carotenoids are another important class of phytochemicals. Although more than 700 types of these pigments have been identified, only about 50 of them are common in our diet, where they are responsible for the yellow, orange, and red colors of apricots, watermelons, sweet potatoes, red peppers, and tomatoes. **(INFOGRAPHIC 9.5)** Some

INFOGRAPHIC 9.5 Examples of Phytochemicals and Their Possible Benefits

Phytochemical	Lycopene	Epigallocatechin gallate (EGCG)	Quercetin	Curcumin
Class/Subclass	Carotenoid	Polyphenol Flavonoid/Flavanol	Polyphenol Flavonoid/Flavanol	Polyphenol Diferuloymethane
Structure				
Excellent Sources	Tomatoes, watermelon, pink grapefruit	White and green tea	Red and yellow onions, hot yellow peppers, kale, capers, cranberries	Turmeric spice
Possible Benefits	Diets high in lycopene may reduce the risk of developing cataracts, and prostate and ovarian cancers.	EGCG is the most abundant flovanoid in green tea. Green tea may have anti-cancer, anti-obesity, anti-atherosclerotic, and anti-diabetic effects.	Quercetin has been shown to have anti-inflammatory effects, and it may reduce the risk of heart disease and cancer.	Curcumin may have antioxidant and anti-inflammatory effects. It may also reduce the risk of cancer, and slow the progression of Alzheimer disease.

Which phytochemical is structurally least like the others?

Photo credits (left to right): Eli Ensor, koosen/Shutterstock, Eli Ensor, D. Shashikant/Shutterstock

Sea Wave/Shutterstock

Flaxseed. *Phytoestrogens are abundant in flaxseed and soy foods and have chemical structures resembling those of estrogen hormones made by the body. They are being studied to better understand their effects in the body from both food and supplement sources.*

INFOGRAPHIC 9.6 Strategies to Consume More Plants (and Phytochemicals)

Use These Tips to Help Consume More Plant-based Foods

- ✓ Plan to eat at least 5 portions of fruits and vegetables every day. Start with breakfast. Aim for 3 servings by lunchtime.
- ✓ Start your morning with fruit in plain yogurt, cereal, or sliced fruit on whole wheat toast.
- ✓ Add nuts to yogurt, cereal, or salads.
- ✓ Scramble your eggs with diced vegetables.
- ✓ Go for color. Prepare tomato-based soup and include a vegetable of every color.
- ✓ Add steamed vegetables or legumes to your favorite pasta.
- ✓ Add vegetables to pizza to increase nutritional punch.
- ✓ Drink black, green, or herbal teas.
- ✓ Add spices to your meals like garlic, basil, oregano, sage, turmeric, thyme, or ginger.
- ✓ Keep frozen vegetables on hand to add to casseroles and soups and to stretch takeout stir fry and pasta dishes.
- ✓ Try soy products like tofu and vegetable protein meat substitutes.
- ✓ Make your sandwiches more interesting with cabbage, carrots, cucumber, peppers, and a rainbow of lettuce colors and textures.

Frequently Consume These Phytochemical-rich Foods

- Apples
- Apricots
- Berries
- Bok choy
- Broccoli
- Brussels sprouts
- Cabbage
- Cantaloupe
- Carrots
- Celery
- Garlic
- Green tea
- Horseradish
- Kale
- Leeks
- Lentils
- Olives
- Onions
- Pears
- Seeds
- Soy nuts
- Spinach
- Tomatoes
- Turnips

Photo credit: Eli Ensor

important examples of carotenoids are beta-carotene, lycopene, lutein, and zeaxanthin. Adequate intakes of lutein and zeaxanthin (high in spinach and kale) from food sources have been shown to be important for eye health, while those with high intakes of dietary lycopene (in tomatoes and watermelon) have been seen to have a lower risk of prostate orovarian cancer. Strategies to help you incorporate more phytochemical-rich foods into your diet are included in INFOGRAPHIC 9.6.

NUTRITIONAL CONSIDERATIONS AND CONCERNS ASSOCIATED WITH PLANT-BASED DIETS

Vegetarians must choose their foods carefully, since some nutrients are more abundant in animal foods, and others are less

INFOGRAPHIC 9.7 Make a Nutrition Plan *Individuals who infrequently consume animal products may need a plan to obtain these important nutrients.*

Achieving adequate intakes of iron, iodine, and omega-3 fatty acids are of concern for all vegetarians. Deficiencies of calcium, vitamin D, vitamin B_{12}, and riboflavin are primarily of concern for those who follow a vegan diet.

Iron

Legumes and nuts are high in iron, but the iron is poorly absorbed. Eat these foods with a source of vitamin C (like peppers and citrus fruits) to improve iron absorption.

Some green-leafy vegetables such as bok choy and broccoli are good sources of iron, which are also reasonably high in vitamin C.

Dried apricots and raisins are good sources of iron.

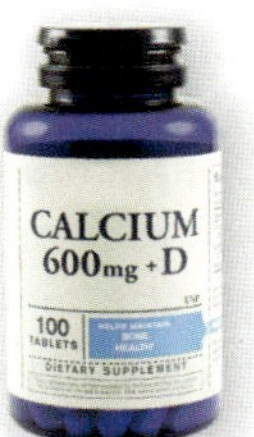

If you take a calcium supplement, do not take it with meals rich in iron, as large doses of calcium will decrease iron absorption.

Vitamin D

If the intake of vitamin D-fortified foods and sun exposure is inadequate to meet needs, a vitamin D supplement should be taken.

Riboflavin

Mushrooms and cooked spinach are naturally good sources, and many breakfast cereals are fortified with high levels of riboflavin.

Vitamin B_{12}

Vitamin B_{12} is found only in foods of animal origin and fortified plant food, including some soy and rice milks, soy-based meat analogs, and some breakfast cereals. If vegans do not consume vitamin B_{12} regularly from fortified foods, a vitamin B_{12} supplement must be taken.

Omega-3 Fatty Acids: EPA and DHA

Dietary supplements containing DHA from microalgae are available, as are soy milk and breakfast bars fortified with DHA. Vegetarians should include good sources of linolenic acid in their diet (flaxseed, walnuts, and soy and canola oils), which can be converted into EPA and DHA.

Iodine

Because vegetarians may be at a higher risk of iodine deficiency than nonvegetarians, when salt is used it should be iodized.

What vegetable is a good source of both iron and calcium? Why is vitamin B_{12} intake not a concern for lacto-ovo-vegetarians?

Photo credits (iron — top left to bottom right): Jamikorn Sooktaramorn/Shutterstock, Eli Ensor, Lightspring/Shutterstock, Datacraft Co Ltd/Getty Images, subjug/Getty Images, David Lee/Alamy, Donna Beeler/Shutterstock; (riboflavin — left to right): Annabelle Breakey/Getty Images, Image Source/Getty Images, ifong/Shutterstock; (iodine): Anthony Pleva/Alamy

bioavailable when consumed from plants. (INFOGRAPHIC 9.7)

Protein

Protein is one important example. Individuals who eat fish, dairy, or eggs typically don't have to worry about getting enough *complete proteins*—proteins that contain all nine essential amino acids in the amount needed to support protein tissue construction. Some plant foods contain all nine essential acids in the ideal proportions, such as soybeans (found in tofu, tempeh, edamame, soy milk, and other soy-based products) and quinoa, the seed of a plant related to spinach. But most plant foods provide *incomplete*

INFOGRAPHIC 9.8 Examples of Foods that Provide Complementary Proteins

Rice, corn, and beans all have at least one essential amino acid that is present in a lower amount than what is needed to support protein synthesis in the body. Because the essential amino acid that is low in both corn and rice is different from the one that is low in beans, combining beans with either corn or rice provides complete protein.

? Why does the protein in corn not complement the protein in rice?

Photo credits (left to right): Takao Onozato/Aflo/Getty Images, Chris Howes/Wild Places Photography/Alamy

proteins because they do not provide all nine amino acids in the amounts needed to synthesize protein in the body. As a result, vegetarians need to eat **complementary protein** foods to provide their bodies with sufficient quantities and proportions of the essential amino acids. **(INFOGRAPHIC 9.8)** Examples of complementary protein foods are beans and grains (such as rice) or beans and most nuts; the two foods contain different amino acids that together provide adequate amounts of all essential amino acids. There is no need to eat the complementary proteins in the same meal, but they should be consumed within the same day as part of an overall balanced and varied diet.

COMPLEMENTARY PROTEINS *two or more incomplete protein sources that together provide adequate amounts and proportions of all the essential amino acids*

Iron

Iron is another nutrient of which vegetarians sometimes consume too little. Although iron is found in plant foods in a form called *non-heme,* it is less bioavailable than the *heme* iron found in animal foods. (See Chapter 14 for full discussion of heme and non-heme iron.) As a result, the recommended non-heme iron intake for vegetarians, particularly vegans, is 80% higher than the iron recommendation for nonvegetarians. Iron-containing plant foods are soybeans, legumes, dried fruit, as well as fortified grains and cereals.

Other nutrients to obtain

Zinc, found in soy products, legumes, grains, and nuts, is less well-absorbed from plant foods than it is from animal foods, though vegetarians typically get enough of this mineral. Calcium, vitamin D, and riboflavin (vitamin B_2) are nutrients of concern for vegetarians. Although intake of these nutrients in lacto-ovo vegetarians is similar to intake in nonvegetarians, vegans may fall below recommended intake levels because they avoid dairy products, which are important sources of these nutrients. Vegans can, however, obtain calcium by eating leafy greens and broccoli, and they can get both calcium and vitamin D from milk alternatives such as soy, rice, and almond milks, as well as some other fortified foods and beverages such as orange juice. Riboflavin is found in high amounts in almonds, and moderately high amounts in mushrooms and cooked spinach, and in lower amounts in whole and enriched grains, but supplementation may sometimes be warranted.

Vitamin B_{12} is only found in foods of animal origin, and because unfortified plant foods contain no B_{12}, vegans must be careful to obtain it from fortified foods such as soy and rice beverages, certain breakfast cereals and meat analogs (which are often soy-based products), as well as specially fortified nutritional yeasts. If vegans can't get enough B_{12} from these sources, a daily vitamin B_{12} supplement may be necessary. Vegetarians who avoid eating seafood must take care to consume enough iodine by selecting iodized forms of table salt, the primary source of iodine in the U.S. diet.

Finally, vegetarian diets are generally rich in omega-6 fatty acids, but they may be marginal in the omega-3s EPA and DHA, which are found primarily in fish (see Chapter 6). Plant foods do provide the omega-3 ALA (alpha-linolenic acid), but less than 10% of ALA in plant foods is converted to EPA and DHA. EPA and DHA can also be found in foods such as soy milk, margarine, and eggs. Vegetarians who avoid animal sources of DHA and EPA should maximize their intake of ALA through walnuts, flaxseeds, soy, and algae.

PLANT-BASED AND VEGETARIAN GUIDELINES

The Academy of Nutrition and Dietetics recommends that vegetarians choose a variety of foods including whole grains, vegetables, fruits, legumes, nuts, seeds, and, if desired, eggs and dairy products (ideally the lower-fat varieties, in moderation). It also helps for vegetarians to choose whole, unrefined foods and to minimize their intake of highly sweetened, fatty, and heavily refined foods, which can be lower in nutrient content. Strict vegetarians such as vegans may need to take supplements or consume fortified food products to ensure that they are getting adequate amounts of every nutrient—particularly vitamin B_{12} and vitamin D, if sunlight exposure is limited. Vegetarians should consult with health care providers or registered dietitians to assist with their dietary planning and for advice regarding dietary supplement needs. There are also many dietary planning resources for vegetarians. The USDA's www.ChooseMyPlate.org, for instance, includes useful tips for vegetarians.

Individuals following a diet free of all animal products can meet vitamin B_{12} needs by eating fortified breakfast cereals, fortified soy milk, and fortified meat substitutes.

Susan Gottberg/Alamy

Children and adolescents can be vegetarians, too, but given their higher nutrient needs to support growth and development, they may be at more risk than adults for nutrient inadequacies. Lacto-ovo-vegetarian kids grow similarly to their nonvegetarian peers, but some studies suggest that vegan children tend to be slightly smaller than nonvegetarian children, although they do tend to be within the normal range for their height and weight. (Poor growth in children

ORGANIC
foods produced through approved methods that generally avoid the use of chemical fertilizers and antibiotics

MyPlate gives guidance for building a vegetarian meal. Simply fill the protein portion of the plate with plant sources of protein including beans, peas, lentils, soy, seeds, and nuts.

mama_mia/Shutterstock

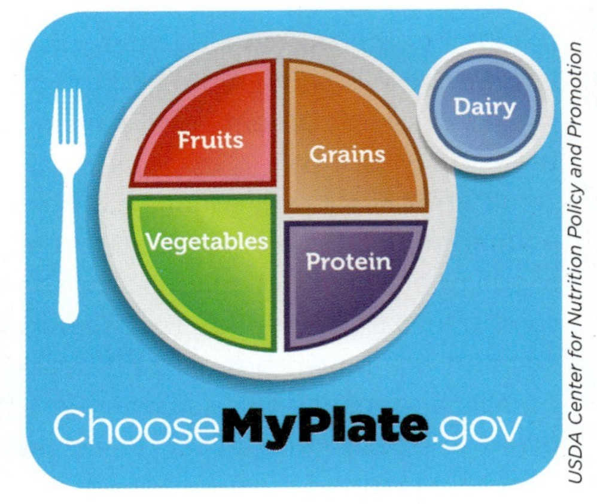

USDA Center for Nutrition Policy and Promotion

has primarily been seen in children whose diets are deficient in protein, calories, and other nutrients.) When adolescents are careful to consume the nutrients they need, plant-based diets can have health advantages: research suggests that vegetarian teens consume more fiber, folate, vitamin A, and vitamin C than nonvegetarians do. They also tend to consume fewer sweets, fast foods, and salty snacks compared with nonvegetarian teens, which can reduce the risk of obesity and chronic disease later in life.

HOW ARE ORGANIC FOODS DIFFERENT THAN CONVENTIONAL FOODS?

Should individuals choose **organic** fruits and vegetables over conventionally produced ones? Organically grown food, which includes both plant and animal products, has been produced through approved methods that integrate cultural, biological, and mechanical practices that foster the recycling of resources, promote ecological balance, and conserve biodiversity. To sell, label, or represent their products as organic, farmers and food producers must follow all of the

Although organic farming practices benefit the environment in many ways, organically farmed food does not have more nutrients than conventionally farmed foods.

Kitra Cahana/Getty Images

USDA specifications for organic regulations. Organically grown and produced foods can be labeled in four ways: *100% Organic*; *Organic* (which means it must be 95% organic); *Made with organic ingredients* (which means it is at least 70% organic); or *Some organic ingredients* (which means it is composed of less than 70% organic ingredients).

Organic diets expose consumers to less antibiotic-resistant bacteria and fewer pesticides. In addition, organic farming has been demonstrated to have a more positive environmental impact than conventional farming does, however, organic farmers are still allowed to use certain pesticides, many of which have not been well-tested. Current evidence does not support any meaningful nutritional benefits associated with eating organic foods compared with conventional foods, and there are no well-powered human studies that directly demonstrate health benefits or disease protection as a result of consuming an organic diet.

THE MEDITERRANEAN DIET

With the news of the recent Spanish clinical trial, many Americans are giving the Mediterranean diet a try now, too. The Mediterranean diet is considered a more "whole diet approach" rather than a focus on certain dietary components, like reducing intake of specific dietary fats; it focuses on consuming a variety of healthy whole foods. **(INFOGRAPHIC 9.9)**

INFOGRAPHIC 9.9 The Mediterranean Diet Pyramid

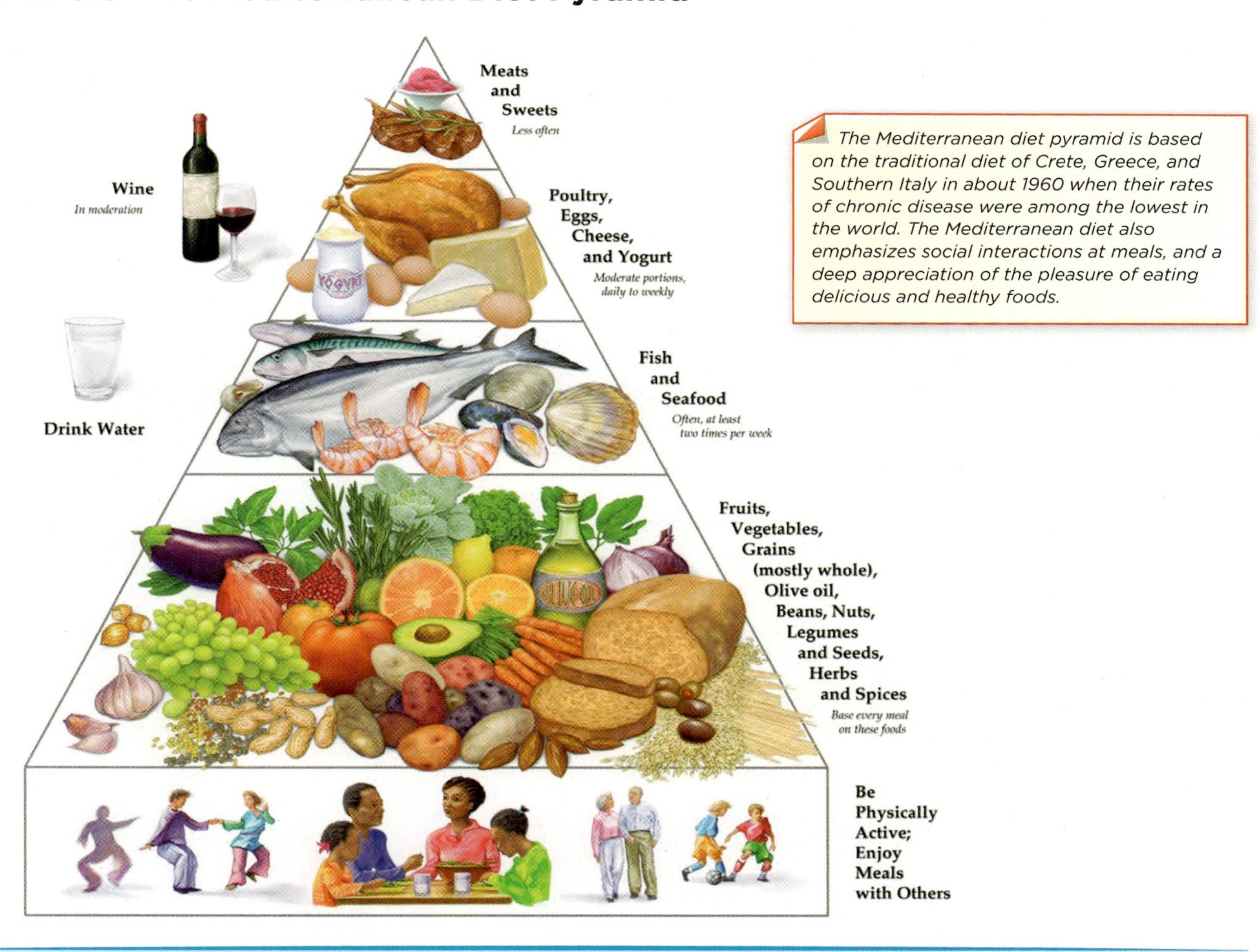

The Mediterranean diet pyramid is based on the traditional diet of Crete, Greece, and Southern Italy in about 1960 when their rates of chronic disease were among the lowest in the world. The Mediterranean diet also emphasizes social interactions at meals, and a deep appreciation of the pleasure of eating delicious and healthy foods.

? *What category of food is eaten daily by most individuals living in the United States but is consumed only occasionally by those who adhere to a Mediterranean diet?*

Photo credit: Oldways Preservation and Exchange Trust/Illustration by George Middleton

There is, of course, no single Mediterranean diet: at least 16 countries border the Mediterranean Sea, and diets differ in these countries and in the regions within them, just as people in California eat differently from people in Alabama. Plus, individuals in Mediterranean countries also have different cultures, ethnic backgrounds, religions, economies, and agricultures, resulting in sometimes radically different food choices.

Nevertheless, most Mediterranean diets share certain features. For instance, more than half of the fat calories in a Mediterranean diet come from monounsaturated fats (mainly from olive oil), which appears to have beneficial blood lipid and other cardioprotective effects. In addition, the Mediterranean dietary pattern emphasizes a high intake of vegetables, legumes, fruits, nuts, whole grains, cheese or yogurt, fish, and healthy oils. Perhaps as a result of the shared features of this diet and the associated lifestyle, the incidence of heart disease in most Mediterranean countries is lower than it is in the United States. Death rates are lower, too. However, it's impossible to know how much of a role diet plays and how much can be attributed to lifestyle, as people living near the Mediterranean Sea lead very different lives in general than Americans do.

In addition to its heart-healthy benefits, the Mediterranean diet is associated with a reduced risk of diabetes, Parkinson's disease, allergies (in children), and cancer. The mechanism by which it reduces cancer risk may have something to do with the fact that the Mediterranean diet provides a healthy ratio of omega-6 to omega-3 fats and that it is high in fiber, antioxidants, and polyphenols. The diet also involves a "Mediterranean way of drinking," which means regular, *moderate* consumption of wine, mainly with food, which has been associated with a reduced risk of chronic disease.

The Mediterranean diet may also reduce the risk of Alzheimer disease. In a 2009 study, researchers at Columbia

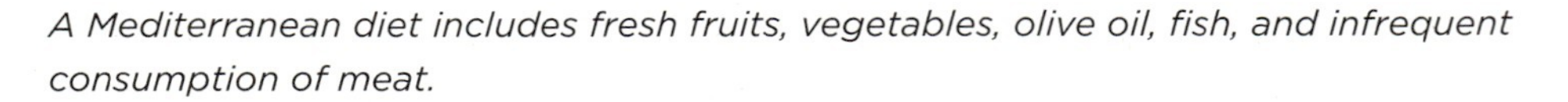

A Mediterranean diet includes fresh fruits, vegetables, olive oil, fish, and infrequent consumption of meat.

Sergey Yechikov/Shutterstock

University interviewed 1,393 healthy New Yorkers older than 65 years about their eating habits once a year for four and a half years. Those who ate in a way that most resembled the Mediterranean diet were 28% less likely to show signs of cognitive decline at the end of the experiment than were those who ate diets high in saturated fat, dairy products, and red meat. The researchers also followed a second group of older adults who already had mild cognitive impairment and reported that those who ate the most Mediterranean-like diets were 48% less likely to develop Alzheimer disease after four years than were their Western diet-eating counterparts. The 2015 Dietary Guidelines for Americans include the Healthy Mediterranean-Style Eating Pattern as an alternative to the Healthy U.S.-Style Eating Pattern (see Chapter 2). According to the Guidelines, it was designed "to more closely reflect eating patterns that have been associated with positive health outcomes in studies of Mediterranean-Style diets." The Pattern contains more fruits and seafood and less dairy than the U.S.-Style, and with the exception of calcium and vitamin D, has similar nutrient content.

The Guidelines state that "the healthfulness of the Pattern was evaluated based on its similarity to food group intakes reported for groups with positive health outcomes in studies rather than on meeting specified nutrient standards."

■ ■ ■

So if you want to stave off your risk of chronic disease, what diet is best? No one yet knows for sure, but it seems that plant-based diets and Mediterranean diets are both excellent choices. Columbia University neurologist Nick Scarmeas, who led the 2009 study, admits that he's partial to the Mediterranean diet, but that's more because of how he was raised than anything else. "I am originally from Greece," he says. "This is the diet my mother cooked."

CHAPTER 9 **BRING IT HOME**

Adopting a vegetarian diet

Imagine your 15-year-old daughter Sophie has decided to avoid all animal foods. You are concerned about potential nutritional inadequacies and how Sophie's diet will affect family food shopping, meals, and budget. Sophie shares that she and her closest friend decided to avoid all animal foods after watching a film on animal rights. Sophie's weight is within a healthy range and she does not presently have any medical conditions that might affect her health or nutritional status. She is on her high school dance team and practices several afternoons a week. You ask Sophie to describe a recent day's intake. Based on her 24-hour recall and what you've learned about vegetarian diets, address the following questions.

Sophie's 24-hour recall:

BREAKFAST
1 cup oatmeal made with water
1 banana, sliced

LUNCH
1 bagel with 2 tablespoons peanut butter
1 ½ ounces potato chips
12 ounces soda

SNACK
2 ounces pretzels
1 apple

DINNER
2 cups pasta
1 cup plain marinara sauce
3 meatless (soy-based) meatballs
2-ounce slice sourdough bread
1 tablespoon olive oil (for bread)
2 cups tossed vegetable salad
2 tablespoons Italian dressing

SNACK
1 cup grape juice
¼ cup soy nuts

Consider

1. Based on Sophie's present dietary choices, what type of vegetarian diet has she decided to follow?
2. What is Sophie's primary motivator for adopting this eating style? What are other reasons individuals decide to adopt predominantly plant-based diets?
3. What are the best sources of protein in Sophie's 24-hour recall? Discuss how individuals avoiding animal foods can obtain all the essential amino acids and provide specific examples.
4. Did Sophie consume appropriate number of foods from each food group? Did she consume an adequate amount of calcium and fiber? Explain.
5. What modifications would you make to this day of meals to maximize nutrient density?
6. Considering Sophie's age and life stage, should she visit a Registered Dietitian for further advice? Why or why not?
7. Identify at least two resources to share with Sophie to assist with dietary planning and ideas for food products and recipes.
9. Sophie expresses the hope that her parents might also consider reducing intake of animal foods in favor of a more plant-based diet. Sophie's father asks if they have to give up all animal foods to adopt a vegetarian diet and if there are actual health benefits if they do so. Discuss.

Take It Further

Record your own 24-hour recall. Examine your day's intake in relation to plant-based food choices. Would you consider your overall intake primarily plant-based? Why or why not? What changes might you make to boost intake of whole plant-based foods? Over the next week, plan to have at least three vegetarian (or vegan) lunches or dinners in place of meals that might contain meats, poultry, or other animal foods. What typical meals might you change and what plant foods would you substitute for animal foods? If you already follow a vegetarian eating style, identify the sources of complete or complementary proteins in several of your meals.

KEY IDEAS

Plant-based diets can vary significantly by including differing types and amounts of animal foods, and they can have varying degrees of healthfulness, depending on the quality and distribution of the nutrients being consumed and the types of food choices being made.

A vegan diet consists only of plant foods, while other types of vegetarian diet plans may include dairy foods and eggs; semi-vegetarian diets may include limited amounts of fish, poultry, and red meats.

Plant-based diets are associated with numerous health benefits. Research suggests that vegetarians have a lower risk of obesity, heart disease, hypertension, cancer, type 2 diabetes, and overall mortality.

Cancer is a disease that results from the uncontrolled growth or division of abnormal cells that invade a part or parts of the body. Cancer develops in a multistep process that includes initiation, promotion, and progression. Dietary choices can both increase and decrease risk of cancer.

Phytochemicals are chemical constituents in plants that have a wide array of effects in the body and are associated with many benefits to human health.

Vegetarian diets that exclude all or most animal foods require planning to meet nutrient needs. Nutrients of potential concern include protein, iron, calcium, vitamin D, riboflavin, vitamin B_{12}, omega-3 fatty acids, and iodine.

Complementary protein foods are incomplete protein plant foods that when consumed together or during the course of a day provide all essential amino acids to help meet protein needs.

Organic is a labeling term that indicates that a food or other agricultural product has been produced through approved methods and follows USDA specifications for organic regulations.

The traditional Mediterranean diet has unique characteristics and multiple health benefits.

NEED TO KNOW

Review Questions

1. Marilyn follows a vegan diet, so which of the following foods would she avoid?
   a. butter
   b. coconut oil
   c. maple syrup
   d. peanut butter
   e. white rice

2. Pete is a quasi-vegetarian, so he includes all of the following in his diet, EXCEPT:
   a. butter.
   b. eggs.
   c. lean beef.
   d. shrimp.
   e. turkey.

3. Studies examining the benefits of vegetarianism cite all of the following, EXCEPT:
   a. a lower risk of cancer.
   b. a lower incidence of iron-deficiency anemia.
   c. a lower risk of obesity.
   d. a lower incidence of high blood pressure.
   e. lower rates of type 2 diabetes.

4. Vegetarians that emphasize whole plant foods tend to have higher intakes of all of the following, EXCEPT:
   a. dietary fiber.
   b. magnesium.
   c. potassium.
   d. riboflavin.
   e. vitamin C.

5. During which stage of cancer development does the body lose control over the mutated cells allowing metastasis to other tissues?
   a. initiation
   b. promotion
   c. progression
   d. proliferation
   e. termination

6. Characteristics of the polyphenols include all of the following, EXCEPT:
   a. they are the most abundant and diverse category of phytochemicals.
   b. they are found in salmon and other cold water fish.
   c. they are found in coffee, tea, and red wine.
   d. they include flavonoids and isoflavones.
   e. many have anti-inflammatory effects.

7. The type of phytochemicals responsible for the yellow, orange, and red colors of fruits and vegetables are:
   a. anthocyanins.
   b. carotenoids.
   c. isoflavones.
   d. lignans.
   e. resveratrol.

8. All of the following are true about complementary protein foods, EXCEPT:
   a. they must be consumed at the same meal or snack to support protein synthesis.
   b. together they can improve overall protein quality.
   c. they have different amino acids that together provide all essential amino acids.
   d. that an example would be refried beans on a corn tortilla.

9. Iron found in plant foods (non-heme iron) is lower in bioavailability than the heme iron found in animal foods. Thus, the U.S. Dietary Reference Intake of non-heme iron is ________ higher for vegetarians than the iron recommendation for nonvegetarians.
   a. no
   b. 20%
   c. 40%
   d. 60%
   e. 80%

10. Jason follows a vegan diet and avoids all animal foods. He consumes a wide variety of vegetables, fruits, grains, beans, and other whole plant-based foods. Despite his nutrient-rich diet, supplementation or use of foods fortified with ________ is recommended.
    a. potassium
    b. protein
    c. extra fiber
    d. vitamin B_{12}
    e. vitamin K

11. Based on several current studies, organic foods do not differ significantly from nonorganic in:
    a. cost.
    b. nutrient content.
    c. pesticide content.
    d. herbicide content.

12. Characteristics of the traditional Mediterranean diet include all of the following, EXCEPT:
    a. an avoidance of alcohol.
    b. an emphasis on healthy fats.
    c. limited consumption of red meat.
    d. limited consumption of refined, processed foods.
    e. small portions of nuts.

Take It Further

Develop a daily menu that excludes all animal foods (vegan), but includes at least two types of complementary protein foods.

Dietary Analysis Using SuperTracker

Evaluating the vegetarian diet

Research has demonstrated many health benefits associated with vegetarian diets, but careful diet planning is critical in maintaining adequate nutrient intake.

1. Log on to the United States Department of Agriculture (USDA) website at www.supertracker.usda.gov. If you have not done so already, you will need to create a profile to get a personalized diet plan. This profile will allow you to save your information and diet intake for future reference. *Do not use the general plan.*
2. Click the Track Food and Activity option.
3. Record your food and beverage intake for one day that most reflects your typical eating patterns. Enter each food and beverage you consumed into the food tracker. Note that there may not always be an exact match to the food or beverage that you consumed, so select the best match available.
4. Once you have entered all of your food and beverage choices into the food tracker, on the right side of the page under the bar graph, you will see Related Links: View by Meal and Nutrient Intake Report. Print these reports and use them to answer the following questions:
    a. Evaluate your intake of dietary protein. Did you meet the targets for protein on this day?
    b. Identify the plant protein foods that you consumed on this day. Identify the animal sources of protein that you consumed on this day.
    c. Did you meet your dietary target for fiber for this day? How would consuming more plant foods, especially plant protein foods, affect fiber intake?
    d. Were you above the daily limits for saturated fat and cholesterol for this day? How could adopting a plant-based diet affect saturated fat and cholesterol intake?
    e. Did you meet your targets for vitamin D, vitamin B_{12}, and omega-3 fatty acids for this day? How might your intake change if you consumed a plant-based diet?
    f. Identify three potential health benefits associated with consuming a plant-based diet.

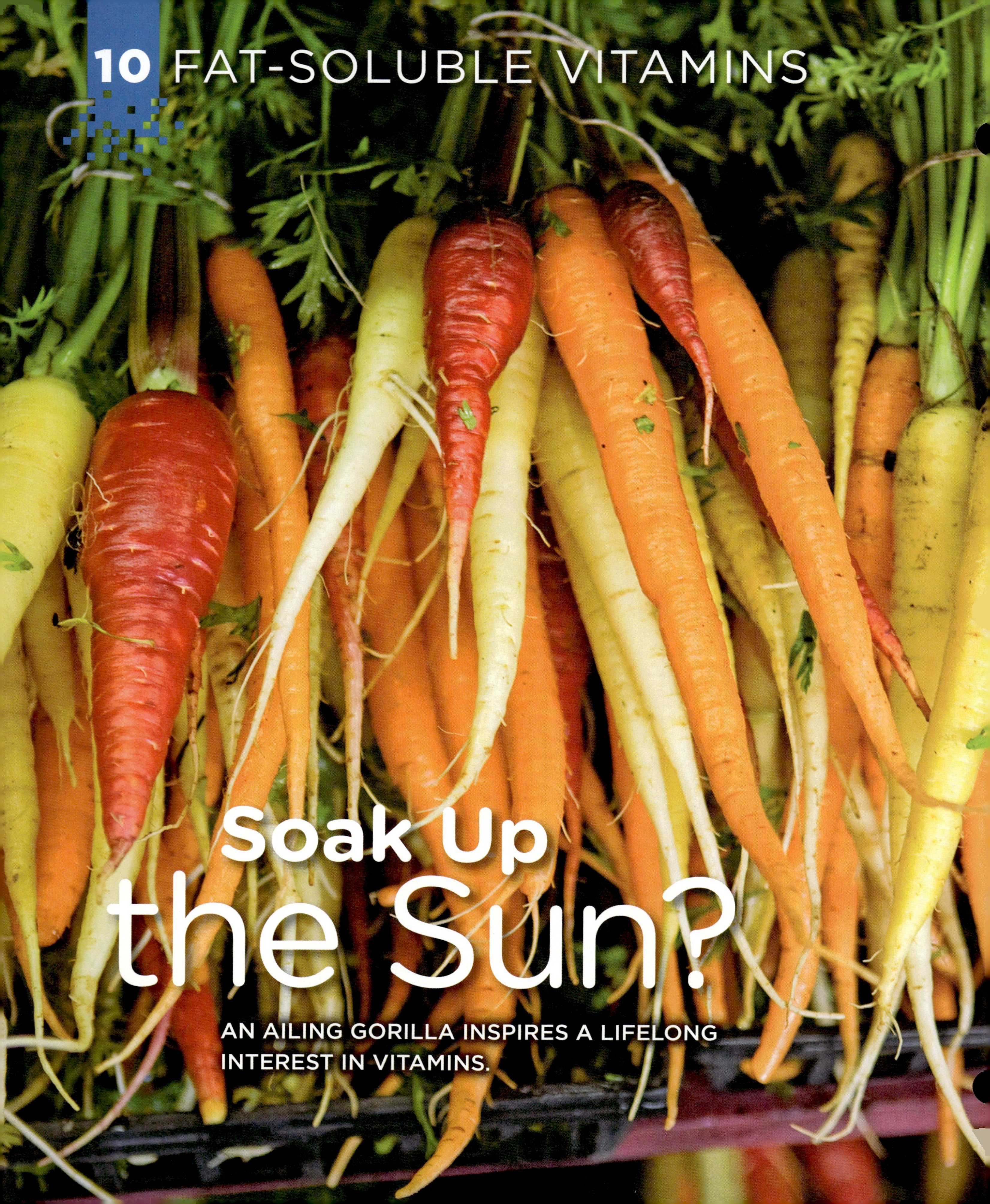

10 FAT-SOLUBLE VITAMINS

Soak Up the Sun?

AN AILING GORILLA INSPIRES A LIFELONG INTEREST IN VITAMINS.

Kate Williams/Getty Images

LEARNING OBJECTIVES

- Identify the fat-soluble vitamins and their primary functions (Infographic 10.1)
- Describe the properties of the fat-soluble vitamins (Infographic 10.2)
- Describe the role of vitamin D in blood calcium regulation (Infographic 10.3)
- Identify excellent dietary sources of the fat-soluble vitamins (Infographics 10.4, 10.5, 10.9, and 10.11)
- Describe the role of vitamin A in healthy vision (Infographic 10.6)
- List sources of free radicals and describe how antioxidants can block oxidative damage (Infographics 10.7 and 10.8)
- Explain how vitamin E works as an antioxidant in the cell membrane (Infographic 10.10)

In 2005, Michael Holick, PhD, MD, received a desperate call. Kimani, a young lowland gorilla—the first to be born at a local zoo—was dying. Could he help?

When Holick, an endocrinologist at Boston University, arrived at Franklin Park Zoo, his heart broke. The seven-month-old gorilla was so weak she could hardly nurse from her mother and her parents were exhibiting obvious signs of stress. After examining her, Holick could see immediately that Kimani had many signs of a disease called **rickets**, in which inadequate intake, or improper absorption or utilization of a few minerals leads to bone malformations and muscle weakness.

In humans, rickets affects bone development in children usually because of an extreme vitamin D

Kimani the lowland gorilla suffered from the deficiency disease rickets, but her condition improved when given supplementary vitamin D.

David L Ryan/The Boston Globe via Getty Images

RICKETS *a condition caused by vitamin D deficiency, characterized by "bowed" legs due to impaired bone mineralization, softening of bones, skeletal malformations, and muscular weakness*

and calcium deficiency. Although it is a relatively rare condition among humans in the United States, here it was in a young gorilla in captivity.

Kimani's doctors had tried giving her a vitamin D supplement, but it hadn't made much of a difference. Holick immediately told them to increase the dosage more than tenfold; this was no time to be conservative. Fortunately, Kimani responded to the treatment and her symptoms significantly improved. Although she will likely never be as tall as other gorillas her age, she eventually recovered from her bout with rickets. At her first birthday party, Holick was the guest of honor. The experience fascinated Holick and launched a lifelong interest in the functions of vitamins.

Vitamins are abundant in whole foods such as fruits and vegetables.

Dasha Petrenko/Shutterstock

INTRODUCING THE VITAMINS

Vitamins are *organic compounds* (compounds containing both carbon-carbon and carbon-hydrogen bonds) that are needed in small quantities for very specific functions, such as the maintenance of regulatory and metabolic processes in the body. Consequently, vitamins (and minerals) are considered ***micro*nutrients**, because they are required in much smaller amounts than *macro*nutrients (carbohydrates, fat, protein, and water). In nearly all cases, human cells cannot synthesize vitamins, so it is essential to obtain them through diet. The few vitamins manufactured by the body (or by intestinal bacteria) are not produced in amounts sufficient to meet our needs. If we don't consume enough of a vitamin, or are unable to adequately absorb or utilize it, we may develop symptoms characteristic of an insufficiency of that vitamin.

In many cases, researchers discovered each of the individual vitamins by working backwards from symptoms or diseases that resulted from deficiencies—such as when people (or lowland gorillas) fell ill because they lacked enough of a particular vitamin. As we'll discuss in Chapter 11, early pioneers in vitamin research recognized that there was a link between these unique and "vital" constituents in foods and the maintenance of health and body functions. The observations, research, and isolation of specific vitamins spanned decades and, even today, the door isn't closed to the possible identification of additional vitamins.

Vitamins and macronutrients are essential nutrients found in foods, but they differ in significant ways. For instance, macronutrients in foods are typically complex structures that the body must break down into individual units before they can be utilized for energy or other functions; vitamins, in contrast, are already individual units. Macronutrients (including water) are also incorporated into structures of the cell—lipids, for example, are incorporated into cell membranes—but vitamins do not have this role in cellular structure. Finally, carbohydrates, lipids, and proteins provide energy; vitamins do not.

BIOAVAILABILITY AND SOLUBILITY

Even though vitamins are not a source of energy, we need them to extract energy from macronutrients, as well as to provide many regulatory and metabolic functions in the body. Vitamins are also necessary for proper cell functioning, development, and growth. The **bioavailability**, or the degree to which nutrients can be absorbed and utilized by the body, is influenced by many factors, including physiological and dietary conditions. The presence of disease, chronic alcohol abuse, and age-related physiological changes decrease bioavailability. The bioavailability of nutrients is also affected by the presence of other nutrients or food components that can enhance or reduce absorption, or by food-handling practices, such as whether you cook a food or eat it raw.

Vitamins are commonly grouped according to their **solubility**—or ability to disperse or dissolve in water. Solubility is an important determinant of how vitamins are absorbed and transported in the body. Solubility also affects where the vitamins are stored in the body and, potentially, the risk of toxicity (adverse effects from getting too much). Of the 14 currently identified vitamins, four dissolve in fat and are stored in body tissue. These **fat-soluble vitamins**, namely vitamins A, D, E, and K, are the subject of this chapter. The B vitamins, choline, and vitamin C—which disperse easily in water-based solutions, such as blood—are called **water-soluble vitamins** and are the subject of Chapter 11.

PROPERTIES OF FAT-SOLUBLE VITAMINS

The four distinct fat-soluble vitamins that are required by the body are vitamins A, D, E, and K. The fat-soluble vitamins have diverse functions in the body. For example, vitamin A is involved in eyesight, while others are involved in such roles as bone growth, blood clotting, and antioxidant activity. (INFOGRAPHIC 10.1)

Because fat-soluble vitamins are not soluble in water, they generally need the presence of dietary fat and the assistance of bile to be absorbed effectively. Once they are absorbed, fat-soluble vitamins, like fats, typically leave the small intestine in chylomicrons via the lymph. They aren't excreted as easily as water-soluble vitamins, which means they are more likely to cause toxicity. But this rarely occurs

VITAMINS
organic compounds that are required in small quantities for specific functions in the body

MICRONUTRIENTS
essential dietary constituents needed in small quantities for human health, such as vitamins and minerals

BIOAVAILABILITY
the degree to which nutrients can be absorbed and utilized by the body

SOLUBILITY
the ability of a substance (solute) to dissolve in a solution (solvent)

FAT-SOLUBLE VITAMINS
essential micronutrients; vitamins A, D, E, and K; soluble in fat, require the presence of bile for absorption, and are stored in body tissue

WATER-SOLUBLE VITAMINS
vitamins that disperse easily in water-based solutions; includes the B vitamins, vitamin C, and the vitamin-like nutrient choline

INFOGRAPHIC **10.1** **Primary Functions of Fat-Soluble Vitamins** *Fat-soluble vitamins are stored in your body and used when your dietary intake falls short. Each fat-soluble vitamin plays an important role in the body. See Appendix 3 for a review of vitamins and minerals by function in the body.*

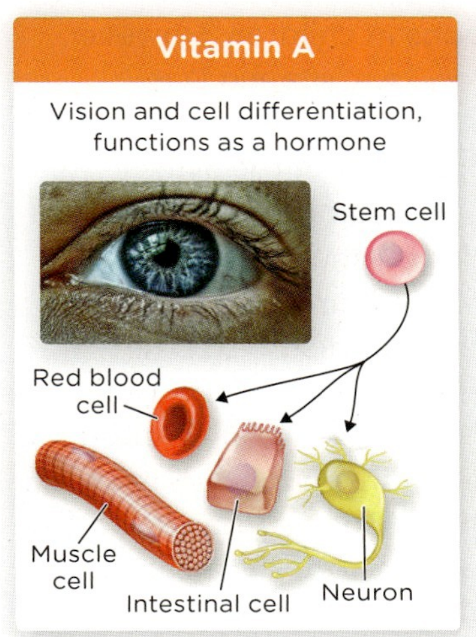

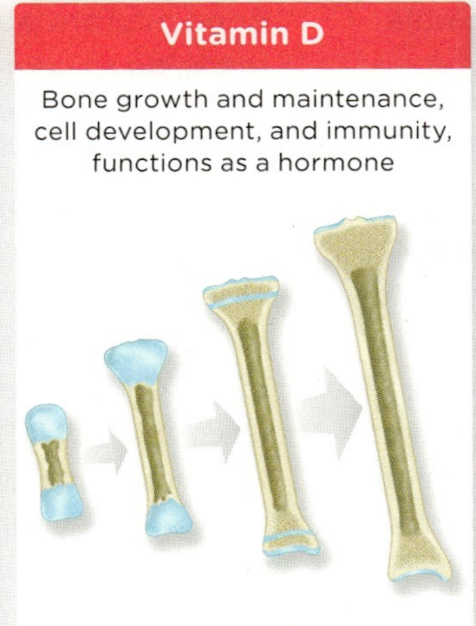

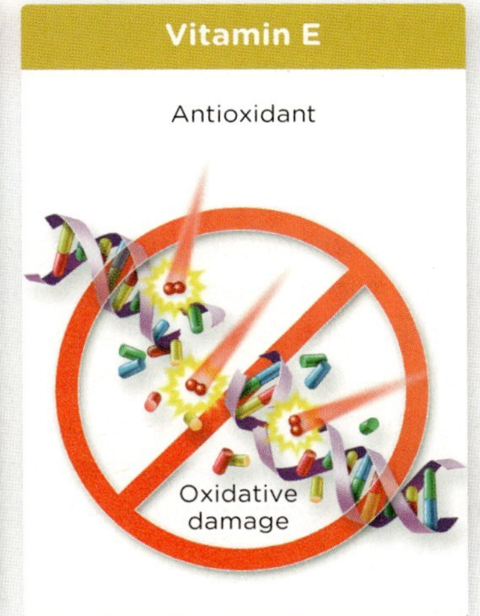

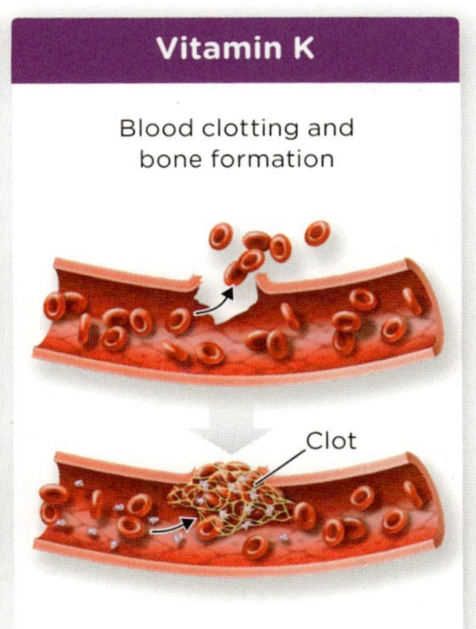

Photo credit: Michael Siward/Getty Images

INFOGRAPHIC **10.2** **Properties of the Fat-Soluble Vitamins**

? *Under what circumstances might a person be at risk from a deficiency of fat-soluble vitamins?*

with the exclusive consumption of vitamins that are naturally present in food—mostly, it is the result of consuming fortified foods and supplements at levels that exceed established Tolerable Upper Intake Levels (ULs).

When we take in more of a fat-soluble vitamin than we need, it—again, like fat—is stored in fatty tissues and the liver, from which it is released when the body needs it. And since the body may have significant stores of fat-soluble vitamins, deficiency symptoms and disease may take longer to appear than with water-soluble vitamins, which are not stored in significant amounts. Although each vitamin has its own distinct functions, metabolism, and mechanism of action, there are some general properties to keep in mind. (INFOGRAPHIC **10.2**)

VITAMIN D

Most people have not considered that a primate could experience a vitamin D deficiency and be stricken with rickets, a condition typically associated with bone malformations in children. However, many people know that vitamin D is often called the "sunshine vitamin," since it can be produced in the skin from cholesterol with exposure to ultraviolet (UV) light. Most of us meet at least some of our vitamin D needs this way, but with limited sun exposure or impaired synthesis it becomes important to consume sufficient vitamin D through our diet. But, whether through synthesis in the skin or through foods, vitamin D must be activated in the kidneys and

Soaking up the sun. *Sunlight can be a significant source of vitamin D. The body makes vitamin D when skin is exposed to the sun's ultraviolet B (UVB) energy.*

Catchlight Visual Services/Alamy

liver to fulfill its biological functions in the body.

■ ■ ■

And that, Holick realized, is ultimately what had sickened Kimani—as an animal in captivity, she hadn't been getting enough sunlight or enough dietary sources of vitamin D, so she required supplements to help her recover. Holick was well aware of the potential benefits of supplemental vitamin D in rectifying deficiency issues, so although Kimani's recovery was gratifying, it was not surprising.

Sources of vitamin D

Although sunlight can be a significant source of vitamin D, the RDAs—established to maintain bone health and normal calcium metabolism in healthy people—are set on the basis of minimal sun exposure. In 2010, the Institute of Medicine increased the RDA for vitamin D for all age groups—it was increased threefold to 600 IU (15 mcg), for example, for adults in the 19 to 50 years age range. However, the recent appreciation of the importance of vitamin D for functions beyond that of bone health has caused some scientists to question whether the new RDAs are actually high enough to provide for optimal health.

It can be difficult to meet those RDAs from food alone, because vitamin D is present in very few foods naturally. Some of the best sources include fatty fish (such as salmon, tuna, and mackerel) and fish liver oils. Beef liver, cheese, eggs, and some mushrooms (particularly those exposed to UV light) contain small amounts of vitamin D. Alternatively, most Americans get dietary vitamin D from fortified foods—most of the U.S. milk supply is fortified with vitamin D (100 IU, or 2.5 mcg, per cup). This is the result of a milk fortification program that was instituted in the 1930s in the United States to reduce the incidence of rickets,

INFOGRAPHIC 10.3 Dietary Sources of Vitamin D *Vitamin D does not occur naturally in many foods besides fish; the most common sources in the U.S. diet are fortified foods.*

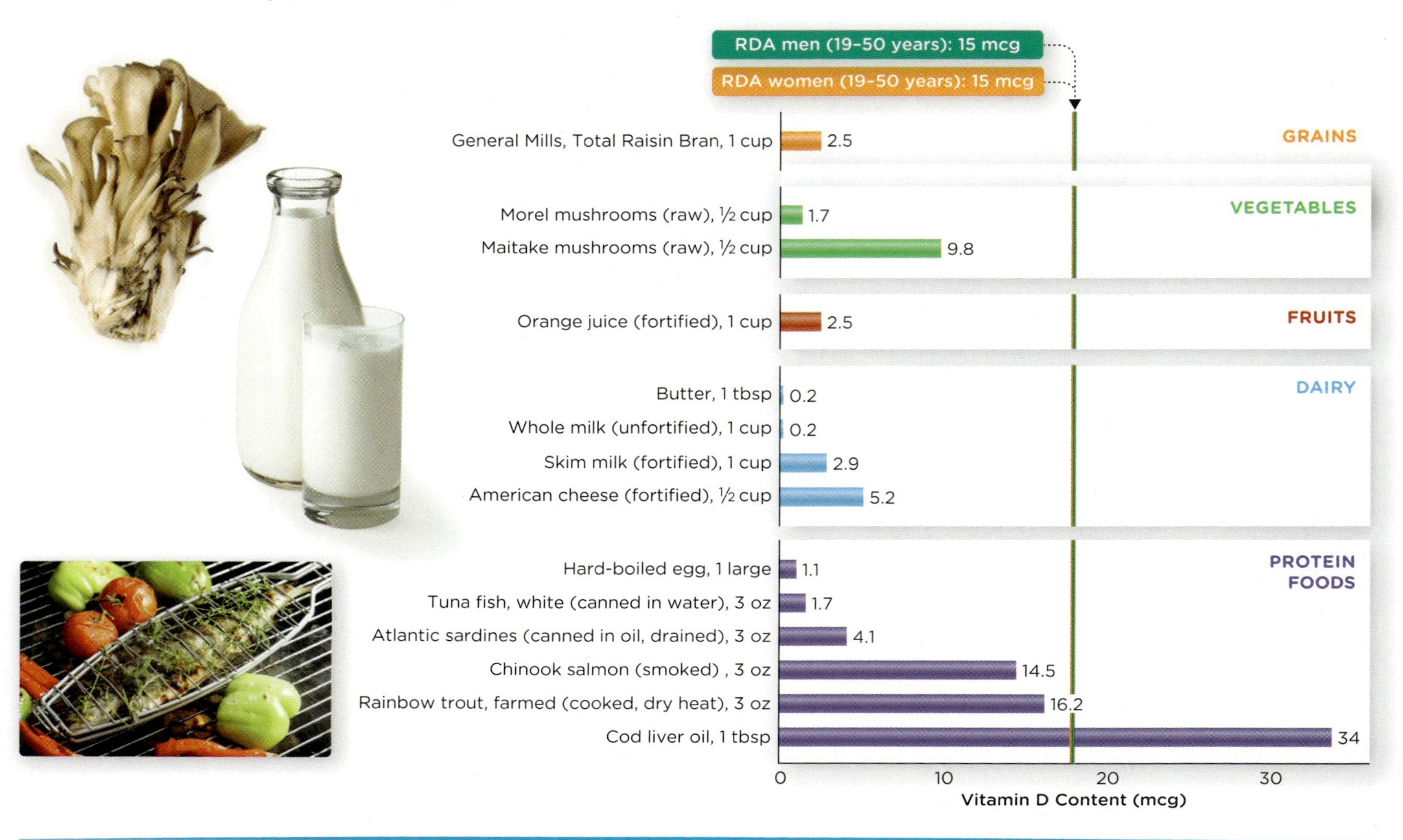

? *Check the food labels at home to determine which foods have been fortified with vitamin D.*

Photo credits (top to bottom): Daniel Bendjy/Getty Images, Paul Johnson/Getty Images, Lehner/Getty Images

a significant public health issue at the time. **(INFOGRAPHIC 10.3)**

Functions of vitamin D in the body

Vitamin D has many important known functions in the body. Mainly, as illustrated by the case of Kimani, vitamin D plays a key role in the growth and maintenance of bone by maintaining blood concentrations of minerals involved in bone development, such as calcium and phosphorus. In addition, vitamin D works in combination with other nutrients and hormones in bone growth and maintenance. But beyond bone health, research continues to explore and expand vitamin D's role in human health.

Vitamin D actually functions as a hormone, since, like hormones, it is made in one part of the body, but carries out its regulatory effects elsewhere. Since it is made from cholesterol, vitamin D is a member of the steroid family of hormones (which includes estrogen, testosterone, and cortisol, among others).

Vitamin D also plays an important role in the regulation of calcium. Most calcium in the body is stored in bones (see Chapter 13), with less than 1% circulating in the blood; however, this circulating calcium is critical for normal muscle and nerve functioning and must be maintained. **(INFOGRAPHIC 10.4)**

The regulation of blood calcium requires the interaction of bones, kidneys, and intestines. For example, when the concentration of calcium in blood falls, vitamin D is converted into its fully active hormone form (*calcitriol*) in the kidneys. This active hormonal

INFOGRAPHIC 10.4 Vitamin D Activation and Calcium Maintenance in the Blood

Vitamin D functions as a hormone to regulate calcium metabolism. Together with parathyroid hormone, vitamin D tightly controls blood (serum) concentrations of calcium.

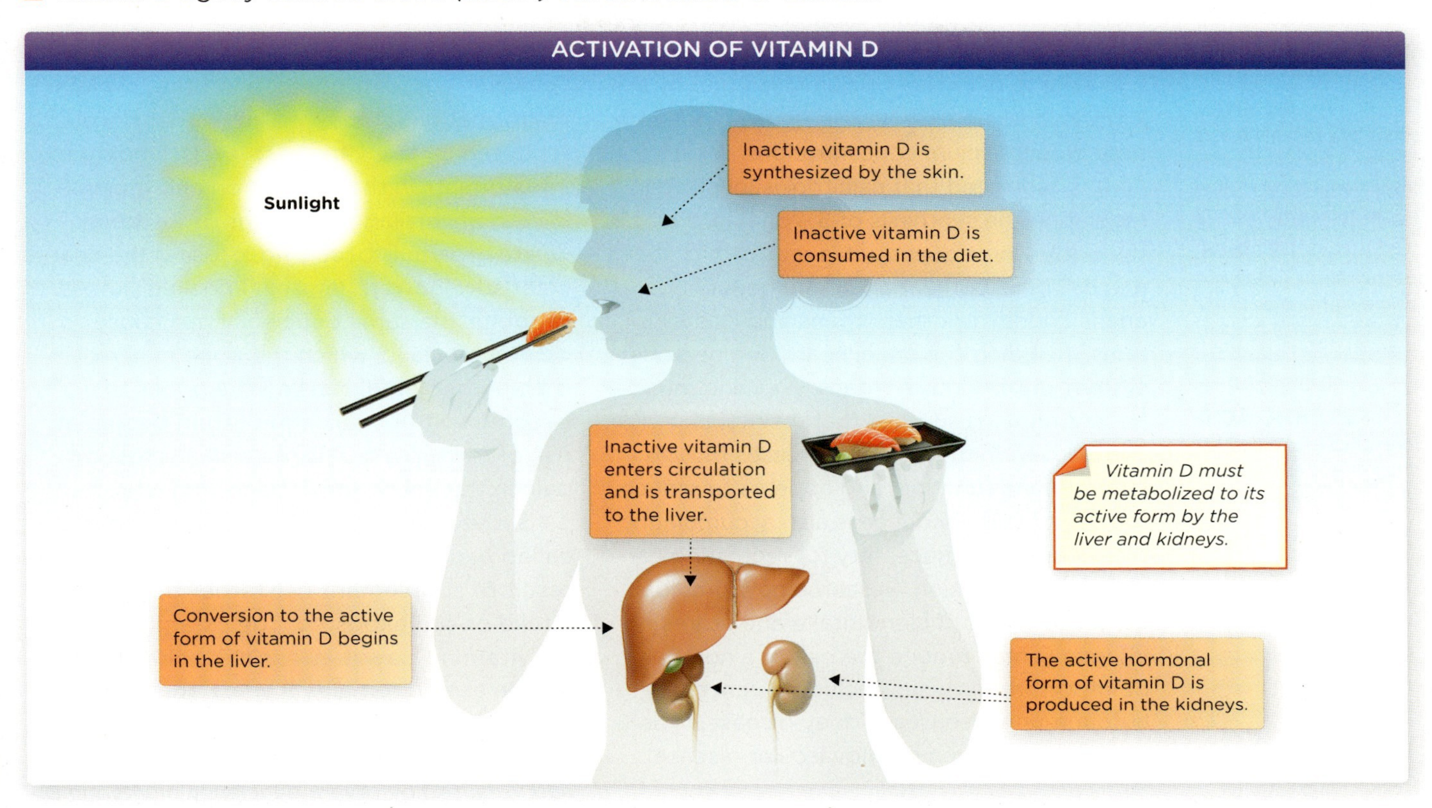

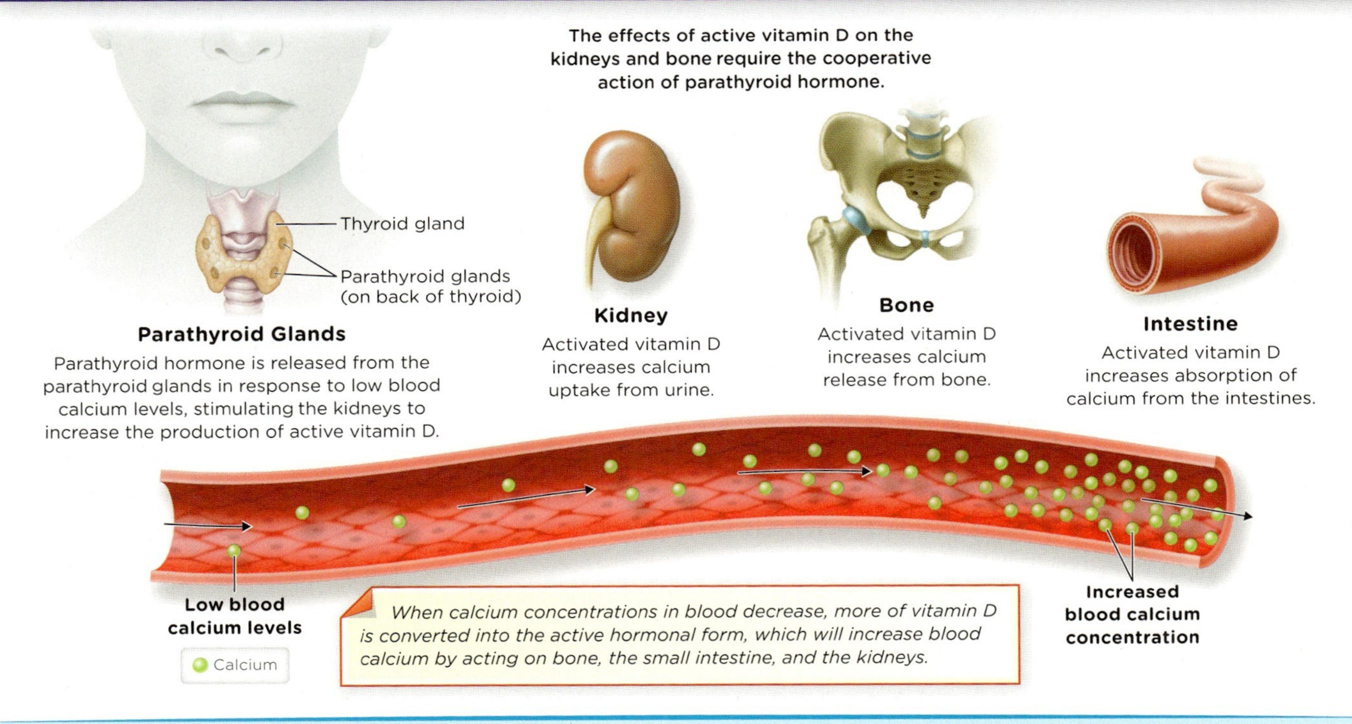

? *Why might the maintenance of appropriate blood calcium concentrations be impaired in someone with advanced liver or kidney disease?*

OSTEOMALACIA
the adult form of rickets, characterized by softening and weakening of bones (brittle bones) accompanied by pain in the pelvis, lower back, and legs

RETINAL
a form of vitamin A that is critical for vision and is derived from the conversion of retinol in the body

RETINOL
the active form of vitamin A in animal and fortified foods; the form of vitamin A stored in the body and transported between tissues

RETINOIC ACID
a form of vitamin A derived from retinal; essential for growth and development

form of vitamin D then increases the absorption of calcium in the gastrointestinal tract—for this reason, you will often find calcium supplements that also contain vitamin D. Activated vitamin D also works together with parathyroid hormone to decrease calcium excretion by the kidneys, and increase its release from bone to raise blood calcium concentrations.

It is estimated that calcitriol is involved in regulating the synthesis of 5% of all proteins in the body. Consequently, it is required for normal cell development and immune function and is critical for other organs and body systems, including the brain, heart, and nervous system, reproductive organs, skin, and muscle. Recent findings suggest that maintaining vitamin D levels in the elderly increases muscle strength and function, and appears to reduce the risk of falls. Evidence is also accumulating that vitamin D is necessary for regulating cardiovascular function. Research now provides evidence that sufficient intake and stores of vitamin D may reduce the risk of a number of important diseases, including cancer, autoimmune diseases, kidney disease, type 2 diabetes, and even cardiovascular disease.

Vitamin D deficiency and toxicity

Vitamin D is listed as a nutrient of concern in the 2015 Dietary Guidelines for Americans. According to data from the 2007-2010 National Health and Nutrition Examination Survey, 94% of all Americans consume less than adequate intakes of vitamin D increasing risk of deficiency particularly in vulnerable populations. People most at risk are those who avoid vitamin D-fortified dairy products (because of allergies, intolerance, or a vegan diet); people with dark skin (since the melanin that darkens skin can interfere with the body's ability to make vitamin D from sunlight); and those with little sun exposure, or who regularly use sunscreen. Although sunscreen can block UV rays and reduce the risk of skin cancer, it also prevents vitamin D synthesis. Other groups at risk include infants who are exclusively breast-fed, since breast milk is not a rich source of vitamin D, and the elderly, who have a reduced ability to synthesize vitamin D when exposed to sunlight. Many elderly individuals are also more likely to stay indoors and therefore have less opportunity to synthesize vitamin D.

Having extremely low levels of vitamin D can, over time, cause serious bone diseases. Starting in the mid-1600s, researchers began describing mysterious symptoms in children that Holick instantly recognized in Kimani—"bowed" legs, soft bones, and other skeletal malformations. Today we know that rickets is caused by low vitamin D levels, leading to impairments in the maturation and mineralization of cartilage in regions of the bone where growth is occurring, causing the characteristic "bowed" legs or "knocked" knees. In adults, the same deficiency can cause **osteomalacia**, in which the bone mineral is being depleted, causing the bones to become soft or weak and putting people at risk of fractures or falls, as well as creating pain in the pelvis, lower back, and legs.

But you can also get too much vitamin D—called *hypervitaminosis D*—which is likely the result of supplement use rather than food intake or sun exposure. The established UL for vitamin D is set at 4,000 IU (100 mcg) for males and females 9 years and older.

Rickets causes softening of bones in children and is caused by severe deficiency of vitamin D.

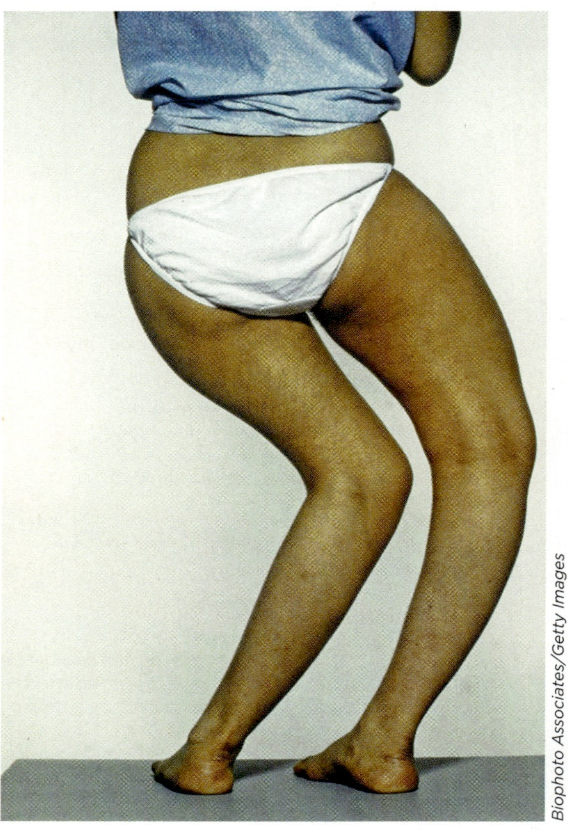

Biophoto Associates/Getty Images

The symptoms of hypervitaminosis D vary, but may include loss of appetite, weight loss, irregular heartbeat, and frequent urination. Having too much vitamin D can also increase levels of calcium in the blood, which leads to calcification of the soft tissue and damage to the heart, blood vessels, and the formation of kidney stones.

VITAMIN A

Despite the recent focus on vitamin D, and what it can or cannot do for human health, the fat-soluble vitamin with the longest history is vitamin A—the first vitamin ever identified (thus the decision to name it with the first letter of the alphabet). Scientists originally called it "fat-soluble A," since it was initially identified in animal fats.

Stored primarily in the liver, vitamin A encompasses a group of fat-soluble compounds composed of *retinoids* and several *carotenoids*. *Retinoid compounds* are **retinal**, **retinol**, and **retinoic acid**. Retinol is also referred to as **preformed** vitamin A because it is already in an active form in foods. Though retinol is the only retinoid present in significant amounts in our diet, once it is absorbed we can readily (and reversibly) convert it to retinal, and a small amount of retinal is irreversibly converted to retinoic acid. In contrast, compounds that have vitamin activity only after conversion to active forms are **provitamins**. For example, provitamin A **carotenoids** are compounds (primarily **beta-carotene**) that have vitamin A activity once they are converted in the body into one of the active forms of the vitamin (retinol, retinal, or retinoic acid).

Sources of vitamin A

In the United States, approximately two-thirds of the vitamin A is consumed as preformed vitamin A (retinol) from fortified foods, and supplements, and foods that naturally contain retinol. For example, preformed vitamin A occurs in animal products; it is highest in liver, but fish, eggs, and dairy foods (containing fat) are also good sources. In the United States and Canada, reduced-fat milk and yogurt must be fortified with vitamin A to make up for the loss of this vitamin during the removal of the fat.

The rest of our vitamin A comes from provitamin A carotenoids, which must be converted into an active form in the body before they can fulfill biological functions. These compounds are the yellow, orange, and red pigments of fruits and vegetables—think sweet potatoes, carrots, cantaloupe, and apricots, as well as dark, leafy greens (which, despite their color, contain a lot of beta-carotene). Although more than 700 carotenoids have been identified, about 50 are present in commonly consumed foods, and of the approximately 12 carotenoids that have been identified in blood and tissues, only three are converted in significant amounts to active vitamin A—with beta-carotene being by far the most abundant.

The bioavailability of vitamin A differs upon the food source. Preformed vitamin A (retinol), found in foods from animal sources, is more easily absorbed than the carotenoids found in foods from plant sources. To account for these differences in bioavailability, the Dietary Reference Intake (DRI) values are expressed as micrograms (mcg; one millionth of a gram) of **retinol activity equivalents**, or RAE. The RDA for vitamin A for men aged 19 to 50 years is thus 900 mcg RAE; for women of the same age, it is 700 mcg RAE. Of the provitamin A carotenoids, beta-carotene is best converted into retinol, but this conversion is never complete.

PREFORMED
vitamins already present in their active form

PROVITAMIN
the inactive form (or precursor) of a vitamin that requires conversion to the active form to fulfill biological functions in the body

CAROTENOIDS
a group of pigments synthesized by plants, algae, and some bacteria and fungi; examples include carotenes and xanthophylls

BETA-CAROTENE
a pigment found in plants, and a precursor to vitamin A

RETINOL ACTIVITY EQUIVALENTS (RAE)
measure of vitamin A activity; accounts for differences in bioavailability between preformed vitamin A (retinol) and provitamin A carotenoids

Food sources of preformed vitamin A and sources of beta-carotene. *The bowl on the left contains liver and the bowl on the right has oily fish. These animal foods contain preformed vitamin A. Papaya, carrots, mango, and dark green leafy vegetables are good sources of beta-carotene.*

Even when beta-carotene is consumed as a supplement in oil (to improve absorption), it takes two micrograms of beta-carotene to provide the equivalent of one microgram of retinol. Because the absorption of carotenoids from foods is even poorer than from supplements, you need to eat 12 micrograms of dietary beta-carotene to reach the equivalent RAE of only one microgram of retinol. Because the release of carotenoids from food is difficult, processing the food by slicing, chopping, and even juicing can improve their bioavailability. The smaller food particles are more completely broken down by mechanical digestion than are larger particles, allowing nutrients to disperse more readily into digestive fluids. Cooking can also often increase nutrient bioavailability because it ruptures plant cells, releasing the nutrients that otherwise might be trapped within those cells and subsequently excreted. Thus, the potential vitamin A is actually higher from sliced and cooked carrots than from raw carrots. **(INFOGRAPHIC 10.5)**

INFOGRAPHIC 10.5 Dietary Sources of Vitamin A *It is estimated that about 40% of individuals living in the United States consume less than the RDA of vitamin A.*

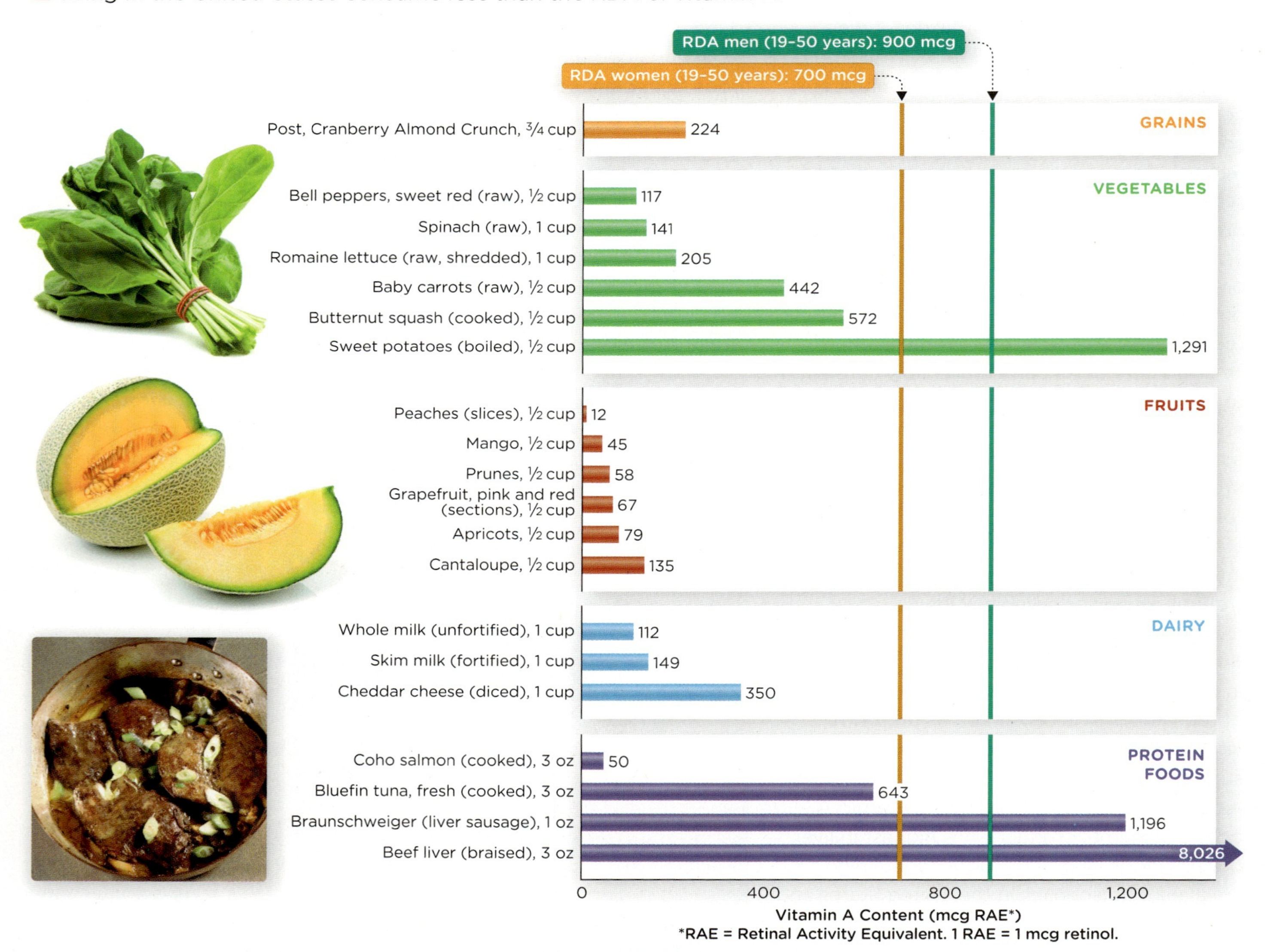

What color are many of the vitamin A–rich vegetables and fruits? Why might beef liver be rich in vitamin A?

Photo credits (top to bottom): Kim Nguyen/Shutterstock, SOMMAI/Shutterstock, John Peacock/Getty Images

Functions of vitamin A in the body

Vitamin A serves many critical biochemical and physiological functions in the body related to vision, cell development, immune function, and growth. Along with these important functions, vitamin A also plays key roles in bone health and reproduction, such as sperm and fetal development. Except for its role in vision, vitamin A almost always functions as a hormone that exerts its effects by controlling the synthesis of numerous proteins encoded in our genes.

The hormonal actions of vitamin A play a key role in normal, healthy cell development, such as in cell differentiation (the process by which cells become progressively more specialized). Vitamin A is particularly important to *epithelial cells*, which form the skin and mucous membranes inside the body, such as those present in our eyes, lungs, and intestines. It also plays a role in the development of immune cells, which affects how well our immune system functions and, therefore, influences our susceptibility to disease. In developing countries, a child's vitamin A deficiency significantly increases his or her risk of death from an infectious disease.

In addition, vitamin A is required by our eyes to convert light into nerve impulses that bring messages to the brain, telling us what we're seeing. Specifically, vitamin A is a key component of **rhodopsin**, the visual pigment that is formed when retinal binds to the protein opsin. Rhodopsin is found in light-sensing cells within the retina at the back of the eye. When rhodopsin absorbs light, retinal changes its shape and is then released, triggering a chain of events that generate a nerve impulse that transmits the visual signal to the brain. When the vitamin A level is low, these light-sensing cells are unable to quickly regenerate rhodopsin, which can make it difficult to see in low light. We also need vitamin A to maintain the health of the cornea, the clear outer covering at the front of the eye. (INFOGRAPHIC 10.6)

RHODOPSIN
a pigment in the retina that absorbs light and triggers nerve impulses to the brain for vision

INFOGRAPHIC 10.6 The Role of Vitamin A in the Detection of Light by the Eye *One of the forms of vitamin A (retinal) is a necessary component of the visual pigment rhodopsin that detects light in our eyes. The cells at the back of the eye (the rods in the retina) that are primarily responsible for allowing us to see in low light are particularly sensitive to a vitamin A deficiency.*

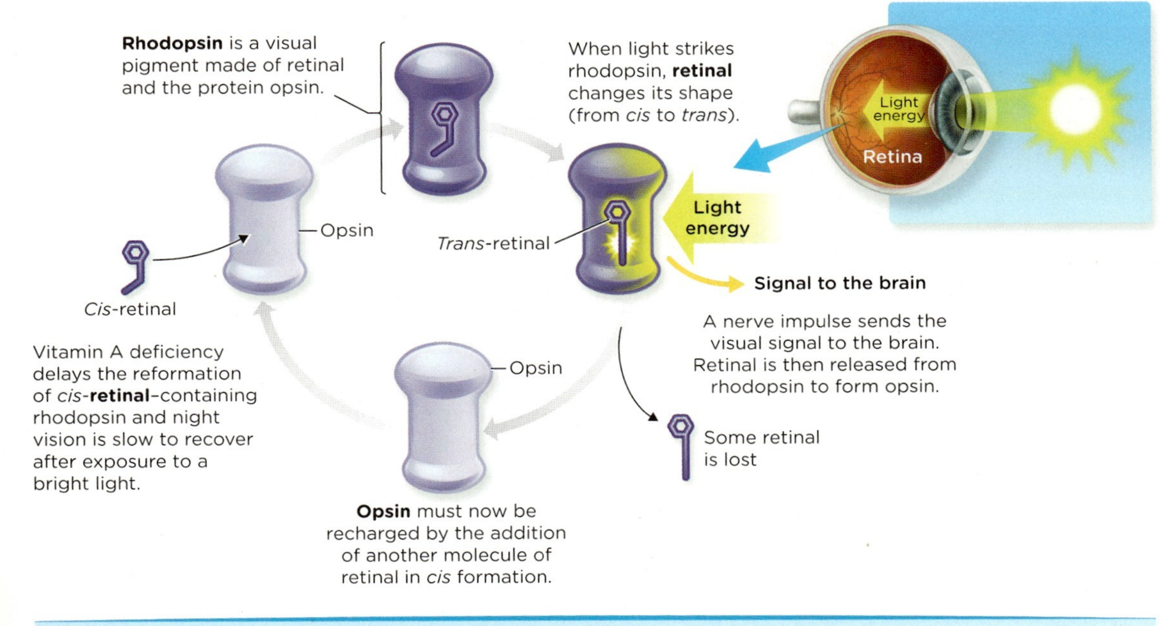

Why is the ability of the eye to detect light decreased when there is a vitamin A deficiency?

INFOGRAPHIC 10.7 Sources of Free Radicals and Their Effects *The free radicals we are exposed to can come from environmental sources or are produced by our own bodies. Although free radicals have necessary functions in our body, high levels can cause damage.*

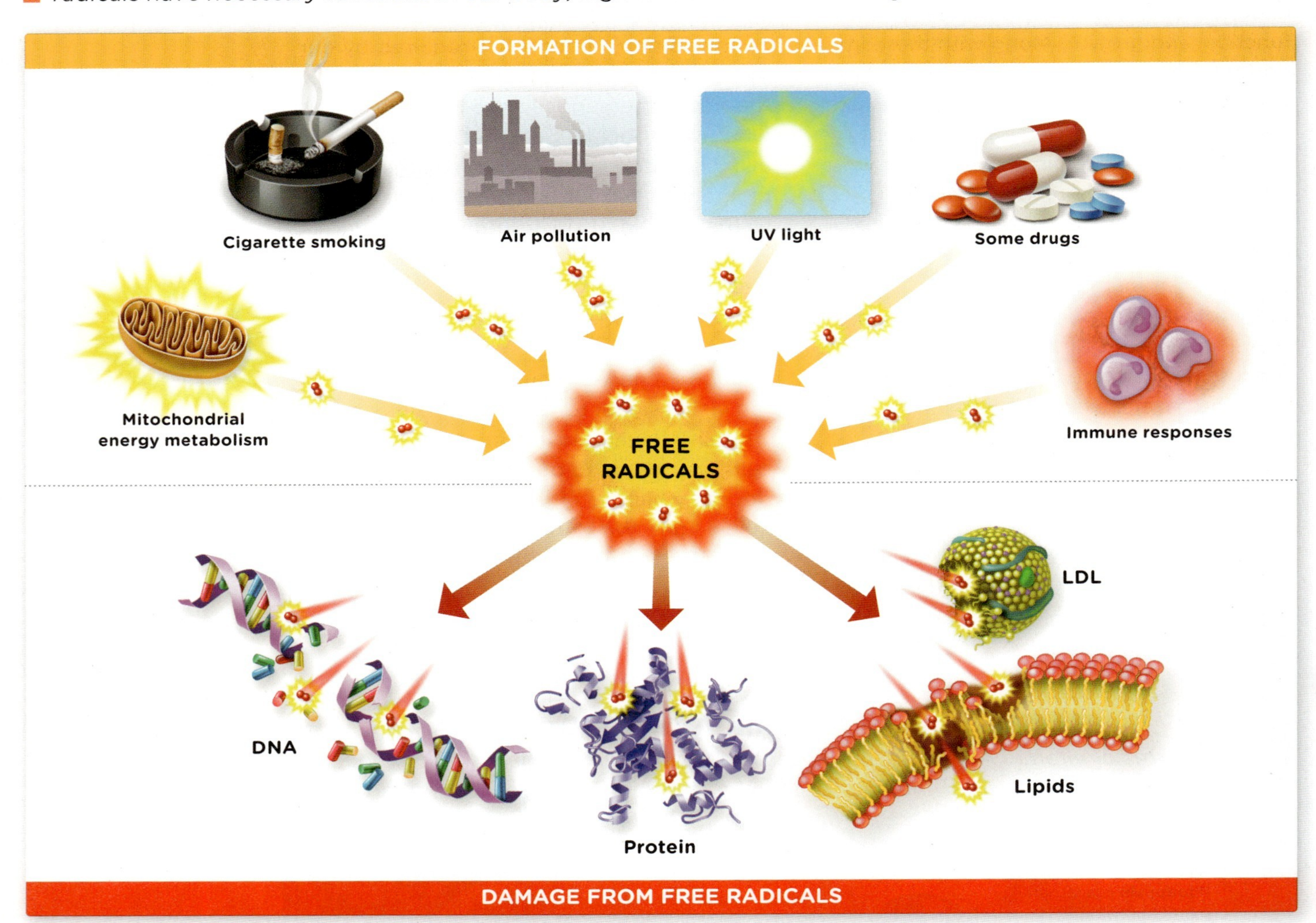

Why might those who suntan regularly have increased skin wrinkling as they age?

ANTIOXIDANT *a substance that prevents damage to cells by inhibiting the oxidation caused by free radicals*

FREE RADICALS *molecules that are naturally formed in the body or present in the environment that have an unpaired electron; at high levels they damage cells, including DNA, through a process called oxidation*

ANTIOXIDANTS PROTECT AGAINST DAMAGE FROM FREE RADICALS

The plant form of vitamin A—beta-carotene—as well as the other dietary carotenoids have important roles as health-promoting phytochemicals. One of the health-promoting functions of carotenoids may be their ability to function as **antioxidants** in the body, which helps protect cells from damage caused by highly unstable molecules called **free radicals**. Free radicals are reactive molecules with unpaired electrons that attempt to pair up with other molecules, atoms, or individual electrons to create a stable state. These can arise from many sources: they are generated during normal energy metabolism and they are present in air pollution and tobacco smoke. **(INFOGRAPHIC 10.7)** Although excess free radicals can have significant deleterious effects, we actually need some of the free radicals generated in the body for several important functions, such as cellular communication and killing bacteria.

At high levels, free radicals can cause damage known as **oxidation**, which can alter DNA and destabilize cell membranes.

Oxidative damage can lead to chronic diseases like cancer and heart disease, so eating a plant-rich diet that contains beta-carotene and other antioxidants may reduce the risk of those diseases. (Taking supplements may not be the answer: Some research has suggested that beta-carotene pills may actually increase the risks of some types of cancer in smokers.) Ample dietary intakes of carotenoids also appear to provide some protection to the skin from the damaging effects of UV-radiation from sunlight. (INFOGRAPHIC 10.8)

Vitamin A deficiency and toxicity

Given its crucial roles in the body, vitamin A deficiencies can have major effects on health. Although uncommon in the United States, people in developing countries develop *hypovitaminosis A* if they have little access to preformed vitamin A and beta-carotene. (Even in the United States, nutrition surveys suggest that most of us could benefit from higher intake closer to the DRI.) The hallmark symptoms of vitamin A deficiency affect the eye. Night blindness occurs first, the result of problems in the synthesis of rhodopsin. The

OXIDATION
a loss of electrons; in the body, this results in damage to cells

INFOGRAPHIC **10.8** **Antioxidants Defend Against Oxidative Damage Caused by Free Radicals** *Free radicals are molecules containing unpaired electrons, which makes them highly reactive. The free radical either causes oxidative damage by reacting with another molecule and chemically modifying it, or it stabilizes itself by stealing an electron from a nearby molecule, which creates a new free radical and begins a chain reaction. Antioxidants are able to stop the chain reaction, by donating an electron.*

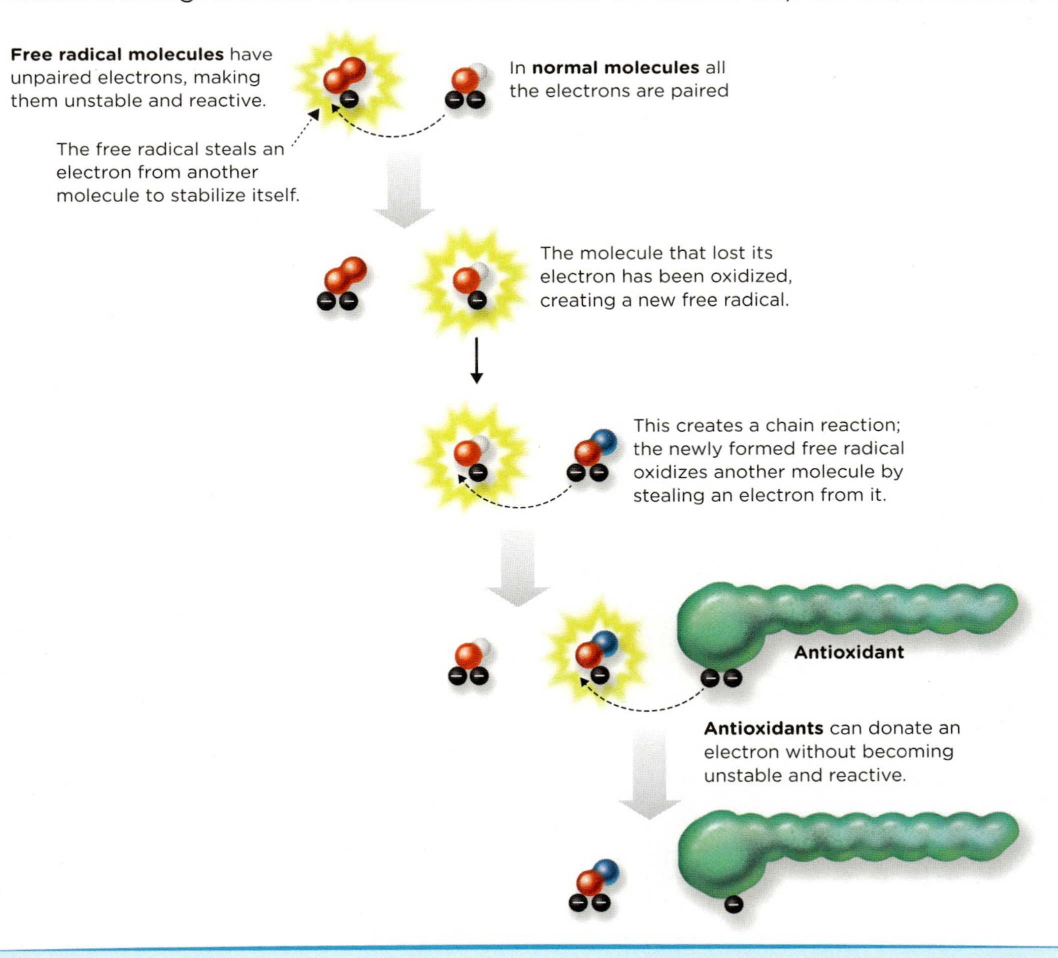

What characteristic do antioxidants possess that allow them to break the chain of oxidative damage?

Vitamin A deficiency can affect the surface of the eye. *A severe deficiency of vitamin A can cause drying and thickening of the corneas of the eye.*

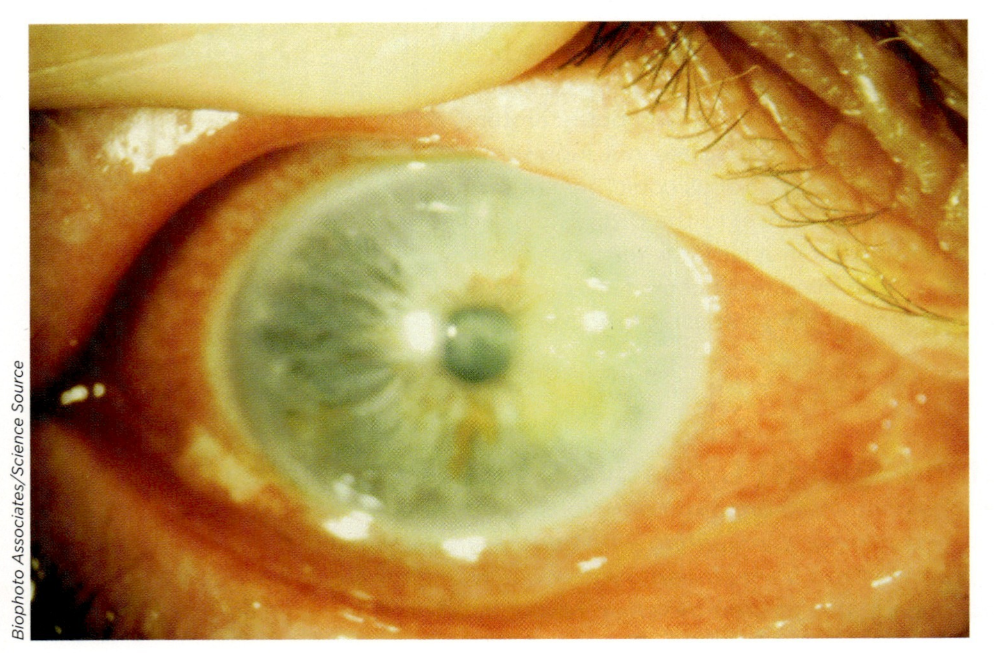

Biophoto Associates/Science Source

Yellowing of the skin. *The hand on the right shows signs of carotenemia, and the hand on the left is normal (for comparison). Carotenemia is elevated blood levels of carotene caused by excessive consumption of foods such as carrots, squash, and sweet potatoes.*

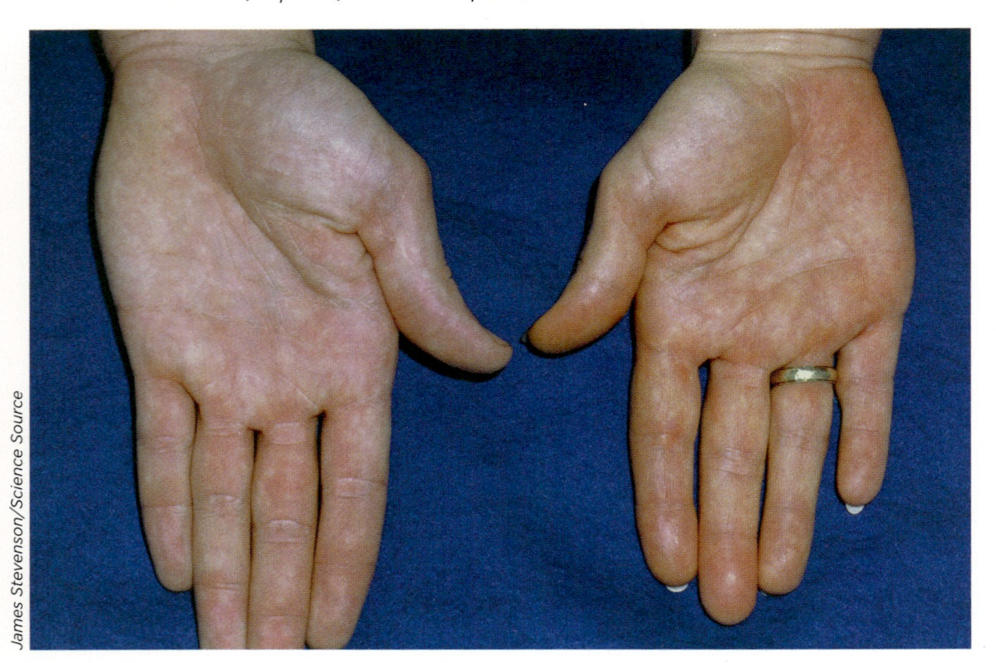

James Stevenson/Science Source

changes in protein synthesis resulting from a deficiency of vitamin A decrease the production of mucus in tears, and then the cornea of the eye begins to dry. If the deficiency continues, ulcers form on the cornea and this can lead to scarring—resulting in permanent blindness.

In fact, vitamin A deficiency is the leading cause of preventable blindness in children worldwide. Other symptoms include impaired immunity (increasing the risk of infections), and rough, dry, or scaly skin, resulting from a loss of moisture in the mucous membranes. This last symptom—known as *keratinization*—occurs as the epithelial cells that cover the surface of our bodies (like the cornea of the eye) and line body cavities cannot develop properly; as a consequence they become filled with a fibrous structural protein called *keratin.* These are the same type of proteins that make up our nails, hair, and even the horns of animals.

When people consume too much vitamin A they develop *hypervitaminosis A*, which mostly occurs after taking supplements of large quantities of vitamin A. Here, the risks are largely defects or weaknesses in bones, leading to *osteoporosis* or fractures. The UL for vitamin A for adults 19 years and older is set at 3,000 mcg RAE. In pregnant women, high amounts of vitamin A can affect fetal development and cause birth defects. But these problems can result from too much vitamin A from animal foods or from dietary supplements—excess beta-carotene from plant sources does not cause toxicity, but may temporarily turn your skin yellow or orange color in color.

VITAMIN E

Vitamin E encompasses a group of fat-soluble compounds found primarily in vegetable oils. The compounds are known as **tocopherols**—a name derived from Greek words related to fertility (*tókos* [birth] and *phérein* [to bear or carry]), since researchers originally discovered that these oil components prevent sterility in rats. However, only alpha-tocopherol supplements have been shown to reverse vitamin E deficiency symptoms in humans. For this reason, it is the only form defined by the

INFOGRAPHIC 10.9 Dietary Sources of Vitamin E *It is found in a variety of foods but the best sources are many nuts and some oils.*

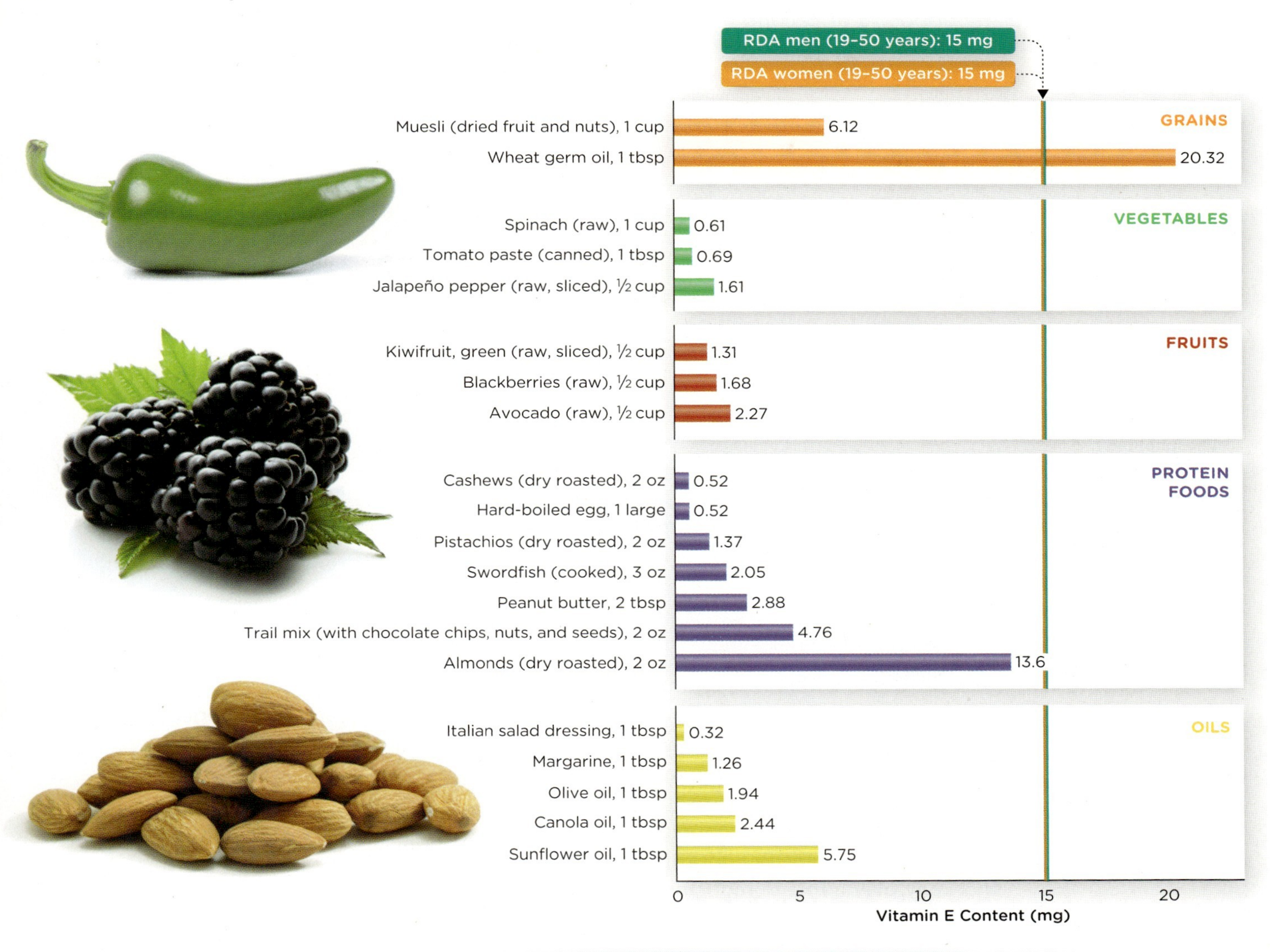

? *What foods that you consume regularly are good sources of vitamin E?*

Photo credits (top to bottom): Barry Sutton/Getty Images, Vitalina Rybakova/Getty Images, Alasdair Thomson/Getty Images

Food and Nutrition Board of the Institute of Medicine as meeting human vitamin E requirements.

Sources of vitamin E

Some foods (such as breakfast cereals) are fortified with vitamin E, but vitamin E is also found naturally in plant-based oils—primarily those from nuts and seeds (safflower and sunflower oils are good sources), and products made with these oils (such as margarine and salad dressings). Other natural sources include wheat germ, whole nuts and seeds, and leafy green vegetables. The RDA for vitamin E for adults 19 years and older is 15 mg (22.4 IU). **(INFOGRAPHIC 10.9)**

Functions of vitamin E in the body

Because of its unique ability to effectively be incorporated into cell membranes,

TOCOPHEROLS
a group of fat-soluble vitamin E molecules

INFOGRAPHIC 10.10 The Antioxidant Functions of Vitamin E *Vitamin E can break the chain of oxidation by donating an electron to free radicals without becoming unstable. Because it is a fat-soluble vitamin it is particularly good at performing this function in cell membranes, and even in lipoproteins like LDL.*

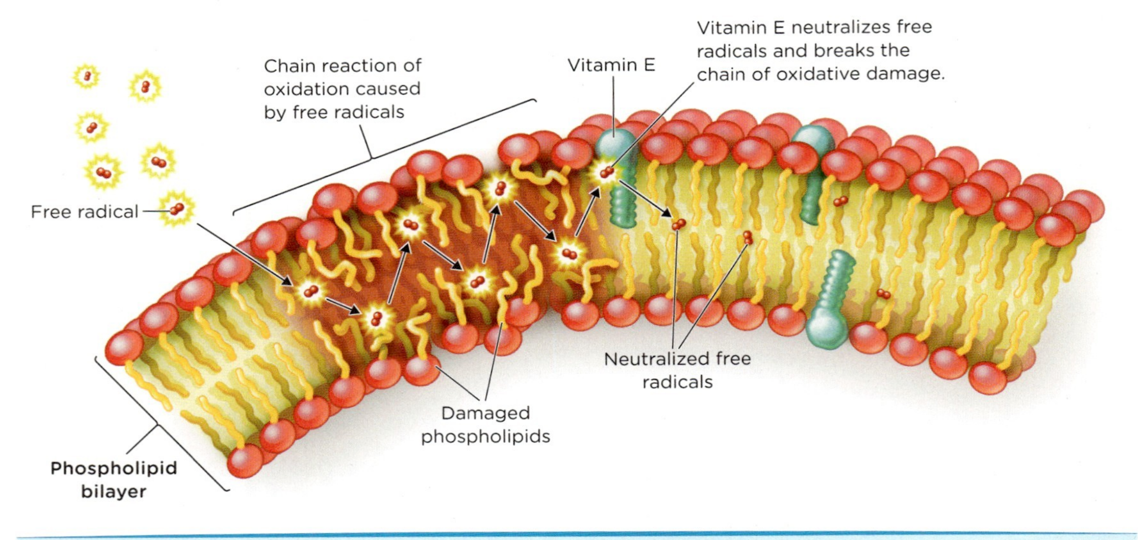

What might be the consequences of free-radical damage to the cell's DNA?

vitamin E acts within membranes as an antioxidant, protecting cells throughout the body from the oxidative damage that results from exposure to free radicals produced in the body or present in the environment. It also helps prevent oxidative changes in low-density lipoproteins (LDLs) that may play a role in reducing plaque formation in blood vessels. **(INFOGRAPHIC 10.10)**

Vitamin E appears to be of particular importance in maintaining healthy immune function by protecting white blood cells from oxidative damage, particularly in aging adults. Adequate intake of vitamin E is also necessary to prevent damage to nervous tissue. Many claims have been made about vitamin E's potential to promote health and prevent and treat disease. Despite its role as an antioxidant, rigorous studies have not produced consistent and convincing evidence that vitamin E supplements (which are almost always alpha-tocopherol) reduce the risk of cancer or heart disease. However, recent research suggests that other forms of vitamin E that are abundant in our diet may reduce our risk of cancer, cardiovascular, and neurodegenerative diseases.

Vitamin E deficiency and toxicity

Nutrition surveys suggest Americans may be getting less vitamin E than they should, but researchers have only rarely documented deficiency in healthy adults, so specific symptoms are unclear. So, too, for toxicity—vitamin E is less likely to be toxic than either vitamin A or vitamin D, and is only observed with very high intakes from supplements, which increase the tendency to bleed. Because adverse effects have only been seen at very high levels, the UL for vitamin E (1,000 mg, or 1,500 IU) is set at more than 60 times the RDA.

VITAMIN K

Another fat-soluble vitamin is vitamin K, an essential nutrient that's found in food. However, a significant amount of our daily

INFOGRAPHIC 10.11 Dietary Sources of Vitamin K *It is found most abundantly in green leafy vegetables.*

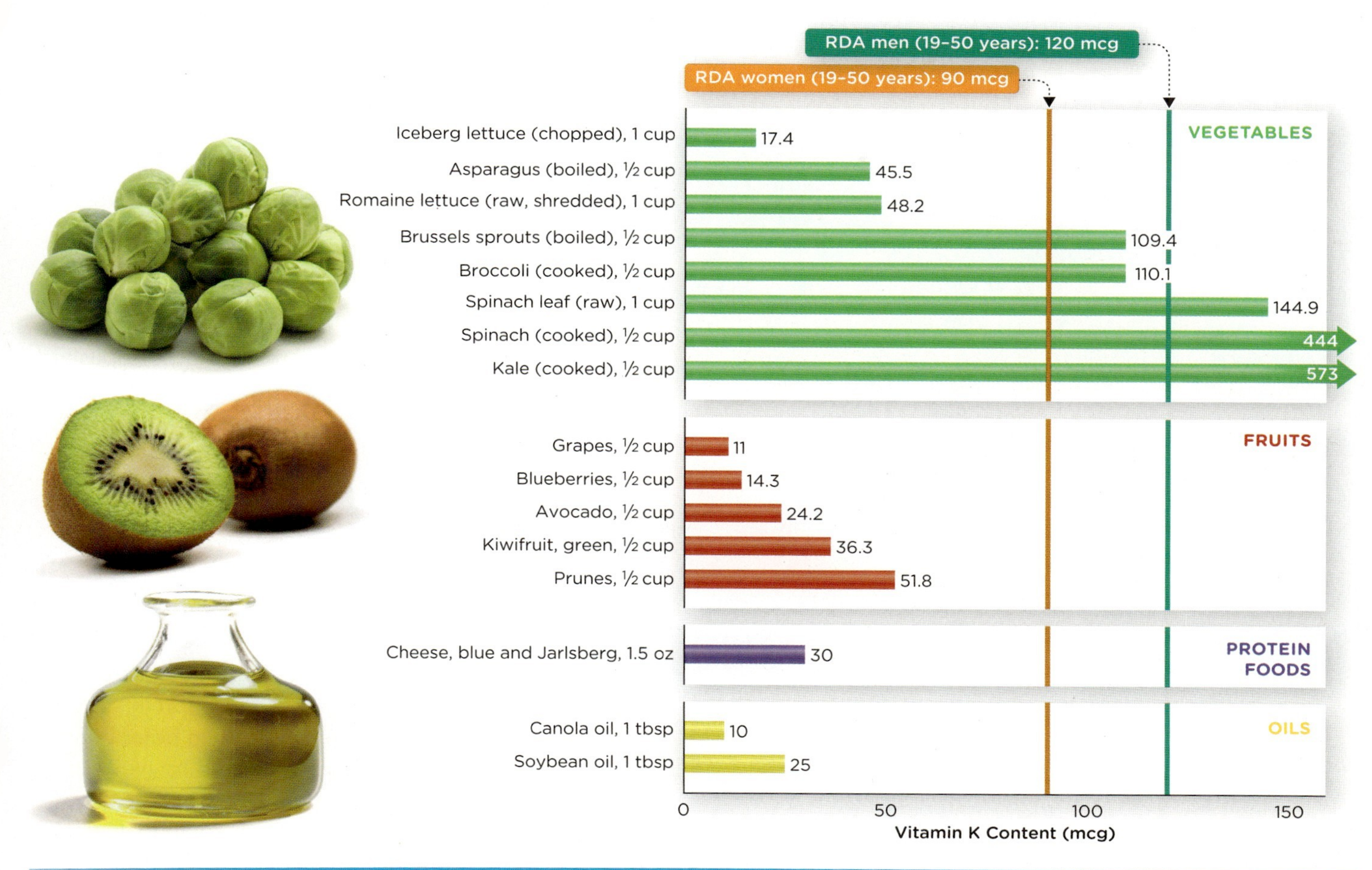

? *What leafy green vegetables do you regularly consume that are high in vitamin K?*

Photo credits (top to bottom): Alasdair Thomson/Getty Images, macxoom/Getty Images, James And James/Getty Images

requirement of vitamin K may be produced by bacteria in the intestine. The most significant dietary sources of vitamin K are green leafy vegetables; some cheeses, fruits, and vegetable oils are also good sources. The requirements for vitamin K are given in Adequate Intake (AI) amounts since current knowledge does not provide sufficient evidence to establish a specific RDA. The AI for men 19 years and older is 120 mcg and for women in the same age range the AI is set at 90 mcg. **(INFOGRAPHIC 10.11)**

Functions of vitamin K in the body

Vitamin K is required to complete the synthesis of several proteins in blood that cause the blood to clot when those proteins are activated. The vitamin also plays a key role in bone metabolism by modifying bone proteins, which allows them to bind calcium and regulate bone formation. Although studies suggest that a diet high in vitamin K is associated with lower risk of hip fractures in aging adults, current evidence is insufficient to recommend supplementation for the prevention of osteoporosis or fractures.

All babies are born with low levels of vitamin K. *Babies generally have enough vitamin K stores to stop bleeding. However, one in 10,000 babies experience vitamin K deficiency bleeding. These babies don't have enough vitamin K to make their blood clot. For this reason, doctors recommend that all babies receive vitamin K at birth.*

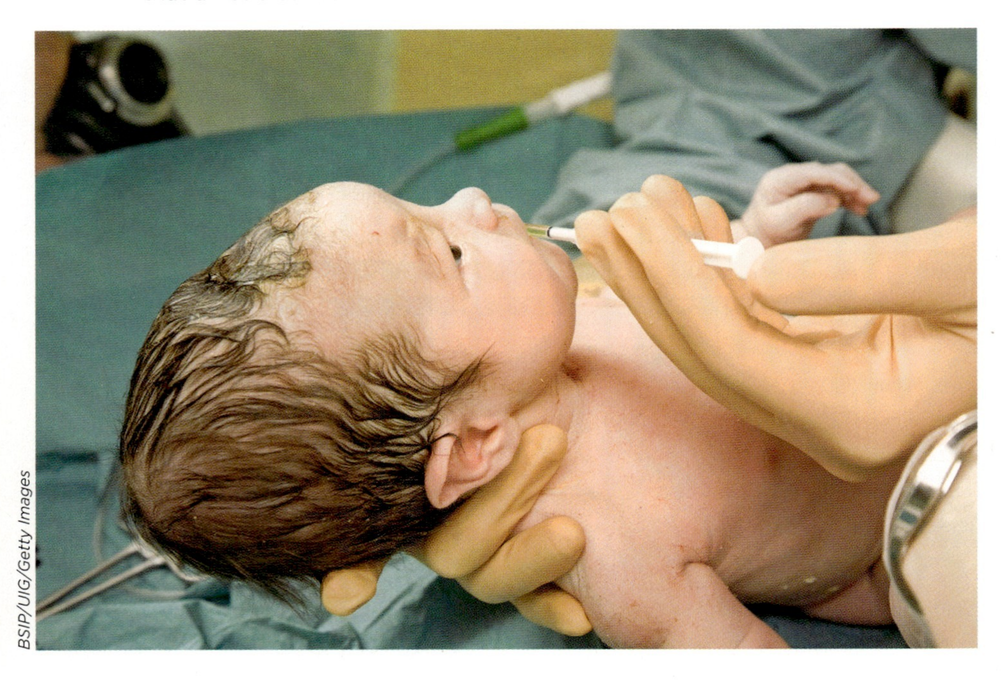

BSIP/UIG/Getty Images

Sun protection is essential to skin cancer prevention. *Ninety percent of non-melanoma skin cancers and 65% of melanomas are associated with exposure to UV radiation from the sun. For this reason, the American Association of Dermatology recommends getting your vitamin D from food to minimize the risk of skin cancer.*

Oleg Mikhaylov/Shutterstock

Vitamin K deficiency and toxicity

Given its key role in blood clotting, vitamin K deficiency can cause uncontrolled bleeding, or *hemorrhaging*. This can result from long-term use of antibiotics that destroy intestinal bacterial, or the use of medications that prevent clotting (*anti-coagulants*) and interfere with the metabolism of vitamin K. Since the early 1960s, newborn babies have received a dose of vitamin K at birth to reduce the risk of hemorrhage because they have limited liver stores of vitamin K and relatively "sterile" intestines.

Toxicity is rare and there is no established UL for vitamin K, but individuals who are prescribed anti-coagulate drugs must strive to keep vitamin K intake at consistent levels from day to day as the vitamin will interfere with those medications.

■ ■ ■

Over the years, Holick hasn't just treated one lowland gorilla—he's administered supplemental vitamin D to twin polar bears who could no longer walk, and revived ailing Komodo dragons at the National Zoo. And his work is hardly restricted to animals—he's treated innumerable people whose health problems he believed stemmed from vitamin D deficiencies.

But it's important to not take concerns over vitamin D deficiency too far, cautions Clifford Rosen, MD, who helped the Institute of Medicine establish the latest DRIs for vitamin D. Vitamin D deficiencies were more of a concern 20 years ago, before people became aware of the vitamin's importance and started adding it to foods, he says. Although millions of Americans still have slightly low levels of vitamin D, that doesn't mean they are at risk of bone weaknesses and other problems, says Rosen. "Most of these people are healthy people, with slightly low vitamin D levels." A diet that meets the DRI of 600 IU per day guarantees that "virtually everybody" gets enough of the fat-soluble vitamin, he says.

Healthy people who consume foods fortified with vitamin D and dairy products do not need vitamin D supplements.

Blend Images/Alamy

Instead, what concerns Rosen is the emphasis on supplementation, which can put people at risk of problems from too much vitamin D. Rosen, Director of Clinical and Translational Research and a Senior Scientist at Maine Medical Center's Research Institute in Scarborough, recently treated a patient with symptoms of hypervitaminosis D, causing her blood calcium levels to spike. "We just lowered her vitamin D and she was fine." He even sees people who take supplements that contain 5,000 IUs to 10,000 IUs. "That's an issue." And there may be little benefits that offset the risks of supplements—a recent review found that taking supplements doesn't decrease the risk of cancer and other health problems linked to low vitamin D.

People who are otherwise healthy don't need to take supplements of vitamin D, says Rosen—simply eat dairy products, get some exercise, and step outdoors occasionally to get a bit of sun. "That should do it." However, the American Association of Dermatology, in its statement on vitamin D, urges Americans to stay out of the sun, because of risk of skin cancers. "Getting vitamin D from a healthy diet, which includes naturally enriched vitamin D foods, fortified foods and beverages, and/or vitamin supplements, and practicing sun protection offer a healthier and safer alternative."

CHAPTER 10 **BRING IT HOME**

Focus on the fat-soluble vitamins

On any given day, you can explore scientific literature or read headlines in the news and find articles reporting the results of research on vitamins and their role in promoting health or preventing disease. To acquaint yourself with some of the newer research and vitamin news, choose one of the fat-soluble vitamins and explore media outlets and the scientific literature for recent reports and studies.

1. Using a popular search engine (for example: Google News), search for recent media articles or reports (in newspapers, on television, or from other popular news providers) that highlight either vitamin A, E, D, or K. List one to two recent headlines and their sources that feature any report, finding, or claim regarding the vitamin you have chosen. (For example: "Vitamin D found to reduce risk of asthma" *USA Today.*)
2. Visit the MEDLINE database at www.pubmed.gov (which is maintained by the U.S. National Library of Medicine at the National Institutes of Health) for access to thousands of abstracts from peer-reviewed medical and scientific journal abstracts. Type your vitamin's name in the search box. Scroll through the first page or two listing of articles, taking note of the variety and focus of research. If you are looking for information on a specific claim, you might search the vitamin and the condition; for example, "vitamin D AND asthma." List the title and source of at least two abstracts.
3. You can access an abstract or summary of the research in PubMed by clicking on the link of the article. Skim the one to two abstracts you listed, taking note of the author's results or conclusions. *Briefly* describe the study findings on the vitamin.
4. Research on diet, including the risks and benefits of vitamins, is abundant and ongoing. *Briefly* discuss new research areas or findings about the vitamin you explored.
5. Imagine yourself a nutrition scientist. What research questions might you propose on the vitamin you've explored?

KEY IDEAS

Vitamins are *organic compounds* (containing both carbon-carbon and carbon-hydrogen bonds) that are needed in small quantities for very specific functions, such as the maintenance of regulatory and metabolic processes in the body.

Most vitamins are in their active form (preformed) in foods, however, in some cases, inactive, precursor forms of vitamins, called provitamins, must be converted to fulfill biological functions in the body.

Inadequate or excessive intake of vitamins can lead to deficiency or toxicity with adverse health effects.

The 14 known vitamins are classified by their solubility as fat-soluble or water-soluble.

The fat-soluble vitamins—A, D, E, and K—have diverse functions in the body.

Fat-soluble vitamins are transported in the body via the lymph and excess intake is stored in fatty tissues and the liver, from which they are released when needed.

The body derives vitamin A from preformed A (retinol) found in animal foods and fortified foods. And, to a lesser degree, from provitamin A carotenoids in plant foods, as conversion to the active form is incomplete.

Vitamin A serves many critical biochemical and physiological functions in the body related to vision, cell development, reproduction, immune function, and growth, among others. The carotenoids have antioxidant properties in the body, which help protect cells from damage caused by free radicals.

Vitamin D, which functions as a hormone in the body, is actually not considered an essential nutrient in the diet because it can be synthesized from cholesterol through exposure of the skin to UV light. Among other important functions, vitamin D plays a key role in the growth and maintenance of bone in part through its role in maintaining calcium concentrations in the blood.

Serving primarily as an antioxidant in the body, vitamin E encompasses a group of fat-soluble compounds called tocopherols found primarily in nuts and vegetable oils. Only alpha-tocopherol meets dietary requirements in humans.

Vitamin K is an essential nutrient found in foods such as leafy greens. A significant amount of our daily requirement may also be produced by bacteria in the intestine.

Vitamin K plays a key role in blood clotting and in bone metabolism.

NEED TO KNOW

Review Questions

1. Characteristics of vitamins include all of the following, EXCEPT they:
   a. are considered micronutrients.
   b. do not provide energy.
   c. contain carbon molecules.
   d. are incorporated into the structure of cells.
   e. are necessary for development and growth.
2. Vitamins are classified according to their:
   a. bioavailability.
   b. function.
   c. molecular structure.
   d. solubility.
3. Inactive, precursor forms of vitamins are termed:
   a. antioxidants.
   b. enzymes.
   c. hormones.
   d. preformed vitamins.
   e. provitamins.
4. Possible toxicity from vitamins is MOST likely to occur with consuming:
   a. large portions of vegetables and fruits.
   b. excess intake of water-soluble vitamins.
   c. fortified foods and dietary supplements.
   d. more than two servings of milk per day.
5. All of the following are TRUE regarding carotenoids, EXCEPT:
   a. only three are converted to active vitamin A, with beta-carotene as the most abundant.
   b. they are primarily provitamins that must be converted before they can be utilized in the body.
   c. they are the yellow, orange, and red pigments in fruits and vegetables.
   d. they supply approximately two-thirds of vitamin A in the U.S. diet.
   e. they function as antioxidants in reducing damage from free radicals.
6. Vitamin A deficiency:
   a. compromises the ability to see in low light.
   b. results in keratinization of epithelial cells.
   c. increases the risk of death from infectious disease in children.
   d. is the leading cause of preventable blindness in children worldwide.
   e. All of the answers are correct.
7. All of the following are TRUE regarding vitamin D, EXCEPT:
   a. it is also known as cholecalciferol.
   b. its synthesis in the body is not diminished by use of sunscreens.
   c. it is important for maintaining calcium concentrations in the blood.
   d. it is only essential in the diet when sun exposure or synthesis is insufficient to meet needs.
   e. it functions as a hormone in the body.
8. The Recommended Dietary Allowances (RDA) for vitamin D:
   a. are established based on minimal sun exposure.
   b. are established based on maximum sun exposure of 4 to 6 hours a day.
   c. are established to maintain calcium concentrations in healthy people.
   d. can be met through the consumption of adequate amounts of fruits and vegetables.
   e. can only be met through the use of dietary supplements.
9. The fat-soluble vitamin that is LEAST likely to be toxic when consumed in amounts above recommended intake levels is:
   a. vitamin A.
   b. vitamin D.
   c. vitamin E.
   d. All of the fat-soluble vitamins have some risk of toxicity.

10. All of the following are characteristics of vitamin E as an antioxidant, EXCEPT that it:
    a. protects against the oxidation of polyunsaturated fatty acids in cell membranes.
    b. boosts the functioning of the immune system by protecting white blood cells.
    c. protects against oxidative damage that results from exposure to free radicals.
    d. protects the fat in low-density lipoproteins from oxidation.
    e. dramatically reduces the risks of heart disease and cancer.

11. Vitamin K has a key role in:
    a. blood clotting.
    b. energy metabolism.
    c. the formation of red blood cells.
    d. nerve impulse transmission.
    e. regulating blood glucose.

12. Which of the following is the BEST food source of vitamin K?
    a. apples
    b. Brussels sprouts
    c. eggs
    d. lean beef
    e. yogurt

Dietary Analysis Using SuperTracker

Analyzing the intake of fat-soluble vitamins using SuperTracker

1. Log onto the United States Department of Agriculture (USDA) website at www.supertracker.usda.gov. If you have not done so already, you will need to create a profile to get a personalized diet plan. This profile will allow you to save your information and diet intake for future reference. *Do not use the general plan.*
2. Click the Track Food and Activity option.
3. Record your food and beverage intake for one day that most reflects your typical eating patterns. Enter each food and beverage you consumed into the food tracker. Note that there may not always be an exact match to the food or beverage that you consumed, so select the best match available.
4. Once you have entered all of your food and beverage choices into the food tracker, on the right side of the page under the bar graph, you will see Related Links: View by Meal and Nutrient Intake Report. Print these reports and use them to answer the following questions:
    a. Did you meet the target recommendations for the fat-soluble vitamins (A, D, E, and K) on the day you selected? If not, which vitamins fall below the target numbers?
    b. For each vitamin target that you missed, list two specific foods you could consume to increase your intake of that vitamin. For vitamins for which you met the target, what foods contributed most to your intake?
    c. Were you above the targets for any of the fat-soluble vitamins? If so, which ones? Are any of these numbers above the toxicity level for that vitamin?
    d. Fat-soluble vitamins require dietary fat to be optimally absorbed. If you ate a tossed salad, what could you include in the salad to help you absorb the fat-soluble vitamins?

11 WATER-SOLUBLE VITAMINS

It's Not a Germ

PIONEERING RESEARCH UNCOVERS VITAMIN DEFICIENCY DISEASES

Marco Veringa/Getty Images

LEARNING OBJECTIVES

- Describe the general properties of the water-soluble vitamins and proper handling techniques to preserve vitamins in foods (Infographics 11.1 and 11.2)
- Describe the possible causes of vitamin deficiencies (Infographic 11.3)
- Describe the role of vitamins as coenzymes (Infographic 11.4)
- List two excellent food sources for each of the water-soluble vitamins (Infographics 11.5–11.9, 11.12, and 11.13)
- Identify the vitamin deficiencies that cause megaloblastic anemia and explain how this anemia occurs (Infographic 11.11)
- Identify vitamins that have neurological functions, act as an antioxidant, or are involved in energy metabolism, red blood cell production, and DNA and RNA synthesis (Infographic 11.14)

In 1914, an esteemed young doctor named Joseph Goldberger was asked by the U.S. Surgeon General to investigate a devastating and mysterious epidemic sweeping across the American South. Known at the time as *mal de rosa,* now more commonly referred to as **pellagra**, the condition was causing horrifying **signs** and **symptoms** among children and adults—scaly skin, mouth sores, diarrhea, confusion, and, ultimately, mental deterioration. (The signs of pellagra are sometimes known as the three Ds: dermatitis, diarrhea, and dementia.) Although statistics are hard to come by, in 1912, South Carolina alone

Dr. Goldberger is sitting at a table surrounded by hospital personnel and patients.

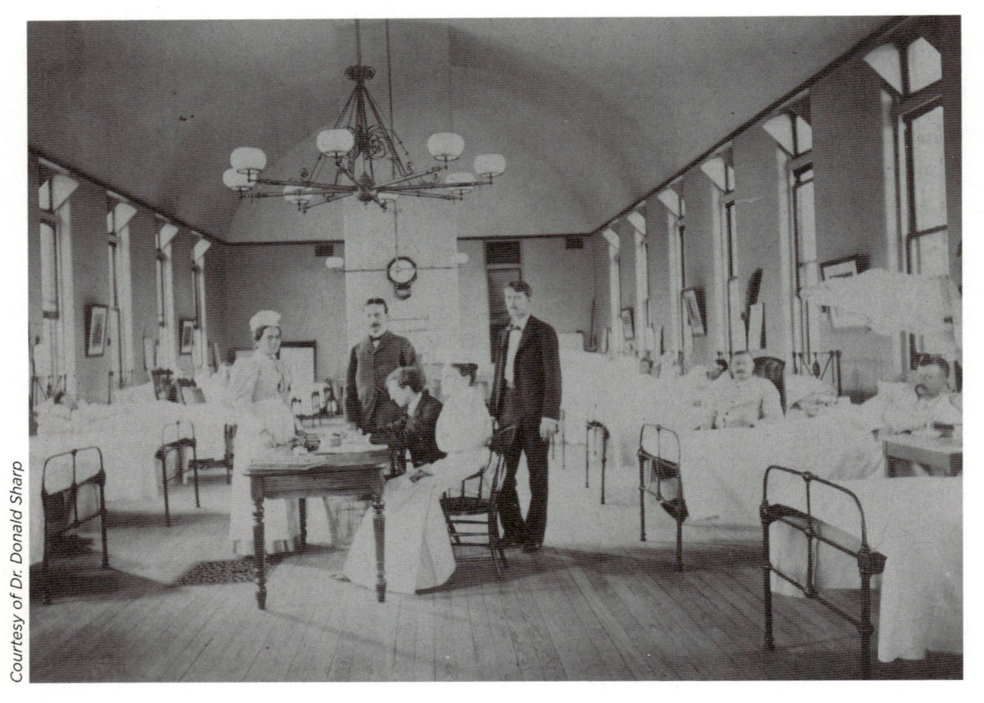

Courtesy of Dr. Donald Sharp

reported 30,000 cases of this mysterious disease and 40% of those who developed it died.

At the time, most doctors believed that the mysterious condition was an infectious disease caused by an as-yet-unidentified microbe. But Goldberger wasn't convinced. He noticed that the condition tended to afflict poor people, not wealthy people, and that it would sicken prisoners in correction facilities but not the guards. A germ, he knew, wouldn't make such a sociological distinction. If not a germ, what was causing this terrible disease, and how? Goldberger's idea, wildly controversial at the time, was that the devastating disease was caused by diet, as he had noticed that the people who were ill tended to eat nutrient-poor diets consisting of cornbread, molasses, and a little pork fat.

In 1915, in what would be considered a highly unethical experiment today, Goldberger experimented on 11 inmates in a Mississippi prison who had volunteered to participate in exchange for a pardon. Most of the inmates in this particular prison ate a well-balanced diet and did not suffer from pellagra. Goldberger fed his volunteers a much less nutritious corn-based diet and watched what happened. Within five months, six of the men showed signs of pellagra, whereas the other inmates in

Pellagra causes thickening, peeling, and discoloration of the skin.

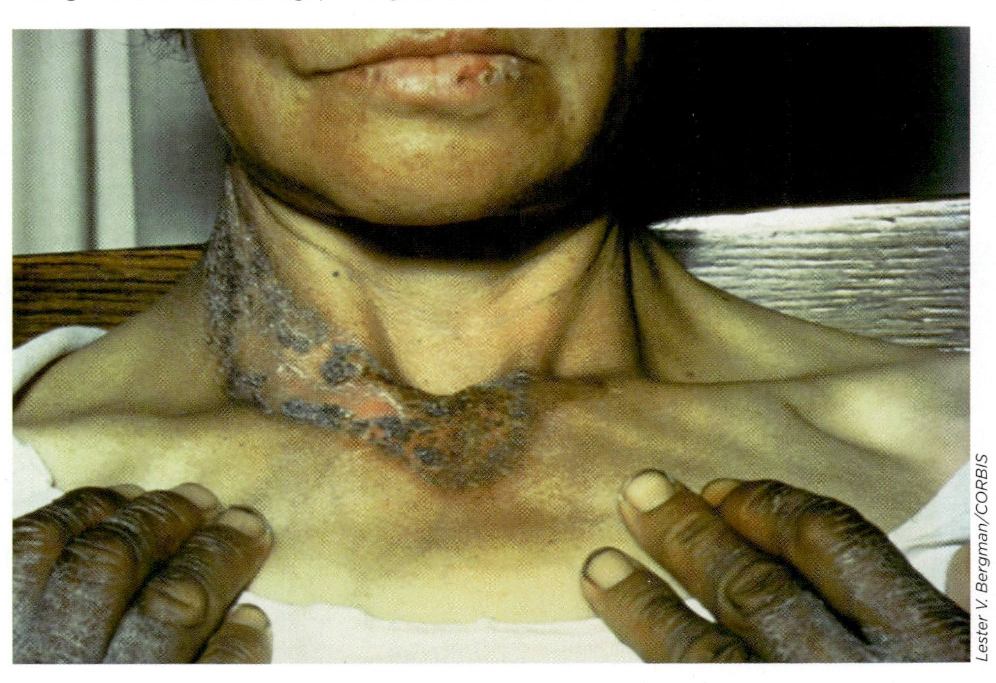

Lester V. Bergman/CORBIS

PELLAGRA
a disease caused by niacin deficiency and characterized by the three Ds—dermatitis, diarrhea, and dementia

SIGNS
are objective evidence of disease that are observed by health care professionals, such as a rash or abnormal blood tests

SYMPTOMS
are subjective evidence of disease that are experienced by the individual that only they can perceive, such as a stomachache or fatigue

the prison, who ate the more balanced diet, did not. In another experiment, Goldberger supplemented the diets of those suffering from pellagra with fresh meat, milk, and vegetables, and they quickly recovered.

The results of Goldberger's experiments seemed to implicate diet in the development of the disease, but the case was not yet closed.

■ ■ ■

Today we understand that the foods we eat contain vitamins, *organic* compounds (compounds containing both carbon-carbon and carbon-hydrogen bonds) that are needed in small quantities for very specific functions, such as the maintenance of regulatory and metabolic processes in the body. If we don't get enough of a vitamin, or are unable to adequately utilize it, we may develop signs and symptoms characteristic of an insufficiency of that vitamin.

The fat-soluble vitamins (A, D, E, and K) were the subject of Chapter 10. The vitamins that disperse easily in water-based solutions such as the blood, called **water-soluble vitamins**, are discussed in this chapter.

CHARACTERISTICS OF THE WATER-SOLUBLE VITAMINS

Water-soluble vitamins include eight **B vitamins**, vitamin C, and the vitamin-like nutrient, choline. The B vitamins function primarily as **coenzymes**, chemical compounds that bind enzymes and are required by the enzymes to carry out their function or activity. As coenzymes, they participate in **energy metabolism**, a series of reactions in the body that results in energy production, as well as myriad other types of reactions. In addition to functioning as a coenzyme in several types of reactions, vitamin C also serves as an antioxidant, protecting cells from free-radical damage. (INFOGRAPHIC 11.1)

WATER-SOLUBLE VITAMINS
vitamins that disperse easily in water-based solutions; include the B vitamins, vitamin C, and the vitamin-like nutrient choline

B VITAMINS
a group of water-soluble vitamins that serve as coenzymes in the conversion of carbohydrates, fat, and protein into energy

COENZYME
a compound that binds to a protein (enzyme) and is required for its function or activity

ENERGY METABOLISM
a series of reactions in the body that result in energy production

INFOGRAPHIC **11.1** **Properties of the Water-Soluble Vitamins**

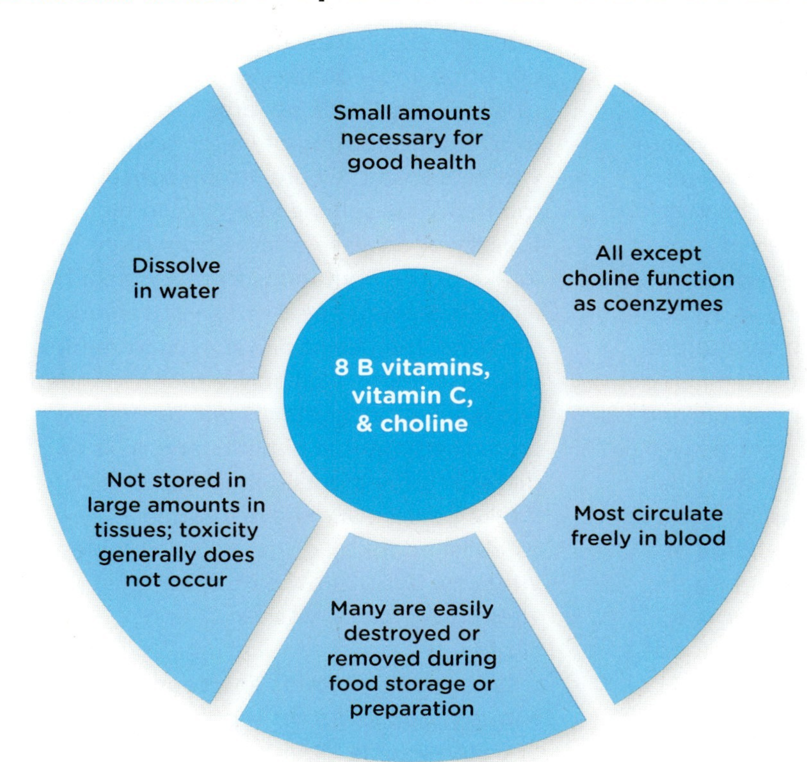

INFOGRAPHIC 11.2 Preserving Vitamins in Foods

- ✓ Cook fresh fruits and vegetables with low heat. Avoid high cooking temperatures and long heat exposure.
- ✓ Leaving the skin on fruits like apples and pears adds to the amount of vitamins already available in the fruit.
- ✓ When fruit has been peeled and air hits the fruit, some vitamins in the fruit start to break down through oxidation.
- ✓ Some vitamins are lost from fruits and vegetables when they are stored, so eat fresh produce as soon as possible after harvesting or purchase.
- ✓ Freezing will slow vitamin degradation in fresh produce but not stop it. Since some vitamins will continue to breakdown even in frozen vegetables, do not store them for extended periods.
- ✓ Refrigerating fruits and vegetables will reduce spoilage and slow the breakdown of some vitamins in some foods.
- ✓ Refrigerate food in airtight moisture-proof containers. The loss of some vitamins is slowed by cold temperatures and less air contact.
- ✓ More nutrients are retained when minimal water is used and cooking time is shorter for food preparation.
- ✓ Keep milk refrigerated and tightly capped, away from strong light. Riboflavin is very sensitive to direct light.
- ✓ Do not soak vegetables for long periods.

Photo credits (left side — top to bottom): Evan Sklar/Getty Images, StudioSmart/Shutterstock; (right side — top to bottom): Vlue/Shutterstock, Yagi Studio/Getty Images

Unlike fat-soluble vitamins, which leave the small intestine in chylomicrons via the lymph and are stored in the body like fat, water-soluble vitamins immediately enter the blood following absorption, where most circulate freely. Because the water-soluble vitamins are not stored in large quantities, we must consume adequate amounts of them consistently. In addition, foods must be handled with care to preserve vitamin content; some vitamins (both fat- and water-soluble) are unstable and can be destroyed by ultraviolet light, as well as by cooking and storage methods. When you boil foods that contain water-soluble vitamins, for instance, a percentage of the vitamins typically leach out into the cooking water. However, steps can be taken to help preserve vitamins in foods. **(INFOGRAPHIC 11.2)**

When we consume more of a water-soluble vitamin than we need, our body typically eliminates the excess through our urine. Individuals are much less likely to experience toxic adverse effects due to overconsumption of water-soluble vitamins than they are if they consume too many fat-soluble vitamins. In fact, toxicity or adverse effects from high intake of the B vitamins from *food* sources alone has almost never been observed. However, the use of dietary supplements can push intake well above recommended intake levels. (Chapter 12 will further explore the potential benefits and risks of dietary supplements.) Tolerable Upper Intake Levels (ULs) have not been established for all of the water-soluble vitamins, but ULs exist for niacin, vitamin B_6, folate, vitamin C, and choline.

In the United States and countries with an adequate food supply, deficiencies of water-soluble vitamins are rare. However,

INFOGRAPHIC 11.3 Possible Causes of Vitamin Deficiencies *Although serious vitamin deficiencies are relatively uncommon in developed countries like the United States, various factors can conspire to cause obvious signs and symptoms of deficiency in some individuals. In contrast, it is likely that a significant number of individuals in the United States have chronically low intake of one or more vitamins resulting in subclinical deficiencies that cause no overt symptoms.*

Cause	Examples
Inadequate intake	Calorie restriction, poverty, anorexia, food fads, difficulty swallowing, dental problems, decreased taste and smell, illness
Decreased absorption	Poor digestion, diarrhea, parasites, intestinal disorders, dietary antivitamin factors, gastrointestinal infections, prescription drugs, alcohol
Decreased utilization in cells	Deficiencies in other nutrients, prescription drugs, alcohol, or infection
Increased requirements	Growth, pregnancy, lactation, chronic illness, infections
Increased breakdown	Prescription drugs, alcohol
Increased losses and excretion	Increased urinary excretion, blood losses (gastric ulcers), parasites, infection

What members of society do you think have the highest risk of vitamin deficiencies?

there are circumstances when the risk of deficiency is higher; when calories are restricted, for example, or under conditions that affect absorption, such as diarrhea; parasitic, bacterial, or viral gastrointestinal infections; and intestinal disorders. **(INFOGRAPHIC 11.3)** In addition, the body's need for most water-soluble (and fat-soluble) vitamins is higher during certain life stages, such as in the elderly, and during pregnancy and lactation, increasing the risk of potential deficiency and the associated consequences. When we suffer from fevers or injuries, and when we recover from surgery, we also need higher-than-normal amounts of water-soluble vitamins.

■ ■ ■

In the early days of the twentieth century, it wasn't easy to accept that a poor diet could cause pellagra's varied signs and symptoms, such as skin lesions, confusion, and diarrhea. Some suspected moldy corn caused the illness, or disease-carrying mosquitos. Even when individuals regained their health when fed a varied and wholesome diet, many doctors thought Goldberger's ideas about pellagra were ridiculous. To prove them wrong, in a final jaw-dropping experiment conducted on April 26, 1916, Goldberger injected five cubic centimeters of blood from someone suffering from pellagra into his assistant, Dr. George Wheeler. Wheeler did the same to Goldberger. They swabbed secretions from the noses and throats of pellagra sufferers and rubbed them onto their own noses and throats. They even swallowed pills containing scabs of pellagra rashes. If pellagra were caused by a germ, they reasoned, they would undoubtedly fall ill. They did not.

Goldberger's uncouth experiments eventually convinced many doctors that pellagra was not an infectious disease, but Goldberger didn't live to figure out exactly what kind of nutritional deficiency caused pellagra. He died of cancer in 1929, leaving it up to someone else to track down the exact culprit.

■ ■ ■

In the early twentieth century, corn was cheap and was an important source of dietary energy and protein for poor people, particularly in rural areas. Niacin (and the niacin precursor tryptophan) has limited bioavailability in corn without first processing it with an alkaline substance such as lime. Pellagra is a disease caused by niacin deficiency.

Firma V/Shutterstock

THE B VITAMINS

You might have wondered why there are so many B vitamins. Initially, the B vitamins were thought to be a single large *B vitamin complex,* but once the structures and functions were better understood, researchers realized they were actually separate vitamins with distinct functions. To better distinguish the compounds, they retained the title "B vitamin" but gave each vitamin a number, and later, a name. So, the B vitamins include thiamin (B_1), riboflavin (B_2), niacin (B_3), pantothenic acid (B_5), pyridoxine (B_6), biotin (B_7), cobalamin (B_{12}), and folate (B_9). Some are commonly called by their chemical name, while others are identified by their numerical designation, particularly when there are multiple chemical forms of the vitamin.

All B vitamins function as coenzymes, with most playing critical roles in energy metabolism. *Energy metabolism* refers to the chemical reactions that break down carbohydrates, fats, and proteins to release energy, and the chemical reactions that use energy to construct molecules and carry out body processes. Coenzymes are not actually part of the enzyme structure; rather, they assist enzymes by accepting and donating hydrogen ions, electrons, and other molecules during reactions. In other words, the B vitamins do not provide energy to the cells, but they play a critical role in energy transformation. **(INFOGRAPHIC 11.4)**

INFOGRAPHIC 11.4 B Vitamins Function as Coenzymes *Coenzymes associate with enzymes to form an active complex that is capable of catalyzing a chemical reaction.*

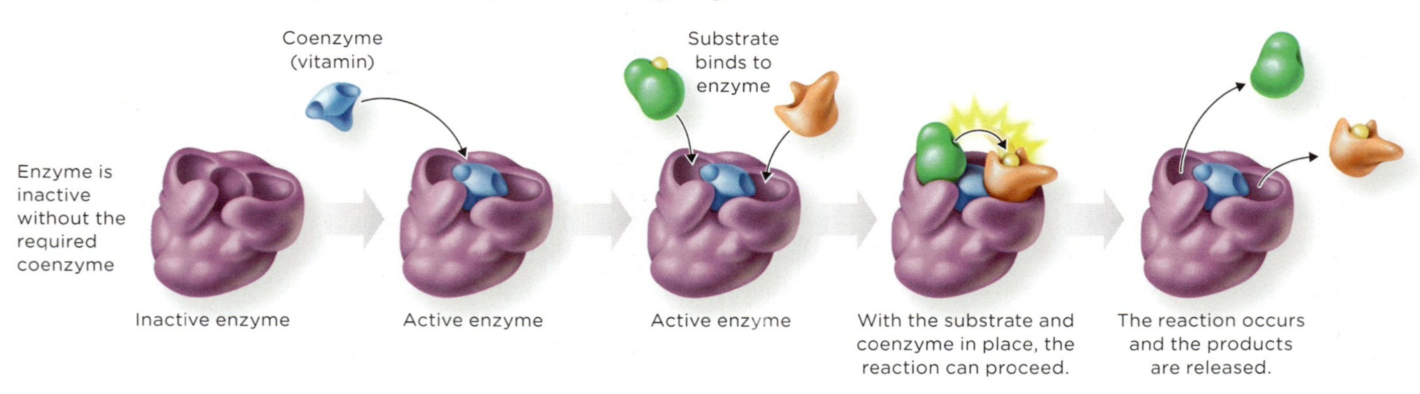

Describe the basic function of a coenzyme.

Beriberi patient on crutches. *Beriberi is a disorder of the nervous system caused by a lack of vitamin* B_1 *(thiamin) in the diet. It can cause weight loss, emotional disturbances, and weakness and pain in the limbs, among other symptoms.*

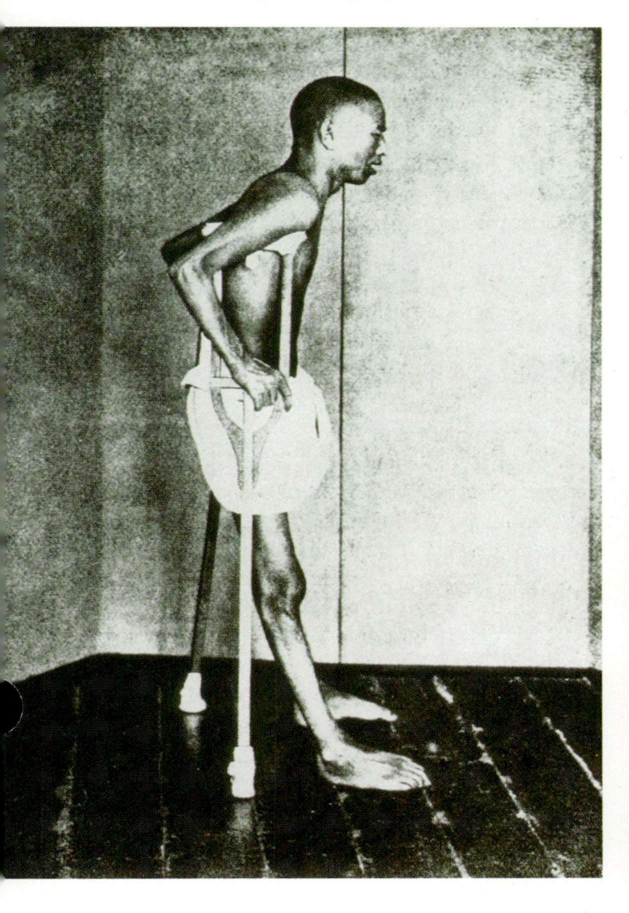

B vitamins also participate in amino acid metabolism; reactions that build and maintain bones, muscles, and red blood cells; and the synthesis of nucleic acids that are needed for DNA synthesis and cell division.

Many of the B vitamins are interrelated and interdependent, in that the successful functioning of one B vitamin often depends on the successful functioning of others. Likewise, deficiencies of B vitamins rarely occur in isolation, because many B vitamins typically occur together in the same types of foods, and chances are that a diet inadequate in one B vitamin would be low in others. However, a deficiency of a single B vitamin wreaked havoc in the not-too-distant past.

■ ■ ■

BERIBERI
a condition characterized by extreme weakness; caused by thiamin deficiency

A painful degenerative disease was sweeping East Asia in the closing decades of the 1800s, causing its victims to experience terrible muscle wasting and eventual heart failure. The disease was known as **beriberi**, and was at one time fairly common among populations whose diet consisted largely of rice.

Technology of the time made it possible to refine white rice cheaply and efficiently.

Rice with and without hulls. *The white "polished" rice on the right has been stripped of the nutrients found in the hulls.*

Tim Gainey/Alamy

Polish-born American biochemist Dr. Casimir Funk, who coined the term "vitamin," is shown in New York City, January 20, 1953.

AP Photo

The hulls of the rice were removed and discarded to improve the storage life of the rice—brown rice would spoil faster than white rice. What the people didn't understand, however, was that removing the rice hull removed many essential nutrients from the rice.

As chance would have it, a Dutch researcher named Christiaan Eijkman noticed that chickens fed a diet of white rice became ill; so ill that they would stumble drunkenly and collapse on their sides. When the brown hulls were added into their diet the chickens recovered. Eijkman surmised incorrectly that there were illness-causing compounds within the white rice, and that something in the rice hulls cured it.

NEUROTRANSMITTERS *chemical substances involved in transmitting signals between nerve cells*

Many years later a Polish-born American biochemist Casimir Funk was attempting to isolate and identify the exact anti-beriberi factor present in the hulls of brown rice. Funk called the factor *vitamine* (from *vital* and *amine*) and the term was later shortened to "vitamin." In fact, in 1912, Funk was the first to propose that beriberi, scurvy, rickets, and pellagra were caused by deficiencies of specific vitamin(e)s.

Thiamin

Funk named his first vitamin B_1; today, it's commonly known as *thiamin*. The coenzyme form of thiamin is needed to provide energy from the breakdown of glucose, fatty acids, and some amino acids—where it participates in reactions that release carbon dioxide (CO_2). It is also required for the production of sugars needed for the synthesis of RNA and DNA. Good food sources of thiamin include pork and fortified grain products. **(INFOGRAPHIC 11.5)**

The disruptions in metabolism that occur with a thiamin deficiency alter the production of several **neurotransmitters**, leading to mental disturbances, such as apathy, irritability, and confusion. Other deficiency symptoms include fatigue and muscle weakness.

Although beriberi is not common in the United States today, another thiamin deficiency, called *Wernicke-Korsakoff syndrome*, can be found in many developed nations. Alcohol abuse is the leading cause of Wernicke-Korsakoff syndrome, with as many as 80% of chronic alcohol abusers showing signs of deficiency, often resulting in severe neurological disturbances. Alcoholics are at high risk of thiamin deficiency because alcohol consumption displaces the intake of nutrient-rich foods, decreases thiamin absorption, increases its excretion in urine, decreases its storage in the liver, and decreases conversion of thiamin into its coenzyme. In fact, chronic alcohol abuse increases the risk of deficiency

INFOGRAPHIC 11.5 Food Sources of Thiamin

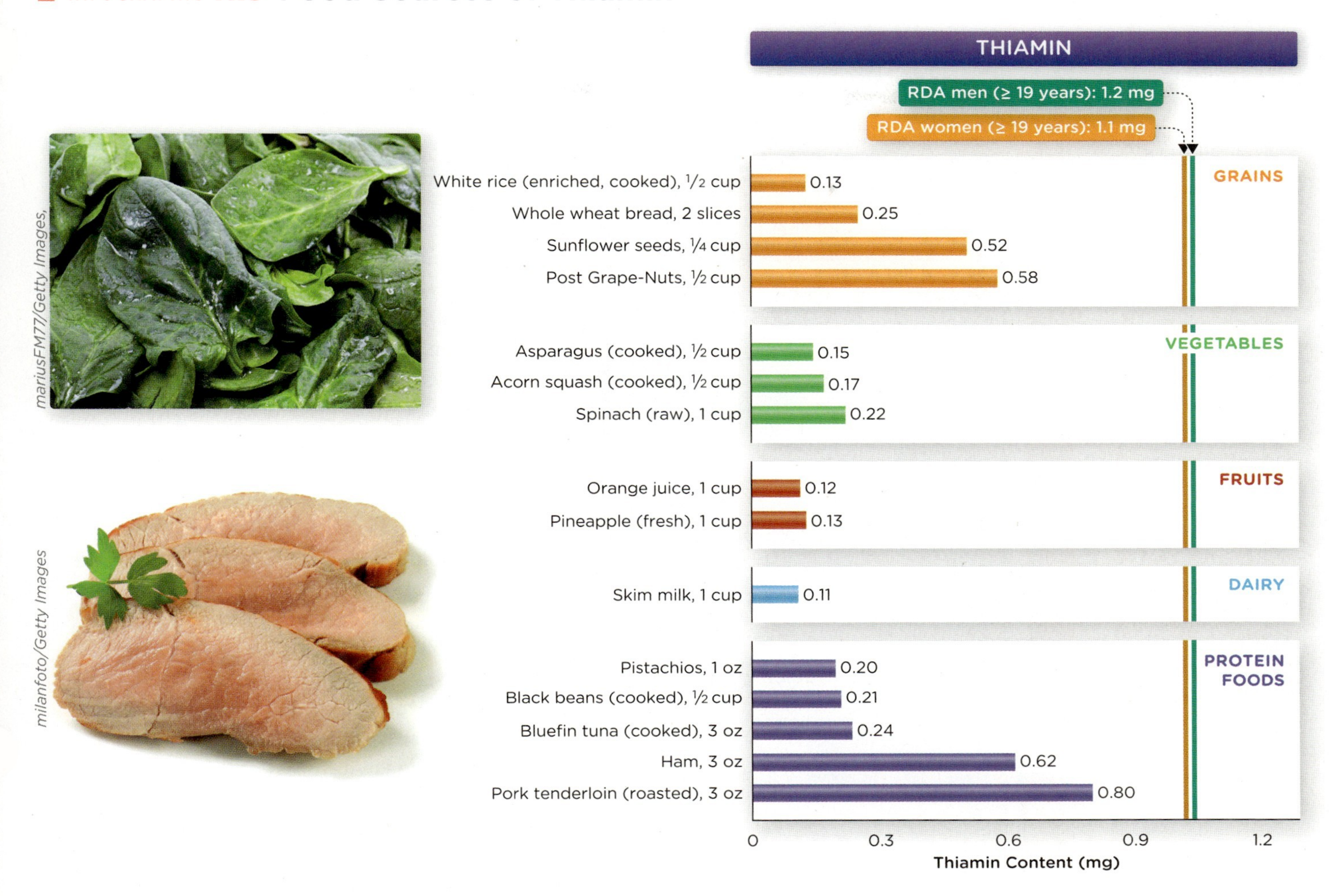

for most of the water-soluble vitamins through several of these mechanisms.

Niacin

Recall that pellagra was sweeping the southern United States in the early part of the twentieth century, and that Dr. Goldberger had determined that eating a diet that contained a variety of foods was curative. It wasn't until 1937, eight years after his death, that the specific curative compound in food was identified. An American biochemist gave a dose of nicotinic acid (niacin) to a dog suffering from a disease known as black tongue, the canine equivalent of pellagra. The dog was cured, and subsequent testing determined that niacin was also an effective treatment for humans with pellagra.

Niacin is rich in meats (especially poultry) and fish, as well as peanuts, mushrooms, and fortified cereals. It can also be synthesized in the body, albeit rather inefficiently, from the amino acid tryptophan found primarily in protein-rich foods. About 1 mg of niacin is made from every 60 mg of tryptophan provided in food, so dietary recommendations are often given in **niacin equivalents (NEs)**. NEs are used to describe

NIACIN EQUIVALENTS (NEs) *the total amount of niacin that is provided by a food from both the preformed vitamin and that which can be synthesized in the body from tryptophan in the food*

Niacin in the ingredients list. *Flour loses nutrients in processing; food manufacturers enrich flour by returning some B vitamins and iron to the product.*

Nutrition Facts
Serving Size 2 whole crackers (27g)
Servings Per Container about 15

Amount Per Serving	
Calories 110	Calories from Fat 10
	% Daily Value*
Total Fat 1g	**2%**
Saturated Fat 0g	**0%**
Trans Fat 0g	
Polyunsaturated Fat 0g	
Monounsaturated Fat 0g	
Cholesterol 0mg	**0%**
Sodium 130mg	**5%**
Total Carbohydrate 22g	**7%**
Dietary Fiber 1g	**4%**
Sugars 7g	
Protein 2g	
Iron 6%	

Not a significant source of vitamin A, vitamin C, and calcium.

*Percent Daily Values are based on a 2,000 calorie diet. Your daily values may be higher or lower depending on your calorie needs:

	Calories:	2,000	2,500
Total Fat	Less than	65g	80g
Saturated Fat	Less than	20g	25g
Cholesterol	Less than	300mg	300mg
Sodium	Less than	2,400mg	2,400mg
Total Carbohydrate		300g	375g
Dietary Fiber		25g	30g

Calories per gram:
Fat 9 • Carbohydrate 4 • Protein 4

INGREDIENTS: ENRICHED **WHEAT** FLOUR (**WHEAT** FLOUR, NIACIN, REDUCED IRON, THIAMIN MONONITRATE, RIBOFLAVIN AND FOLIC ACID), GRAHAM FLOUR, SUGAR, HIGH FRUCTOSE CORN SYRUP, HONEY, CONTAINS TWO PERCENT OR LESS OF: VEGETABLE OIL (CONTAINS ONE OR MORE OF THE FOLLOWING OILS: INTERESTERIFIED **SOYBEAN**, CANOLA) WITH CITRIC ACID AND TBHQ ADDED TO PRESERVE FRESHNESS, MOLASSES, SODIUM BICARBONATE, SALT, **SOY** LECITHIN (AN EMULSIFIER), AMMONIUM BICARBONATE, MALTED CEREAL SYRUP, NATURAL AND ARTIFICIAL FLAVORS, SODIUM SULFITE AND ENZYMES.
CONTAINS: SOYBEAN, WHEAT.

Martin Shields/Alamy

FORTIFICATION
the addition of vitamins and/or minerals to a food product

ENRICHMENT
a process used to replace some of the B vitamins (and iron) that are extracted from grains when they are refined

the contribution to dietary intake of the preformed niacin and that which can be synthesized from the tryptophan provided by foods.

Therapeutically, doctors sometimes prescribe niacin in high doses to help lower "bad" LDL cholesterol and increase "good" HDL cholesterol. However, as with most vitamins, taking high doses creates side effects and patients must be monitored to ensure the potential benefits outweigh the risks.

Typically, the first sign of pellagra is fatigue, because niacin plays such an important role in the conversion of the macronutrients in food into energy. In addition to being involved in energy metabolism, niacin is also required for the synthesis of glucose, fatty acids, cholesterol, and steroid hormones. It also plays critically important roles in DNA repair, cell signaling, and the regulation of gene expression.

To reduce the incidence of pellagra, the U.S. government instituted niacin **fortification** by adding niacin to grain products in 1938. Although manufacturers today fortify foods for various reasons, the original goal of food fortification was to correct identified nutrient deficiencies by adding specific vitamins and/or minerals to foods. Refined grain products may also be **enriched** with niacin, meaning that some of the B vitamins (and iron) that were extracted from the grains as part of the refining process were returned. Today, bread products and fortified cereals are a primary source of niacin in our diets, along with protein-rich foods, particularly poultry. Though bread products are not particularly high in niacin, they contribute significantly to our niacin intake because they are so abundant in our diet. **(INFOGRAPHIC 11.6)**

There is no evidence that consuming too much niacin from food leads to adverse effects. However, people who take high doses of niacin in supplements may suffer side effects such as flushing and gastrointestinal distress. Individuals who are prescribed niacin to control their cholesterol also must be carefully monitored to ensure that excessive amounts of the vitamin do not cause liver problems or glucose intolerance.

INFOGRAPHIC 11.6 Food Sources of Niacin

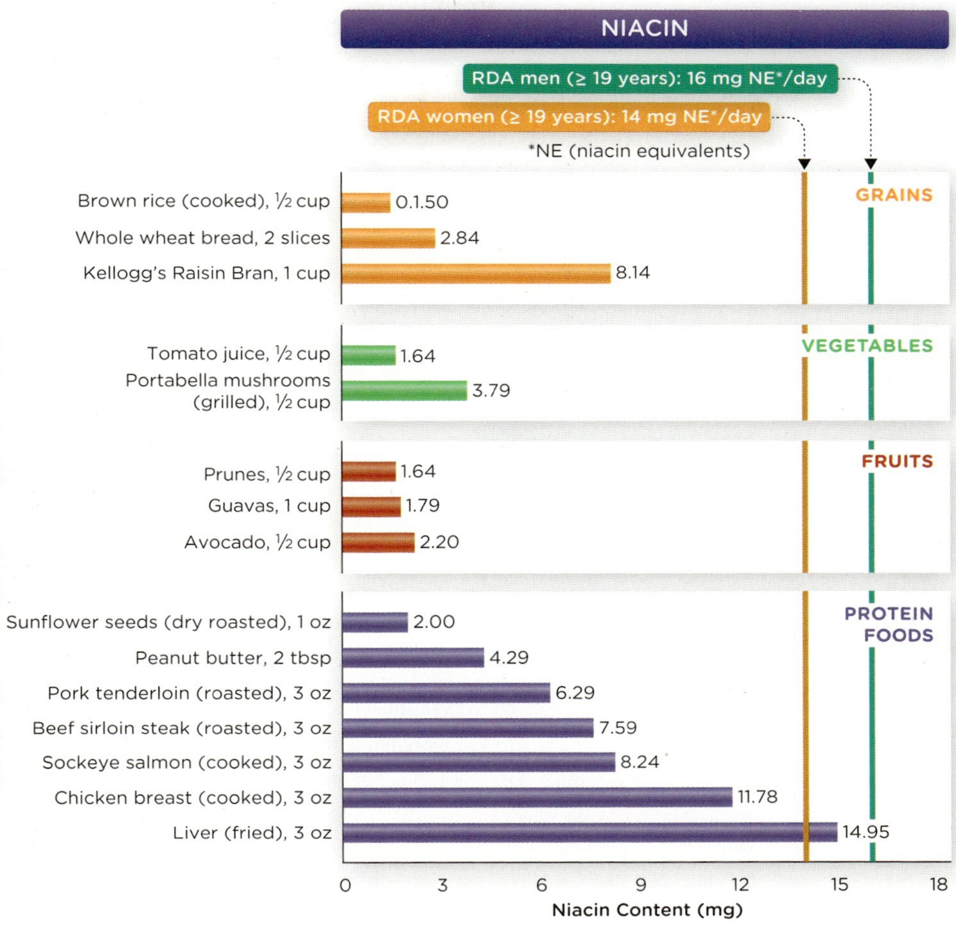

Riboflavin

Riboflavin was the second B vitamin to be isolated and its function is similar to that of niacin: It's important for the metabolism of proteins, lipids, and carbohydrates. Milk and other dairy products are particularly good sources of riboflavin, and it is because riboflavin is destroyed by ultraviolet light that milk is kept in paper cartons or opaque plastic containers. A riboflavin deficiency (known as *ariboflavinosis*) is characterized by cracks and redness on the lips and corners of the mouth, swelling of tissues in the mouth, and a sore throat. Deficiency may be seen with chronic alcohol abuse or malabsorptive conditions and seldom occurs by itself; it typically occurs along with deficiencies in other B vitamins. No toxicity with riboflavin consumption from food or supplements has been observed. (INFOGRAPHIC 11.7)

Vitamin B_6

Vitamin B_6—sometimes referred to by one of its chemical names, *pyridoxine*—functions as a coenzyme in the release of glucose from stored glycogen and in amino acid

INFOGRAPHIC **11.7** **Food Sources of Riboflavin**

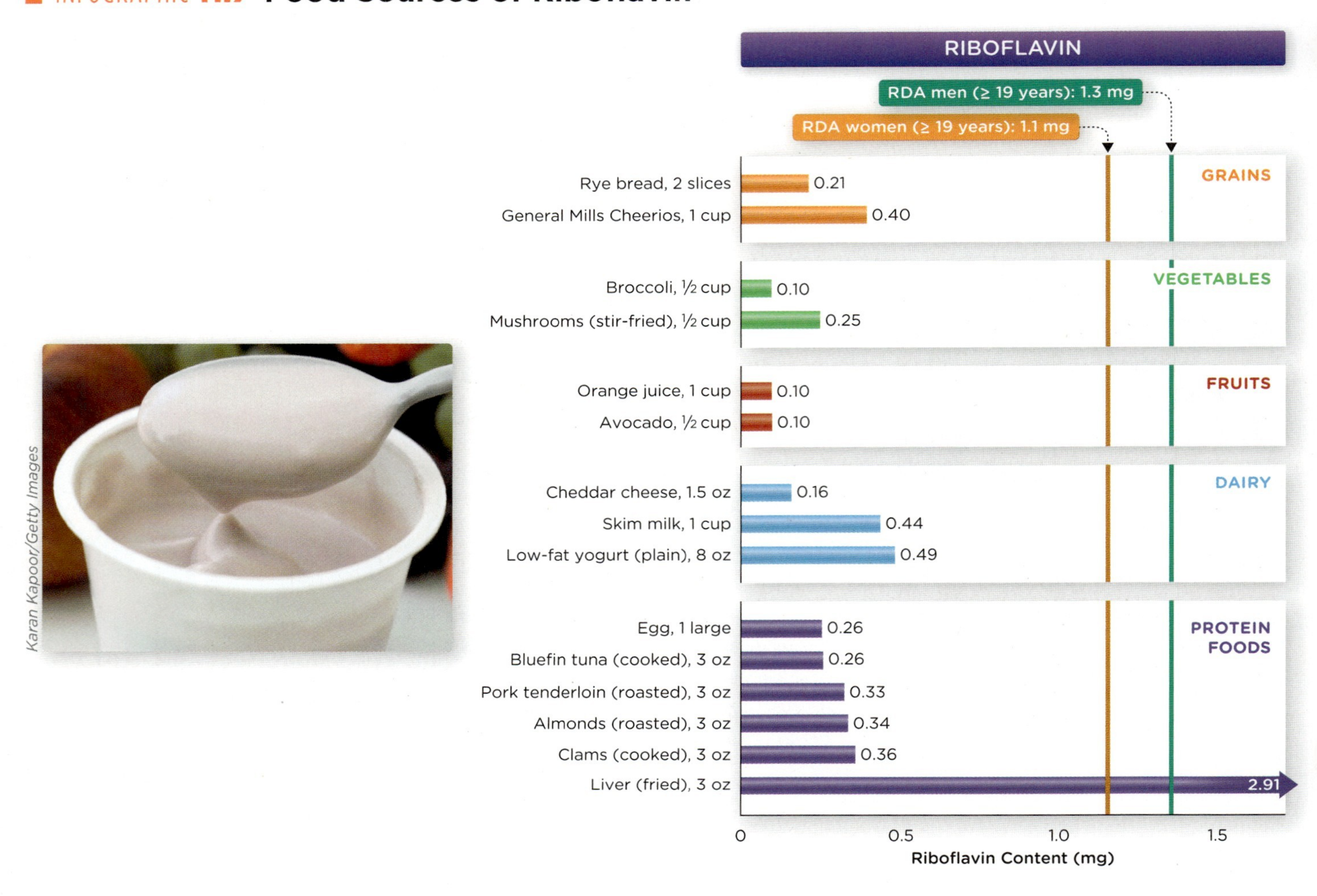

metabolism. It is required for the conversion of the amino acid tryptophan to niacin. The body also uses vitamin B_6 for the production of hemoglobin (found in red blood cells) and neurotransmitters.

Deficiency in vitamin B_6 can cause various signs and symptoms, including anemia, impaired immune function, weakness, dermatitis, and neurological disorders such as confusion and convulsions. Poor vitamin B_6 status has also been associated with an increased risk of cardiovascular disease. Adverse and sometimes irreversible neurological effects such as pain and numbness in the extremities can occur in people who take vitamin B_6 supplements in doses that result from intakes above the UL. **(INFOGRAPHIC 11.8)**

Pantothenic acid

Another B vitamin, *pantothenic acid*, has critical functions in energy metabolism and is also required for the synthesis of fatty acids, cholesterol, steroid hormones, and two neurotransmitters. Named for the Greek word *pantothen*, meaning "from all sides," deficiencies of pantothenic acid are rare because of its widespread occurrence in virtually all foods.

Folate

Folate is an important vitamin particularly for developing fetuses. Toward the end of World War II, Nazis blockaded western Netherlands, limiting millions of people's access to food and causing a terrible six months of starvation known as the Dutch Hunger Winter. The story of the Dutch

INFOGRAPHIC 11.8 Food Sources of Vitamin B_6 (Pyridoxine)

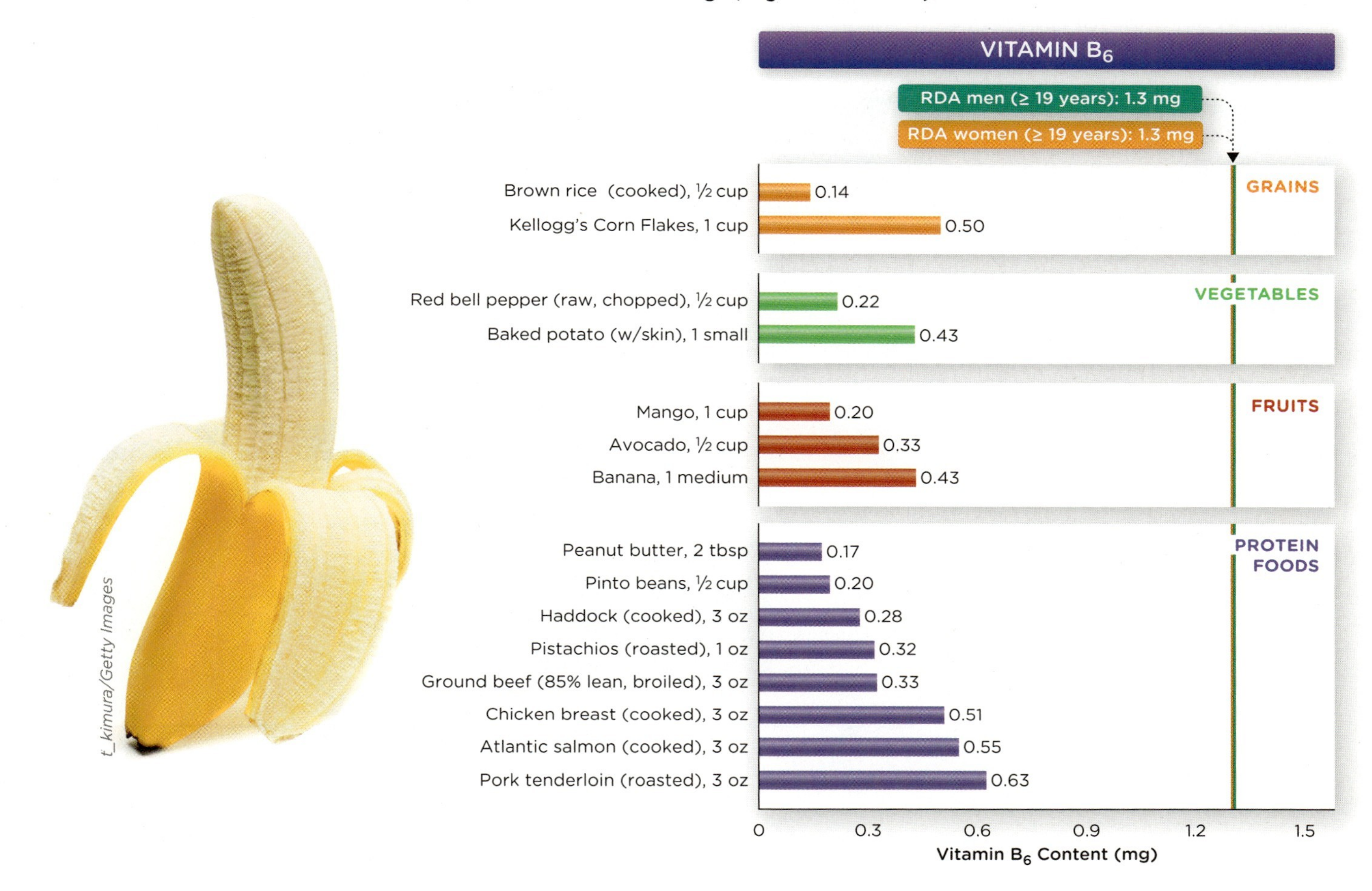

struggle was featured in Chapter 1 of this book. As tragic as the famine was—it is believed to have killed more than 20,000 Dutch people—the event also gave scientists a rare opportunity to study the biological consequences of nutrient deficiency, particularly during pregnancy. Babies conceived during the famine, when mothers were unable to eat folate-rich foods, were twice as likely as children conceived before or after to have spinal defects like spina bifida (see Chapter 18). Research suggests that folate may be among the most important nutrients for pregnant women during early pregnancy, because the nervous system requires folate to carry out a number of fundamental metabolic processes.

Folate—its synthetic form, folic acid, is used in supplements—acts as a coenzyme in the metabolism of certain amino acids and production of nucleic acids that are required for DNA and RNA synthesis, so it is essential for normal cell division and development. A calculation known as the **dietary folate equivalent (DFE)** is used to consider the difference in bioavailability between the different types of folate. Not only is the folate found naturally in foods absorbed half as well as the folic acid added to foods, it is also degraded much more readily. **(INFOGRAPHIC 11.9)**

Like the Dutch women who were pregnant during the Hunger Winter, expectant mothers today who don't consume enough folate are at an increased risk of giving birth to children with abnormalities of the spinal cord and brain known as neural tube defects. These children are also more likely to develop schizophrenia, autism, attention problems, and language delays.

DIETARY FOLATE EQUIVALENT (DFE) *a system established to account for the differences in bioavailability between folic acid in dietary supplements or foods, and folate found naturally in food*

INFOGRAPHIC 11.9 Understanding Folate Equivalents *Dietary folate equivalent (DFE) units are used to account for differences in the efficiency of absorption of naturally occurring food folate and synthetic folic acid from dietary supplements or fortified foods.*

Folate Equivalents		
Source	**Conversion**	**Reason**
Food folate	1 mcg = 1 mcg DFE	50% of food folate is absorbed. Folate from food is set as the standard for comparison.
Folic acid from supplements	1 mcg = 2 mcg DFE	100% of folic acid taken as a supplement without foods is absorbed, so two times as much is absorbed than an equal mass of food folate.
Folic acid from fortified foods	1 mcg = 1.7 mcg DFE	When folic acid is consumed with food, either as a supplement or in fortified foods, there is a slight decrease in absorption compared with its consumption without food.

EXCELLENT SOURCES OF DIETARY FOLATE EQUIVALENTS

? *If a peanut butter sandwich provides 25 mcg of folate and 30 micrograms of folic acid, how many mcg DFE units does this provide? Identify the source of the folic acid and the folate in this sandwich.*

Photo credits (left to right): AN NGUYEN/Shutterstock, Kelly Cline/Getty Images, FotografiaBasica/Getty Images

MEGALOBLASTIC ANEMIA *a type of anemia characterized by larger-than-normal red blood cells; usually caused by folate or vitamin B_{12} deficiency*

Folate deficiency also causes **megaloblastic anemia**, a form of blood abnormality characterized by large, immature, and sometimes irregularly shaped red blood cells. Folate deficiencies can also cause poor growth and gastrointestinal tract disturbances because the rate of DNA synthesis is highest during growth or in rapidly dividing cells, such as those found in the brush border of the small intestine. Some research suggests that folate may reduce the risk of heart disease because a high level of the amino acid homocysteine, which folate converts (along with vitamin B_{12}) into another amino acid (methionine), has been linked with an increased risk of heart disease. A 2012 meta-analysis of 14 epidemiologic studies found that for every 200 extra micrograms of folate consumed per day, an individual's risk of developing heart disease dropped by 12%. However, because randomized controlled trials have not found folic acid supplements to be effective at reducing the risk of heart disease, they are not recommended by the American Heart Association. **(INFOGRAPHIC 11.10)**

The RDA for folate for men and women older than 19 years is 400 mcg. The consumption of at least 400 mcg of folic acid daily from supplements, fortified foods, or both, in addition to consuming food folate from a varied diet prior to conception has been shown to prevent approximately two-thirds of all cases of neural tube defects, the most common of all birth defects, in newborns. Once a woman becomes pregnant, her RDA of folate goes up to 600 mcg, which she can get from supplements or fortified foods,

INFOGRAPHIC 11.10 Megaloblastic Anemia *A deficiency in either folate or vitamin B_{12} disrupts DNA synthesis and therefore impairs cell division of red blood cells and other rapidly dividing cells in the body.*

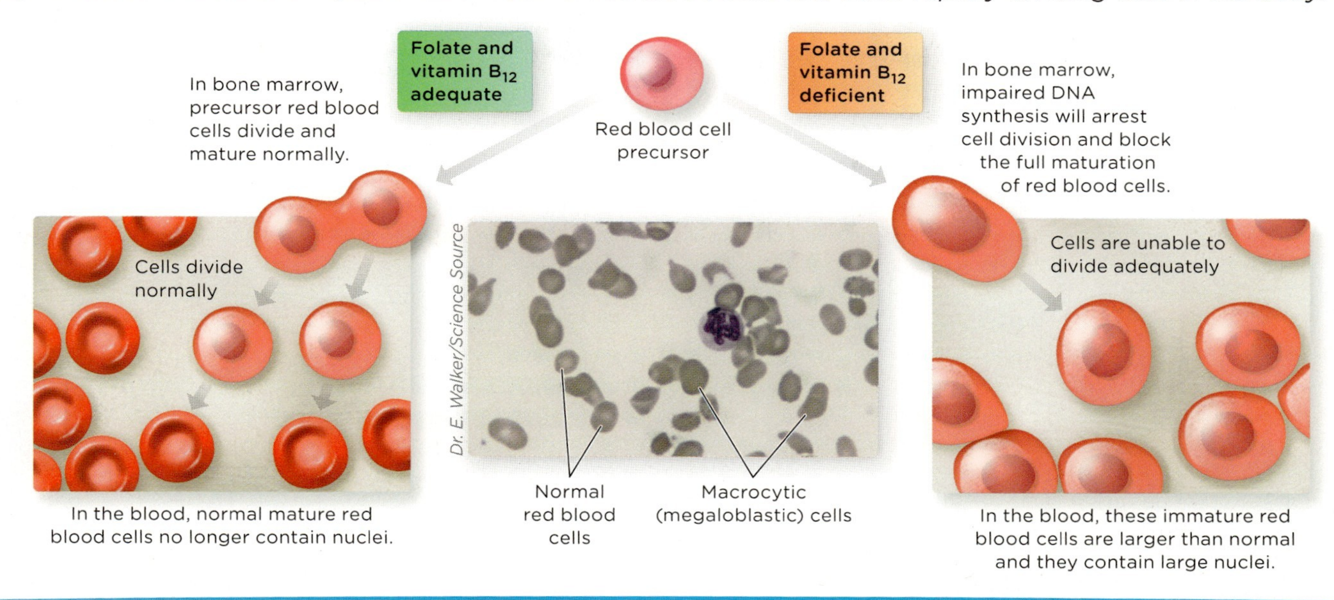

What cellular structure is present in megaloblastic red blood cells that is absent in mature red blood cells?

such as enriched cereals and bread products, in addition to foods that contain naturally occurring folate. Foods with plentiful naturally occurring folate include green leafy vegetables, avocado, broccoli, and beans. **(INFOGRAPHIC 11.11)**

People who abuse alcohol, have malabsorptive conditions, or take certain medications (such as high doses of anti-inflammatory drugs or anticonvulsants) are at an increased risk of developing folate deficiencies. The UL for folate applies to synthetic forms found in supplements (folic acid) and is set at 1,000 mcg per day. No adverse effects are known to occur with the amounts of folate found naturally in foods, but getting too much folic acid from fortified foods or supplements can mask vitamin B_{12} deficiency symptoms.

Vitamin B_{12}

Vitamin B_{12} is unique among the vitamins for several reasons: It has the largest and most complex structure and it contains a mineral (cobalt; for this reason it is known as *cobalamin*). Unlike other water-soluble vitamins, it is stored in significant quantities in the liver. Because vitamin B_{12} found naturally in foods is bound to food proteins, gastric acid from the stomach is required for it to be released from those proteins. It must then bind to a protein produced in the stomach called intrinsic factor. The cobalamin-intrinsic factor complex is then absorbed in the small intestine.

B_{12} acts as a coenzyme in only two reactions, one of which is important in deriving energy from several amino acids, and the other is the aforementioned reaction involving folate and the conversion of homocysteine to methionine. This last reaction is also indispensable for activating folate; therefore, without B_{12}, folate becomes trapped in an unusable form. For this reason B_{12} is also required for DNA synthesis and cell division.

INFOGRAPHIC **11.11** **Food Sources of Folate**

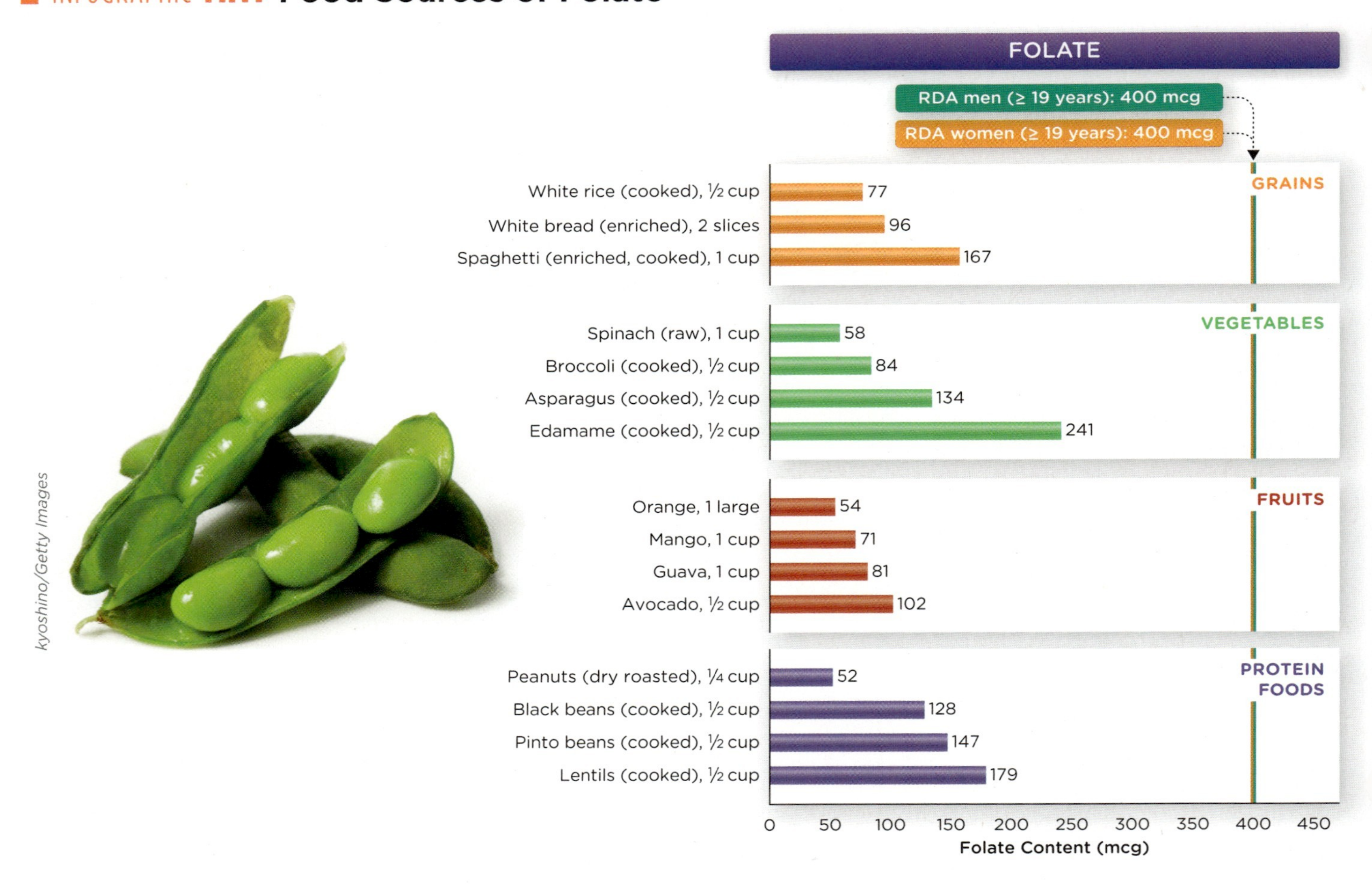

Because the metabolism of folate and vitamin B_{12} are closely linked, a deficiency in B_{12} produces the same megaloblastic anemia that is seen with a folate deficiency, causing increased fatigue during physical activity.

As a B_{12} deficiency continues, it often causes a tingling or lack of sensation in the legs and arms, and may progress to include cognitive impairment and problems with motor control. The risk of developing a deficiency of vitamin B_{12} increases as we age. Diminished or compromised food intake, sometimes seen in the aging population, decreases the amount of B_{12} available through the diet. In addition, between 10% and 30% of older people don't properly absorb the B_{12} that is found naturally in food because of a common condition that reduces the production of both gastric acid and intrinsic factor by the stomach. For this reason, individuals older than 50 years are advised to mainly meet their B_{12} RDA by consuming foods fortified with vitamin B_{12} or by taking supplements containing B_{12}, because this form of B_{12} is not bound to food proteins and therefore does not rely on gastric acid for absorption. A 2013 study noted that new diagnostic tests have revealed a "surprisingly high prevalence" of a subtle form of subclinical B_{12} deficiency among the elderly. Low vitamin B_{12} status is associated with more rapid cognitive decline as we age, adding more importance to adequate B_{12} intake.

Get your vitamin B_{12}. *Some people don't consume enough vitamin B_{12} to meet their needs, while others can't absorb enough, no matter how much they take in. As a result, vitamin B_{12} deficiency is relatively common, especially among older people.*

Animal foods such as meat, poultry, fish, and dairy are the only natural sources of vitamin B_{12}; some grain and soy products are fortified with B_{12}. Vegans (those who consume no animal products, including dairy and eggs) must either take supplements or regularly consume fortified food products to meet their RDA. **(INFOGRAPHIC 11.12)**

Gastric bypass patients are also at risk for B_{12} deficiency because less intrinsic

INFOGRAPHIC 11.12 Food Sources of Vitamin B_{12} (Cobalamin)

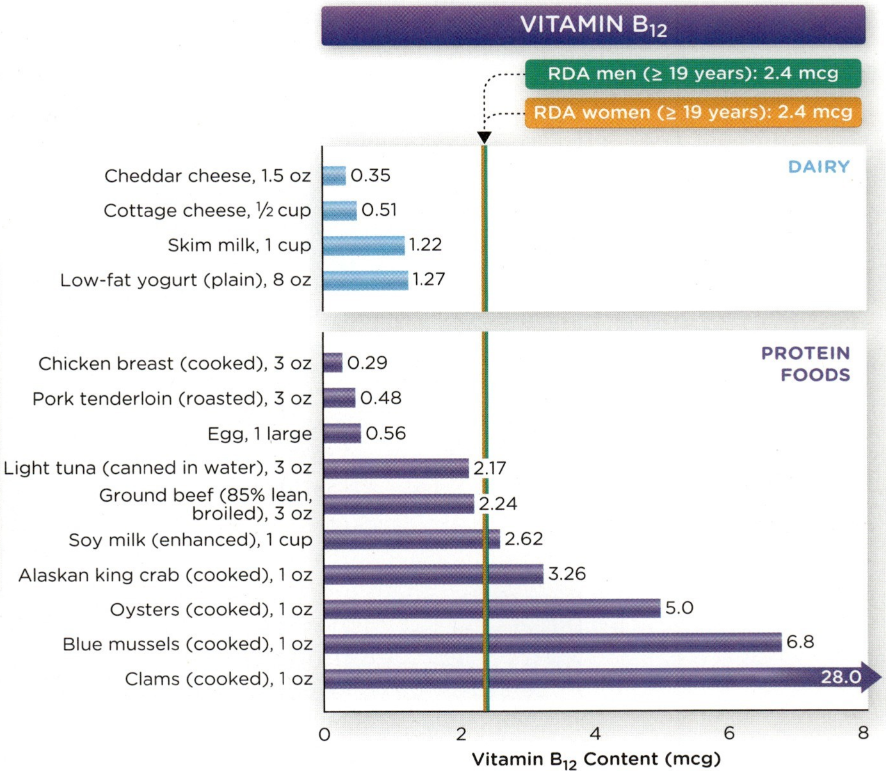

PERNICIOUS ANEMIA *a condition caused by a failure to produce intrinsic factor, resulting in vitamin B_{12} malabsorption*

SCURVY *a disease caused by a deficiency of vitamin C resulting in bleeding gums, bruising, joint pain, and impaired wound healing*

factor is produced by the stomach after the surgery, and less B_{12} is released in the stomach as much of the stomach is "bypassed." Other groups at risk for developing B_{12} deficiencies include those with **pernicious anemia**—a condition caused by the failure to produce intrinsic factor (as may often be the case in the elderly), resulting in vitamin B_{12} malabsorption and megaloblastic anemia. In these cases, treatment may require periodic injections of vitamin B_{12}, or the oral administration of very high daily doses (generally 1 mg per day) of the vitamin. Very high doses can result in absorption by passive diffusion of a small amount in the absence of intrinsic factor.

CHOLINE

Choline is the most recent compound to be added to the list of essential nutrients. Though not a B vitamin, it is a water-soluble compound and its function in the body is intertwined with that of folate and vitamin B_{12}. In addition, it forms the critically important neurotransmitter, acetylcholine, and it is part of two of the most abundant phospholipids in cell membranes. The primary sign of a choline deficiency is liver damage. Though we are capable of synthesizing choline, it is considered an essential nutrient because many people, particularly men, require dietary sources to meet their needs. Excessive intake of choline, typically only through supplementation, causes a fishy body odor and a slight drop in blood pressure.

VITAMIN C

In the eighteenth and nineteenth centuries, as many as two million sailors were afflicted by a disease known as **scurvy**, a condition that causes anemia, bleeding

Scurvy causes the subcutaneous bleeding (bleeding under the skin) visible on the legs of this person suffering from a vitamin C deficiency.

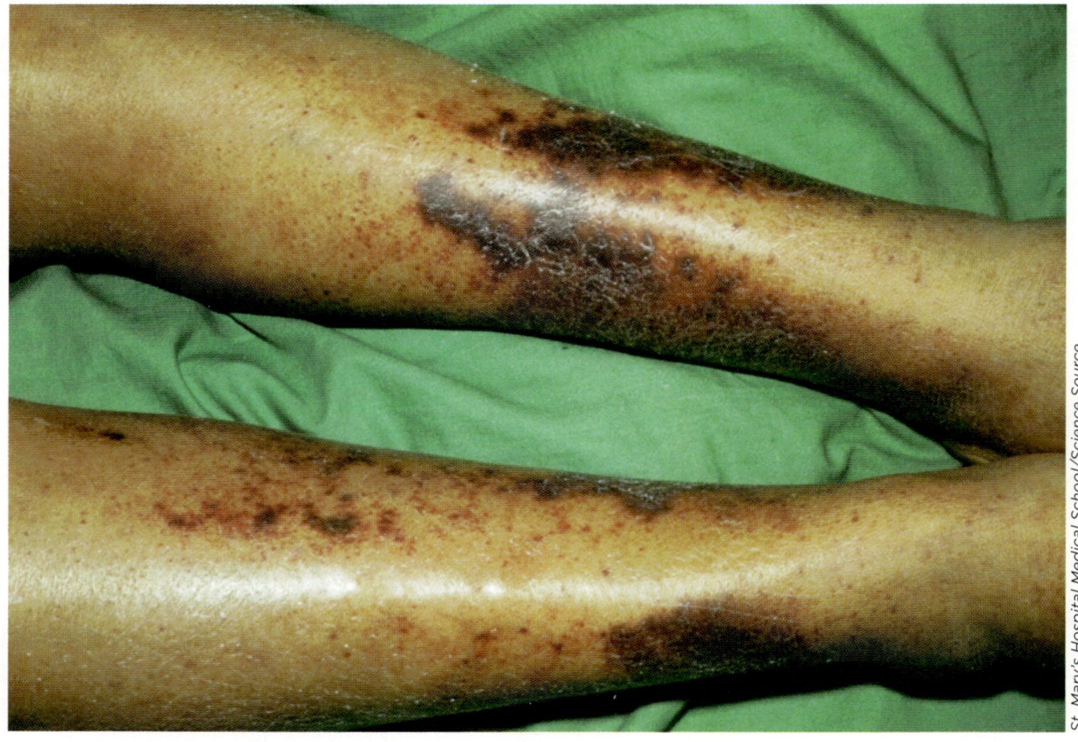

St. Mary's Hospital Medical School/Science Source

Citrus fruits are an excellent source of vitamin C.

lacaosa/Getty Images

gums, weakness, fatigue, joint pain, fragile bones, bruising, impaired wound healing, and immunity problems. In 1747, Scottish physician James Lind conducted a trial of six different treatments for 12 sailors with scurvy and found that only lemons and oranges were effective at curing the condition. These citrus fruits are rich in vitamin C; a deficiency of this water-soluble vitamin was the cause of the malady.

Vitamin C, also known as *ascorbic acid*, acts as a coenzyme in biological reactions and aids in hormone production. It's also involved in the synthesis of collagen, which is used to build bone, teeth, scar tissue, and arterial walls, and it enhances iron absorption. The majority of the signs associated with scurvy are caused by the inability to synthesize collagen appropriately, which causes blood vessels to leak, wounds to heal poorly, and gums to bleed. In addition, vitamin C functions as an **antioxidant**, a substance that prevents damage to cells by inhibiting the oxidation caused by free radicals, which can damage DNA and tissues.

Despite the old wives' tales, vitamin C is not a cure for the common cold: A 2013 evaluation of published research on vitamin C concluded that "no consistent effect of vitamin C was seen on the duration or severity of colds in the therapeutic trials," meaning that there is likely little benefit to taking vitamin C supplements once the symptoms of a cold are detected. In contrast, *regular* vitamin C supplementation may result in a very slight decrease in the duration of a cold by about 8% in children and 14% in adults. In individuals experiencing short-term physical stress, like running a marathon, regular vitamin C supplementation can reduce the incidence of colds, by as much as 50%. As with some of the B vitamins, vitamin C may have a role in reducing the risk of heart disease.

ANTIOXIDANT
a substance that prevents damage to cells by inhibiting the oxidation caused by free radicals

INFOGRAPHIC **11.13** **Food Sources of Vitamin C (Ascorbic Acid)**

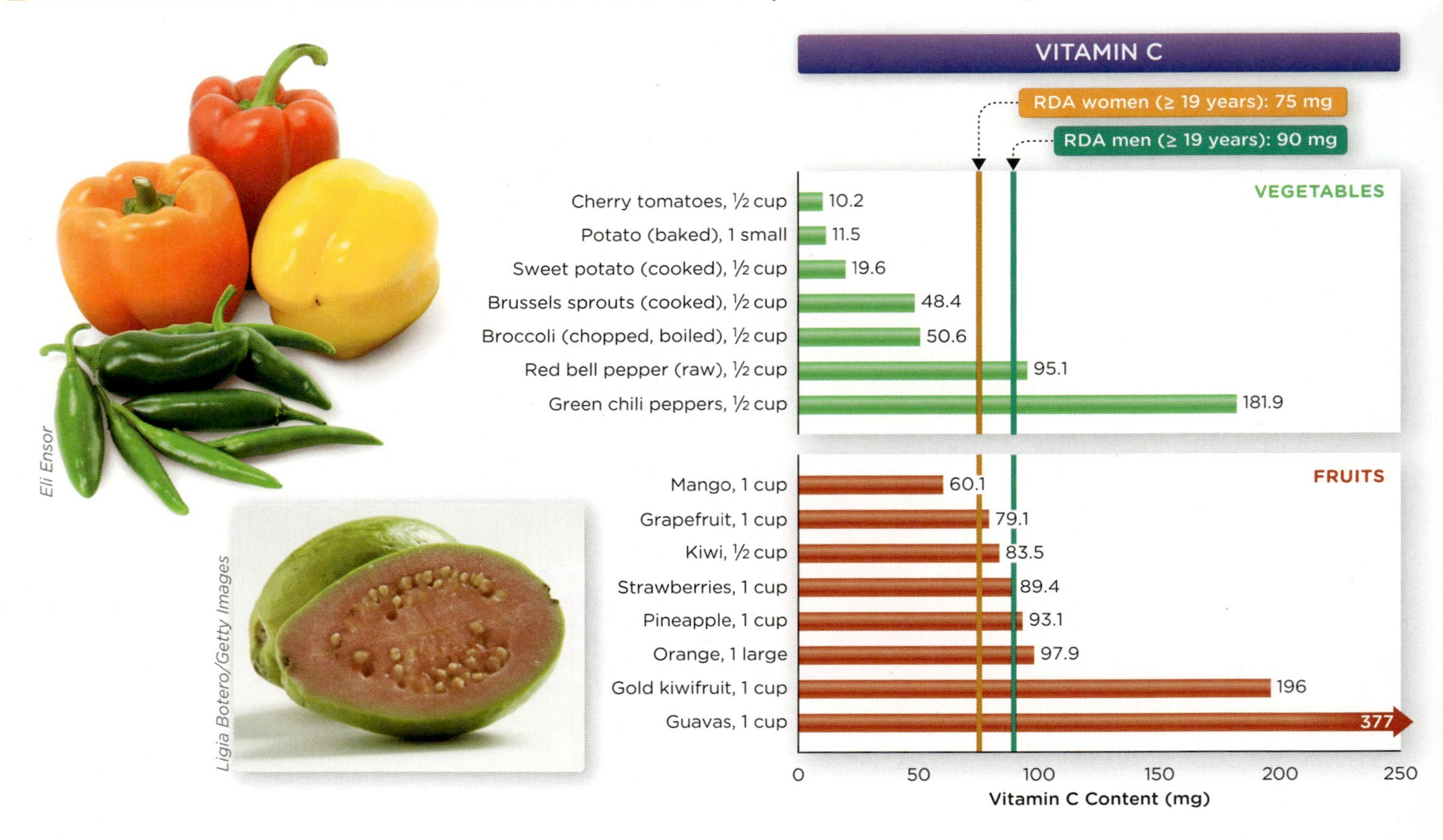

Vitamin C is naturally found in citrus fruits, tomatoes, tomato juice, potatoes, Brussels sprouts, cauliflower, broccoli, strawberries, cabbage, and squash (INFOGRAPHIC **11.13**). However, it is readily destroyed by storage and cooking. Potatoes and apples lose about 50% of their vitamin C content in 4 to 5 months of storage.

Consumption of vitamin C in excess of the UL, set at 2,000 mg for both men and women, may cause diarrhea and bloating, and it may increase the risk of kidney stones in people with kidney disease. The RDA is 90 mg for men 19 and older and 75 mg for women 19 and older. Because smokers are exposed to greater oxidative stress, and they have lower concentrations of vitamin C in the body, they are advised to increase their intake by another 35 mg over the RDA.

INFOGRAPHIC **11.14** shows a list of all of the water-soluble vitamins and their functions.

■ ■ ■

Scurvy, beriberi, and pellagra are diseases largely of the past, because of the work of the physicians, physiologists, and chemists who were pioneers in vitamin research. Although our understanding of vitamins and other nutrients in foods is ever-growing, micronutrient deficiency still haunts people in developing nations, people in war-torn regions, and refugees from conflict. Relief organizations do their best to find ways to distribute food and supplementary nutrients to support those populations. However, as we shall see in Chapter 13—iron, iodine, and zinc deficiencies are still common.

INFOGRAPHIC 11.14 Functions of the Water-Soluble Vitamins

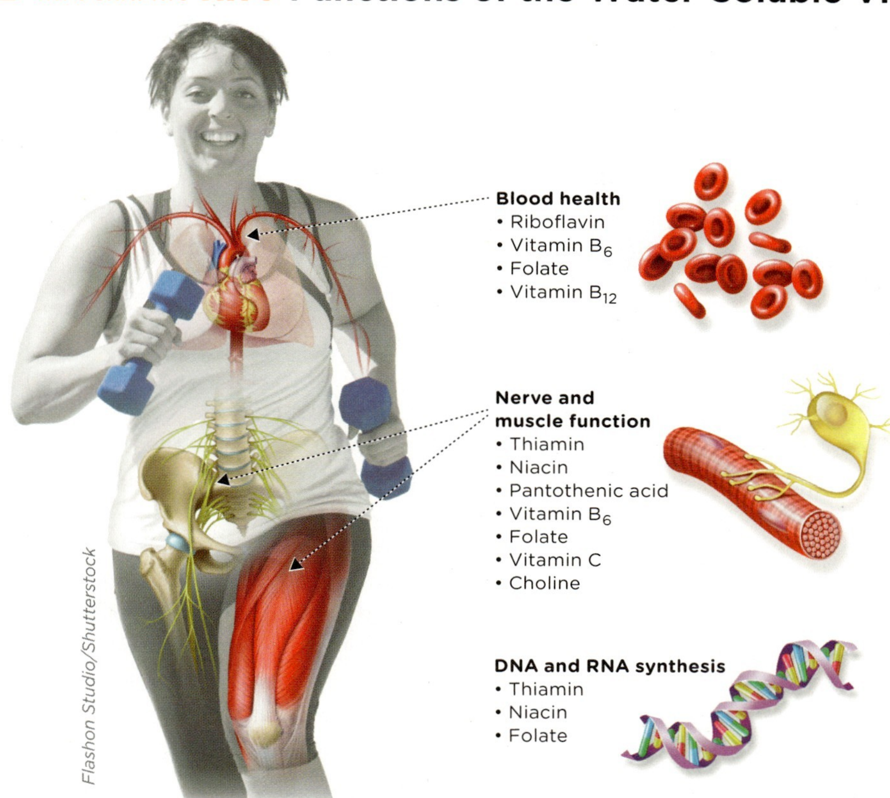

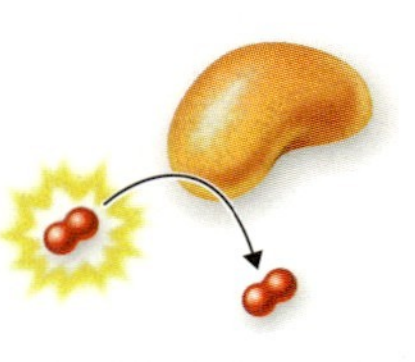

Summary of the Water-Soluble Vitamins

Vitamin	Excellent Food Sources	Deficiency Signs and Symptoms	Toxicity/Adult UL
Thiamin	Pork, fortified and enriched cereal products, seeds and some nuts, fish	**Beriberi:** Mental disturbances, muscle weakness, impaired cardiac function	None
Riboflavin	Dairy products, fortified cereals, liver, almonds, clams, pork	**Ariboflavinosis:** Inflammation of the mouth and tongue, cracks in the corner of the mouth	None
Niacin	Poultry, meat, fish, fortified cereals, peanuts, mushrooms	**Pellagra:** Diarrhea, dermatitis with sun exposure, delirium	None
Pantothenic acid	Widely distributed in foods; highest in liver, fortified cereals, poultry, potatoes, and mushrooms	Neurological disturbances, irritability and restlessness, fatigue, and gastrointestinal disturbances	Flushing of the skin, liver toxicity **45 mg** from supplements
Vitamin B_6	Meat, fish, potatoes, bananas	Dermatitis, anemia, depression, confusion, and weakness	Numbness in hands and feet resulting from nerve damage **100 mg**
Biotin	Liver, eggs, meat, fish	Dermatitis, hair loss, depression, and neurological abnormalities	None
Folate	Legumes, green leafy vegetables, liver, fortified grains and cereals	Megaloblastic anemia, neural tube defects, malabsorption and diarrhea	Masks a B_{12} deficiency **1,000 mcg** from fortified foods and supplements
Vitamin B_{12}	Fish and shellfish, liver, beef	Megaloblastic anemia, nerve damage, neurological disorders	None
Vitamin C	Guava, citrus fruits, peppers, kiwi, broccoli	**Scurvy:** Bleeding gums, loose teeth, pinpoint bruising, poor wound healing, severe joint pain	Diarrhea and gastrointestinal disturbances **2 g**
Choline	Liver, eggs, meat	Liver damage	Fishy body odor, excess perspiration, decreased blood pressure **3.5 g**

CHAPTER 11 **BRING IT HOME**

Evaluating vitamin B and vitamin C dietary supplements

You may have considered purchasing and taking a dietary supplement after hearing or reading about its potential benefits to your health, physical performance, energy level, or appearance. Follow the steps below to explore some resources that can help you learn more about one of the B vitamins or vitamin C as a dietary supplement. This isn't a comprehensive or conclusive review; no definite conclusion can be made from looking at a few studies, but you will learn what questions to consider when evaluating dietary supplements and which online resources have more extensive information to guide you.

Choose one of the B vitamins (thiamin, riboflavin, niacin, pantothenic acid, biotin, folic acid, vitamin B_6, or vitamin B_{12}) or choline or vitamin C to investigate. Ideally, you will have access to the vitamin supplement's bottle or package. If you don't have a supplement bottle, you can also find specific product information and label information by visiting supplement manufacturer websites.

1. Which vitamin supplement did you chose to evaluate?
2. Are there any warnings or precautions for use of this product listed on the bottle or package?
3. Does the label include any health-related claims? Are these FDA-approved health claims or structure/function claims? Do you trust the claims? Why or why not?
4. Does the label of your supplement include any designations or claims regarding its purity, safety, quality, or effectiveness? Does the label have a symbol that the product has been evaluated for quality assurance from an independent laboratory?
5. Why do you think someone might choose to take this supplement or product? What benefit do you think people expect when they purchase and use this supplement?
6. Visit www.pubmed.gov (U.S. National Library of Medicine, National Institutes of Health) for access to thousands of abstracts from peer-reviewed medical and scientific journals. Type your vitamin's name in the search box. If you are looking for information on a specific claim, you might search the vitamin and the condition—for example "folate AND birth defects." Read at least two medical abstracts regarding the vitamin you've chosen, taking special note of the "Results" and "Conclusions" sections in the abstract, if they exist. From your brief review, in your opinion, is there sufficient medical evidence to justify the supplement's claims?
7. Explore at least two of the websites listed below for additional information on your dietary supplement. List three things you learned about the vitamin supplement.

   National Institutes of Health
   dietary-supplements.info.nih.gov

   U.S. Office of Dietary Supplements
   ods.od.nih.gov/index.aspx

   National Center for Complementary and Alternative Medicine www.nccam.nih.gov

   Medline Plus www.nlm.nih.gov/medlineplus/druginformation.html

   Dietary Supplement Information Bureau
   www.supplementinfo.org

   Natural Medicines Comprehensive Database
   www.naturaldatabase.com
8. Based on your review of the vitamin supplement label and readings/research above, what recommendations and/or reservations do you have regarding the use of this vitamin as a dietary supplement?

KEY IDEAS

The essential water-soluble vitamins disperse easily in water-based solutions and include the B vitamins, vitamin C, and the vitamin-like nutrient, choline.

Water-soluble vitamins are not stored in large quantities in the body and must be consistently consumed in adequate amounts to meet the body's needs and prevent deficiencies.

Dietary deficiency diseases of the water-soluble vitamins result from inadequate intake or conditions that decrease absorption with subsequent low blood levels. Although deficiencies of many of the B vitamins may have similar signs and symptoms and often occur simultaneously, unique deficiency diseases have been identified for many of the B vitamins and vitamin C.

Although water-soluble vitamins rarely reach toxicity levels from food alone, excess intake through dietary supplements and/or fortified foods may exceed the established Tolerable Upper Intake Level (UL) for some vitamins.

The B vitamins include thiamin (B_1), riboflavin (B_2), niacin (B_3), pantothenic acid (B_5), pyridoxine (B_6), biotin (B_7), cobalamin (B_{12}), and folate (B_9).

The B vitamins function primarily as coenzymes, chemical compounds that bind enzymes and are required for their function or activity. As coenzymes, they participate in energy metabolism—transforming carbohydrate, protein, and fat into energy—as well as a myriad of other types of reactions.

Megaloblastic anemia—characterized by large, immature, and sometimes irregularly shaped red blood cells—results from folate deficiency, but is also seen with vitamin B_{12} deficiency.

Choline, the most recently identified essential nutrient, is a water-soluble compound frequently grouped with the B vitamins because of its intertwined role with folate and vitamin B_{12}.

Vitamin C is a water-soluble vitamin that functions as a coenzyme in biological reactions, aids in hormone production, serves as an antioxidant, and is involved in the synthesis of collagen.

NEED TO KNOW

Review Questions

1. Characteristics of water-soluble vitamins include all of the following, EXCEPT:
   a. they all function as coenzymes.
   b. they are stored in muscle and liver.
   c. most circulate freely in the blood.
   d. they dissolve or disperse in water.
   e. they are easily destroyed or removed during food preparation.

2. The B vitamins include all of the following, EXCEPT:
   a. biotin.
   b. folate.
   c. niacin.
   d. pantothenic acid.
   e. tocopherol.

3. Chemical compounds that bind enzymes and are required for their function are termed:
   a. catalysts.
   b. coenzymes.
   c. hormones.
   d. lipoproteins.
   e. precursors.

4. The deficiency disease associated with the B vitamin thiamin is:
   a. beriberi.
   b. pernicious anemia.
   c. megaloblastic anemia.
   d. pellagra.
   e. rickets.

5. The vitamin that is sometimes given as a prescription to help lower LDL cholesterol and raise HDL cholesterol is:
   a. riboflavin.
   b. niacin.
   c. pyridoxine.
   d. thiamin.
   e. vitamin E.

6. The primary source of riboflavin in the U.S. diet is:
   a. citrus fruits.
   b. grains.
   c. milk.
   d. nuts.
   e. shellfish.

7. A chemical name for vitamin B_6 is:
   a. cobalamin.
   b. linolenic acid.
   c. pantothenic acid.
   d. pyridoxine.
   e. tocopherol.

8. Absorption of vitamin B_{12} requires:
   a. the presence of vitamin C in the same meal.
   b. the presence of bile in the small intestine.
   c. binding to a protein in the stomach called intrinsic factor.
   d. incorporation into chylomicrons in the blood.

9. When consumed in adequate amounts before conception and early in pregnancy, _____ has been shown to prevent as many as two-thirds of all cases of _____.
   a. vitamin A; immune-related disorders
   b. folate; neural tube defects in newborns
   c. vitamin D; premature bone loss
   d. vitamin B_6; morning sickness in women

10. Functions of vitamin C include all of the following, EXCEPT:
    a. it acts as a coenzyme.
    b. it aids in hormone production.
    c. it serves as an antioxidant.
    d. it is a structural component of cell membranes.
    e. it synthesizes collagen.

11. The absorption of dietary iron is enhanced by the presence of:
    a. choline.
    b. fiber.
    c. pyridoxine.
    d. vitamin C.
    e. water.

Dietary Analysis Using SuperTracker

Analyzing intake of water-soluble vitamins

The water-soluble vitamins include the B vitamins and vitamin C (and the nutrient choline). Because these vitamins are generally not stored in the body, deficiency conditions of the water-soluble vitamins develop more quickly than for the fat-soluble vitamins. Using SuperTracker, identify the sources of water-soluble vitamins in your diet and compare the amount you get with the current dietary recommendations.

1. Log on to the United States Department of Agriculture (USDA) website at www.supertracker.usda.gov. If you have not done so already, you will need to create a profile to get a personalized diet plan. This profile will allow you to save your information and diet intake for future reference. *Do not use the general plan.*
2. Click the Track Food and Activity option.
3. Record your food and beverage intake for one day that most reflects your typical eating patterns. Enter each food and beverage you consumed into the food tracker. Note that there may not always be an exact match to the food or beverage that you consumed, so select the best match available.
4. Once you have entered all of your food and beverage choices into the food tracker, on the right side of the page under the bar graph, you will see Related Links: View by Meal and Nutrient Intake Report. Print these reports and use them to answer the following questions:
   a. Did you meet the targets for the water-soluble vitamins on the day you selected? If not, which vitamin intakes are below the target numbers?
   b. For each vitamin target that you missed, list two specific foods you could consume to increase your intake of that vitamin.
   c. Was your intake above the targets for any of the water-soluble vitamins? If so, which ones? Are there any issues with toxicity that are associated with those vitamins?
   d. Discuss how the enrichment of grains in the United States has helped eradicate certain water-soluble vitamin deficiencies. What deficiencies are rarely seen in the United States as a result of this policy?
   e. Many foods and beverages are fortified with water-soluble vitamins. Discuss the pros and cons of this practice.

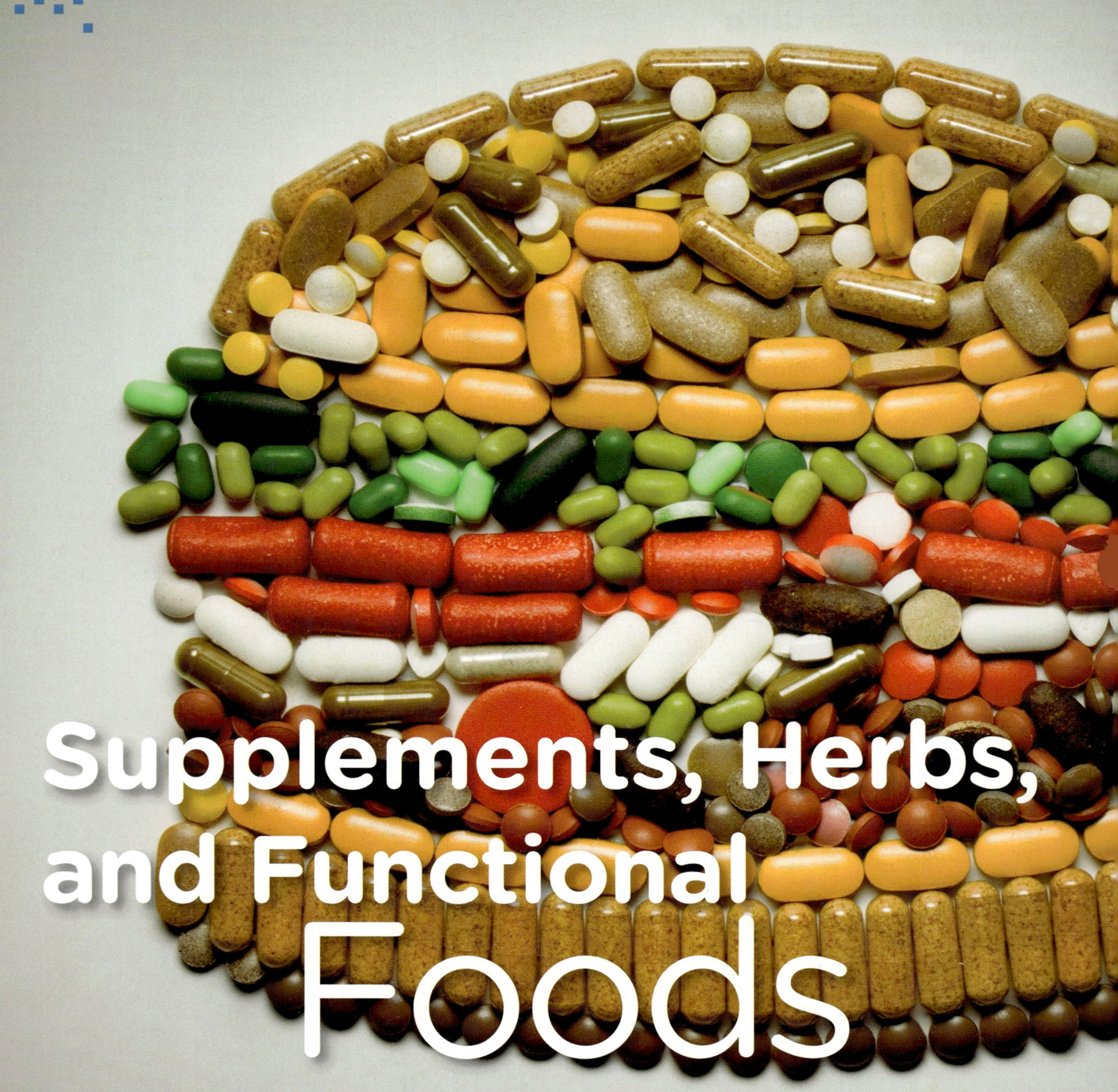

Supplements, Herbs, and Functional Foods

SURPRISING STUDIES ON THE VALUE OF VITAMIN SUPPLEMENTS

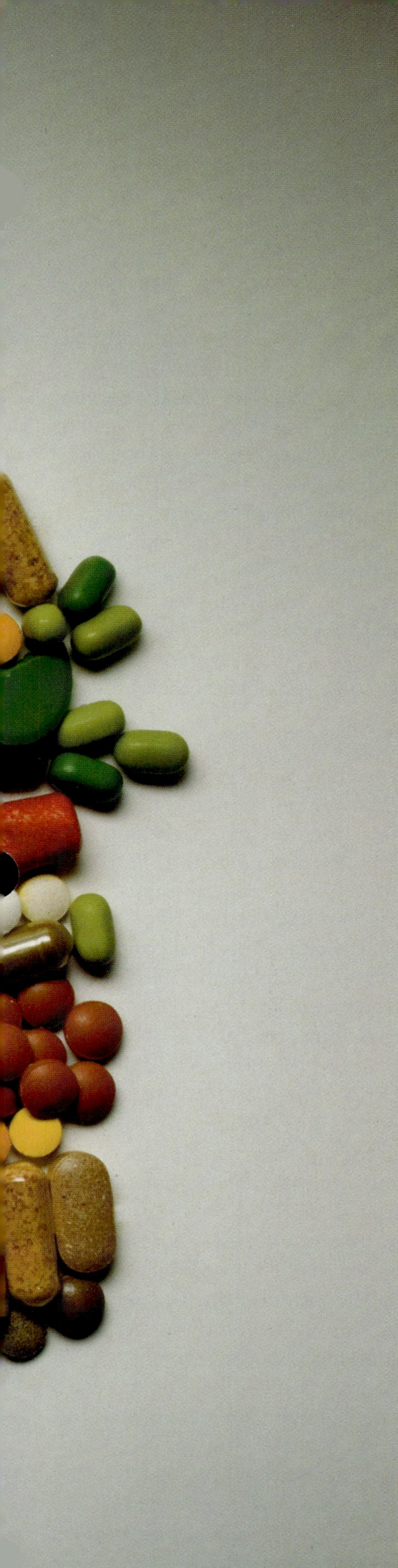

Jasper White/Getty Images

LEARNING OBJECTIVES

- Identify at least three situations or conditions for which specific supplementation may be warranted (Infographic 12.2)
- Provide an overview of the regulatory policies in the United States for dietary supplements compared with those for prescription or conventional drugs (Infographic 12.3)
- Describe the types of information provided on a Supplement Facts Panel (Infographic 12.4)
- Describe how approved health claims differ from structure/function claims for dietary supplements and provide at least two examples of each (Infographics 12.5 and 12.6)
- Provide an example of an herbal supplement, and explain its possible benefits and adverse effects (Infographic 12.9)
- Describe what might make a food "functional" and how these foods might affect health, dietary quality, and overall nutrient intake (Infographic 12.10)
- Characterize the difference between probiotics and prebiotics, and give examples of each (Infographic 12.11)

Gilbert Omenn swallowed hard. It was April 1993, and he had just been called into an emergency breakfast meeting with two scientists from the National Cancer Institute (NCI), a division of the United States National Institutes of Health. "We have a very serious problem. We would like to tell you all about it," the NCI scientists told Omenn, who at the time was Dean of the University of Washington's School of Public Health and Community Medicine. "On one critical condition: You must not tell anybody else."

Omenn didn't like secrets, but he agreed—not least because he wanted to know if their concerns

Gilbert Omenn, Professor of Internal Medicine, Human Genetics, and Public Health at the University of Michigan. *Omenn was principal investigator of the Beta-Carotene and Retinol Efficacy Trial (CARET) which explored potential lung cancer and heart disease preventatives.*

University of Michigan Health System

had anything to do with the large NCI-funded clinical trial he was leading called Beta-Carotene and Retinol Efficacy Trial (CARET). He was testing whether large doses of beta-carotene plus vitamin A could prevent lung cancer in more than 18,000 people who either currently smoked, had smoked in the past, or had been exposed to asbestos. As it turned out, the NCI scientists were worried about a different ongoing antioxidant trial being conducted by researchers at the NCI and in Finland. The scientists pulled out some of the preliminary Finnish trial data and laid it out in front of him.

Omenn was looking at the lung cancer rates of people in the Finnish trial who had taken beta-carotene plus either vitamin E supplements or sugar pills. "What it showed was a difference in cancer incidence in the two treatment groups of the study that received beta-carotene, starting a year or two after the beginning, getting bigger and bigger," Omenn recalls. That made sense: Beta-carotene was probably preventing cancer, but the placebo wasn't. But when Omenn saw the data labels, he froze. The study participants who were developing cancer were the ones taking the beta-carotene—not the ones taking placebos.

"This was a big shock," Omenn recalled—so much so that he didn't believe it. The labels must have accidentally been switched, he said. The NCI scientists assured him they weren't.

In closed meetings that Omenn attended over the course of the next few weeks, the cancer agency confirmed that beta-carotene had increased lung cancer risk by 18% compared with placebo in the Finnish trial, although vitamin E seemed to have no effect. All the while, Omenn's own trial continued. A year later, in late spring 1994, the NCI gave Omenn the green light to tell his colleagues and his 18,314 trial participants about the Finnish findings. Only 606 of his participants chose to drop out of the study. By January 1996, Omenn had enough data to crunch numbers, and what he discovered was distressing: The combination of beta-carotene and vitamin A was increasing his participants' risk of lung cancer by a whopping 28% compared with sugar pills. For the safety of everyone enrolled, Omenn immediately stopped the trial, 21 months ahead of schedule. Looked at another way, Omenn's results, which were published in the *Journal of the National Cancer Institute* later that year, suggested that the people had a 1 in 1,000 increased chance of developing cancer each year as a direct

Does vitamin A supplementation prevent cancer? *Observational studies had shown that people eating more fruits and vegetables, which are rich in beta carotene (that the body can convert into retinol, or vitamin A), had lower rates of lung cancer. The Beta-Carotene and Retinol Efficacy Trial (CARET) tested the combination of beta-carotene and vitamin A supplements in men and women at high risk of developing lung cancer. The CARET intervention was stopped 21 months early because of clear evidence of no benefit and substantial evidence of possible harm.*

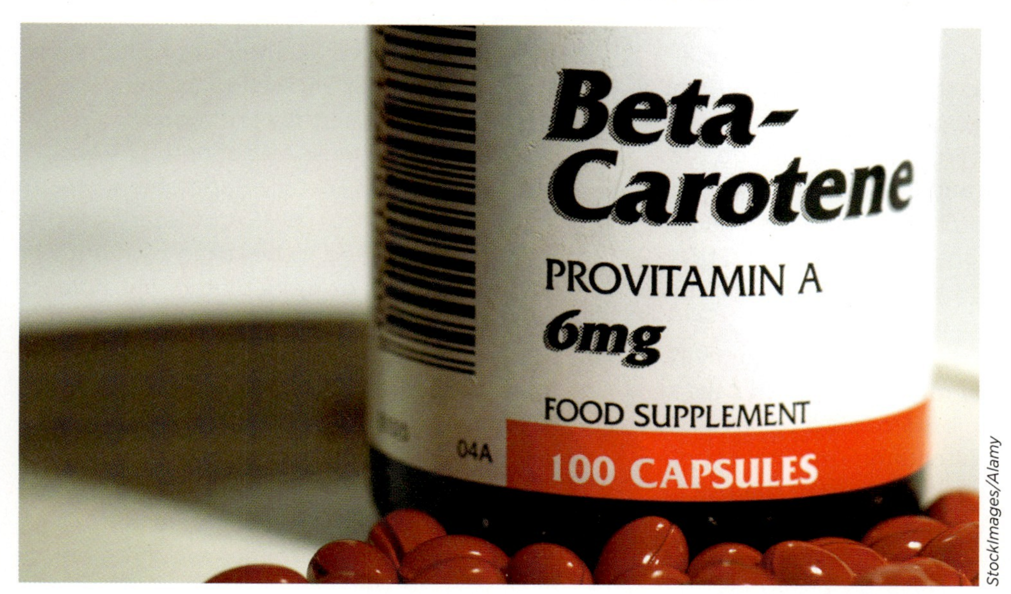

StockImages/Alamy

result of taking the beta-carotene/vitamin A supplements.

■ ■ ■

Chances are, if you opened the medicine cabinet in your bathroom or the cabinets in your kitchen you would find a vitamin bottle, or a drink powder, or perhaps a nutrition bar of some sort. Dietary supplements are a more than 30 billion dollar a year industry, more than 65,000 varieties are sold in the United States today, and more than half of all Americans take supplements in an attempt to stay healthy. The most common supplements are multivitamin and multi-mineral supplements, followed by calcium, omega-3 fatty acid supplements (fish oil), and vitamin D. Yet fewer than a quarter of Americans who take supplements do so at the recommendation of a qualified health care provider.

Dietary supplements are a big business in the U.S. *According to NHANES data, the percentage of the U.S. population who used at least one dietary supplement increased from 42% in 1988–1994 to 53% in 2003–2006.*

Keith Brofsky/Getty Images

DIETARY SUPPLEMENT
a food or substance that supplements the diet and contains one or more dietary ingredients (including vitamins, minerals, herbs, amino acids, and certain other substances) or their constituents

Supplements are beneficial for some individuals, particularly those who cannot meet their nutritional requirements because of disease, increased need, or restricted diets. But many people take supplements despite the fact that they are already in good health and likely get adequate nutrients through their diet. Women are more likely to take supplements to keep their bones healthy, while men tend to pop pills in the hope of preventing heart problems. But do these supplements actually help? A growing body of research, including the results of those National Cancer Institute trials from the 1990s, suggests that high doses of single vitamins, as well as lower doses of multivitamins, may be only mildly helpful at best and may pose health risks at worst. The majority of studies have found that multivitamin/mineral supplement use does not decrease the risk of death or chronic disease, and several studies have even reported an increased risk of death associated with their use. People who take supplements tend to eat well already, so those who might benefit the most from supplements are, ironically, often the ones who are least likely to take them.

WHAT ARE DIETARY SUPPLEMENTS?

Dietary supplements can come in the form of pills, capsules, tablets, liquids, powders, and bars. This diverse array of products are grouped together and called "dietary supplements" because they are regulated by a common set of rules. As defined by Congress in the 1994 Dietary Supplement Health and Education Act (DSHEA), dietary supplements are intended to be taken by mouth and contain one or more dietary ingredients or their constituents: vitamins; minerals; herbs or other botanicals; amino acids; or other dietary substances, such as enzymes. Essentially any substance that is found in any food, even if present only in minute quantities, can be extracted or concentrated and then sold as a dietary supplement. Therefore, these substances are often being consumed at levels far exceeding those ever achieved from the intake of foods. **(INFOGRAPHIC 12.1)**

INFOGRAPHIC **12.1** **What Are Dietary Supplements?** *Dietary supplements are a diverse array of products meant to "supplement the diet." They include vitamins, minerals, herbals, botanicals, amino acids, and enzymes.*

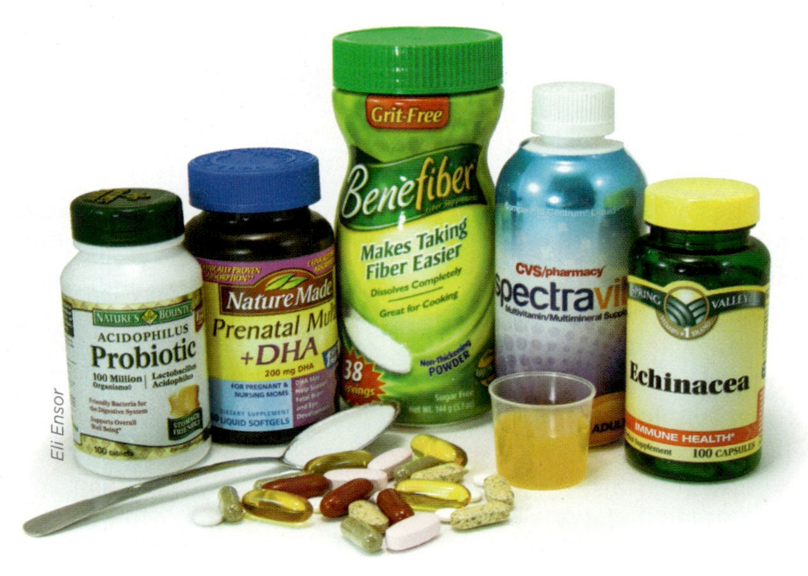

Women who are pregnant, or plan to become pregnant, are advised to take a folic acid supplement to help prevent neural tube defects.

Who benefits from taking supplements? For the average person, the Academy of Nutrition and Dietetics recommends that the best nutrition-based strategy for health is to eat a wide variety of foods. However, some populations may benefit from using dietary supplements. For example, women who have heavy menstrual bleeding may need supplemental iron, and people who have had procedures that interfere with nutrient absorption, such as gastric bypass surgery, may need particular nutrient supplements. **INFOGRAPHIC 12.2** lists additional circumstances that might require the use of a dietary supplement.

REGULATION OF DIETARY SUPPLEMENTS

The FDA regulates both dietary supplement products and the ingredients found within them, and it does so under a different set of regulations than those governing "conventional" foods and drugs. Unlike drug regulation, the FDA does not approve dietary supplements for their effectiveness or safety before they are made available to consumers. Dietary supplements do not undergo the rigorous testing for effectiveness, interaction, or safety requirements that prescription and over-the-counter drugs do. In fact, as DSHEA dictates, the *manufacturer* of a dietary supplement or ingredient is responsible for ensuring that the product is safe, unadulterated, produced with good manufacturing practices, and properly and truthfully marked with a label that identifies the product as a dietary supplement and includes specific information about the supplement and its use.

Although dietary *ingredients* found in supplements are federally regulated, ingredients and additives that were already in the food supply prior to when DSHEA went into effect on October 15, 1994, were grandfathered in as **"generally recognized as safe" (GRAS)**. GRAS substances don't need FDA approval before being marketed, so many

GENERALLY RECOGNIZED AS SAFE (GRAS)
any substance intentionally added to food that is generally recognized, among qualified experts, as having been adequately shown to be safe under the conditions of its intended use; not subject to premarket review and approval by the FDA

The FDA does not approve dietary supplements for their effectiveness or safety. *The manufacturer of a dietary supplement or ingredient is responsible for ensuring product safety.*

H. Mark Weidman Photography/Alamy

INFOGRAPHIC 12.2 Circumstances that May Warrant Nutrient Supplementation

Nutrient supplements may be useful in some circumstances, but they cannot replace a healthy diet.

Population Group *Do you fit into one of these categories?*	**Dietary Concerns**
Infants and children	Breastfed children and any child consuming less than 1 qt/day of vitamin D-fortified milk should receive a vitamin D supplement.
Women who may become pregnant	Supplemental folic acid reduces the occurrence of neural tube defects.
Pregnant women	A folic acid supplement is recommended during pregnancy. A multivitamin/mineral (MVM) supplement is recommended for anemia, women carrying multiple fetuses, or women consuming little or no animal proteins.
Vegans	The only source of B_{12} is animal proteins and fortified foods, so vegans who eat no animal products may need a supplement as well as supplementary calcium, iron, and zinc.
Those who do not consume dairy products	Because milk and other dairy products are an important source of vitamin D and calcium, a supplement providing these nutrients may improve bone health.
Adults older than 50 years	B_{12} and vitamin D supplements are recommended because B_{12} absorption tends to decline with age and older individuals synthesize less vitamin D when exposed to UV light.
Those with dark skin	Vitamin D supplements are recommended because skin pigments block UV light and decrease the synthesis of vitamin D.
Individuals on restricted diets	Those with low food intake or limited food choices may benefit from an MVM supplement.
Smokers, alcohol-dependent individuals, and those taking some medications	Nutrient absorption, utilization, and excretion can be affected by prescription or recreational drug use, therefore an MVM supplement may be warranted.
Women who are pregnant; women with heavy menstrual periods. Individuals who frequently donate blood, as well as those with some stomach and intestinal conditions (food sensitivity, hookworms).	Iron supplementation may be necessary.

J Am Diet Assoc. 109: 2073–2085, 2009.

Identify at least three underlying reasons why supplements might be necessary in these circumstances.

long-standing supplement ingredients have not actually endured FDA scrutiny for safety.

For new dietary ingredients, DSHEA requires that manufacturers must notify the FDA 75 days before the product is to be introduced and provide the agency with evidence that the supplement is "reasonably expected to be safe." Unfortunately, it is common for supplement distributors and manufacturers to ignore this requirement, as well as other regulations.

Because the FDA does not regulate dietary supplements as rigorously as it does drugs, supplements are sometimes sold contaminated with banned substances or prescription drugs. In 2010, for instance, the agency warned consumers not to take the Chinese weight-loss supplement Fruita Planta, because FDA testing had revealed it contained sibutramine, a drug that had been withdrawn from the United States market earlier that year for safety reasons. And in a 2013 study, Pieter Cohen, an assistant professor of medicine at Harvard Medical

INFOGRAPHIC 12.3 FDA Regulations Governing the Introduction of New Dietary Ingredients in Supplements *Many dietary ingredients legally present in supplements have not been reliably demonstrated to be safe. The 2011 Food Safety Modernization Act expands the authority of the FDA to oversee new dietary ingredients, but has not yet been fully implemented.*

Current Regulations	Proposed New Regulations
• Manufacturers must submit to the FDA information regarding the "safety and efficacy" of a dietary supplement containing a new dietary ingredient. • The manufacturer needs only to demonstrate that the new dietary ingredient can "reasonably be expected to be safe" by providing *some* evidence of safety but not actually proving safety. • The FDA does not approve new dietary ingredients or supplements—they are free to be marketed 75 days after the information was submitted to the FDA.	• Safety must be established by a documented history of use at the same or higher dosage, frequency, and duration of use. • Laboratory or animal studies would be required for products marketed for consumption at higher than historical intakes. • No human studies are required, even for substances without any documented historical use. • The company is not obligated to report results showing adverse effects. They need only to submit positive results supporting their claim of safety.

New Eng J Med. 366: 289-391, 2012.

Give an example of "some evidence of safety" that would NOT actually prove that the ingredient is safe.

School who studies dietary supplement safety, and his colleagues found traces of a methamphetamine-like substance in a popular workout supplement.

The FDA is responsible for taking action against unsafe or improperly labeled dietary supplements after they go to market, but this is not easy to do: The agency has to *prove* that the product is unsafe to restrict its use or remove it, and "there's no effective system that the FDA has to identify these supplements, so hazardous supplements stay on the marketplace for years," Cohen says. Even when the FDA announces that a supplement is unsafe, it may not cease being sold. In a 2011 study, Cohen and his colleagues found that after the FDA recalled a popular weight-loss supplement called Pai You Guo in 2009, women were able to continue buying it for years. This all said, the 2011 Food Safety Modernization Act (discussed in Chapter 20) expanded some of the FDA's authority over supplement manufacturers. **(INFOGRAPHIC 12.3)**

UNDERSTANDING SUPPLEMENT LABELS

The FDA requires that dietary supplement manufacturers list certain details about their products on product labels. The general information required on the package includes the name of the product; the word "supplement" or a statement that the product is a supplement; the quantity of the package contents; the name and location of the manufacturer, packer, or distributor; and directions for using the product.

In addition to the general information, a supplement must also have what is called a **Supplement Facts Panel**. This panel must include information on serving size and amount of product per serving size (by weight), the percent of Daily Value (%DV) that a particular ingredient or nutrient provides per serving (if this is known), and a list of the product's dietary ingredients. **(INFOGRAPHIC 12.4)** If a dietary ingredient is a botanical, the panel must list the scientific name of the plant or the common name that has been standardized in

SUPPLEMENT FACTS PANEL
package label that must indicate that the product is a supplement, not a conventional food, and must include serving size and amount of product per serving size, the percent of Daily Value that a particular ingredient or nutrient provides per serving, and a list of the product's dietary ingredients

INFOGRAPHIC 12.4 Dietary Supplement Labeling Requirements *One of the most significant differences between the "Facts Panel" for a supplement versus a food is that substances and ingredients that do not have recommendations are allowed on the supplement panel.*

SUPPLEMENT FACTS PANEL

Supplement Facts

Serving Size: One Capsule
Servings Per Container: 60

	Amount Per Serving	% Daily Value
Vitamin C	5mg	8
Thiamin	500mcg	33
Niacin	5mg	25
Vitamin B_6	5mg	250
Magnesium	35mg	9
Valerian Root (Valeriana Officinalis)	100mg	††
Chamomile Flowers (Matricaria recutita L.)	75mg	††
Passion Flower (Passiflora Incarnata)	65mg	††
Melatonin	250mcg	††

††Daily Value Not Established

Inactive Ingredients: Plant-Derived Cellulose (capsule), Rice Flour, Magnesium Stearate

SERVING SIZE must be provided at the top of the panel.

NUTRITION LABELING

NAME AND AMOUNTS OF DIETARY INGREDIENTS WITH DAILY VALUES must be listed first.

% DAILY VALUE must be given for those nutrients with recommendations.

DIETARY INGREDIENTS WITHOUT DAILY VALUES are listed next. For botanical ingredients, either the standard common name or the scientific name of the plant must be provided. This label provides both names.

SUPPLEMENTS CONTAINING PLANT MATERIAL must indicate the part of the plant that is used.

INGREDIENTS NOT LISTED ABOVE are listed here in descending order of predominance (by weight). These are compounds used in the manufacture of the supplement, such as binders, colors, fillers, flavors, and sweeteners.

Botanicals: *a plant or plant part valued for its health-promoting properties, flavor, and/or scent. Herbs are a subset of botanicals.*

Maintains healthy neurotransmitter levels †
Fall asleep faster and sleep longer †

Sixty (60) Capsules

Distributed by:
Betamed, Inc.
PO BOX 555, 67851
Call toll-free to reorder
(888) 555-5555

†These statements have not been evaluated by the Food and Drug Administration. This product is not intended to diagnose, treat, cure, or prevent any disease.

THE STATEMENT OF IDENTITY must include the name of the product and identify it as a dietary supplement.

CONTACT INFORMATION (domestic address or telephone number) must be provided for consumers to report adverse effects.

STRUCTURE/FUNCTION CLAIMS are allowed on supplement labels, but they must be accompanied by this disclaimer (which is not required on foods).

Must all ingredients be declared on the label of a dietary supplement?

the reference book *Herbs of Commerce* (2000 edition). The panel must also include name of the plant part that has been used. If the dietary ingredient is a proprietary blend, meaning a blend that is exclusive to the manufacturer, the Supplement Facts Panel must list the total weight of the blend and its components in descending order of predominance by weight.

Supplements must also include another ingredients panel. This lists all nondietary components found in the product such as fillers; water; artificial colors; sweeteners; flavors; and processing aids, such as binders, gelatin, and stabilizers. These ingredients are listed by common name or proprietary blend in descending order of predominance by weight. The ingredients listed in this panel may include the sources of the dietary ingredients if they are not identified in the Supplement Facts Panel—for instance, a label might list rose hips as the source of vitamin C.

Finally, supplement labels may also contain cautionary statements about potential side effects, but if a supplement does not have a cautionary statement, it does not mean that the product is completely safe. Unlike conventional drugs, supplement manufacturers do not have to list known adverse effects on their labels.

Health claims

Supplement labels can also include *health claims* that describe a relationship between a dietary supplement ingredient and a reduced risk of a disease or condition. The FDA must pre-approve these claims based on Significant Scientific Agreement (SSA) about the publicly available scientific evidence. A supplement containing calcium and vitamin D can legally claim that it reduces the risk of osteoporosis, for instance, and a folic acid supplement can say that it may prevent fetal neural tube defects. **(INFOGRAPHIC 12.5)** The FDA also allows the use of *qualified* health claims for conventional foods and dietary supplements when the evidence linking a food, food component, or supplement to a reduced risk of a disease is emerging but not well enough established to meet the SSA standard for a true health claim. For example, the number of studies demonstrating a beneficial effect may be limited, or the results of studies may be inconsistent. Qualifying language is included to indicate that the evidence supporting the relationship is limited—for instance, you might read on a label that "supportive but not conclusive research shows that consumption of EPA and DHA omega-3 fatty acids may reduce the risk of coronary heart disease." But even qualified claims have to be approved by the FDA based on the quality and strength of the scientific evidence.

Structure/function claims

Supplement manufacturers can't make claims that their product treats, prevents, or cures disease unless it has been approved for a health claim or qualified health claim. However, companies can make a *structure/function claim* on the label about how that product could affect the body's structure or function. **(INFOGRAPHIC 12.6)** "Calcium builds strong bones" is an example of a structure-related claim, whereas "fiber maintains bowel regularity" or "antioxidants maintain cell integrity" are function-related claims. Alternatively, these claims may state that consuming a nutrient or dietary ingredient may improve general well-being or describe a benefit related to a nutrient deficiency disease. Labels containing these claims must state in a disclaimer that the FDA has not evaluated the claim, and that the product is not intended to diagnose, treat, cure, or prevent any disease. Still, "more times than not, the label is going to be misleading, and the claims are going to be overly positive," Cohen says.

Supplement quality

How do we know if supplements on the market are pure and of high quality? The FDA does not monitor supplements for quality assurance, potency, purity, or efficacy—there are far too many supplements for the agency to handle—but the FDA does track

INFOGRAPHIC 12.5 FDA-Approved Health Claims and Qualified Health Claims *When there is significant scientific agreement that evidence supports a link between a diet or nutrient and a disease, the FDA establishes approved health claims. When the evidence is not as strong, the FDA allows qualified health claims.*

Selected Approved Health Claims

Nutrient and Disease	Claim Statement
Calcium and Osteoporosis	Adequate calcium throughout life, as part of a well-balanced diet, may reduce the risk of osteoporosis.
Dietary Fat and Cancer	Development of cancer depends on many factors. A diet low in total fat may reduce the risk of some cancers.
Sodium and Hypertension	Diets low in sodium may reduce the risk of high blood pressure, a disease associated with many factors.
Dietary Saturated Fat and Cholesterol, and Risk of Coronary Heart Disease	Although many factors affect heart disease, diets low in saturated fat and cholesterol may reduce the risk of this disease.
Fiber-Containing Grain Products, Fruits, and Vegetables and Cancer	Low-fat diets rich in fiber-containing grain products, fruits, and vegetables may reduce the risk of some types of cancer, a disease associated with many factors.
Fruits, Vegetables, and Grain Products that Contain Fiber, Particularly Soluble Fiber, and Risk of Coronary Heart Disease	Diets low in saturated fat and cholesterol and rich in fruits, vegetables, and grain products that contain some types of dietary fiber, particularly soluble fiber, may reduce the risk of heart disease, a disease associated with many factors.
Folate and Neural Tube Defects	Healthful diets with adequate folate may reduce a woman's risk of having a child with a brain or spinal cord defect.
Potassium and the Risk of High Blood Pressure and Stroke	Diets containing foods that are a good source of potassium and that are low in sodium may reduce the risk of high blood pressure and stroke.
Whole Grain Foods and Risk of Heart Disease and Certain Cancers	Diets rich in whole grain foods and other plant foods and low in total fat, saturated fat, and cholesterol may reduce the risk of heart disease and some cancers.

Selected Qualified Health Claims

Nutrient and Disease	Eligible Foods	Claim Statement
Green Tea and Cancer	Green tea and conventional foods and dietary supplements that contain green tea	Green tea may reduce the risk of breast or prostate cancer. The FDA has concluded that there is very little scientific evidence for this claim.
Selenium and Cancer	Dietary supplements containing selenium	Selenium may produce anticarcinogenic effects in the body. Some scientific evidence suggests that consumption of selenium may produce anticarcinogenic effects in the body. However, the FDA has determined that this evidence is limited and not conclusive.
Antioxidant Vitamins and Cancer	Dietary supplements containing vitamin E and/or vitamin C	Vitamin C may reduce the risk of gastric cancer. The FDA has concluded that there is very little scientific evidence for this claim. Vitamin E may reduce the risk of colorectal cancer. The FDA has concluded that there is very little scientific evidence for this claim.
Omega-3 Fatty Acids and Coronary Heart Disease	Conventional foods and dietary supplements that contain EPA and DHA omega-3 fatty acids	Supportive but not conclusive research shows that consumption of EPA and DHA omega-3 fatty acids may reduce the risk of coronary heart disease.

Which one of the sample qualified health claims has the strongest supporting evidence?

■ INFOGRAPHIC 12.6 Examples of Structure/Function Claims on Dietary Supplements

What is the difference between the statements "calcium builds strong bones" and "calcium reduces the risk of osteoporosis"?

reports of illness, injury, or reactions that might occur in consumers after taking supplements. Supplement manufacturers are now required to report serious harmful effects to the FDA, too. Supplements may be labeled as "pure," "natural," or "quality-assured," but, because the FDA does not regulate these terms, these claims may not be true. Supplements that claim to be "all natural," for instance, are not always better or safer than refined or synthetic substances, because natural and synthetic forms generally have the same chemical structure and do not differ in terms of how they are

Meaningless words. *Supplements may be labeled with words that are meant to make a product seem trustworthy such as "quality," "pure," or "natural," but these terms have no legal meaning.*

Vadim Petrov/Shutterstock

Look at the label. *Independent labs such as United States Pharmacopeial Convention (USP) and NSF International (NSF) an evaluate and set standards for dietary supplements.*

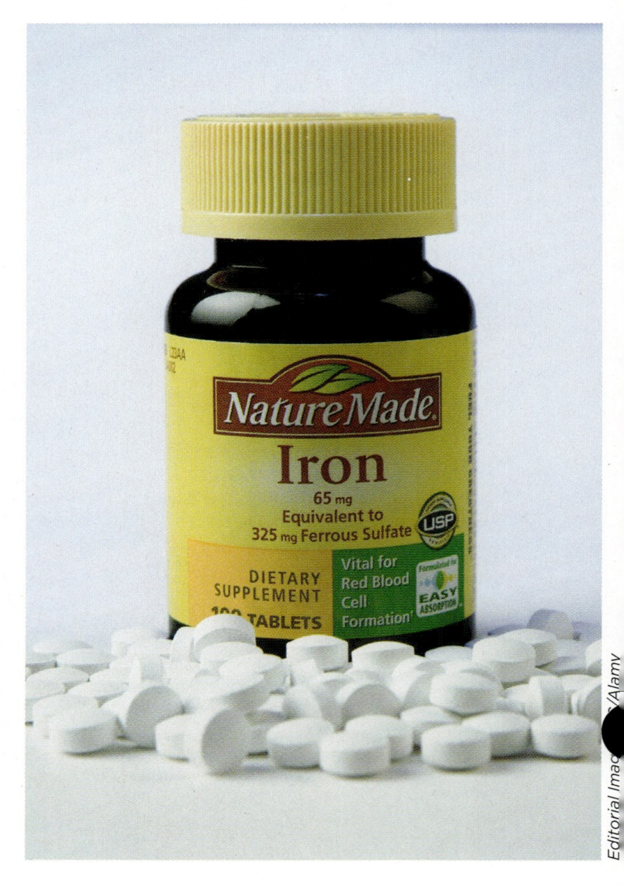

Editorial Images/Alamy

absorbed or used by the body. They may, however, differ in price—the "natural" forms are often costlier.

Consumers do have some ways of gauging supplement quality, though. Independent labs test supplements that manufacturers voluntarily submit; some labs also do product reviews. Organizations such as United States Pharmacopeial Convention (USP)—a scientific nonprofit organization that sets standards for the identity, strength, quality, and purity of medicines, food ingredients, and dietary supplements distributed and consumed worldwide—provide seals of quality that companies can display on their products. Another independent lab that provides quality labels is NSF International (NSF)

ARE DIETARY SUPPLEMENTS HARMFUL?

Of course, purity is not the only important quality. The question that is much harder to answer is: Will this supplement actually improve the health of the general population? The National Cancer Institute studies led by Omenn and his colleagues suggest that some supplements are not only ineffective, but can actually be harmful. A small subset of studies, however, suggests that supplements may be

INFOGRAPHIC 12.7 Effects of Long-term Use of Multivitamin/Mineral Supplements on Mortality

Neither the long-term use of full-spectrum multivitamin/mineral supplements, nor the administration of more narrowly formulated vitamin/mineral supplements decrease the risk of death.

Studies Examining Multivitamin/Mineral Supplements

- Of eight observational studies that examined the effect of long-term multivitamin/mineral (MVM) supplements on mortality, six studies found no effect of MVM use, while two studies observed an increased risk of death in those taking MVM supplements.
- In a recent randomized clinical trial, approximately 10 years of MVM administration caused no decrease in deaths from cardiovascular disease or cancer. MVM supplements did not affect the occurrence of CVD, though there was a slight decrease in the occurrence of all cancers.
- An earlier randomized clinical trial found no decrease in mortality following the administration of an MVM supplement to a poorly nourished population for six years, nor was there any affect on mortality 20 years later.
- Analysis of data pooled from 78 clinical trials involving nearly 300,000 participants found no benefit of antioxidant vitamins and minerals on the risk of death; however, consuming supplements of beta-carotene and vitamin E were found to increase the risk of death.
- The analysis of pooled data from 21 studies that had administered supplements containing three or more vitamins and minerals to more than 90,000 participants found no effect on mortality risk.

From the research findings summarized here, identify perhaps the best reason to NOT use vitamin/mineral supplements.

beneficial for preventing certain conditions. (INFOGRAPHIC 12.7)

For instance, in 1997, researchers at Harvard University and its affiliated hospitals and schools began a clinical trial as part of the Physicians Health Study II to test whether moderate doses of multivitamins might prevent cancer or heart disease. "When you think about individual vitamin supplements tested in trials, there's been an emphasis on high doses," explains Howard Sesso, one of the study's leaders and an associate professor of epidemiology at the Harvard School of Public Health. "We thought it would be interesting to have the opposite approach: It might be more appropriate to test a standard common multivitamin that has all the essential vitamins and minerals in combination, but in lower, more usual, doses that you would get in your diet."

To do this, the Harvard researchers tracked the health of nearly 15,000 physicians older than 50 years who had been randomly assigned to take either a standard multivitamin or a sugar pill every day for an average of 11 years. Although the multivitamin takers did not end up with a reduced risk for heart disease, they were 8% less likely to develop cancer during the follow-up period than the doctors who did not take multivitamins. A separate arm of the study that evaluated the effects of taking 400 mg vitamin C every day and 400 IU vitamin E every other day did not, however, find that the two vitamins reduced cancer risk. It seemed to be the combination of many vitamins, at lower doses, that made a positive difference.

Sesso and his colleagues were very pleased with their study results, "But what we were immediately trying to think through was: How do you explain the findings?" Indeed, one major question is exactly *how* vitamins may protect against cancer. Another question is whether a person is better off getting vitamins through whole foods or through pills. "I think a lot of people would argue that the natural approach through food is the way to go," Sesso says, but more research is needed to prove this point.

Supplements seem to pose the highest risks when they are what are known as "high-potency supplements," which include one or more nutrients or ingredients in amounts significantly in excess of

INFOGRAPHIC 12.8 Tips for Choosing a Multivitamin Supplement *If you are in a category that might benefit from taking a multivitamin supplement (see Infographic 12.2), these guidelines can help you.*

Tips for Choosing a Multivitamin Supplement

- **Read the label carefully**. Examine which nutrients are included and the amounts contained within each serving. In general, choose a supplement that provides 100% of the Daily Value (DV) for most of the vitamins and minerals in that supplement. Some nutrients, like calcium and magnesium, are rarely included at 100% because the pill would be too large to swallow.
- **Look for quality products.** The initials USP stand for U.S. Pharmacopeial Convention, and NSF stands for NSF International; both are reputable organizations that test dietary supplements for quality.
- **Look for the expiration date.** Select products that will have a long shelf life.
- **Consider formulas for men, women, and age groups.** Choose a multivitamin designed for your age and sex so that the nutrients included will be right for you.
- **Don't overdo it.** Avoid multivitamins that exceed 100% of daily recommended values.

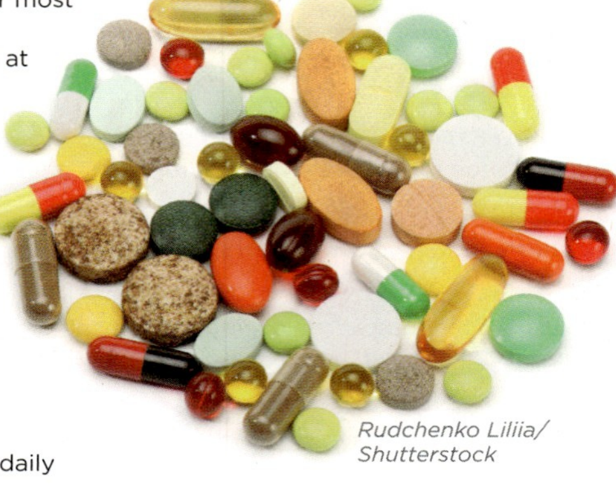

Rudchenko Liliia/ Shutterstock

SPECIAL CAVEATS

- **Beware of interactions.** Taking a combination of supplements together with medications could produce adverse effects. For example: Coumadin (a prescription drug), ginkgo biloba (an herbal supplement), aspirin (an OTC drug) and vitamin E (a vitamin supplement) can each thin the blood, and taking any of these products together can increase the potential for internal bleeding. The herbal supplement St. John's wort may also reduce the effectiveness of prescription drugs for heart disease, depression, seizures, certain cancers, and oral contraceptives.
- **Some supplements may interfere with surgeries.** Before elective surgery you may be asked to stop taking vitamins, minerals, or herbal supplements to avoid potentially dangerous supplement/drug interactions—such as changes in heart rate, blood pressure, and increased bleeding—that could adversely affect the outcome of your surgery.

recommendations. These supplements include the mega-doses of beta-carotene (30 mg) and vitamin A (25,000 IU) that the participants were taking as part of Omenn's CARET trial. Sometimes, these supplements also exceed the established Tolerable Upper Intake Limit (UL), the maximum level of daily intake that is likely to cause no risk of adverse health effects to almost all individuals in the population. No laws establish or cap potency for any supplements, except potassium. But high doses of some supplements can cause fatigue, diarrhea, hair loss, kidney stones, liver and nerve damage, and birth defects. They can lead to nutrient imbalances or interactions, reducing the absorption and utilization of other nutrients. And taking high doses of many minerals—as well as the fat-soluble vitamins A, D, and E—can be toxic. Even high doses of the water-soluble vitamin B_6 are toxic, potentially causing permanent nerve damage. It's important to remember that even if you don't exceed the UL for a nutrient through supplement use, you may exceed it when you combine the amount you're getting from a supplement with the amount you're getting from food. (INFOGRAPHIC 12.8)

HERBAL SUPPLEMENT *a type of dietary supplement that includes plants (botanicals), singly or in combination; typically dried preparations of flowers, leaves, roots, bark, and seeds*

PROS AND CONS OF HERBAL SUPPLEMENTS

Botanical supplements are valued for their medicinal or therapeutic properties to treat disease or maintain health. Botanicals include any supplement that is derived from plants, and may include liquid extracts, oils, or herbs. **Herbal supplements** are a subset of botanicals that are typically dried preparations of flowers, leaves, roots, bark, or seeds. They are less popular in the United States than vitamin and mineral supplements, but more than one-fifth of U.S. adults take them. There are 550 primary herbs with 1,800 names, but examples of some of the most common herbs sold in this country include echinacea, flaxseed, ginseng, ginkgo, saw palmetto, St. John's wort, black cohosh, milk thistle, and garlic. **(INFOGRAPHIC 12.9)** From a medicinal perspective, herbs are less potent crude drugs and can have druglike effects, yet they do not undergo the same stringent approval process as drugs do. Sometimes, herbal supplements can contain biologically active ingredients and toxins in addition to their active "useful" components. Even though herbal supplements are often considered natural, they can still cause drug interactions and serious adverse effects and even exacerbate medical conditions; there is no legal definition for the term "natural," and it certainly does not mean a product is safe or effective.

INFOGRAPHIC 12.9 Possible Adverse Effects and Benefits Associated with the Use of Herbal Supplements

Potentially Effective Botanical Supplements and Possible Adverse Effects

Supplement	Possible Benefits	Adverse Effects
Senna	Laxative	Liver failure with excessively high doses
Licorice root	Protection against liver damage, anti-ulcer effects	Hypertension
Hawthorn	Cardiovascular benefits	None
Ginger	Reduce nausea and vomiting	None
Garlic	Reduction of hypertension and cardiovascular benefits	Decreased Clotting
Black cohosh	Relief of menopausal symptoms	Possible liver injury with long-term use
Holy basil	Anti-inflammatory, anticarcinogenic effects	None
Fenugreek	Lowers blood glucose and improves insulin sensitivity	Diarrhea, low blood glucose
St. John's wort	Treatment of mild to moderate depression	Hypertension
French maritime pine bark (Pycnogenol)	Antioxidant, decreased hypertension, improved cardiovascular function	May cause mild dizziness, nausea, headache

A recent study found that bottles labeled as St. John's wort from two manufacturers contained none of the medicinal herb; one contained only rice powder and the other contained senna.

? *What adverse effects might you experience if you had purchased the bottle containing senna?*

Photo credits (top to bottom): wasanajai/Shutterstock, limpido/Shutterstock, jizni/Shutterstock, Edward Westmacott/Shutterstock, Timmary/Shutterstock, Neil Fletcher and Matthew Ward/Getty Images, Swapan Photography/Shutterstock, picturepartners/Shutterstock, Scisetti Alfio/Shutterstock, Vahan Abrahamyan/Shutterstock

Even though herbal supplements are called "natural" they can cause drug interactions and exacerbate medical conditions.

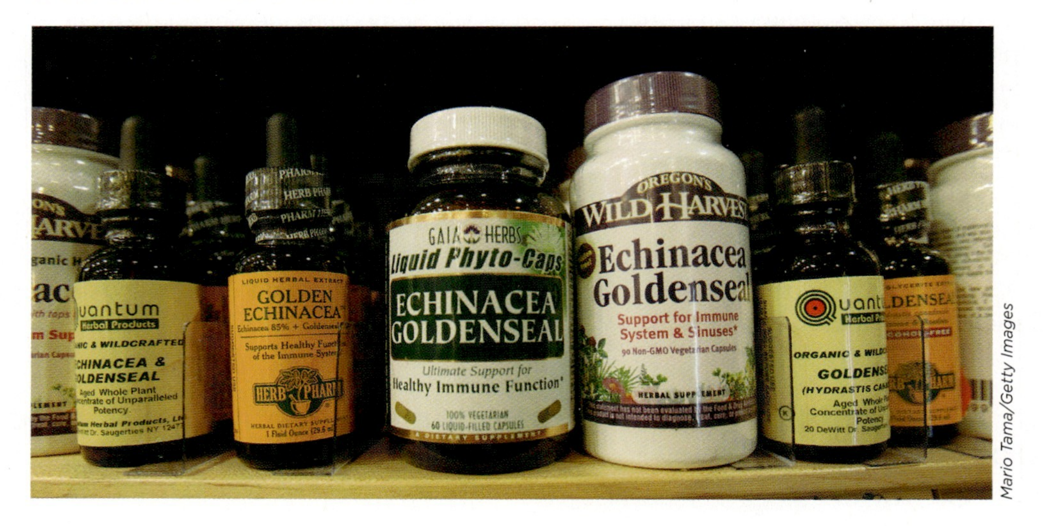

Mario Tama/Getty Images

FUNCTIONAL FOODS

Food manufacturers may opt to add additional nutrients to food products to boost nutritional value and potential health benefits. These products fall under a class of foods called **functional foods** (also called *nutraceuticals*). Functional foods contain nutrients or other constituents, such as phytochemicals, that may enhance their contribution to health and disease prevention beyond their basic nutritional content. Functional foods may be whole foods or processed foods that have been touted to decrease the risk of cancer, heart disease, diabetes, or obesity, or to slow the aging process. Some of these claims may be overhyped, but some may be accurate.

Functional foods include whole foods, like vegetables, berries, and fish; enriched grains;

Supplements for athletes. *Energy gels, bars and powders are used by athletes to improve performance, or to last through a grueling workout. See Chapter 16 for a complete discussion of this topic.*

B. Christopher/Alamy

FUNCTIONAL FOOD *a food that provides additional health benefits beyond basic nutrition that may reduce disease risk or promote good health*

fortified juices and milks; and enhanced foods, such as sports bars and sports drinks designed to help you run longer or play sports harder. Sports supplements, known as ergogenic aids, are discussed in detail in Chapter 16.

The Academy of Nutrition and Dietetics noted, in its 2013 position statement on functional foods, that "all food is essentially functional at some level as it provides energy and nutrients needed to sustain life. However, there is growing evidence that some food components, not considered nutrients in the traditional sense, may provide positive health benefits." **(INFOGRAPHIC 12.10)**

Nutrient-dense plant-based functional foods may, for instance, contain disease-fighting phytochemicals, or phytonutrients, which are biologically active constituents in foods. More than 2,000 phytochemicals have been found in plant-based foods, and many have antioxidant or hormonelike actions and may help to reduce the risk of certain types of cancer and other chronic diseases. Or, functional foods might contain *prebiotics* (nondigestible carbohydrates broken down by colon bacteria) or *probiotics* (live beneficial bacteria found in fermented foods). Probiotics can restore or maintain a healthy balance of beneficial bacteria, while prebiotics feed and cultivate them. Probiotics and prebiotics can reduce the risk of illness, because in order to establish infections, "bad" bacteria must adhere to the lining of the gut. As "good" bacteria increases in the gut in response to

INFOGRAPHIC 12.10 Examples of Functional Foods *Functional foods have health benefits beyond those provided by the vitamins and minerals they traditionally contain.*

Orange juice with added calcium to improve bone health.

Blueberries and blackberries contain high amounts of anthocyanins that promote cardiovascular health.

Yogurt with live bacterial cultures to promote gastrointestinal health.

Margarine with plant sterols added to lower blood cholesterol.

Tomatoes are naturally an excellent source of the phytochemicals lutein and lycopene that are associated with a reduced risk of eye disease and some cancers, respectively.

Why is the orange juice pictured here considered a functional food?

Photo credits (all photos): Eli Ensor

INFOGRAPHIC 12.11 Prebiotics and Probiotics *Prebiotics are compounds that promote the growth of beneficial bacteria in our gut. Probiotic bacteria may have many possible health benefits.*

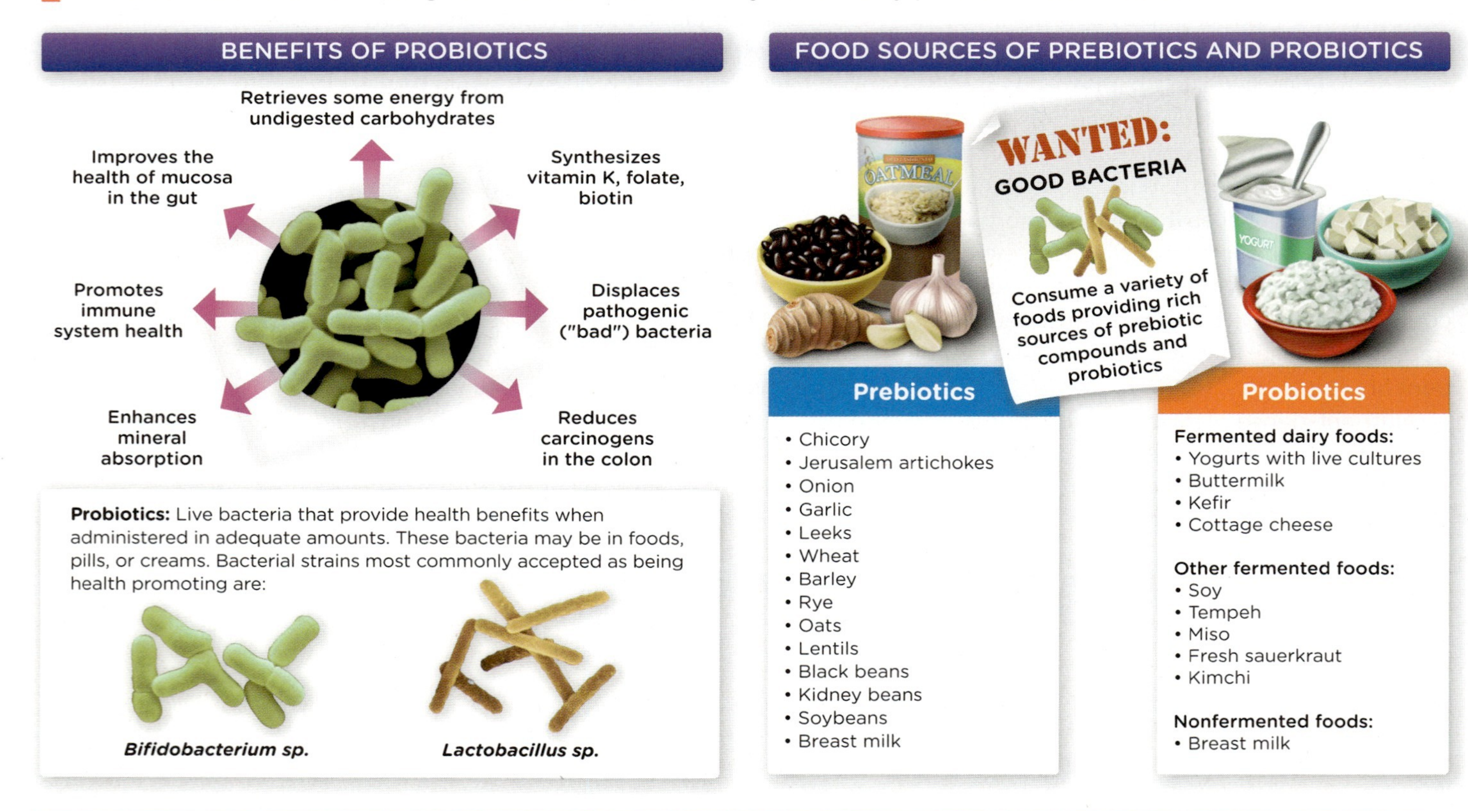

What types of grain products are most likely to have prebiotic properties?

Photo credit (all photos): Scimat/Science Source

the ingestion of probiotics (or in response to prebiotics), they take up more and more real estate, making it difficult for the bad bacteria to find a home. In addition, as the number of good bacteria increases, they grab the majority of the available nutrients, leaving little for the unwanted microbes. **(INFOGRAPHIC 12.11)**

The United States currently has no statutory legal definition for functional foods, nor any specific regulatory policies for them, so they can be categorized as conventional foods, dietary supplements, or medical foods. Functional foods are common, representing the largest percent of new food products introduced to the market over the past decade. Many packaged functional foods include health claims on their labels that are regulated by the FDA, and most are high in nutrients in relation to calories, so they can contribute to dietary quality, optimal health, and disease prevention when consumed as part of a varied, balanced diet and healthy lifestyle.

LET THE BUYER BEWARE

Dietary supplements can be important and potentially life-saving products for individuals who cannot get enough nutrients through foods—for instance, vegans, whose only source of vitamin B_{12} is fortified foods, may not get enough through their diet alone. But because supplements are not as tightly regulated as drugs, and because in some cases scientists do not fully understand their potential effects, supplements do not come without risks. "The great majority of the time people are taking supplements and it's just a waste of money," Cohen says. "And a small percentage of the time, they're taking something that's actually dangerous."

CHAPTER 12 **BRING IT HOME**

Functional foods: What's in this food?

Food manufacturers are expanding the range of nutrients they add to foods as well as the types of foods to which they are added. The increasing presence of fortified and enhanced foods in supermarkets has the potential to affect the nutrient intake and health of individuals. What role do functional foods play in your diet?

1. Choose a functional food product (one that has one or more added nutrients or dietary constituents) that you typically consume.
2. What nutrients or dietary constituents have been added to this product?
3. Are any health claims on the food label? If so, list the claim.
4. Would you consider the added nutrients to be ones that most people need more of in their diets?
5. Why do you use this particular food product? Do you purchase it for the taste, the added nutrients, the potential health benefits, or other reasons?
6. Is this particular type of product available without the added nutrients or extra fortification? For example, if you chose orange juice with added calcium, the answer would be "yes," this product is available as orange juice without added calcium. Do you know if there a cost difference between the regular product and the functional food version? Are you willing to pay more for the enhanced product?
7. What role do you think functional foods should play in the food market and in the overall population's diet?
8. Do you see any potential for misuse or excessive nutrient intake through the use of this or similar products? What if an individual is also taking a dietary supplement that includes the same nutrient?

KEY IDEAS

Dietary supplements, including vitamins, minerals, herbs, botanicals, amino acids, and enzymes, are meant to "supplement the diet," not replace nutrients that are best obtained through a varied and balanced diet.

Under the 1994 Dietary Supplement Health and Education Act (DSHEA), and more recently the 2011 Food Safety Modernization Act, the United States Food and Drug Administration (FDA) is responsible for regulating the sale, labeling, and manufacturing of dietary supplements, as well as approving any health claims made for the supplements.

Current regulations do not require dietary supplements, including botanical supplements, to undergo the same rigorous testing for effectiveness, interaction, or safety requirements as conventional drugs.

Supplement manufacturers are responsible for—and accountable to the FDA for—ensuring that supplements are safe, unadulterated, and produced with good manufacturing practices.

Manufacturers must provide specific product information and ingredients on the Supplement Facts Panel.

Supplements may also include health claims on the label that describe a relationship between a dietary supplement ingredient and a reduced risk of a disease or condition. These claims must be pre-approved for use on supplement labels by the FDA using Significant Scientific Agreement (SSA).

Manufacturers are also able to make certain claims about how a product affects the body's structure or function. These structure/function claims must include a disclaimer that the claim has not been evaluated by the FDA.

Current evidence does not demonstrate that dietary supplements provide significant health benefits. They may, in some cases, increase the risk of disease and mortality. However, there are some individuals and groups that may benefit from using supplemental nutrients to help meet their nutritional needs.

The use of high-potency supplements may result in nutrient intakes above the Tolerable Upper Intake Level and cause adverse effects.

Some foods may have a positive effect on health beyond that of basic nutrition. These are called functional foods, and the group includes whole foods, as well as fortified food products.

NEED TO KNOW

Review Questions

1. In the United States, nutrition surveys indicate that the people who are most likely to use dietary supplements:
   a. usually are in poor health with diets deficient in multiple nutrients.
   b. tend to already have healthful diets adequate in most nutrients.
   c. do so based primarily on the advice of their health care provider.
   d. experience immediate and significant health benefits.
2. Which of the following would NOT meet the criteria for being called a "dietary supplement" in the United States?
   a. omega-3 fish oil capsule
   b. multivitamin and mineral tablet
   c. chewable children's multivitamin "gummy"
   d. vitamin B_{12} injection or shot
3. Dietary supplements in the United States are monitored by the FDA for:
   a. purity.
   b. potency.
   c. effectiveness.
   d. reports of illness, reactions, or harmful effects.
4. According to the Dietary Supplement Health and Education Act, what is considered a "new" dietary ingredient?
   a. a dietary ingredient not sold in the United States in a dietary supplement before 1994
   b. a dietary ingredient that has not been evaluated by the FDA for safety and efficacy
   c. any type of ingredient or supplement other than a vitamin or a mineral
   d. vitamins or minerals added to processed foods for fortification purposes
5. Dietary supplements that have GRAS status:
   a. require the FDA's approval before they can be marketed and sold.
   b. are generally recognized as safe by the FDA.
   c. have undergone rigorous testing by the FDA for safety.
   d. have been banned for use in the United States.
6. Which of the following is an example of an acceptable structure/function claim that might appear on a dietary supplement label?
   a. helps maintain cardiovascular health
   b. prevents heart disease
   c. lowers cholesterol to prevent heart disease
   d. alleviates chest pain in individuals with heart disease
7. Synthetic vitamins are:
   a. tested for safety and efficacy by the FDA.
   b. significantly less effective than natural vitamins.
   c. generally chemically identical to vitamins labeled "natural."
   d. always classified as a new dietary ingredient by the FDA.
8. The vitamins and minerals found in high-potency supplements:
   a. are proven to improve health and promote longevity.
   b. cannot by law exceed the percent Daily Value.
   c. may exceed the Tolerance Upper Intake Level.
   d. will not result in toxicity because any excess is excreted through the urine.
9. Herbal supplements include all of the following characteristics, EXCEPT:
   a. the same approval process as drugs.
   b. a label that clearly states the scientific or standardized name of the plant used to make the supplement.
   c. possible toxins in addition to active components.
   d. being medicinally considered crude drugs.
10. All of the following are true in regard to functional foods, EXCEPT:
    a. they represent the largest percentage of new products introduced to the food marketplace in the United States.
    b. they are generally high in nutrients in relation to calories.
    c. they are only whole foods; processed or packaged foods would not meet the criteria.
    d. if packaged, they often include a health claim or nutrient content claim on their label.

11. Probiotics can be defined as:
    a. foods with added nutrients to replace vitamins lost in processing.
    b. the addition of the B-vitamin biotin to reduce the risk for certain cancers.
    c. nondigestible carbohydrates broken down by colon bacteria.
    d. live, beneficial bacteria found in fermented foods like yogurt.

Take It Further

A friend shows you the vast array of dietary supplements that he takes each morning in hopes of improving his fitness, boosting his energy, and increasing his immunity. Your friend is in good health, at a healthy weight, and seems to follow a varied and balanced diet. What are three considerations or cautions that come to mind regarding the use of dietary supplements in this scenario? Do you feel supplementation is warranted? Why or why not?

Dietary Analysis Using SuperTracker

Does a multivitamin or mineral supplement fit in your diet?

One key aspect of achieving a healthy diet is for your average intake of vitamins and minerals to meet the RDAs/AIs over the course of several days. In this exercise you will evaluate your diet to determine if functional foods or dietary supplements are necessary for you to meet the recommended intakes of one or more vitamins or minerals.

1. Log onto the United States Department of Agriculture (USDA) website at www.supertracker.usda.gov. If you have not done so already, you will need to create a profile to get a personalized diet plan. This profile will allow you to save your information and diet intake for future reference. Do not use the general plan.
2. Click the Track Food and Activity option.
3. Record your food and beverage intake for one day that most reflects your typical eating patterns. Enter each food and beverage you consume into the food tracker. Note that there may not always be an exact match to the food or beverage that you consume, so select the best match available.
4. Once you have entered all of your food and beverage choices into the food tracker, on the right side of the page under the bar graph, you will see Related Links: View by Meal and Nutrient Intake Report. Print these reports and use them to answer the following questions:
    a. On the nutrient intake report, identify any vitamins or mineral levels that are below your Recommended Dietary Allowance/Adequate Intake (RDA/AI).
    b. For each vitamin and mineral identified above, list *one* food that you enjoy eating that is also a good food source of that nutrient.
    c. On the nutrient intake report, identify any vitamins or mineral levels that are above 100% of your RDA/AI.
    d. Do you think you could change your diet enough to meet the recommendations identified in the nutrient intake report? Are there any functional foods that you could consume to increase your intake of the nutrients that fall below the recommended levels?
    e. What are the shortcomings of having analyzed your diet for only one day?
    f. Do you currently take a vitamin or mineral supplement? Considering the results of the nutrient intake report, do you think you need to be taking that supplement? Explain.
    g. If you do not take any supplements, do you think you should be? Explain.
    h. Discuss what you can look for on a supplement label that can provide assurance about the quality of the supplement.

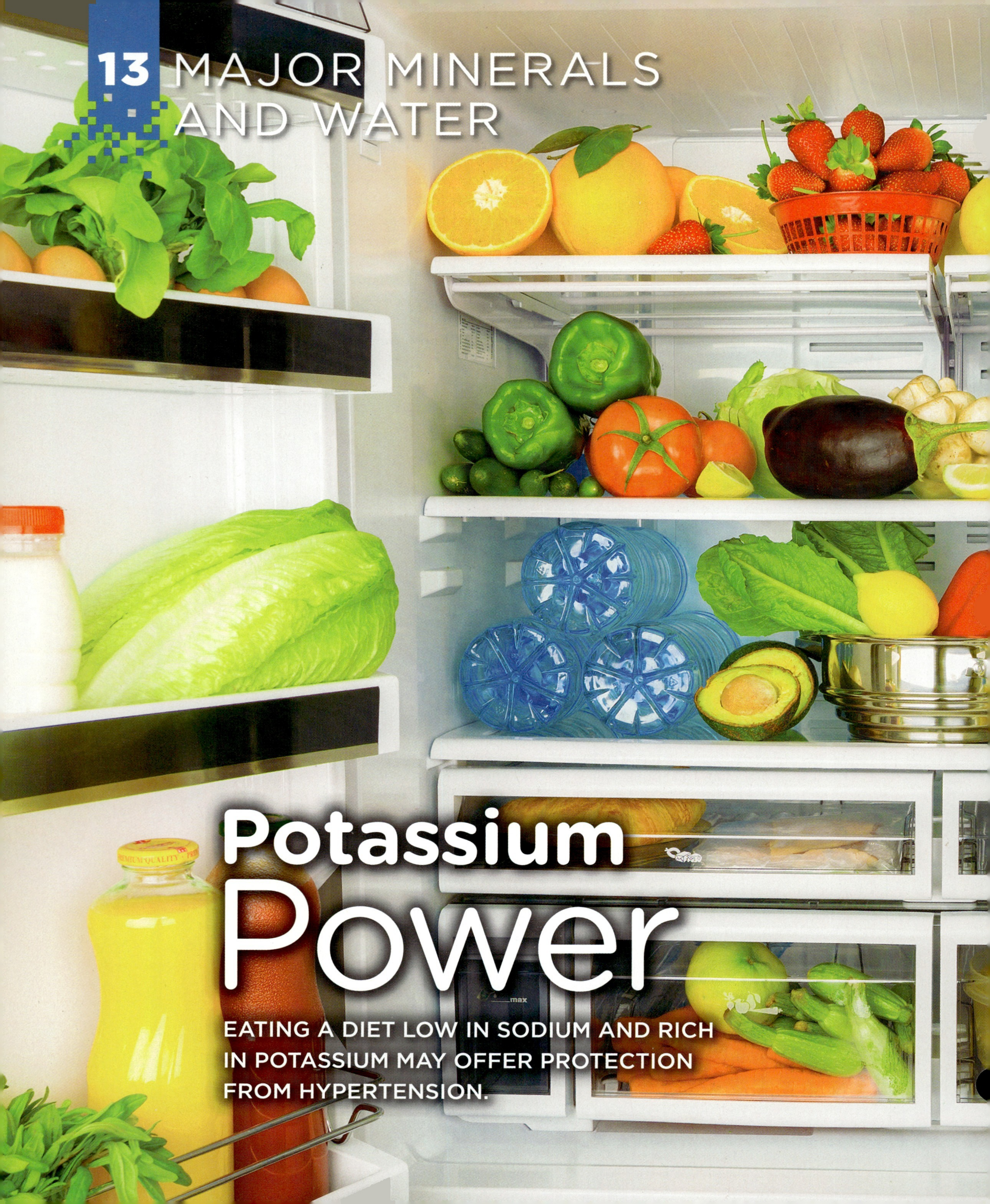

Potassium Power

EATING A DIET LOW IN SODIUM AND RICH IN POTASSIUM MAY OFFER PROTECTION FROM HYPERTENSION.

Anna Omelchenko/Shutterstock

LEARNING OBJECTIVES

- Identify the major minerals that have structural functions in the body, as well as the specific structures into which they are incorporated (Infographic 13.1)
- Discuss the role of calcium in bone health and the consequences of inadequate intake during the growing years and as we age (Infographics 13.2 and 13.4)
- Describe factors that may influence the bioavailability of minerals in the body
- Identify the major minerals that serve as electrolytes, and describe their individual roles in fluid balance (Infographics 13.6 and 13.10)
- Describe the effect that changes in sodium and potassium intake have on the risk of hypertension
- Identify primary sources of calcium, sodium, potassium, and magnesium in the diet (Infographics 13.3, 13.5, 13.7, and 13.9)
- Identify at least five functions of water in the body
- Identify the sources of water for the body, and the ways water is lost from the body (Infographic 13.11)
- Describe how antidiuretic hormone regulates water balance (Infographic 13.12)

In the 1960s, Lewis Dahl, a physician at the Brookhaven National Laboratory on Long Island, New York, spent a lot of time with rats. He was trying to understand the link between sodium and blood pressure, having noticed from national statistics that populations that eat low-salt diets, such as Eskimos, have a lower prevalence of high blood pressure than do populations that eat high-salt diets, such as the Japanese. To further explore this relationship, Dahl fed rats in his lab chronically high-sodium diets and observed them over time. He found that three-quarters of them developed high blood pressure (hypertension). He fed

13 MAJOR MINERALS AND WATER

Lewis K. Dahl, of Brookhaven National Laboratory. *Dahl is known for his pioneering work on the interactions of salt, the kidney, and hypertension.*

Courtesy of Brookhaven National Laboratory

MINERALS *inorganic individual chemical elements; there are 16 minerals considered essential in human nutrition with diverse regulatory and structural functions*

WATER *an essential nutrient central to all body functions*

MAJOR MINERALS *minerals with a daily requirement of 100 milligrams or more: sodium, potassium, chloride, calcium, magnesium, phosphorus, and sulfur*

TRACE MINERALS *minerals with a daily requirement of less than 100 milligrams: iron, zinc, copper, iodine, selenium, molybdenum, fluoride, manganese, chromium*

IONS *elements that carry a positive or negative charge*

other rats low-sodium diets, and they rarely developed high blood pressure. As Dahl put it in a 1961 paper, "the evidence is unequivocal that, as the amount of salt ingested daily is increased, both incidence and severity of hypertension will be increased." But then Dahl discovered something else interesting: When he fed his rats extra potassium along with the extra sodium, their blood pressure did not rise nearly as much as it did when he fed them sodium alone. In other words, sodium and potassium seemed to work against each other when it came to determining blood pressure—potassium protected against hypertension, while sodium increased the risk.

We often hear that we should eat less sodium, but we don't often hear that we need more potassium. Yet as it turns out, only 3% of American adults have usual intakes of potassium that exceed the Adequate Intake (AI) levels set by the Institute of Medicine. Indeed, so many of us do not get enough of the mineral that in 2010, the Dietary Guidelines for Americans identified potassium as a "nutrient of public concern." Add to this the fact that we're all eating way too much sodium—on average, we ingest 50% more than the government-recommended amounts—it's no wonder that one in every three Americans has hypertension. According to a 2011 study published by researchers at the U.S. Centers for Disease Control and Prevention, a high consumption of sodium and low consumption of potassium work in concert to threaten health, increasing the risk of death from any cause by 50% and doubling the risk of death from heart attacks.

Sodium and potassium are two of the 16 **minerals** that are essential for human health and nutrition. Minerals are solid, stable inorganic elements. Unlike vitamins, minerals can't be broken down into smaller constituents or destroyed by heat, light, cooking, or digestion. As the examples of sodium and potassium illustrate, both underconsumption and overconsumption of minerals can have a huge impact on human health and can have a range of physiological effects. Minerals have a variety of important jobs in the body, such as maintaining proper fluid balance and bone growth and maintenance.

Water is also an essential nutrient for the human body and it maintains health in multiple ways. Of the six categories of nutrients—the others being carbohydrates, lipids, proteins, vitamins, and minerals—water is arguably the most critical and indispensable. Because water is central to all our body functions, we can survive only a few days without it, while we can survive weeks without food. The primary ingredient in our bodies, after all, is water. The first part of this chapter is devoted to the topic of the major minerals. The second part discusses the role of water in health.

OVERVIEW OF THE MINERALS

Essential minerals are categorized according to how much we need each day: **major minerals** (sometimes called "macro" minerals) have a daily requirement of 100 milligrams or more, and **trace minerals** (sometimes called "micro" minerals) have a daily requirement of less than 100 milligrams.

The major minerals include *sodium, potassium, chloride, calcium, magnesium, phosphorus,* and *sulfur;* the trace minerals are *iron, zinc, copper, iodine, selenium, molybdenum, fluoride, manganese,* and *chromium* (see Chapter 14). Overall, minerals make up about 4% of our body weight, with major minerals composing more and trace minerals composing less. But don't let the names fool you: Both major and trace minerals play equally critical roles in human health. **(INFOGRAPHIC 13.1)**

Minerals have diverse regulatory and structural functions. Many work in partnership with other minerals. They often function as parts of enzymes and hormones, participate in chemical reactions, transmit nerve impulses, maintain fluid balance, and support the immune system. They can also work with enzymes as *cofactors*, inorganic substances that facilitate and catalyze chemical reactions. Like many of the B vitamins, minerals also play roles in energy metabolism, the chemical reactions that release energy from food. Minerals also help build and maintain structural components in the body such as bones, teeth, cell membranes, and connective tissue.

Minerals are **ions**, elements with a positive or negative charge due to their unequal numbers of electrons and protons. The charges of mineral ions, whether positive or negative, allow them to participate in chemical reactions and bond with other molecules.

Minerals can't be synthesized by the body, so they must be ingested through diet. Yet they also have a narrow range of safe intake, so there's often a fine line between getting too little and too much of a mineral. We couldn't live without any sodium, for instance, but we often get too much of it along with too little potassium, which increases the risk for high blood pressure and chronic disease. Intake of minerals above the recommended levels can also have other adverse effects including gastrointestinal problems. With the possible exception of sodium, we are at greatest risk of consuming too much of a particular mineral when we take dietary supplements, as overconsumption rarely occurs through food alone.

INFOGRAPHIC 13.1 Total Body Content of the Major Minerals and Their Functions *See Appendix 3 for a review of vitamins and minerals by function in the body.*

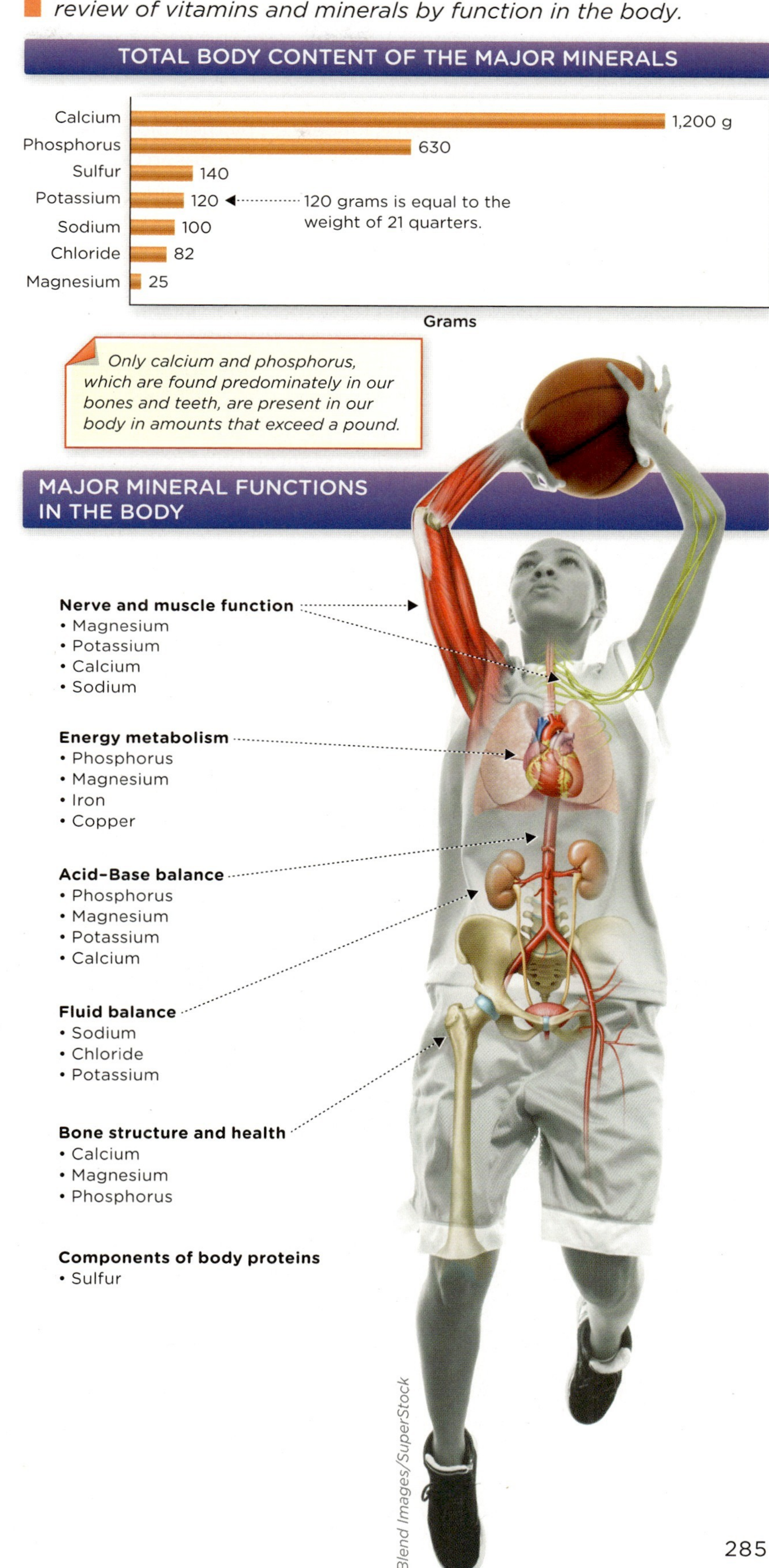

MINERAL ABSORPTION AND BIOAVAILABILITY

Humans absorb most minerals in the small intestine, and both active and passive processes transport them into the blood for distribution and use throughout the body. Mineral absorption is influenced by their *bioavailability*, the ability of nutrients to be absorbed and utilized in the body, which is itself affected by various factors. One such factor is nutritional status. When a person is deficient in a mineral or the needs for that mineral have increased, the absorption and bioavailability of that mineral tends to increase. For instance, during pregnancy, a woman's body absorbs calcium more efficiently than usual. Likewise, people absorb less of minerals if they already get enough. It's important to note, however, that although absorption of minerals slows during overconsumption, it doesn't stop—so it is still possible to consume excess, or even toxic, amounts of minerals. When mineral levels in the body increase, the excess is typically excreted in the urine.

The presence of dietary elements called *binders* can also influence mineral absorption. Binders such as phytates, found in whole cereal grains; oxalates, found in spinach and some vegetables; and polyphenols, present in coffee and tea, can limit bioavailability by chemically binding to minerals. Minerals can also compete with one another for absorption. For example, excess iron consumption from supplements can interfere with zinc absorption. Finally, the presence of certain vitamins can also affect bioavailability. Vitamin D, for instance, actually boosts absorption of calcium and phosphorus.

MINERALS IN OUR FOOD

The best way to get essential minerals is to eat an adequate, varied, and balanced diet. Tap water, too, can be a source of some essential minerals: "hard" water, which is

Whole cereal grains contain binders, such as phytates, which can influence the absorption of minerals. A varied diet rich in foods such as vegetables and fruits, animal and vegetable protein, can ensure adequate intake of minerals.

Dennis Gottlieb/Getty Images

more commonly found in the United States, typically contains more calcium and magnesium, whereas "soft" water contains more sodium, which is added in proportion to the "hardness" of the water source. Mineral content in tap water, whether hard or soft, can vary regionally and according to treatment methods. Tap water may contribute as little as 6% of overall intake of calcium or magnesium, but can add around 30% with consumption of approximately two liters a day in some areas. Likewise, consuming two liters of "soft" tap water generally contributes rather insignificant amounts of sodium to overall intake, but in some areas it can add up to almost half of the daily sodium recommendations. To find what minerals are present in significant amounts in your water, refer to the annual municipal water report for the area in which you live.

Minerals in food come from the soil. *Minerals are dissolved in water and are absorbed through the roots of the plant.*

Minerals in plants reflect the mineral content of the soil in which they are grown. Because of this, mineral content can vary drastically depending on the region in which the plants are grown and farming practices used to grow them. Animal foods are typically a better source of minerals, in part because animals eat plants and concentrate their nutrients. (As a result, vegans and vegetarians may be at a higher risk for deficiencies in the minerals that are typically concentrated in animal foods, such as calcium, iron, and zinc.)

Processing and refining foods can sometimes boost, but more often reduce, mineral content. For instance, when cereal grains are refined and milled, they lose a portion of many important minerals and only iron is added back through enrichment. To maximize mineral intake then, individuals should emphasize eating whole, unprocessed foods. And although cooking itself does not degrade or break down minerals, boiling foods can result in the leaching (or loss) of a varying amount of minerals into the cooking water—thus the recommendation to use cooking water in sauces or soups, whenever possible.

CALCIUM, MAGNESIUM, AND PHOSPHORUS HAVE DIVERSE STRUCTURAL ROLES IN THE BODY

Bones are the structural component of the body that shield our brain and organs from injury and make it possible to move. Minerals make up approximately two-thirds of the mass of the skeletal system and are involved in growth and maintenance of cellular membranes and connective tissues. Three major minerals of particular importance for bone formation and maintenance are calcium, phosphorus, and magnesium; with crystals of primarily calcium and phosphorus laid down on a matrix of proteins (predominantly collagen, a structural protein found in connective tissue) during bone formation. (INFOGRAPHIC 13.2)

Functions and sources of calcium

Calcium (Ca) is the most abundant mineral in the body, with 99% found in bones and teeth, where it provides an essential structural component for their formation. Bones provides

INFOGRAPHIC 13.2 The Bone Workers: Calcium, Phosphorus, and Magnesium Are Involved in Maintenance of Bone *Accounting for 98% of the body's mineral content by weight, calcium, phosphorus, and magnesium play key roles in the development and maintenance of bone and other calcified tissues.*

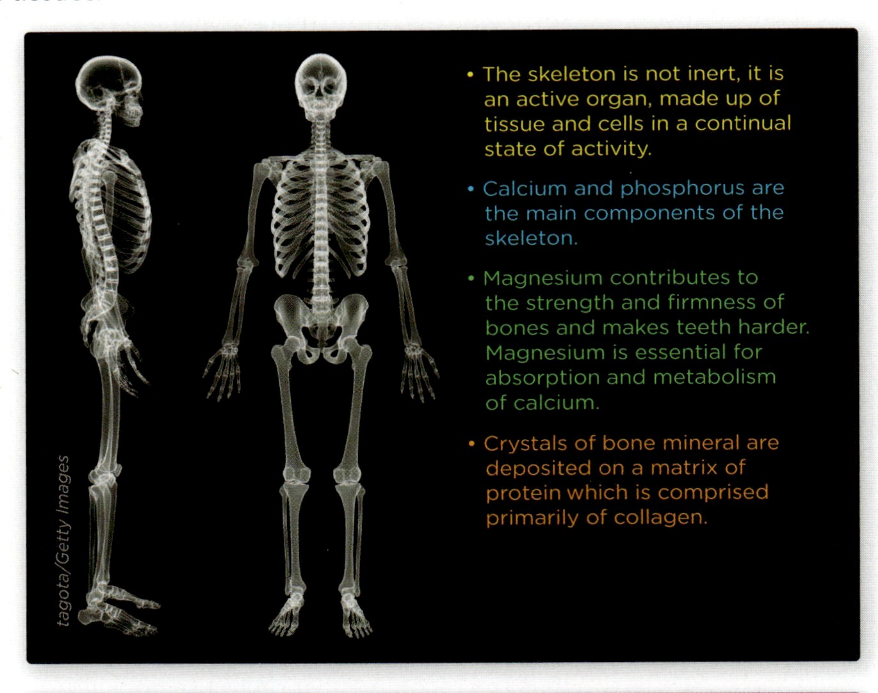

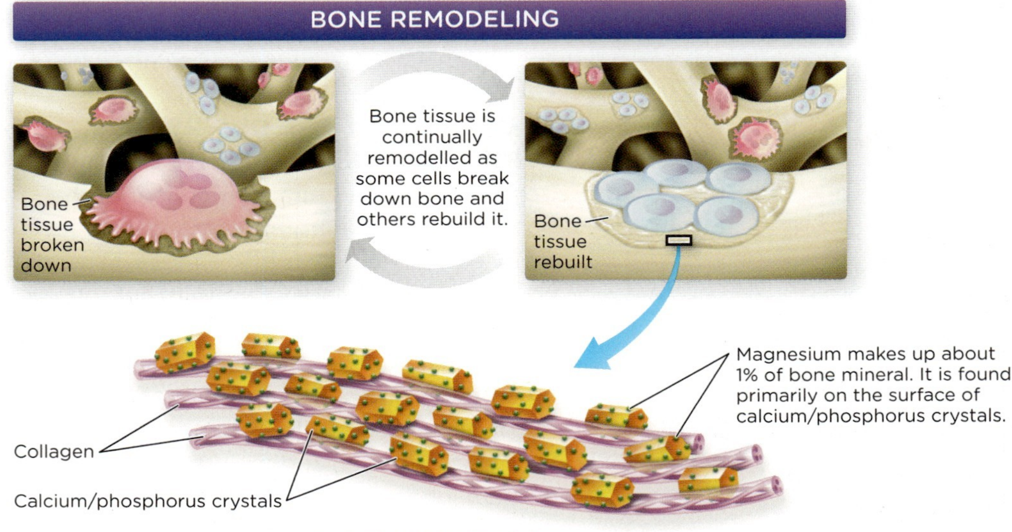

a reservoir of calcium that can be tapped to supply calcium to body fluids when its concentration in blood decreases. The other 1% is located in the body cells and fluids, where it is necessary for many essential functions such as blood clotting, hormone secretion, muscle contraction, and nerve transmission.

Because so many critical body functions depend on calcium, its concentration in blood is tightly regulated so that it remains nearly constant regardless of dietary calcium intake. When calcium in blood falls even slightly, it will be released from bone to maintain steady blood calcium levels. The mechanism by which the body maintains calcium levels in the blood is known as **calcium homeostasis**. When blood calcium levels fall, the parathyroid gland releases **parathyroid hormone (PTH)**, which

CALCIUM HOMEOSTASIS *the balance between the actions of parathyroid hormone, vitamin D, and the kidneys to tightly control serum calcium levels*

stimulates the production of the active form of vitamin D (calcitriol) and thereby increases calcium absorption from the intestine. (Refer back to Infographic 10.4 to review this process.) PTH and activated vitamin D work together to mobilize calcium from the bone and decrease calcium excretion from the kidneys.

Calcium plays an indispensable role in bone and tooth formation. In fact, bone is constantly being broken down and rebuilt in a process known as remodeling. Bone remodeling is necessary not only to maintain blood calcium levels, but it is also required during bone growth in the young, to allow bone to adapt to strain, and to repair the microscopic damage that occurs daily.

The Institute of Medicine has set the AI for calcium at 1,000 milligrams per day for men and women aged 19 to 50 years; the Tolerable Upper Intake Level (UL) is set at 2,500 milligrams. Calcium-rich foods include milk, yogurt, and cheese, as well as some legumes, and certain dark-green leafy vegetables, such as Chinese cabbage, kale, and broccoli. Because these vegetables are low in oxalates the bioavailability of calcium is high: about 50% is absorbed. In contrast, though spinach is high in calcium, it is also high in oxalates that bind calcium and inhibit its absorption. Consequently, the bioavailability of calcium in spinach is low: only about 5% is absorbed. Calcium is also added to some grains, juices, tofu, and cereals. Milk, an excellent natural source of calcium, is usually also fortified with vitamin D, which works with calcium to promote bone health. Most studies indicate that calcium from food is better absorbed than calcium from supplements. This is likely due to improved absorption with meals and the tendency of people to consume smaller amounts of calcium more frequently, which likely improves absorption efficiency. **(INFOGRAPHIC 13.3)**

PARATHYROID HORMONE (PTH) *a hormone released from the parathyroid gland in response to low serum calcium levels*

BONE REMODELING *the process of continuous bone breakdown and rebuilding, which is required for bone maintenance and repair*

INFOGRAPHIC **13.3** **Dietary Sources of Calcium** *Milk and many other dairy products are an important source of calcium. Sardines eaten with the bones are also a very good source.*

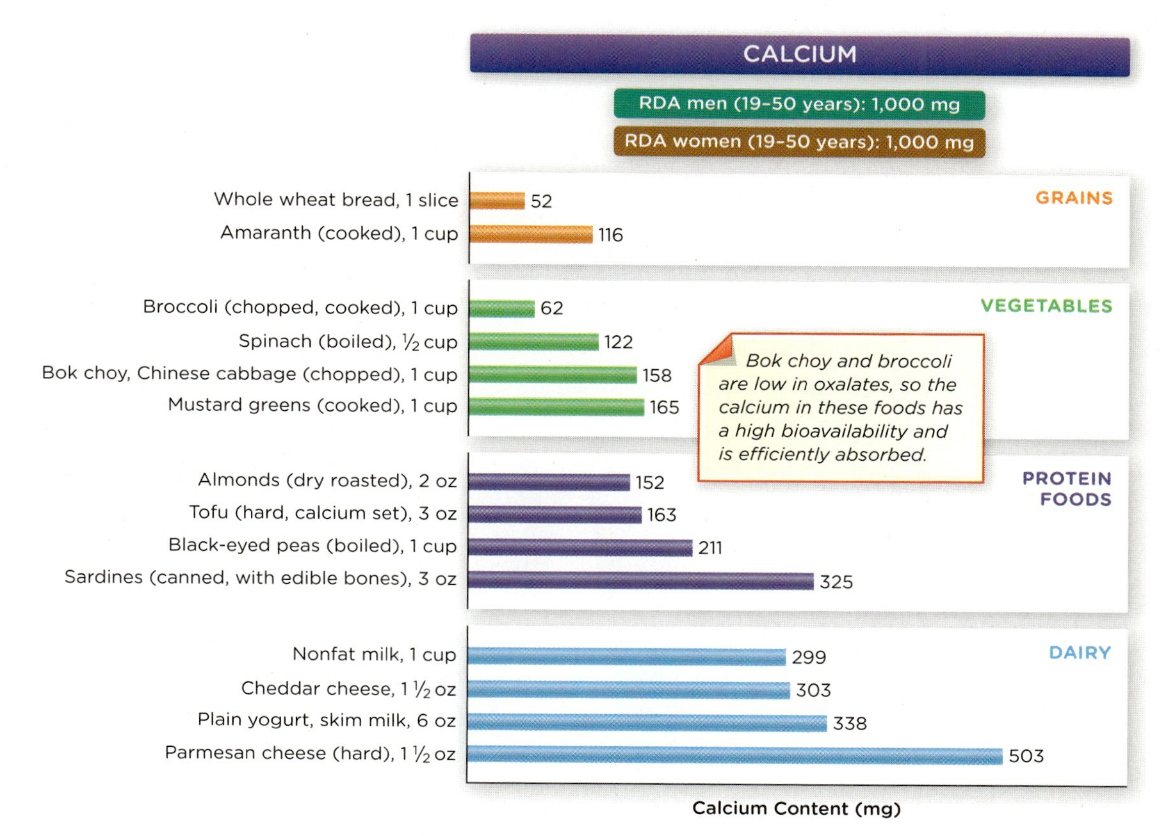

? *Identify several foods that together would allow you to meet your RDA for calcium without consuming more than a single serving of any one food.*

Photo credits (top to bottom): BIWA/Gallery Stock, Datacraft Co Ltd/Getty Images, Maximilian Stock Ltd./Getty Images

OSTEOPENIA
a condition characterized by low bone mineral density

OSTEOPOROSIS
a bone disease in which the bone density and total mass are decreased, leading to porous bones, increased fragility, and susceptibility to fractures

HYPERCALCEMIA
a high level of calcium in the blood

Calcium and bone health

With age, the balance of calcium release and deposition in bone changes. During years of growth, such as childhood, more calcium is added to bone in relation to the amount lost, but as we get older, bone breakdown often exceeds formation. Peak bone mass is established at around age 30, so it is important, during the formative years of bone development, to consume adequate amounts of calcium and vitamin D. If intake is low or absorption is impaired, bone loss occurs because the body uses the calcium in bone to maintain blood levels and support calcium-dependent bodily functions.

Most people realize that adequate calcium status is important for optimal bone health. However, a significant number of Americans have low bone mass. Although some bone loss is a normal consequence of aging, bone loss accelerates in postmenopausal women because of low levels of the hormone estrogen. This reduced bone mass, or bone density, along with reduced mineral content, can lead to a condition called **osteopenia**. When osteopenia becomes severe, and bone loss worsens to cause bones to be fragile and porous, a person develops **osteoporosis**, or "porous bones," and the risk of bone fractures is dramatically increased.

Osteoporosis afflicts more than 10 million Americans; approximately one-half of all women and one-quarter of all men older than 50 years of age experience osteoporosis-related bone fracture. This is particularly worrisome as the risk of mortality increases by as much as four times in the first four months following an osteoporosis-related fracture. Risk factors for osteoporosis include advanced age as well as a history of inactivity, smoking, excess alcohol consumption, and a family history of osteoporosis. To reduce their risk, individuals should maintain a healthy diet with adequate intake of calcium and vitamins D and K throughout life. Vitamin K is needed for the functioning of several proteins that are involved in regulating bone formation. Some studies have found that the risk of bone fractures decrease as intake of vitamin K increases. It is also important to participate in regular weight-bearing exercises, such as walking, running, or tennis, as well as perform resistance exercises to maintain bone health and reduce risk of osteoporosis. (INFOGRAPHIC **13.4**)

An important age for bone building. *A person with bone mass as a young adult will be less likely to develop osteoporosis later in life.*

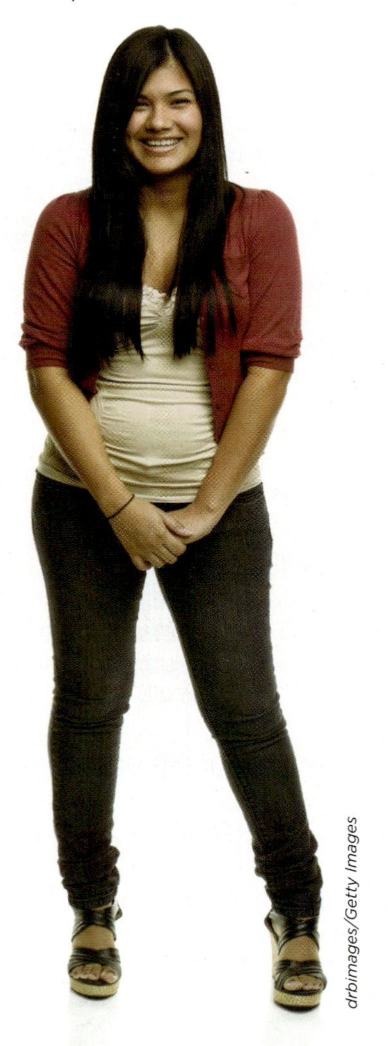

drbimages/Getty Images

Calcium supplementation

Although it is tempting to think that supplements are a "sure thing" to confirm that you are getting enough calcium, supplementation can potentially push intake to higher-than-recommended levels, especially when added to daily food sources that are fortified with calcium. Studies show that more than 60% of women older than 60 years take calcium supplements, which could push their intake close to the UL. High dietary intakes may cause constipation and can interfere with iron absorption. Although rare in healthy people, excess intake may contribute to **hypercalcemia**, or excess calcium in the blood.

INFOGRAPHIC 13.4 Osteoporosis and Bone Mass *Achieving a higher peak bone density decreases the risk of osteoporosis as one ages.*

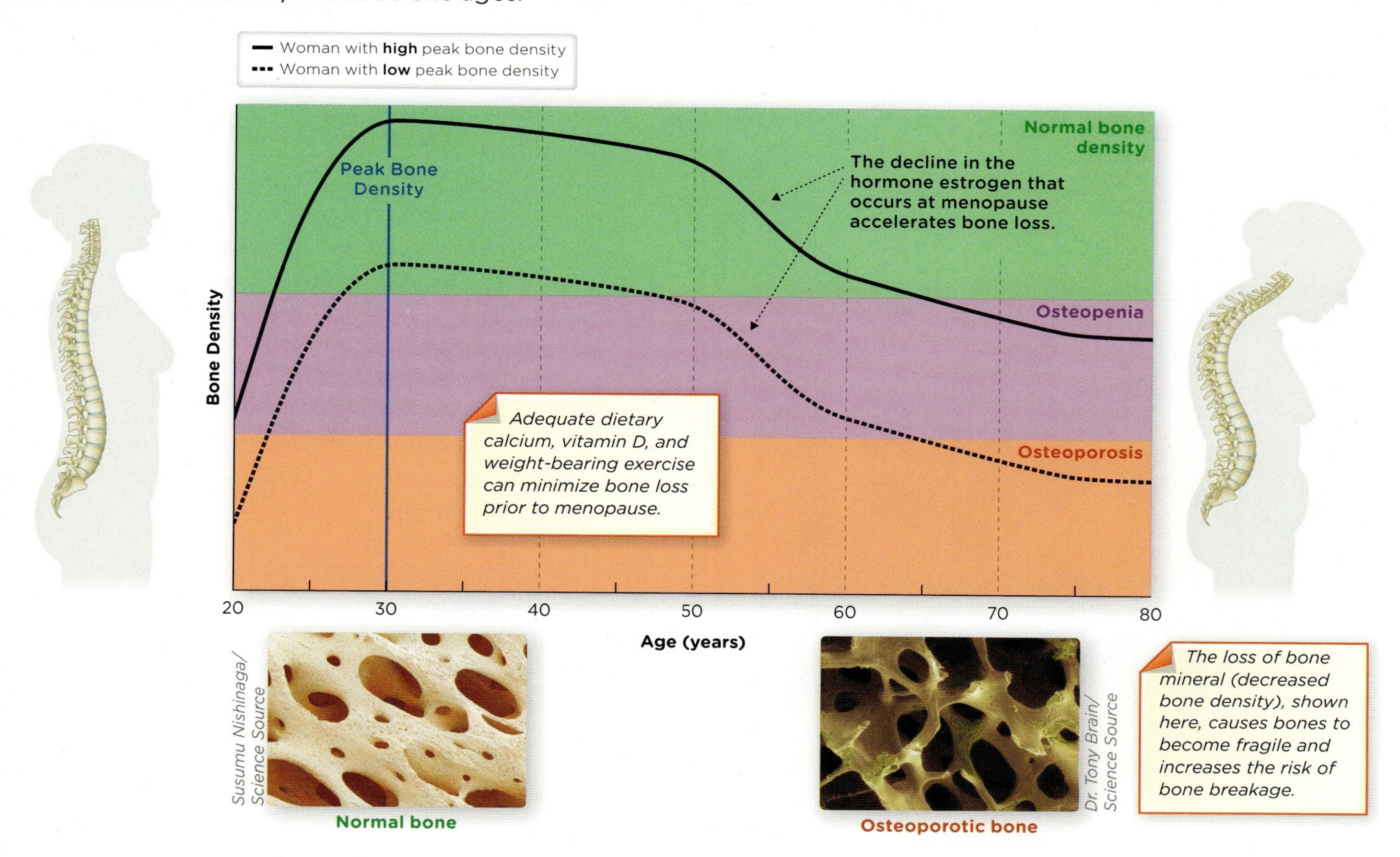

Contrary to what is commonly believed, there is little evidence that high calcium intake promotes the formation of kidney stones.

Functions of magnesium

Magnesium (Mg) is a cofactor in more than 300 chemical reactions in the body. An adult body contains about 25 grams of magnesium, most of it—about 50% to 60%—is found in the bones; the rest is in the cells and fluids of the body. Magnesium plays a role in the transport of ions across the cell membrane, a process that is important to muscle contraction, nerve impulse conduction, and maintaining the rhythm of the heart. Magnesium is involved in extracting energy from carbohydrates, fats, and protein, as well as using that energy to perform work; and it plays a role in protein production. Magnesium is also necessary to convert vitamin D into its active form (calcitriol) to increase calcium absorption and thus plays an important role in bone health.

Approximately 60% of American adults do not consume the recommended intake level of magnesium, but outright deficiency symptoms are rare, because the kidneys limit excretion when intake is low and the body may absorb more. However, low intakes of magnesium are a risk factor for osteoporosis. Marginal or moderate magnesium deficiencies may also increase the risk of atherosclerosis, cancer, diabetes, and hypertension. There is ongoing research about the role of magnesium in preventing and managing these disorders. Excess consumption from the diet is also rare, but toxicity can occur from supplement misuse. (INFOGRAPHIC 13.5)

Phosphorus

Phosphorus (P) is the second most abundant mineral in the body and is present in every cell of the body. It too plays a critical role in bone health and is an essential component of bone and cartilage, phospholipids, DNA, and RNA. It is also involved in energy metabolism,

INFOGRAPHIC **13.5** **Dietary Sources of Magnesium** *Magnesium is present in all food groups in small quantities. Nuts, whole grains, and leafy greens are good sources.*

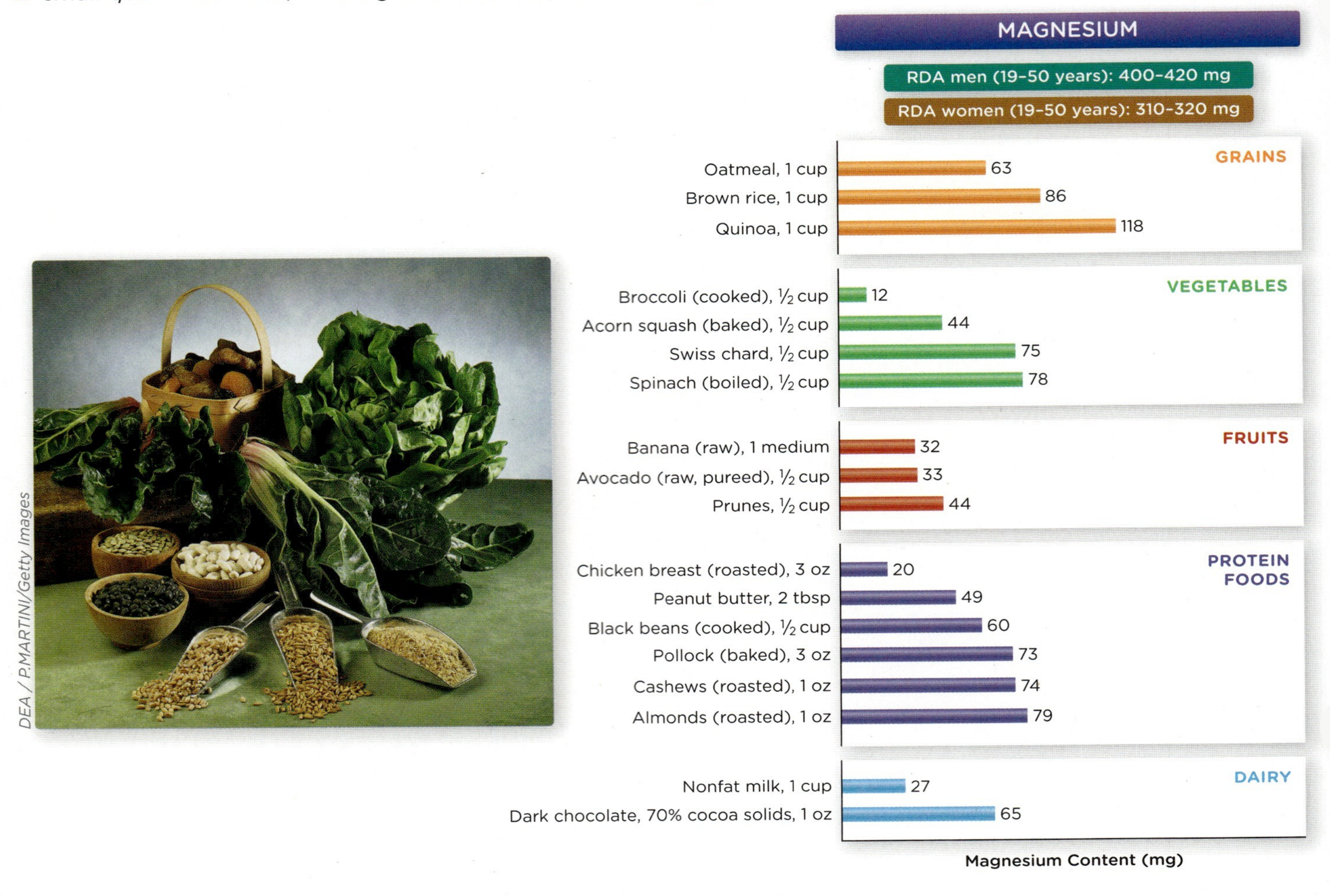

and a multitude of enzymes and other proteins depend upon phosphorus to regulate their activity. Phosphorus is important in the maintenance of proper *acid-base balance* in the body. Phosphorus deficiency is rare, in part because the mineral is found in most protein-rich foods, such as meats and dairy, and because many food additives also contain phosphorus. In fact, there is some concern that Americans may be chronically overconsuming phosphorus, a potential problem considering that increased blood levels of the mineral have been associated with cardiovascular and other types of chronic diseases, particularly in people with kidney disease.

ELECTROLYTES *electrically charged minerals (ions) dissolved in body fluids that balance the fluid outside the cells with the fluid inside the cells*

SULFUR IS A COMPONENT OF PROTEINS

Sulfur (S) is a mineral that occurs in our diet as a component of other compounds. It is present in the vitamins thiamin and biotin; and it is present in two of the amino acids (cysteine and methionine) that are found in our body's proteins, as well as those that we eat. When we need to synthesize sulfur-containing compounds, the ultimate source of that sulfur is almost invariantly one of those amino acids. Because sulfur is a component of all proteins, deficiency is virtually unknown. Toxicity is also rare because the body can excrete excess sulfur in the urine.

SODIUM, POTASSIUM, AND CHLORIDE MAINTAIN FLUID BALANCE IN THE BODY

The minerals sodium, potassium, and chloride maintain fluid balance in the body, transmit nerve impulses, and help muscles contract. When dissolved in bodily fluids, such as blood and urine and the fluids inside and outside our cells, these electrically charged minerals are known as **electrolytes**. Maintaining the right balance of electrolytes is necessary for the transport of nutrients, muscle

INFOGRAPHIC 13.6 Distribution of Body Fluids *Sodium (Na^+), potassium (K^+), and chloride (Cl^-) are electrolytes that maintain fluid balance in the body. The majority of sodium and chloride ions are found in extracellular fluid, and potassium ions are highest in intracellular fluid.*

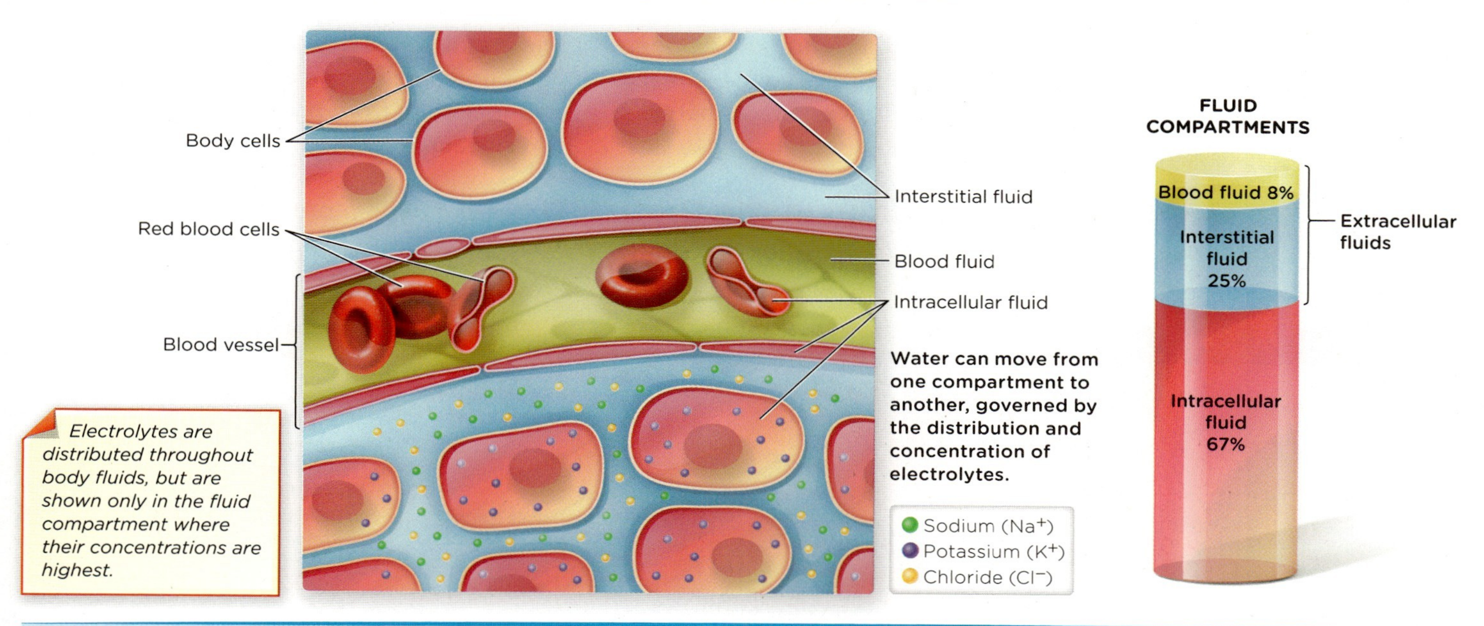

Extracellular fluids include which two fluid compartments?

contractions, and other processes. With the help of the kidneys, electrolytes balance the fluid outside the cells with the fluid inside the cells. Sodium and chloride are primarily found outside cells and potassium is typically found inside cells. (INFOGRAPHIC 13.6)

Sources and functions of sodium

Sodium (Na) is a mineral we are all familiar with, in part because it is widely consumed as table salt, which is otherwise known as sodium chloride (NaCl). Sodium is used extensively in food processing and is often added to foods during cooking and at mealtimes to improve taste. The sodium content of packaged foods is a mandatory component of the Nutrition Facts Panel on food labels.

An essential nutrient, sodium is the primary electrolyte responsible for maintaining fluid balance between cells and throughout the body. It plays a crucial role in regulating blood pressure, and is the major positively charged ion in extracellular fluids such as blood. It is also required to transmit nerve impulses and contract muscles, and is involved in the active transport of a number of nutrients. For instance, sodium is required to move glucose and amino acids into cells. Its role in the active transport of a variety of compounds also means that it is indirectly needed to maintain the body's acid-base balance and a stable pH.

Sodium intake recommendations

Overconsumption of sodium is common and increases the risk of **hypertension**, defined as having a blood pressure of at least 140/90 mmHg most of the time. On average, as Dahl and others have found, blood pressure rises progressively with intakes of sodium above 2,300 milligrams per day. Hypertension is a major risk factor for cardiovascular disease and stroke.

The Institute of Medicine has set the AI level for sodium at 1,500 milligrams per day for men and women aged 19 to 50 years, noting that needs can be increased in endurance athletes who perspire a lot and people who work in hot and humid conditions. The UL for adults 19 years and older is 2,300 milligrams daily. The Dietary Guidelines for Americans recommend limiting sodium intake to 2,300 milligrams per day for adults and children ages 14 and older. The recommended limits align with the UL for sodium for all age groups. No more than 1500 milligrams of sodium per day is advised for adults who would benefit from

HYPERTENSION *blood pressure of at least 140/90 mmHg most of the time; a major risk factor for cardiovascular disease and stroke*

blood pressure lowering, specifically those with prehypertension or hypertension. Although reducing sodium intake to this level has been demonstrated to reduce the risk of hypertension there is currently only moderate evidence that it actually reduces the risk of heart disease and stroke. In fact, there is some evidence that low sodium intake (less than 2,300 mg) may actually be associated with adverse health effects in some subgroups.

Currently, an estimated 90% of Americans exceed sodium recommendation and average about 3,400 milligrams a day, but that number varies by age and sex. We get most of our sodium from packaged processed and restaurant foods, with only about 10%

INFOGRAPHIC **13.7** **Sources of Sodium in the Diet** *Sodium is abundant in processed foods and in foods from restaurants. The Nutrition Facts Panel on commercial food and beverage packaging is a good source of information about the sodium content.*

MOST SODIUM COMES FROM PROCESSED AND RESTAURANT FOODS

Source: Sources of Sodium in the Diets of the U.S. Populatioin Ages 2 Years and Older, NHANES 2005–2006.

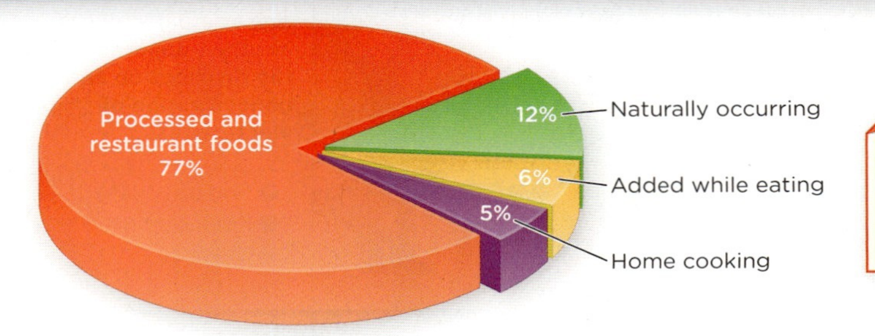

When purchasing processed foods, select "reduced sodium" or "low sodium" varieties and watch the serving sizes.

FINDING SODIUM IN FOODS

Eli Ensor

INGREDIENTS: WHEAT FLOUR, WATER, SHREDDED LOW-MOISTURE PART-SKIM MOZZARELLA CHEESE (PART-SKIM MILK, CHEESE CULTURE, SALT, ENZYMES), TOMATO PASTE, PEPPERONI MADE WITH PORK, CHICKEN, AND BEEF (PORK, MECHANICALLY SEPARATED CHICKEN, BEEF, SALT, CONTAINS 2% OR LESS OF SPICES, DEXTROSE, PORK STOCK, LACTIC ACID STARTER CULTURE, OLEORESIN OF PAPRIKA, FLAVORING, SODIUM ASCORBATE, SODIUM NITRITE, BHA, BHT, CITRIC ACID. MAY ALSO CONTAIN PAPRIKA, NATURAL SMOKE FLAVOR), VEGETABLE OIL (CORN OIL OR SOYBEAN OIL), CONTAINS LESS THAN 2% OF SUGAR, SALT, YEAST, FLAVOR, SPICE, GARLIC (12% PEPPERONI)

Nutrition Facts

Serving Size ¼ pizza (146g)
Servings Per Container 4

Amount Per Serving

Calories 360	Calories From Fat 160
	% Daily Value*
Total Fat 18g	**28%**
Saturated Fat 7g	**35%**
Trans Fat 0g	
Cholesterol 35mg	**12%**
Sodium 770mg	**32%**
Total Carbohydrate 35g	**12%**
Dietary Fiber 2g	**8%**
Sugars 4g	
Protein 16g	**22%**

Vitamin A 8% • Vitamin C 2%
Calcium 25% • Iron 15%

*Percent Daily Values are based on a 2,000 calorie diet. Your daily values may be higher or lower depending on your calorie needs.

	Calories:	2,000	2,500
Total Fat	Less than	65g	80g
Sat Fat	Less than	20g	25g
Cholesterol	Less than	300mg	300mg
Sodium	Less than	2,400mg	2,400mg
Potassium		3,500mg	3,500mg
Total Carb		300g	375g
Fiber		25g	30g
Protein		50g	65g

*Calories per gram:
Fat 9 • Carbohydrate 4 • Protein 4

Tips to Manage Sodium Intake

- ✓ Even if a food doesn't taste "salty," sodium may be present in large quantities. Check the Nutrition Facts Panel for accurate information.
- ✓ If you eat more than one serving of a food, your sodium intake will increase.
- ✓ The Nutrition Facts Panel lists the **Percent Daily Value (%DV)** of sodium in **one serving** of a food.
- ✓ The %DV for sodium is based on 100% of the recommended amount of sodium, which is **less than 2,400 milligrams (mg) per day.**
- ✓ **Remember:**
  - **5%DV (120 mg) or less of sodium per serving is LOW.**
  - **20%DV (480 mg) or more of sodium per serving is HIGH.**
- ✓ The American Heart Association has designated six commonly eaten foods that are loaded with sodium as the "salty six." These include **bread, cold cuts, pizza, poultry, soup, and sandwiches.**

How many times does the word "sodium" or "salt" appear in the ingredients list?

coming from salt added at the table and another 10% coming from "fresh" unprocessed foods. (INFOGRAPHIC 13.7)

DASH diet

To help Americans reach these lower sodium goals—as well as boost intake of potassium, calcium, and magnesium—the American Heart Association and the American College of Cardiology both recommend the **DASH diet**, which emphasizes "fruits, vegetables, whole grains, low-fat dairy products, poultry, fish and nuts." The DASH eating plan has been found to be an effective approach in helping lower blood pressure and reduce the risk of heart disease. (INFOGRAPHIC 13.8)

DASH DIET
Dietary Approaches to Stop Hypertension; a food plan that moderates sodium intake while increasing the intake of foods rich in potassium, calcium, and magnesium to assist with blood pressure control

INFOGRAPHIC **13.8** **DASH Diet** *Dietary Approaches to Stop Hypertension. The DASH eating plan shown below is based on a 2,000 kcal/day diet. The DASH diet reduces sodium in the diet, and encourages a variety of foods rich in nutrients such as potassium, calcium, and magnesium.*

GRAINS
6–8 Daily Servings

Serving Sizes:
- 1 slice bread, ½ cup cooked rice, pasta, or cereal
- 1 oz dry cereal

Whole grains are recommended for most servings.

FRUITS
4–5 Daily Servings

Serving Sizes:
- 1 medium fruit
- ¼ cup dried fruit
- ½ cup fresh, frozen, or canned fruit
- ½ cup fruit juice

VEGETABLES
4–5 Daily Servings

Serving Sizes:
- 1 cup raw leafy vegetables
- ½ cup cut-up cooked or raw vegetables
- ½ cup vegetable juice

FAT-FREE OR LOW-FAT MILK PRODUCTS
2–3 Daily Servings

Serving Sizes:
- 1 cup milk or yogurt
- 1 ½ oz cheese

LEAN MEATS, POULTRY, AND FISH
6 or Fewer Daily Servings

Serving Sizes:
- 1 oz cooked meat, poultry, or fish
- 1 egg

NUTS, SEEDS, AND LEGUMES
4–5 Servings per Week

Serving Sizes:
- ⅓ cup or 1 ½ oz nuts
- 2 tbsp peanut butter
- 2 tbsp or ½ oz seeds
- ½ cup cooked legumes

FATS AND OILS
2–3 Daily Servings

Serving Sizes:
- 1 tsp soft margarine
- 1 tsp vegetable oil
- 1 tbsp mayonnaise
- 2 tbsp salad dressing

SWEETS AND ADDED SUGARS
5 or Fewer Servings per Week

Serving Sizes:
- 1 tbsp sugar
- 1 tbsp jelly or jam
- ½ cup sorbet or gelatin
- 1 cup lemonade

Photo credits (left side—top to bottom): Camera Press Ltd/Alamy, Brett Stevens/Getty Images, Multi-bits/Getty Images, Carlos Gawronski/Getty Images; (right side—top to bottom): Jill Chen/Getty Images, Paul Kennedy/Getty Images, Maximilian Stock Ltd./Getty Images, Kate Sears/Getty Images

HYPOKALEMIA decreased blood levels of potassium

Potassium

Potassium (K), the primary electrolyte within cells, works together with sodium (and chloride) to maintain fluid balance. As Dahl found back in the 1960s, the level of intake of these minerals affects blood pressure. A diet that emphasizes foods rich in potassium (the DASH diet, for example) may help mitigate some of the effects of excess sodium by increasing sodium excretion in the urine. Potassium may also help relax blood vessel walls, which can also lower blood pressure. Found in a wide range of minimally processed foods, particularly fruits and vegetables, potassium also functions as a co-factor for certain enzymes, helps nerves transmit and muscles contract, plays a role in nutrient transport, and helps to maintain the electrical activity of the heart to sustain a steady heartbeat. **(INFOGRAPHIC 13.9)**

INFOGRAPHIC 13.9 Major Sources of Dietary Potassium

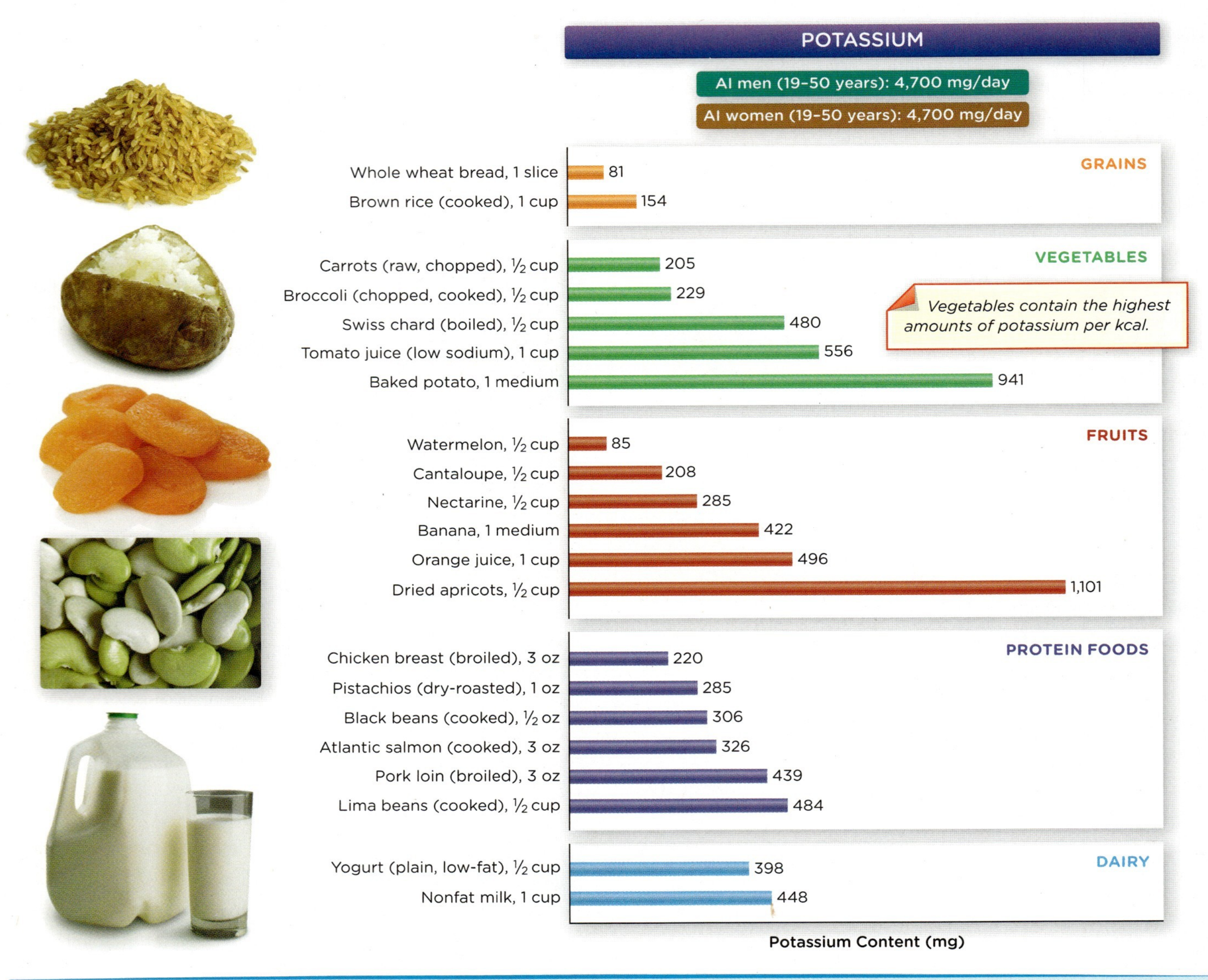

? *Vegetables have the highest potassium contents per kcal, but potatoes have a higher content per serving. Considering the concept of nutrient density explain why this is so.*

Photo credits (top to bottom): Burwell and Burwell Photography/Getty Images, John Scott/Getty Images, ferhat mat/Getty Images, Brian Yarvin/Getty Images, Tetra Images/Getty Images

Potassium deficiency, when serious—such as when there is excessive fluid loss through vomiting, diarrhea, use of diuretics, or kidney disorders—can lead to **hypokalemia**, a disorder characterized by fatigue, muscle weakness, abnormal heart rhythms, increased calcium excretion, and reduced insulin production. Milder forms of potassium deficiency—which are common in the United States, as most Americans only get about half the potassium they should—can increase the risk of hypertension, stroke, heart attacks, and other health problems. Toxicity associated with excess consumption of potassium is extremely rare and would likely only result from dietary supplement misuse. However despite the prevalence of low intake, multivitamin mineral supplements in the United States contain only about 2% of the AI for potassium because of the risk of side effects, in particular low heart rate and abnormal heart rhythm. Potassium supplementation may sometimes be warranted, but should be prescribed and monitored by a health care provider because of potential side effects.

Water stop. *Staying properly hydrated is important during marathon racing for both safety and performance.*

Dream Pictures/Ostrow/Getty Images

Chloride

Chloride (Cl) is the third electrolyte involved in fluid balance. Like sodium, it is primarily found in extracellular fluids such as blood. Since chloride composes half of the mineral composition of table salt (NaCl), our chloride intake correlates directly to our sodium intake. In addition to its role in fluid balance, chloride also forms half of hydrochloric acid (HCl), found in the stomach's gastric juices. Because sodium is abundant in the American diet, chloride deficiencies are extremely rare. Overconsumption of chloride, like sodium, increases blood pressure.

■ ■ ■

On the morning of April 19, 2004, 27-year-old Mark Robinson was in Boston preparing to run his first marathon. It was going to be unexpectedly hot—forecasts said it might reach 90 degrees—so, concerned about dehydration, Robinson drank more

Mark Robinson survived a dangerous case of hyponatremia by taking in too much water during his first marathon attempt.

Jodi Hilton/The New York Times/Redux

than a gallon of water prior to the race. During the race, he stopped at every water station to drink a few cups, too.

Suddenly, at mile 19, Robinson started feeling nauseated and his legs began to cramp. But he kept going, assuming it was just typical end-of-marathon fatigue. By mile 23, though, Robinson could no longer run. He walked the rest of the way. After he crossed the finish line, he drank two more quarts of water, but he continued to feel terrible. He was vomiting, having diarrhea, and he felt disoriented. Suddenly, Robinson screamed, jumped up in the air, and fell squarely on his shoulder, breaking it. His mother called 911 and a helicopter took him to Boston Medical Center, where he was given a saline solution intravenously to treat what the doctors believed was extreme dehydration.

It was not dehydration. Robinson was actually dangerously overhydrated. Because he had consumed so much water before, during, and after the race, Robinson's blood was diluted and contained low levels of sodium. The extra fluid he consumed was also causing his brain to swell. Robinson wound up in a coma and was placed on a life-support machine, but regained consciousness four days later and, thankfully, fully recovered. (Nutrient and fluid needs in the athlete will be addressed in Chapter 16)

Cooling off at the water fountain on a warm day. *One important role of water is to help regulate body temperature.*

DreamPictures/Shannon Faulk/Getty Images

WATER

The body maintains fluid balance by moving electrolytes to where more water is needed. However, the level of electrolytes in your body can become too low or too high. That can happen when the amount of water in your body changes, upsetting fluid balance and causing dehydration or even overhydration.

Approximately 60% to 70% of our bodies are made up of water, but the ratio of water to other substances in our tissues varies, depending primarily on the ratio of fat mass to lean body mass (body composition), which is strongly influenced by age and sex. The body of an adult man is approximately 60% water, and an adult woman is about 55% water—with water composing about 75% of the mass of muscle, and 15% of adipose tissue mass. As people age, they typically lose muscle mass and gain fat, which will decrease the percent of body mass that is water. Organs have different water percentages, too: The brain and heart are made of approximately 73% water, but bone is only 10% water.

Roles of water in the body

Water has many important roles in the body. It helps regulate body temperature within a very narrow range, which is important because even slight variations can affect body functions and damage organs.

(Think about how a fever of 102°F—which is slightly higher than three degrees above normal—makes you feel.) The temperature of water rises slowly, so, because there is so much water in your body, it provides an important "brake" for unwanted body temperature fluctuations.

Water transports nutrients and oxygen to tissues in your body. It is also present in the mucus and salivary juices of the digestive system, which help to move food through the digestive tract. Water also lubricates joints and mucous membranes in our noses, eyes, and the gastrointestinal tract. Since water can't be compressed, it helps to protect delicate tissues like the brain, eyes, and spinal cord against injury and shock.

Water acts as a **solvent**—a liquid substance that is capable of dissolving another substance, and participates in the body's biochemical reactions. For instance, without water our bodies can't break down proteins, carbohydrates, or fats to extract energy. Water also helps to remove waste from the body via urination, perspiration, and bowel movements.

Water balance in the body

Approximately two-thirds of the water in our bodies is *intracellular*, or found inside cells. The rest is *extracellular*, found outside cells, primarily in the blood and the fluid that surrounds our cells (*interstitial fluid*). Water may move from outside to inside cells and vice versa, in a process called **osmosis**. The direction of water movement depends on the concentration of dissolved substances in intracellular and extracellular fluids called **solutes**. Some of the most important solutes are the electrolytes sodium, potassium, and chloride. Water moves across a membrane toward areas where there is a greater concentration of solutes, and, therefore a lower concentration of water molecules. **(INFOGRAPHIC 13.10)**

SOLVENT
a liquid substance that is capable of dissolving another substance

OSMOSIS
a process by which water passes between intracellular and extracellular spaces (or compartments) through cellular membranes

SOLUTE
a dissolved substance

INFOGRAPHIC **13.10 Osmosis** *A process by which water (but not solutes) passes through a selectively permeable membrane, from a less concentrated area into a more concentrated one.*

Osmosis occurs when solutions with different solute concentrations are separated by a selectively permeable membrane—in this case a membrane that lets water but not the solutes pass through the membrane.

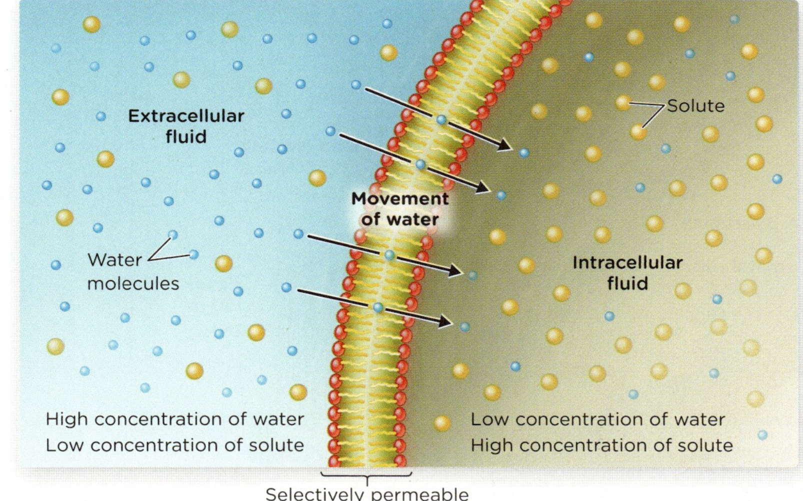

In this example, the solution in the cell has more solute particles in the cell than outside the cell (when comparing equal volumes), so the concentration of solute particles is higher inside the cell than outside it.

Since water can pass across the membrane, but the solutes cannot, water will be drawn into the cell toward the higher concentration of solute particles in an attempt to equalize the concentration of solutes on both sides of the membrane.

? *As water moves into the cell what will happen to the size of the cell?*

RESPIRATION
the process of transporting oxygen from the air to the cells within tissues (inhalation) and transporting carbon dioxide from cells to the air (exhalation)

In the case of our marathoner, as he was rapidly drinking the low-solute water, his cells were absorbing dangerous amounts of water. In essence, the water passed, via osmosis, from his blood into his cells to try to equalize the concentration of solute particles between the intracellular and extracellular compartments. The sudden flood of water into the cells of his brain (the neurons in particular) caused his brain to swell and press against the inside of his skull, which prevented blood, and thus essential oxygen, from reaching his brain.

One of the reasons we need to drink fluids is because we lose water throughout the day and our bodies can't store extra to fill the void. But how much water we need to drink depends on many factors. We lose, on average, about 350 ml, or 1½ cups, of water each day through **respiration**, the process of transporting oxygen from the air to the cells within tissues (inhalation) and, conversely, transporting carbon dioxide to the air (exhalation). When we breathe, water evaporates from the lungs and the skin. How much water we lose while respiring depends on the air temperature and our body temperature, how active we are, how humid the air is, and how big our bodies are. The more we respire—for instance, when we exercise and breathe hard—the more water we lose. **(INFOGRAPHIC 13.11)**

We lose 140 ml to 150 ml of water, or about two-thirds of a cup, through our bowel movements, too; the amount is small because most of the water in our stools is re-absorbed

INFOGRAPHIC **13.11** **Water Balance** *To remain in water balance, fluid input must equal its output. This figure shows the approximate amounts of water sources and losses for an average 20-year-old man who is sedentary and does not have extensive water loss from sweating.*

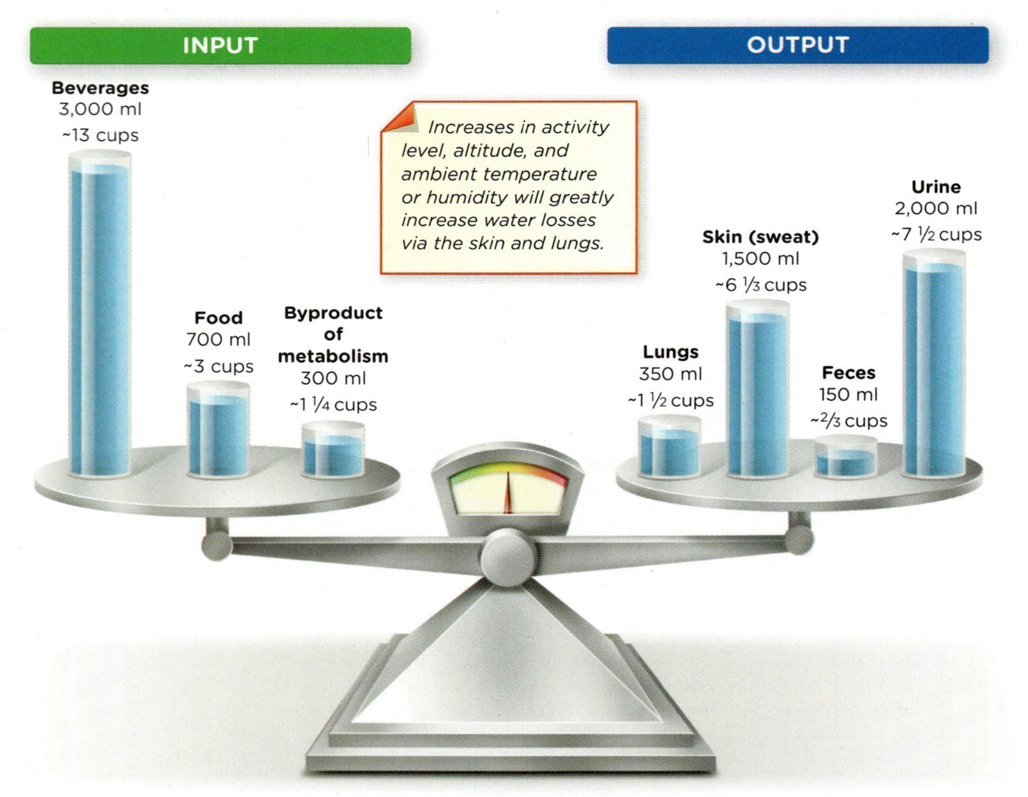

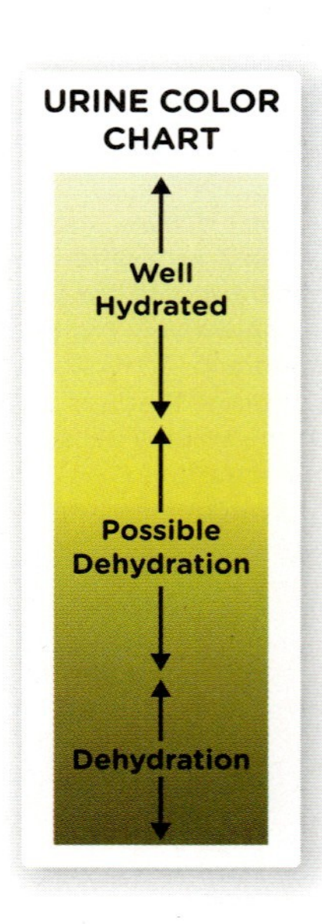

If someone increases their fluid intake without increasing physical activity, through what route would this excess fluid be lost?

into the body before the stool is eliminated. Diarrhea, however, can dramatically increase water loss, which can be dangerous: Diarrhea-associated dehydration is a major cause of child death in areas of the world that do not have access to clean water, as contaminated water can cause diarrheal illnesses. When we suffer diarrhea, we lose from 3,300 ml to 7,300 ml of water a day—as much as 70 times what we would lose through normal bowel movements.

Our water needs can fluctuate because of other factors, as well. We need more water when we are ill, because we respire more with a fever, and our body perspires to cool itself; vomiting also increases water loss. We also need more water when we eat high-protein diets, in part because water is necessary to remove the byproducts of protein metabolism that accumulate in the blood. High-fiber diets increase our water needs, too, because fiber increases the water content of our stools, making it easier for us to pass them. Water needs are increased when we drink alcohol, take certain medications, and spend time at high altitudes.

The body achieves water balance by attempting to ensure that the amount of water we consume in food and drinks and the amount we produce through metabolism equals the amount of water we excrete. We can thank our kidneys and our brain for achieving this balance. The kidneys conserve water by reducing urine volume when necessary, and they excrete excess water by increasing the volume of urine and making it more dilute when we have excess fluid.

When the concentration of solutes in the blood increases, or when blood volume drops, the brain responds by stimulating the pituitary to release **antidiuretic hormone (ADH)**. ADH tells the kidneys to conserve water to bring more water back into the bloodstream, which will also lower the concentration of solutes. When the brain detects that water volume has increased again, it decreases the production of ADH so that the kidneys stop conserving water. Without ADH, urine becomes very dilute. **(INFOGRAPHIC 13.12)**

Thirst is typically a powerful and rapid barometer that tells us when we need to drink (and conversely, lack of thirst tells us we don't need to drink). The same factors (an increase in solutes in the blood or a

ANTIDIURETIC HORMONE (ADH)
a peptide hormone produced by the hypothalamus, with the primary function of decreasing the amount of water excreted by the kidneys; stored and released in the posterior pituitary gland

INFOGRAPHIC 13.12 Antidiuretic Hormone and Water Balance *Antidiuretic hormone (ADH) increases total body water content by reducing water excretion.*

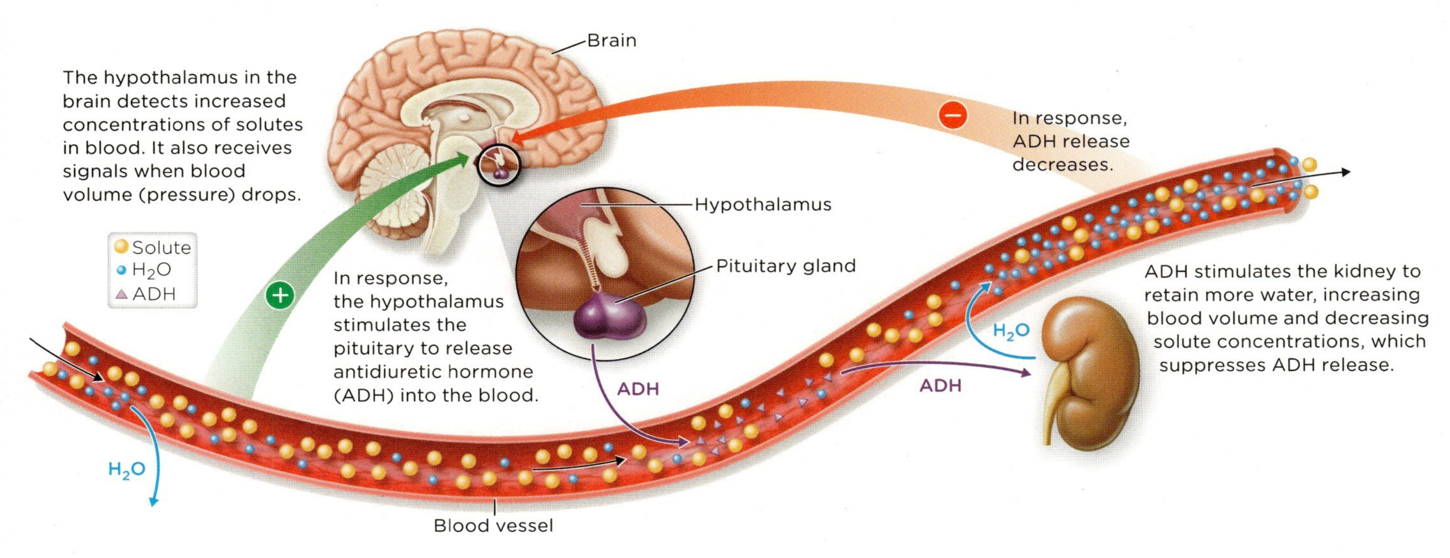

? *Why will intense exercise stimulate ADH secretion?*

drop in blood volume) that stimulate the release of ADH also stimulate the brain to generate the sensation of thirst. So, in the end, ADH and thirst work in concert to increase body fluids and decrease the concentration of solutes in blood. However, in those over the age of 65, the thirst sensation is diminished when they become dehydrated or experience increased solute concentrations in blood, which slows the restoration of fluid balance.

WATER INTAKE RECOMMENDATIONS

Because people vary so much in terms of how much water they need each day, there is not an established Recommended Daily Allowance (RDA) or Estimated Average Requirements (EAR) for water. However, in 2004, the Institute of Medicine set an AI value of 3.7 liters (about 15½ cups) of total water intake per day for men living in temperate climates and 2.7 liters (about 11½ cups) per day for women living in temperate climates. **(INFOGRAPHIC 13.13)**

INFOGRAPHIC 13.13 Adequate Intake of Fluids, and Water Content of Selected Foods *These recommendations are for individuals aged 19 to 30 years living in moderate climates. Higher intakes of total water are required for those who are physically active or exposed to hot environments.*

Adequate Daily Intake of Fluids

Adequate Intake (AI)	Men	Women
Total water intake Food, beverages, and water	3.7 liters (~15.5 cups)	2.7 liters (~11.5 cups)
Fluid intake Beverages and water	3.0 liters (~13 cups)	2.2 liters (~9 cups)

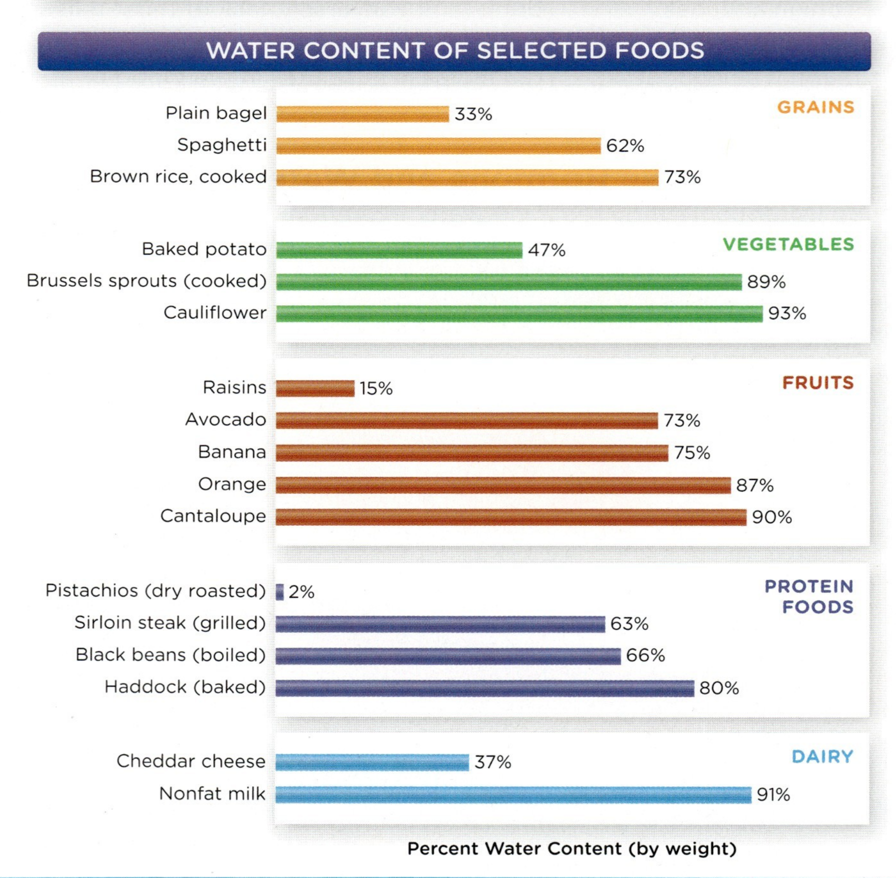

Examine the AI's for water. How much water are men and women expected to obtain from foods?

Tap or bottled? *The Food and Drug Administration (FDA) oversees bottled water, while the Environmental Protection Agency (EPA) regulates tap water, so the safety of both sources is comparable. Some people may opt for tap water because of its lower cost, and to reduce waste from plastic bottles.*

AVAVA/Shutterstock

These totals include water from a combination of food, drinking water, and other beverages. Current consumption patterns indicate that fluids provide about 80% of our total water intake, with plain water making up about 45% of those fluids and other beverages making up the rest. Therefore, the AI for fluids is 3.0 liters for men and 2.2 liters for women. (Both tap and bottled water are considered safe to drink; the EPA regulates the former, while the FDA regulates the latter, and they have similar standards.) The body also produces a small amount of water as a byproduct of metabolic reactions, which also helps meet our total water needs.

■ ■ ■

During his marathon, Robinson continued to drink even when he was not thirsty, causing a major water imbalance. But another factor, out of his control, may have contributed to his imbalance, too. Some athletes release excess ADH when exercising vigorously for long periods, making the kidneys retain water. This hormonal reaction likely worsened Robinson's water imbalance. He then started to feel the common symptoms of **hyponatremia**, a condition in which sodium levels in the blood become low because of increased blood volume.

Unfortunately, the symptoms of hyponatremia sometimes mimic the symptoms of **dehydration**, or deficiency of water, which is in part why Robinson was given saline at the hospital—the doctors there assumed he was dehydrated from his run. There are a variety of symptoms associated with dehydration, including nausea, dizziness, elevated temperature, and concentrated urine. Dehydration is typically caused by excessive sweating, excessive urine output (because of diuretic use or uncontrolled diabetes), fever, vomiting, or diarrhea. It is also a widespread problem in developing countries, where safe, drinkable water can be hard to find. Although 70% of Earth is composed of water, only 3% of it is drinkable.

HYPONATREMIA
a condition characterized by a low serum sodium concentration and clinical signs of confusion, nausea, vomiting, bloating, and swelling around brain; may be seen in athletes who consume excess water with no sodium

DEHYDRATION
water deficiency resulting from fluid losses that exceed intake

To distinguish between hyponatremia and dehydration, some city marathons now test blood sodium concentrations in unwell athletes before deciding whether to treat for dehydration (which might involve saline) or for hyponatremia (which involves a hypertonic, or high-sodium solution). Other marathons have scales available so that runners can weigh themselves; weight gain from excess fluid during a marathon is a common sign of hyponatremia.

Water, one of the six essential classes of nutrients, is vital for life, but any time we overconsume a nutrient, even water, we may encounter trouble. The seven essential *major* minerals are also vital—in some ways they are just as important to our health as our own tissues; our cells and organs could not do their jobs properly without them. But as we will see in the next chapter, the nine *trace* minerals, though less abundant in our bodies, are just as important to our survival.

CHAPTER 13 **BRING IT HOME**

Focus on major minerals

All the major minerals have vital roles in supporting body processes, maintaining health, and potentially preventing diseases. In the United States and most developed nations, people consume too much sodium and too little potassium in comparison with recommended intake levels. In addition, for much of the population, women in particular, calcium intake is suboptimal. These disproportionate or inadequate intake amounts contribute to the development of multiple health conditions. To explore how an average day's intake of sodium, potassium, and calcium stacks up against expert recommendations, you will use the USDA National Nutrient Database for a closer look at these three major minerals by analyzing a typical American day's diet described in the table below.

1. Navigate to the USDA National Nutrient Database http://ndb.nal.usda.gov/ndb/nutrients/index.
2. On the drop-down menus for First Nutrient, Second Nutrient, and Third Nutrient choose sodium (Na), potassium (K), and calcium (Ca).
3. Under Food Subset select All Foods.
4. Using the "Food Groups" drop-down menu to limit search items, complete the table below. As you look up each item, fill in the corresponding column in the table with the milligrams of sodium, potassium, and calcium from the information provided. Take note of portion sizes: For some food items, you will need to multiply the mineral content accordingly. (Tip: Click the back arrow between each food item to return to Nutrient Database.)
5. Total each column.

FOOD GROUP	FOOD ITEM	MEASURE	SODIUM (mg)	POTASSIUM (mg)	CALCIUM (mg)
Baked Products	Bagels, plain, enriched	1 mini (2½″ diam)			
Dairy and Egg Products	Cheese, cream	1 tbsp			
Fruits and Fruit Juices	Tangerines (mandarin oranges), raw	1 cup			
Fast Foods	McDonald's Big Mac	1			
Fast Foods	McDonald's French Fries	1 small			
Beverages	Carbonated beverage, low calorie, cola	12 oz			
Snacks	General Mills Chex Mix, traditional flavor	2 oz			
Dairy and Egg Products	Yogurt, vanilla, low-fat, 11 grams protein per 8 ounce	One 6-oz container			
Restaurant Foods	Chinese, Orange Chicken	1 order			
TOTALS					
Recommended Intake (RDA/AI)			*1,500 mg*	*4,700 mg*	*1,000 mg*
Difference (+/−)					

6. Compare the totals for each of these major minerals with the RDA/AI for men and women 19 years and older (provided in the table). Determine the difference and record a "+" (exceeded recommended intake) or a "−" (fell short of recommended intake).
7. What was the total sodium intake for this sample day? ______________________

   What was the difference from the recommended intake? ______________

   From the food lists under Food Groups, identify lower sodium alternatives for one or more of the food items in the scenario (provide food item, serving size, and milligrams of sodium). In addition, what are two ways that sodium intake could be reduced to better meet recommended intake levels?

   ____
   ____
   ____
8. What was the total potassium intake for this day? ______________________

   What was the difference from the recommended intake? ______________

   From the food lists under Food Groups, identify higher potassium alternatives for one or more of the food items in the scenario (provide food item, serving size, and milligrams of potassium). In addition, what are two ways to boost potassium intake to better meet recommended intake levels?

   ____
   ____
   ____
9. What was the total calcium intake for this day?

   ____

   What was the difference from the recommended intake? ______________

   From the food lists under Food Groups, identify higher calcium alternatives for one or more of the food items in the scenario (provide food item, serving size, and milligrams of calcium). In addition, what are two ways to boost calcium intake to better meet recommended intake levels?

   ____
   ____
   ____

For help and information on using the USDA National Nutrient Database, go to http://ndb.nal.usda.gov/ndb/help/index/nutrients.

KEY IDEAS

- The 16 essential minerals in the diet are inorganic elements supplied through food.
- Minerals in the diet are classified as major or trace based on the amount required per day. The major minerals include sodium, potassium, chloride, calcium, magnesium, phosphorus, and sulfur.
- Minerals have diverse regulatory and structural functions in the human body, including participation in chemical reactions, transmission of nerve impulses, maintenance of fluid balance, and support of the immune system.
- The bioavailability of the minerals we consume is affected by various factors, including nutritional status as well as the presence of binders, other minerals, and certain vitamins.
- Sodium, potassium, and chloride are electrolytes that function in fluid balance, transmission of nerve impulses, and muscles contraction.
- Dietary intakes of sodium and potassium affect our risk of hypertension.
- The Dietary Approaches to Stop Hypertension (DASH) diet includes strategies to reduce sodium intake and increase intake of foods rich in potassium, calcium, and magnesium.
- Calcium levels in the blood are regulated through a process called calcium homeostasis.
- Adequate calcium intake across the lifespan is important in the maintenance of bone density and the prevention of osteoporosis.
- Among its essential roles in human health, magnesium serves as a cofactor in more than 300 chemical reactions in the body. Phosphorus (P) is the second most abundant mineral in bone and throughout the body. Sulfur (S) is an essential mineral present primarily in two amino acids found in dietary proteins.
- Water is an essential nutrient with critical roles in the body. Among its many functions, water helps regulate body temperature, transports nutrients and oxygen to tissues, lubricates joints and mucous membranes, cushions tissues, and removes waste.
- Water makes up the greatest percentage of our body, both by weight and by volume, and is present in varying amounts in every tissue and organ.
- To maintain water balance in the body, intake of fluid from beverages, food, and water produced by metabolism must equal water output from urine, respiration, perspiration, and stools.
- Consuming inadequate or excess amounts of water can lead to health issues.

NEED TO KNOW

Review Questions

1. All of the following are true regarding essential minerals in the diet, EXCEPT that they:
   a. are classified as major or trace based on daily requirement.
   b. make up about 4% of our body weight.
   c. are easily destroyed by heat, light, cooking, or digestion.
   d. have an electric charge and are considered ions.
   e. function as part of enzymes and hormones.

2. The bioavailability of minerals:
   a. is unaffected by dietary intake or nutritional status.
   b. determines their caloric contribution to the diet.
   c. is always enhanced when minerals are consumed together.
   d. can be influenced by the presence of certain vitamins.
   e. is improved when coffee or tea is consumed at each meal.

3. The 2015 Dietary Guidelines for Americans recommend that adults 51 years and limit their sodium intake to _____ milligrams a day.
   a. 1,500
   b. 2,300
   c. 3,000
   d. 3,400
   e. 4,000

4. The primary focus of the DASH diet is to help control:
   a. high cholesterol levels.
   b. type 2 diabetes.
   c. blood triglycerides.
   d. high blood pressure.
   e. body weight.

5. The primary intracellular electrolyte in the body is:
   a. calcium.
   b. chloride.
   c. potassium.
   d. sodium.
   e. zinc.

6. All of the following are true regarding calcium blood levels in the body, EXCEPT that they are:
   a. primarily determined by dietary calcium intake through the diet.
   b. supplied primarily from calcium in bone.
   c. regulated through a process called calcium homeostasis.
   d. influenced by calcium absorption and excretion.
   e. increased in response to the release of parathyroid hormone.

7. Which of the following minerals does NOT serve as a structural component of bone and teeth?
   a. calcium
   b. chloride
   c. magnesium
   d. phosphorus

8. Water plays an important role in:
   a. preventing obesity in the U.S. population.
   b. providing a source of energy.
   c. providing a good source of potassium in the diet.
   d. making up 50% of the body in men.
   e. regulating body temperature.

9. Daily water requirements are:
   a. the same for all adults: eight glasses per day.
   b. determined by losses through urine, respiration, perspiration, and feces.
   c. established by specific RDAs or AIs for all ages and both sexes.
   d. met only by drinking water, not through foods or other beverages.
   e. the same for men and women if they live in temperate climates.

10. All of the following are true regarding antidiuretic hormone (ADH), EXCEPT that it:
    a. is released by the pituitary gland.
    b. is released when blood volume falls.
    c. prompts the kidneys to conserve water.
    d. functions to release water and dilute urine.
    e. is decreased when blood volume increases or normalizes.

11. A condition characterized by low sodium levels in the blood, increased blood volume, and symptoms that may include nausea and confusion is termed:
    a. ketoacidosis.
    b. hypokalemia.
    c. hyponatremia.
    d. dehydration.
    e. homeostasis.

Take It Further

Describe a scenario in which water balance might be disrupted by water losses exceeding water intake. What are some of the signs of dehydration indicating that this is happening?

Dietary Analysis Using SuperTracker

Analyzing the intake of major minerals and water

Major minerals have many regulatory and structural functions in the human body. They also play important roles in the prevention of some chronic diseases. Using SuperTracker, you will identify your intake the major minerals and water and compare them with the current dietary recommendations.

1. Log on to the United States Department of Agriculture (USDA) website at www.supertracker.usda.gov. If you have not done so already, you will need to create a profile to get a personalized diet plan. This profile will allow you to save your information and diet intake for future reference. *Do not use the general plan.*
2. Click the Track Food and Activity option.
3. Record your food and beverage intake for one day that most reflects your typical eating patterns. Enter each food and beverage you consumed into the food tracker. Note that there may not always be an exact match to the food or beverage that you consumed, so select the best match available.
4. Once you have entered all of your food and beverage choices into the food tracker, on the right side of the page under the bar graph, you will see Related Links: View by Meal and Nutrient Intake Report. Print these reports and use them to answer the following questions:
    a. Did you meet the target recommendations for the major minerals for the day you selected? If not, what minerals fall below the target numbers?
    b. For each major mineral target that you missed, list one food that you could consume to increase your intake of that mineral.
    c. Were you above the targets for any of the major minerals? If so, which ones? Are any of these numbers above the toxicity level for that mineral?
    d. Evaluate your intake of sodium. Were you above the UL? If so, identify the foods that contributed the most sodium to your diet for this day.
    e. Evaluate your intake of potassium. Did you meet the target intake for this day? If not, what foods could you include in your diet to increase your intake of this mineral?
    f. Evaluate your intake of water. Did you meet the recommendations? If not, what dietary modifications could you make?
    g. Evaluate your risk for osteoporosis. Did you meet the target amounts for calcium and vitamin D and K? Do you think you need to take supplements for these nutrients? How might you alter food choices to meet your needs? Explain your answer.

14 TRACE MINERALS
Small Quantities with Big
Importance
HOW LACK OF TRACE MINERALS CAN WREAK
HAVOC ON YOUR HEALTH

Tim Gainey/Alamy

LEARNING OBJECTIVES

- Describe the major functions of copper, fluoride, iodine, iron, selenium, and zinc in the body (Infographics 14.1 and 14.12)
- Identify the general properties of trace minerals (Infographic 14.2)
- Describe how the use of iodine and the production of thyroid hormone are controlled (Infographic 14.3)
- Describe the symptoms of deficiency and toxicity for copper, fluoride, iodine, iron, selenium, and zinc (Infographics 14.4 and 14.12)
- Identify at least two dietary sources of copper, iodine, iron, selenium, and zinc (Infographics 14.5 and 14.8–14.12)
- Discuss the dietary sources of heme and non-heme iron and the factors that affect their absorption (Infographics 14.6–14.8)

In the days that led up to World War I, a great swath of the United States was known as "the goiter belt," because alarming numbers of men, women, and children in the area were showing signs of **goiter**, an enlargement of the thyroid gland. During the draft for World War I, the Michigan State Department of Public Health found that the prevalence of goiter reached as high as 64.4% in some parts of Michigan. Many of the military recruits were deemed too unhealthy to serve in the war effort.

The soils of the Great Lakes, Appalachians, and Northwestern regions of the United States are poor in the trace mineral *iodine.* The body needs iodine, an element that is transported in the body as *iodide,* to make thyroid hormones, which play key roles in many essential metabolic processes. Most of Earth's

An abnormal enlargement of the thyroid gland is called a goiter. *The thyroid is a butterfly-shaped gland located at the base of the neck. It secretes several hormones that affect metabolism, growth, and development.*

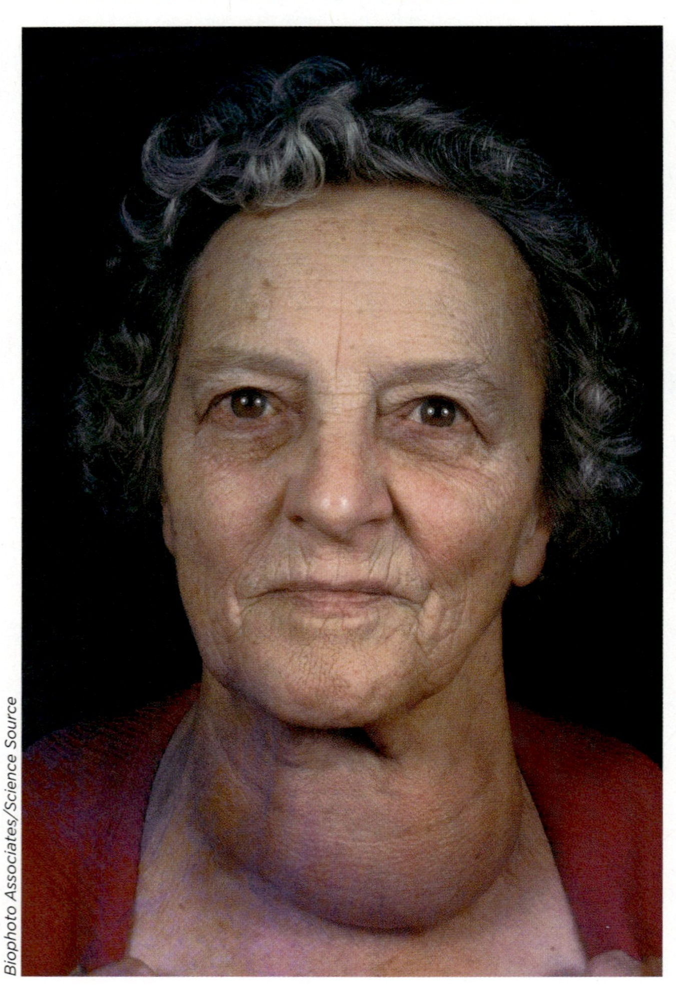

Biophoto Associates/Science Source

iodine is found in oceans and iodine content in the soil varies by location, which affects the iodine content of crops. In some regions of the world, particularly mountainous areas, or flooded river valleys, iodine-deficient soil is common, increasing the risk of iodine deficiency among people who consume foods primarily from those areas.

Since the days of World War I, many countries have implemented programs to add iodine to table salt and have dramatically reduced the prevalence of iodine deficiency worldwide. However, iodine deficiency is still "one of the most important public health issues globally," according to a 2012 study published by researchers at the Boston Medical Center. In the United States, where much of our table salt is iodized, most people appear to consume sufficient iodine to offset deficiency, although certain groups, such as pregnant women, may be at risk of suboptimal intake and iodine status.

GOITER
enlargement of the thyroid gland, most often caused by lack of iodine in the diet

TRACE MINERALS
minerals with a daily requirement of less than 100 milligrams (sometimes called microminerals)

CO-FACTOR
nonprotein inorganic substances that enable enzymes to carry out chemical reactions

■ ■ ■

INTRODUCING THE TRACE MINERALS

Iodine is categorized as one of the **trace minerals**, because the body requires significantly less of them to meet daily needs than the major minerals. (Trace minerals are also found in smaller amounts in the body.) Minerals with a daily requirement of 100 milligrams or more are considered major (or "macro") minerals, while those with a daily requirement less than 100 milligrams are considered trace (or "micro") minerals. In addition to iodine, the other essential trace minerals are iron, zinc, copper, selenium, molybdenum, manganese, and chromium. Fluoride is another important trace mineral that is required for optimal health, but because it is not required to sustain life it is not considered an essential nutrient. (INFOGRAPHIC 14.1)

Although the body doesn't need large amounts of the trace minerals, they are crucial for health. Trace minerals often act as **co-factors** by binding to enzymes, enabling them to carry out their chemical reactions. Trace minerals are also incorporated into other types of proteins in addition to enzymes; iron, for example, is a component of the oxygen-carrying protein in red blood cells called *hemoglobin*. Trace minerals also serve other roles. Fluoride, for example, has an important structural function in that it hardens the enamel on our teeth, reducing the occurrence of dental caries.

Trace minerals are found in plant-based and animal-based foods, but their bioavailability—the extent to which a mineral can be used by the body—can be influenced by many

INFOGRAPHIC 14.1 Trace Minerals in the Body and Their Functions

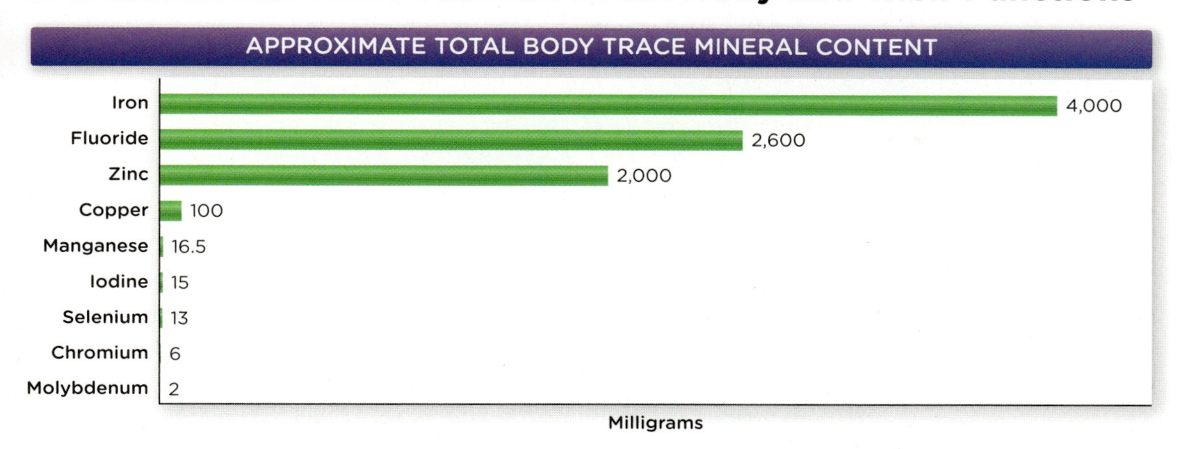

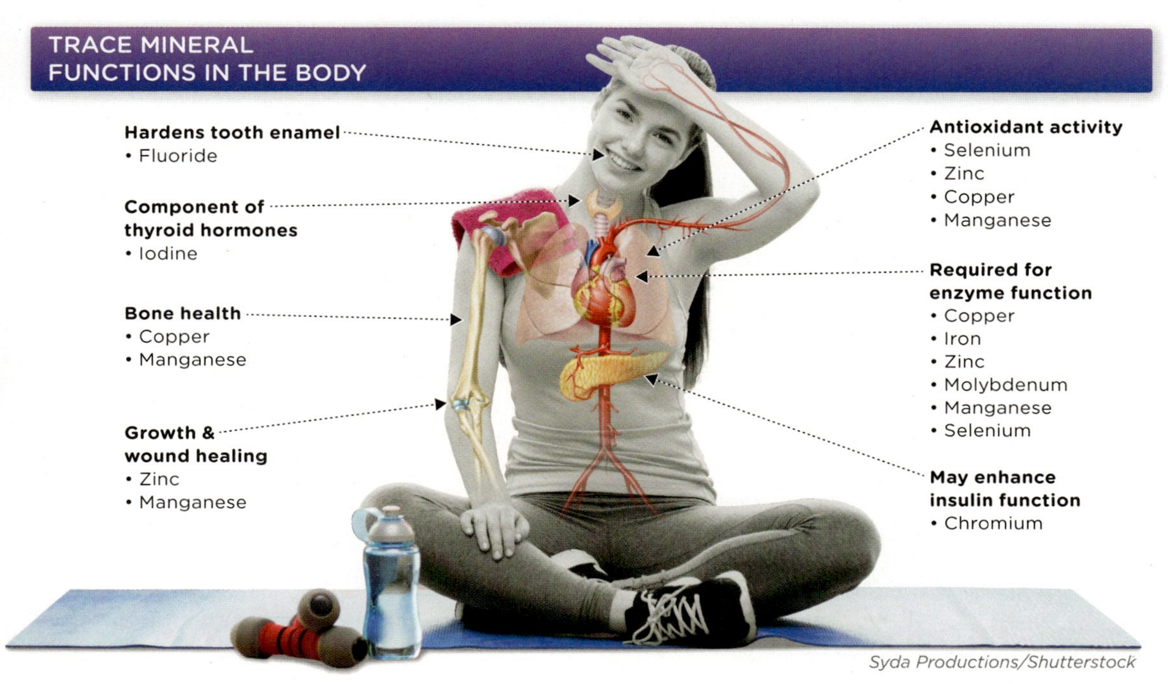

Syda Productions/Shutterstock

factors, including the form of the food, our nutrition status, our age, and pregnancy status. The safe range of intake for trace minerals is also narrow compared with the major minerals, since they are required in such small amounts. Trace mineral deficiencies can cause vague or varied symptoms that are difficult to diagnose; but, sometimes, inadequacies have severe and even deadly consequences. Pregnant women and children in rapid growth stages are especially affected by inadequate intakes of trace minerals. As with the vitamins and major minerals, it is rare for individuals to overconsume trace minerals through food alone, but intake above the Tolerable Upper Intake Level (UL) is possible through supplementation, resulting in potentially dangerous adverse effects. For example, excess selenium can result in hair loss and brittle nails and in extreme cases, respiratory distress along with kidney and heart failure. In general, by eating a balanced and varied diet, we can get all the trace

INFOGRAPHIC 14.2 General Properties of the Trace Minerals

To combat goiters, the Morton Salt Company fortified its salt with iodide. Iodide is the ion and transport form of iodine in the body.

Mary Evans Picture Library/Alamy

minerals we need without excess intake. (INFOGRAPHIC 14.2)

■ ■ ■

Although the problems with iodine deficiency are most pronounced in developing countries, the United States and other westernized nations are not exempt. In 2006, nearly half of all pregnant American women's diets were estimated to be deficient in the mineral according to World Health Organization intake recommendations. These low intakes are a major problem considering that iodine deficiency during pregnancy can lower fetal IQ to the point of causing mental retardation.

IODINE: SOURCES AND FUNCTIONS

The iodine content of food depends on the iodine content of the environment it comes from, which varies from region to region. The highest *naturally* occurring concentrations of

INFOGRAPHIC 14.3 Regulation of Thyroid Hormone Production

NORMAL IODINE LEVELS

Low thyroid hormone levels cause the pituitary to secrete TSH.

Pituitary gland

TSH (thyroid-stimulating hormone)

The pituitary detects increased thyroid hormone levels and it decreases the release of TSH.

Iodine

Thyroid gland

Thyroid hormone

TSH stimulates the thyroid gland to take up more iodine and produce and release more thyroid hormone.

IODINE DEFICIENCY

Low thyroid hormone levels

Pituitary gland

TSH (thyroid-stimulating hormone)

Goiter

Iodine

Enlarged thyroid gland

Thyroid hormone

Goiter

Biophoto Associates/Science Source

When iodine in the body is low, thyroid hormone production cannot increase in response to increasing TSH levels. Consequently, TSH levels stay elevated, stimulating the thyroid gland to enlarge.

? *Graves' disease is caused by the body's immune system producing antibodies that the thyroid gland detects as TSH. Why does this cause the thyroid gland to enlarge as well as produce excessive amounts of thyroid hormone?*

iodine are in foods from the sea, such as fish, shellfish, and plants (e.g., seaweed, algae), because the ocean contains considerable iodine. In part because the problem of goiter was so widespread, the U.S. salt company Morton began to add iodide to its salt in 1924, and the U.S. Food and Drug Administration recommended that the product be labeled with the following message: "This salt provides iodide, a necessary nutrient."

Although iodine is added to most table salt in the United States, only 15% of daily salt intake comes from iodized table salt. The majority of salt intake in the United States comes from processed foods, but food manufacturers often use noniodized salt. Milk and processed grain products provide the majority of total iodine intake in the United States (about 70%). Iodine in milk originates primarily from the fortification of animal feed and the use of iodine-containing sanitizers during milk collection, while its presence in processed grain products is largely due to the use of iodine-containing food additives (dough conditioners).

Iodine is an essential component of thyroid hormone, thus it is required for normal function of the **thyroid gland**. The pituitary gland in the brain secretes **thyroid-stimulating hormone (TSH)**, which regulates the thyroid by increasing the uptake of iodine from the blood, stimulating thyroid hormone production and release. The thyroid hormones regulate energy metabolism and protein synthesis and play critical roles in the development of the fetal skeleton and brain, which is why iodine deficiency is especially problematic during pregnancy. (INFOGRAPHIC 14.3)

When individuals do not get enough iodine, the thyroid gland cannot produce adequate levels of thyroid hormones. This can lead to goiter or hypothyroidism

THYROID GLAND
a gland located in the neck that releases the iodine-containing thyroid hormones, involved in metabolism

THYROID-STIMULATING HORMONE
a hormone released from the anterior pituitary that stimulates the thyroid gland to produce and secrete thyroid hormones

INFOGRAPHIC 14.4 The Adverse Effects of Iodine Deficiency

THE ADVERSE EFFECTS OF IODINE DEFICIENCY

SEVERE DEFICIENCY

MENTAL RETARDATION, CRETINISM

POOR GROWTH, STUNTING

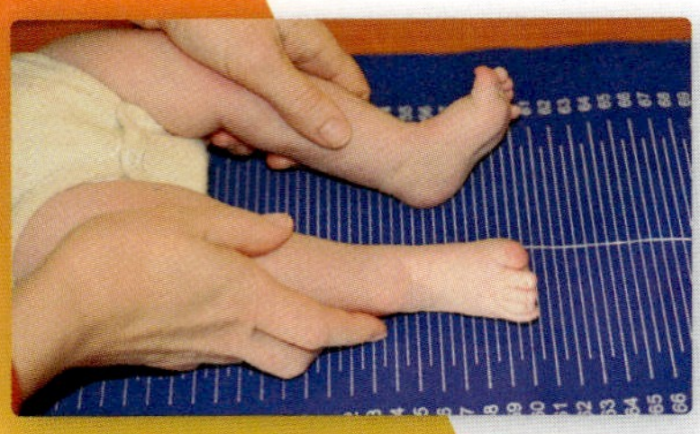

INCREASED INFANT MORTALITY

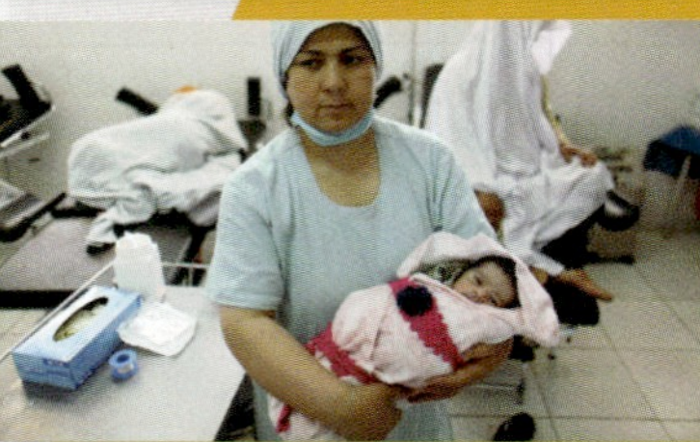

GOITER AND HYPOTHYROIDISM

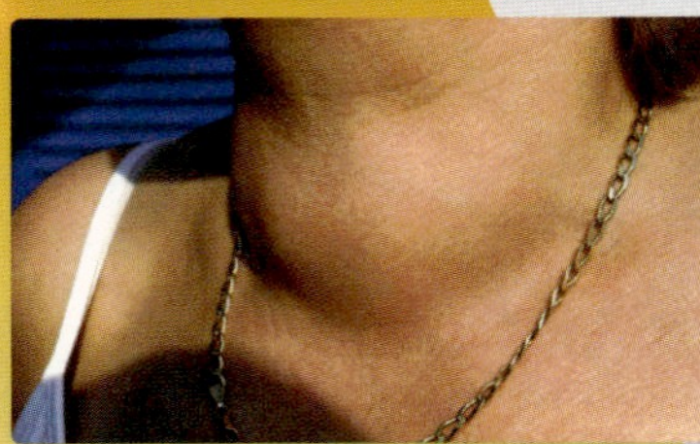

NODULAR GOITER AND HYPERTHYROIDISM

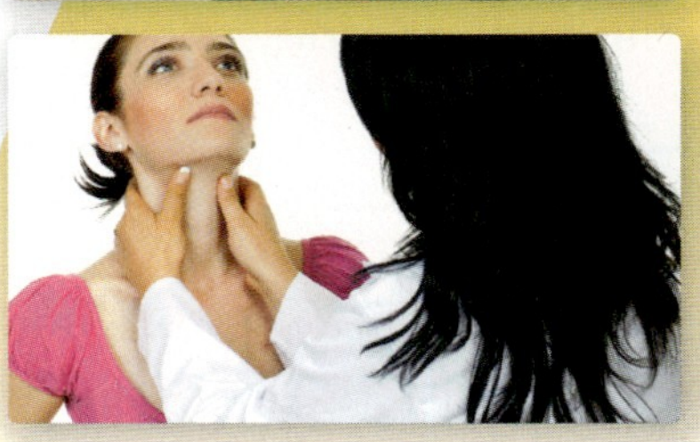

LOWER INTELLIGENCE, POOR EDUCABILITY

MILD DEFICIENCY

Photo credits (left—top to bottom): CNImaging/Newscom, Horizons WWP/Alamy, Blaj Gabriel/Shutterstock; (right—top to bottom): Rafael Ben-Ari/Alamy, Chris Pancewicz/Alamy, Zurijeta/Shutterstock

(underactive thyroid), which slows the metabolic rate. Goiter can occur when the thyroid gland—an endocrine gland located at the front of the neck—is not able to make enough thyroid hormone to meet the body's needs. The TSH levels rise and the thyroid gland attempts to make thyroid hormones, but it does not receive enough iodine. The thyroid gland grows and expands, trying to do its job, forming a lump in the neck that can grow as large as a grapefruit.

CRETINISM *a condition characterized by arrested mental and physical development; can be caused by severe iodine deprivation during fetal growth*

The milder iodine deficiencies that have developed in the United States are most common among pregnant women, as the RDA for iodine during pregnancy is 50% higher than it is for nonpregnant women. Lack of iodine can adversely affect brain development and growth of the developing fetus. If a pregnant woman is extremely deficient in iodine, then **cretinism** can develop in her child, characterized by mental retardation, deafness and muteness, stunted growth, delayed sexual maturation, and other abnormalities. Less severe iodine deficiencies can affect neurological development in young children, resulting in below average intelligence. (INFOGRAPHIC 14.4)

INFOGRAPHIC **14.5** **Selected Sources of the Mineral Iodine**

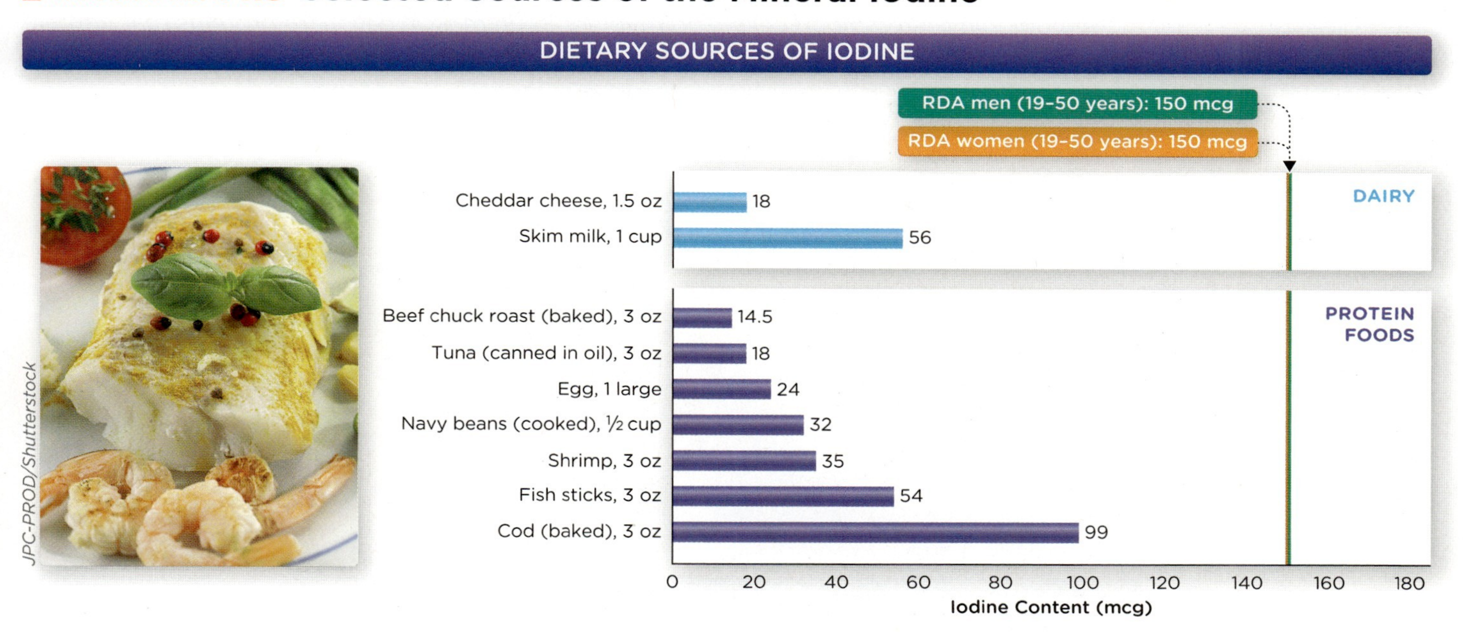

Iodine recommended intakes

Because of these problems, iodine supplements are sometimes recommended for pregnant women and women of child-bearing age (unfortunately many prenatal vitamins do not contain supplemental iodine). The RDA for iodine for men and women 19 years and older is 150 micrograms (the UL is 1,100 micrograms). Excessive intake can elevate levels of TSH, disrupting thyroid function and sometimes causing some of the same symptoms as iodine deficiency: goiter, elevated TSH levels, and hypothyroidism. (INFOGRAPHIC **14.5**)

■ ■ ■

Today, there is no longer a "goiter belt" in the United States thanks to iodized salt; however a major public health concern worldwide is iron deficiency, which is generally caused by a person's increased demand for iron (because of blood loss or growth) or through insufficient intake or absorption. The World Health Organization considers iron deficiency to be the number one nutritional disorder in the world, affecting developing and industrialized nations.

IRON

As you saw in Infographic 14.1, *iron* (Fe) is the most abundant of the trace minerals in our body. Iron is a crucial component of hundreds of enzymes and other proteins in our body.

Most iron in our body occurs as **heme iron**. Heme iron is a critical part of the protein *hemoglobin.* In the body, red blood cells pick up oxygen from the lungs and release it into the tissues. Red blood cells are able to pick up and transport oxygen because they contain a protein within them called **hemoglobin**. Hemoglobin is made up of four units, each unit contains one heme group (an iron atom surrounded by a ring-shaped structure) and one protein chain. These heme groups contain positively charged iron atoms that can bind to oxygen molecules and transport them to various areas of the body. The structure of hemoglobin allows it to be loaded with oxygen in the lungs, so it is sometimes referred to as an oxygen transport protein. In muscles, the iron-containing oxygen storage protein is **myoglobin**, which is similar in structure to hemoglobin but has only one heme unit and one protein chain.

HEME IRON
iron derived from hemoglobin; the most bioavailable form of dietary iron found in meat, fish and poultry

HEMOGLOBIN
a protein in red blood cells that contains iron and carries oxygen to tissues

MYOGLOBIN
a protein that functions to provide oxygen to muscles; contains less iron than hemoglobin

INFOGRAPHIC **14.6 Hemoglobin and Myoglobin Contain Heme Iron**

Hemoglobin (in red blood cells) and myoglobin (in muscle cells) are proteins that contain iron, which binds oxygen.

HEMOGLOBIN (FOUND IN RED BLOOD CELLS)

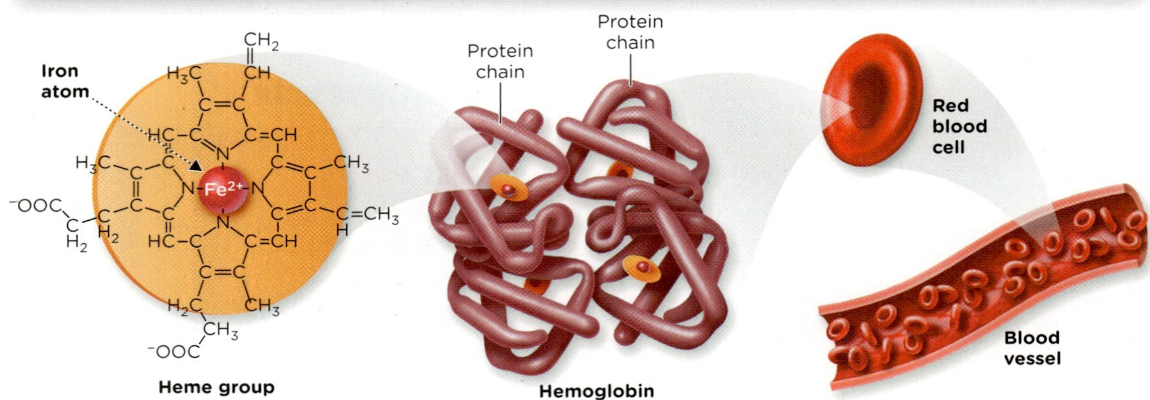

MYOGLOBIN (FOUND IN MUSCLE CELLS)

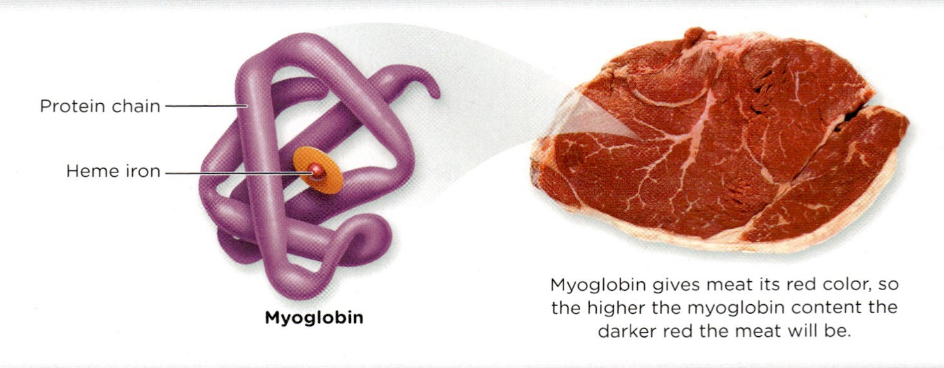

Myoglobin gives meat its red color, so the higher the myoglobin content the darker red the meat will be.

OXYGEN BINDS TO HEME IN HEMOGLOBIN AND MYOGLOBIN

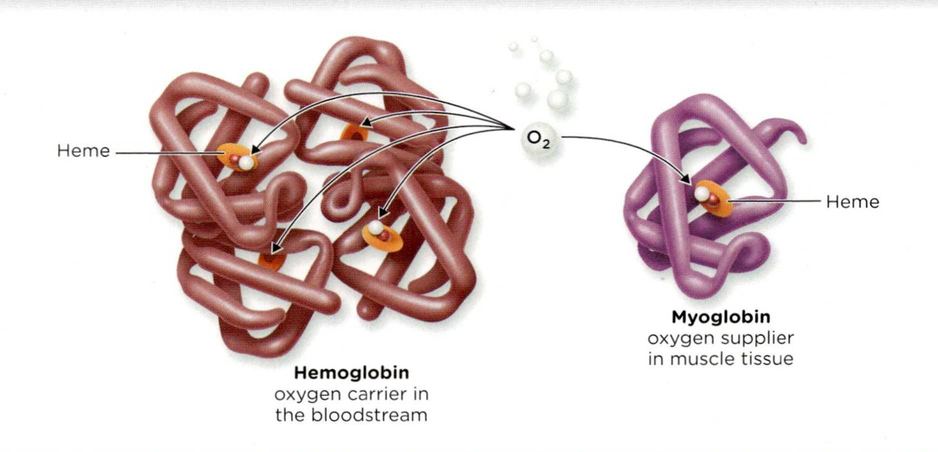

What part of a chicken would likely have the most heme iron?

Photo credit: Michael Krinke/Getty Images

Myoglobin transports and stores oxygen in muscle cells and helps to coordinate the supply of oxygen to the demand of working muscles. **(INFOGRAPHIC 14.6)**

In addition, iron is also an integral component of many enzymes required for a host of crucial processes in the body, including energy metabolism, protection against antioxidants, the immune response, and DNA synthesis. Because of its role in DNA synthesis, iron is required for a wide variety of critically important functions, including reproduction, growth, and healing.

Although most iron is in heme form in the body, **non-heme iron** is also important because it assists enzymes in their functioning. For example, the enzymes that bind iodine to form thyroid hormone, and metabolize beta-carotene to vitamin A, both use non-heme iron as a co-factor.

Food sources offer iron in both the heme and non-heme forms. Dietary heme iron (from hemoglobin and myoglobin in animals) is present in red meats, poultry, and fish. Non-heme iron is also found in meat and fish; additionally, it is the only form of iron found in plant foods, such as lentils and beans, dried fruits, and grain products—particularly those grain products that have been fortified with iron. Although non-heme iron makes up the majority of the iron that we consume, it is not as well-absorbed by the body as heme iron.

Iron absorption

Non-heme iron is strongly influenced by absorption enhancers and inhibitors, while the absorption of heme iron is little affected. For example, non-heme iron absorption is enhanced when consumed with vitamin C or with meat, fish, or poultry. On the other hand, soy protein, phytates (compounds found in whole grains and legumes), and polyphenols (such as tannins in tea and red wine) have been shown to inhibit non-heme iron absorption by binding to iron in the gastrointestinal tract. Calcium is the only dietary factor that may reduce the absorption of both heme and non-heme iron. Because calcium and iron compete with each other for absorption, when large amounts of calcium (such as from calcium supplements) are consumed with iron at the same time or during the same meal, iron absorption is decreased. However, studies have found that calcium has little

NON-HEME IRON
iron derived from plant foods such as lentils and beans, but it is also present in significant quantities in meat and fish; less bio-available than heme iron

Iron absorption is improved when a food source containing non-heme iron (like beans) is accompanied by a food rich in vitamin C (such as tomatoes and peppers).

denio109/Shutterstock

INFOGRAPHIC 14.7 Non-heme Iron Absorption Enhancers and Inhibitors

ENHANCERS

- MFP (meat, fish, poultry) factor
- Vitamin C

INHIBITORS

- Phytate
- Soy protein
- Compounds in tea and red wine (polyphenols)
- High calcium intake in the same meal with the iron

Photo credits (top—left to right): sasaken/Shutterstock, Milleflore Images/Shutterstock; (bottom—left to right): aboikis/Shutterstock, Liv friis-larsen/Shutterstock, naito8/Shutterstock, Barnaby Chambers/Shutterstock, Donna Beeler/Shutterstock

long-term effect on iron absorption and status. (INFOGRAPHIC 14.7)

Iron stores in the body are regulated only by controlling its absorption in the small intestine, because, unlike other minerals, iron cannot be excreted in urine or bile. In healthy individuals about 10% to 15% of dietary iron is absorbed. However, when iron stores are low, iron is absorbed more efficiently, and when its stores are high, iron is absorbed less efficiently.

IRON-DEFICIENCY ANEMIA *a condition characterized by fatigue, decreased immune function, and impaired development, due to reduced levels of iron containing hemoglobin in red blood cells and, therefore, decreased oxygen transport around the body*

The body also carefully conserves iron so that daily losses are minimized. While we do lose iron through blood loss, the shedding of cells from the gastrointestinal tract and skin, and small losses in sweat, the body holds on to as much as possible. For example, when red blood cells die, iron is recycled and incorporated into new red blood cells. Iron balance in the body is achieved by carefully regulating iron absorption, storage, release, and transport. When imbalances develop, dangerous deficiencies and toxicities can occur.

Iron deficiency anemia

As many as 30% of the world's people may suffer from **iron deficiency anemia** (also known as microcytic hypochromic anemia), a serious condition that develops gradually when a person's iron intake does not meet his or her daily needs. "Anemia" is a condition in which the oxygen carrying capacity of red blood cells is inadequate and "iron deficiency" is a cause of the type of anemia that results from a shortage of oxygen carrying hemoglobin. When iron intake is low over time, iron stores can become depleted; if iron stores become completely depleted, blood levels of iron fall, and hemoglobin synthesis is impaired.

The left image shows normal red blood cells, and the right image shows iron-deficient red blood cells. The red blood cells in iron-deficiency anemia are smaller and paler than normal red blood cells.

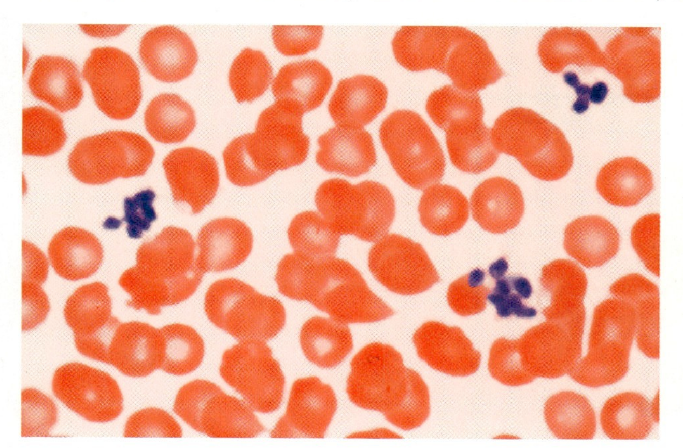

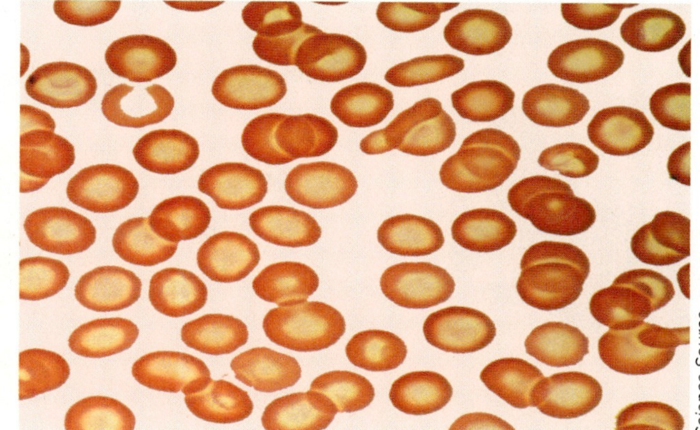

Science Source

The decrease in hemoglobin synthesis results in less hemoglobin incorporation into red blood cells, and iron deficiency anemia develops.

Iron deficiency anemia has many causes. The most common cause is blood loss. The combination of blood loss through menstruation and a limited or restricted diet that may minimize sources of heme iron can put premenopausal women at risk of iron-deficiency anemia. It is estimated that 15% of women aged 20 to 49 years may have iron-deficiency anemia. It can also result from insufficient dietary iron intake and is more common among vegans, who avoid foods of animal origin, and thus consume no heme iron through food consumption. The condition can also develop when the mineral cannot be properly absorbed because of disease or the presence of dietary components that inhibit non-heme iron absorption. Infants and young children need higher-than-normal amounts of iron because they are rapidly growing; iron needs are also increased during pregnancy. Iron lost through menstruation, frequent blood donations, or any other form of blood loss must be replaced. Adult men and postmenopausal women are at lower risk for iron deficiency anemia, because they lose very little iron through blood loss.

People who suffer from iron deficiency anemia can have many symptoms. They can feel tired and out of breath, perform poorly at work or at school, and have slow cognitive

As many as 25% of college age women have suboptimal iron status. In addition, athletes who train on a regular basis are estimated to have a 30% higher average iron requirement due to increased iron losses. Consuming foods rich in iron and use of grain products fortified with iron can help meet daily needs.

Getty Images

INFOGRAPHIC 14.8 Selected Sources of the Mineral Iron

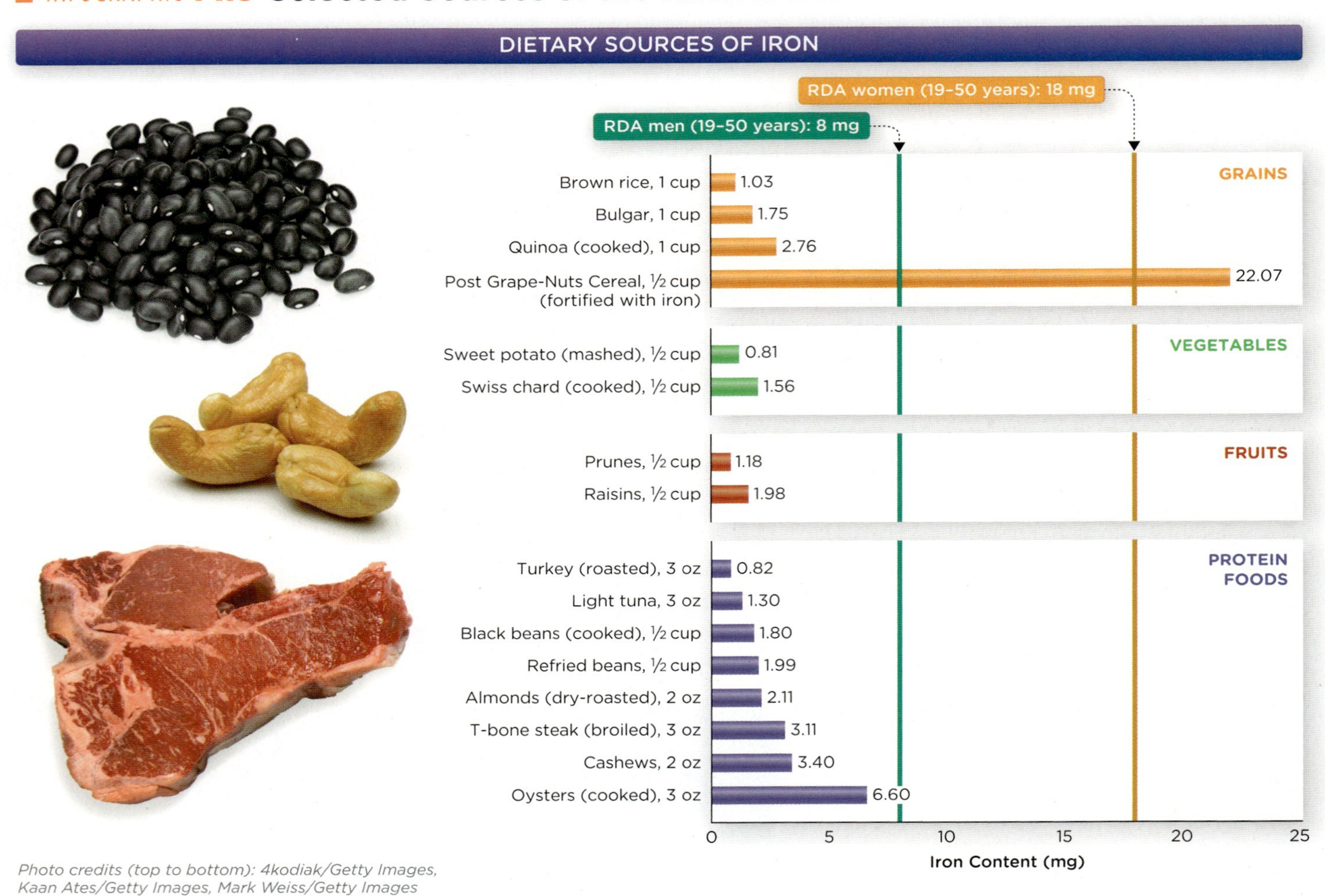

Photo credits (top to bottom): 4kodiak/Getty Images, Kaan Ates/Getty Images, Mark Weiss/Getty Images

and social development during childhood. They often have trouble maintaining their body temperature and are more susceptible to infections because their immune systems are not working properly. Iron deficiency anemia during pregnancy increases the risk of preterm birth and low birth weight, increasing the infants' risk of health problems later. Individuals thought to be suffering from this form of anemia should be evaluated and managed by a health care provider, who will try to determine the cause of the condition and may prescribe iron supplements.

Iron intake recommendations

The RDA for iron for men 19 years and older is 8 mg; for women aged 19 to 50 years, it is 18 mg. However, these recommendations assume that 75% of iron consumption is heme iron, so for vegans and vegetarians, who primarily ingest non-heme iron, intake recommendations are approximately doubled. Athletes who train intensively on a regular basis are estimated to have a 30% higher average iron requirement because of increased iron losses. National nutrition surveys suggest that men of all racial and ethnic groups consume the recommended amounts of iron, but intakes are generally lower than recommended in women of childbearing age and young children. **(INFOGRAPHIC 14.8)**

The UL for iron for men and women 19 years and older is 45 mg, and higher intakes of iron can cause gastrointestinal distress. Individuals who take high doses of iron to prevent or treat iron deficiency anemia also sometimes suffer from side effects, including constipation, nausea, vomiting, and diarrhea, especially when the supplements are taken on an empty stomach. Because very little iron is excreted from the body, iron toxicity can occur when intake is too high, causing symptoms such as apathy, fatigue, liver damage, and immune problems.

In children, iron poisoning is a major cause of unintentional poisoning death, causing symptoms such as nausea, vomiting, diarrhea, constipation, rapid heartbeat, dizziness, shock, and confusion (which explains why you will not find iron in "gummy" multivitamin/mineral supplements; children might mistake them for candy and consume too many with potentially fatal results). Men are at a higher risk for iron toxicity than are women because men don't experience monthly blood loss. Individuals with hereditary hemachromatosis, sometimes called iron overload disease, are also at high risk of iron toxicity because of increased absorption and high iron stores.

■ ■ ■

In 1958, a young physician named Ananda Prasad evaluated a patient in Iran who suffered from severe iron deficiency and its associated symptoms. In addition, the severely stunted patient appeared to be about eight years old and had not gone through puberty, although his chronological and bone age indicated he was closer to 21 years old. This was not an isolated case; the same condition was so prevalent in Iran that it was considered an epidemic.

Dr. Prasad studied the curious problem, administering iron to the patients with iron deficiency. However, Prasad found it difficult to assign all the health problems to iron deficiency, since growth retardation and the lack of secondary sex characteristics are not typically linked to iron deficiency.

Seeking clues to effectively treat the patients, Prasad went to Egypt to study rural farmers with similar signs of illness, taking careful inventory of their diets. Soon, Prasad began to target the mineral zinc, which, at that time, was not thought to be of importance to human health. He knew that in the developed world, zinc is found in a variety of food sources, such fish, red meat, and dairy products. However, the diets of the individuals he studied from the developing world relied heavily on breads and grains, which contain phytates, which are substances that bind zinc and iron and prevent both minerals from being properly absorbed, accounting for both the iron deficiency and the zinc deficiency.

In 1961, Dr. Prasad published an article in the *American Journal of Medicine*, suggesting for the first time that zinc deficiency could account for human growth retardation. By 1963, Dr. Prasad started administering zinc through clinical trials, and his participants began growing taller and developing secondary sex characteristics.

ZINC

Zinc (Zn) is required for the function of perhaps more proteins in the body than any other mineral. Research indicates that zinc binds to about 10% (~2,800) of all proteins in the body, including more than 900 enzymes. Zinc functions as a co-factor for enzymes that participate in most major metabolic pathways. The binding of zinc

A mother gives a baby supplemental zinc. *Zinc supplementation has been a successful treatment for childhood diarrhea in developing nations. In addition, the prevention of zinc deficiency through zinc supplementation also reduces the risk of death from infectious disease and improves the growth and development of children.*

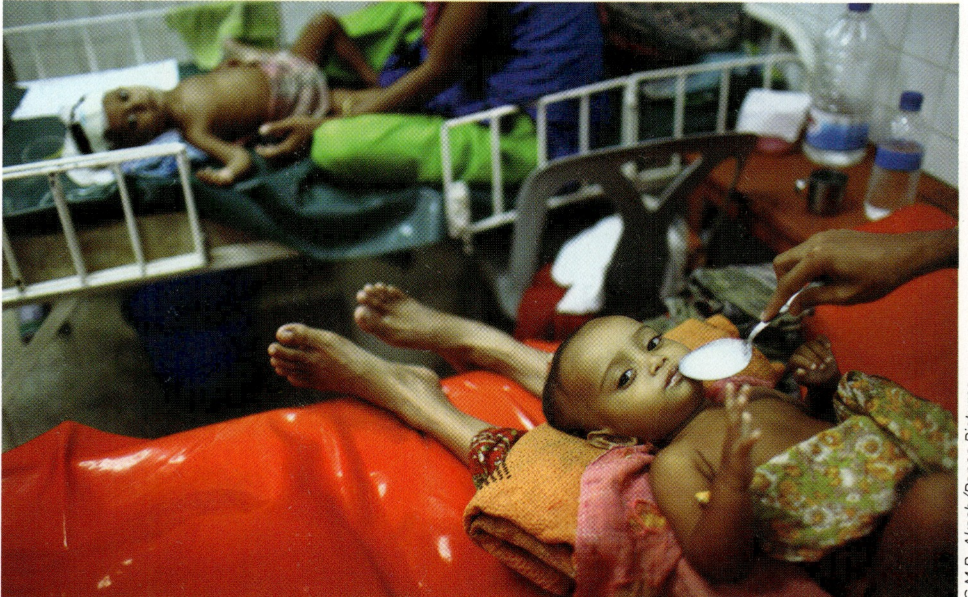

G.M.B. Akash/Panos Pictures

to proteins also plays critically important structural roles by allowing proteins to achieve and maintain their appropriate shapes. The prolific presence of zinc in so many enzyme systems and regulatory proteins means that it is required for virtually every essential process in the body, including the regulation of protein synthesis, reproduction, cell division, growth and development, immune responses, and neurological functions.

Although zinc deficiencies serious enough to produce readily identifiable symptoms are uncommon in the United States, it is estimated that about 12% of the population is at risk of deficiency. Some groups, including alcoholics, vegetarians, and the elderly, are particularly at risk of a zinc deficiency. Alcohol reduces zinc absorption and increases zinc excretion in the urine, while phytates in whole grains and legumes, which are dietary staples among vegetarians, inhibit the mineral's absorption. Zinc status in the elderly may be compromised because of reduced food intake and impaired absorption resulting from low gastric acid production.

Seniors may experience impaired absorption of zinc because of reduced production of gastric acid. In addition, zinc absorption is inhibited by phytates in whole grain cereals.

Hill Street Studios/age fotostock

Mild to moderate zinc deficiency can cause impaired immune function, appetite loss and weight loss, delayed sexual maturation, and slowed growth. Severe zinc deficiency results in hair loss, diarrhea, infertility in men, and impaired neurological and behavioral functions. Symptoms of zinc deficiency can mimic symptoms of other nutrient deficiencies and can occur along other deficiencies in part because zinc plays an important role in the proper functioning of many other nutrients.

The UL for zinc intake for men and women 19 years and older is 40 mg. Short-term symptoms of excessive zinc intake include nausea, vomiting, abdominal cramps, and diarrhea. Chronic, longer-term effects may include copper deficiency (excessive zinc reduces the absorption of copper), altered iron function, reduced immune function, and lowered levels of high-density lipoproteins. **(INFOGRAPHIC 14.9)**

The recommended intake of zinc is 11 mg for men 19 years and older and 8 mg for women in the same age group. Since the body cannot store zinc, a regular daily intake is required to maintain adequate zinc status, although absorption does increase in the small intestine when intake is low. Because vegetarians absorb less zinc than nonvegetarians do, the Dietary Reference Intakes recommend that vegetarians and vegans consume twice as much zinc as nonvegetarians. Zinc is found in many foods but is most concentrated in meats, poultry, and certain types of seafood, such as oysters, which contain more zinc per serving than any other food. Fortified cereals, beans, and nuts also provide zinc, as do certain brands of cold lozenges and some over-the-counter drugs sold as cold remedies.

COPPER

Copper (Cu) is a mineral perhaps more familiar to us for its use in cookware, wire, and even jewelry than as a nutrient. However, copper functions as a co-factor in oxygen-dependent enzymes in the body. Though there are only about a dozen copper-containing human enzymes, they participate in a variety of critical physiological processes, including but not limited to, energy metabolism, formation

INFOGRAPHIC 14.9 Selected Sources of the Mineral Zinc

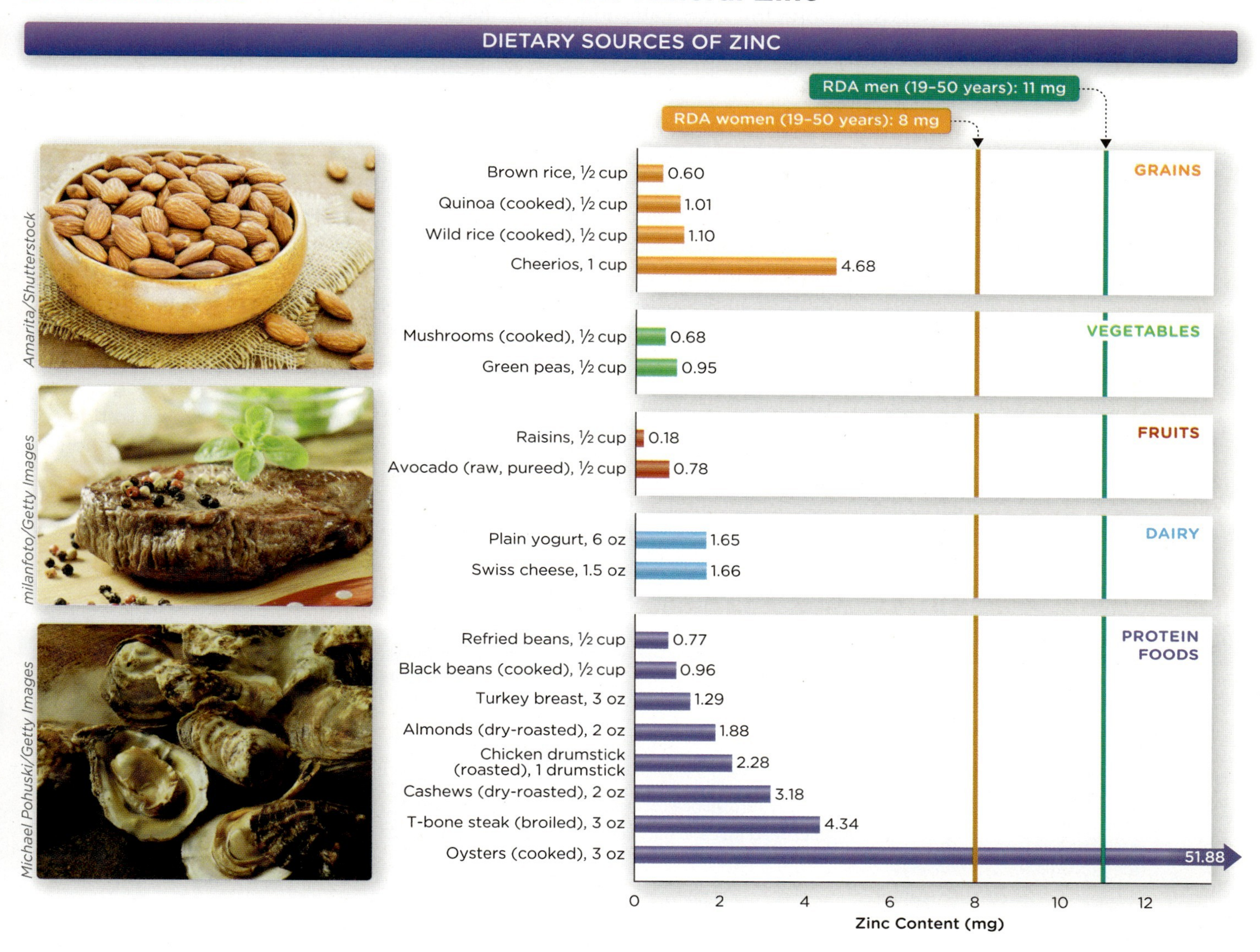

of connective tissue, regulation of iron storage and transport, and antioxidant functions. Copper may play a role in slowing the progression of age-related macular degeneration, which is responsible for causing severe vision loss in the elderly.

Deficiencies of copper from low dietary intakes are rare in the United States as most of the population meets or exceeds the RDA for copper. However, deficiencies that are secondary to other factors are more common. Copper deficiencies are somewhat common in people who have gastric bypass surgery to treat obesity, as well as in those with other conditions that cause nutrient malabsorption. Excessive zinc intake decreases copper absorption and is another cause of copper deficiency. One of the most common copper deficiency symptoms highlights another important mineral-mineral interaction (in addition to that of copper and zinc): A copper deficiency often results in anemia because copper-containing enzymes are required to release iron from stores and for its incorporation into its heme form. Symptoms of copper deficiency include impaired immune response and osteoporosis.

The RDA for copper is 900 micrograms per day in adults. Approximately 55% to 75% of dietary copper is absorbed, which is much more efficient than other trace minerals. Although copper is found in a wide variety of foods, it is highest in organ meats, shellfish, nuts, seeds, mushrooms, chocolate, and legumes. **(INFOGRAPHIC 14.10)**

SELENIUM

The important role of the trace mineral *selenium* (Se) in the function of certain enzymes wasn't established until the early 1970s. Currently, about two dozen selenium-containing human proteins have been identified. Selenium functions as a co-factor for several antioxidant enzymes and with other selenium-dependent enzymes in the activation of thyroid hormone, making it essential for normal growth, development, and metabolism.

A selenium deficiency by itself seldom causes obvious symptoms or illness. However, additional stresses, such as viral infections, chemical exposure, or low intakes of other antioxidant nutrients, together with low selenium intake can cause clinical illness. For example, a selenium deficiency increases the likelihood of developing a vitamin E deficiency when intake of the vitamin are low. This occurs because selenium functions as a cofactor for an important antioxidant enzyme system and a deficiency of selenium increases oxidative stress and places a greater demand on the antioxidant function of vitamin E. Low-selenium status increases the risk of a particular form of heart disease and may increase the risk of some cancers. Inadequate selenium may also decrease immune function.

The RDA for selenium is 55 micrograms per day for adults. The highest concentrations of selenium are present in organ meats and seafood. The most common sources in the diet are meat and cereals. The selenium content

INFOGRAPHIC 14.10 Selected Sources of the Mineral Copper

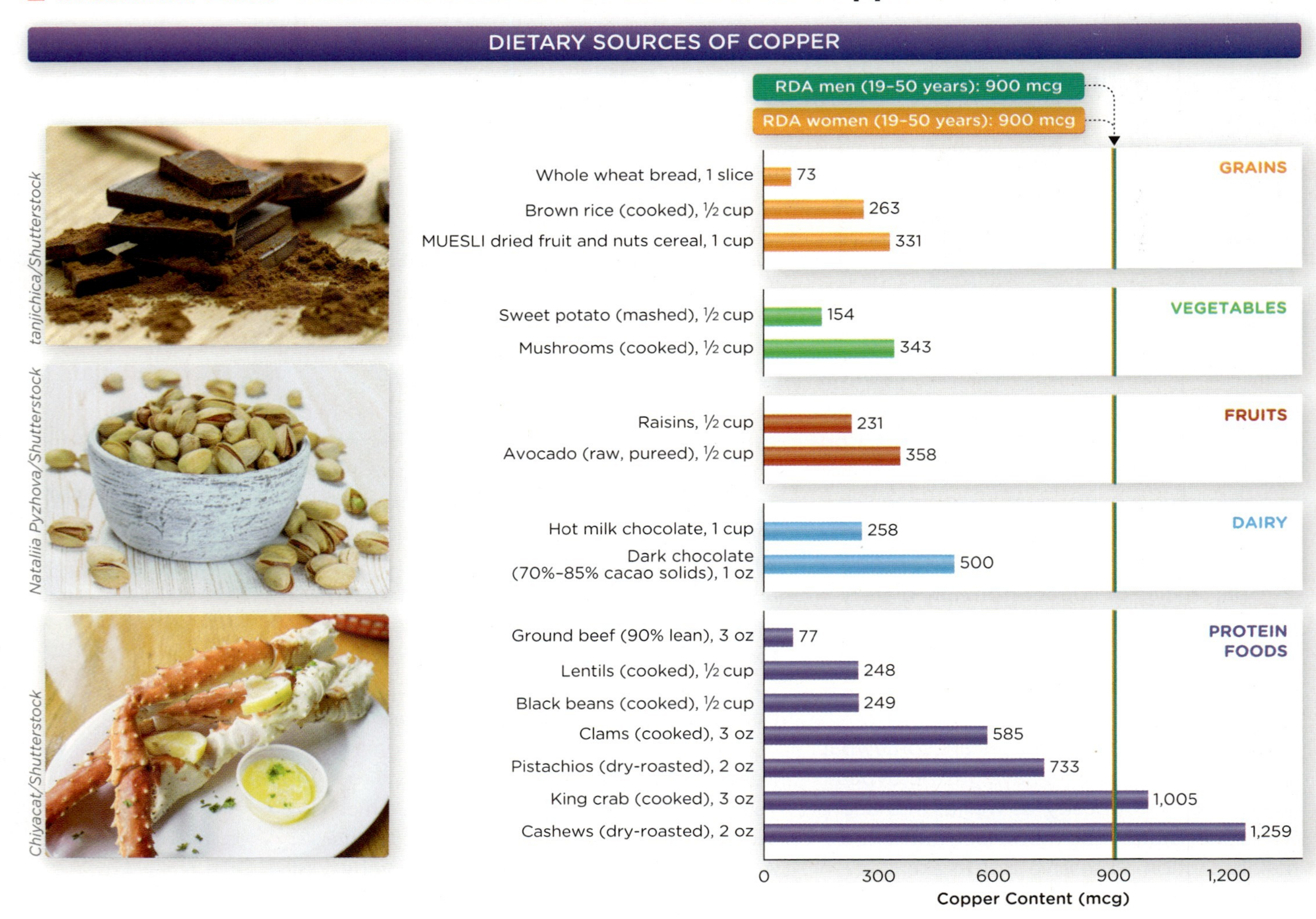

of meats is fairly consistent, but the selenium content of plant foods varies widely depending on the content of the mineral in the soil where the plant was grown. Dietary selenium is generally well-absorbed, and unlike most other minerals, its absorption is not regulated in relation to nutritional status. **(INFOGRAPHIC 14.11)**

The UL for selenium is 400 micrograms per day in adults. Although a single high dose of selenium may be fatal, selenium toxicity generally occurs with long-term exposure to moderately high levels of intake either through supplementation or, in rare instances, through frequent consumption of selenium-rich foods. For example, a single one-ounce serving of about six Brazil nuts contains 544 micrograms of selenium, which is about 25% above the UL. The most common toxicity symptoms are hair and nail loss, and brittleness.

FLUORIDE

Fluoride has a well-established role in the prevention of dental caries through the hardening and maintenance of tooth enamel. Fluoride is also important in stabilizing the structure of bone. As fluoride is not required for growth, for reproduction, or to sustain life, it is not by definition considered an essential nutrient. However, the decline in incidence of tooth decay over the past 70 years has been largely attributed to the addition of fluoride to public water supplies.

Around the same time as other fortification initiatives in the United States, such as the iodization of salt to prevent goiter and the addition of vitamin D to milk to prevent rickets, the fluoridation of drinking water began in the 1940s to help prevent dental caries. It is estimated that 60% to 70% of the

INFOGRAPHIC 14.11 Selected Sources of the Mineral Selenium

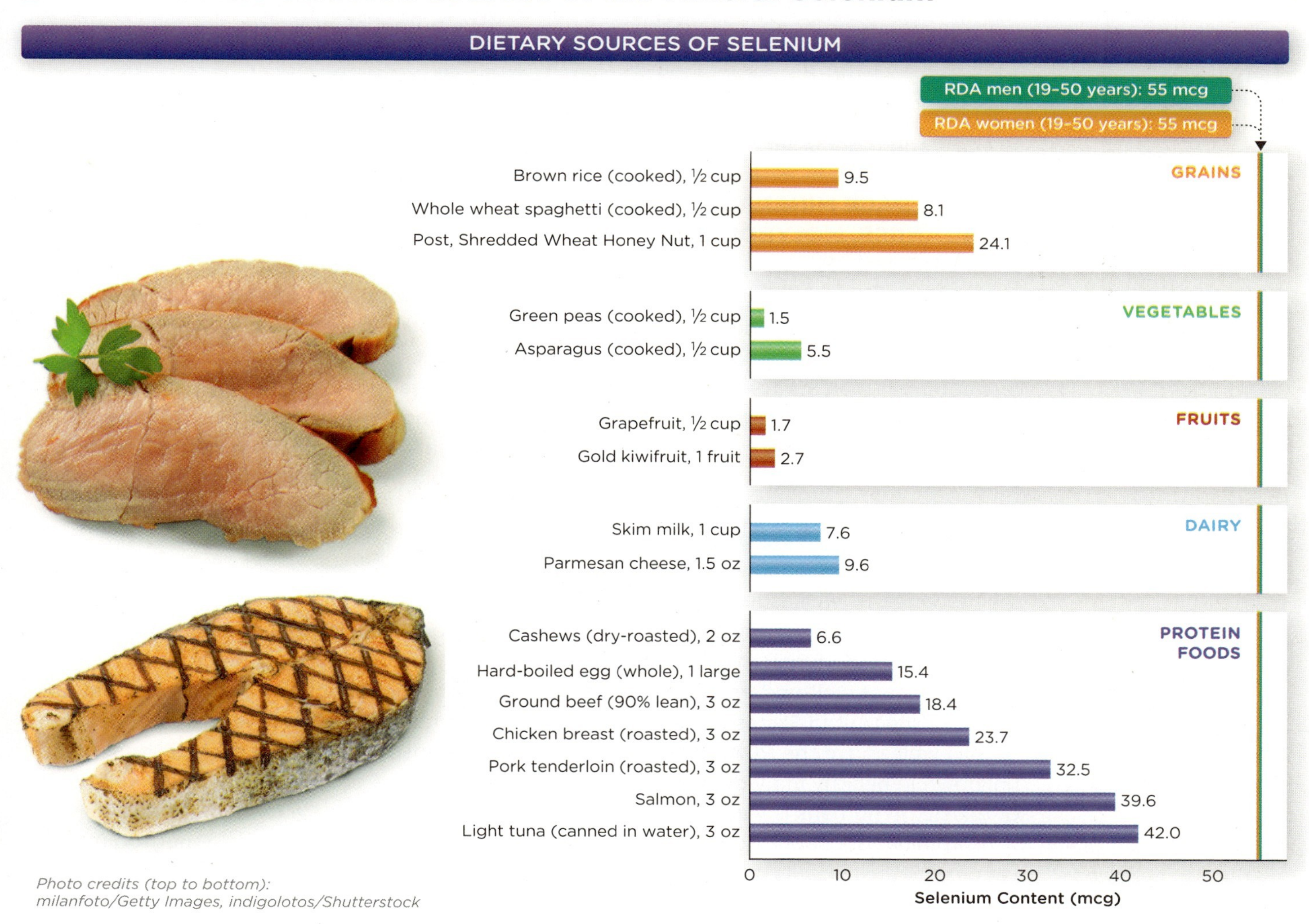

Photo credits (top to bottom): milanfoto/Getty Images, indigolotos/Shutterstock

population is served by community water systems that receive fluoridated water. The Environmental Protection Agency (EPA) is responsible for safeguarding our public drinking water, which includes establishing ranges for fluoride that reduce rates of dental caries while not exceeding upper thresholds.

In addition to fluoridated water, we obtain fluoride through fluoridated dental products, beverages made with fluoridated water, and to a lesser extent, marine fish. The AI for fluoride is set at 4 mg for men 19 years and older, and at 3 mg for women in the same age range. Inadequate fluoride, not surprisingly, results in an increased risk of dental caries. Fluoride may cause health issues if drinking water exceeds the standards set by the EPA. Tooth and skeletal fluorosis is characterized initially by small opaque flecks on the teeth and then by stains or pits in the teeth with longer-term exposure. The EPA warns that excessive intake over a lifetime may have adverse effects on bone, among them an increased likelihood of fractures because bones become brittle as they become excessively dense. The UL for men and women 19 years and older is set at 10 mg.

OTHER TRACE MINERALS: MANGANESE, MOLYBDENUM, AND CHROMIUM

Despite being one of the most poorly absorbed of the trace minerals (generally less than 5% is absorbed), *manganese* deficiencies are far less common than are toxicities. It functions as a co-factor for enzymes involved in antioxidant functions, and is involved in the metabolism of carbohydrates, cholesterol, and amino acids. The role of manganese in amino acid metabolism makes it important for the synthesis of proteins needed for bone growth and maintenance, as well as for wound healing.

Molybdenum is a co-factor for only four enzymes, and the function of one of these enzymes is not well-understood. The average intake of the United States population is usually well-above the RDA; deficiencies are unknown in healthy individuals and the risk of toxicity in humans is very low.

Chromium is a popular dietary supplement, yet despite the hype, its proposed benefits have not been supported by the results of intervention trials. There is essentially no credible evidence that it promotes weight loss, or that it improves muscle mass with resistance training. Perhaps even more surprising, recent studies have failed to conclusively demonstrate that chromium can enhance the ability of insulin to control blood glucose, and no progress has been made in identifying the means by which chromium might exert its biological affects. Also, chromium deficiency in humans has been observed only in hospitalized patients who had been receiving 100% of all nutrients intravenously for an extended period. For these reasons, some experts now question chromium's designation as an essential nutrient.

ULTRATRACE MINERALS
minerals that are found in the body, but not considered "essential" at this time

ULTRATRACE MINERALS

In addition to trace minerals, individuals may also need to ensure that they are getting enough of the **ultratrace minerals**, which have potential intake requirements of less than 1 microgram per day. The ultratrace minerals include *arsenic* (As), *boron* (B), *nickel* (N), *silicon* (Si), and *vanadium* (V). Although scientists have not yet identified specific biological or physiological functions for ultratrace minerals, and these minerals are not currently classified as essential nutrients, animal data suggest that they may be important for human nutrition. Daily recommendations for the intake of ultratrace minerals are considered ND, or "not determinable," because so little data exists; as a result, food is considered the only safe source of these nutrients to avoid possible overconsumption.

■ ■ ■

All of the trace minerals found in your body could fit in the palm of your hand. Yet their contributions are crucial to your health. Their functions are as important as the functions of the major minerals calcium and phosphorus, which each account for more than a pound of your body weight. The key is to consume a range of foods to meet our need for trace minerals—not too much and not too little.

INFOGRAPHIC **14.12** **Summary of the Trace Minerals** *See Appendix 3 to review vitamins and minerals organized by functions in the body.*

Mineral	Major Functions	RDA or AI* for Adults, per Day	Food Sources	Deficiency Signs and Symptoms	Toxicity/Adult UL
Iron	Oxygen transport and storage, energy metabolism, detoxification, thyroid hormone synthesis.	**Male:** 8 mg **Female:** 18 mg	Meat, fish, poultry, legumes, nuts, enriched grains.	Anemia, decreased thyroid hormone production, impaired immune function, growth, and intellectual development.	Nausea, vomiting, abdominal pain, coma, heart, kidney, liver failure. **45 mg**
Zinc	Needed for the regulation of protein synthesis, antioxidant functions, growth, development, reproduction and immune and neurological functions. Present in nearly 1,000 enzymes in nearly all major metabolic pathways.	**Male:** 11 mg **Female:** 8 mg	Oysters, meat, poultry, legumes, nuts, whole grains.	Impaired immune function, delayed sexual maturation, slow growth, hair loss, neurological impairment.	Nausea, vomiting, diarrhea, decreased immune function, copper deficiency. **40 mg**
Copper	Co-factor for about a dozen enzymes involved in antioxidant functions, energy metabolism, iron transport, healthy connective tissue including bone, and synthesis of hormones and neurotransmitters.	**Male:** 900 mcg **Female:** 900 mcg	Organ meats, shellfish, nuts, seeds, chocolate, and legumes.	Anemia, poor immune function, osteoporosis, and poor growth.	***Rare:*** Very high doses cause nausea, vomiting, and diarrhea. **10 mg**
Iodine	Required for the synthesis of thyroid hormone, which regulates protein synthesis and metabolic rate.	**Male:** 150 mcg **Female:** 150 mcg	Seafood, dairy products, iodized salt.	Inadequate thyroid hormone production, goiter, impaired brain function and growth, and reduced work capacity.	Goiter, and impaired thyroid function. **1,100 mcg**
Selenium	Co-factor for about two dozen enzymes. Important antioxidant functions and the production of active thyroid hormone.	**Male:** 55 mcg **Female:** 55 mcg	Brazil nuts, seafood, meat, whole grains.	Generally does not occur without additional stress factors. Can cause cancer and heart disease.	Hair and nail brittleness and loss. **400 mcg**
Chromium	May enhance insulin action.	**Male (19–50):** 35 mcg* **Male (>50):** 30 mcg* **Female (19–50):** 25 mcg* **Female (>50):** 20 mcg*	Broccoli, grape juice, processed meats, whole grains, and spices.	Not observed in healthy individuals.	***Rare:*** kidney failure and liver disease possible **No UL**
Fluoride	Hardens tooth enamel.	**Male:** 3 mg* **Female:** 2 mg*	Treated drinking water, tea, grape juice, and sardines.	Increased risk of dental caries.	Nausea, vomiting, and diarrhea, staining of teeth. **10 mg**
Manganese	Co-factor for enzymes involved in antioxidant functions, energy metabolism, bone development, and neurotransmitter production.	**Male:** 2.3 mg* **Female:** 1.8 mg*	Grain products, tea, coffee, nuts, legumes, spinach.	***Rare:*** Poor growth and dermatitis. Low status may increase the risk of diabetes, osteoporosis, and atherosclerosis.	Neurological abnormalities. **11 mg**
Molybdenum	Co-factor for only four enzymes that metabolize some amino acids, nucleic acids, as well as one group of other related compounds.	**Male:** 45 mcg **Female:** 45 mcg	Legumes, whole grains, and nuts.	Not observed.	Unknown **2,000 mcg**

*Adequate Intake (AI)

CHAPTER 14 **BRING IT HOME**

Iron across the life span

Among the essential vitamins and minerals, the variability in iron requirements between sexes and across the lifespan can be significant. The different RDAs of the Dietary Reference Intakes (DRIs) reflect the critical role of iron in oxygen transport, enzyme function, and growth and development.

Using the Interactive DRI for Health Professionals (http://fnic.nal.usda.gov/fnic/interactiveDRI/) complete the following chart. You can also use the DRI chart in the appendix of this book.

Sex	Age	Height	Weight	Activity Level	RDA Iron (mg)
Male	9	4 feet 2 inches	62 pounds	Low active	
Female (Not Pregnant or Lactating)	9	4 feet	62 pounds	Low active	
Male	29	5 feet 10 inches	165 pounds	Low active	
Female (Not Pregnant or Lactating)	29	5 feet 7 inches	135 pounds	Low active	
Female (Pregnant, 25 Weeks)	29	5 feet 7 inches	150 pounds	Low active	
Female (Lactating)	29	5 feet 7 inches	145 pounds	Low active	
Male	59	5 feet 10 inches	170 pounds	Low active	
Female (Not Pregnant or Lactating)	59	5 feet 7 inches	140 pounds	Low active	

After completing the chart with the RDA for iron for each of the individuals listed address the following questions.

1. Do the requirements for iron differ for boys and girls at age 9?

   YES NO

2. Do the requirements for iron differ for men and women at age 29?

   YES NO

   a. What factors might influence the difference in iron needs?

   b. How might iron requirements change if either the man or woman followed a vegan diet?

3. How do iron requirements for women aged 19 to 50 years change with pregnancy and lactation?
4. Why do you think the RDA for a woman at age 59 is actually lower than for a man of the same age?
5. Using the sample day's intake below:

   a. Identify two "good" sources of heme and non-heme iron.

   1. 2.

   1. 2.

   b. Underline the dietary components that might enhance the iron absorption and *circle* those that might inhibit iron absorption or bioavailability.

Sample Day's Intake

BREAKFAST
1 cup whole-grain fortified breakfast cereal
1 cup low-fat milk
2 tablespoons raisins
1 cup coffee

LUNCH
2 ounces sliced turkey breast
1 ounce Swiss cheese
2 slices whole wheat bread
1 tablespoon mayonnaise
2 sliced tomatoes
½ cup baby carrots

SNACK
2 tablespoons peanut butter
4 graham cracker squares
1 apple
1 cup hot tea

DINNER
3 ounces lean beef steak, broiled
1 medium potato, baked
1 tablespoon sour cream
2 cups raw spinach with 2 tablespoons slivered almonds and ½ cup fresh orange segments
2 tablespoons salad dressing
1 ounce French bread

KEY IDEAS

Trace minerals are essential nutrients that are required in very small amounts in human nutrition. These include chromium, copper, fluoride, iodine, iron, manganese, molybdenum, selenium, and zinc.

Trace minerals have vital roles in the body such as acting as co-factors in numerous chemical reactions.

Trace minerals are found in both plant and animal foods, but their actual bioavailability is influenced by many factors.

Trace mineral deficiencies can have varied, yet serious, consequences, particularly of concern in pregnant women and growing children.

Chromium may enhance the action of insulin in maintaining blood glucose levels.

Copper is a component of certain enzymes and has critical roles in the body including energy metabolism, iron transport and storage, and antioxidant function.

Fluoride is important for hardening tooth enamel and the prevention of dental caries.

Iodine is a component of thyroid hormones that helps regulate energy metabolism.

Iron is a component of many enzymes, as well as proteins, that transport oxygen in the body.

Heme iron is the predominant form of iron in the body and is found only in animal foods, while non-heme iron, found in both plant and animal foods, makes up the majority of iron we consume. Non-heme iron is not as well absorbed as heme iron.

Iron deficiency is the most prevalent nutritional disorder in the United States and globally. It is caused not only by insufficient intake, but also by blood loss or impaired absorption.

Selenium is a component of certain human proteins and functions as a co-factor in certain enzyme systems.

Zinc is present in many enzyme systems and regulatory proteins making it essential for all body processes.

Trace minerals are found across all food groups, thus sufficient intake can generally be achieved by consuming a varied and balanced diet.

Ultratrace minerals are not currently classified as essential nutrients, but may have important roles in the body. They include arsenic, boron, nickel, silicon, and vanadium.

NEED TO KNOW

Review Questions

1. Trace minerals differ from major minerals in all of the following ways, EXCEPT:
   a. their daily requirement is less than 100 milligrams.
   b. their safe range of intake is more narrow.
   c. they are found in smaller amounts in the body.
   d. they are less essential in human health.
   e. they include iodine and iron.

2. All of the following are general properties of trace minerals, EXCEPT:
   a. they are found only in foods of animal origin.
   b. many are co-factors for enzymes.
   c. very little digestion is required.
   d. they are absorbed primarily in the small intestine.
   e. they circulate freely in the blood.

3. The majority of iodine intake in the United States comes from:
   a. fruits and vegetables.
   b. iodized salt.
   c. milk and processed grains.
   d. nuts and vegetable oils.

4. Which trace mineral is an essential component of thyroid hormones?
   a. chromium
   b. iodine
   c. iron
   d. selenium
   e. zinc

5. According to the World Health Organization, the number one nutritional disorder worldwide is:
   a. beriberi.
   b. goiter.
   c. iron deficiency anemia.
   d. pellagra.
   e. zinc deficiency.

6. In contrast to heme iron, non-heme iron:
   a. is found only in animal foods.
   b. makes up the majority of iron we consume.
   c. is better absorbed by the body.
   d. is less influenced by dietary components that inhibit absorption.
   e. is the most prominent form of iron in the body.

7. Of the trace minerals, only _____ cannot be excreted in the urine or bile.
   a. chromium
   b. copper
   c. iodine
   d. iron
   e. zinc

8. Which trace mineral has been demonstrated to play a role in slowing progression of age-related macular degeneration, which can lead to loss of vision?
   a. chromium
   b. copper
   c. manganese
   d. selenium
   e. zinc

9. Which trace mineral is required for the function of perhaps more proteins in the body than any other mineral?
   a. copper
   b. chromium
   c. iron
   d. selenium
   e. zinc

10. Which of the following is NOT an ultratrace mineral?
   a. boron
   b. manganese
   c. nickel
   d. silicon
   e. vanadium

Dietary Analysis Using SuperTracker

Discovering the trace minerals in your meals

1. Log on to the United States Department of Agriculture (USDA) website at www.supertracker.usda.gov. If you have not done so already, you will need to create a profile to get a personalized diet plan. This profile will allow you to save your information and diet intake for future reference. *Do not use the general plan.*
2. Click the Track Food and Activity option.
3. Record your food and beverage intake for one day that most reflects your typical eating patterns. Enter each food and beverage you consumed into the food tracker. Note that there may not always be an exact match to the food or beverage that you consumed, so select the best match available.
4. Once you have entered all of your food and beverage choices into the food tracker, on the right side of the page under the bar graph, you will see Related Links: View by Meal and Nutrient Intake Report. Print these reports and use them to answer the following questions:

a. Did you meet the target recommendations for the trace minerals (iron, iodine, chromium, fluoride, copper, zinc, selenium, molybdenum, and manganese) for the day you selected? If not, what minerals fall below the target numbers?

b. For each trace mineral target that you missed, list two specific foods you could consume to increase your intake of that mineral.

c. Were you above the targets for any of the trace minerals? If so, which ones? Are any of these numbers above the toxicity level for that mineral?

d. Iron deficiency anemia is the number one nutritional deficiency in the United States. What groups are most at risk for this deficiency? If you developed iron deficiency anemia, what physical symptoms would you be likely to experience?

15 ENERGY BALANCE AND OBESITY
The Sitting Disease
UNDERSTANDING THE CAUSES AND CONSEQUENCES OF OBESITY

Alija/Getty Images

LEARNING OBJECTIVES

- List the health consequences associated with obesity and describe how excess body fat increases the risk of developing these conditions (Infographics 15.1 and 15.3)
- Describe the concept of energy balance (Infographics 15.2 and 15.4)
- Identify and describe the components of total energy expenditure (Infographic 15.5)
- Describe factors that affect basal metabolic rate (Infographic 15.6)
- Describe activities that are examples of nonexercise activity thermogenesis (NEAT) (Infographic 15.7)
- Describe factors that contribute to the development of obesity (Infographic 15.8)
- Use body mass index and waist circumference to evaluate someone's risk of chronic disease (Infographics 15.9–15.11)
- Describe methods for determining body composition (Infographic 15.12)
- Discuss strategies for successful weight loss and maintenance (Infographics 15.13 and 15.15)

When James Levine was 11 years old, he began a science experiment in his bedroom. His test subjects? Snails.

In glass-walled fish tanks he built himself, Levine collected pond snails from nearby Regents Park in his native London. He then methodically monitored and recorded their movements. "My idea was that every snail has a built-in hard-wired style of movement," explains Levine. "One snail will do swirly-whirly-whirly, while another snail will always move in a straight line."

Dr. James Levine and his treadmill station. *Levine, a professor of medicine at the Mayo Clinic and Arizona State University, is an expert on the physiology of weight gain and loss.*

AP Photo/Jim Mone

To test this hypothesis, every night, between the hours of 9 PM and 5 AM, he'd wake up hourly with an alarm clock and mark on the glass where the snail had moved. At the end of the night, he'd trace each snail's journey.

"Not surprisingly, I was constantly asleep at school," Levine confesses.

By the time he finished his experiment, Levine had 270 snail tracings. What he discovered was that snails didn't quite sort the way he thought. But they did have stereotypical styles of movement. "Joanna [he gave each snail a name] always does ziggidy-zaggidy-ziggidy-zaggidy. And John always moves in a smooth way."

The snail experiment had a lasting impact on Levine, who eventually went on to medical and graduate schools. Now, 35 years later, Levine has focused his love of experimentation on another slow-moving creature: the human couch potato.

OBESITY
excess amount of body fat that adversely affects health

OVERWEIGHT
a moderate amount of excess body fat or an excess amount of body weight from muscles, bone, fat, and water

Levine, a professor of medicine at the Mayo Clinic and Arizona State University, is an expert on the physiology of weight gain and loss, with a special focus on **obesity**. Obesity—having excess body fat—has been called American's number one health crisis. Obesity has been linked to a whole host of health problems, including heart disease, diabetes, stroke, Alzheimer disease, hypertension, and even certain cancers. It is also a major killer: In the United States, only tobacco use causes more premature deaths per year. **(INFOGRAPHIC 15.1)**

Alarmingly, rates of obesity have skyrocketed over the past four decades, leading many to refer to an obesity "epidemic." As of 2014, more than one-third of U.S. adults were obese, and more than two-thirds were **overweight** or obese, and the prevalence of obesity in children is increasing at a similar rate. The United States is not alone: Global obesity rates have almost doubled since 1980 and the World Health Organization estimates that more than 10% of the world adult population is obese.

Obesity is a complex disease that is influenced by multiple factors; genetics, environment, and behavior are the main causes. Genetics is seen as the primary factor determining an individual's susceptibility to obesity. However, the rapid rise in the prevalence of obesity in the last few decades is largely attributed to changing environmental factors, because human genes have essentially remained the same over this time. For this reason it is said that, "Genetics loads the gun, and environment pulls the trigger," because the environments that we live and work in can strongly affect our behaviors. For example, the *built environment*—our surroundings that are designed by humans—can strongly affect how likely we are to engage in physical activity. Also, the social and physical factors that influence the foods we eat (the *food environment*) can make it challenging for us to make healthy food choices. Consequently, the dramatic rise in obesity is believed to result from environmental changes that promote low levels of physical activity and the consumption of energy-dense foods that are particularly likely to promote weight gain in genetically susceptible individuals.

Although almost everyone agrees that obesity seriously compromises our health

INFOGRAPHIC 15.1 Prevalence and Health Consequences of Obesity in the United States

TRENDS IN THE PREVALENCE OF OBESITY AND OVERWEIGHT IN THE UNITED STATES

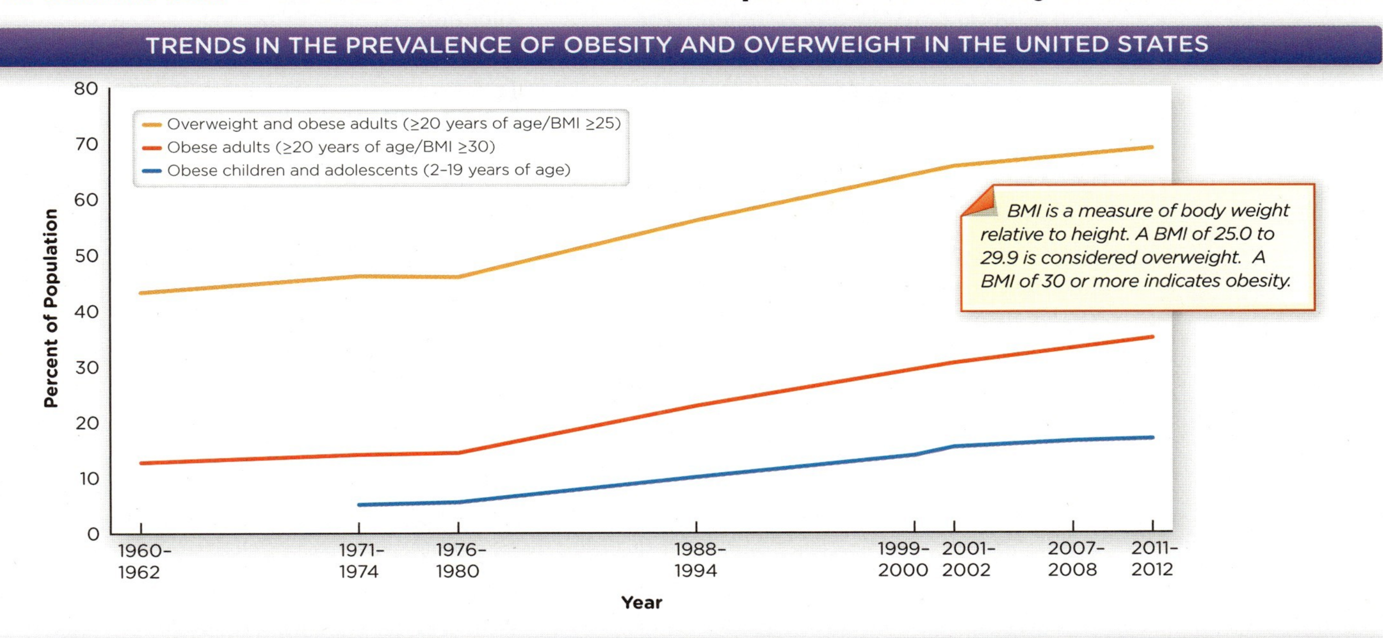

HEALTH CONSEQUENCES OF OBESITY

Alzheimer disease, stroke

Heart disease, hypertension

Fatty liver disease, gallstones

Type 2 diabetes

Arthritis

Gout

Sleep apnea

Gastroesophageal reflux

Increased risk of some cancers (particularly colon, esophagus, uterus, breast, pancreas, kidney, gallbladder)

Infertility in women, increased risk of miscarriage and birth defects

Erectile dysfunction in men

Obesity increases the risk of a number of chronic diseases, and ultimately increases the risk of premature death. One study found that the risk of death was increased two- to threefold in healthy, middle-aged individuals who have never smoked.

http://www.nejm.org/doi/full/10.1056/NEJMoa055643#t=articleResults

? *Looking at the graph, how would you describe the change in the number of overweight (nonobese) adults?*

and longevity, scientists are divided about just what is causing our waistlines to expand at an ever-alarming pace. Is it increased food intake—linked perhaps to larger portion sizes in restaurants and in supermarkets? Or is it decreased energy expenditure—a result of our increasingly sedentary lifestyle? Arguments have been made on both sides of the coin, but no strong consensus has emerged. Now, with the help of some clever experiments using sophisticated undergarments, Levine thinks he has found the answer.

ENERGY IN, ENERGY OUT

To understand our obesity crisis, says Levine, you have to realize that it didn't happen overnight. "Obesity doesn't occur over minutes and hours," says Levine. "Obesity occurs over years, decades, and generations."

What that means is that even small changes in the way we eat or the way we move can seriously add up. Even something as simple as an extra spoonful or two at each meal or the amount of time we spend sitting or standing—if multiplied consistently across time—can profoundly affect how much energy we store or expend.

Energy, defined as the capacity to do work, is required to perform all the various functions that are necessary to sustain life, from breathing to moving to digesting food to maintaining a constant body temperature. Humans and other animals obtain energy through the breakdown of carbohydrates, fats, proteins, and alcohol in food and beverages. The energy contained in the chemical bonds of these molecules is released by the chemical reactions of metabolism and captured in a form that can be used to do the body's work. (To learn more about the cell's energy supply, see Chapter 16.)

One way scientists measure energy is in units called calories. A **calorie** is defined as the energy required to raise 1 g of water 1°C. The energy in food is commonly measured in units of kilocalories (1,000 calories). A kilocalorie (kcal) is the energy required to raise 1 kg of water 1°C. All food labels in the United States report the energy in foods in kilocalories, although these are usually just written as "calories."

Obesity results from a chronic imbalance of energy intake and expenditure. According to the laws of thermodynamics, energy is neither created nor destroyed but merely changes form. This principle, known as the conservation of energy, means that when we consume more energy than we expend, the excess has to go *somewhere*. Most often, it goes to our hips, thighs, and bellies as fat.

To be in energy balance means that the amount of energy we take in ("energy in") equals the amount of energy we use ("energy out"). When this occurs, our body weight is stable. Any increase in our body weight indicates that "energy in" is greater than "energy out." Fundamentally, the only way to gain weight is to consume calories in excess of what is expended (**positive energy balance**), and the only way to lose weight is to consume fewer calories than expended (**negative energy balance**). Obesity results from chronic positive energy balance. (INFOGRAPHIC 15.2)

This discussion reveals something unique about energy nutrients compared with other nutrients: body weight provides us with an easily monitored indicator of adequacy, excess, or insufficiency.

Just why is having excess body fat bad for you? Scientists used to explain the negative consequences of obesity in terms of the added strain put on a person's heart, lungs, and other organs. But that's no longer thought to be the primary problem. Although it is true that some problems, such as sleep apnea, can result from being physically large, most of the negative health consequences of obesity are believed to stem from the biochemical effects of fat in our bodies. Fat is not just a blob of inert yellow goop hanging out in our bodies; rather, it's dynamic tissue with an active life all its own. It secretes an abundance of different hormones that exert effects on other tissues in the body. These hormones cause a low-grade chronic inflammatory state

ENERGY
the capacity to do work

CALORIE
a unit of measure defined as the energy required to raise 1 g of water 1°C

POSITIVE ENERGY BALANCE
the state when energy intake exceeds energy expenditure that leads to weight gain

NEGATIVE ENERGY BALANCE
the state when energy intake is less than energy expenditure, resulting in weight loss

INFOGRAPHIC 15.2 Energy Balance

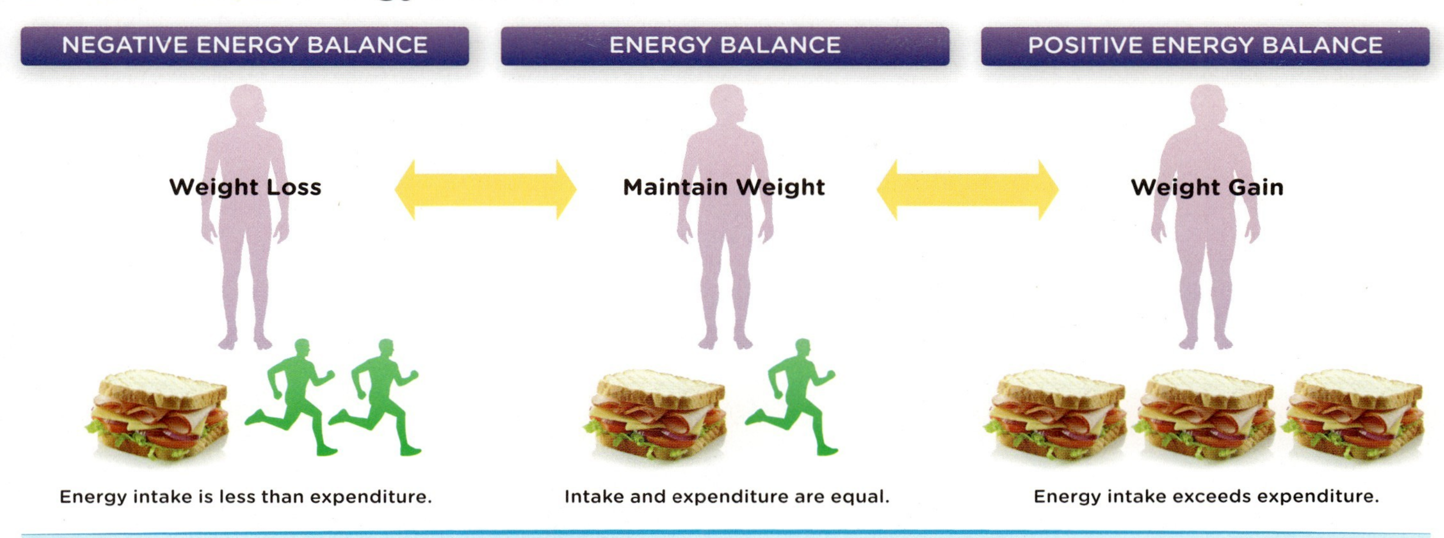

? *What are two strategies to avoid or reverse weight gain?*

Photo credit: baibaz/Shutterstock

throughout the body that is believed to be the primary means by which obesity causes health problems and contributes to the development of many chronic diseases. (INFOGRAPHIC 15.3)

THE BIOLOGY OF HUNGER

Given energy's central role in sustaining life, it is not surprising that obtaining and storing energy has a complex physiological regulation. Our bodies will let us know when we are hungry, and when we have eaten enough (if we pay attention). Through this complex physiological control system, which involves a constant dialogue between our brains and our gastrointestinal tract, we are able to obtain and maintain sufficient energy stores to power our activities.

There are two different ways of regulating energy balance and food intake, a short-term system and a long-term system. The short-term system is mediated by hormones and stomach pressure and is responsible for triggering hunger and satiety before and after individual meals. The long-term system, mediated by a different set of hormones, adjusts food intake and energy expenditure to maintain adequate fat stores.

Hormones participate in energy balance

When we haven't eaten for a while, our stomach begins to grumble. That grumbling is a sign that a hormone called **ghrelin** is racing into action. Ghrelin, nicknamed the "hunger hormone," is a 28-amino acid peptide hormone that is produced primarily in the stomach. It is the only hormone that has been found to increase hunger. Circulating ghrelin levels in the blood surge just before meals and decrease after eating. Ghrelin stimulates hunger by activating specific neurons in the brain. Ghrelin secretion decreases only once nutrients from the meal are absorbed into the blood. Its secretion is most effectively inhibited by carbohydrates, and then by proteins; fat is least effective.

Satiation is the process that leads to the termination of a meal and refers to the sense of fullness that we feel while eating. **Satiety** is the effect that the meal has on our interest in food after a meal; it operates in the interval between meals and affects when we feel hungry again. The primary factors affecting these two processes are gastric distention—how much our stomach has expanded to let food in—and the release of hormones produced by specialized cells in the gastrointestinal tract. Nerves in the stomach sense its expansion

GHRELIN *a hormone that stimulates hunger before meals*

SATIATION *the sense of fullness while eating that leads to the termination of a meal*

SATIETY *the effect that a meal has on our interest in food after and between meals and when we feel hungry again*

INFOGRAPHIC 15.3 Obesity, Low-Grade Inflammation, and Chronic Disease

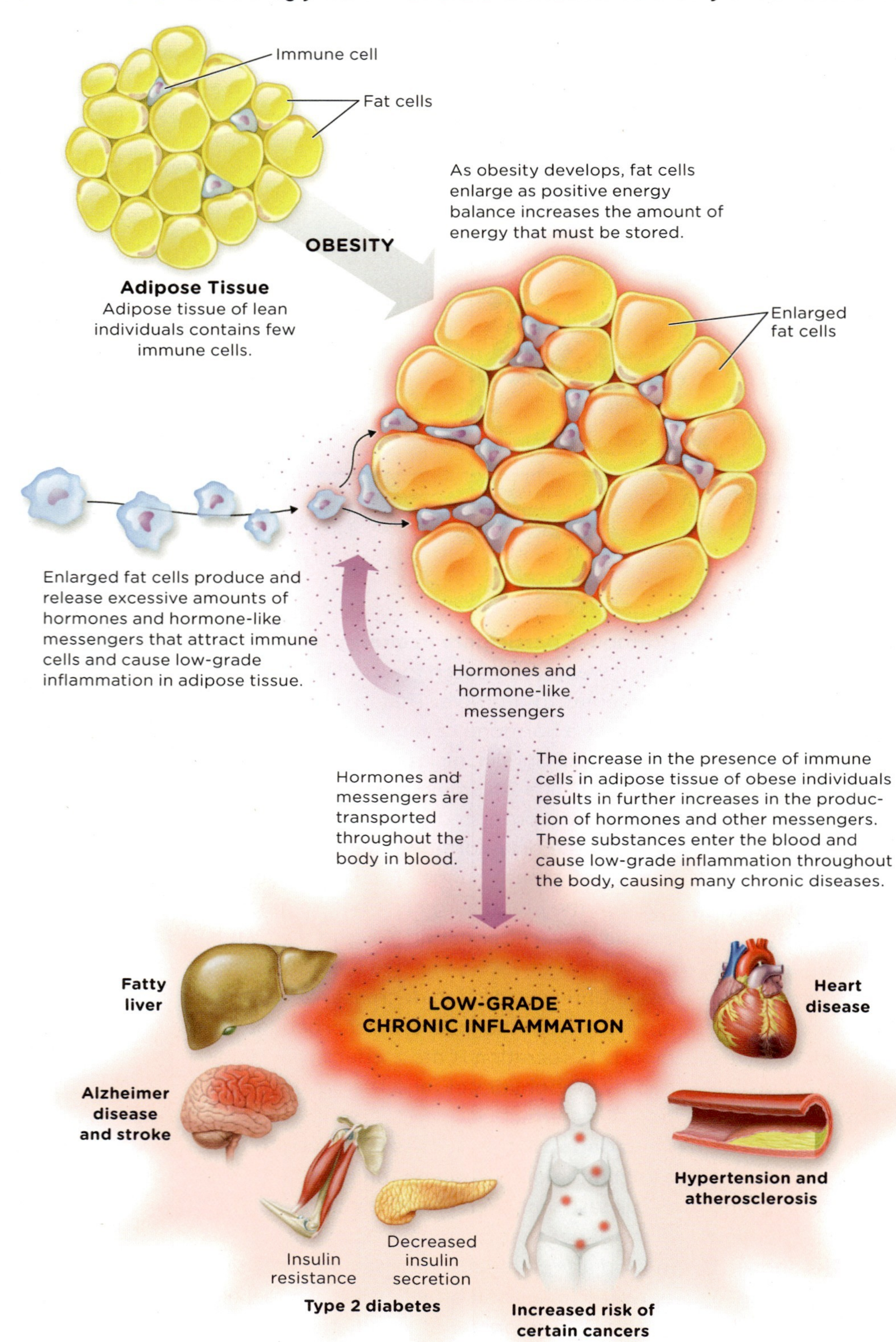

? *What directly stimulates the development of the low-grade inflammation that is responsible for many of the chronic diseases that are associated with obesity?*

and relay signals to the brain to communicate a sense of fullness. At the same time, several gut peptide hormones are produced by specialized cells in the small intestine in response to the detection of nutrients in the gut.

Importantly, calories in beverages appear to bypass mechanisms of satiation. A number of studies have demonstrated that soft-drink calories are less satiating than calories from solid foods.

Over the long term, energy balance is affected by a hormone called **leptin**. Leptin is produced primarily by adipose (fat) tissue. Its circulating concentration is closely associated with total body fat. When fat stores increase, more leptin is produced. The leptin level increase in the blood acts on the brain to suppress hunger, and increases energy expenditure to avoid excess weight gain. **(INFOGRAPHIC 15.4)**

If leptin suppresses hunger, and fat people have more leptin, then why do people ever become obese? It turns out that obese individuals have higher levels of circulating leptin than lean individuals, but they seem to be resistant to the effects of leptin. Thus, they don't experience the same suppressed hunger or increased expenditure of energy. Some evidence suggests that a high-fructose diet may contribute to leptin resistance in humans.

Hunger and appetite

In addition to the internal cues that regulate energy balance, there are also many environmental cues that influence food intake and energy expenditure. It is critical to differentiate between **hunger** and **appetite**. Hunger is the biological impulse that drives us to seek out food and consume it to meet our energy needs. Appetite, however, is often viewed as the liking and wanting of food for reasons other than, or in addition to, hunger. Appetite is often a product of sensory stimuli (the sight or smell of appealing foods) and the perceived pleasure we will experience as we satisfy our appetite and eat the desired food.

Most of us live in an environment in which we are surrounded by a plethora of palatable foods that are readily accessible. Hunger is certainly not the only reason (and perhaps not the dominant one) that motivates us to eat on many occasions. (See Chapter 19 to learn more about determinants of eating behavior.) We often eat because it is pleasurable to do so. Furthermore, food

LEPTIN
a hormone produced by adipose tissue that plays a role in body fat regulation and long-term energy balance

HUNGER
the biological impulse that drives us to seek out and consume food to meet our energy needs

APPETITE
a desire for food for reasons other than, or in addition to, hunger

INFOGRAPHIC **15.4** **Hormones Help Regulate Energy Balance and Food Intake**

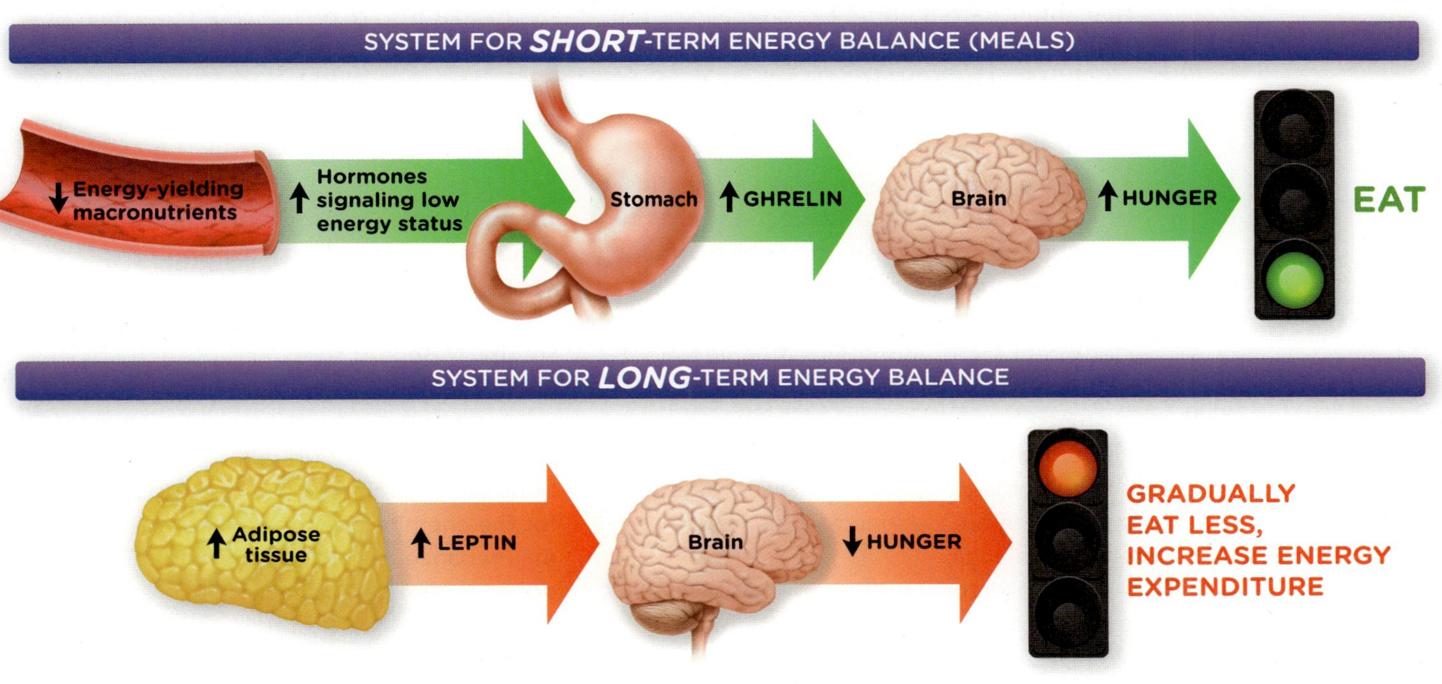

Cheap, energy-dense foods. *The easy accessibility of highly palatable, energy-dense foods contributes to the obesity problem in the United States.*

Tim Boyle/Getty Images

manufacturers spend billions of dollars in research designing foods that will trigger our "bliss point"—with just the right balance of salt, fat, and sugar that we find nearly irresistible. Similarly, large sums of money are spent marketing these products to us. We are often surrounded by cues (such as advertisements) that stimulate our appetite and make us not only want to eat, but to eat specific foods—and more of them as portion sizes increase. In the end, we often eat when we are not hungry, causing us to consume more calories than we expend, and therefore, to gain weight. This is often referred to as the "toxic food environment" because the easy accessibility of highly palatable, energy-dense foods, and the resulting overconsumption of those foods, is a significant factor in the obesity epidemic.

We also eat for many other reasons, such as in response to stress, boredom, and difficult emotions. All of these reasons for eating can, to some degree, override the satiety signals that are designed to keep us in energy balance.

So is overconsumption of food the cause of our obesity epidemic? Are we simply being goaded into eating more food than we used to, and paying the price in wider waistlines? Levine and his colleagues think that there is more to this story than just an increase in energy intake.

TOTAL ENERGY EXPENDITURE (TEE) *the total amount of energy expended through basal metabolism, thermic effect of food, and activity energy expenditure*

BASAL METABOLISM *the energy expenditure required to maintain the essential functions that sustain life*

A NEAT CAUSE OF WEIGHT GAIN

We all know people who seem to be able to eat anything they wish and never gain weight. Likewise, there are those who "merely have to look at food" to put on pounds. It turns out there is scientific support for these subjective observations. Studies that have looked at weight gain in response to overfeeding have shown that people vary greatly in how much fat they accumulate. The biological mechanism that allows some individuals to resist weight gain more than others, however, has not been identified.

In 1999, Levine set out to solve the mystery. He and a team of researchers at the Mayo Clinic in Rochester, Minnesota, recruited 16 nonobese adults (12 men and 4 women from 25 to 36 years of age). These individuals underwent measures of body composition and energy expenditure before and after eight weeks of supervised overfeeding by 1,000 kcal per day.

The researchers found that fat gain varied tenfold among individuals in the study, ranging from a gain of only 0.36 kg to a gain of 4.23 kg (1 kg = 2.2 lb). Why was the weight gain so disparate? One obvious possibility is each person expended a different amount of energy.

Understanding energy expenditure

A person's **total energy expenditure (TEE)** can be divided into three main components: (1) basal metabolism, (2) the thermic effect of food, and (3) activity energy expenditure.

Basal metabolism is the energy expenditure required to maintain the essential functions that sustain life. This energy is required for the chemical reactions in our cells, the maintenance of muscle tone, and the work done by our heart, lungs, brain, liver, and kidneys—with much of this work depending on the active transport of electrolytes and other nutrients in our cells. In research

INFOGRAPHIC 15.5 Components of Total Energy Expenditure in a Typical Sedentary Individual

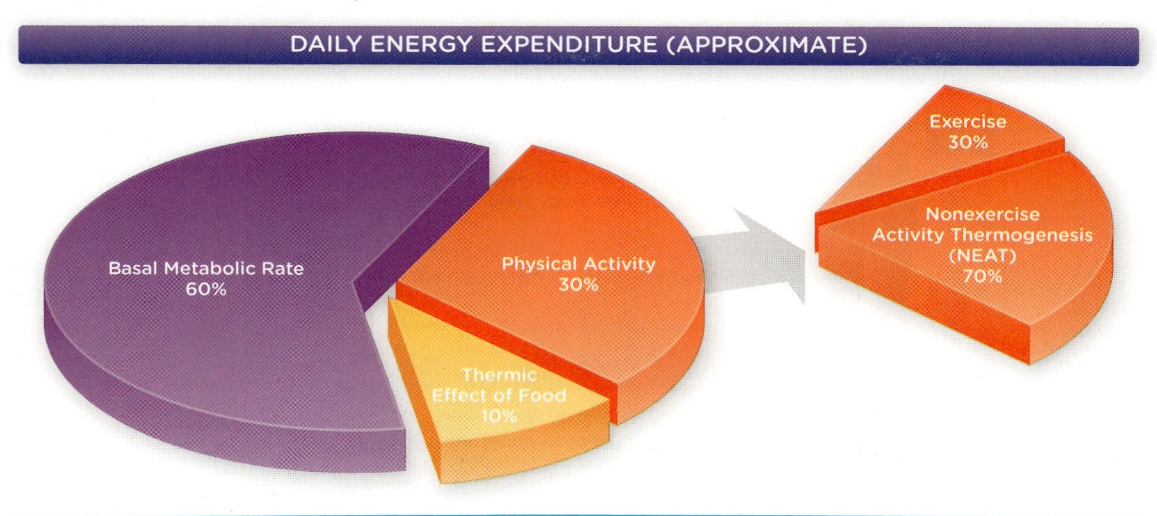

If you increase your energy expenditure by 500 kcal per day, how many more calories would you need to consume to stay in energy balance?

studies, basal metabolism is measured while the person is lying completely still but awake, at a comfortable temperature, following an overnight fast, and without any physical activity for the preceding eight hours. For most individuals, basal metabolism is the largest component of total daily energy expenditure, accounting for about 60% of TEE in a typical sedentary individual.

Thermic effect of food (TEF) is the energy needed to digest, absorb, and metabolize nutrients in our food. TEF is generally equivalent to 10% of the energy content of the food ingested and does not vary greatly between people. **Activity energy expenditure (AEE)** is the amount of energy individuals expend in physical activity per day, both planned and spontaneous. It includes all the energy expended in the contraction of skeletal muscles to move our body and to maintain posture (sitting or standing versus lying down). AEE is the most variable component of TEE. (INFOGRAPHIC 15.5)

Fat-free mass and basal metabolic rate

Though people can and do vary in the amount of energy expended by their basal metabolism, having a "low metabolism" is not a widespread cause of obesity. Nearly all of the variation in basal metabolism from person to person is accounted for by individual differences in fat-free mass (FFM), which is total body mass minus the fat mass (adipose tissue). The greater a person's FFM, the higher his or her **basal metabolic rate (BMR)** will be. Although increasing fat mass will also increase BMR to some degree, adipose tissue is far less metabolically active than other tissues. Our organs, such as the brain, kidney, heart, and liver, have the highest metabolic activity, but even at rest, skeletal muscle is about three times more metabolically active than adipose tissue. What this means is that a lean individual expends more energy at rest than someone of the same weight who has more body fat. Proportionally higher fat mass and lower FFM in women nearly completely accounts for the lower BMR in women than in men. People also tend to have higher fat mass as they get older, which accounts for the decline in BMR that generally occurs with age. Some external factors, such as how much caffeine we consume and whether we smoke cigarettes can also affect BMR. (INFOGRAPHIC 15.6)

THERMIC EFFECT OF FOOD (TEF) *the energy needed to digest, absorb, and metabolize nutrients in food*

ACTIVITY ENERGY EXPENDITURE (AEE) *the amount of energy expended in physical activity per day*

BASAL METABOLIC RATE (BMR) *the amount of energy expended in basal metabolism over a fixed period, typically expressed as kcal per day*

INFOGRAPHIC 15.6 Factors that Affect Basal Metabolic Rate

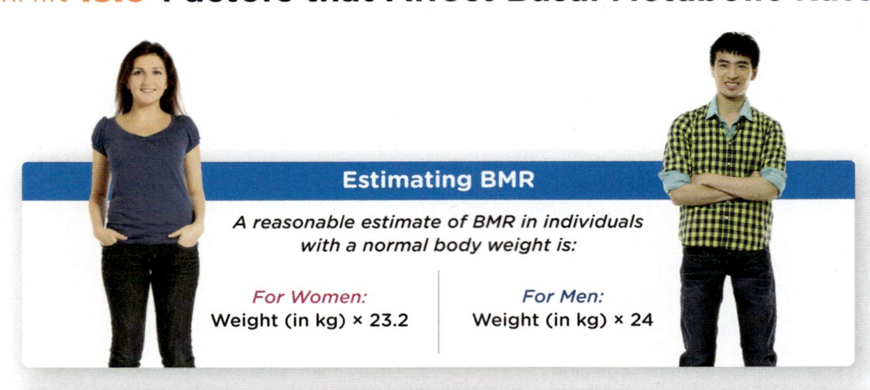

Factor	Effect on BMR
Fat-Free Mass	Fat-free mass (FFM) is body weight minus the weight of adipose tissue. FFM is the single most important factor affecting BMR because adipose tissue is metabolically much less active than most other tissues. Our organs have the highest metabolic activity while we are at rest, with the liver, brain, heart, and kidneys accounting for about two-thirds of our BMR. Though at-rest skeletal muscle is far less metabolically active than our major organs, it is still about three times more so than adipose tissue.
Age	BMR decreases by about 1%–2% per decade after young adulthood. The vast majority of this decrease is due to the decreased mass of both our organs and skeletal muscle.
Sex	A woman will generally have a lower BMR than a man with the same body weight because FFM generally makes up a lower percent of a woman's total body weight than a man's. Women also tend to be smaller than men.
Growth	The energy demands of growth increase BMR from infancy through adolescence, but only during the first 6 months of life do the energy demands of growth cause a significant increase in BMR.
Hormones	Elevated levels of thyroid hormone and epinephrine (released in response to stress) increase BMR.
Starvation	Fasting or low-calorie diets decrease BMR, to conserve energy when it is scarce.
Illness	A fever, burns, and trauma will increase BMR.
Pregnancy and Lactation	During pregnancy, BMR increases because of the increased work required to support the maternal circulation, respiration, and kidney function, as well as to support the increased tissue mass. During lactation, BMR increases to meet the energy demands of milk production.
Ethnicity	The BMR of Whites may be 5%–10% higher than that of other ethnic groups.
Caffeine	Caffeine consumption increases BMR. The amount of caffeine equal to that of a typical soft drink has been shown to raise BMR by about 5%, while an amount approximately equivalent to that in three to four 8-ounce cups of coffee has been shown to raise BMR by about 10%. Those who regularly consume high amounts of caffeine may develop tolerance to its effects.
Smoking	Nicotine tends to increase BMR.
Sleep	Sleeping metabolism is about 10% less than basal metabolic rate.

What two factors affect BMR primarily because they strongly influence a person's FFM?

Photo credits (left to right): kaarsten/Shutterstock, East/Shutterstock

In Levine's study, he found only very minor increases in BMR and TEF in his study participants, which was not enough to account for the tenfold variance in fat gain among them. By contrast, levels of physical activity varied markedly between study participants. Interestingly, intentional exercise was not the crucial difference in activity level.

Levine designed his study in such a way that exercise was kept at a constant and minimum level across all study participants. Therefore, the differences in physical activity were principally non-exercise-related. Levine has a name for this type of activity: he calls it **NEAT**, short for nonexercise activity thermogenesis.

NEAT includes all the activities of daily living, such as household chores, yard work, shopping, occupational activity, walking the dog, and playing a musical instrument. It also includes the energy expended to maintain posture and spontaneous movements such as fidgeting, pacing, and even chewing gum. In other words, NEAT encompasses all the activities we do as part of daily living, separate from planned, intentional exercise like going for a run, taking an aerobics class, or hopping on an exercise bike.

In Levine's study, NEAT varied greatly among individuals (fewer than 100 kcal to more than 700 kcal per day). More important, changes in NEAT were inversely correlated with changes in weight gain. In other words, NEAT proved to be the principal mediator of resistance to fat gain in these individuals. (INFOGRAPHIC 15.7)

Next, Levine wanted to know if NEAT plays a role in obesity. Do lean and obese people differ in their levels of NEAT, for example? To get at this question, Levine and his colleagues recruited 20 healthy volunteers who were self-proclaimed "couch potatoes." Ten participants (five women and five men) were lean and 10 participants (also five women and five men) were mildly obese. The volunteers agreed to have all of their movements measured for 10 days. They were instructed to continue their normal daily activities and not to adopt new exercise regimens.

To measure NEAT, Levine and colleagues devised a novel way to track activity levels in test volunteers. They built a special kind of underwear outfitted with electronic sensors that detect movement. The undergarments were built to allow

NEAT (NONEXERCISE ACTIVITY THERMOGENESIS)
the energy expended for everything we do that is not sleeping, eating, or sportslike exercise

Yard work counts. *NEAT includes all the activities of daily living.*

Jodi Jacobson/Getty Images

INFOGRAPHIC 15.7 NEAT and Its Impact on the Risk of Obesity

Total daily energy expenditure can be increased significantly when we minimize the time spent quietly sitting and find ways to keep moving, which decreases our risk of weight gain.

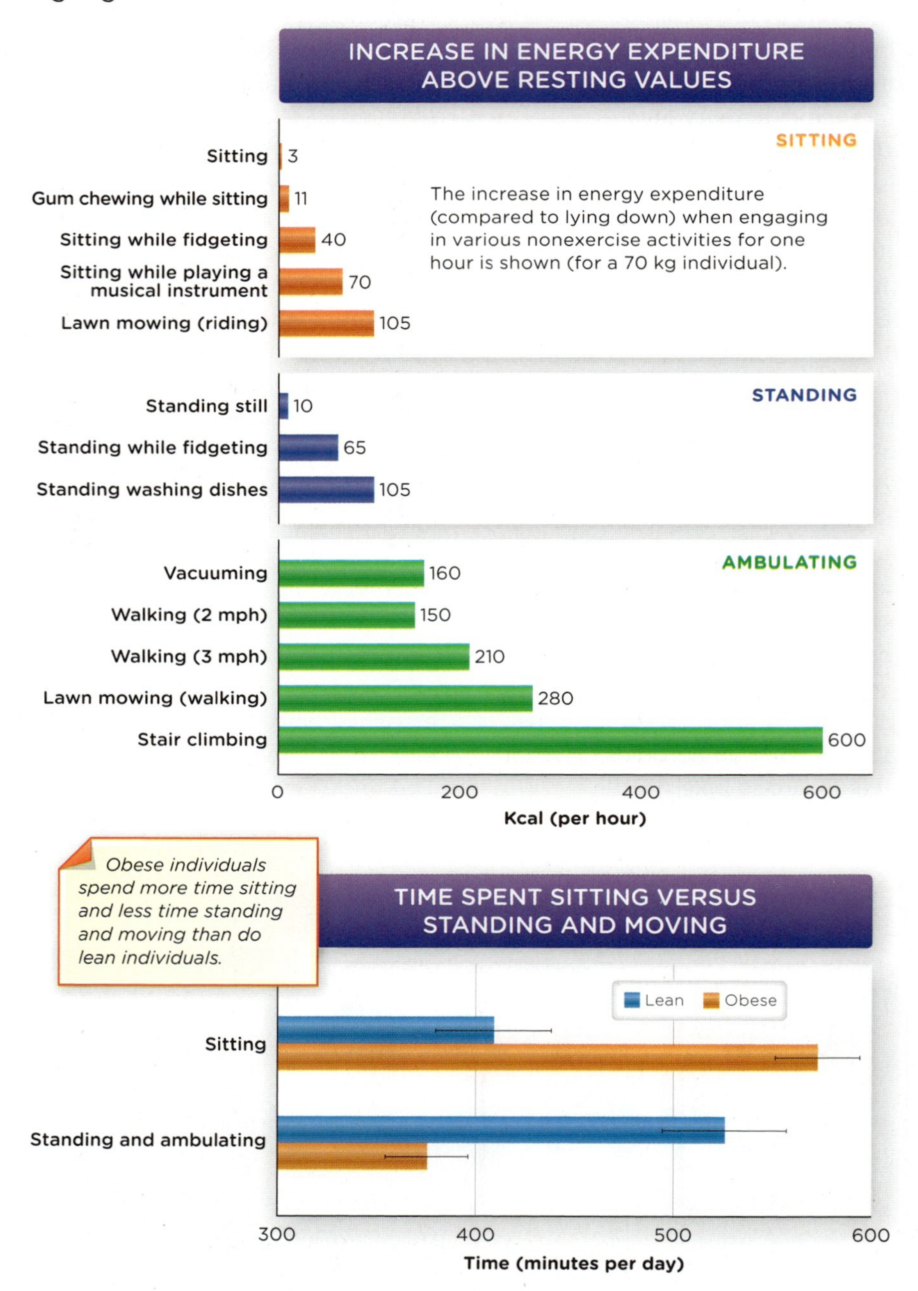

Describe three or four changes that you can make to your daily routine to increase your nonexercise activity thermogenesis.

How many additional calories would you expend by climbing stairs for one minute compared with riding in an elevator for the same time?

people to wear them essentially all the time—even while going to the bathroom and having sex. "We literally have snapshots as to how real people live their lives every half second of every day for days and days and days on end," says Levine.

From those measurements, they can then calculate how many calories a person expends per day. Over the 10-day period, Levine's team collected 25 million data points on NEAT for each individual.

The bottom line? Obese people sit on average 2.25 hours longer than their leaner counterparts. Moreover, this tendency seems to be innate: Lean individuals sit the same amount of time even after they are forced to gain weight, and obese individuals sit the same amount of time even after they are forced to lose it.

Modern conveniences and technologies encourage inactivity.

DCPhoto/Alamy

What does all this have to do with obesity? Levine thinks that it's not only that people are eating more than they did a few decades ago, but also that we have been "seduced" (his word) by our environment into expending less energy by sitting more.

LIFESTYLE AND ENERGY BALANCE

One hundred fifty years ago, 90% of the world's population lived in agricultural regions. Much like our distant evolutionary ancestors, they walked to work, performed manual labor, and walked home at the end of the day. They manually prepared their food and washed their clothes. Physical work—and thus caloric expenditure—was required to get the job of living done.

Today, in developed countries, most people live in cities and work behind a computer. A modern person sits during his drive to work, sits all day at work, sits to drive home, and sits in the evening watching television, surfing the Internet, or playing video games. "In a mere 150 years," says Levine, "*Homo sapiens* has become addicted to the chair."

Notably, in adopting this sedentary lifestyle, Levine estimates that humans have decreased their NEAT by approximately 1,500 kcal per day. At the same time, we have continued to eat similar amounts of—if not more—food. As a result, we are experiencing a population-wide positive energy balance, which Levine believes has led to our current obesity epidemic. **(INFOGRAPHIC 15.8)**

Of course, not everyone who lives in our modern society becomes obese. Why is that? Levine would say that is because some people naturally have higher NEAT than others. These high-NEAT individuals are the toe-tappers and fidgeters among us who just can't seem to sit still. Fortunately, our NEAT quotient is made up of more than our propensity to fidget, and some of it is within our conscious control.

"The brain is just like a muscle," says Levine. "If you exercise a muscle, it gets bigger and stronger, and if you exercise and activate your brain, your brain becomes NEATER."

INFOGRAPHIC 15.8 Factors Contributing to the Development of Obesity

Food Environment
The availability of cheap, palatable, energy-dense foods appears to be a major factor contributing to the rise in obesity in the United States.

Portion Sizes
When we are served larger portion sizes, we tend to eat more.

Fewer Home-Cooked Meals
Those who frequently eat at restaurants or purchase take-out meals are at a higher risk of obesity.

Sleep Deprivation
Sleep deprivation has been shown to increase food consumption.

Pregnancy
Some women have difficulty losing the weight they gained during pregnancy.

Emotions
Boredom, anger, or stress can result in overeating.

Sedentary Lifestyle
Decreased energy output promotes a positive energy balance and weight gain.

Television
Besides contributing significantly to the time we spend sitting each day, television viewing negatively affects our diet, often causing us to eat more at meals, snack more often, and choose less healthy foods.

Communities
The environments we live and work in can promote weight gain by creating barriers to physical activity if there are few sidewalks, trails, parks, or recreation centers, or if we feel unsafe.

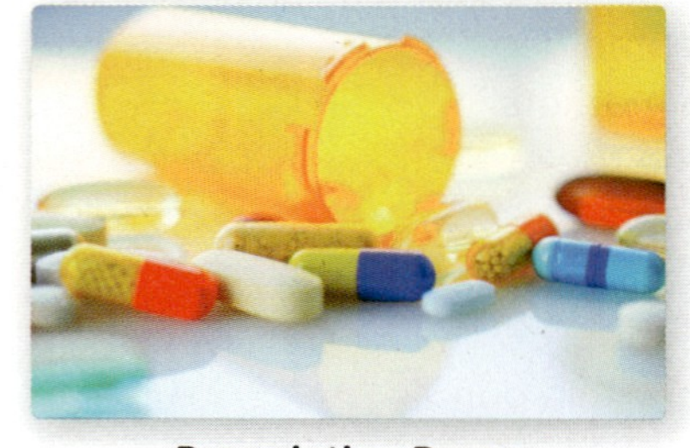

Prescription Drugs
Some prescription drugs can cause weight gain, perhaps by increasing appetite, or decreasing metabolism or physical activity.

Smoking Cessation
Individuals often eat more and gain weight once they have quit smoking.

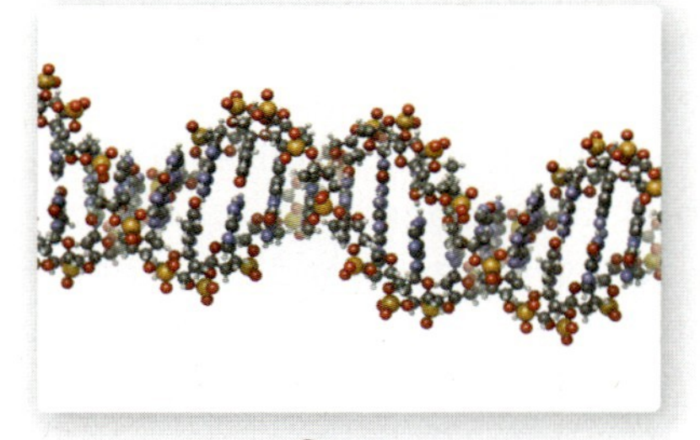

Genes
Our genetic makeup can potentially affect how well we regulate food intake, and how easily we gain weight when we overeat.

What factors might pose significant challenges for you as you seek to avoid weight gain?

Photo credits (top left to bottom right): Roberto Machado Noa/LightRocket via Getty Images, Carl Pendle/Getty Images, Jose Luis Pelaez, Inc./Getty Images, Sacha Bloor/Getty Images, Kohei Hara/Getty Images, Jamie Grill/Getty Images, Image Source/Getty Images, maxriesgo/Shutterstock, South West Images Scotland/Alamy, Tetra Images/Getty Images, Marcela Barsse/Getty Images, molekuul.be/Shutterstock

INFOGRAPHIC 15.9 BMI and Risk of Death among Healthy Women Who Never Smoked *Mortality increases with increasing BMI, but being underweight is also associated with increased mortality. Underweight may reflect a genetic predisposition toward thinness, but can indicate prolonged illness or nutritional inadequacy. Risks may include increased bone loss, compromised immune function, and disrupted hormonal regulation.*

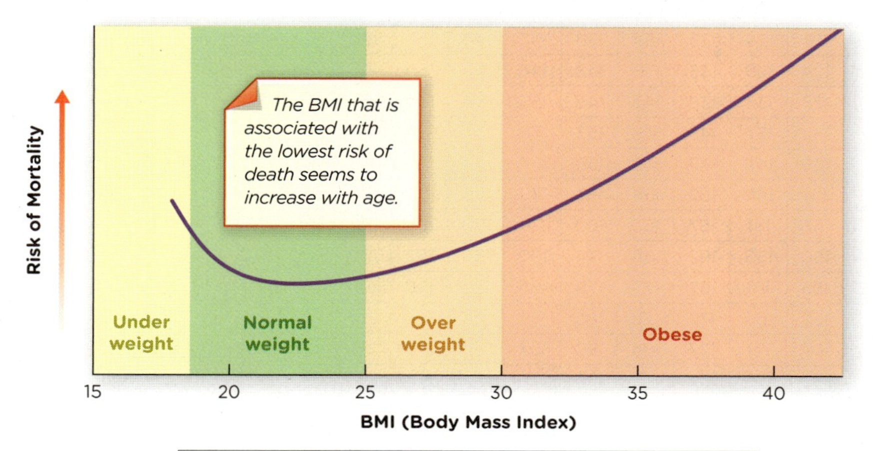

Adapted from: Berrington de Gonzalez A, et al. Body-mass index and mortality among 1.46 million white adults. N Engl J Med 2010;363:2211-9.

What is a healthy body weight?

Although it is difficult to determine an ideal weight or body fat level, we do know that health risks are associated with body weight status. It is possible, for example, for a thin person to be less healthy than an overweight person (especially if the thin person is a smoker). **(INFOGRAPHIC 15.9)** Some research even suggests that being slightly overweight is associated with higher longevity than being underweight. Nevertheless, the link between obesity and health risks is irrefutable, and health professionals need a guide for action.

Body mass index

The most common tool for assessing body fat is the **body mass index (BMI)**. BMI is calculated from a person's weight and height and provides an indirect estimate of body fat. "Underweight," "normal," "overweight," and "obese" are all labels for ranges of weight on the BMI scale. The greater the BMI, the higher the risk of obesity-associated diseases such as coronary heart disease, hypertension, stroke, and type 2 diabetes. **(INFOGRAPHIC 15.10)**

Although easy to use, BMI has limitations. For example, BMI does not distinguish between excess fat, muscle, and bone mass, and may thus overestimate body fat in muscular individuals, such as highly trained athletes who have increased muscle mass.

Waist circumference

Another way to assess whether a person's body fat level carries health risks is to measure waist circumference, which is an indicator of *abdominal obesity*. Waist circumference indicates body fat distribution—in particular, the presence of excess **visceral fat** located in the abdominal area that surrounds the body's internal organs, which has been shown to be an independent health risk. The greater the waist circumference, the greater the risk of cardiovascular disease, type 2 diabetes, and premature all-cause mortality. Even individuals who are of normal weight (BMI less than 25),

BODY MASS INDEX (BMI)
an indirect measure of body fat calculated from a person's weight and height

VISCERAL FAT
fat in the abdominal area that surrounds the body's internal organs; has been shown to be an independent health risk

INFOGRAPHIC 15.10 **Determine Your Body Mass Index** *To find your BMI, use the table or one of the equations provided below.*

	HEALTHY WEIGHT						OVERWEIGHT					OBESE					
BMI	**18.5**	**20**	**21**	**22**	**23**	**24**	**25**	**26**	**27**	**28**	**29**	**30**	**31**	**32**	**33**	**34**	**35**
Height	**Weight (in pounds)**																
4'10"	88	98	100	105	110	155	119	124	129	134	138	143	148	153	158	162	167
4'11"	91	99	104	109	114	119	124	128	133	138	143	148	153	158	163	168	173
5'0"	95	102	107	112	118	123	128	133	138	143	148	153	158	164	169	174	179
5'1"	98	106	111	116	121	127	132	137	143	148	153	158	164	169	174	180	185
5'2"	101	109	115	120	125	131	136	142	147	153	158	164	169	175	180	186	191
5'3"	104	113	118	124	130	135	141	146	152	158	163	169	175	180	186	192	197
5'4"	108	116	122	128	134	140	145	151	157	163	169	174	180	186	192	198	203
5'5"	111	120	126	132	138	144	150	156	162	168	174	180	186	192	198	204	210
5'6"	114	124	130	136	142	148	155	161	167	173	179	185	192	198	204	210	216
5'7"	118	127	134	140	147	153	159	166	172	178	185	191	198	204	210	217	223
5'8"	121	131	138	144	151	158	164	171	177	184	190	197	203	210	217	223	230
5'9"	125	135	142	149	155	162	169	176	182	189	196	203	209	216	223	230	237
5'10"	129	139	146	153	160	167	174	181	188	195	202	209	216	223	230	236	243
5'11"	132	143	150	157	165	172	179	186	193	200	207	215	222	229	236	243	250
6'0"	136	147	155	162	169	177	184	191	199	206	213	221	228	235	243	250	258
6'1"	140	151	159	166	174	182	189	197	204	212	219	227	234	242	250	257	265
6'2"	144	155	163	171	179	187	194	202	210	218	225	233	241	249	256	264	272
6'3"	148	160	168	176	184	192	200	208	216	224	232	240	247	255	263	271	279

Find your height in the column on the left, and then move across that row until you find the weight closest to your body weight. Your BMI is at the top of the column where you found your body weight.

BMI does not account for differences in body composition. In very muscular athletes, BMI will overestimate body fat content because their muscle mass contributes an unusually high percent to their total body mass.

CALCULATING BODY MASS INDEX (BMI)

Calculating BMI (kg/m²)

$$\textbf{BMI} = \frac{\text{body weight (kg)}}{\text{height}^2\ (\text{m}^2)}$$

EXAMPLE:

For someone 143 lb and 5'4" in height:

*Convert weight in pounds to kilograms by dividing by 2.2. (143 lb ÷ 2.2 lb/kg = **65 kg**)*

*Convert height in inches to meters by multiplying by 2.54 (to convert to centimeters) and then dividing by 100 (to convert to meters). (64 in × 2.54 cm/in = 162 cm = **1.626 m**)*

$$\textbf{BMI} = \frac{65}{1.626^2}$$

$$= \frac{65}{2.66}$$

$$= 24.6\ \text{kg/m}^2$$

Alternative Method

$$\textbf{BMI} = \frac{\text{body weight (lb)}}{\text{height}^2\ (\text{in}^2)} \times 703$$

EXAMPLE:

For someone 143 lb and 5'4" in height:

$$\textbf{BMI} = \frac{143}{64^2} \times 703$$

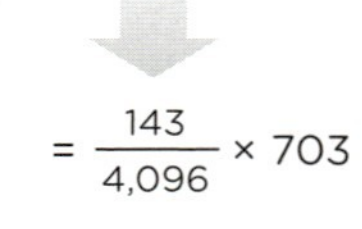

$$= \frac{143}{4{,}096} \times 703$$

$$= 24.5$$

BMI may underestimate body fat content in the elderly, many of whom have lost considerable amounts of muscle. In this case, body fat will constitute a higher percent of their total body mass than would be predicted by BMI.

Calculate your BMI using one of the methods above.

Photo credits (top to bottom): Predrag Vuckovic/Getty Images, Dean Mitchell/Getty Images

but with a large waist are at an increased risk of a number of chronic diseases. Considering waist circumference in addition to BMI is a better predictor of health risk than BMI alone. Abdominal obesity is considered present when waist circumference exceeds 35 inches for women and 40 inches for men. **(INFOGRAPHIC 15.11)**

INFOGRAPHIC 15.11 Abdominal Obesity and Waist Circumference

The distribution of body fat is an important risk factor for obesity-related diseases.

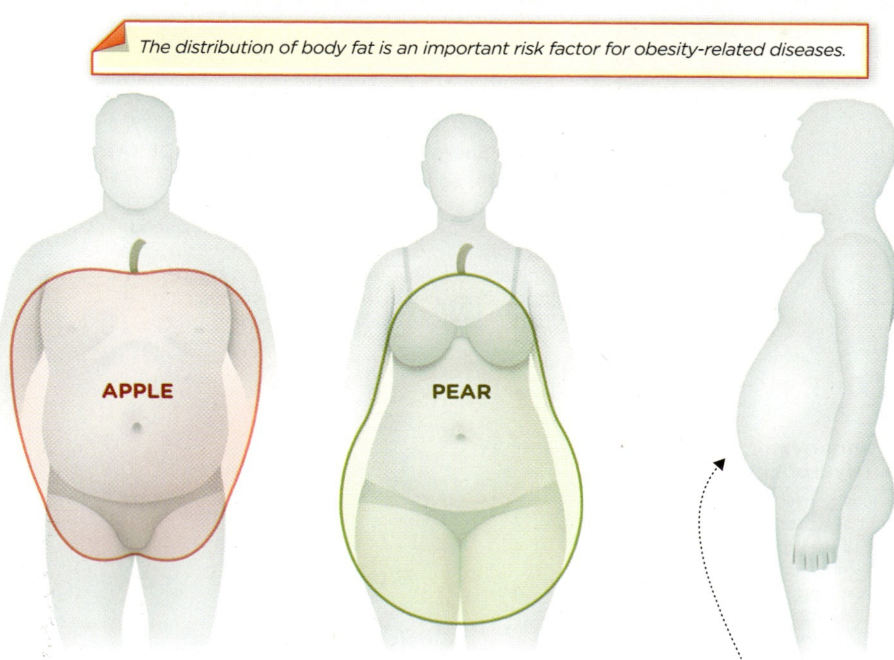

Those who accumulate fat in the abdominal region (abdominal obesity) and have an apple-shaped body have a higher risk of chronic disease than those who accumulate fat in their hips and thighs and have a pear-shaped body. The apple body shape is more common in men, while the pear body shape is more common in women.

Normal-weight individuals (BMI ≤ 25) can also have abdominal obesity and be at an increased risk of chronic disease.

How to Measure Waist Circumference

- At your side, locate the top of your hip bone.
- Place the measuring tape just above your hip bone and circle your body with it.
- Make sure the tape is snug but does not push into the skin.
- Ensure that the tape is parallel to the floor.
- Measure your waist circumference after breathing out normally.

Measuring waist circumference can easily assess the presence of abdominal obesity.

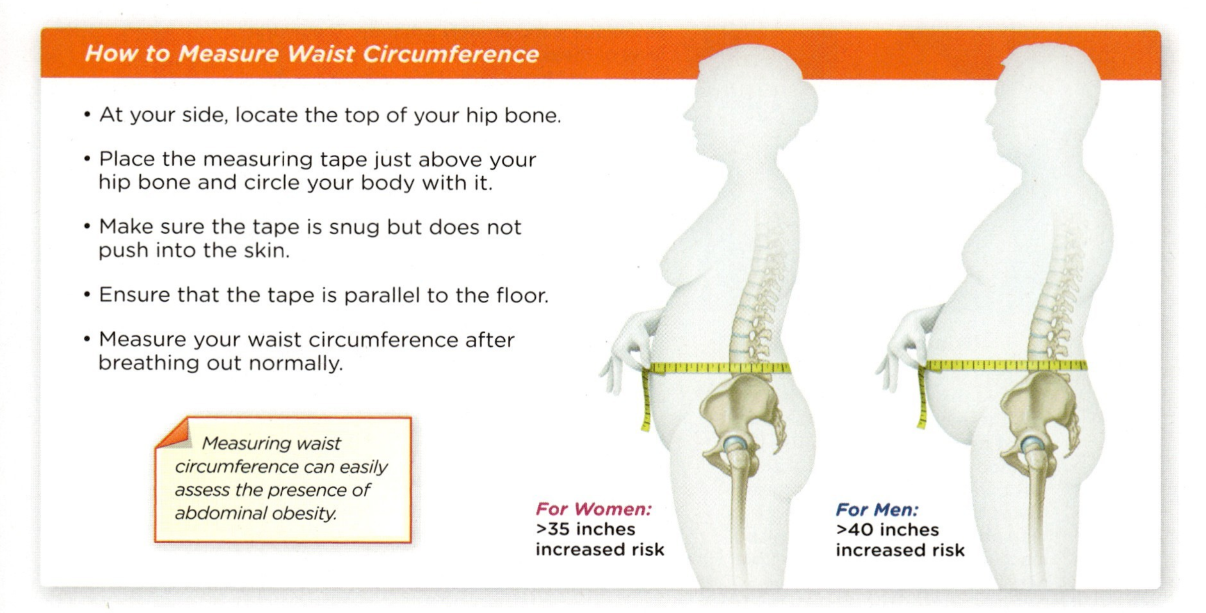

Why is abdominal obesity of more concern than excess fat in the hips and thighs?

INFOGRAPHIC 15.12 Methods for Determining Body Composition

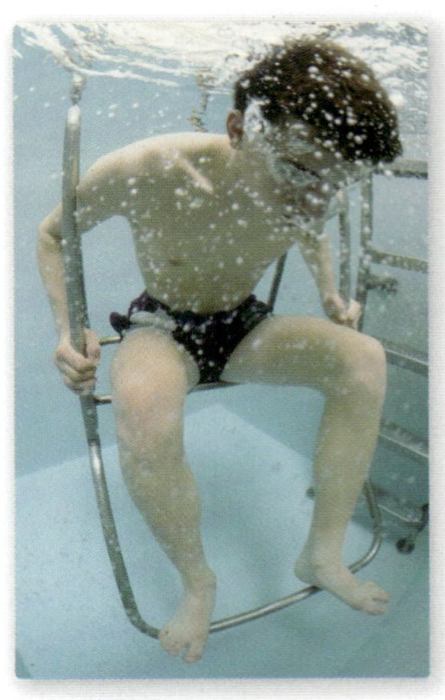

Underwater Weighing (Hydrodensitometry)

Fat is less dense, and therefore more buoyant in water than is muscle. The density of the body is calculated from the difference in the person's body weight on land from that while completely submerged (after exhaling all the air possible from their lungs). This method is not suitable for children, the elderly, or disabled individuals. Accuracy is fairly good, but depends on the ability to exhale all air from the lungs.

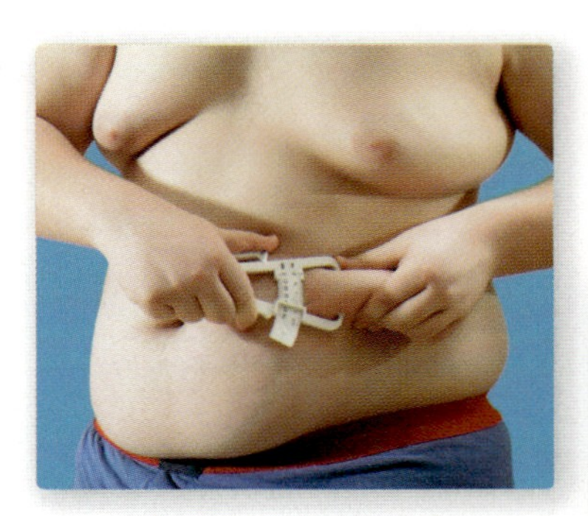

Skinfold Thickness

Calipers are used to measure the skinfold thickness at multiple sites, and these measurements are used to estimate body fat. Because this method only measures fat that is just beneath the skin it underestimates body fat content in those with abdominal obesity. It is also inaccurate in those who are either very lean or very obese.

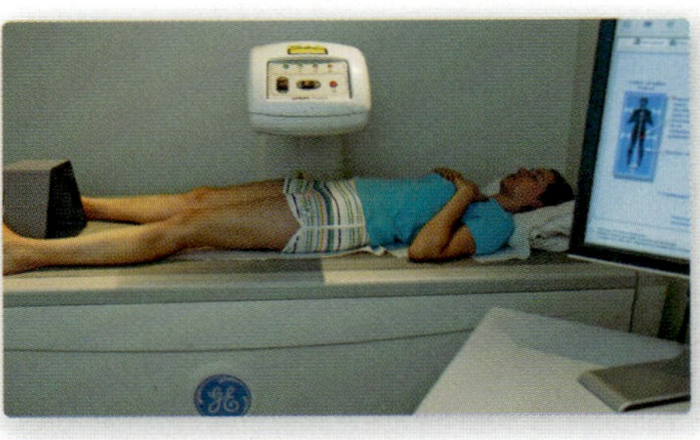

Dual-Energy X-ray Absorptiometry (DEXA)

The differential absorption of x-rays of two different energy levels are used to calculate bone mineral mass, lean soft-tissue mass, and fat mass. It is fairly accurate but the results are influenced by hydration status and the person's body shape. Also, the tables cannot accommodate those who are severely obese.

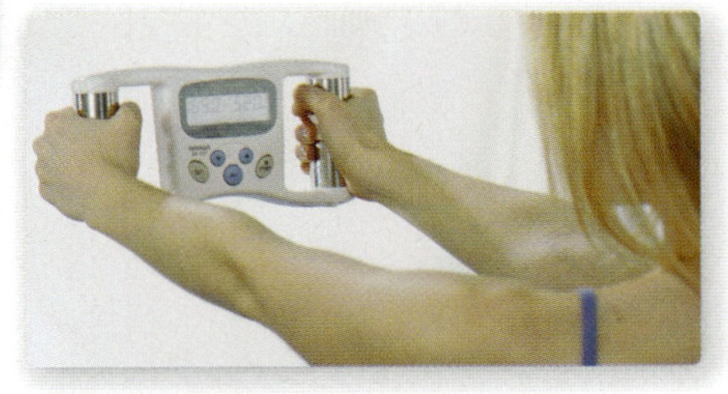

Bioelectrical Impedance

This method exploits the fact that tissues conduct a current differently depending largely upon their water content. Because adipose tissue contains little water, current flows poorly through it. Those with a high percent of body fat will have low total body water and display a higher level of resistance (or impedance). This technique is moderately accurate. The hydration level of the subject will significantly alter the results, and bioelectrical impedance poorly predicts fat content in severely obese subjects.

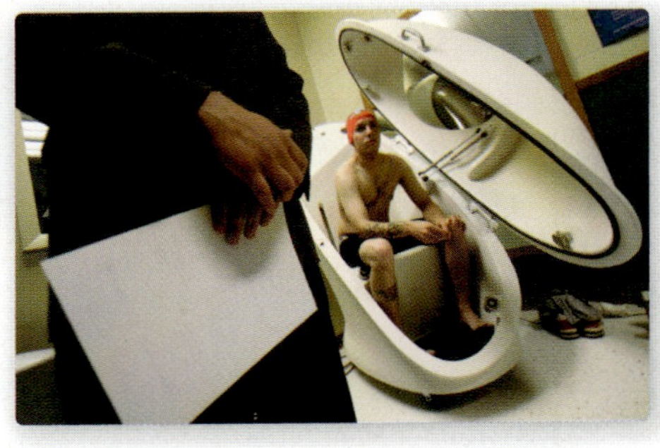

Air Displacement

Air displacement is used to determine the person's body volume, and this is used to calculate body density. From body density, body composition can be estimated. Accuracy is fairly good.

If you want your body composition assessed, which method would you choose and why?

Photo credits (left—top to bottom): David Madison/Getty Images, Steven Puetzer/Getty Images; (right—top to bottom): BSIP/UIG Via Getty Images, May/Science Source, AP Photo/Jeff Adkins

Analyzing body composition

BMI and waist circumference provide useful estimates of body fat in most people, but they are both indirect measures; they do not actually measure an individual's body composition in terms of fat, muscle, bone, and so forth. A variety of additional methods that better determine body composition are available to researchers and health professionals. These include measuring skinfold thickness with calipers, underwater weighing, and air displacement. All methods estimate percent body fat, but vary in terms of their sophistication, cost, and accuracy. Levine and his colleagues, for example, used a highly regarded research method called dual-energy x-ray absorptiometry (DEXA), which uses a small dose of radiation in the form of x-rays to distinguish between fat, muscle, and bone.

One method of estimating body composition that has become accessible to the general public is bioelectrical impedance, which is the method used by home bathroom scales or handheld devices that measure body fat content. It is interesting to note that although scientists and health care providers agree that excess body fat is risky, there is little scientific consensus and no established standards on exactly how much excess body fat would identify someone as obese or at increased health risk. (INFOGRAPHIC 15.12)

For the average person, BMI is still the most accessible indicator of health risk, since it only requires knowing his or her height and weight. BMI is also what health professionals (and insurance companies) typically use to help determine what intervention may be warranted for overweight or obese patients.

WEIGHT LOSS RECOMMENDATIONS

Weight loss is recommended for anyone with a BMI of 30 or higher or those who are overweight and have two or more risk factors, or have a large waist circumference. Examples of risk factors include cardiovascular disease or a family history of it, smoking, hypertension, age (men: 45 years or older; women: 55 years or older, or postmenopausal), diabetes, and physical inactivity.

Surgical approaches to weight loss

Dietary modifications and increased physical activity are generally recommended to help people achieve and maintain a healthier body weight. But for very obese individuals, additional treatment options or interventions may be warranted. There are a few FDA approved antiobesity drugs that can be prescribed as an adjunct to diet, exercise, and behavior therapy for these individuals. For those with extreme obesity (BMI of 40 or above or a BMI of 35 or above with additional risk factors), weight loss surgery may be recommended. At present, in the case of extreme obesity, weight loss surgery (also known as bariatric surgery) is the most effective treatment to yield significant weight loss and reduction of weight-related disorders. Although not without significant risk, these procedures dramatically reduce stomach capacity, limit food intake, and increase satiety. Losses of 50% of excess body weight are not unusual. (INFOGRAPHIC 15.13)

A modest reduction in body weight can result in significant improvements in health for people who are overweight or obese. Losing 5% to 10% of body weight (and maintaining that lower weight) can reduce the risk of chronic diseases (cardiovascular disease and diabetes, for example) and premature all-cause mortality by about 50%.

Dietary and lifestyle approaches to weight loss

For those with less severe weight problems, dietary and lifestyle approaches to weight loss may be enough. There are numerous dietary approaches to weight loss, as shown by the number of "diet" books published each year, magazine articles promising an end to the battle of the bulge, and advertisements for commercial weight-loss programs and products. Low-fat diets, low-carbohydrate diets, meal-replacement plans, and others

INFOGRAPHIC 15.13 Surgical Treatments for Obesity

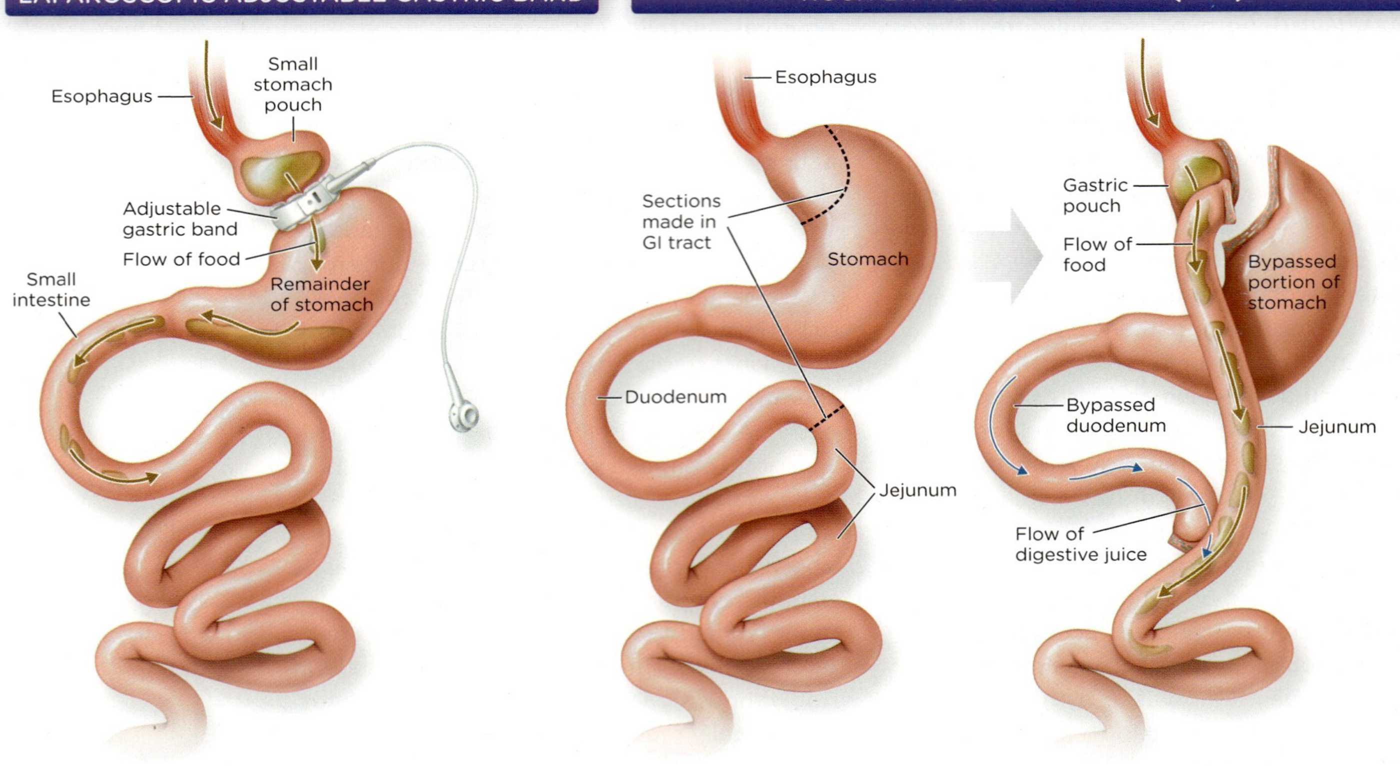

With gastric banding, an adjustable band is placed around the upper portion of the stomach, creating a small stomach pouch and a narrow opening through which food can flow. This delays gastric emptying and increases satiety with smaller amounts of food intake.

Gastric bypass surgery is effective at inducing weight loss because it both reduces food intake and decreases nutrient absorption. Food intake is reduced because of the drastically smaller volume of the newly created stomach pouch. Nutrient absorption is decreased because a portion of the small intestine (the duodenum) is bypassed.

Which one of these surgeries do you expect has the greatest potential to create problems in the future?

Which one of these surgeries do you expect will result in the most rapid rate of weight loss? Explain your answer.

vie for the attention of Americans who are considering or actively pursuing weight loss. Studies demonstrate that most "diets" result in a reduction in calorie intake and that ultimately they vary little in terms of weight loss. Of course, some approaches are more nutritionally sound and conducive to sustaining changes and keeping weight off.

Healthy weight loss plans maximize nutrient density while reducing calorie intake; increase physical activity; incorporate behavioral strategies to enhance compliance; address individual health concerns; and consider not only dropping pounds, but the maintenance of a lower body weight. One effective and healthful dietary approach seems to be a moderately low-fat diet (less than 30% of total calories from fat), coupled with a moderately high protein intake (25% to 30% of total calories). Including significant amounts of nonstarchy vegetables, fruit, minimally processed whole grains

Coming up with a strategy. *Healthy eating plans maximize nutrient density, reduce calorie intake, and increase physical activity.*

Westend61 GmbH/Alamy

that are high in fiber, and lean sources of protein will decrease the energy density of the diet. This will allow for the consumption of greater quantities of food while keeping energy intake low, which may increase the perception of satiation while adhering to a reduced-calorie diet.

A common misconception about weight loss is that a person can simply reduce his or her food intake by a certain amount of calories and expect to lose a proportional amount of weight (assuming that a pound of fat contains approximately 3,500 kcal). For example, if a 700 kcal per day decrease in energy intake is made today, and that new lower-calorie intake is maintained for 100 days, many would predict a body weight loss of 20 pounds (70,000 kcal/3,500 kcal/lb of fat). In reality, only about half that weight would be lost in that time because a lighter body expends fewer calories while resting and while moving. As weight loss continues, the magnitude of the calorie deficit will be consistently shrinking (along with the shrinking body), and the rate of weight loss will continually slow. **(INFOGRAPHIC 15.14)**

Weight-loss maintenance

The likelihood of regaining weight lost through any diet or program is well-known and discouraging. However, the National Weight Control Registry (NWCR) provides hope and strategies through the findings from data collected from more than 10,000 people who have successfully maintained long-term weight loss. According to the NWCR, most "successful losers" share common strategies that include maintaining a lower-fat, reduced-calorie eating plan; eating breakfast; weighing themselves at least once a week; watching fewer than 10 hours of television per week; and exercising on average about one hour per day.

A MORATORIUM ON THE CHAIR

When we first caught up with Levine to discuss NEAT, we found him in transit, walking to his office in downtown Phoenix, Arizona. Levine walks *a lot*. He routinely conducts meetings, interviews, and many other tasks on the go.

INFOGRAPHIC 15.14 Determining Your Estimated Energy Requirement (EER)

1. Estimate your physical activity level.

Physical Activity (PA) Factors			
	Male	**Female**	**Physical Activity**
Sedentary	1.0	1.0	Typical daily living activities
Low active	1.11	1.12	Typical daily living activities + 30–60 minutes moderate activity (Equivalent to walking ~2–3 miles daily)
Active	1.25	1.27	Typical daily living activities + ≥ 60 minutes moderate activity (Equivalent to walking ~4–6 miles daily)
Very active	1.48	1.45	Typical daily living activities + ≥ 60 minutes moderate activity, plus an additional 60 minutes of vigorous, or 120 minutes of moderate, activity (Equivalent to walking ~10–20 miles daily)

Moderate activity is the equivalent of walking at 3–4 mph.

2. Convert your body weight in pounds to kilograms by dividing by 2.2.

3. Convert your height in inches to meters by multiplying by 2.54 (to convert to centimeters) and then dividing by 100 to convert to meters.

4. Enter your values into the appropriate equation below.

For men 19 years and older:
EER = (662 − (9.53 x **age**)) + **PA** x (15.91 x **weight, kg**) + (539.6 x **height, m**)

For women 19 years and older:
EER = (354 − (6.91 x **age**)) + **PA** x (9.36 x **weight, kg**) + (726 x **height, m**)

EXAMPLE:

For a 20-year-old woman who is 67 inches in height, weighs 143 pounds, and engages in 45 minutes of moderate physical activity per day in addition to normal activities of daily living:

1. PA factor = **1.12**
2. 143 lb ÷ 2.2 = **65 kg**
3. 67 inches x 2.54 ÷ 100 = **1.7 m**
4. **EER** = (354 − (6.91 x **20**)) + **1.12** x (9.36 x **65**) + (726 x **1.7**)

= (354 − (138.82)) + 1.12 x (608.4) + 1,234.2

= 215.18 + 681.4 + 1,234.2

= **2,130.78 kcal/day**

Calculate your EER. How much would your EER change if you went up one physical activity level? How much would your EER change if you gained or lost 20 pounds?

Photo credit: Eugenio Marongiu/Shutterstock

"Contrary to popular belief, I do sit from time to time," says Levine, but he makes a habit of doing most things standing if he can. Asked just how bad sitting is for you, Levine rattles off a list of 16 associated health risks including obesity, diabetes, hypertension, high cholesterol, cardiovascular disease, depression, swollen ankles, joint problems, back pain, depression, and cancer. Even one's creativity, he suggests, may be dulled from sitting too much.

That's why Levine has called for a "moratorium on the chair." He believes it is time to fundamentally redesign our environments so that higher NEAT is the norm. Toward that end, he is working with the mayor of Phoenix on initiatives to encourage more walking among commuters, and he also works with businesses and school systems to make workplaces and classrooms more active—or, as Levine puts it, "NEATER."

"What's really cool about NEAT is that everyone can do it," says Levine. "You can have somebody who's 150 pounds going for a 'walk and talk' meeting with someone who's 350 pounds. You can promote NEAT in all people without having to change their clothes."

Though it's easy to get discouraged about our modern obesity epidemic—as discouraged, perhaps, as a person who tries desperately, and fails, to lose weight—Levine stresses that obesity has crept up on us slowly. "The obesity epidemic has occurred over about four generations," he notes. If little steps got us into this mess, he notes, then little steps may get us out—provided they are done consistently.

The degree of positive energy balance that has produced our obesity epidemic has been termed our "energy gap." Addressing the energy gap means identifying the change in energy expenditure relative to energy intake necessary to restore energy balance. It turns out that the amount is less than you might think.

By some estimates, a lifestyle change that reduced energy intake or increased energy expenditure by 100 kcal per day would completely abolish the energy gap, and hence weight gain, for most of the

Get out of the chair!

Westend61 GmbH/Alamy

Taking a walk with a friend is a pleasant way to increase energy expenditure.

Lumi Images/Hudolin-Kurtagic/Getty Images

INFOGRAPHIC 15.15 Strategies for Successful Weight Loss *Successful weight loss involves approaches that address diet, physical activity, and behavior modification to help with adherence to the diet and activity recommendations.*

- ✓ Ideally, work with a trained healthcare professional, such as a registered dietitian, behavioral psychologist, or other trained weight loss counselor.
- ✓ Develop social networks that encourage healthy eating and physical activity patterns.
- ✓ Set realistic and sustainable goals for weight loss and changes in diet and activity level.
- ✓ Monitor intake with a food log or diet tracking application.
- ✓ Emphasize nutrient-dense foods while reducing empty calorie and energy-dense foods.

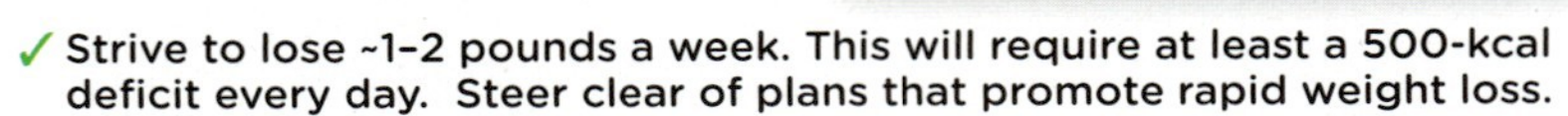

- ✓ Strive to lose ~1–2 pounds a week. This will require at least a 500-kcal deficit every day. Steer clear of plans that promote rapid weight loss.
- ✓ For women this often means eating 1,200 to 1,500 kcal/day, and 1,500 to 1,800 kcal/day for men.
- ✓ Avoid very low-calorie diets providing less than 800–1,000 kcal/day unless medically warranted and under a doctor's supervision.
- ✓ Increase physical activity within daily routine and through intentional exercise. Reduce time spent sitting.
- ✓ Develop a healthy eating plan that you can maintain even after you reach your target weight, which generally means consuming food from each food group.

Photo credit: Frances L Fruit/Shutterstock

population. Walking just one extra mile each day would increase one's energy expenditure by about 50 to 100 kcal per day, depending on body weight. You can estimate the energy you expend in walking by using this relationship: For every mile you walk (between 2 and 4 miles per hour) the energy you expend (in kilocalories) is essentially equal to your body weight in kilograms. For example, if you weigh 150 pounds (68 kg), you will expend approximately 68 kcal for every mile you walk. Similarly, taking two or three fewer bites of food at each meal would reduce energy intake by 100 kcal per day. The problem, of course, is that our current environment strongly discourages both of these things. **(INFOGRAPHIC 15.15)**

■ ■ ■

James Levine once referred to NEAT as the "crouching tiger, hidden dragon" of societal weight gain. What he meant was that the behaviors that lead to obesity may be both sneakier and more deadly than we ever imagined. "Whoever thinks about the amount of time they spend sitting?" Levine asks. "No one does."

CHAPTER 15 **BRING IT HOME**

Review of popular weight-loss diets

Nutrition surveys demonstrate that more than half of American adults are concerned about their weight and are actively trying to lose unwanted pounds. Many turn to popular "diet" books and programs for guidance and support. Evaluating the safety and effectiveness of these books or programs can be a challenge for both consumers and health professionals. However, by examining the claims, components, and credibility of these popular approaches, you can help determine if a book or program is a sound and safe approach to healthy weight loss and maintenance. Using book reviews written by registered dietitians at the website for the Academy of Nutrition and Dietetics, we address the following questions to explore considerations for evaluating popular weight loss diets or programs.

Visit the website for the Academy of Nutrition and Dietetics Consumer Diet and Lifestyle Book Reviews at www.eatright.org/dietreviews. Scroll through the Latest Reviews or the extensive alphabetical listing of diet- and health-related books by title to choose two diet books. Click on the link for each diet book and read the review, taking note of factors that the reviewer considers in evaluating the author's claims, nutritional aspects and other components of the plan that influence compliance, weight loss and maintenance, and overall safety and efficacy of the plan or program.

Name of first book you chose to review:

____

1. Briefly describe the book's claims. After reading the review, do you feel the claims are realistic, grounded in science, and achievable? Explain.
2. After reading the reviewer's synopsis of the diet plan, how do you think it differs from other diet plans? What makes it unique or potentially appealing to dieters?
3. After reading the reviewer's evaluation of the book's nutritional pros and cons, are there any specific types or groups of foods that are restricted? What were the reviewer's primary nutritional concerns about the diet, if any?
4. After reading the reviewer's Bottom Line, would you recommend this book or program to an overweight or obese individual? Why or why not?
5. Do you feel this book's plan or program is conducive to long-term weight management? Why or why not?

Name of the second book you chose to review:

____

1. Briefly describe the book's claims. After reading the review, do you feel the claims are realistic, grounded in science, and achievable? Explain.
2. After reading the reviewer's synopsis of the diet plan, how do you think it differs from other diet plans? What makes it unique or potentially appealing to dieters?
3. After reading the reviewer's evaluation of the book's nutritional pros and cons, are there any specific types or groups of foods that are restricted? What were the reviewer's primary nutritional concerns about the diet, if any?
4. After reading the reviewer's Bottom Line, would you recommend this book or program to an overweight or obese individual? Why or why not?
5. Do you feel this book's plan or program is conducive to long-term weight management? Why or why not?

Consider your answers and what you've learned from this chapter, and list at least five characteristics of a healthy, safe, and potentially effective weight-loss and weight-loss maintenance program.

1.
2.
3.
4.
5.

This website may also help you to complete this activity: *US News and World Reports* Best Commercial Diet Plans: http://health.usnews.com/best-diet/best-commercial-diets

KEY IDEAS

- Obesity is at epidemic proportions in the United States, with more than two-thirds of adults classified as overweight or obese.
- Energy, the capacity to do work, is required to perform all functions necessary to sustain life and is obtained through the breakdown of carbohydrates, fats, proteins, and alcohol in food.
- A calorie, a unit of measure, is defined as the energy required to raise 1 g of water 1°C. The energy in food is commonly measured in units of kilocalories (1,000 calories). A kilocalorie (kcal) is the energy required to raise 1 kg of water 1°C.
- Energy balance is a reflection of energy intake versus expenditure. Although there are many factors that contribute to the development of obesity, fundamentally it results from chronic positive energy balance. Negative energy balance is necessary for weight loss.
- Obesity increases the risk of multiple chronic diseases and premature death. Hormonal changes associated with excess body fat cause a low-grade chronic inflammation throughout the body that has adverse health effects.
- Energy balance and food intake are regulated through a short-term system and a long-term system. The short-term system is mediated by hormones and stomach pressure and is responsible for triggering hunger and satiety during individual meals. The long-term system, mediated by a different set of hormones, adjusts food intake and energy expenditure to maintain adequate fat stores.
- Satiation is the sense of fullness we feel while eating and leads to the termination of a meal. Satiety is the effect that the meal has on our interest in food and hunger levels after and between meals.
- The hormone ghrelin stimulates hunger by activating specific neurons in the brain. Ghrelin levels in the blood increase just before meals and decrease after eating.
- The hormone leptin is produced primarily by adipose (fat) tissue and has a role in long-term energy balance. Its circulating concentration is closely associated with total body fat.
- Hunger is the biological impulse that drives us to seek out food and consume it to meet our energy needs. Appetite is a desire for or liking of food for reasons other than, or in addition to, hunger.
- Total energy expenditure (TEE) is composed of basal metabolism, the thermic effect of food, and activity energy expenditure.
- Basal metabolism is the energy expenditure required to maintain the essential functions that sustain life. It accounts for about 60% of TEE in a typical sedentary individual, with most variation from person to person accounted for by differences in fat-free mass (FFM).
- Thermic effect of food (TEF) is the energy needed to digest, absorb, and metabolize nutrients in our food. TEF is generally equivalent to 10% of the energy content of the food ingested and does not vary greatly between people.

Activity energy expenditure is the amount of energy individuals expend in physical activity per day and is the most variable component of TEE.

Body mass index (BMI) is an indirect measure of body fat calculated from a person's weight and height. "Underweight," "normal," "overweight," and "obese" are all labels for ranges of body weight on the BMI scale.

The greater the BMI, the higher the risk of premature mortality and the risk of obesity-associated diseases. A BMI in the underweight range is also associated with increased premature mortality.

Waist circumference indicates body fat distribution and the presence of excess visceral fat, which has been shown to be an independent health risk.

Body composition can be measured in a variety of ways, including skinfold thickness with calipers, underwater weighing, air displacement, bioelectrical impedance, and dual-energy x-ray absorptiometry.

Weight loss, including dietary modifications and increased physical activity, is recommended for anyone with a BMI of 30 or higher or those who are overweight and have two or more risk factors, or have a large waist circumference. For those with extreme obesity (a BMI of at least 40 or a BMI of *at least* 35 with additional risk factors), weight loss (bariatric) surgery may be recommended.

Modest weight loss and maintenance (5% to 10% of body weight) can improve health, as well as reduce the risk of chronic diseases and premature death.

Healthy weight-loss plans maximize nutrient density while reducing calorie intake; increase physical activity; incorporate behavioral strategies to enhance compliance; address individual health concerns; and consider maintenance of a lower body weight.

NEED TO KNOW

Review Questions

1. A calorie is defined as:
   a. the energy required to raise 1 mg of water 1°C.
   b. the energy required to raise 1 g of water 1°C.
   c. a molecule that provides energy to cells.
   d. a byproduct of carbohydrate and fat metabolism.

2. Current evidence suggests that a primary cause of adverse health effects related to obesity is:
   a. excess body fat in the hip and thigh area.
   b. long-term negative energy balance.
   c. low-grade chronic inflammation throughout the body.
   d. changes in the acid-base balance of the blood.

3. Satiation can be defined as:
   a. the sense or feeling of fullness while eating that leads to termination of the meal.
   b. the effect of a meal on level of hunger and desire to eat after or between meals.
   c. the saturation of adipose tissue with the hormone ghrelin.
   d. an appetite for foods high in fat, particularly saturated fat.

4. All of the following are TRUE with regard to the hormone leptin, EXCEPT:
   a. it is primarily produced by adipose tissue.
   b. its circulating concentration is closely associated with total body fat.
   c. increased levels of leptin in the blood act in the brain to suppress hunger.
   d. it functions primarily in short-term energy balance.

5. The primary contributor to an individual's total energy expenditure is:
   a. basal metabolism.
   b. thermic effect of food.
   c. activity energy expenditure.
   d. meal timing and composition.

6. The most variable component of an individual's total energy expenditure is:
   a. basal metabolism.
   b. thermic effect of food.
   c. activity energy expenditure.
   d. meal timing and composition.

7. The thermic effect of food is generally equivalent to _____ of the energy content of food ingested.
   a. 10%
   b. 15%
   c. 20%
   d. 25%
   e. 30%

8. All of the following are TRUE with regard to body mass index (BMI), EXCEPT:
   a. that it is a measure of an adult's weight in relation to his or her height.
   b. that it provides an indirect measure of body fat.
   c. that there is an increased risk of obesity-related diseases and conditions with increasing BMI.
   d. measurement requires the use of skinfold calipers or other body composition assessment tools.
   e. that it may overestimate body fat in athletes and highly trained individuals.

9. A BMI of 25 to 29.9 would classify a person as:
   a. underweight.
   b. normal weight.
   c. overweight.
   d. obese.
   e. a candidate for weight-loss surgery.

10. Waist circumference can be an indication of "risky" abdominal obesity. A waist circumference of _____ inches for a man would indicate excess abdominal fat and increased health risk.
    a. 27
    b. 32
    c. 37
    d. 42

11. All of the following are TRUE with regard to weight loss (bariatric) surgery, EXCEPT:
    a. it dramatically reduces food intake and limits stomach capacity.
    b. it does not have any effect on satiety.
    c. it is reserved for individuals with BMIs of *at least* 40 or *at least* 35 with additional risk factors.
    d. it produces significant weight loss in obese patients.

Dietary Analysis Using SuperTracker

Using SuperTracker to Assess Energy Intake

Obesity—having too much body fat—has been linked to a number of health problems, such as heart disease, diabetes, stroke, and many others. The rate of obesity in the United States has continued to increase over the last four decades. Using SuperTracker, you will examine your energy intake and assess your body weight status. You will also identify appropriate diet changes.

1. Log onto the United States Department of Agriculture (USDA) website at www.supertracker.usda.gov. If you have not done so already, you will need to create a profile to get a personalized diet plan. This profile will allow you to save your information and diet intake for future reference. *Do not use the general plan.*
2. Click the Track Food and Activity option. Record three days of food intake.
3. After you have entered your food for each day, you can then use the "My Reports" feature to analyze your intake. The "Food Groups and Calories" report shows your average intake for each food group. It also shows average calorie intake. The "Nutrients Report" shows an average intake of specific nutrients (such as sodium, calcium, and vitamin D), as well as the foods you consumed that provide the highest or lowest amount of each nutrient. Use the Nutrients Report to answer the following questions.
    a. Did you consume an appropriate number of calories for the three days you selected? If not, were you over or under the target number?
    b. Do you feel that these three days are a typical intake pattern for you? Explain your answer.
    c. According to the body mass index chart, is your body weight within a healthy range? If not, are you overweight or underweight?
    d. Identify three health risks associated with being overweight and obese.
    e. Based on your results, what dietary changes do you believe you should make to be more in line with the SuperTracker recommendations?

Eating to Win

RESEARCH SUGGESTS THAT WHEN ATHLETES EAT MAY BE JUST AS IMPORTANT AS WHAT THEY EAT

Chris Keane/Reuters/Corbis

LEARNING OBJECTIVES

- Identify five health-related benefits of regular physical activity (Infographic 16.1)
- Describe the four components of physical fitness (Infographic 16.2)
- Identify the components of the anaerobic and aerobic energy systems and rank these four components according to the rate at which they produce adenosine triphosphate (ATP) (Infographic 16.3)
- Describe how exercise intensity and duration affects the use of carbohydrates and fat as fuel (Infographics 16.5 and 16.6)
- Define carbohydrate loading and describe when the practice may be beneficial (Infographics 16.8 and 16.9)
- Explain why adequate dietary carbohydrates are necessary for endurance training to be effective (Infographics 16.8 and 16.10)
- Identify factors that affect fluid needs of athletes and discuss strategies to optimize hydration during and after exercise (Infographic 16.11)

Madison Hubbell is not a typical 22-year-old woman. Every day, she ice skates for four or more hours and then sweats through 90 minutes of intense cardiovascular exercise or specialized lessons in ballroom dancing. Hubbell is one of the world's top ice dancers—she and her partner Zachary Donohue finished fourth at the 2014 U.S. Figure Skating Championships and first at the Four Continents Figure Skating Championship in 2014.

Ice dancing differs from traditional pairs skating in that, rather than skating in unison and having to perform lifts, spins, and throws, ice dancers focus on rhythm, musical interpretation, and precise steps.

Madison Hubbell and her partner, Zachary Donohue, of the United States skate their free dance program at Skate America 2013 in Detroit, Michigan.

Rebecca Cook/Reuters/Corbis

PHYSICAL FITNESS *the ability to perform moderate to vigorous activity without undue fatigue*

PHYSICAL ACTIVITY *any bodily movement produced by skeletal muscles that causes energy expenditure*

EXERCISE *intentional physical exertion*

RESISTANCE ACTIVITIES *activities performed against an opposing force that increase muscle strength, improve body composition, and promote healthy bone density*

CARDIORESPIRATORY ACTIVITIES *activities that increase the heart rate and promote increased use of oxygen to improve overall body condition and endurance*

NEUROMOTOR ACTIVITIES *activities that incorporate motor skills such as balance, coordination, and agility (also known as functional fitness)*

And even though ice dancing is considered a winter sport, Hubbell doesn't get a summer vacation. "In our off-season, we train more," she says.

As a result of such constant intense training, competitive athletes like Hubbell have achieved extremely high levels of **physical fitness**, the ability to perform moderate to vigorous activity without undue fatigue. That's not to say that Hubbell doesn't get tired—running through an ice dancing routine leaves her exhausted every time, but that's in part because she strives to improve her performance with each run-through, which requires her to exert herself physically at very high levels each time. She aims for peak physical performance.

COMPONENTS OF FITNESS

Physical activity is any bodily movement produced by skeletal muscles—whether it's running, playing a sport, brushing our teeth, vacuuming, or walking to school. Physical activity can be spontaneous and integrated into our activities of daily living (nonexercise activity thermogenesis, as discussed in Chapter 15), or it can be a planned, discrete, and intentional bout of activity called **exercise**. Generally, more physical activity, either through activities of daily living or planned exercise results in a higher level of physical fitness, as well as better physical and mental health. If exercise were a life-saving drug, some say, it would be the most beneficial, effective, and affordable drug in the world. In the short term, exercise helps people lose body fat, build muscle mass, and become physically fit, while over the long term, research suggests that regular exercise reduces the risk of heart disease, stroke, type 2 diabetes, osteoporosis, and certain cancers; ultimately, it lowers a person's risk of premature death. Regular exercise has also been demonstrated to help manage stress while reducing anxiety and symptoms of depression. Recent studies have demonstrated that moderate-intensity activity, like jogging for as little as 10 minutes per day (most days of the week), can significantly reduce the risk of CVD and all-cause mortality, with additional exercise bringing greater benefits. While the optimal exercise dose has yet to be defined, new evidence also suggests that the benefits of exercise may diminish when duration or intensity are excessive. (INFOGRAPHIC 16.1)

Even though Hubbell's passion is ice dancing, her training regimen encompasses many types of physical exercise, including the four types of exercise recommended

INFOGRAPHIC 16.1 Benefits of Regular Exercise *Exercise decreases the overall risk of death. The majority of benefits are associated with cardiorespiratory fitness, the ability of the circulatory and respiratory systems to supply oxygen to muscles during sustained physical activity.*

My Good Images/Shutterstock

by the American College of Sports Medicine for promoting maximal health and preventing diseases and injuries: **resistance**, **cardiorespiratory**, **neuromotor**, and **flexibility**. (INFOGRAPHIC 16.2) All of these types of exercise are recommended for adults.

To succeed as a competitive ice dancer, however, Hubbell has to do more than just train vigorously. She also has to eat carefully to achieve and maintain a body composition that is optimal for her sport. Hubbell already had a very healthy diet—often,

FLEXIBILITY ACTIVITIES
activities that promote the ability to move joints through their whole span of movement

INFOGRAPHIC 16.2 Types of Exercise and Physical Activity Recommendations for Adults *A fitness program should regularly include all four types of exercise. For those with low baseline physical fitness, aerobic exercise for a shorter time, or at a lower intensity than recommended, can still result in improved fitness.*

Morsa Images/Getty Images

RESISTANCE
Strength Training

Recommendations
Train each major muscle group (legs, hips, back, abdomen, chest, shoulders, and arms) 2–3 days per week.

Michael DeYoung/Getty Images

CARDIORESPIRATORY
Endurance, or Aerobic Training

Recommendations
At least 30 minutes of moderate-intensity exercise 5 or more days per week for at least 150 min per week. OR at least 25 minutes of vigorous-intensity exercise 3 or more days per week for at least 75 minutes per week.

Adrianna Williams/Getty Images

NEUROMOTOR
Balance, Agility, and Coordination

Recommendations
≥ 2–3 days per week, ≥ 20–30 minutes per day.

Take A Pix Media/Getty Images, Michael DeLeon/Getty Images, Vasko Miokovic Photography/Getty Images, Jeremy Ric[illegible]tty Images.

FLEXIBILITY
Stretching

Recommendations
≥ 2–3 days per week. Stretch to the point of feeling slight discomfort or tightness and hold the stretch for 10–30 seconds.

the qualities that make a person athletic, such as self-awareness and self-discipline, also induce them to make healthy food choices—but according to her coaches, she still needed to make changes. They believed Hubbell could find some competitive advantage by slightly reducing her percentage of body fat, but she needed to do this without compromising her strength or health.

To get sound nutrition guidance, she asked around for nutritionist recommendations. Several colleagues pointed her to Dan Benardot, PhD, a sports nutritionist and registered dietitian who runs the Laboratory for Elite Athlete Performance at Georgia State University in Atlanta. Benardot has spent his career studying what foods athletes should eat and when they should eat them. So, Hubbell paid Benardot a visit in Atlanta. Benardot was indeed the right person to go to: He had been in charge of nutritional strategy for the U.S. Women's Gymnastics Team from 1992 to 1996.

For Hubbell to meet her goals, Benardot recommended that she should eat small meals frequently throughout the day—not more than 600 calories in one sitting. He also told Hubbell to ignore the advice she had heard about avoiding carbohydrates. "He said, 'That's completely backwards—you have to eat a lot of carbs,'" Hubbell recalls.

Eating at regular intervals throughout the day will allow athletes to avoid extended periods of calorie deficit that can lead to **catabolism**, which is the breakdown of large molecules (protein, fat, and glycogen) into smaller ones, causing a loss of skeletal muscle mass. When sufficient glucose is not available, the body will accelerate the breakdown of muscle proteins to provide the amino acids that are required to synthesize the needed glucose, which leads to a reduction in lean body mass—something most athletes want to avoid. This is exactly the opposite of **anabolism**, the process by which the body synthesizes protein, fat, and glycogen; and builds muscle mass and lean tissue. To build muscle and increase the proportion of lean tissue to fat tissue, both athletes and nonathletes must consume adequate calories to prevent catabolic processes from exceeding anabolic processes over the course of the day, while engaging in activities (primarily resistance training) that promote muscle growth.

FUELING THE BODY

When a person exercises, muscles experience an increased demand for energy to contract. That energy comes from one of two main sources: carbohydrates in the form of glucose and fats in the form of fatty acids. These **energy substrates** provide the fuel that humans need to keep moving.

Both glucose and fatty acids are rich in chemical energy, stored in the chemical bonds holding the molecules together. Before that energy can be tapped by the body, however, it must be converted to a form that cells can use to perform work. That usable form is called **adenosine triphosphate (ATP)**. Commonly referred to as the cell's energy currency, ATP stores chemical energy in the bonds of its three phosphate groups. When our cells need energy, they typically break the bond between the last two phosphates, releasing the stored energy and forming adenosine diphosphate (ADP). (The "di" in diphosphate means "two," as in two phosphates; the "tri" in "triphosphate" refers to its three phosphates.)

You can think of the energy in glucose and fat as the value of a gold brick: It's worth a lot of money, but you couldn't buy even a cup of coffee with it. ATP, however, is like bills and coins—it's the energy your cells can actually spend.

ATP is produced in the body by three separate energy systems, two of which are **anaerobic** (not requiring oxygen), and one of which is **aerobic** (requiring oxygen). The anaerobic energy systems reside in the **cytosol**, the intracellular fluid of the cell outside the **mitochondria**. The aerobic energy system resides in mitochondria, small organelles within cells that are often called the powerhouses of our cells because this is where the vast majority of all ATP is produced. The production of ATP in mitochondria is completely dependent on the availability and use of oxygen. These three energy systems differ in the speed at which they replenish ATP for use by cells and, in the case of the aerobic energy system, the fuel that is burned. (INFOGRAPHIC 16.3)

Anaerobic energy systems

Why have three separate ways to make ATP? These systems provide overlapping coverage to replenish ATP over the short, medium, and long terms. The most important source of

CATABOLISM
the breakdown of body muscle and tissue

ANABOLISM
the process of building up body muscle and tissue

ENERGY SUBSTRATES
macronutrients used by the body to provide energy

ADENOSINE TRIPHOSPHATE (ATP)
the primary energy currency of our cells

ANAEROBIC
occurring in the absence of oxygen

AEROBIC
occurring in the presence of oxygen

CYTOSOL
fluid inside cells

MITOCHONDRIA
an organelle in the cytoplasm of cells that functions in aerobic energy production

INFOGRAPHIC 16.3 ATP-Producing Energy Systems—An Overview

Adenosine triphosphate (ATP) has high energy content and is often referred to as the energy currency of cells. ATP is replenished by three energy systems, and the speed of its production depends upon the system being used, and in the case of the aerobic energy system, the fuel being burned.

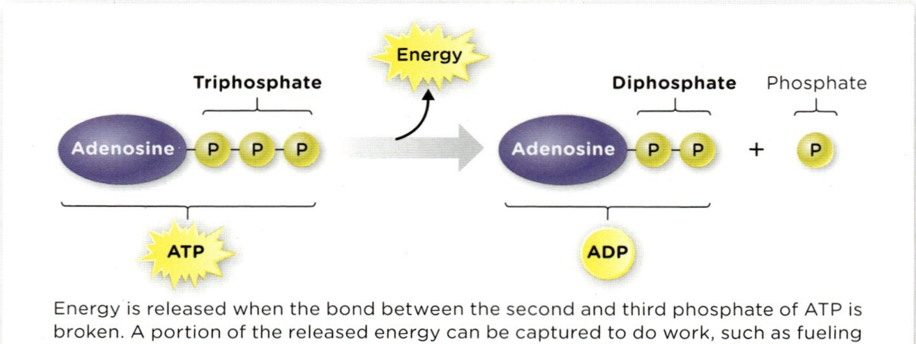

Energy is released when the bond between the second and third phosphate of ATP is broken. A portion of the released energy can be captured to do work, such as fueling muscle contractions. ADP is then used to reform ATP by the three energy systems.

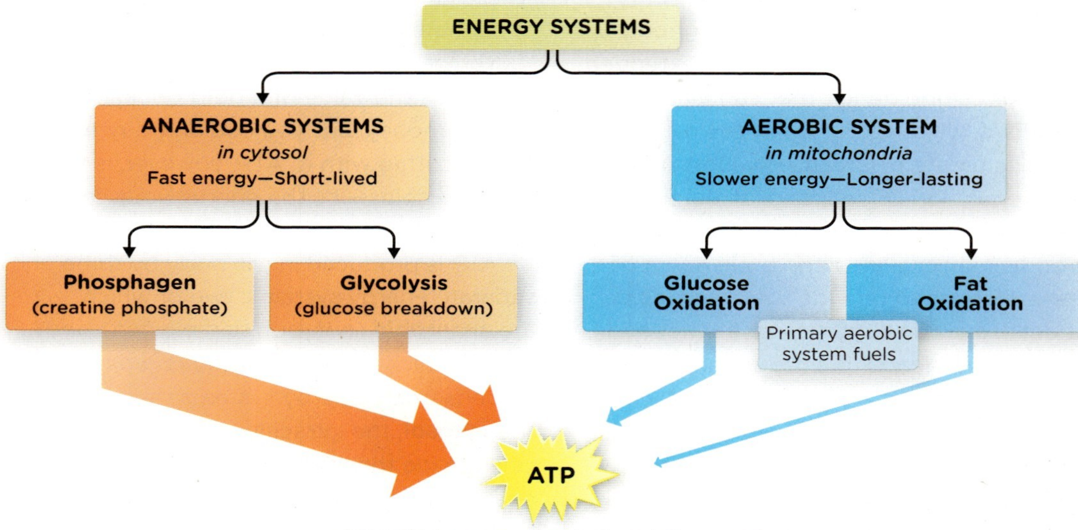

Arrow thickness represents the relative speed of ATP production of each energy system.

What fuel can provide ATP both anaerobically and aerobically? Why would someone run slower if they are relying on fat as the primary fuel?

CREATINE PHOSPHATE (PHOSPHOCREATINE)
a compound that readily transfers its phosphate and stored bond energy to ADP to replenish ATP

GLYCOLYSIS
a series of metabolic reactions in the breakdown of glucose to provide energy

energy for short bursts of highly intense exercise, such as the 15-second sprints that Hubbell does while training, is the anaerobic *phosphagen system*. The amount of ATP stored in resting muscle is limited and depleted after only a few seconds of such vigorous exercise. To continue to meet energy demands and produce more ATP, muscle contains **creatine phosphate** (also known as phosphocreatine), which can readily transfer its phosphate and stored bond energy to ADP to quickly make more ATP.

The other significant source of energy for high-intensity exercise is **glycolysis**, an anaerobic energy system that breaks down glucose (obtained from the blood

Because the phosphagen system and glycolysis (the anaerobic energy systems) are the only processes that produce ATP fast enough to meet the demands of high-intensity exercise, they are the dominant sources of ATP during resistance training (weight lifting) or short sprints, such as the race to the waterfront during the starting moments of a triathlon.

ZUMA Press, Inc./Alamy

or muscle glycogen) into a three-carbon molecule called **pyruvate**, producing ATP in the process. Like the phosphagen system, glycolysis occurs in cytosol, the intracellular fluid of the cell outside the mitochondria.

When energy demands are high, glycolysis produces pyruvate at a rate faster than it can be used as a fuel for aerobic metabolism in mitochondria. As pyruvate accumulates in the muscle, it must be converted to **lactate** (or lactic acid) for glycolysis to continue. Lactate then enters the blood to be delivered to the liver where it can be recycled into glucose and returned to contracting muscle for use as an energy source. The heart, brain, and nonworking muscles can also take up lactate and use it directly as a source of energy.

Lactate gets a lot of bad press, but it turns out that essentially none of it is true. It is often blamed for causing muscle fatigue or the muscle "burn" associated with intense exercise, but in reality neither lactate nor muscle becoming more acidic are significant contributors to fatigue or muscle soreness.

Slightly longer bursts of exercise lasting a few minutes, such as Hubbell's short ice dance routines (lasting 2 minutes and 50 seconds), rely heavily on glycolysis, but the aerobic energy system will also begin to come into play. During the first minute or so of this routine, the dominant means of creating ATP will be through the phosphagen system and glycolysis. As the program continues, however, the aerobic energy system begins to dominate. And this is the case for recreational exercisers, as well: The first minute of a jog or run will rely primarily on anaerobic energy systems, and then as the run continues the oxygen-dependent aerobic energy system begins to take over.

Aerobic energy system

Aerobic metabolism is a process that occurs within a muscle cell's mitochondria, where

PYRUVATE
a compound produced from the breakdown of glucose; lactate is produced from pyruvate and is produced faster than it can be metabolized in mitochondria

LACTATE (LACTIC ACID)
a molecule formed by glycolysis when the energy demands of skeletal muscle are high

INFOGRAPHIC 16.4 Energy System Fuels *Creatine phosphate, glucose, and fatty acids are the primary fuels supplying energy to produce ATP for muscle contractions.*

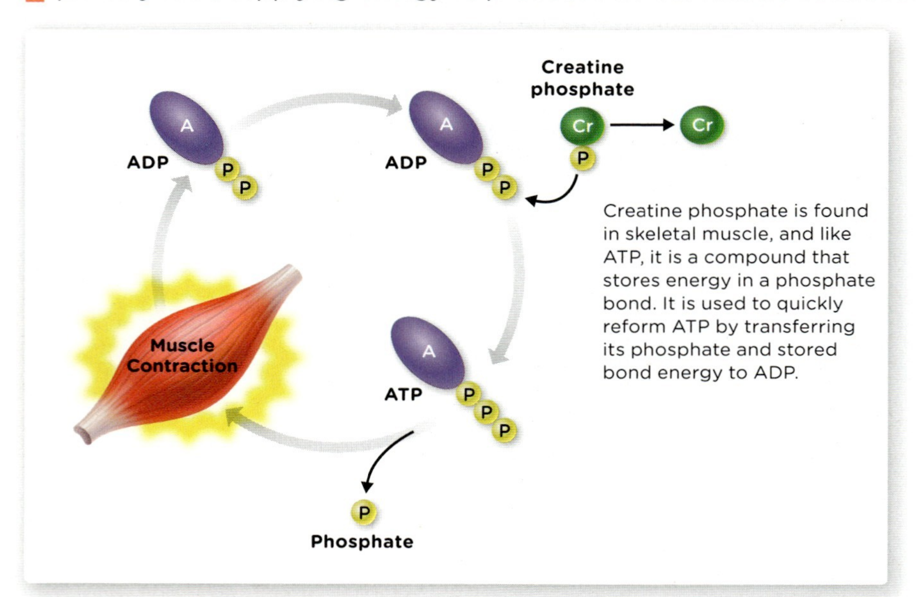

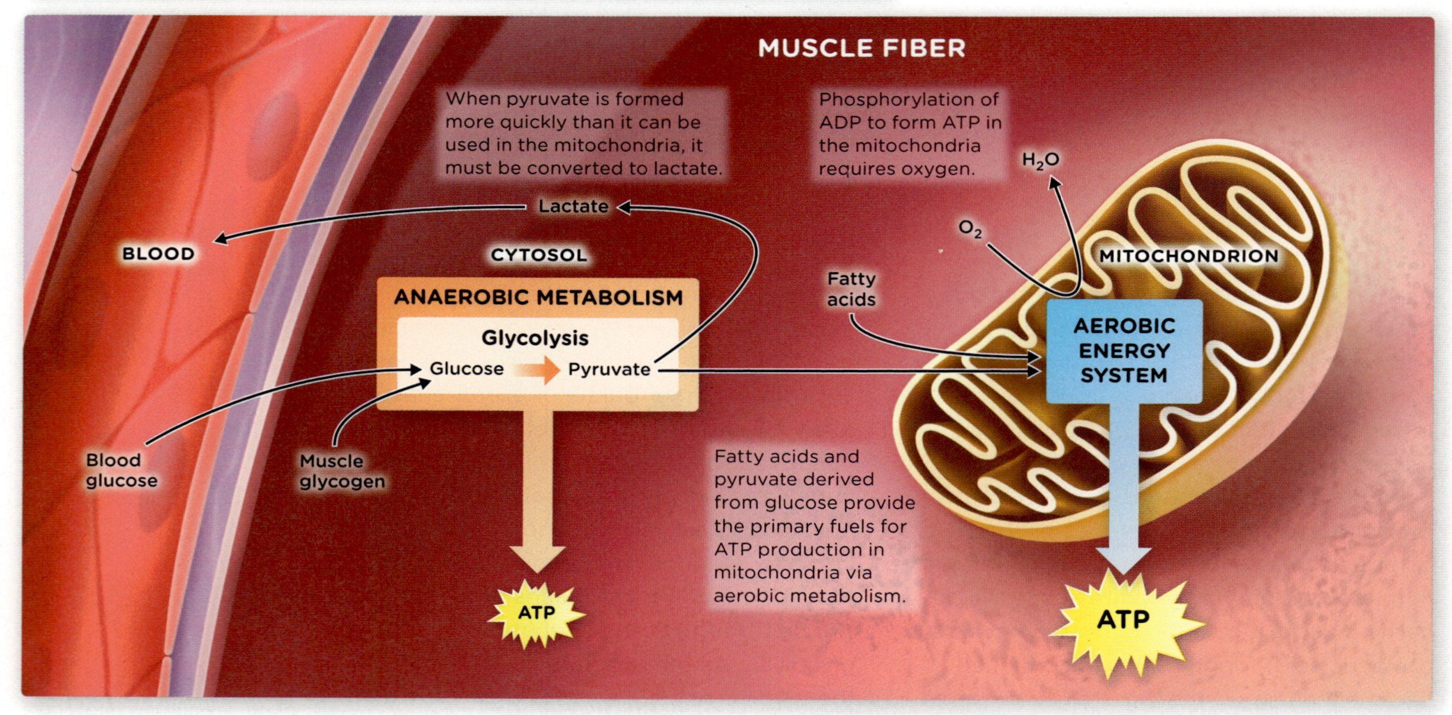

What fuels supply energy for anaerobic metabolism? What fuels supply energy for aerobic metabolism?

oxygen is required to drive ATP production. **(INFOGRAPHIC 16.4)**

Despite producing ATP more slowly than anaerobic energy systems, longer-lasting, lower-intensity activities such as long-distance running or cycling rely almost entirely on the aerobic energy system in the mitochondria. Mitochondria and the aerobic energy system are so critically important for endurance exercise performance that the primary adaptation to endurance training within skeletal muscle

INFOGRAPHIC 16.5 Influence of Exercise Intensity on Energy System Use *Relative energy system contribution to total energy expenditure for any given duration of maximal intensity exercise (e.g., the point at 3 minutes assumes that someone ran as far as possible in that time). Even for very brief, high-intensity exercise, aerobic systems make some contribution to ATP production, however, as exercise duration increases, so does the contribution aerobic metabolism makes to total energy supply.*

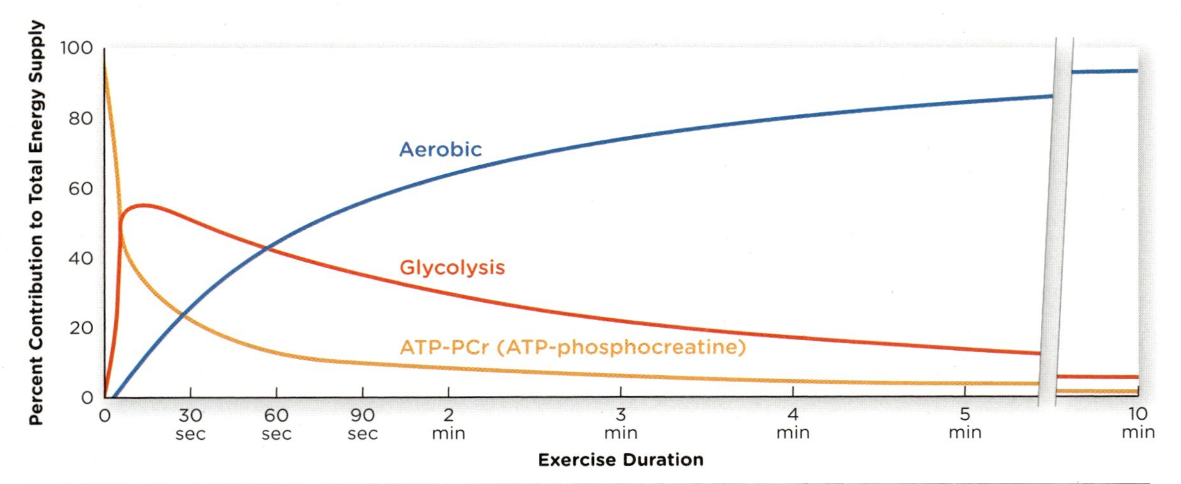

Sources: Gastin, P.B. Energy system interaction and relative contribution during maximal exercise. Sports Med. 2001;31:725–741.
Dawson, B, & Goodman, C. Energy system contribution to 1500- and 3000-metre track running, Journal of Sports Sciences 2005;23: 993–1002.

At approximately what duration of maximal exercise intensity does aerobic metabolism begin contributing at least 50% of total ATP production? Approximately what percent of ATP production is provided by aerobic metabolism during a 15-second maximal sprint?

is to increase the number of mitochondria. Aerobic metabolism is relied on heavily during endurance exercise because it can be maintained for much longer periods than anaerobic metabolism (with much higher yields of ATP). The primary fuels for the aerobic energy system are pyruvate generated by glycolysis in the cytosol, as well as fatty acids that are released from triglycerides stored both in adipose tissue and skeletal muscle. **(INFOGRAPHIC 16.5)**

Ultimately, the energy contributions made through anaerobic and aerobic pathways combine to provide muscles with enough ATP to meet demands. The percentage of each depends on the intensity and duration of the activity.

Both glucose and fats are burned via the aerobic energy system to make ATP. The

The increase in exercise intensity that often occurs while running an uphill portion of a race increases the use of carbohydrate (glucose and glycogen) and decreases the use of fat for energy.

H. Mark Weidman Photography/Alamy

INFOGRAPHIC 16.6 Fat Utilization and Endurance Exercise Intensity *Although the percent of energy provided by fat is highest with low exercise intensity, the total amount of fat used during exercise is highest during moderately intense exercise.*

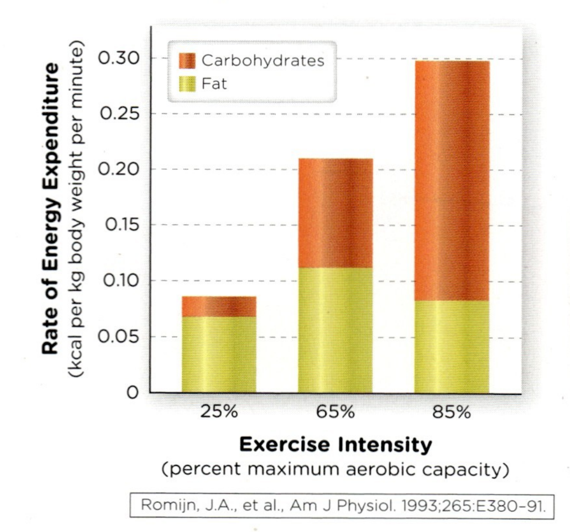

Romijn, J.A., et al., Am J Physiol. 1993;265:E380–91.

During moderate-intensity and high-intensity exercise, the rate of energy expenditure is so high that the total amount of fat used as fuel exceeds what is used during low-intensity exercise, even though the percent of total energy provided by fat decreases (as carbohydrate use increases) with increasing exercise intensity (as shown below).

Total calories and calories from fat expended by a 70-kg individual exercising for 30 minutes at three different intensities (25%, 65%, and 85% of maximum aerobic capacity).

Exercise intensity	Total kcal expended	Total kcal from fat	Percent kcal from fat
25%	181	143	79%
65%	441	235	53%
85%	626	174	28%

? *Explain why total fat use during exercise at 65% of maximum oxygen consumption is higher than it is at 25%, despite the fact that fat provides a lower percent of total energy at the higher exercise intensity. Approximately how many kcal would an individual weighing 70 kg expend in 30 minutes of exercise at 85% of their VO_2 max?*

relative contributions of glucose and fat depend on the intensity and duration of the activity. The lower the intensity of an activity is, the more likely it is to burn fat. The use of fat as an energy source decreases proportionally as energy intensity increases, until at maximal intensity the energy contribution of fat is negligible. This does not mean that exercising at low intensity is better for losing weight. As the *percentage* of fat's contribution to energy use decreases with increasing exercise intensity, the *total amount* of energy being expended is increasing even faster.

Experts point out that if low-intensity exercise truly maximized fat utilization, then staying in bed would be the best way to lose weight since an even higher percent of energy use in muscle is derived from fat when one is completely at rest. Total fat use during both moderate-intensity and high-intensity exercise exceeds that of low-intensity exercise of the same duration because energy use is so much higher. Furthermore, the higher overall calorie expenditure will potentially lead to a greater energy deficit and more effectively promote gradual weight loss. **(INFOGRAPHIC 16.6)**

MAXIMAL OXYGEN CONSUMPTION (VO_{2max}) *the maximum amount of oxygen that can be used at the peak of intense aerobic exercise*

MEASURES OF EXERCISE INTENSITY

A high-intensity exercise for one person may be low-intensity for another; it depends on the person's fitness and aerobic capacity. Someone in poor shape may find brisk walking difficult, whereas Hubbell would barely break a sweat on a power walk. Many different scales are used to estimate exercise intensity in an individual, including the Borg Rating of Perceived Exertion Scale and a measure called **maximal oxygen consumption (VO_{2max})**, which is the maximum amount of oxygen that can be used at the peak of intense aerobic exercise.

The harder an athlete works during an activity, the more oxygen is consumed and the faster her heart beats until she reaches her VO_{2max}, or her maximum heart rate. Exercise intensity is often estimated as a percent of one's VO_{2max}, or as a percent of one's maximum heart rate.

INFOGRAPHIC 16.7 Estimating Exercise Intensity *Several methods can be used to estimate the intensity of aerobic activity. Two common methods are the Borg Rating of Perceived Exertion Scale, and heart rate as a percent of maximum heart rate.*

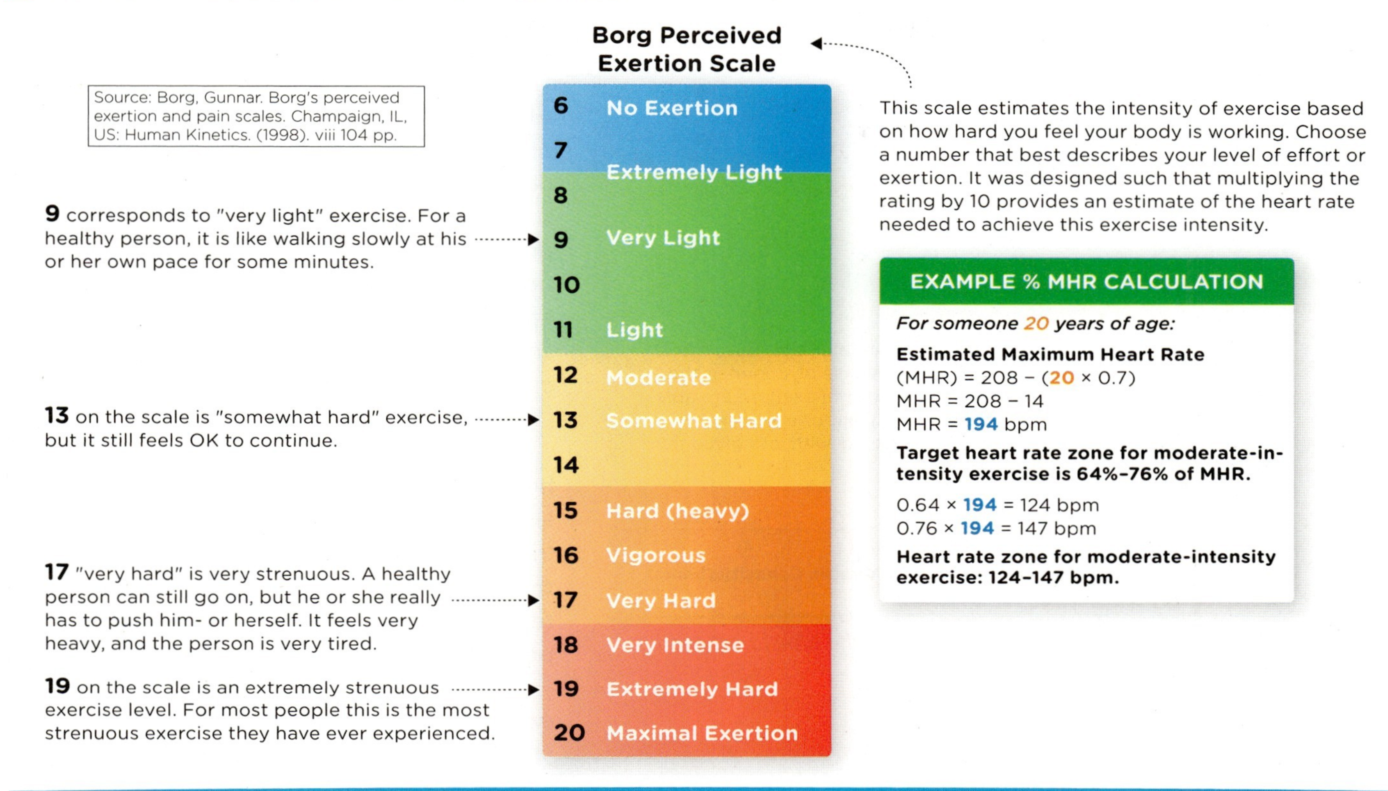

Vigorous-intensity exercise is defined as an exercise intensity resulting in a sustained heart rate that is 77% to 95% of maximum heart rate. Calculate the heart rate range that corresponds to vigorous-intensity exercise for a 30-year-old individual.

The accuracy of determining exercise intensity using heart rate depends on the accuracy of estimating maximum heart rate (MHR). One of the best methods for estimating maximum heart rate is MHR = 208 − (age × 0.7). However, this is only an estimate of MHR, and it does not work well for everyone. For example, older individuals who have remained fit throughout their adult life typically have a MHR well-above what is predicted by this equation. For an average 22-year-old, vigorous-intensity exercise—such as a running a 5K race—would equate to exercising at 77% to 95% of her maximum heart rate, which occurs at heart rates between 148 and 183 beats per minute. Sedentary 22-year-olds may not have the fitness to run a 5K race and could see a similar increase in heart rate by brisk walking the same distance. Increased fitness enables a person to engage in higher-intensity activity more comfortably. **(INFOGRAPHIC 16.7)**

DIETARY CARBOHYDRATES FOR ENDURANCE EXERCISE

Because glucose is such an important source of energy for muscle contraction, the body does its best to store glucose in a form that is easily accessible. Dietary carbohydrates are stored in the liver and muscle as **glycogen**, a polymer of glucose (see Infographic 4.2). At any one time, a body's stores of glycogen can fuel up to 2,000 calories of activity—about 1.5 to 2.5 hours of high-intensity exercise. Athletes who run out of muscle glycogen during training or competition experience sudden, serious fatigue known as "hitting the wall."

GLYCOGEN
storage form of carbohydrate in the muscle and liver

INFOGRAPHIC 16.8 Muscle Glycogen, Dietary Carbohydrates, and Endurance *The depletion of muscle glycogen during aerobic exercise is the most significant factor leading to exhaustion.*

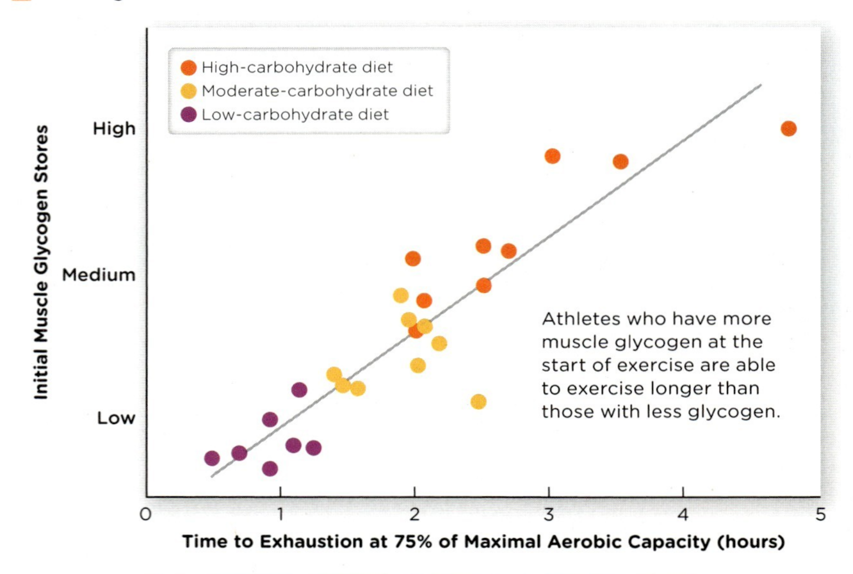

? *Explain why a high-carbohydrate diet increases the time to exhaustion during intense aerobic exercise.*

To delay or prevent this from happening, athletes who are training on consecutive days must consume enough carbohydrates each day to allow glycogen stores to be replenished before their next training bout. If glycogen stores are not replenished at the beginning of a training session, exercise performance will suffer, while perception of the effort required—the discomfort—will increase. Therefore, exercising in a glycogen-depleted state not only feels terrible, it also decreases training effectiveness. Because exercise of longer duration and higher intensity uses more glycogen, total daily carbohydrate intake should increase as training volume (total amount of work done) and intensity increase to ensure that glycogen stores are replenished. **(INFOGRAPHIC 16.8)**

To increase muscle glycogen stores even more, and further delay fatigue during competition, endurance athletes often use a technique called *carbohydrate loading*. The resulting increase in muscle glycogen stores can increase time to exhaustion during intense exercise by 90 minutes or more. To "carbo load," athletes will typically consume about 10 grams to 12 grams of carbohydrate per kilogram of body weight each day, for 1 to 2 days prior to competition, while reducing training volume and intensity. At 1 to 2 hours before exercise, they will often consume meals or snacks that are rich in carbohydrates to top off glycogen stores and to provide additional carbohydrates to be used during the upcoming bout of exercise. As the time between eating and beginning exercise increases so should the amount of carbohydrate consumed before an event or training bout. **(INFOGRAPHIC 16.9)**

Eating carbs *during* exercise can also be helpful. During competitions, for instance, Hubbell often eats carb-heavy snack bars. Ingesting carbs during intense exercise that lasts more than an hour has consistently been shown to improve exercise performance.

For exercise sessions lasting 1 to 2 hours, experts recommend that carbohydrates should be consumed at a rate of at least 30 grams per hour. As the exercise duration increases, so also should the amount of carbohydrates ingested. When the carbohydrates being consumed comprises glucose alone (dextrose, maltodextrin, and/or starch), the maximum rate of ingestion should be limited to 72 grams per hour, because glucose cannot be absorbed from the small intestine and utilized faster than this. This rate of carbohydrate intake should be sufficient for exercise sessions lasting up to about 2.5 hours.

High rates of carbohydrate ingestion during exercise can lead to discomfort in some people; however, research has shown that athletes can become accustomed to this practice when they regularly consume carbohydrates during training, along with a high-carbohydrate diet. This emphasizes the importance of practicing any nutrition support strategy during training, and not trying out a new strategy for the first time during competition. Carbohydrates in the form of beverages, gels, or bars are all equally good; however, when consuming gels or bars one

INFOGRAPHIC 16.9 Carbohydrate Loading to Increase Glycogen Stores *When dietary carbohydrate intake is high (10-12 g/kg body weight/day), and training intensity and duration is decreased, glycogen stores in skeletal muscle will increase significantly, and this will improve endurance performance by increasing the time to exhaustion.*

SAMPLE MEAL PLAN FOR A 70-KG ATHLETE

	Kcal	Carbs(g)
BREAKFAST		
Oatmeal, 1 cup	196	28
Blueberries, 1/4 cup	21	5
Brown sugar, 2 tsp	35	9
Milk, 1/2 cup	42	5
Orange juice, 1 cup	117	28
Whole wheat toast	128	24
Raspberry jam, 1 tbsp	50	13
Total	**589**	**112**
MID-MORNING SNACK		
Large banana	121	31
Breakfast bar (BelVita)	190	33
Total	**311**	**64**
LUNCH		
Burrito	577	89
Pulled pork, 3 ounces	*(99)*	*(0)*
Black beans, 1/2 cup	*(114)*	*(20)*
Brown rice, 3/4 cup	*(164)*	*(34)*
Whole wheat tortilla	*(200)*	*(35)*
Large apple	116	14
Large oatmeal raisin cookie	390	62
Total	**1,083**	**165**
DURING TRAINING AND POST TRAINING		
Sports drink, 32 ounces	240	60
Cinnamon-raisin bagel	320	65
Total	**560**	**125**
DINNER		
Green peas, 1 cup	134	25
Herb chicken tortellini, 1 1/2 cup	495	78
Roasted garlic sauce, 3/4 cup	105	21
Italian bread, 2 large slices	162	30
Skim milk, 1 1/2 cup	125	18
Fig bars, 2	110	22
Total	**1,131**	**194**
EVENING SNACK		
Large peach	68	17
Reduced-fat vanilla Ice cream, 1 cup	200	30
Total	**268**	**47**
DAILY TOTAL	**3,942**	**707**

Dionisvera/Shutterstock

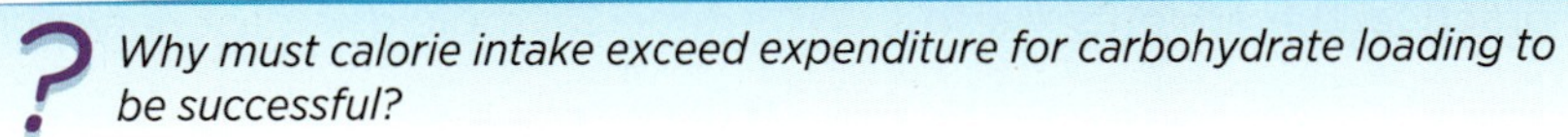

Why must calorie intake exceed expenditure for carbohydrate loading to be successful?

Photo credit: Eli Ensor unless credited otherwise.

INFOGRAPHIC 16.10 Recommendations for Daily Carbohydrate Intake During Endurance Training *These are general recommendations that may need to be adjusted for individual athletes.*

Objective	Carbohydrate Intake	Comments
Daily Carbohydrate Intake:	Total daily carbohydrate intake increases with increasing training volume (intensity and/or duration) to provide adequate fuel during exercise and to allow for the replenishment of glycogen stores during recovery.	
Moderate Intensity about 1 hour per day	5–7 g per kg* body weight (b.w.) per day	
Moderate to High Intensity 1–3 hours per day	7–10 g per kg b.w. per day	
Extreme Intensity 4–6 hours per day	10–12 g per kg b.w. per day	
Carbohydrate Loading	10–12 g per kg b.w. per day	Consume for 1–2 days prior to competition while reducing training volume and intensity.
Pre-exercise Meals 1–4 hours prior to exercise	1–4 g per kg b.w. at 1–4 hours prior to exercise or competition	When eating 2 hours prior to exercise, consume 2 g per kg b.w.; at 3 hours, 3 g per kg b.w., etc.
During Intense Exercise:		
Lasting 1–2 hours	30–60 g per hour	A sports drink should not be more concentrated than an 8% solution (80 g/L, or about 9.5 kcal/oz.) as this will slow gastric emptying when the carbohydrate source provides only glucose.
Lasting 2–2.5 hours	60 g per hour	
Lasting ≥2.5 hours	Up to 90 g per hour	Must consume a mix of glucose and fructose.

* 1 kg = 2.2 lb

Modified from: Burke, L.M., et al., Carbohydrates for training and competition. J Sports Sci. 2011;29:S17–27.
Jeukendrup, A. The new carbohydrate intake recommendations. Nestle Nutr Inst Workshop Ser. 2013;75:63–71.

How many grams of carbohydrate should a 154-pound athlete consume two hours prior to competition? What is the lowest amount of kcal from carbohydrates that should be consumed per day by a 132-pound athlete who trains four hours a day?

must be mindful to consume adequate water. (INFOGRAPHIC 16.10)

U.S. GYMNASTICS TEAM IN TRAINING

Historically, athletes and their coaches have thought about energy needs very differently. When Benardot was first asked to advise the U.S. gymnastics team in 1992, for instance, he observed one of their practices to see if he could identify any problems. He saw that the gymnasts trained every afternoon for five hours straight, but they were never allowed to eat or drink anything during that period. About halfway through the practice, the athletes started getting grumpy and didn't perform as well, which Benardot attributed to the fact that their blood sugar levels had dropped very low. That evening there was a team meeting, and Benardot told the coaches, "If you really want to get the most out of these practices, you need to feed these gymnasts every two or three hours." As Benardot recalls, the coaches looked at him as if he had just been dropped in from Mars. "They said, 'Are you kidding? This is gymnastics; we don't have to eat during gymnastics.'" Benardot didn't give up, however, and eventually, the team's head technical coach, a three-time Olympic gymnast named Muriel

Dan Benardot, PhD, RD, LD, FACSM, provides nutrition counseling to Olympic-level athletes to help them improve performance in sport.

Robin Bernadot

Grossfeld, agreed to try Benardot's nutritional approach. Once.

In his trial run, Benardot walked out onto the gymnastics floor exactly halfway through an afternoon practice with a tray of fruit and juice. The gymnasts looked at him curiously and apprehensively. "This is a breach of protocol," one said. But Dominique Dawes, the team captain, called her team members over and told them to start eating. They did, nervously but happily. Exactly 15 minutes later, Grossfeld took gymnast Amanda Borden aside, who had been struggling to perfect her vault routine, and asked her to try it one more time. "I'm very certain," Benardot recalls, "that Muriel was thinking that Amanda would start projectile vomiting—that she was trying to make a point that gymnasts really can't snack. But Amanda did the vault perfectly for the first time."

The success convinced Grossman that snacking might be a good idea, and the team began taking regular food breaks during their practices. Four years later, at the 1996 Olympics in Atlanta, the team won the first gold medal ever awarded to a U.S. Women's Gymnastics Team.

FEMALE ATHLETE TRIAD

Extended periods of low energy intake also place female athletes at risk for the **female athlete triad**, which refers to the interrelationships among low energy intake, menstrual dysfunction (called **amenorrhea**), and bone loss (osteoporosis). Inadequate energy intake may result from eating disorders (such as anorexia nervosa or bulimia nervosa), intentional caloric restriction in an attempt to reduce body fat or weight, or because she is unintentionally eating inadequate calories to meet the increased energy needs of an athlete. Doing what the U.S. Gymnastics Team used to do—not eating during their long training hours—may make it difficult to consume adequate energy and essential nutrients during the remaining portion of the day. When energy intake is inadequate for an extended time, the processes that are necessary to sustain reproductive and bone health are impaired,

FEMALE ATHLETE TRIAD
a condition recognized in female athletes, characterized by interrelated energy restriction, menstrual dysfunction, and bone loss (osteoporosis)

AMENORRHEA
an abnormal absence of menstruation

Getting stronger. *Muscles grow when they are challenged by strength training, and properly fueled by a balanced diet.*

Bud Force/Getty Images

and this can have life-long consequences that are not reversed even when energy balance is later restored.

BODY BUILDING

Exercise—particularly resistance training—leads to increases in muscle mass and strength through the growth of muscle fibers. Muscle fibers are largely made of protein. To repair and build new muscle fibers, the body needs amino acids for the assembly of new protein. A large industry exists around the promotion of amino acid and protein powders marketed to athletes and bodybuilders, based on the notion that added protein in the diet will help to build additional muscle. This is false. Amino acids and protein powders alone do not cause muscle growth; protein intake has to be combined with exercise—particularly strength training—and sufficient calorie intake to increase muscle mass.

Most people, including athletes, get plenty of protein from food—adult men in the United States tend to consume about 99 grams of protein per day, about 75% more than the recommended daily allowance (RDA) for protein, whereas women eat around 68 grams, about 40% more than their RDA. Some strength athletes, such as competitive weight lifters, may benefit from protein intakes that are 50% to 100% over the recommended intake, but most individuals already consume this amount. (For more on the protein needs of athletes refer to Chapter 8.) Furthermore, consuming more protein than the body needs will cause an increase in urea production and increased amounts of urine as the kidneys excrete excess nitrogen, which may make it more difficult to stay well hydrated.

What about vitamins and minerals? As with protein powders, supplemental vitamins and minerals have few advantages. Athletes who eat an adequate, balanced, and varied diet are typically able to meet their vitamin and mineral needs. There is limited evidence that athletes may require slightly more riboflavin and vitamin B_6 than sedentary individuals. However, the increased energy expenditure that accompanies training generally results in an increase in food intake and an adequate

Female athletes have special nutritional concerns. *Although a healthy diet is important for all athletes, female athletes need sufficient calories and adequate nutrients, including iron (to reduce the risk of iron-deficiency anemia) and calcium (for bone health).*

Jonathan Larsen/Diadem Images/Alamy

intake of these vitamins when sound nutritional principles inform the athlete's food choices. The exception to this may be in situations when athletes are restricting calories during training to reduce body fat. In this circumstance, a multivitamin supplement may be useful. There is more extensive evidence that some athletes—female endurance athletes, especially—are at risk for iron deficiencies. Iron is a component of hemoglobin, the protein that carries oxygen to muscles. Because of possible depletion of iron stores with intense training, loss of iron from menstruation, as well as inadequate intake of foods rich in iron, many female athletes may have suboptimal iron status and require supplementation. (Refer to Chapter 14 for more information.)

HYDRATING THE ATHLETE

The human body is 70% water, and water is essential for proper cell function. Maintaining adequate hydration by drinking water is important because we all lose water through the normal bodily processes of urination, perspiration, exhalation, and defecation, and because **dehydration** can have dangerous consequences. Athletes lose and need more fluids than the average person does, and even in cool conditions fluid losses through sweating can be quite high during intense exercise. Exercising in hot or humid conditions will greatly exacerbate these losses, and everyone needs to be concerned with staying hydrated when engaging in any type of outdoor physical activity (such as yard work or hiking) in these conditions. Without enough fluids in our body, we can become nauseous, dizzy, lightheaded, and confused.

Sweating rates increase with body weight, exercise intensity, hot or humid conditions, and aerobic exercise training. When engaging in intense exercise at even moderate temperatures, it is impossible to stay hydrated as sweat losses exceed the rate at which water can be absorbed from the gut. The goal of fluid consumption during exercise is to limit fluid loss to 2% of body weight, as fluid losses beyond this compromise aerobic exercise performance. Since it is often impossible to stay fully hydrated during intense exercise, it is of utmost importance to begin every training session or competitive event fully hydrated. To monitor their hydration status, athletes should weigh themselves immediately before and after training to assess how much body weight was lost during the training bout. One pound of weight loss is the equivalent to losing about 16 ounces of water. Individuals can also look at the color

DEHYDRATION
water deficiency caused by insufficient intake or excessive loss of fluids

INFOGRAPHIC 16.11 Fluid Losses and Recommendations for Fluid Replacement *Sweating rates vary among individuals and are strongly influenced by exercise intensity and environmental conditions. In most conditions, exercise duration must exceed one hour to experience water loss sufficient to impair performance.*

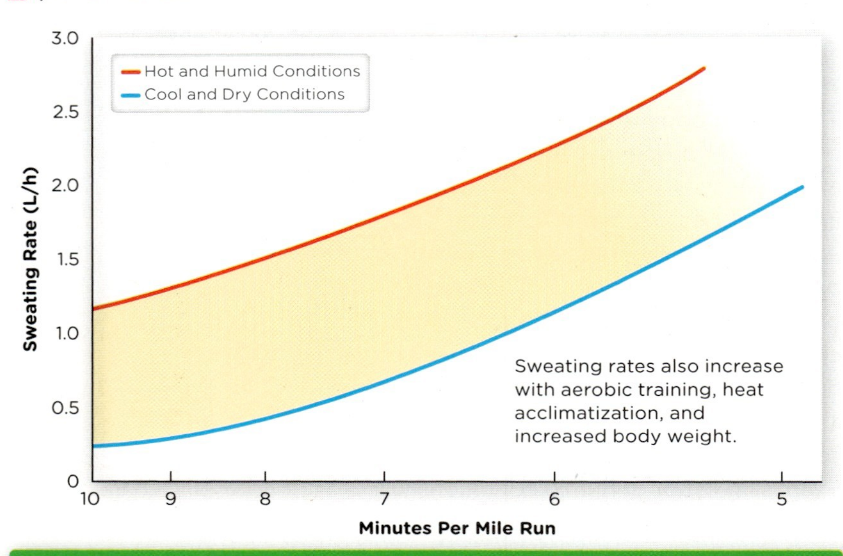

Recommendations to Avoid Fluid Loss

BEFORE EXERCISE:

- ✓ Drink 5–7 ml/kg body weight 4 hours prior to exercise (~½ – ⅓ cup per 25 lb of body weight).

DURING EXERCISE:

- ✓ Drink to avoid a water deficit > 2% of body weight.
- ✓ Start with ½ to 1 cup every 15 minutes if sweating rate is not known.
- ✓ Sports beverages should contain no more than 8% carbohydrates (when provided as glucose only), or about 9.5 kcal per ounce.

AFTER EXERCISE:

- ✓ For every pound of body weight lost during exercise, consume 16–24 ounces of fluid (2–3 cups).
- ✓ If not eating following exercise, the rehydration beverage should contain electrolytes (sodium).

Figure from: Sawka M, Pandolf K. Effects of body water loss on physiological function and exercise performance. In: Gisolfi C, Lamb D, editors. Perspectives in exercise science and sports medicine. Vol. 3. Fluid homeostasis during exercise. Carmel (IN): Benchmark Press Inc., 1990: 1–38

If an athlete lost three pounds of body weight during training, how much fluid should they consume after exercise?

of their urine; it should be pale yellow, not bright or dark yellow. **(INFOGRAPHIC 16.11)**

ERGOGENIC AIDS *substances or treatments believed to improve athletic performance*

Drinking water too quickly or too frequently while exercising, however, can be dangerous and lead to *hyponatremia,* a condition in which there is a deficiency of sodium in the blood, a potentially fatal situation given the importance of sodium to normal cell function. Hyponatremia generally occurs only in individuals exercising for extended periods at low intensity and in mild conditions while consuming only water. However, it may also occur when athletes attempt to rehydrate following intensive, prolonged exercise (like running a marathon) by consuming only water and without eating any food.

An effective way to stay hydrated during or after prolonged exercise, without becoming hyponatremic, is to consume sports drinks, which hydrate as well as replenish sodium and other electrolytes that may be lost through sweating. Hubbell, for instance, consumes a sports drink during each training session. Sports drinks also typically contain carbohydrates, which provide fuel for muscle contractions during exercise, and help to replenish glycogen stores during recovery. Sports beverages should contain no more than 8% carbohydrates (when provided as glucose only), or approximately 9.5 kcal per ounce. For the recreational athlete and those who engage in moderate exercise, drinking water is sufficient to stay hydrated.

SPORTS SUPPLEMENTS

In competitive sports, the pressure to outperform other athletes can be fierce. Not surprisingly, many athletes resort to **ergogenic aids** to improve performance. Ergogenic aids are any physical, mechanical, nutritional, psychological, or pharmacological substances or treatments that either directly improve physiological variables associated with exercise performance or remove subjective restraints that may limit physiological capacity. Purported ergogenic aids abound, but research is limited regarding how well they actually improve performance and long-term fitness. Hubbell, for one, doesn't use them.

Two ergogenic aids that have been well studied and shown to be efficacious

Exercise benefits all. *Moderate-intensity activity includes walking briskly, pushing a lawn mower, or ballroom dancing. Vigorous-intensity activity includes jogging, racewalking, running, swimming laps, playing basketball, biking faster than 10 miles per hour, or hiking uphill.*

imageBROKER/Alamy

Christopher Futcher/Getty Images

Radius Images/Alamy

are caffeine and creatine monohydrate. **Caffeine**, a central nervous system stimulant, has been repeatedly demonstrated to improve performance in almost all high-intensity activities lasting more than one minute, including long-distance endurance events. Studies have shown that supplementing with creatine monohydrate increases skeletal muscle creatine phosphate (recall its role in ATP production) by 10% to 20%. Benefits to performance are commonly seen with short duration, anaerobic, intermittent exercise.

CAFFEINE *a central nervous system stimulant*

PHYSICAL FITNESS: NOT JUST FOR ATHLETES

In the United States, too few individuals get the recommended physical activity they need to stay healthy. It is estimated that in 2010 approximately one-third of all U.S. adults (18 years and older) participated in absolutely no leisure-time physical activity.

Part of the problem is that pervasive technologies such as escalators, elevators, cars, and moving sidewalks have eliminated the need for much activity or exertion. In addition, people spend time watching television and playing video games instead of, say, playing touch football or taking walks, and office jobs have replaced some manual labor. Americans sit, on average, for four hours a day, but a 2012 study reported that if Americans reduced this to less than three hours a day, they would live, on average, two years longer.

The U.S. Department of Health and Human Services recommends that adults participate in at least 2 hours and 30 minutes of moderate-intensity exercise per week plus muscle strengthening activities, or 75 minutes of high-intensity exercise per week plus muscle strengthening activities, or an equivalent combination of the two. Relatively modest amounts of physical activity will improve the fitness of inactive people, but people who want to reap substantial health benefits such as weight maintenance, or reducing the risk of diabetes or heart disease, may need to participate in more than 30 minutes of moderate activity most days of the week. To achieve significant weight loss, more than

45 minutes most days of the week would be needed.

The American College of Sports Medicine aligns with the United States Department of Health and Human Services to convey the message that *any* amount of exercise is better than none. Even modest amounts of exercise can reduce the risk of cardiovascular disease and type 2 diabetes. With an environment that fosters a sedentary lifestyle, pursuits of opportunities to move more must be intentional. Most people are not professional athletes for whom exercise is inherent in their jobs. Scheduling walks or runs with friends, taking the stairs whenever possible, enrolling in exercise or dance classes, participating in recreational sports, tracking and boosting the number of steps you take per day by wearing a pedometer or downloading a tracking app on your phone, or *just sitting less* can all help increase physical activity and personal fitness.

So has Benardot's strategy for eating to win made a difference for Hubbell? Yes—in several important ways. Seven months after she weighed in at 22% body fat, she was tested again. She had lost three pounds and had built two-and-a-half pounds of muscle, giving her a body fat percentage of a little more than 15%—a 32% drop. "And it was crazy how much different I felt," she says. Training sessions were still hard, of course, but she felt stronger and better able to tackle the challenges that her routines presented. "As hard as the sport is, your muscles can do it, your body can do it—it's just whether you're feeding them and making sure they have the right things to use for energy," she says.

CHAPTER 16 **BRING IT HOME**

Weekly exercise log

Use this form to track your activity for a week as you consider incorporating the four components of fitness as set forth by the American College of Sports Medicine. Record a brief description of your activities along with the length of time spent when indicated. At the end of the week total the amount of time and/or episodes at the bottom of the log.

The American College of Sports Medicine Recommendation			
Cardiorespiratory (Aerobic)	**Resistance**	**Flexibility**	**Neuromotor**
Activities that increase the heart rate and promote increased use of oxygen to improve overall body condition and endurance	*Activities performed against an opposing force that increase muscle strength, improve body composition, and promote healthy bone density*	*Activities that promote the ability to move joints through their whole span of movement*	*Activities that incorporate motor skills such as balance, coordination, and agility (also known as functional fitness)*
30–60 minutes of moderate-intensity exercise (5 days per week) OR 20 or more minutes of vigorous-intensity exercise (3 or more days for total of ≥ 75 minutes per week)	2–3 days per week train each major muscle group Major Muscle groups: legs, hips, back, abs, chest, shoulders, arms	2–3 days per week do flexibility exercises to improve range of motion	20–30 minutes of exercises involving motor skills—balance, agility, coordination, gait (2–3 days per week)
Effort Scale: 6–20 (see Borg Perceived Exertion Scale Infographic 16.7); 6 = No Exertion; 20 = Maximal Exertion			
Moderate Intensity (64%–76% of Maximal Heart Rate) Relatively moderate-intensity activity is a level of perceived effort of about 7 to 8. • Walking briskly (3 miles per hour or faster, but not race-walking) • Water aerobics • Bicycling around 10 miles per hour • Tennis (doubles) • Ballroom dancing • General gardening		Vigorous Intensity (77%–95% of Maximal Heart Rate) Relatively vigorous-intensity activity is a 12 or 14 on this scale. • Race-walking, jogging, or running • Swimming laps • Tennis (singles) • Aerobic dancing • Vigorous cycling • Jumping rope • Heavy gardening (continuous digging or hoeing, with heart rate increases) • Hiking uphill or with heavy backpack	

Type of Exercises					
	Aerobic Exercise		Resistance Training	Flexibility	Neuromotor
Date	**Description & Intensity**	**Duration**	**Description**	**Description**	**Description**
Monday					
Tuesday					
Wednesday					
Thursday					
Friday					
Saturday					
Sunday					
	My total number of minutes or hours this week				
	American College of Sports Medicine Recommendation	~150 minutes Moderate intensity OR ~75 minutes Vigorous intensity	2–3 days	2–3 days	50–75 minutes

Take It Further

Comment on how your documented activity compared with recommendations. List ways you (or your less-active friends) could achieve the recommended levels of activity for each of the four types of exercise: aerobic, resistance training, flexibility, and neuromotor.

KEY IDEAS

- Physical fitness is the ability to perform moderate to vigorous activity without undue fatigue and can be achieved through regular exercise or intentional physical exertion.
- The four types of exercise recommended by the American College of Sports Medicine for promoting maximal health and preventing diseases and injuries include resistance, cardiorespiratory, neuromotor, and flexibility.
- The increased energy demands of exercise are fueled by energy substrates in the form of glucose and fats in the form of fatty acids.
- Both glucose and fatty acids are rich in chemical energy that must be converted into a form that cells can use, namely ATP (adenosine triphosphate).
- ATP, the primary energy currency of our cells, is produced in the body by three separate energy systems, two of which are anaerobic (not requiring oxygen) and occur in the cytosol of the cell, and one of which is aerobic (requiring oxygen) and occurs in the mitochondria.
- The energy contributions made through anaerobic and aerobic pathways combine to provide muscles with enough ATP to meet demands. The percentage, or relative contribution, of each depends on the intensity and duration of the activity.
- The anaerobic energy system supplies ATP quickly for high-intensity exercise, but only for a short time and relies heavily on carbohydrates as fuel.
- Longer-lasting, lower-intensity activities rely almost entirely on the aerobic energy system, which produces ATP more slowly and utilizes a higher percentage of fat as a fuel source.
- Several methods can be used to estimate the intensity of aerobic activity. Two common methods are the Borg Rating of Perceived Exertion Scale, and heart rate as a percent of maximum heart rate.
- The depletion of muscle glycogen during aerobic exercise is the most significant factor leading to exhaustion. Carbohydrate loading increases muscle glycogen stores, which can help athletes sustain and recover from high-intensity endurance exercise.
- Protein supplements are generally not warranted and alone do not cause muscle growth; protein intake must be combined with exercise—particularly strength training—and sufficient calorie intake to increase muscle mass.
- The goal of hydration during exercise is to limit fluid loss to less than 2% of body weight.
- Overhydrating through excess water consumption while exercising can be dangerous and lead to hyponatremia (low levels of sodium in the blood).
- Individuals can reduce their risk of chronic diseases and weight gain through moderate exercise for at least 30 minutes a day, most days of the week.

NEED TO KNOW

Review Questions

1. The four types of exercise specifically recommended by the American College of Sports Medicine include all of the following, EXCEPT:
   a. calisthenics.
   b. cardiorespiratory.
   c. flexibility.
   d. neuromotor.
   e. resistance.

2. All of the following are TRUE with regard to ATP, EXCEPT that it:
   a. is often referred to as the energy currency of cells.
   b. stores energy in the bonds of its three phosphate groups.
   c. can be produced both aerobically and anaerobically.
   d. cannot be produced without the presence of oxygen.
   e. is the abbreviation for the chemical compound adenosine triphosphate.

3. The aerobic energy system occurs in what part of the cell?
   a. cellular membrane
   b. cytosol
   c. mitochondria
   d. nucleus
   e. ribosome

4. Anaerobic energy systems:
   a. provide energy for low-intensity exercise.
   b. provide energy for short, intense exertion.
   c. provide unlimited energy.
   d. require oxygen.
   e. utilize fatty acids as primary fuel source.

5. More carbohydrate is used as fuel when someone is:
   a. walking.
   b. jogging.
   c. running.
   d. sprinting.

6. All of the following are TRUE with regard to carbohydrate loading, EXCEPT that it:
   a. can increase time to exhaustion during intense exercise by 90 minutes or more.
   b. helps prevent depletion of glycogen during activities of long duration.
   c. is intended to increase glycogen stored in muscle.
   d. promotes unlimited glycogen stores.

7. Female endurance athletes are at higher risk for being deficient in which of the following nutrients?
   a. biotin
   b. chromium
   c. iron
   d. sodium
   e. zinc

8. Female athletes who restrict energy intake and experience menstrual dysfunction are at higher risk for the long-term complication of:
   a. high blood pressure.
   b. neural tube defects.
   c. osteoporosis.
   d. type 2 diabetes.

9. What would you tell a friend who inquires if amino acid and protein supplements will help build muscle?
   a. Amino acids and protein powders alone do not increase muscle mass.
   b. Protein intake has to be combined with exercise and sufficient calorie intake to increase muscle mass.
   c. Excess protein intake contributes to energy needs or may be stored as fat.
   d. It is likely he or she is already consuming sufficient protein.
   e. All of the above.

10. All of the following are TRUE with regard to hydration for athletes, EXCEPT:
    a. It is impossible to consume too much water—the more the better.
    b. You should begin intense activities fully hydrated.
    c. Urine color can be an indication of hydration status.
    d. Comparing body weight before and after exercise is a good indication of hydration status.
    e. Sports drinks may help hydrate and prevent hyponatremia during and following prolonged exercise.
11. According to the U.S. Department of Health and Human Services Physical Activity Guidelines, adults gain substantial health benefits from _____ a week of moderate-intensity aerobic physical activity or _____ a week of vigorous activity.
    a. 2 hours; 1 hour
    b. 2.5 hours; 1 hour and 15 minutes
    c. 3 hours; 2 hours
    d. 4.5 hours; 2 hours and 20 minutes
    e. 5 hours; 3 hours

Analyzing Exercise Patterns Using SuperTracker

Understanding your level of physical activity

Regular physical activity has many health benefits, yet the majority of Americans do not get the amount of regular physical activity recommended by health experts. Using SuperTracker, you will be able to compare your exercise patterns with current recommendations. You will also be able to identify the appropriate dietary changes needed as you become physically active.

1. Log onto the United States Department of Agriculture (USDA) website at www.supertracker.usda.gov. If you have not done so already, you will need to create a profile to get a personalized diet plan. This profile will allow you to save your information and diet intake for future reference. Do not use the general plan.
2. Click the Track Food and Activity option. Record three days of activity.
3. After you have entered your activity for each day, you can then use the "My Reports" feature to analyze your activity.
    a. Did you meet the weekly aerobic activity target for the week? Discuss how your activity compares with the recommendations.
    b. Did you meet the weekly muscle-strengthening activity target for the week? Discuss how your activity compares with the recommendations.
    c. List two changes you can make to meet your activity recommendations.
    d. Discuss the dietary recommendations for an athlete. How are the recommendations for carbohydrate and protein different for an athlete than for a person who is not physically active?
    e. Why might an iron supplement be recommended for a female athlete?
    f. In your opinion, how can health experts motivate Americans to increase their level of physical activity? What motivates you to increase your physical activity?

17 NUTRITION FOR PREGNANCY, BREASTFEEDING, AND INFANCY

Nourishing Mother and Baby

NEW RESEARCH SUGGESTS THAT DELAYING CORD CLAMPING AFTER BIRTH PRESERVES IRON STATUS.

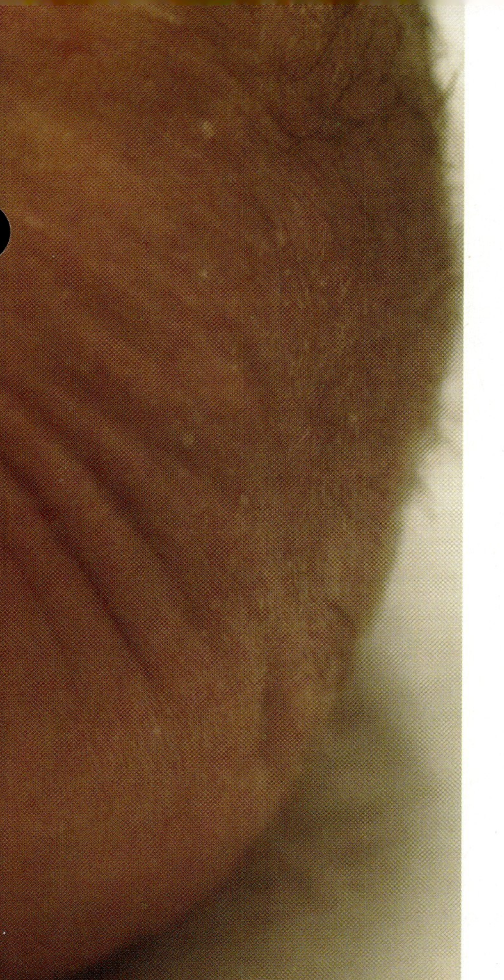

Gallery Stock

LEARNING OBJECTIVES

- Discuss recommendations and rationale for appropriate weight gain during pregnancy (Infographic 17.3)
- Define small for gestational age (SGA) and describe consequences of a small-for-gestational-age birth weight (Infographic 17.4)
- Compare energy and nutrient needs during pregnancy and lactation and compare to recommendations for non-pregnant women (Infographic 17.5)
- Identify foods and beverages that should be avoided during pregnancy and explain why they should be avoided (Infographic 17.6)
- Identify at least five benefits infants derive from breastfeeding (Infographic 17.8)
- Describe appropriate growth patterns in the first two years of life (Infographic 17.10)
- Discuss timing and rationale for the introduction of solid foods into an infant's diet (Infographic 17.11)

Few sights are more distressing to a midwife than that of an unresponsive newborn. But this was the situation Judith Mercer, a certified nurse-midwife, was facing when she assisted in a home birth in 1979. As she stared at the pale and lifeless little boy, she was terrified that she wouldn't be able to resuscitate him. She tried to stimulate him with her fingers and he did not respond. She looked down at the umbilical cord, which connected the baby directly to his mother's placenta, and saw that it was pulsating—still delivering blood to the boy's body. His color was ever-so-slowly returning. A minute-and-a-half later, the little boy opened his eyes and took a quiet breath. He did not even cry. The still-attached umbilical cord,

Cutting the umbilical cord. *Babies are born attached to their mothers by the umbilical cord, which delivers nutrients and oxygen throughout pregnancy. After birth, the cord is clamped near the baby's navel, and is then cut. Delaying cord clamping by a minute or two may improve a baby's iron stores.*

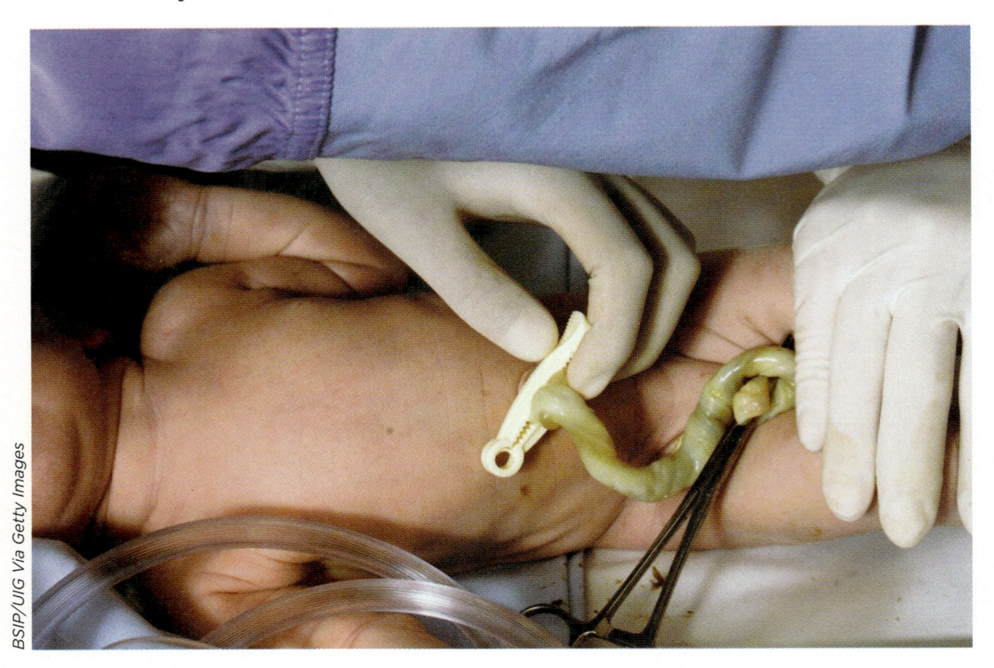

BSIP/UIG Via Getty Images

Mercer realized, had seemingly saved this boy's life by pumping blood and oxygen—life, basically—from his mother's body to his own.

Mercer was awed by what she had seen. She vowed, right then and there, to study what she had witnessed and understand why and how the umbilical cord might have saved that baby boy. "That baby affected the whole second half of my lifetime," she says. Mercer went back to school, got a PhD, and as a clinical professor of nursing at the University of Rhode Island, she has dedicated her life to studying the benefits of what is known as *delayed cord clamping*. Usually, within seconds of a birth, obstetrical care providers clamp the umbilical cord connecting mother to baby, severing the blood flow between them. As Mercer's research has shown, however, delaying the clamping of the umbilical cord by several minutes can provide benefits. Not only does the cord continue to provide blood and oxygen to the newborn in the event of a trauma, but it also increases the baby's stores of iron—crucial for the development of

Judith Mercer, D.N.Sc., C.N.M., F.A.C.N.M. Professor Emerita, University of Rhode Island. *Dr. Mercer is the principal investigator studying the protective effects of delayed cord clamping in very low-birth-weight infants.*

Joe Giblin/Giblin & Company Photography

certain types of brain cells—yet a nutrient in that some babies may become deficient of, as breast milk is not naturally rich in iron.

In March 2013, the Cochrane Collaboration, a nonprofit international consortium funded in part by the U.S. Department of Health and Human Services, analyzed the results of 15 clinical trials—many conducted by Mercer—and concluded that delayed clamping increases birth weight (by increasing the newborn's blood volume) and halves the risk that babies will be iron-deficient when they are three or six months old. (Iron deficiency is a risk factor for future cognitive problems.) Yet few healthcare providers who deliver babies today delay the clamping of umbilical cords, in part because it takes extra time and in part because delaying clamping slightly increases the risk for newborn jaundice, a condition that arises when a baby's liver has trouble processing a byproduct of red blood cell breakdown called bilirubin so that it can be excreted. Jaundice, however, can usually be treated with ultraviolet light therapy in the hospital and rarely causes complications.

Ultimately, Mercer hopes that with additional research—she is now starting a clinical trial to see whether delayed cord clamping improves scores on developmental tests at age two years—she can prove that the practice is both safe and beneficial.

■ ■ ■

Although Mercer's research points to the importance of the **umbilical cord** after birth, its primary purpose is to supply nutrients and oxygen to a developing baby during **pregnancy**, the period from fertilization to birth. Pregnancy begins when a woman's egg is fertilized by a sperm, forming a zygote that develops into an embryo and then a fetus.

Starting about five weeks after fertilization, the umbilical cord develops and then supplies the fetus with nutrients and oxygen and removes waste until it is clamped after birth. The fetus is carried in a fluid-filled amniotic sac in the muscular organ known as the **uterus**. The umbilical cord connects the fetus to the **placenta**, which is attached to the uterus. (INFOGRAPHIC **17.1**)

UMBILICAL CORD
a ropelike structure that supplies the fetus with nutrients and oxygen and removes waste

PREGNANCY
the condition of being pregnant, encompassing the time from fertilization through birth

UTERUS
a muscular organ that holds the developing fetus

PLACENTA
organ within the uterus that allows for exchange between maternal and fetal circulations, via the umbilical cord

■ INFOGRAPHIC **17.1** **The Placenta** *The placenta connects the uterus to the fetus by the umbilical cord.*

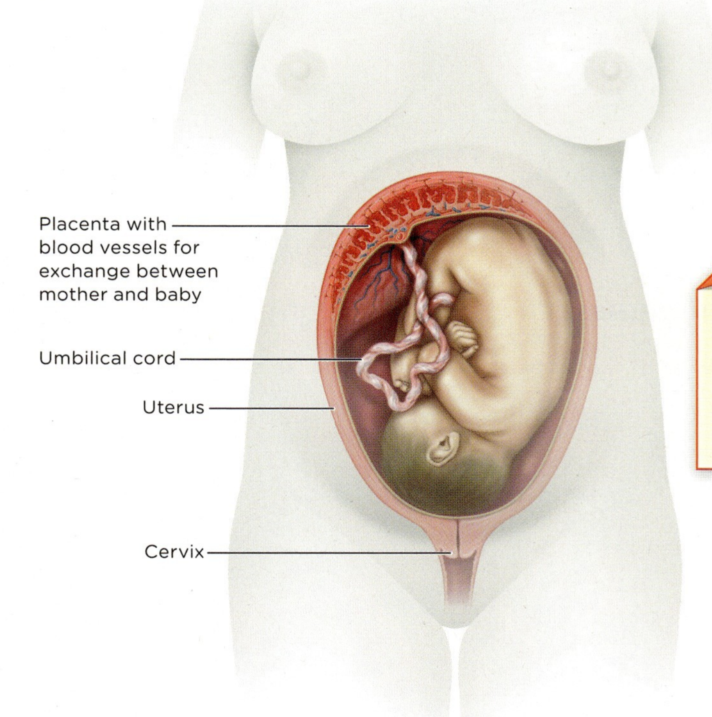

The placenta is a structure that forms during pregnancy within the uterus of a woman. It provides oxygen and nutrients from the mother to a growing baby, and removes waste products from the baby's blood through the umbilical cord.

EMBRYO
the developing human during the first two to eight weeks of gestation

FETUS
the developing human from eight weeks of gestation until birth

GESTATION
the time during which the embryo develops in the uterus—from fertilization to birth

TRIMESTER
one-third of the normal gestation period of a pregnancy

Between two and eight weeks after fertilization, as the organs and vital systems begin to develop, a developing human is referred to as an **embryo**. At the tenth week of pregnancy (eight weeks after fertilization), the developing human is called a **fetus**, and its organs mature as it puts on significant amounts of weight (from less than one ounce to between about seven and eight pounds at birth).

The entire period of development from fertilization to birth is called **gestation**. Full-term pregnancies last between 38 and 42 weeks, and are, on average, about 40 weeks long, calculated from the first day of the woman's last menstrual period. Pregnancy is split into three periods called **trimesters**: weeks 1 to 13 are considered the *first trimester*; weeks 14 to 27 compose the *second trimester*; and weeks 28 to 40 make up the *third trimester*. (INFOGRAPHIC 17.2)

CHANGING NUTRITIONAL NEEDS

Once a woman is pregnant, her body makes gradual yet significant changes to support the growth of the fetus, and this shift alters her nutritional needs. One key change is that her heart works harder and pumps more blood throughout her body—her blood volume typically increases by about 50%—and her breathing rate increases. Additionally, gastrointestinal motility decreases, slowing the passage of food and potentially causing constipation, a common complaint during pregnancy.

Metabolic rate is also affected, and subsequent energy demands increase between 5% and 20% due to the oxygen demands of the developing fetus and maternal support tissues. A pregnant woman's energy needs thus increase to compensate. Because a fetus prefers glucose as a primary fuel source, the mother's body provides glucose to her developing baby from the

INFOGRAPHIC **17.2** **Pregnancy Timeline** *The entire period of prenatal development from fertilization to birth is called gestation. Because doctors date pregnancy from the first day of the woman's last menstrual period, which is typically two weeks before fertilization, fertilization is shown at week two.*

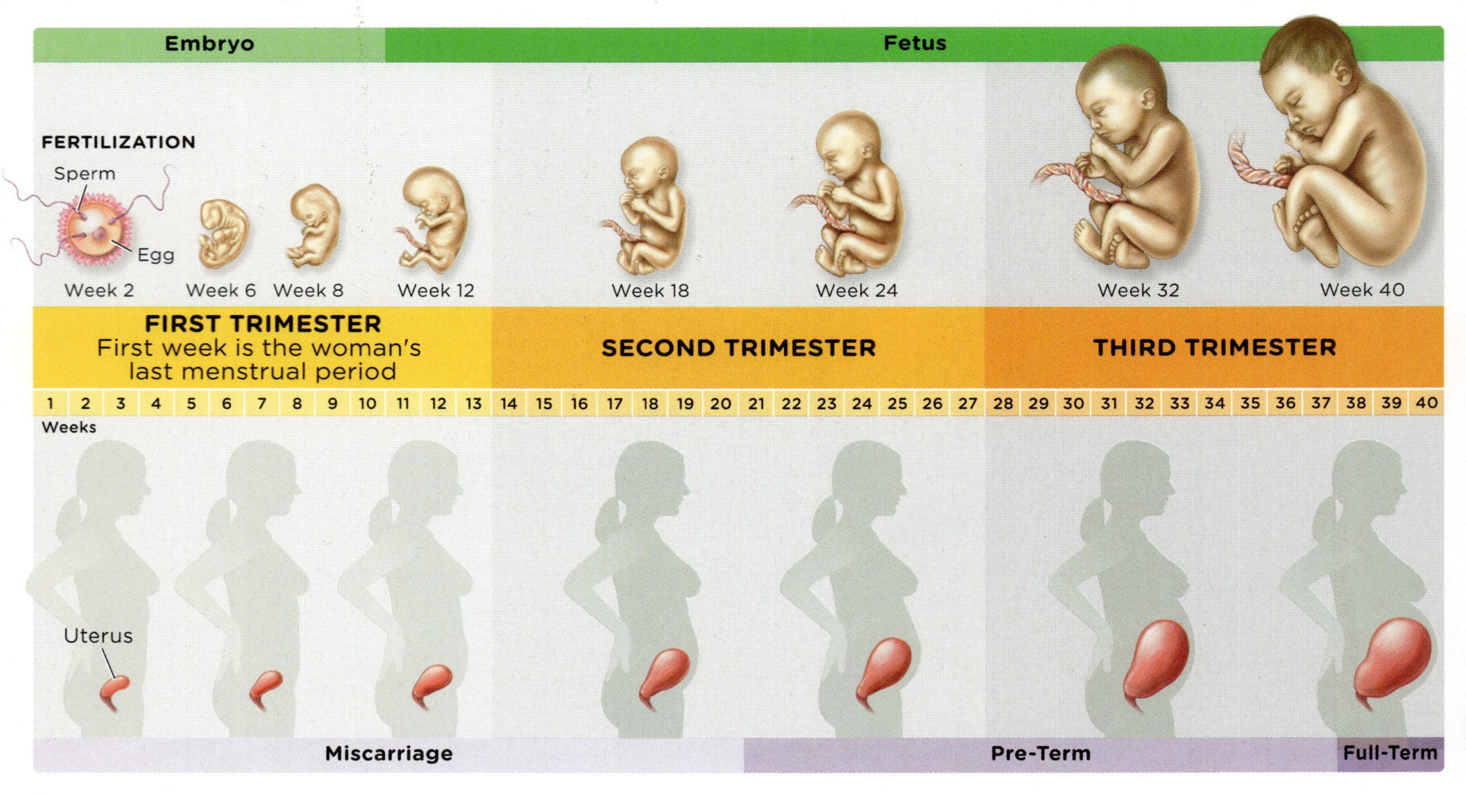

food she digests and uses predominantly fatty acids to fuel her own needs.

Women need extra food and nutrients to help their babies grow. However, more than half of all women are overweight (defined as a body mass index [BMI] over 25) or obese (BMI over 30) when they become pregnant. Still others gain too much weight during pregnancy—both of which can put the mother and the baby's health at risk, according to recent research.

Appropriate weight gain in pregnancy

In 2009, the Institute of Medicine re-examined the guidelines on weight gain and recommended that women should ideally not only have a healthy BMI (see Chapter 15) when they conceive, but that they should gain weight within a certain range that depends on whether they go into pregnancy underweight, at normal weight, overweight, or obese. For example, a woman with a normal or healthy BMI before pregnancy would be given a recommendation to gain 25 to 35 pounds while a woman with a BMI in the obese range would be given a recommendation to limit weight gain to between 11 and 20 pounds. **(INFOGRAPHIC 17.3)**

Underweight women and those who fail to gain sufficient weight during pregnancy can also potentially compromise the health of their babies by increasing the risk that they will be born preterm, meaning younger than 37 weeks, or born too small for their age (**small for gestational age**, or **SGA**). SGA is defined as a birth weight that is below the 10th percentile of gestational age. SGA infants are not only at a higher risk of stillbirth and dying, they also have a heightened risk for medical conditions such as hypertension, diabetes, and heart disease as adults. With new and advanced medical technologies, babies born prematurely or SGA, are more likely to survive—and thrive—than in years past. **(INFOGRAPHIC 17.4)**

Factors influencing birth weight

Because of the risk of not gaining enough weight, even overweight or obese pregnant women should not "diet"; doing so increases the risk of delivering a low-birth-weight baby who has not received sufficient nutrition. Exercise is a different story: Unless otherwise advised by a health care provider, pregnant women can and should stay active throughout their pregnancies. The four main factors that influence birth weight are: (1) the duration of pregnancy; (2) the weight status of the mother before conception; (3) the amount of weight the mother gained during pregnancy; and (4) whether the mother smoked during pregnancy.

Nutrition before conception

Eating well to sustain the fetus involves establishing a healthy lifestyle far before conception occurs. A woman who is physically

SMALL FOR GESTATIONAL AGE (SGA)
a birth weight that is below the 10th percentile of gestational age

Activity is beneficial. *Daily exercise during pregnancy may lift mood, increase energy, reduce backache, and prevent excess weight gain and gestational diabetes.*

Blend Images/Alamy

INFOGRAPHIC 17.3 Weight Gain During Pregnancy *Gaining an appropriate amount of body weight during pregnancy is important for the health of both the mother and her baby.*

DISTRIBUTION OF PREGNANCY WEIGHT GAIN: TOTAL 29 LBS

Lost at or Soon After Birth	Total of 29 lbs Gained	Retained at Birth
Baby 7.5		Fat 7.5
Body water 3.5		Uterus 2
Blood 3.5		Breasts 1.5
Amniotic fluid 2		
Placenta & umbilical cord 1.5		

Recommended Weight Gain Based on Prepregnancy BMI

Prepregnancy BMI	Total Weight Gain Range (lbs)	Rate of Weight Gain 2nd and 3rd Trimesters. Average per Week (lbs)
Underweight (<18.5)	28–40	1
Normal weight (18.5–24.9)	25–35	1
Overweight (25.0–29.9)	15–25	0.6
Obese (≥30.0)	11–20	0.5

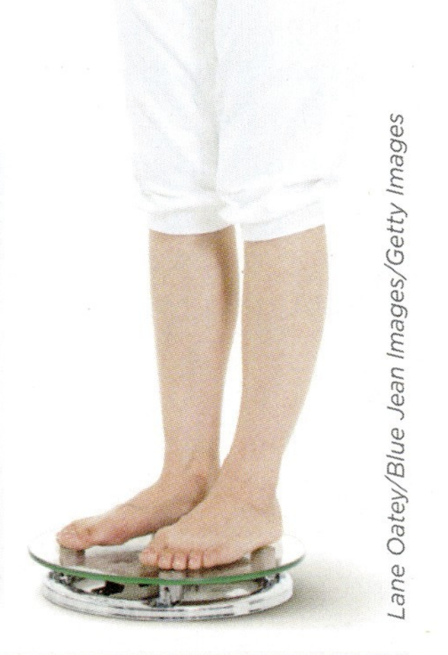

Lane Oatey/Blue Jean Images/Getty Images

Consequences of Inadequate Weight Gain

For Mom — Increased risk of:
- Unsuccessful breastfeeding

For Baby — Increased risk of:
- Preterm birth, and/or small for gestational age
- Stillbirth and infant mortality in underweight or normal-weight women
- Impaired neurological development

Consequences of Excessive Weight Gain

For Mom — Increased risk of:
- Pregnancy-related hypertension
- Cesarean delivery
- Weight retention
- Gestational diabetes

For Baby — Increased risk of:
- Large for gestational age
- Preterm birth
- Increased infant mortality in women with BMI ≥ 25
- Childhood leukemia
- Obesity, and breast and ovarian cancer as adults

FERTILITY *the capability to produce offspring*

active, eats well, makes responsible choices by not smoking or using harmful substances, and gets regular medical care is much more likely than a less healthy woman to be **fertile**, that is, to have the ability to produce offspring and have a healthy pregnancy.

ENERGY AND NUTRIENT NEEDS DURING PREGNANCY

Typically, a pregnant woman only needs to start consuming more calories after the first trimester, because early in pregnancy, the developing fetus is comparatively small in relation to

INFOGRAPHIC 17.4 Classification of Infants by Birth Weight and Gestational Age

Infants are considered to be appropriate for gestational age when their birth weight falls within the 10th to 90th percentile range. Infants born SGA are at increased risk of health problems at birth and as adults.

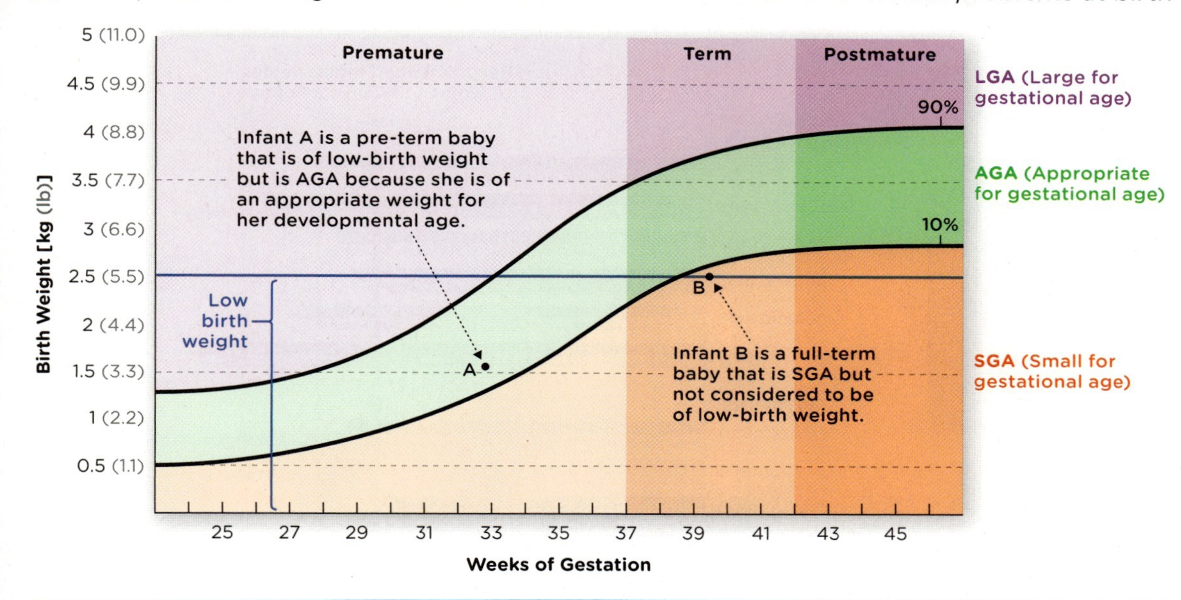

What is the approximate weight range of an average infant that is born at 36 weeks of gestation?

the mother's body mass. In general, in the second and third trimesters, pregnant women need between 2,200 and 2,900 total calories a day; the exact number depends on prepregnancy weight and the mother's activity level. Women should increase their caloric intake gradually using nutrient-dense foods—ideally to meet increased energy demands they should eat on average an extra 340 calories per day in the second trimester and an extra 450 kcal per day in the third trimester. (Underweight women need more calories since they should gain more weight; overweight women need fewer. Physically active women require more calories to offset expenditure.)

One excellent resource for pregnancy meal planning is the Health & Nutrition Information for Pregnant & Breastfeeding Women on the United States Department of Agriculture (USDA) ChooseMyPlate.org website, which provides a personalized daily food plan based on age, height, weight, physical activity level, and stage of pregnancy or breastfeeding status. That all said, a better gauge of appropriate intake is to monitor weight gain during pregnancy than to count daily calories. (INFOGRAPHIC 17.5)

Pregnant women need more calories to support their growing baby and they need more nutrients for themselves, too, all of which are best supplied through a wide variety of nutrient-dense foods. That's because all energy and nutrient needs for a baby's growth and development come from what the mother eats. Even if she eats enough calories, a woman who does not consume adequate nutrients during pregnancy—or who consumes excess amounts—can put her baby at risk for birth defects and other anomalies, particularly if these discrepancies occur during periods of intense or accelerated fetal development known as **critical periods**. Critical periods are genetically determined pathways that direct the development of specific cell types, organs, and tissues. Disruption of growth or development during a critical period (from nutritional problem or toxic substance) may be irreversible.

Nutrients needed in increased amounts

Ultimately, pregnant women need only 15% more total calories than nonpregnant women, but about 50% more of some nutrients such as protein, folate, zinc, iodine, and iron.

CRITICAL PERIODS *developmental events occurring during the first trimester of pregnancy in which cells differentiate and organs and vital systems begin to develop*

INFOGRAPHIC 17.5 Recommended Energy and Nutrient Intakes During Pregnancy and Lactation *This graph compares the recommended intakes for women ages 19 to 30 years old who are in the third trimester of pregnancy and in the first six months of lactation as compared to their recommended intake when not pregnant or lactating.*

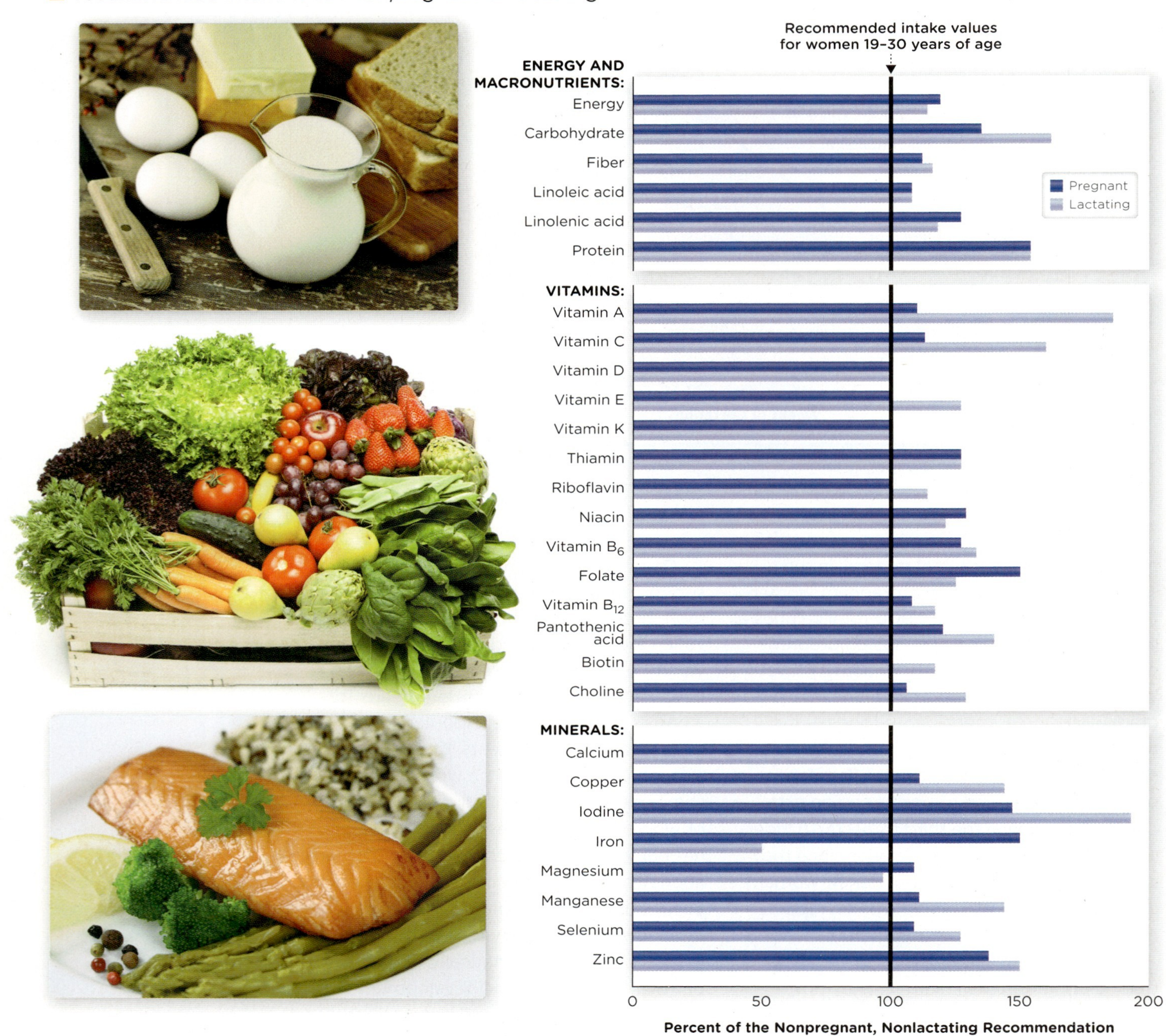

? *What nutrients have higher recommendations during lactation than during pregnancy?*

Photo credits (top to bottom): Svetlana Foote/Shutterstock, Cristian Baitg/Getty Images, Markus Mainka/Shutterstock

Pregnant women can meet most of their nutritional requirements through food; in fact, the only nutrient for which a supplement is universally recommended is iron. However, prenatal multivitamin-mineral supplements are sometimes recommended for women who do not consume an adequate diet or those who have high-risk pregnancies. Women who are vegans should also supplement with additional vitamin D, iron, and vitamin B_{12}. When choosing supplements, women should follow the advice of their doctor or midwife and not buy products that contain more than the recommended intake of any nutrient.

Folate

Folate (the synthetic form of which is called folic acid), a water-soluble B vitamin, is one of the key nutrients mothers-to-be need before and during pregnancy, as it is crucial for the development of new cells. Folate deficiency is associated with fetal growth problems, birth defects, schizophrenia, and autism; in addition, women who don't get enough folate during the first 12 weeks of pregnancy are up to 70% more likely to give birth to babies with **neural tube defects**, such as spina bifida.

NEURAL TUBE DEFECTS *malformation of the spine during early development*

The U.S. Centers for Disease Control and Prevention (CDC) recommends that all women between the ages of 15 and 45 years—even if they are not pregnant—consume a daily dose of 400 micrograms of folic acid in a multivitamin or the equivalent through folate-rich foods. This recommendation applies to all women of child-bearing age because half of U.S. pregnancies are unplanned, and birth defects from folate deficiency develop as early as three to four weeks into pregnancy, before most women even know they are pregnant. During pregnancy, women should consume the equivalent of 600 micrograms of folate daily from foods or supplements. To help women meet these requirements, in 1998, the U.S. Food and Drug Administration began requiring food manufacturers to add folic acid to common grain products such as breads, cereals, flours, and pastas.

Vitamin A

Another nutrient of particular concern during pregnancy is vitamin A, but expectant women

Spina bifida is linked to insufficient folic acid in a mother's diet. *Spina bifida is a neural tube defect—a disorder involving incomplete development of the brain, spinal cord, and/or their protective coverings. It occurs when the fetus's spine fails to close properly during the first month of pregnancy.*

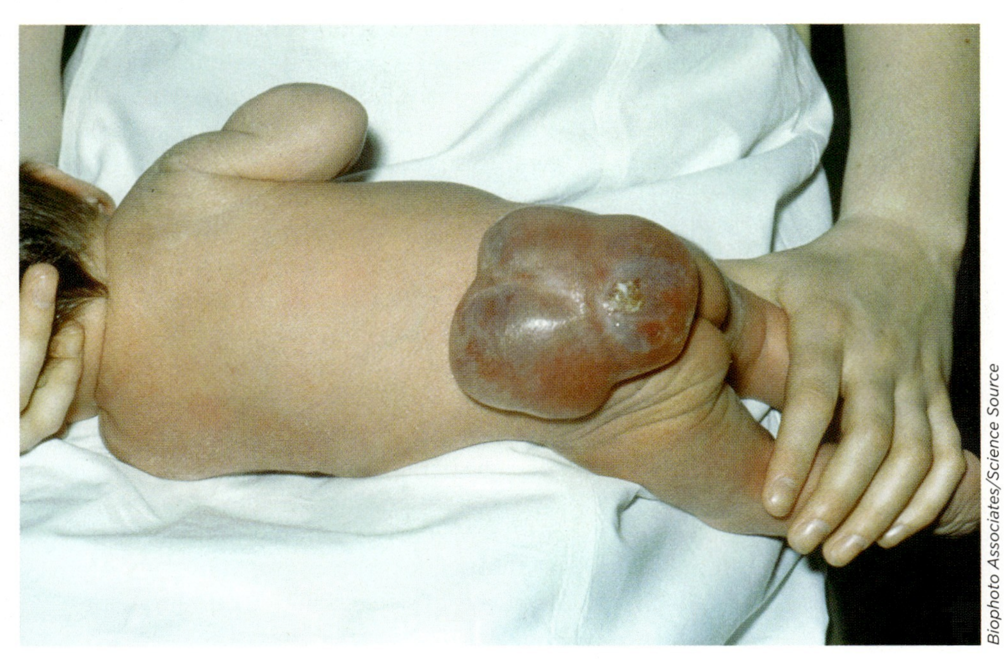

Biophoto Associates/Science Source

need to be careful about how much of it they consume, because although too little vitamin A can cause developmental problems, too much can also cause birth defects such as facial and heart deformities. Women can thus keep the risk of vitamin A toxicity low by meeting their needs through the consumption of bright orange, deep yellow, and light red vegetables and fruits that are rich in the vitamin A precursor, beta carotene. See Chapter 10 for more information on food sources of vitamin A.

The Institute of Medicine recommends that pregnant women between the ages of 19 and 50 years consume 770 micrograms of vitamin A per day, not exceed the Tolerable Upper Intake Level of 3,000 micrograms per day, and stop taking medications that contain vitamin A. The precursor to vitamin A, beta-carotene, however, does not seem to have the same potential adverse effects.

Iron

Iron deficiency is the most common deficiency in pregnant women. Iron is used to make hemoglobin, the molecule that transports oxygen through blood, and pregnant women vastly increase their production of hemoglobin to supply oxygen to their fetuses and to help build a fetal blood supply. The daily recommended intake of iron for pregnant women is 27 milligrams per day, compared with only 15 to 18 milligrams for nonpregnant women, yet national surveys have reported that pregnant women generally consume only 15 milligrams per day—just over half of the recommended amount. To ensure adequate iron intake, women should consume fortified cereals and green leafy vegetables, both good sources of non-heme iron; particularly if consumed with vitamin C-rich foods such as citrus fruits. Red meats are a good source of readily absorbable heme iron.

It is difficult for women to meet their increased iron needs from food alone, so supplements of 30 milligrams of iron are typically recommended during the second and third trimesters. Women who don't take supplements are at an increased risk of suffering from iron-deficiency anemia and are also at risk for giving birth to low-birth-weight babies who become iron-deficient in their first year of life.

Other nutrients of importance

Many other nutrients are important for a healthy pregnancy. There's calcium, which is crucial for the formation of healthy bones, and although absorption is enhanced during pregnancy, the recommendation for pregnant women is still 1,000 milligrams daily. Vitamin D helps to incorporate calcium into bones and also appears to play a role in programming genes in ways that could reduce the risk of chronic diseases; pregnant women should consume 600 IUs (15 micrograms) daily. Iodine is required for normal brain development and growth and recent studies indicate that approximately one-third of pregnant women in the United States are marginally deficient in iodine. The Institute of Medicine recommends 220 micrograms per day for pregnant women.

Omega-3 fatty acids

The omega-3 fatty acids, such as EPA and DHA, are critical for the development of the fetal brain. During pregnancy, the placenta transfers nutrients from the mother to the fetus. The amount of omega-3 fatty acids in the fetus is correlated with the amount consumed by the mother, underscoring the importance of adequate nutrition. Although the U.S. government has not made official recommendations about omega-3 intake during pregnancy, several organizations recommend that pregnant women consume 300 milligrams of DHA daily—yet most women do not meet this recommendation. Cold-water fish are a great source of these omega-3 fatty acids and thus recent recommendations *encourage* pregnant women to consume 8 to 12 ounces of fish each week. But because of concerns about mercury contamination in some types of fish, pregnant women are told to choose low-mercury fish and avoid fish that have potentially high levels of mercury, such as shark, swordfish, king mackerel, and tilefish. Women who consume too much mercury during pregnancy are more likely to give birth to babies with central nervous system defects and slow cognitive development.

However, pregnant women may safely enjoy salmon, cod, shrimp, sardines, anchovies, and trout, and are advised to limit albacore tuna and tuna steak to six ounces a week.

A HEALTHY PREGNANCY

Some women have trouble getting the nutrients they need because they suffer from **morning sickness**, triggered by hormonal changes that can cause nausea and vomiting in the first trimester and may sometimes cause unique food aversions. They are also at a heightened risk for **pica**, an eating disorder that causes individuals to want to ingest nonfood substances such as clay, paint chips, paste, plaster, dirt, or hair. Pica may be a sign of anemia.

With new diagnostic criteria and increasing rates of obesity, it is estimated that as many as 18% of pregnant women may develop **gestational diabetes** (see Chapter 5), which is characterized by elevated levels of blood glucose that pass to the fetus, causing it to be large for gestational age, increasing the risk of complications. Although gestational diabetes generally resolves after pregnancy, it increases the risk that the mother will develop type 2 diabetes later in life.

Pregnant women should also avoid alcohol. **Fetal alcohol syndrome** is a group of conditions causing physical, behavioral, and learning problems in children whose mothers drank heavily during pregnancy. Alcohol consumed by a woman during pregnancy travels through her blood and into the baby's blood, tissues, and organs, where it breaks down slowly, exposing the fetus to the alcohol for long periods. The CDC says that there is no safe level of alcohol consumption during pregnancy.

Although caffeine is not nearly as dangerous as alcohol, a health care provider might recommend limiting the amount of caffeine consumed because it crosses the placenta into the bloodstream of a developing baby. Although most studies report no adverse effects of maternal caffeine consumption on the infant, there is some evidence that high intakes of caffeine are associated with an increased likelihood of SGA births, miscarriage, and childhood acute leukemia. The World Health Organization recommends a maternal caffeine intake of below 300 mg per day and the American College of Obstetricians and Gynecologists recommends less than 200 mg per day. Brewed coffee contains about 135 mg per 8-ounce cup and caffeine-containing soft drinks generally contain between 40 mg to 50 mg per 12 ounce can.

Smoking—although not a nutritional issue per se—should be discontinued during pregnancy because it increases the risk of miscarriage, puts the baby at risk of premature and SGA birth, and infant mortality.

Individuals exposed to alcohol in the womb may suffer lifelong effects. *The effects of alcohol on a developing fetus can have profound consequences, including growth problems, brain damage, learning disabilities, physical abnormalities, and behavior issues.*

Stuart Wong/KRT/Newscom

MORNING SICKNESS
nausea and vomiting often experienced during the first trimester of pregnancy

PICA
an eating disorder characterized by the desire to eat nonfood substances such as dirt, clay, or paint chips

GESTATIONAL DIABETES
pregnancy-induced elevated blood glucose levels

FETAL ALCOHOL SYNDROME
mental and physical defects observed in infants born to mothers who consumed a significant amount of alcohol during pregnancy

INFOGRAPHIC 17.6 Food Safety During Pregnancy *During pregnancy a woman's immune system is suppressed, placing her at increased risk of contracting a food-borne illness. In some cases, these infections and toxins in foods can place either mom or the fetus at risk of severe complications.*

	Foods and Beverages to Avoid	Why?	Possible Consequences
	Unpasteurized milk and cheeses made from it, including: brie, feta, camembert, blue-veined cheese, and queso bianco and fresco cheeses. Refrigerated, smoked fish, meat spreads, and pâtés.	May contain the bacteria *Listeria*	Pregnant women are about 20 times more likely to be infected than those who are not pregnant. Infection can cause premature birth, miscarriage, stillbirth, or severe illness in the newborn.
	Undercooked chicken and poultry, in particular, but other meats and seafoods as well. Avoid contaminating other uncooked foods with drippings from raw poultry.	Nearly all raw poultry is contaminated with the bacteria *Campylobacter jejuni*. Other raw meats and seafood may also be contaminated.	*C. jejuni* infections may spread to the fetus and cause abortion, stillbirth, or early neonatal death.
	Marlin, shark, tilefish, swordfish, king mackerel. In general, avoid large predatory fish that are typically contaminated with higher levels of mercury.	These fish are high in mercury.	Too much mercury can damage the developing brain and nervous system of the fetus.
	Alcoholic beverages	Fetal alcohol syndrome is caused by a woman drinking during pregnancy. There is no safe level of alcohol consumption or safe time to drink during pregnancy.	Fetal alcohol syndrome can cause fetal death; abnormal facial features; neurological impairment, and heart, kidney, bone, or hearing defects.
	Foods to Be Cautious With	**Why?**	**Solution**
	Raw, fresh produce	May be contaminated with *Listeria* (as well as other bacteria)	Wash produce well before eating, even if you are going to peel it before it is eaten.
	Deli meats, hot dogs	May be contaminated with *Listeria* (as well as other bacteria)	Heat deli meats and hot dogs until they are steaming hot (165°F) before eating.
	Albacore tuna	Contains moderate amounts of mercury.	Eat no more than six ounces per week during pregnancy.

Photo credits (top to bottom): Will Heap/Getty Images, Radius Images / Alamy, Foodcollection/Getty Images, Steve Wisbauer/Getty Images, Olesya Feketa/Shutterstock, Carlos Gawronski/Getty Images, Davies and Starr/Getty Images, Foodcollection RF/Getty Images

FOOD SAFETY

Pregnant women need to be careful to avoid microbial-contaminated foods. Hormonal changes during pregnancy suppress the immune system of the mother. While such changes are necessary for the survival of the fetus, they increase the chance of food-borne infections. (See Chapter 20 for specific information on food safety.) Pregnant women are about 20 times more likely to develop **listeriosis** than a nonpregnant individual. Listeriosis is a serious infection caused by eating food tainted with the bacterium *Listeria monocytogenes*, which can cause premature birth, miscarriage, fetal death, and newborn illness, as the bacteria can cross the placenta. To minimize their risk of listeriosis, expecting women should not eat hot dogs or other luncheon meats unless they have been heated to steaming hot. They should also avoid unpasteurized cheeses, such as brie, blue cheeses, camembert, and some fetas; uncooked refrigerated smoked seafood; refrigerated meat spreads and pâtés; and unpasteurized milks and juices. **(INFOGRAPHIC 17.6)**

POSTNATAL NUTRITION

Breastfeeding

Once babies are born, of course, they still need adequate energy and nutrients. One of the best ways to provide this essential nutrition is through breastfeeding (**lactation**), which has become more common in recent years. Seventy-seven percent of U.S. newborns are breastfed by their mothers, and nearly 50% are breastfed until six months of age—up from just 35% in 2000. The American Academy of Pediatrics recommends that women exclusively breastfeed their babies through the age of six months, at which point they should continue breastfeeding—ideally to the one-year mark or beyond—while also introducing complementary solid foods.

The human breast is a gland made up of connective and fatty tissues that support and protect the milk-producing areas of the breast and give it shape. Shortly after birth, milk is produced in clusters of small sacs called *alveoli*. The alveoli group together and form *lobules*. Milk produced in the lobules travels through ducts, which eventually exit the skin in the nipple. The dark area of skin surrounding the nipple is called the areola. **(INFOGRAPHIC 17.7)**

The human breast does not store a large amount of milk. In fact, most milk is produced during nursing, based on need. In other words, the volume of milk produced varies with demand—the volume can go down if the mother does not nurse or pump, and the volume may increase if the baby feeds a lot. When the infant sucks on

LISTERIOSIS
a serious infection caused by eating food tainted with the bacterium Listeria monocytogenes, *which can cause premature birth, miscarriage, fetal death, and newborn illness, as the bacteria can cross the placenta*

LACTATION
production and secretion of milk from the mammary glands

INFOGRAPHIC 17.7 The Interior of the Breast *Breast milk is produced in the alveoli, travels through the ducts and exits the breast through the nipple.*

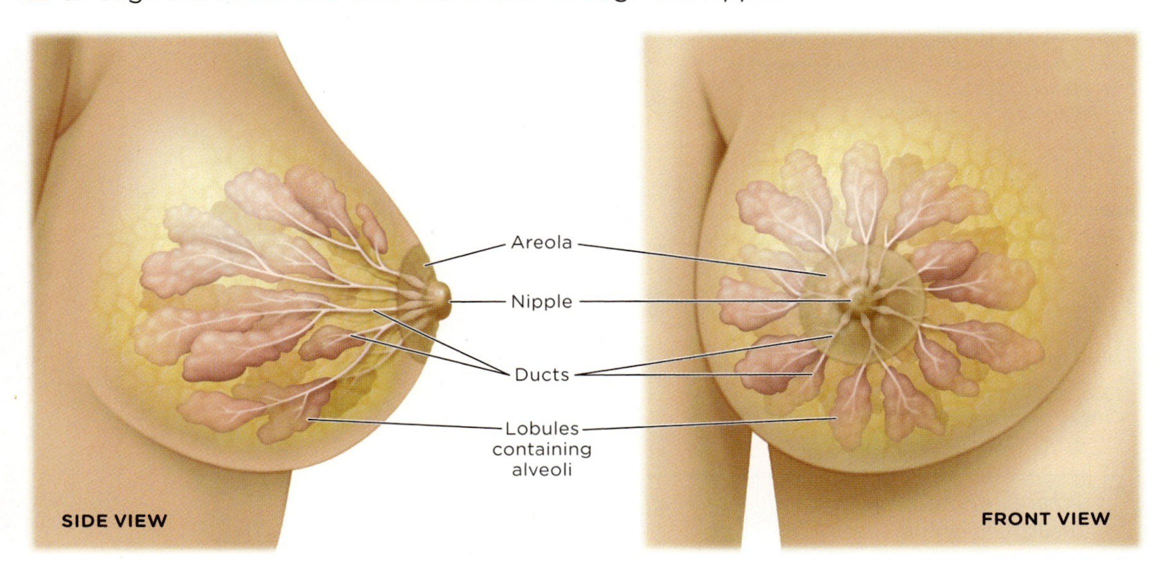

the nipple, the hormones prolactin and oxytocin are released. Prolactin stimulates milk production while oxytocin stimulates contraction (or the "let-down reflex"), which shortens and widens the milk ducts ejecting milk through the duct system and out of the nipples.

Composition of breast milk

The composition of breast milk varies depending on the stage of lactation, milk volume, feeding frequency, and differences between women. However, once lactation is established, the primary components are water (87%), carbohydrates, fats, proteins, vitamins, and minerals. At the beginning of the feeding, the milk contains lactose and proteins, but little fat; it is called *foremilk*. The end of the feeding produces *hindmilk*. The hindmilk contains more fat, the main source of energy for the infant. Both are nutritious, but hindmilk has more calories and babies will be more satiated if they get both foremilk and hindmilk during feedings.

Breast milk is the ideal infant food. It provides babies with many crucial nutrients, as well as antibodies that enhance their ability to fight infections and strengthen their immune systems. Breast milk is rich in vitamins, full of essential fatty acids for brain development, contains the appropriate balance of proteins and minerals to enhance digestion, and promotes infant oral motor development. Additionally, breastfeeding also reduces the risk of diarrhea and vomiting in infants and mitigates their risk of becoming obese later in life. There is also limited evidence that it may reduce the risk for chronic diseases, such as type 2 diabetes and heart disease, later in life.

INFOGRAPHIC **17.8 Benefits of Breastfeeding for the Infant and Mother** *There are huge benefits to breastfeeding that affect both the infant and mother.*

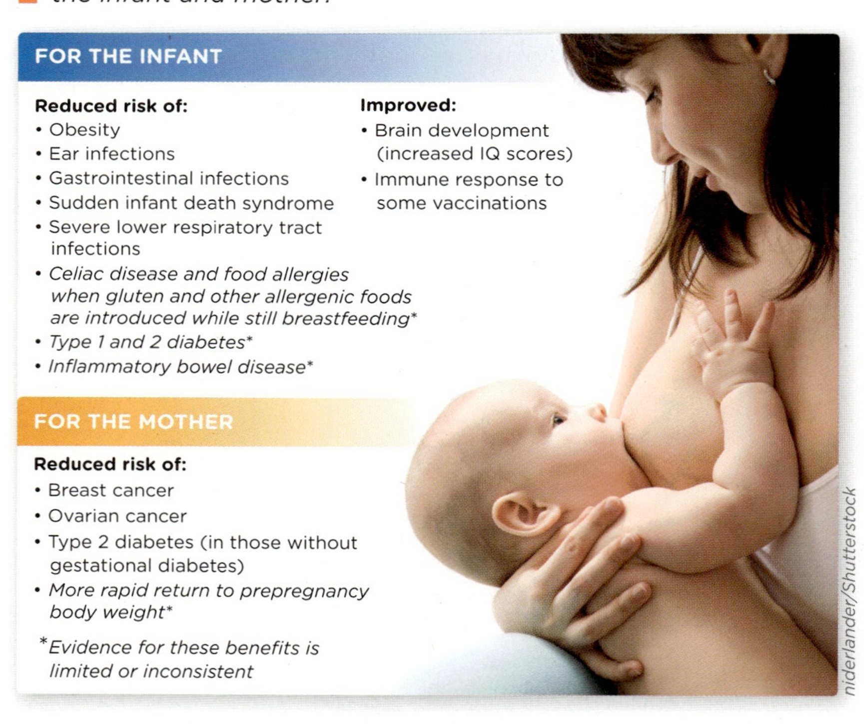

What are some factors that influence whether a woman might breastfeed or bottle feed?

There are also documented benefits to breastfeeding, including reductions in incidence of upper respiratory infections in the first two years of life, childhood leukemia, sudden infant death syndrome, and orthodontic problems. Evidence also exists to support higher IQ scores among children who were breastfed for at least six months. Plus, breastfeeding is convenient and less expensive than formula feeding and it plays an essential role in promoting mother–infant bonding. There are added benefits for mothers who nurse that include a reduced risk of ovarian and breast cancer, less risk of type 2 diabetes, and possibly a faster return to pre-pregnancy body weight. **(INFOGRAPHIC 17.8)**

NUTRITION DURING LACTATION

Women who are breastfeeding need to replenish their nutrient stores and remain healthy to produce enough nutrient-rich milk. Although lactation increases energy use by about 600 kcal a day during the first six months of breastfeeding, it is recommended that intake be increased by only 330 kcal a day, as the remaining energy will come from fat stored during pregnancy. During the second six months of lactation a woman's body weight is generally stable and the recommended increase in energy intake is equal to the energy present in the milk produced each day (400 kcal). As with pregnancy, energy needs should be met by consuming nutrient dense foods.

There are a few physical conditions that preclude new mothers from breastfeeding their babies. These conditions include HIV/AIDS, herpes simplex or chicken pox lesions on one or both nipples, alcohol or drug addiction, nipples that have been removed and replaced, or cancer treated with cytotoxic drugs. Otherwise, most chronic conditions can be managed with drugs that are safe to take while breastfeeding. Despite the encouraging news that more women are breastfeeding, there are challenges that make breastfeeding difficult for some. **(INFOGRAPHIC 17.9)**

INFOGRAPHIC 17.9 Potential Barriers to Breastfeeding

MECKY/Getty Image

UNCERTAINTY

Lack of knowledge about how to breastfeed—how to hold and position a baby at the breast, how to achieve an effective latch, and other breastfeeding techniques.

Confusing and contradictory advice received from friends and family.

Concern about insufficient milk supply.

DISCOMFORT

Frequently cited problems with breastfeeding include sore nipples, engorged breasts, leaking milk, pain, and failure to latch on by the infant.

EMBARRASSMENT

Embarrassment remains a barrier to breastfeeding and is related to disapproval when breastfeeding in public. Women may find themselves excluded from social interactions when they are breastfeeding because others are reluctant to be in the same room while they breastfeed.

WORK ISSUES

Returning to work can be a significant barrier to breastfeeding. Women often face inflexibility in their work hours and locations and a lack of privacy for breastfeeding or expressing milk, and milk storage problems.

ATTITUDES

Negative attitudes of family and friends can pose a barrier to breastfeeding.

Fathers may oppose breastfeeding because of concerns about bonding with an infant they were unable to feed, and how the mother would be able to accomplish household responsibilities if she breastfed.

What ideas can you suggest to overcome each of these issues?

FORMULA FEEDING

Nutritionally, breastfeeding is the best option for babies, but it's not right or possible for every mother. Bottle feeding commercially prepared infant formulas can be a nutritious adjunct to, or replacement for, breast milk. Manufactured under sterile conditions, commercial formulas are designed to duplicate breast milk using sophisticated combination of proteins, sugars, fats, and vitamins.

Although formula provides the basic nutrients an infant needs, it lacks the antibodies and many of the other components that only breast milk contains. Still, some individuals may feel more comfortable with formula feeding. It allows a mother to know exactly how much food the baby is getting, and there's no need to worry about the mother's diet or how medications might affect the breast milk. However, formula feeding can be expensive, and formulas should never be diluted to stretch out a supply. Anyone thinking of formula feeding an infant should check with a physician about selecting an appropriate formula.

NUTRITION FOR THE GROWING CHILD

Babies don't get all of their nutrients through breast milk alone. Parents are advised to supplement their breastfed babies with 400 IU of vitamin D each day starting from the first few days of life. Breastfed babies of vegan or vitamin B_{12}-deficient mothers may also require vitamin B_{12} supplements. All newborns are also typically given an injection of vitamin K in the hospital after birth. Babies have very little vitamin K, an important blood-clotting vitamin, in their bodies because it crosses the placenta to the developing baby poorly. In addition, though it has recently been recognized that the unborn infant's gut may contain low levels of certain bacteria that synthesize vitamin K, very little of the vitamin is absorbed from the colon where these bacteria reside.

Infants vary considerably in terms of their growth, development, nutritional needs, and feeding patterns. During the first two to six weeks of their lives, they primarily feed, sleep, and grow. By late infancy, newborn reflexes have gone away and a baby has mastered certain physical tasks that allow him to progress from a diet of exclusive

INFOGRAPHIC 17.10 The Growing Infant *Infants change dramatically in the first 24 months of life.*

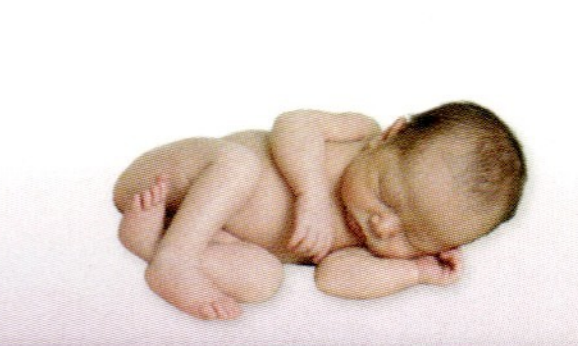

NEWBORN
Immediately after birth, an infant normally loses about 5% to 10% of his or her birth weight. However, by about two weeks of age an infant begins to gain weight and grow steadily.

0–6 MONTHS
From birth to six months of age, a baby may grow ½ to 1 inch a month. A healthy baby will double his or her birth weight between four and six months.

6–12 MONTHS
From six to twelve months of age, growth slows somewhat. The baby will grow about ⅜ inch a month and triple birth weight by about 12 months.

1–2 YEARS
Between one to two years of age, a toddler will gain only about five pounds.

Photo credits (left to right): Ahturner/Shutterstock, Tetra Images/Getty Images, KidStock/Getty Images, jo unruh/Getty Images

Baby's first solid foods. *Around six months of age, babies may be ready for their first taste of solid food.*

Food collection/Getty Images

breast milk or formula to foods with an increasingly wide variety of flavors and textures. The ways parents feed their babies not only nourish them, but also help to promote motor development and establish key feeding skills, healthy habits, and strong family relationships.

Babies grow at different rates because of differing genetic factors and prenatal history, but growth is also an important indicator of adequate nutrition. Inadequate intake of calories or essential nutrients can significantly affect mental and physical development, particularly during the critical periods from birth to one year. Infants should double their birth weight by four to six months and typically triple it by their first birthday. They also grow in length by approximately 50% in their first year. **(INFOGRAPHIC 17.10)**

Babies who are fed on demand typically consume enough and grow appropriately, and they can eat quite a lot—about five times more in relation to their body weight than adults eat. Parents should not limit the fat intake of their children in the first two years of their life, because fat is essential for meeting caloric needs, facilitating brain development, and helping the baby's body absorb fat-soluble vitamins.

Introduction of solid food

Solid foods—which the American Academy of Pediatrics advises should be introduced around the age of six months and the World Health Organization advises should be introduced between the ages of four and six months—are intended to complement, but not replace, breast milk or formula. Infants who are exclusively breastfed for around six months have consistently been shown to have a reduced risk of gastrointestinal tract infections and grow and develop as well as infants who have solid foods introduced before six months.

According to the American Academy of Pediatrics, a baby may be ready to eat solid foods when she is able to hold up her head and sit in a high chair or feeding seat with good head control. The tongue thrusting reflex that allows a baby to suck efficiently should be diminished enough so that the infant can move food from the spoon to the back of his mouth for swallowing. Typically infants double their birth weight at around four months and weigh 13 pounds or more and may, at that point, be big enough to try solid foods. Finally, babies may indicate a readiness for solid foods when they

INFOGRAPHIC 17.11 Introducing Solid Foods *The American Academy of Pediatrics recommends breastfeeding as the sole source of nutrition for the first six months of life. As solid foods are added to the baby's diet, breastfeeding should continue until at least 12 months.*

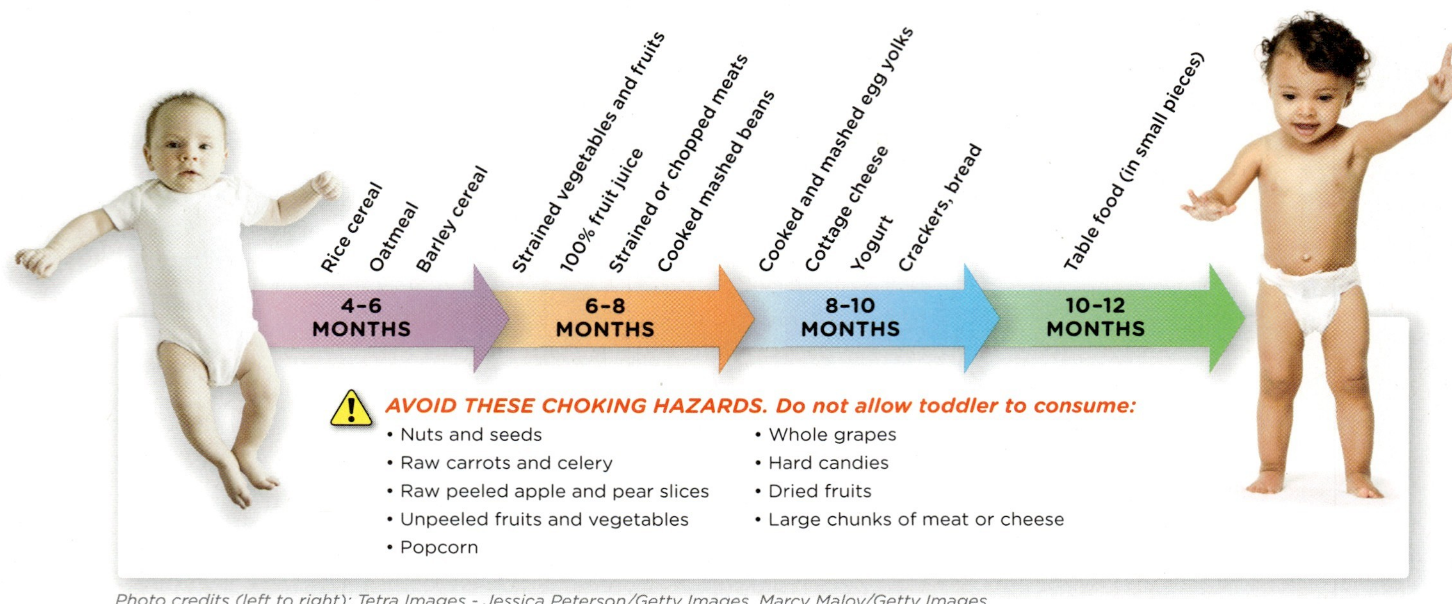

Photo credits (left to right): Tetra Images - Jessica Peterson/Getty Images, Marcy Maloy/Getty Images

show an interest in what the family is eating, and can signify when they are satiated, for instance by turning their head when they are full or by refusing to open their mouth. Babies who are fed solid foods before they are ready may be at risk for becoming overweight, developing food allergies, and suffering upset stomachs. (See Chapter 18 for information about food allergies). **(INFOGRAPHIC 17.11)**

Iron needs in infancy

Following a healthy pregnancy, most—but not all—infants have iron stores to help meet their needs around four months of age, after which the risk for iron deficiency increases. Iron is necessary for the formation of brain cells, so iron deficiency can hinder psychomotor development. Research suggests that treating young children once they have iron deficiencies is not enough—it does not prevent cognitive problems—so it may be more important to *prevent* iron deficiency from developing in the first place. Babies at the highest risk for iron deficiency are those who were born prematurely, who are fed formula that is not fortified with iron, and who are exclusively breastfed without iron supplements. Because of the risks, doctors recommend that parents begin feeding their infants iron supplements after the age of four months if babies are not being fed iron-fortified complementary foods, such as rice cereal.

IRON DEFICIENCY AND CORD CLAMPING

Because iron deficiency during pregnancy is so common and so difficult to prevent without supplementation, the case for delayed cord clamping is strong. To date, six studies have confirmed that when doctors or midwives delay the clamping of the umbilical cord by two to three minutes, an infant's risk of iron deficiency at the age of four to six months significantly drops. It's possible, then, that by allowing this blood and iron to pass from mother to child right after birth—by letting a mother give her baby one more gift before physically separating them—"we can affect their health throughout their lifespan," Judith Mercer says. Yet obstetrical care providers say that the practice is still rare, and that umbilical cords are typically clamped immediately after birth unless a parent specifically requests otherwise.

Mercer is convinced of the universal benefits of delayed cord clamping. And the practice is not only gaining acceptance but gaining endorsement by some obstetric health organizations due, in part, to her work and advocacy. It could be a while before all obstetrical practitioners change their routine practice, however, so Mercer plans to continue designing, conducting, and publishing studies in the hopes of turning the medical tide. She will also have to prove that the benefits of delayed cord clamping outweigh any potential risks.

CHAPTER 17 **BRING IT HOME**

Healthy pregnancy

Bethany is 29 years old, five foot seven, and reports a prepregnancy weight of 140 pounds (BMI = 22 kg/m^2). She is in her first trimester of pregnancy at 11 weeks and is in good health. If Bethany were your friend, you would recommend she discuss these questions with a healthcare provider. However, as a way to review key concepts in this chapter, can you answer Bethany's questions?

1. How much weight am I supposed to gain?
2. Is it okay that I've gained only two pounds so far?
3. What will compose the weight gain? How much will be fat?
4. How many kcal do I need?
5. How much protein do I need during pregnancy?
6. My doctor (or midwife) asked me if I am taking folic acid supplements. I told him only the amount in a regular multivitamin pill. Do I need extra amounts while I'm pregnant?
7. Do I need to take an iron supplement? If so, how much?
8. Before I was pregnant, I walked every day for 45 minutes. How much exercise should I get while pregnant?
9. I'm considering breastfeeding, but wonder if just using formula would be more convenient. What are the pros and cons of each?

Take It Further

Using Bethany's information, visit http://www.choosemyplate.gov/pregnancy-breastfeeding.html to create a *Daily Food Plan for Moms*. For this activity, consider that Bethany has delivered a healthy baby and is breastfeeding—complete the chart below for Bethany to refer to in planning her daily intake while breastfeeding. Can you plan a menu for Bethany that meets the recommended food group servings while she is breastfeeding? For ideas, go to https://www.supertracker.usda.gov/generalplan.aspx.

Daily recommendations from ChooseMyPlate.gov	Prepregnancy	Pregnant (second trimester)	Pregnant (third trimester)	Breastfeeding
kcal	2,200	2,600	2,800	
Grains/whole grains (1-ounce servings)	7	9	10	
Vegetables (cups)	3	3 1/2	3 1/2	
Fruits (cups)	2	2	2 1/2	
Dairy (cups)	3	3	3	
Protein (ounces)	6	6 1/2	7	
Oils (teaspoons)	6	8	8	
Discretionary kcal	≤ 266	≤ 362	≤ 395	

KEY IDEAS

During the 38 to 42 weeks of pregnancy, the developing fetus is supplied with nutrients and oxygen from the mother through the umbilical cord, connecting the fetus to the placenta, which is attached to the uterus.

A woman's nutrient and energy needs change during pregnancy to support growth of the developing fetus and maternal support tissues.

A woman's energy needs gradually increase as pregnancy advances, with an average increase of 15%. Nutrient needs increase by as much as 50% for some vitamins and minerals.

The Institute of Medicine provides guidelines for appropriate weight gain during pregnancy based on BMI at conception to reduce complications during pregnancy and at delivery, as well as to promote the birth of a healthy birth weight baby.

Birth weight is influenced by several factors. Infants who are born preterm or are small for gestational age have increased risk of death, complications, and certain medical conditions as adults.

Insufficient intake of essential nutrients can increase the risk of birth defects, particularly during critical periods of accelerated fetal development.

Folate plays an important role in the prevention of neural tube defects.

Iron needs are increased by 50% during pregnancy, which often warrants supplementation to prevent iron-deficiency anemia that affects oxygen supply and iron status of the fetus.

To reduce risks to the developing fetus, pregnant women are advised to avoid alcohol, limit their intake of certain fish because of potential mercury contamination, and avoid microbial-contaminated foods.

The American Academy of Pediatrics recommends that mothers exclusively breastfeed their babies through six months of age, and then continue breastfeeding (ideally to one year or beyond), while also introducing complementary solid foods according to the infant's developmental readiness.

Breastfeeding supplies the growing infant with essential nutrients in appropriate proportions, provides antibodies to strengthen immunity, and has many additional benefits for both the mother and baby.

Breast milk may not supply sufficient vitamin D, therefore supplementation for breastfed infants is recommended. Infants who are exclusively breastfed after six months of age may require iron supplementation.

Women who are breastfeeding have increased energy and nutrient needs for milk production, as well as to replenish nutrient stores.

Infants develop rapidly in the first year of life and require a consistent supply of energy and essential nutrients. Growth is an important indicator of adequate nutrition.

NEED TO KNOW

Review Questions

1. Physiological changes associated with pregnancy include all of the following, EXCEPT:
   a. decreased gastrointestinal motility.
   b. decreased breathing rate.
   c. increased blood volume.
   d. increased calcium absorption.
   e. increased cardiac output.

2. During pregnancy, there is a maternal adaptation to utilize ________ as primary fuel.
   a. fat
   b. glucose
   c. glycogen
   d. protein

3. Marilyn weighed 140 pounds prior to pregnancy and her BMI was 22. According to healthy weight-gain recommendations, she should anticipate weighing ________ by the end of her pregnancy.
   a. 150 to 160 pounds
   b. 155 to 165 pounds
   c. 160 to 170 pounds
   d. 165 to 175 pounds
   e. 170 to 180 pounds

4. Babies that are considered small for gestational age are defined as having a birth weight that is:
   a. between one and three pounds.
   b. less than 10% of maternal weight gain.
   c. less than 10th percentile of gestational age.
   d. less than 50th percentile of gestational age.
   e. lower than 50% of average U.S. birth weight.

5. Typically, calorie needs during the first trimester of pregnancy:
   a. do not increase over prepregnancy requirements.
   b. increase by 30% over prepregnancy requirements.
   c. increase by 340 kcal over prepregnancy requirements.
   d. increase by 450 kcal over prepregnancy requirements.
   e. are the highest compared with other trimesters because of rapid fetal development.

6. Which of the following about folate is NOT correct?
   a. Adequate intake before and during pregnancy decreases the incidence of neural tube defects.
   b. Between the ages of 15 and 45 years, all women should consume 400 micrograms through folate-rich foods or take folic acid supplements.
   c. During pregnancy, the daily recommended intake is 600 micrograms.
   d. Needs in pregnancy can be met without supplementation through diet and consumption of folate-rich foods.
   e. Supplementation is most critical during the second trimester when folate-related birth defects develop.

7. The only nutrient for which supplementation during pregnancy is universally recommended is:
   a. calcium.
   b. fluoride.
   c. iron.
   d. vitamin B_{12}.
   e. vitamin C.

8. Justin weighed 7 pounds, 3 ounces at birth and was 20 inches in length. What is his anticipated weight and height at one year?
   a. 14 pounds, 6 ounces; 30 inches
   b. 14 pounds, 6 ounces; 40 inches
   c. 21 pounds, 9 ounces; 30 inches
   d. 21 pounds, 9 ounces; 40 inches

9. The benefits of breastfeeding include:
   a. reduced incidence of diarrhea and vomiting in infants.
   b. reduced incidence of childhood leukemia and juvenile diabetes.
   c. enhanced infant oral motor development and digestion.
   d. increased mother–infant bonding.
   e. all of the above.

10. Nutritional recommendations for infants include all of the following, EXCEPT:
    a. exclusive breastfeeding until at least six months of age.
    b. vitamin B_{12} supplementation for all breastfed babies.
    c. vitamin D supplementation for breastfed babies after the first few days of life.
    d. iron supplementation if exclusively breastfed after four months.
    e. solid foods can be introduced to complement breast milk after six months.

Take It Further

Describe developmental signs that an infant is ready for the introduction of solid foods.

Dietary Analysis Using SuperTracker

Using SuperTracker to Understand Nutrition Requirements During Pregnancy

1. Log on to the United States Department of Agriculture (USDA) website at www.choosemyplate.gov. At the top of the page click on Audience and then select Adults, Moms-to-Be.
2. In the left navigation bar select "Making Healthy Choices in Each Food Group." List the five food groups and provide two foods found in each of the food groups. What are two key nutrients found in each group?
3. Next select the "Nutritional Needs during Pregnancy" link.
    a. List the major foods that are considered empty calories.
    b. Explain the guidelines about alcohol consumption and pregnancy.
4. Next select the "Dietary Supplements" link. Click the "Why take a prenatal supplement?" link. What two nutrients are discussed? Why are they so important?
5. Finally, click on "Food Safety for Pregnant and Breastfeeding Women."
    a. Explain recommendations for fish intake.
    b. What is listeriosis? Why is this pathogen dangerous during pregnancy?
    c. Why is food safety especially important for pregnant women?

Food Allergies and Intolerances

AS PARADOXICAL AS IT MIGHT SEEM, COULD DIRT AND GERMS MAKE CHILDREN HEALTHIER FOR LIFE?

D. Hurst/Alamy

LEARNING OBJECTIVES

- Identify four primary objectives for sound nutritional guidance for children (Infographic 18.7)
- Provide an overview of patterns of growth and development from preschool-aged children to early adolescence (puberty) (Infographic 18.1)
- Describe how body mass index is used to assess if children are at a healthy weight for their age (Infographic 18.1)
- Identify at least three nutritional challenges for children and adolescents (Infographic 18.2)
- Describe how changes governing the types of foods that are offered at schools are expected to improve the nutrition profile of school meals (Infographic 18.4)
- Describe how parents can use MyPlate to help in planning a healthy diet for their children (Infographic 18.7)
- Describe at least three ways parents can foster positive eating habits and food choices (Infographic 18.7)
- Discuss the consequences of childhood obesity on future health (Infographic 18.9)
- Describe how food allergies develop, and identify four food allergens that are common among children (Infographic 18.10)

After World War II, a portion of Eastern Finland fell under Russian control. Today known as Russian Karelia, this area is inhabited by Russians as well as former Finns, and to scientists it has presented a fascinating opportunity to research environmental effects on health. "These two populations share partly the same ancestry, but differ in many lifestyle-associated factors," explains Anita Kondrashova, a virologist at the University of Tampere in Finland. Among other things, standards of living and hygiene

Finland and Russian Karelia share a border, but the residents of Karelia suffer fewer autoimmune diseases than their Finnish neighbors.

Michael Schmeling/Alamy

are significantly poorer in Russian Karelia than in Finland. But, surprisingly, some microbial infections are much more common in more affluent Finland.

Thanks to a collaborative effort between Russia and Finland, scientists—including Kondrashova—have, since 1999, been collecting samples and information from children living in both areas to study a controversial idea known as the *hygiene hypothesis*. It posits that reduced exposure to childhood infections in developed countries can help explain drastic increases in the rates of allergies and other immune disorders, such asthma, type 1 diabetes, celiac disease, and multiple sclerosis, in these regions. Whereas only about 2% to 3% of individuals in developing countries report suffering from asthma, for instance, 20% to 30% of those in developed countries do.

Kondrashova and her colleagues have shown that these same health disparities exist between Finland and Russian Karelia. Rates of autoimmune diseases are significantly lower on the Russian side; in fact, rates of type 1 diabetes are six times lower in Russian Karelia than in Finland, even among the Karelian residents who are genetically closest to the Finns. Celiac disease is less common in Russian Karelia than in neighboring Finland, too.

The hygiene hypothesis, which Kondrashova's findings support, is one of several theories that could explain the rising prevalence of allergies, including food allergies, in industrialized countries. Food allergies are thought to affect nearly 5% of U.S. adults and 8% of U.S. children and are becoming more commonplace every year. A 2008 study by researchers at the United States Centers for Disease Control and Prevention (CDC) reported an 18% increase in food allergies between 1997 and 2007.

Because food allergies limit certain food choices, they may make it more difficult for some kids to get the nutrients they need, particularly in children who have allergies to cow's milk and in those who are eliminating multiple foods. But potential nutritional shortfalls can be prevented with proper dietary guidance. According to the Academy of Nutrition and Dietetics, all children—even those with food allergies—should be able to obtain optimal physical and cognitive development, a healthy body weight, and minimize their risk of chronic diseases, such as obesity, heart disease, type 2 diabetes, stroke, cancer, and osteoporosis, through appropriate and enjoyable eating patterns and food choices balanced with regular physical activity.

GROWTH AND DEVELOPMENT IN CHILDHOOD

One way for parents and pediatricians to check that children are consuming an adequate diet is by assessing if their rate of growth is appropriate when compared with growth rates expected for U.S. children and adolescents. The CDC's **growth charts** serve as references for this comparison. These charts include sex-specific plots of weight-for-age, height-for-age, and body mass index (BMI)-for-age curves for children and adolescents from ages 2 to 20 years. The growth charts present these data as a series of percentile curves that allow one to assess if body size and growth of children are appropriate for their age, or if the child is at risk of undernutrition or obesity. A child whose body weight or height falls at or under the fifth percentile for their age is considered at risk of undernutrition, while those who exceed the 85th or 95th percentile of BMI-for-age are considered to be at risk of being overweight and obese, respectively. To learn more about undernutrition see Chapter 20. **(INFOGRAPHIC 18.1)**

Growth is extremely rapid in infancy, but it slows in preschool children aged 2 to 5 years. Still, it continues at an impressive rate with an average weight gain of 4.5 to 6.6 pounds and an average growth of 3 to 4 inches each year. School-aged children 6 to 11 years still continue to grow at a steady pace and have occasional **growth spurts**, periods of accelerated physical development with associated changes in height and weight. Children can vary drastically in height, weight, and build during the school years because of genetics, nutrition, and exercise patterns. That said, on average, children in this age range grow a little more than 2 inches and gain 6.5 pounds per year.

The end of the elementary school years generally marks the beginning of **puberty**, a dynamic time of development with periodic growth spurts that result in changes in body

GROWTH CHARTS
a series of percentile curves illustrating the distribution of selected body measurements (weight, height, body mass index) in children

GROWTH SPURTS
periods of accelerated physical development in children and adolescents with associated changes in height and weight

PUBERTY
dynamic period of development with periodic growth spurts that result in changes in body size, shape, composition, and sex-specific maturation

Children have occasional growth spurts, periods of accelerated physical development with associated changes in height and weight. The boys in this photo are brothers ages 14 to 4.

Erin Patrice O'Brien/Getty Images

INFOGRAPHIC 18.1 Tracking Growth for Children Aged 2 to 20 Years *Examples of charts used to track height and weight-for-age percentiles, and body mass index-for-age percentiles. A range of clinical growth charts for boys and girls is available at CDC.gov/growthcharts.*

HEIGHT/WEIGHT-FOR-AGE PERCENTILES: BOYS, 2 TO 20 YEARS

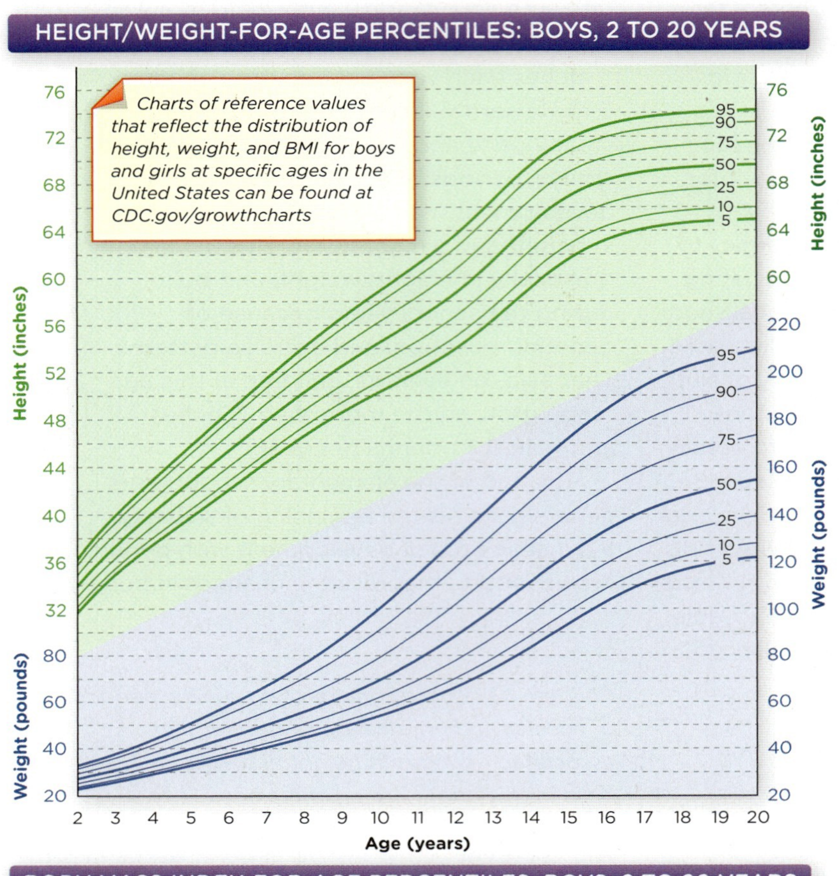

BODY MASS INDEX-FOR-AGE PERCENTILES: BOYS, 2 TO 20 YEARS

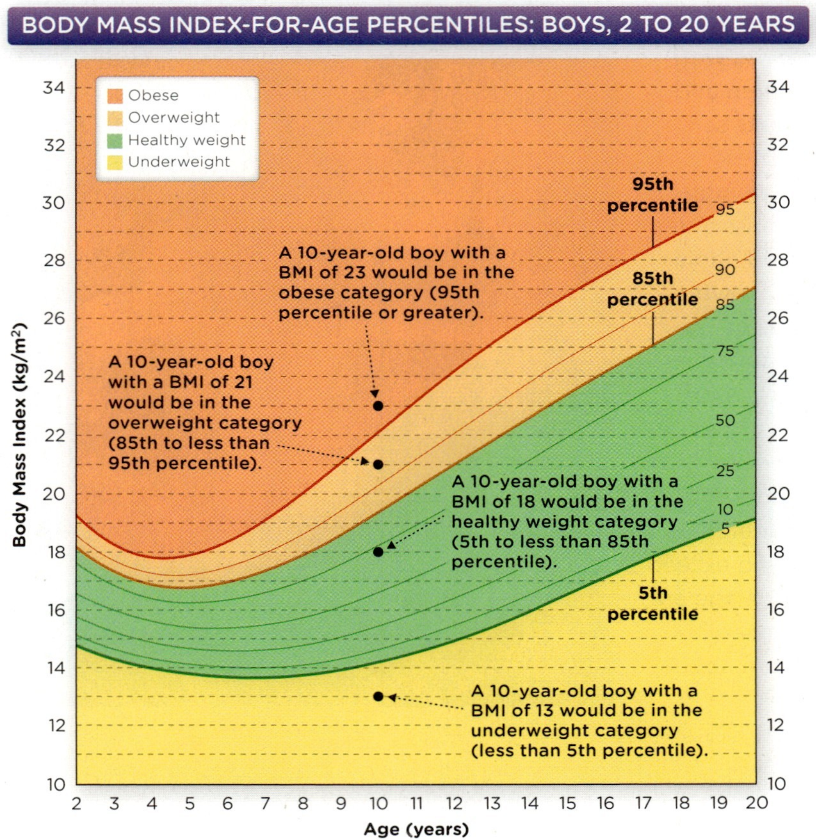

size, shape, composition, and sex-specific maturation. Growth and development vary widely during puberty, but the period is almost always associated with significant weight gain—in fact, 50% of adult body weight is typically gained during puberty. In boys, puberty typically begins at age 12 or 13 with an average of 2.5 inches of growth in height and a weight gain of a little more than 11 pounds a year for a span of about 4 years. In girls, puberty usually begins earlier, at age 10 or 11, and girls gain an average of about 2 inches in height and 9 pounds in fat and lean mass each year. Boys and girls often have different nutritional needs during puberty in part because of their different growth trajectories.

WHAT ARE CHILDREN EATING?

Studies suggest that, on average, U.S. children overconsume energy-dense, nutrient-poor foods at the expense of nutrient-dense foods. For instance, up to 90% of children aged 4 to 13 years do not consume the recommended number of servings of fruits and vegetables, and fewer than 20% consume two or more of the recommended servings of whole grains per day. **(INFOGRAPHIC 18.2)**

Nutrition surveys have found that children and adolescents between the ages of 2 and 18 years get approximately one-third of the total calories they consume, far exceeding recommended allowances, in the form of empty calories. Solid fats and added sugars,

INFOGRAPHIC 18.2 How Well Do Children Aged 2 to 17 Years Meet the Recommendations of the Dietary Guidelines for Americans?

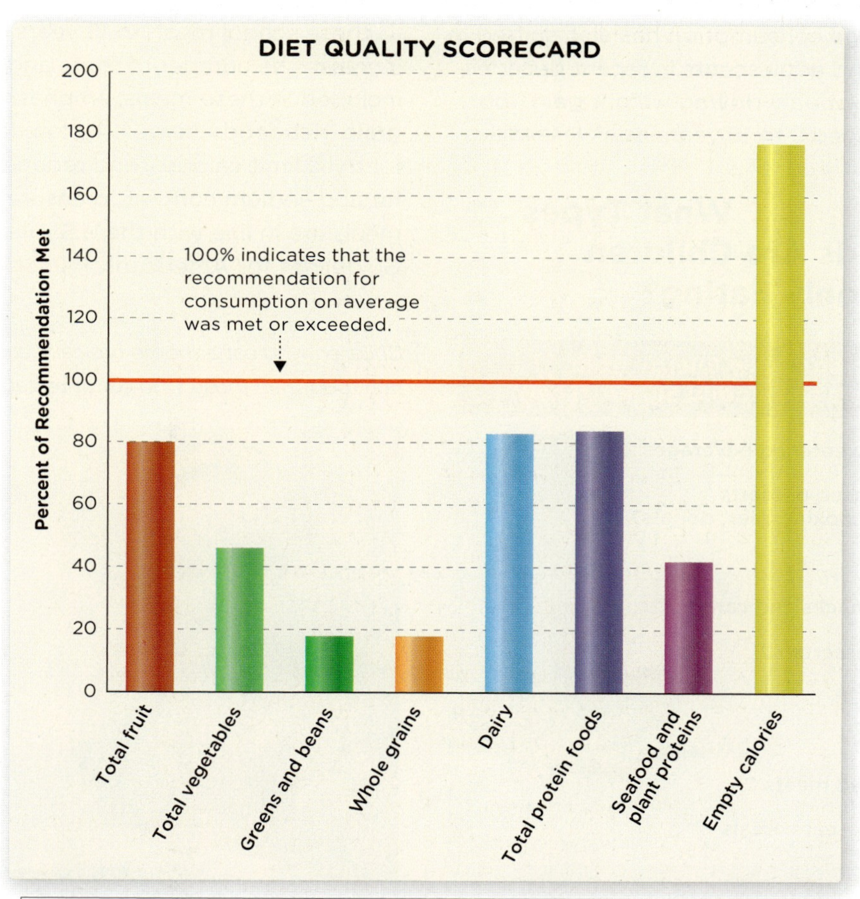

Table Source: National Center for Health Statistics, National Health and Nutrition Examination Survey, 2003-2004, 2005-2006, 2007-2008 and U.S. Department of Agriculture, Center for Nutrition Policy and Promotion, Healthy Eating Index- 2010. http://www.childstats.gov/americaschildren/health6.asp

Did the overall diet quality match up to recommendations? In what areas did it fall short?

NATIONAL SCHOOL LUNCH PROGRAM *federally assisted school meal program that provides nutritionally balanced, low-cost (or free) lunches to children each school day*

which provide a significant amount of these empty calories, are found in abundance in foods like soda, fruit drinks, grain desserts, pizza, whole milk, and many packaged snacks. **(INFOGRAPHIC 18.3)**

Why are children eating so poorly? One problem is that kids are now eating more meals away from home, particularly at fast-food restaurants, which tend to serve energy-dense and nutrient-poor foods in large portion sizes. Studies suggest that children who eat the most fast foods have diets with higher total intakes of total energy, saturated fat, and sodium. Children tend to eat less when they can serve themselves, as they might do at home; children eating at home also tend to eat more nutrient-rich foods. Family meal-times at home can also do a lot to affect nutrition beliefs and attitudes, which typically translates to healthier food choices, reducing kids' future risk for overweight and obesity.

Beverage consumption has also shifted in children and adolescents in recent decades, which is probably driving weight gain, too. Children aged 2 to 18 years drink less milk than did the same age group in the past but more soft drinks. On average, children consume close to 10% of their total calorie intake from sugar-sweetened beverages, and half of all Americans, including kids, consume sugar drinks on any given day. This high sugar consumption could be a primary reason that a whopping 42% of children aged 2 to 11 years have had dental caries in their primary teeth.

One piece of good news, however, is that more and more children—30 million kids in 2013—are participating in the **National School Lunch Program**, a federally assisted meal program operating in public and non-profit private schools and residential child care institutions that provides nutritionally balanced, low-cost (or free) lunches to children each school day. In January 2012, in part as a result of the 2010 passage of the Healthy, Hunger-Free Kids Act, the USDA issued new standards—the first major update to these school meals in 15 years—that increase the number of fruits and vegetables included in these meals, emphasize whole grain-rich foods, serve only low-fat and non-fat milk, limit calories, and reduce saturated fat and sodium content to ensure that the meals are in line with the U.S. Dietary Guidelines for Americans. **(INFOGRAPHIC 18.4)**

INFOGRAPHIC 18.3 What Types of Foods Are Children Commonly Eating?

Average Daily Energy from Solid Fats and Added Sugars (kcal)

1. **Sugar-sweetened beverages** 99
2. **Grain-based desserts (cakes, cookies, pies, donuts)** 78
3. **Milk** 44
4. **Sweet snacks and candy** 32
5. **Dairy desserts** 32
6. **Cheese** 28
7. **Pizza** 26
8. **Processed meats** 25
9. **Ready-to-eat cereals** 17

These top nine sources of solid fats and added sugars contribute 60% of total solid fat and added sugar intake in children and adolescents in the United States (2009–2010).

? *What type of milk is not a source of solid fats and added sugars?*

Photo credit: Eli Ensor

Children who eat at home typically eat more nutrient-rich food than those who eat at restaurants.

Bon Appetit/Alamy

INFOGRAPHIC 18.4 Foods at School and the Healthy, Hunger-Free Kids Act

New nutrition standards affecting lunches and snacks sold during the school day are aimed at ensuring that kids are being fed healthy foods while they are at school.

NEW GUIDELINES FOR SCHOOL LUNCHES

MILK
- Only fat-free (flavored or unflavored) milk or unflavored 1% milk

PROTEIN FOODS
- Tofu may count as meat alternative
- Allows meat/meat alternative in place of grains, after meeting minimum of one grain serving daily

GRAINS
- All grains must be "whole grain" rich

Mike Flippo/Shutterstock

FRUITS
- ½ to 1 cup of fruits per day (depending on age group)
- No added sugar in fruits
- Juice must be 100% juice, no sugar added

VEGETABLES
- ¾ to 1 cup of vegetables per day (depending on age group)
- Weekly requirement for vegetable subgroups—dark green, red/orange, starchy, etc.

- Zero grams trans fat per serving
- Sets a minimum and maximum range of calories per meal
- Sodium to decrease over time

NEW GUIDELINES FOR SNACK FOODS SOLD AT SCHOOL

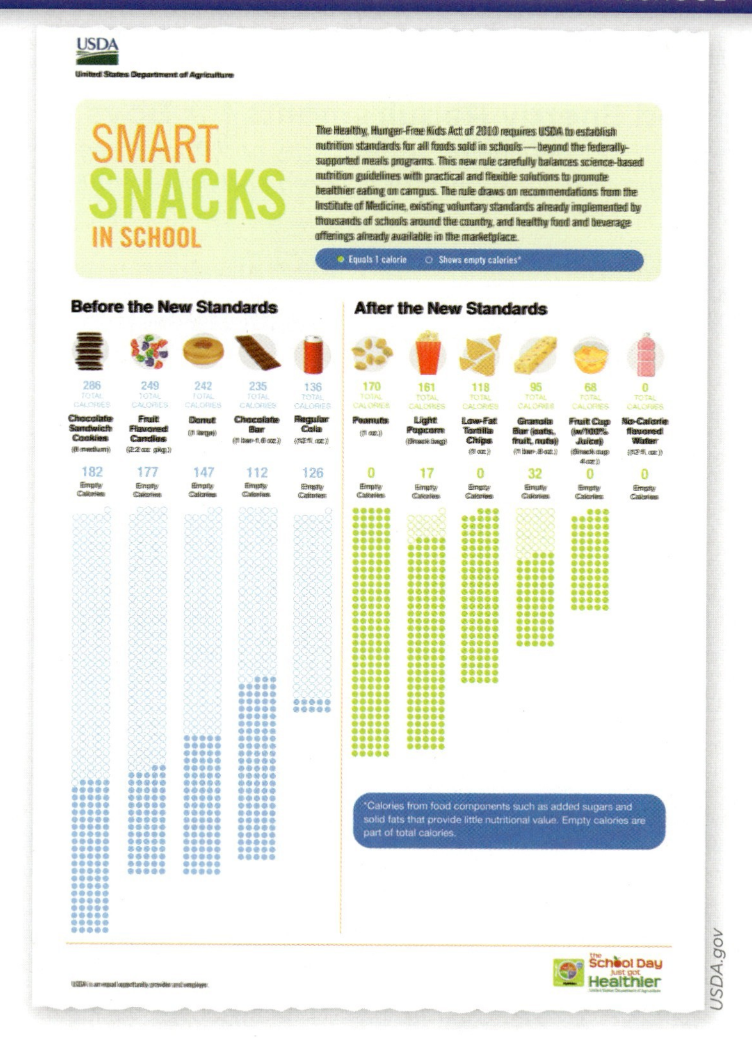

USDA.gov

NUTRITIONAL RECOMMENDATIONS FOR CHILDREN

Just as adults do, children have specific nutrient requirements set through the Dietary Reference Intakes (DRIs) and Accepted Macronutrient Distribution Ranges (AMDRs), as well as recommendations for food choice through the U.S. Dietary Guidelines for Americans and MyPlate.

After one year of age, children learn to feed themselves and consume new and different foods. Although their growth is slower than it was during infancy, it continues steadily through the toddler and preschool years and in periodic "spurts" during the elementary school years and early adolescence. Some parents get concerned because their children's interest in food declines during periods of slower growth, but the trend usually reverses itself during accelerated growth periods. Energy requirements in children, as in adults, are based on age, sex, and activity level. **(INFOGRAPHIC 18.5)**

Children of all ages need adequate amounts of all essential vitamins and minerals, so it's important that they eat varied and balanced diets that emphasize nutrient-dense foods. Although children's nutrient needs are generally lower than those of adults (because of their smaller size), they are no less vital to optimal growth, development, and health. Recommended nutrient intakes do not differ for boys and girls until about age nine years, when maturation and sexual development influence dietary needs. By adolescence, micronutrient needs are similar to those in adulthood.

According to the AMDRs set by the Institute of Medicine (IOM), carbohydrates should be children's primary source of energy, composing 45% to 65% of total calories (the same range recommended for adults). Children need sufficient protein, too—for growth,

INFOGRAPHIC **18.5** **Estimated Energy Requirements for Boys and Girls** *Total daily estimated energy requirements (EER) depend on sex and age, and from three years of age and up EER also depend on activity levels. The EER are shown for children and adolescents with average levels of physical activity.*

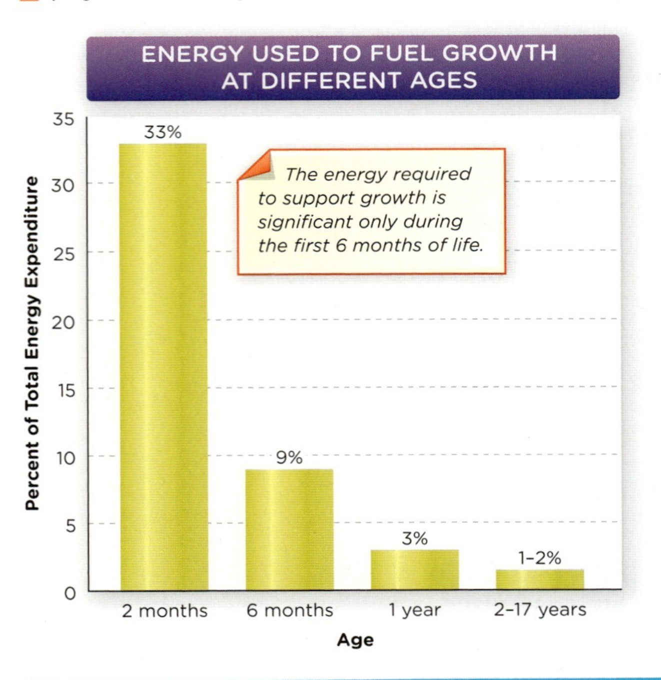

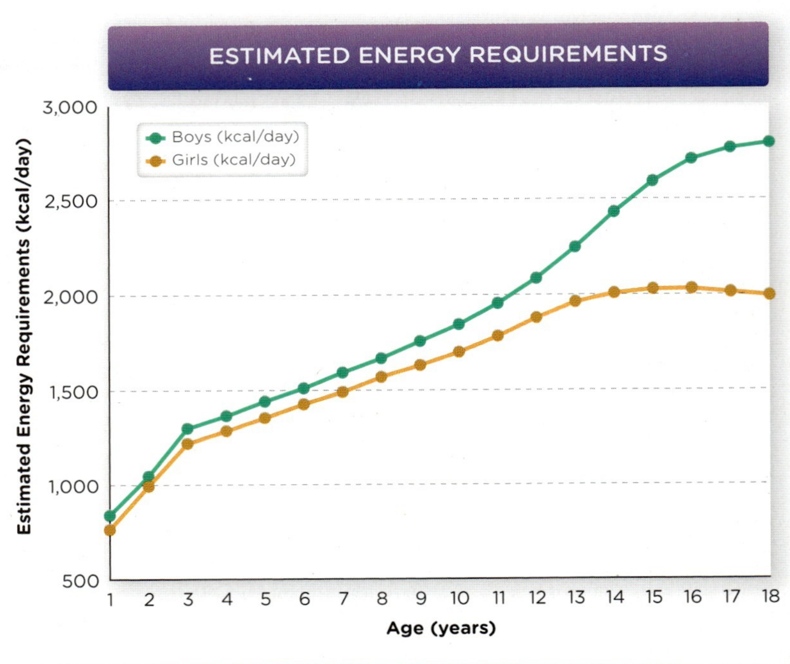

At what ages are the energy requirements for boys and girls very similar? At age 18, how many more kcals do boys require than girls?

INFOGRAPHIC 18.6 AMDRs for Children *Recommended percent of calories that each of the macronutrients should contribute to the total daily energy intake of children.*

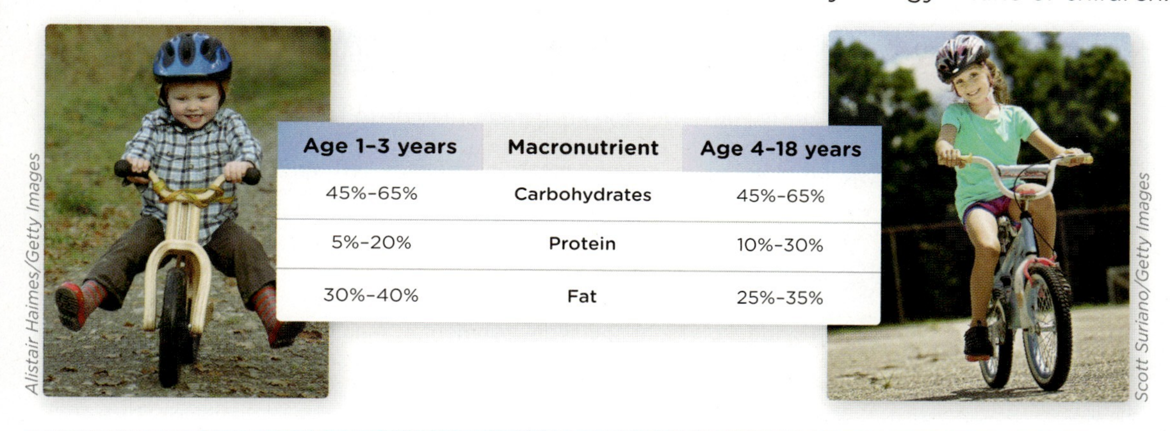

Age 1–3 years	Macronutrient	Age 4–18 years
45%–65%	Carbohydrates	45%–65%
5%–20%	Protein	10%–30%
30%–40%	Fat	25%–35%

Alistair Haimes/Getty Images

Scott Suriano/Getty Images

What AMDRs for children have an upper or lower cut-off that falls outside the intake recommended for adults?

tissue maintenance and repair, and other essential functions. For children aged 1 to 3 years, protein should make up 5% to 20% of total calories, and fat should make up 30% to 40% of their total calories. **(INFOGRAPHIC 18.6)** For older children aged 4 to 18 years, protein should compose 10% to 30% of total calories, and fat should make up 25% to 35% of total calories. These amounts reflect the transition from the higher fat intakes of infancy to the lower recommended fat intakes of adulthood. They also account for the fact that diets too low in fat may not provide sufficient intake of essential fatty acids as well as certain vitamins. Dietary fat also provides energy, which may be particularly important for younger children during the transition from a primarily breast milk or formula-based diet to a mixed diet.

The U.S. Dietary Guidelines for Americans also provide dietary recommendations for the general public 2 years and older, focusing on three main areas of emphasis. The first is the importance of balancing calories with physical activity to manage weight. The second is that individuals should eat more vegetables, whole grains, fat-free and low-fat dairy products, and seafood. The third is that individuals should consume fewer foods high in sodium, saturated fats, trans fats, cholesterol, added sugars, and refined grains.

MyPlate.gov offers daily food plans and science-based advice, too, to help kids and their parents build healthy meals and maintain or reach a healthy weight. Among other things, MyPlate can guide children toward age-appropriate portion sizes. For children up to age 8 years, for instance, one tablespoon per year of age is considered an appropriate serving size for many foods—so it might be appropriate to give a 3-year-old three tablespoons of mashed potatoes rather than an adult serving of ½ cup or more. **(INFOGRAPHIC 18.7)**

Parents also play an important role in shaping their children's eating behaviors by establishing the eating environment and modeling food-related behaviors and attitudes. Decades of research have consistently shown that children's eating patterns are strongly influenced by their physical and social environment—and the factors that influence this environment, such as parental education, time constraints, and ethnicity—and kids, like adults, are likely to eat foods that are available and easily accessible.

Young children especially depend on their parents to provide appropriate nourishment, so early parental influences can play an important role in determining a child's relationship with food later in life. As children grow older, they take more responsibility for feeding themselves and making decisions about

INFOGRAPHIC 18.7 MyPlate Food Plan Recommendations *Sample food plans are shown. For more food plans go to: http://www.choosemyplate.gov/kids/.*

Role of Parents in Shaping Their Child's Diet

- Consistently offer and encourage children to taste healthy foods.
- Provide regularly timed meals in a healthy setting.
- Model healthy food choices.
- Allow the child to control their energy intake.
- Provide a pleasant social environment at mealtime.
- Provide access to healthy foods and snacks.

The goals of sound nutritional guidance for children include assuring appropriate physical and cognitive development, achieving a healthy body weight, and minimizing the risk of chronic diseases.

	GRAINS	VEGETABLES	FRUITS	DAIRY	PROTEIN FOODS
	Make half your grains whole	**Vary your veggies**	**Focus on fruits**	**Get your calcium-rich foods**	**Go lean with protein**
2-year-old 1,000 kcal/day	3 ounces	1 cup	1 cup	2 cups	2 ounces
3-4-year-old 1,400 kcal/day	5 ounces	1 ½ cups	1 ½ cups	2 cups	4 ounces
6-8-year-old 1,800 kcal/day	6 ounces	2 ½ cups	1 ½ cups	2 ½ cups	5 ounces
12-17-year-old* 2,200 kcal/day	7 ounces	3 cups	2 cups	3 cups	6 ounces
12-17-year-old* 2,600 kcal/day	9 ounces	3 ½ cups	2 cups	3 cups	6 ½ ounces
12-17-year-old* 3,000 kcal/day	10 ounces	4 cups	2 ½ cups	3 cups	7 ounces

*Energy needs will differ significantly between individuals in this age group, depending on body size and activity levels.

? What recommendation do you feel children and adolescents typically have the most difficulty achieving?

Photo credits: Center for Nutrition Policy Promotion/USDA

Children learn food habits from their parents.
Parents model food choices and practices surrounding food.

Dragon Images/Shutterstock

food choices, but their tendencies are still strongly shaped by parental influence. For instance, parents can affect children's dietary practices by determining what foods are offered and when, the timing and location of meals, and the environment in which they are provided. They model food choice and intake and socialization practices surrounding food. Also, how parents interact with their children in relation to eating during mealtimes can influence how children relate to food in general, which can have an impact not only on nutritional choices but also on lifelong food preferences and eating habits. Studies demonstrate that although specific parenting styles are not strongly linked to negative eating behaviors and nutrient-poor food choices, a balance between parental authority and permissiveness within an appropriate eating environment helps guide children toward making healthier choices on their own as personal responsibility and independence increase.

To encourage healthy eating habits, parents should provide their kids with a variety of nutritious foods and encourage—but not force or bribe—their children to taste them. Food preferences are in large part learned through repeated exposure. The more a parent offers a food, the more likely a child will accept and try it; and the more a child tries it, the more likely he or she is to like it. Studies suggest that children may need to try a new food at least 10 times before they develop a taste for it. Involving children in food shopping and meal preparation also helps to expose them to different foods, educate them about nutrition, and engage them in the process of feeding the family. Ultimately, nutritionists suggest that parents should consider themselves responsible for the quality (type) of food and the frequency of feeding (when food is offered), but children as young as age two years should be permitted to gauge the quantity they consume. Serving children adult-size portions or forcing them to "clean" their plates can result in excess intake and make it difficult for children to determine when they have had enough to satisfy their hunger and avoid habitual overeating.

Sometimes, children develop particular ways of eating—they might only eat sandwiches if they're cut into triangles, for

Peanut butter food jag. *As they strive for independence, some children may insist on eating the same food day after day.*

Thomas M Perkins/Shutterstock

instance, or they will only eat one food item meal after meal. These behaviors, called **food jags**, are developmentally "normal" as children strive for more independence and control. Kids generally outgrow them with patience and guidance.

In addition to a varied and balanced diet, children should also engage in regular physical activity. Yet research suggests that less than half of all U.S. children meet the physical activity guidelines set by the U.S. Dietary Guidelines for Americans. These recommend that children and adolescents older than six years participate in 60 minutes or more of developmentally appropriate and enjoyable physical activity per day. Younger children should play actively several times a day. It's fine for kids to be active for short bursts of time rather than sustained periods, as long as these bursts add up to meet recommendations.

Unfortunately, children who don't meet the guidelines also tend to have lower diet quality—they eat fewer fruits and vegetables and more energy-dense snacks, drinks, and fast foods—which compounds their risk of obesity and chronic disease.

FOOD JAGS *developmentally "normal" habits or rituals formed by children as they strive for more independence and control*

NUTRIENTS OF CONCERN IN CHILDHOOD

American children consume low levels of several important nutrients, deemed *select nutrients of concern*. One is *calcium*. Most older children and adolescents in the United States do not meet recommended calcium intakes, which is problematic as calcium is necessary for bone health and the development of peak bone mass, both of which prevent fractures and osteoporosis later in life. Optimal calcium intake is particularly important during adolescence, when most bone mineralization occurs and calcium requirements are higher than during any other period of life. One way to help children obtain a range of healthful nutrients is to provide nutrient-dense snack options in a pleasant eating environment. (INFOGRAPHIC 18.8)

Many children under the age of three years also have inadequate dietary intakes of *iron*, because they often consume a lot of cow's milk, which is low in iron, in place of iron-rich foods. This can lead to iron-deficiency anemia, a form of anemia that can, during infancy and childhood, affect short-term and long-term neurological development.

Iron is also important for adolescents and older teens as it supports accelerated growth and, in females, helps to replace iron lost through menstruation. The DRIs recommend that adolescent boys (14 to 18 years of age) consume 11 mg of iron per day, more than the 8 mg recommended for younger boys and men older than 19. In teenage girls, iron needs increase from 8 mg to 15 mg during adolescence. Because few teenage girls consume enough iron to meet this increased need, approximately 10% of adolescent girls are iron deficient.

Fiber is another nutrient of concern, as 9 of every 10 U.S. children—particularly those with low-income and minority backgrounds—fail to meet the IOM's recommendations for fiber intake. Recent surveys suggest that children and adolescents only consume about half of the fiber they should; teens are particularly deficient. Fiber plays an important role in the health of children and adults and has been shown to reduce the risk of several

INFOGRAPHIC 18.8 Building Healthy Eating Habits *By making wise snacking selections, children can consume fiber, calcium, iron, and vitamin D.*

MAKE THE EATING ENVIRONMENT PLEASANT

- ✓ Children can help plan meals, shop for ingredients, and prepare the food.
- ✓ Turn off electronic devices and televisions, and avoid other distractions at mealtimes.
- ✓ Resist eating in the car to make mealtimes and snacks less rushed.
- ✓ Eat meals together as a family when possible.
- ✓ When you introduce a new food, serve it with food your child already likes.
- ✓ Include a vegetable or fruit at each meal and snack.

CHOOSE NUTRIENT-RICH SNACKS

GRAINS

- Tortilla filled with vegetables, salsa, and low-fat cheese
- Popcorn
- Rice cakes with peanut butter and sliced fruit
- Whole grain cereal (fiber) with low-fat vitamin D-fortified milk or soymilk
- Whole grain bread and peanut butter or other nut butter
- Iron-fortified cereal with strawberries (iron plus vitamin C)*
- Oatmeal (iron) with a small glass of orange juice fortified with vitamins C and D
- Whole wheat toast (iron) with grapefruit (vitamin C) on the side

VEGETABLES

- Cut up vegetables with low-fat dressing or yogurt
- Celery sticks with peanut butter and raisins
- Baked potato with salsa and sprinkled cheese
- Tossed salad with a small amount of dressing, preferably a vinaigrette
- Medium baked sweet potato with peel (fiber)
- Pureed vegetables added to soups and sauces
- Steamed vegetables mixed into pasta or rice dishes

FRUITS

- Whole fruit (bananas, apples, oranges, strawberries)
- Unsweetened applesauce
- Frozen grapes
- Frozen bananas blended with yogurt or soymilk and peanut butter
- Berries dipped in vitamin D-fortified yogurt (fiber plus calcium)
- Tortilla with nut butter and sliced fruit

PROTEIN

- Edamame or soy nuts (fiber)
- Baked tortilla chips and black beans with salsa
- Deviled eggs made with hummus
- Scrambled eggs with red and yellow peppers and ham (iron plus vitamin C)
- Nuts (fiber)
- Low-fat cheese in a tortilla with sliced red pepper or crunchy cucumber
- Peanut butter and jelly on bread enriched with iron and an orange on the side (iron plus vitamin C)
- Cottage cheese on a pear half (fiber plus calcium)

BEVERAGES

- Water
- Seltzer
- Low-fat milk, or milk alternatives such as soy, rice, or almond milk
- 100% orange or grapefruit juice fortified with vitamins C and D

*Vitamin C helps with the absorption of iron

Photo credits (top to bottom): Dragon Images/Shutterstock, Andi Berger/Shutterstock, margouillat photo/Shutterstock, MR.TEERASAK KHEMNGERN/Shutterstock, kyoshino/Getty Images, Nata-Lia/Shutterstock

chronic diseases. It also alleviates constipation, which many children experience.

Adequate intake of *vitamin D* is crucial for skeletal health and optimal bone growth and development in children, but few children and adolescents get enough of this vitamin either. This is in part because of the low levels of vitamin D naturally found in most foods. Milk is almost always fortified with vitamin D, but as noted, many children opt for soda or other beverages at mealtimes. Also sunlight exposure, which can help meet all or part of vitamin D needs, has decreased in recent decades as children spend more time indoors. The IOM recently increased its recommended intake of vitamin D by 50% to 15 micrograms (600 IU) for children and adolescents. Health care providers may recommend supplementation for some children to meet these recommended levels.

CHILDHOOD OBESITY

Approximately 17% of American children and adolescents are currently obese, a prevalence that has tripled since 1980. Its effects are vast. Obese children are more likely than normal-weight children to have high blood pressure and high cholesterol, type 2 diabetes, asthma, joint problems, fatty liver disease, and psychological problems. Obese children are at increased risk of being bullied and often endure negative stereotyping and comments from their peers. Moreover, obese children often become obese adults and are at a higher risk of a number of serious chronic diseases. **(INFOGRAPHIC 18.9)**

The causes of childhood obesity are varied. Diet is, of course, one of them—kids are eating more energy-dense foods, bigger portion sizes, and more meals away from home. Most children are also too sedentary. The Dietary Guidelines for Americans note that screen time, and in particular TV viewing, is associated with an increased risk of overweight and obesity. As a result, the American Academy of Pediatrics recommends that children and adolescents spend no more than one to two hours a day watching television, playing electronic games, or using the computer (other than for homework). Yet children tend to spend far more time than is recommended in these sedentary pursuits. A recent study found that children 8 to 10 years of age spend an average of 7.5 hours a day using entertainment media, including 4.5 hours of watching TV. In addition to being a sedentary behavior, television viewing also appears to have a negative impact on diet quality. Increased

INFOGRAPHIC 18.9 Childhood BMI and the Risk of Adult Obesity

Overweight and obese children are at an increased risk of becoming obese as adults.

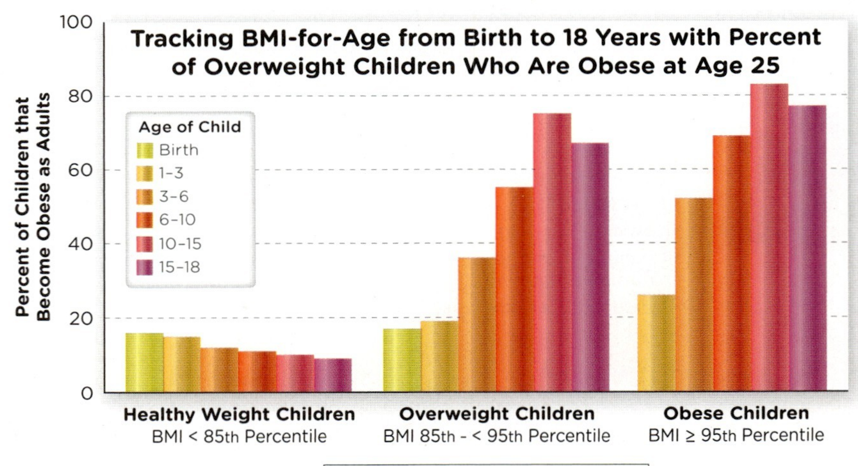

Describe the interaction that occurs between excess body fat and a child's age, and their risk of being obese as an adult.

Fighting against childhood obesity, First Lady Michelle Obama encourages children to exercise through the Let's Move! program.

Win McNamee/Getty Images

duration of television viewing by adolescents in grades 7 through 10 was associated with decreased fruit and vegetable intake, and higher intakes of candy and fast food.

National, state, and private programs are actively working to reduce and prevent the prevalence of childhood obesity through education, community-based programs, health-related legislation, food industry changes, and other initiatives. These include Let's Move!, a comprehensive initiative launched by First Lady Michelle Obama. Another is the Alliance for a Healthier Generation, founded by the American Heart Association and the Clinton Foundation. And the Robert Wood Johnson Foundation has noted recent progress in reducing rates of childhood obesity. For example, the obesity rate among children and young adults ages 2 to 19 has leveled off.

FOOD ALLERGIES

Studies demonstrate that 1 in every 13 U.S. children has a **food allergy**, a reproducible, adverse reaction to a food that is caused by a type of immune reaction to an **allergen** in that food. Usually, a **food allergen** is a small fragment of protein from the food that in susceptible individuals is identified by the body as harmful and that elicits an allergic reaction by the immune system. Interestingly, the first exposure to a food does not cause an allergic response, but it sensitizes susceptible individuals to the food by causing the immune system to produce **antibodies** (**immunoglobulin E** [or IgE]) that are specific to the food allergen. When the food is consumed again the food allergen binds to these antibodies, which stimulates immune cells to release histamine and other chemical substances that trigger an allergic reaction. (INFOGRAPHIC 18.10)

The most common food allergies are to cow's milk, eggs, peanuts, tree nuts, wheat, soy, fish, and crustacean shellfish. For individuals allergic to one or more of these foods, consuming even minute amounts can be immediately life-threatening. Food allergens can cause gastrointestinal reactions including "itchy" mouth, nausea, vomiting, abdominal cramps, and diarrhea; skin reactions such as hives, swelling, itchiness, and flushing; respiratory problems like wheezing, coughing, and runny or itchy nose; and most severely, they can cause **anaphylaxis** or **anaphylactic shock**, a massive immune reaction that can result in death if not treated with injected **epinephrine** (adrenaline). Although symptoms typically appear within two hours of exposure to a food allergen, allergic reactions can also be delayed. Younger children with delayed allergic reactions most commonly experience

FOOD ALLERGY
a reproducible, adverse reaction to a food that is caused by activation of an immune response involving the production of antibodies

ALLERGEN
an antigen that, when exposed to the body, stimulates an abnormal immune response

FOOD ALLERGEN
a substance in food (usually protein) that the body identifies as harmful and that elicits an allergic reaction from the immune system

ANTIBODIES
proteins found in blood that are produced in response to foreign substances, such as bacteria, viruses, and allergens, that have entered the body

IMMUNOGLOBULIN E
a class of antibody released in response to an allergen

ANAPHYLAXIS
a condition caused by decreased oxygen supply to the heart and other body tissues, and by vasodilation as a result of a heightened immune response to an allergen

ANAPHYLACTIC SHOCK
a massive immune response that occurs in oversensitive individuals and can result in death

EPINEPHRINE (ADRENALINE)
a hormone that can be administered as an injection to treat the potentially life-threatening symptoms of anaphylaxis

INFOGRAPHIC 18.10 **The Allergic Reaction Process**

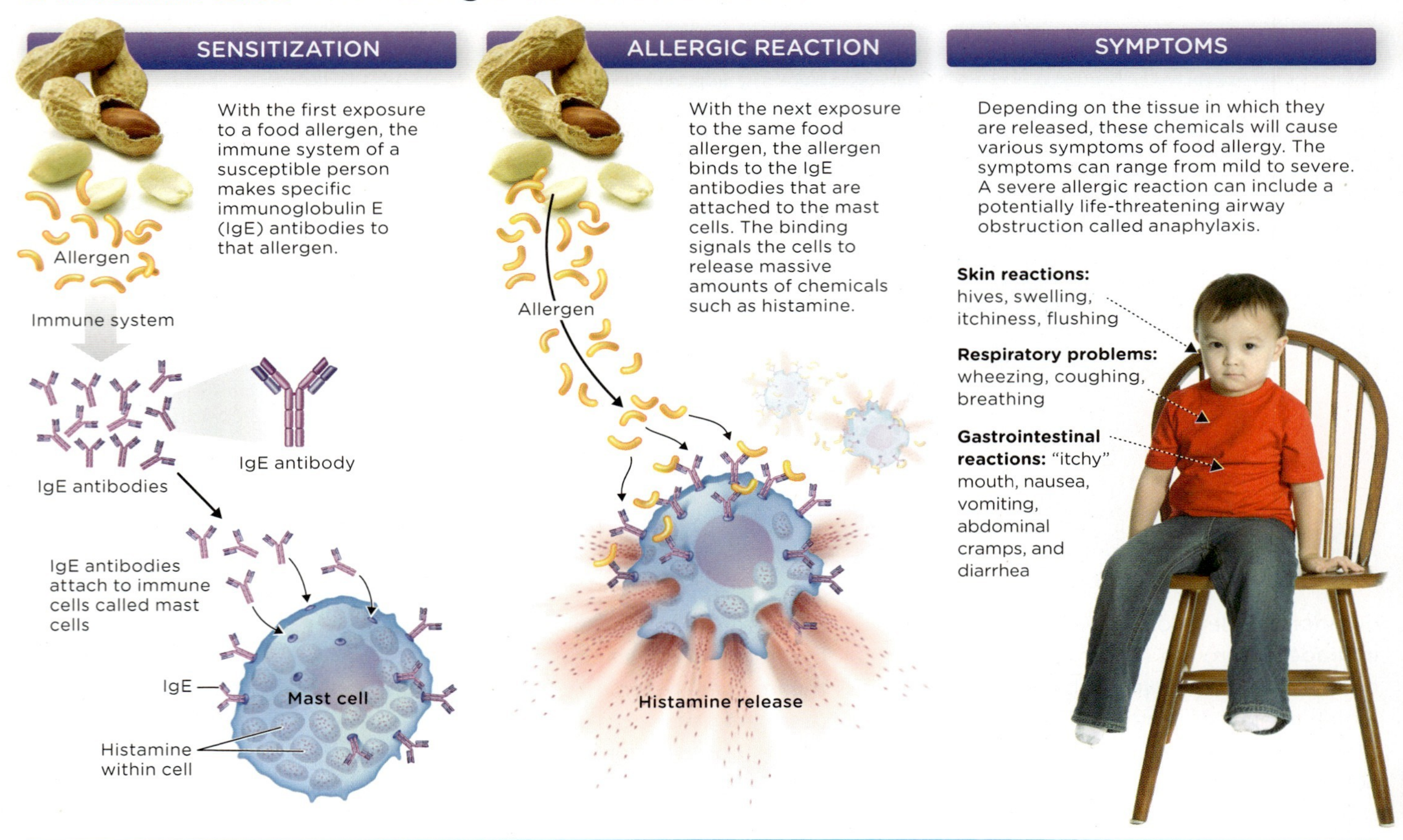

? *Why doesn't first exposure cause any symptoms in those people who develop an allergy?*

Photo credits (left to right): RedHelga/Getty Images, Nina Shannon/Getty Images

heartburn, vomiting, bloody stools, and poor growth. **(INFOGRAPHIC 18.11)**

Although most adverse reactions or sensitivities to food are attributed to allergies, many food-induced symptoms are not allergic in nature and instead may be caused by a **food intolerance**, a reproducible adverse reaction to a food that is not a direct result of an antibody (IgE) dependent **immune response**. Food intolerances can induce allergy like gastrointestinal and respiratory symptoms and often appear within a few hours of eating the food, but they are not life-threatening and are the result of different bodily mechanisms. Lactose intolerance, for example, is caused by reduced levels of the digestive enzyme lactase in the gastrointestinal tract, but there are other potential causes for food intolerance. Celiac disease is also classified as a food intolerance and not a food allergy because it does not involve the production of allergen-specific antibodies. Rather it is caused when immune cells directly attack the lining of the small intestine in the presence of gluten, a protein found in wheat, rye, barley, and many processed foods.

FOOD INTOLERANCE *a reproducible adverse reaction to a food that is not a direct result of an antibody dependent immune response; not usually life-threatening*

IMMUNE RESPONSE *how the body recognizes and protects itself against invading pathogens and foreign substances*

How can parents distinguish between food allergies and intolerances? In general, the best way to diagnose mild food allergies is through immunologist-led, double-blind, placebo-controlled oral food challenges in which kids are exposed to suspected food allergens and to "fake" allergens at alternating times, without knowing which is which, while being closely monitored. In a 2013 study, Dutch researchers administered this kind of food challenge to 116 infants and toddlers believed to have cow's milk allergies and found that only one-third of them were truly allergic. On top of that, some allergies naturally resolve over time. About 80% of children, for instance, outgrow cow's milk allergies, and most also outgrow egg allergies. This could be because the gastrointestinal tract matures to become less permeable with age, making it less likely that an allergen will be able to penetrate the lining of the intestine and initiate an immune response. Other scientists contend that food allergy risks go down with increased exposures to allergens.

Some food allergies are diagnosed in children and persist into adulthood. Allergies to peanuts and tree nuts are good examples of this category. There are some allergies, however, that tend to be diagnosed more commonly in adults. For example, estimates of fish allergy appear to be more than twice as high in adults than in children and allergy to crustaceans was five times more common in adults. Furthermore, food allergies can develop at any point in a person's life.

Why are allergies becoming more common? Some contend that heightened awareness among the public and health care providers translates to increased recognition and diagnosis of allergies. Another theory that was already mentioned was the hygiene hypothesis—the idea that less exposure to dirt and germs interferes with the normal maturation and regulation

INFOGRAPHIC 18.11 The Big Eight Food Allergens

Although many foods can cause allergic reactions, the CDC lists just eight foods that cause 90% of these reactions.

Photo credits (clockwise starting with eggs): Nattika/Shutterstock, RedHelga/Getty Images, YinYang/Getty Images, Filip Krstic/Shutterstock, Africa Studio/Shutterstock, ajafoto/Getty Images, margouillat photo/Shutterstock, Paul Johnson/Getty Images

Read food labels for a list of food allergens.

of the immune system. Changes in food manufacturing and processing could also play a role, in that foods are prepared differently now than they used to be. For instance, peanuts used to be boiled, but now they are roasted. Some scientists also theorize that inadequate intake of certain nutrients, like vitamin D, omega-3 fatty acids, and folate, may contribute to the rise of food allergy in children. Finally, the timing at which foods are introduced to kids could be important: Many parents delay the introduction of allergenic foods to their children, but research suggests that delaying introduction beyond a certain point may actually be detrimental.

Although there are no food allergy "cures," it is now getting easier for families to avoid foods that elicit allergic reactions. As a result of the Food Allergen Labeling and Consumer Protection Act of 2004, food manufacturers

and packagers have to declare the presence of major food allergens on food packages.

The Academy of Nutrition and Dietetics suggests that parents read food labels carefully; that they educate family members, caregivers, and teachers about allergy severity; and that they teach children about their allergies at a young age. It may also help for parents of children with food allergies to consult with a registered dietitian to develop a healthy eating plan that avoids allergens.

The Academy of Nutrition and Dietetics has published tips for parents who want to minimize the risk of food allergies in their children. They suggest exclusive breastfeeding for at least four months during infancy, which decreases the risk of atopic dermatitis, cow's milk allergy, and wheezing compared with the use of cow's milk–based formula. (Soy formula does not seem to affect allergy risk.) The organization also recommends against introducing solid foods before four to six months of age. That said, avoiding common allergens such as fish, eggs, and peanuts during pregnancy, breastfeeding, or beyond the age of four to six months in infants does not seem to be protective. In fact, there is growing evidence that the introduction of potentially allergenic foods while the infant is still receiving the majority of calories from breast milk reduces the risk of subsequent allergies.

In light of the growing evidence in support for the hygiene hypothesis, can parents do anything else, lifestyle-wise, to protect their kids against allergies and other immune conditions? No one advocates that kids should be raised in unsanitary conditions, but there may be other ways to help "prime" kids' immune systems to make them healthier and less susceptible to immune problems. "Be closer to nature," Kondrashova suggests—don't avoid contact with pets, grass, soil, and farms. In other words, let kids be kids. Don't deny them the fun of getting dirty once in a while. As paradoxical as it might seem, dirt and germs could, ultimately, make children healthier for life.

CHAPTER 18 **BRING IT HOME**

Preventing childhood obesity

Childhood obesity is a major public health concern in the United States. It not only affects the well-being of a child, physically and psychologically, but can increase the risk of weight-related conditions and chronic disease.

Visit the websites for the following childhood obesity prevention programs and initiatives. For each, describe the founding organization or initiative and the overall mission and scope.

Let's Move!—www.letsmove.gov

Alliance for a Healthier Generation—www.healthiergeneration.org

Robert Wood Johnson Foundation—http://www.rwjf.org/en/about-rwjf/program-areas/childhood-obesity.html

More info at: CDC Childhood Obesity Strategies and Solutions—http://www.cdc.gov/obesity/childhood/solutions.html

Imagine yourself in each of the following scenarios. From your review of the childhood obesity prevention programs, along with your own ideas, what are two to three specific strategies that these individuals might put into practice to promote healthy eating and exercise habits in children?

Parent of an overweight child

Middle-school food service manager

Elementary school physical education teacher

Owner of a convenience store in an urban neighborhood

Registered dietitian hired as a consultant by a restaurant chain

Vice president of marketing for a supermarket corporation

City planner

KEY IDEAS

Sound nutritional guidance for children encompasses four primary objectives, which include promoting optimal physical and cognitive development, a healthy body weight, enjoyment of food, and reduced risk of chronic disease.

Children grow and develop at varying rates throughout childhood and adolescence.

Growth standards for children and adolescents aged 2 to 20 years are based on BMI-for-age, calculated using height and weight plotted on the Centers for Disease Control growth charts according to a child's sex and age.

Puberty represents a dynamic period of development with periodic growth spurts that result in changes in height, weight, body composition, nutritional requirements, and sex-specific maturation. Growth and development during puberty varies widely among all children.

Nutrition surveys and studies suggest that, overall, U.S. children are consuming more energy-dense, nutrient-poor foods at the expense of nutrient-dense foods.

The National School Lunch Program is a federally assisted meal program operating in public and nonprofit private schools and residential child care institutions. Each school day, it provides children with low-cost or free meals that are consistent with the Dietary Guidelines for Americans.

Children have specific nutrient requirements set through the Dietary Reference Intakes and Accepted Macronutrient Distribution Ranges, as well as recommendations for food choice through the U.S. Dietary Guidelines for Americans and MyPlate.

Children of all ages need adequate amounts of all essential nutrients obtained through a varied and balanced diet to support optimal physical and cognitive development.

Parents influence children's dietary practices in many ways. They establish the eating environment and model health behaviors and attitudes for their children.

Several nutrients have been identified that are consumed in amounts low enough to be of concern among all or subgroups of children. These include calcium, iron, fiber, and vitamin D.

Childhood obesity can have a harmful effect on the body in a variety of ways, such as increasing the risk of obesity in adulthood.

National, state, and private programs are actively engaged in efforts to reduce and prevent the prevalence of childhood obesity through education, community-based programs, health-related legislation, food industry changes, and other initiatives that affect the health and weight status of children.

Food allergies are reproducible, adverse reactions to a food that are caused by a type of immune reaction to an allergen in food and they can be immediately life-threatening. Common food allergies include cow's milk, eggs, peanuts, tree nuts, wheat, soy, fish, and crustacean shellfish.

Food intolerances are reproducible, adverse reactions to a food that are not a direct result of an immune response. Food intolerances can produce reactions similar to food allergies, but are not life-threatening and are produced by different mechanisms in the body.

NEED TO KNOW

Review Questions

1. School-aged children 6 to 11 years old:
   a. experience the greatest increases in height and weight during childhood and adolescence.
   b. grow at a steady pace with occasional growth spurts.
   c. are all quite similar in height and body weight.
   d. would be considered at risk of obesity with a BMI-for-age above the 50th percentile.
   e. have the same nutritional requirements as adolescents.

2. Studies suggest that children who consume the most total energy, saturated fat, and sodium:
   a. determine their own serving sizes by serving themselves.
   b. eat most of their meals with their family.
   c. eat most of their meals at home.
   d. eat most of their meals away from home.
   e. watch the least amount of television.

3. All of the following reflect updates in the National School Lunch Program, EXCEPT:
   a. emphasis on whole grains over refined grains.
   b. increased offerings of vegetables and fruits.
   c. reduced saturated and trans fat.
   d. restriction of total fat to less than 20% of total calories.
   e. reduced sodium content.

4. The Acceptable Macronutrient Distribution Range for dietary fat for children aged 1 to 3 years is _____ of total calories.
   a. 20% to 35%
   b. 20% to 40%
   c. 25% to 40%
   d. 30% to 40%
   e. 35% to 50%

5. MyPlate.gov for children offers guidance toward healthy food choices, as well as age-appropriate serving sizes. An appropriate serving size of mashed potatoes for a 5-year-old would be:
   a. 5 teaspoons.
   b. 5 tablespoons.
   c. ¼ cup.
   d. ½ cup.

6. To encourage healthy eating habits in children, which of the following strategies is NOT recommended for parents or caregivers?
   a. Deny dessert or special treats if a child does not eat all foods offered at a meal.
   b. Repeatedly offer foods even if the child does not initially accept or try a food.
   c. Involve children in shopping for and preparing food.
   d. Model positive food choices and attitude toward nutrition.
   e. Determine which foods are offered and when they are offered.

7. All of the following are true in regards to food jags, EXCEPT that they:
   a. are generally outgrown.
   b. require immediate parental intervention to stop these behaviors.
   c. considered developmentally normal.
   d. can represent a child seeking more independence and control.
   e. are habits, rituals, or particular ways of eating.

8. All of the following have been identified as potential nutrients of concern during childhood, EXCEPT:
   a. calcium.
   b. dietary fiber.
   c. iron.
   d. omega-6 fatty acids.
   e. vitamin D.

9. To help reduce risk of obesity, the American Academy of Pediatrics recommends that "screen time," whether watching television, playing electronic games, or using the computer, should be:
   a. allowed only on the weekends.
   b. minimized to no more than three hours per week.
   c. restricted to less than one hour a day.
   d. limited to one to two hours a day.
   e. limited to three to four hours a day.

10. Food intolerances are reproducible, adverse reactions to food that:
    a. are life-threatening.
    b. involve the immune system.
    c. do not involve the immune system.
    d. are the same as food allergies.

11. The Academy of Nutrition and Dietetics provides tips to help families avoid foods that elicit allergic reactions, including:
    a. read food labels and ingredient listings carefully.
    b. educate family members, caregivers, and teachers about food allergies and allergens.
    c. teach children about their allergies at a young age.
    d. consult a registered dietitian nutritionist to develop a healthy eating plan that avoids allergens.
    e. All of the answers are correct.

Diet Analysis Using Choosemyplate

Healthy eating habits for children and teens

Log onto the United States Department of Agriculture website at www.choosemyplate.gov. At the top of the page click on "Audience" and then "Children," and then "Preschoolers."

1. Click on "Daily Food Plan" in the left hand navigation bar. Select the 1000 calorie food plan.
   a. How many cups of fruits and vegetables should a child consume each day when following this plan?
   b. List the three recommendations provided about protein foods.
   c. How much physical activity is recommended for children ages 2 to 5 years old?
2. Click on the "Cope with a Picky Eater" "Picky eating" link. Identify two ways that parents can encourage picky eaters to try new foods.
3. Return to the main page and click on the "Kids" link. Next, open the link for "Parents & Educators" and select "The School Day Just Got Healthier" link. Identify three changes that have been initiated in the school lunch program to make the meals healthier. Do you think these changes will encourage healthier eating habits in children? Explain your answer.
4. Next, click on the "Cut Back on Your Kid's Sweet Treats" link. Choose two tips that you feel are relevant to the adolescent. Discuss how these tips can be useful to this age group.
5. Now, go back main page and click on "Interactive Tools" and "Daily Food Plans" and open the plan for 2,000 calories for 18+ year olds. Looking at the recommendations, do you believe most 18-year-olds consume a diet similar to the recommendations? Explain your answer.

19 THE COLLEGE YEARS

Determinants of Eating Behavior, Disordered Eating, and Alcohol

HOW FOOD CHOICES ARE INFLUENCED BY SOCIAL NORMS

Ariel Skelley/Getty Images

LEARNING OBJECTIVES

- Identify at least five determinants of eating behavior and describe how each determinant might influence healthy food choices (Infographic 19.1)
- Discuss factors that may contribute to dietary and weight changes during the college years (Infographic 19.2)
- Describe characteristics of and differences among anorexia nervosa, bulimia nervosa, and binge eating disorder (Infographics 19.3 and 19.4)
- Discuss possible psychological and physical indicators of an eating disorder (Infographics 19.3 and 19.4)
- Define moderate alcohol consumption and what is considered a standard drink (Infographic 19.5)
- Describe how alcohol is absorbed and metabolized in the body (Infographic 19.6)
- Discuss the definitions of and potential consequences of heavy drinking and binge drinking (Infographics 19.8, 19.9, and 19.10)
- Identify the symptoms of alcohol use disorder (Infographic 19.11)

When you find yourself snacking on empty-calorie foods, grabbing that candy bar in the checkout aisle, or eating more than you probably should, do you ever wonder why you're doing it? Why people eat the way they do has long intrigued Brian Wansink, director of the Food and Brand lab at Cornell University. Wansink has spent decades studying this question. Although we might think we choose certain foods because they're good for us, or because they are delicious, or because we're having biological "cravings," Wansink has found that many of our food choices are actually

Brian Wansink, PhD is the John Dyson Professor of Consumer Behavior at Cornell University. He is the author of over 150 academic articles and books, including the best-selling Mindless Eating: Why We Eat More Than We Think *(2006). Wansink's academic research is on topics such as eating behavior, behavioral economics, and behavior change.*

Jason Koski/Cornell University

determined by social, environmental, and cognitive factors that have little to do with our physical hunger or need for sustenance. We eat differently—and even taste things differently—depending on whom we're with, the size of our plate, the music we're listening to, the foods we see, the mood we're in, and even the color of the room in which we dine. As Wansink wrote in his book *Mindless Eating*, "everyone—every single one of us—eats how much we eat largely because of what's around us."

In one study, for instance, researchers found that people who were offered three different flavors of yogurt ate 23% more than those who were offered one flavor. That's because we tend to eat more when we're offered more variety. Often cited rat studies have shown that rodents consume more food and gain more weight when offered "cafeteria-style" diets rather than just plain rat chow. And the more people we dine with, the more we eat—studies have shown that if you dine with one other person, you'll eat about 35% more than you would dining alone, whereas when you eat with seven other people, you'll eat as much as 96% more than you would alone. You also eat more when you're served more. In an amusing study conducted in 2000, Wansink invited people to a free movie matinee and gave them either a medium-sized or large-sized bucket of stale popcorn. Despite the fact that the popcorn had been popped *five days earlier*—in other words, it wasn't very appetizing—those who had been given the large size ate 53% more than those given the medium.

FRESHMAN 15 *popular term that perpetuates a misconception regarding a pattern of weight gain associated with the first year of college; in reality, weight gain averages three to four pounds*

It's perhaps not surprising, then, that when adolescents make the big switch from living at home to residing at college—where they have new and different food options, different social circumstances, demanding schedules, and a new dining environment—many students experience changes in how and what they eat. Although the infamous **Freshman 15**—the term used to describe the weight gain associated with the first year of college—is a myth, weight gain does occur in the first year of college but studies show that it averages to only about three or four pounds. This weight gain may be due in part to so many changed environmental factors, making us more likely to overeat.

Wansink's research has found, too, that as college semesters move forward, students tend to eat less healthfully over time, filling their trays with potato chips, curly fries, and chicken fingers instead of healthier options like grilled chicken or fish, vegetables, whole grains, yogurt, and fresh fruit. He speculates that as students get more stressed throughout the semester, they make poorer and more impulsive food choices—stress, in other words, trumps willpower and good intentions.

■ ■ ■

The college years are exciting and challenging times for young adults as they experience more independence, responsibility, and new situations. They are also a time of food-related transition, as college students change their eating patterns, gain more responsibility over their choices, and become more influenced by their peers and changes in their food

Experiencing a new food environment. *The college cafeteria provides the opportunity to sample new foods.*

Justin Sullivan/Getty Images

environment. Unfortunately, these changes tend to make college students eat more poorly. Most teens and college students do not consume the balance of nutrients recommended by dietary guidelines, and this in turn can undermine their physical and psychological health and development, including their academic performance.

GROWTH, DEVELOPMENT, AND NUTRIENT REQUIREMENTS IN LATE ADOLESCENCE

Adolescence (ages 13 to 17 years) and its extension into late adolescence/young adulthood (ages 18 to 21 years) are crucial times for good nutrition, because they are important years for growth and optimal development. Approximately half of adult bone mass is obtained by age 18, with boys gaining more bone mass and size compared with girls. By early adulthood, individuals have also become sexually mature—which means that while men have gained more lean mass, women have accumulated more fat mass, which is important for reproductive health.

Yet adolescents tend to eat more fast food, processed food, and sugar-sweetened beverages than do other age groups, and they eat fewer vegetables, fruits, whole grains, and milk than recommended. Their energy-rich, nutrient-poor diets lead many adolescents to have insufficient intake of several nutrients, including calcium, iron, and fiber. Teens and young adults also love to consume snacks, which while not inherently nutritionally detrimental, these snacks often contain unhealthy fats and added sugars—dietary constituents that the Dietary Guidelines for Americans suggest limiting.

Nutrient requirements for adolescents aged 14 to 18 years are often extrapolated from adult DRIs but may be adjusted to reflect the lower average body weight of the younger age group. Most Recommended Dietary Allowances (RDAs) and Adequate Intakes (AIs) increase at 19 years of age to match recommended intake levels for adults up to age 50 or older. However, the recommendations for phosphorus, calcium, and iron—nutrients important for growth and development—are higher for 14- to 18-year-olds. The recommendations are sex-specific, too, to account for the unique physiological changes and needs of boys and girls.

Influenced by peers. *Adolescents tend to eat more fast food, processed food and sugar-sweetened beverages than do other age groups.*

Nico Kai/Getty Images

CALCIUM

One nutrient that is often underconsumed by adolescents is calcium, which helps support bone growth and mineralization. The RDA for this mineral is set at 1,300 mg for 14- to 18-year-olds, decreasing to 1,000 mg at age 19. However, nutritional surveys indicate that many girls and young women do not meet this requirement. For example, one cup of milk provides 300 mg of calcium, but most young adults consume just one cup of milk a day, as they replace the milk frequently consumed in childhood with sugar-sweetened beverages such as sodas. And intake of other calcium-rich foods, like leafy greens, may be lacking as well.

IRON

By the start of college, growth has ceased for most older teens, which explains why the RDA for iron decreases from 11 mg per day for boys 14 to 18 years old to 8 mg per day for men 19 years and older. It's important to note, however, that the RDA for iron increases to 18 mg per day for women 19 to 50 years of age from the 15 mg per day recommended for adolescent girls.

Iron needs are higher for women than men because of the need to replace iron lost through menstruation. Although nutrition surveys demonstrate that men typically consume iron in excess of their RDA, most women do not meet their RDA for iron. It is estimated that approximately 10% of women 19 to 49 years of age are iron deficient; however, several studies have found that iron deficiency rates are even higher for college-age women. (See Chapter 14 for more about iron.) Suboptimal iron status may result not only in iron-deficiency anemia, but it has also been shown to impair cognitive function.

DETERMINANTS OF EATING BEHAVIOR

As adolescents mature, their food decisions become much more their own, as parents have less of an influence on choices. This new independence, combined with the social and environmental changes, can erode healthy eating habits. Several factors come into play.

Personal food preferences and familiarity with foods start to dictate choice.

Nutrition knowledge and awareness can also have a significant influence on food choice. For example, studies show that student intakes of vegetables, fruits, and whole grains and their overall diet quality improve after they take nutrition courses.

Economic factors such as the price of food and an individual's budget affect food selection.

In addition, *social and cultural influences* also have a significant effect on food choice and eating behavior.

Yogurt for calcium. *One cup of low-fat fruit yogurt has as much as 37% DV for calcium, based on energy and nutrient recommendations for a 2,000-calorie diet.*

A 2014 review of literature in the *Journal of the Academy of Nutrition and Dietetics* concluded that the "I'll have what she or he is having" phenomenon is particularly common in college-aged students: individuals conform to social norms in deciding what and even how much to eat, and, as mentioned before, they tend to eat more when they have more company and variety, as well as bigger serving sizes.

For college students, too, the structure of the meal plan and students' schedules affect how they eat, and the campus environment dictates access to and availability of healthy options: if a dining facility or vending machine provides mostly nutrient-poor choices, that's what students are likely going to eat. Even the design of a campus cafeteria can make a difference. If the salad bar is farther away from the cafeteria entrance making it less visible and convenient, for instance, people are often less likely to opt for salad. Likewise, if candy and potato chips require people to go through separate checkout lines, they are more likely to forgo them. **(INFOGRAPHIC 19.1)**

EATING CHALLENGES ON CAMPUS

The Freshman 15 is a popular term used to describe the pattern of weight gain associated with the first year of college. Although studies suggest that as many as 75% of college freshmen do gain weight, the average gain tends to be between three and four pounds and occurs

INFOGRAPHIC 19.1 Factors That Influence Food Choices *Hunger is the key determinant for eating, and taste and pleasure are perhaps the most powerful factors that affect the foods we choose to eat, but many other factors also strongly influence our food choices.*

What factors besides hunger, taste, and pleasure do you think have the strongest influence on your food choices?

primarily during the first three to nine months of college. It's unclear whether this pattern of weight gain is similar among men and women, or whether it continues into sophomore year or stabilizes. Most studies suggest that students do, unfortunately, tend to retain the weight they gain, which can set them up for a future of unhealthy weight control efforts with eating habits and exercise patterns that may actually increase their risk of being obese or overweight. An estimated 35% of college students are overweight or obese, although prevalence appears to differ geographically, racially, and differs between private and public universities.

Why does this college weight gain occur? There are many possible reasons, including the changes in eating patterns, food choices, living environments, and stress levels previously mentioned. Some experts also speculate that college students have a lower level of physical activity in college than in high school. Yet even when students are aware of the risk of weight gain in college and wish to avoid it, they typically don't. In a 2012 study, researchers surveyed more than 200 college freshmen women. The researchers reported that all of the women had heard of the dreaded Freshman 15, reporting "intense fears about gaining weight." They also believed that the weight gain was inevitable because of the increased availability of vending machines, fast food, alcohol, buffet-style cafeterias, and "food independence," the fact that they had more choice over what they ate than they used to.

Unrealistic fears about the Freshman 15 can, in some students, exacerbate food-related anxieties and increase the risk of eating disorders, particularly among students with preoccupations about body image.

To help college-aged students get the nutrients they need, the USDA developed MyPlate on Campus (choosemyplate.gov/MyPlateOnCampus), an initiative designed to encourage college and university students to promote healthy eating on campus. (INFOGRAPHIC 19.2)

EATING DISORDERS

When young adults move away from home for the first time, the transition can bring up unpredictable emotions. Indeed, transitions and emotional problems that surface during the college years increase the risk of **eating disorders**, which are characterized by an unhealthy and abnormal relationship with food and weight that threaten health and interfere with many areas of a person's life. Eating disorders often arise in response to life stressors or traumatic events as strategies to cope with overwhelming emotions, pressures, or situations.

Eating disorders are associated with emotional and psychological issues such as depression, anxiety, perfectionism, and low self-esteem. For some people, focusing on eating and weight becomes an outlet and a way to manage difficult emotions and demands that feel overwhelming or out of control. The social and cultural environment of college life, with its real or perceived emphasis on appearance and body weight, can add to feelings of insecurity and low self-esteem, potentially leading to unhealthy eating and exercise behaviors.

According the National Eating Disorder Association, the prevalence of eating disorders among college students has grown to 10% to 20% of women and 4% to 10% of men, a much higher prevalence than in any other age group. More than 11 million Americans are estimated to struggle with **anorexia nervosa** or **bulimia nervosa**, and an even higher number with **binge eating disorder**.

Types of eating disorders

For each of the eating disorders discussed, the characteristics described are the diagnostic criteria established by the American Psychiatric Association, and are the traits that are most consistently seen in those with the each specific disorder. However, it is important to recognize that there can be significant overlap in the traits displayed by individuals with different eating disorders. For example, those with bulimia nervosa may also exhibit a distorted **body image**.

Anorexia nervosa, which primarily affects adolescent girls and young women, is characterized by a distorted body image, excessive dietary restriction that leads to severe weight loss, and a pathological fear of weight gain. Approximately 0.5% to 1% of older adolescent girls or adult women

EATING DISORDER
an unhealthy and abnormal relationship with food and body weight that threatens health and interferes with many areas of life

ANOREXIA NERVOSA
a condition characterized by distorted body image and excessive dieting that leads to severe weight loss with a pathological fear of becoming fat; primarily affects adolescent girls and young women

BULIMIA NERVOSA
a condition characterized by frequent episodes of binge eating (at least once per week), followed by purging behaviors to avoid weight gain

BINGE EATING DISORDER
a condition characterized by recurring episodes of eating significantly more food in a short period than most people would eat under similar circumstances, accompanied by feelings of lack of control

BODY IMAGE
the way a person perceives their body size, shape, or overall appearance

INFOGRAPHIC **19.2** **Campus Dining Tips** *The dining hall provides many opportunities to develop healthy eating habits.*

SELECTION STRATEGIES:

✓ Fill your plate with fruits and vegetables.

✓ Take fruit for dessert.

✓ Try new foods, and eat a wide variety of foods. Eating a variety of foods is one way to be sure to meet your nutrient needs.

✓ Salads are a great source of nutrients and fiber, but avoid bacon bits and heavy dressings.

✓ Enjoy a slice of pizza, but pair it with a colorful salad, or a side of vegetables.

✓ Broth-based soups are a great way to fill up, warm up, and can prevent overeating more energy-dense foods.

wavebreakmedia/Shutterstock

BEHAVIOR STRATEGIES:

✓ Dine before you become overhungry. It's easy to overfill your plate when you are starving, and we tend to eat whatever is in front of us.

✓ Use smaller plates to avoid taking too much. Finishing a full plate makes us feel full. The larger the plate, the more we'll need to eat before we see that the plate is emptying and the visual cue tells us to slow down.

✓ Relax and eat slowly. Have a conversation during a meal, or put the fork down between bites.

✓ When we enter an unlimited buffet line, its sheer abundance prompts us to eat more. Reduce the influence of this prompt by thinking about portion control and smaller servings before you enter the line.

✓ Sit as far away from the source of food as possible. The farther you are from the buffet, the less likely you will be to get up and have seconds (or thirds).

✓ Eat with likeminded friends. Surround yourself with friends who are conscientious about nutrition; you'll be less likely to pick up or join in a poor eating routine.

✓ Make desserts a big deal. Save a dessert for a special occasion, and enjoy it carefully and with all of your attention.

✓ Eat and go. The dining hall is a great place for conversation, but hanging out there increases the temptation to go back and get more food. Finish your meal, grab an apple for later, and relocate.

meet the criteria for the diagnosis of anorexia nervosa, but prevalence is higher among college women, and there has been a recent increase in prevalence among young women aged 15 to 19. Men account for 5% to 15% of patients with anorexia. (INFOGRAPHIC **19.3**)

Bulimia nervosa is characterized by frequent (at least once a week) episodes of binge eating followed by inappropriate purging behaviors, such as self-induced vomiting, to avoid weight gain. Vomiting is the most common type of purging in bulimia, but excessive exercise, use of laxatives, or fasting may also follow a binge. Approximately 1% to 2% of late adolescent girls and adult women meet criteria for the diagnosis of bulimia nervosa, but the disorder is likely more common among college women. Men account for 5% to 15% of patients with bulimia.

Binge eating disorder is characterized by recurring episodes (at least once a week over three months) of eating significantly more food in a short period than most people would eat under similar circumstances, accompanied by feelings of lack of control. Someone with binge eating disorder may overeat, even when he or she is not hungry. The person may have feelings of guilt, embarrassment, or disgust, and may binge

INFOGRAPHIC 19.3 Diagnostic Criteria and Warning Signs of Anorexia Nervosa and Bulimia Nervosa

ANOREXIA NERVOSA

To receive a diagnosis of anorexia nervosa a person must display:

1. Persistent restriction of energy intake leading to significantly low body weight for age, height, and sex.
2. Either an intense fear of gaining weight or of becoming fat, or persistent behavior that interferes with weight gain (even though significantly low weight).
3. Body image problems, undue influence of body shape and weight on self-evaluation, or persistent lack of recognition of the seriousness of the current low body weight.

⚠ WARNING SIGNS

- Dramatic weight loss
- Preoccupation with weight, calories, food, fat grams, dieting
- Frequent comments about feeling "fat" or overweight despite weight loss
- Anxiety about gaining weight or being "fat"
- Self-esteem overly related to body image
- Denial of hunger
- Consistent excuses to avoid eating
- Withdrawal from friends and activities
- Excessive and compulsive exercise

- **Approximately 90%–95% of anorexia nervosa sufferers are girls and women.**
- **Between 0.5% and 1% of American women suffer from anorexia nervosa.**
- **Between 5% and 20% of individuals struggling with anorexia nervosa will die.**

BULIMIA NERVOSA

To receive a diagnosis of bulimia nervosa a person must display:

1. Recurrent episodes of binge eating characterized by both of the following:
   a. Eating in a discrete amount of time (within a two-hour period) large amounts of food.
   b. Sense of lack of control over eating during an episode.
2. Purging—vomiting, excessive exercise, or laxative abuse—to prevent weight gain.
3. The binge eating and purging both occur, on average, at least once a week for three months.
4. Self-evaluation is unduly influenced by body shape and weight.
5. The disturbance does not occur exclusively during episodes of anorexia nervosa.

⚠ WARNING SIGNS

- Evidence of binge eating
- Evidence of purging behaviors
  - Visits to the bathroom after meals
  - Abuse of laxatives, diet pills, and/or diuretics
  - Excessive, rigid exercise regimen
  - Calluses on the back of the hands and knuckles from self-induced vomiting
  - Discoloration or staining of the teeth
  - Severe dehydration
  - Sore throat
- Self-esteem overly related to body image
- Withdrawal from usual friends and activities

- **Bulimia nervosa affects 1%–2% of adolescent and young adult women.**
- **Approximately 80% of bulimia nervosa patients are girls and women.**
- **People struggling with bulimia nervosa usually appear to be of average body weight.**

For additional information call the National Eating Disorders Association Information and Referral Helpline at 1-800-931-2237. Or visit one of these websites:

- National Eating Disorders Association: http://www.nationaleatingdisorders.org
- The Alliance for Eating Disorders Awareness: http://www.allianceforeatingdisorders.com/
- National Institutes of Health, National Institute of Mental Health: http://www.nimh.nih.gov/health/publications/eating-disorders/index.shtml

Do you recognize any of these warning signs in yourself or your friends or family?

alone to hide the behavior. This disorder is associated with depression and marked distress. It differs from bulimia nervosa in that there is typically no compensatory behavior. Men make up an estimated 40% of patients with binge eating disorder. (INFOGRAPHIC 19.4)

Newly recognized eating disorders

According to new diagnostic criteria, two new categories of eating disorders include other specified feeding or eating disorder (OSFED) and unspecified feeding or eating disorder (UFED). These new categories are intended to more appropriately recognize and categorize disordered eating conditions that do not meet all the diagnostic criteria for anorexia nervosa, bulimia nervosa, or binge eating disorder. For example, under the new diagnostic guidelines, someone with anorexic features but without the low body weight might receive a diagnosis of OSFED and be eligible for early treatment. These new categories do not necessarily reflect less severe eating disorders—they simply characterize a different set of symptoms and characteristics.

Although not an officially recognized eating disorder, orthorexia nervosa is a term that has been coined to describe those who have a "fixation on righteous eating." Individuals can become obsessed with eating "right" and often have rigid eating styles and exercise patterns that affect the quality of their life and relationships, as well as their health. Interestingly, orthorexia appears to be more prevalent among athletes as well as nutrition and dietetics students and professionals.

Causes of eating disorders

The causes of eating disorders are many, varied, and complex. Genetics and predisposition may play a role, as do psychological and personality issues such as coping skills, perfectionism, trauma, and family and social issues. Social norms and cultural environmental factors can also promote unrealistic ideals. The underlying causes or predisposing factors differ for the different eating disorders. For example, individuals who develop anorexia nervosa are often perfectionists, while those with bulimia nervosa may struggle with impulsive behaviors and mood regulation issues. Other common triggers for eating disorders appear to be lack of self-esteem, past dieting behavior, and the cultural emphasis on thinness.

Eating disorders can also have many symptoms. Depending on the disorder, individuals may show a preoccupation with food and weight; changes in eating or exercise behavior; a distorted body image; abnormal weight loss or weight fluctuations; abuse of laxatives, diuretics, or diet pills; and social isolation and withdrawal. They may also choose to eat alone or be secretive about food and eating. Emotional and psychological signs vary, but they can include difficulty concentrating, poor coping skills, depression, and irregular moods.

INFOGRAPHIC 19.4 Diagnostic Criteria and Characteristics of Binge Eating Disorder (BED)

BED is the most common eating disorder, affecting approximately 3% of adults at some point in their lifetime.

BINGE EATING DISORDER (BED)

To receive a diagnosis of binge eating disorder a person must display:

1. Recurrent episodes of binge eating. An episode of binge eating is characterized by both of the following:
   a. Eating, in a discrete period of time an amount of food that is definitely larger than most people would eat in a similar period of time under similar circumstances
   b. A sense of lack of control over eating during the episode
2. The binge-eating episodes are associated with three (or more) of the following:
   a. Eating much more rapidly than normal
   b. Eating until feeling uncomfortably full
   c. Eating large amounts of food when not feeling physically hungry
   d. Eating alone because of feeling embarrassed by how much one is eating
   e. Feeling disgusted with oneself, depressed, or very guilty after overeating
3. Marked distress regarding binge eating is present.
4. The binge eating occurs, on average, at least once a week for three months.
5. The binge eating is not associated with the recurrent use of inappropriate compensatory behavior (purging).

BED affects individuals of every age, race, and socioeconomic status.

Those with BED can be of normal weight or heavier than average weight.

BED is often associated with symptoms of depression.

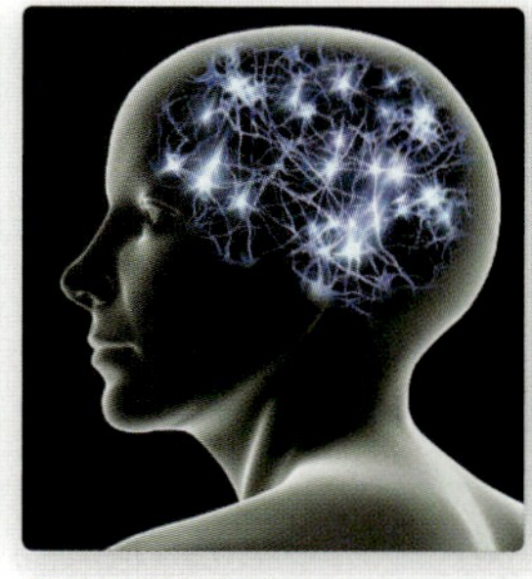

Biological factors can make someone more susceptible to developing BED.

It is common for people with BED to hide their behavior by eating in secret.

Do you recognize any of these warning signs in yourself or your friends or family?

Photo credits (top—left to right): altrendo images/Getty Images, FredFroese/Getty Images; (bottom—left to right): Leanne Surfleet - www.flickr.com/photos/leannesurfleet/Getty Images, Pasieka/Science Source, Image Source/Getty Images
Disclaimer: Licensed Material is being used for illustrative purposes only; persons depicted in the licensed material are models.

Counseling is helpful. *Eating disorders can be effectively treated through psychological intervention such as cognitive behavioral therapy.*

Tetra Images/Getty Images

Eating disorders can have profound health and lifestyle consequences. They can affect quality of life, relationships, and the ability to function day to day. In anorexia, physical manifestations include severe weight loss resulting in muscle wasting and depletion of fat stores. Hormonal and immune functions are impaired. Individuals suffering with anorexia nervosa often experience dehydration, fatigue, low body temperature, hair loss, demineralization of bone, and, in women, cessation of menstruation. Other symptoms, which include low blood pressure and loss of cardiac muscle, can have dire consequences. In bulimia, electrolyte imbalances can result in life-threatening cardiac arrhythmias and death. Self-induced vomiting can lead to dehydration, teeth erosion, and esophageal damage. Binge eating disorder often leads to or perpetuates obesity, which in turn increases risk of associated chronic diseases and overall mortality.

With prompt, aggressive, appropriate, multidisciplinary, and compassionate intervention and ongoing support, it is possible for individuals to overcome eating disorders and normalize their eating and exercise behaviors. According to the American Psychological Association, family therapy, cognitive-behavioral therapy, and interpersonal psychotherapy can help individuals overcome eating disorders and help restore healthy attitudes and behaviors toward eating, exercise, and body weight.

ALCOHOL
a potentially intoxicating ingredient found in beer, wine, and liquor; can be considered a food because it provides energy, but also has druglike effects via the central nervous system

BINGE DRINKING
a pattern of alcohol consumption that brings blood alcohol concentration to 0.08% or higher, corresponding to about five or more drinks on a single occasion for men or four or more for women, generally within about two hours

HEAVY DRINKING
Typically defined as consuming 15 or more standard alcohol drinks per week for men and 8 or more per week for women

ALCOHOL

Alcohol use and overuse is another issue that is often associated with the college years. People consume alcoholic drinks for many reasons—to socialize, celebrate, relax, escape, and as part of cultural and religious practices. According to surveys, 56% of the U.S. adult population reports having consumed alcohol in the past 30 days. Among college students aged 18 to 22 years, the percentage is slightly higher: about 60% of college students have reported drinking alcohol. However, the prevalence of both **binge drinking** and **heavy drinking** in college students is far higher than the general adult population. According to the National Institute on Alcohol Abuse and Alcoholism (NIAAA), "abusive and underage college drinking are significant public health problems, and they exact an enormous toll on the intellectual and social lives of students on campuses across the United States."

Alcohol is the common name for ethanol, a potentially intoxicating ingredient found in beer, wine, and liquor. Alcohol is not a nutrient, but does provide energy in the form of calories

(7 kcal per gram). It has potent druglike effects, acting as a central nervous system depressant. Microscopic yeast obtain energy from simple sugars in grains or fruit via a metabolic pathway called **fermentation**, which produces alcohol and carbon dioxide. The carbon dioxide may be released into the air, or the carbon dioxide may be trapped in the beverage causing it to be carbonated, as when beer is brewed. Beer is made by the fermentation of grains; wine is made from grapes or other fruit. Gin, rum, vodka, and whiskey are distilled, which involves heating a fermented mixture and then cooling it to condense the alcohol content into a more concentrated liquid form.

A standard alcoholic drink is comprised of 14 grams (0.6 fluid ounces) of pure alcohol (ethanol), which can typically be found in 12 ounces of beer, 5 ounces of wine, or 1.5 ounces (a "shot") of 80-proof distilled spirits or liquor. The alcohol content of liquor is half the proof, so 80 proof means 40% alcohol. While defining a "standard" drink is useful for establishing and following health guidelines, in practice, the alcohol content in a typical serving varies significantly. Many cocktails may contain the alcohol equivalent of three or more standard drinks. (INFOGRAPHIC 19.5)

The metabolism of alcohol

Alcohol is readily absorbed into the bloodstream through diffusion and then is transported to the body's cells and tissues and dispersed throughout the water-containing portions of the body. About one-fifth of all alcohol consumed is absorbed through the stomach; the rest is absorbed in the small intestine. When consumed in moderate amounts, alcohol is metabolized primarily in the liver by a two-step process to form acetate. In the first step, the enzyme alcohol dehydrogenase converts alcohol to **acetaldehyde**,

FERMENTATION
production of alcohol (and carbon dioxide) through the action of yeast on the simple sugars in grains or fruit

ACETALDEHYDE
a product of the first step of alcohol metabolism in the liver; a highly reactive and toxic compound that can damage cellular components

INFOGRAPHIC **19.5** **Alcohol Equivalents** *A standard-sized drink provides 0.6 fluid ounces of pure alcohol.*

Moderate drinking is defined as one drink per day for a woman and no more than two drinks per day for a man.

You can calculate the alcohol in a beverage. A handy rule of thumb is to divide 60 by the percent alcohol content (listed on the product label) to determine the number of ounces that will supply 14 grams of alcohol—that of a "standard" drink.
Example: For a beer that is 7.5% alcohol, 60/7.5 = 8 ounces of this beer will contain 14 grams of alcohol. Vodka 80 proof (40% alcohol): 60/40 = 1.5 ounces of vodka supply 14 grams of alcohol.

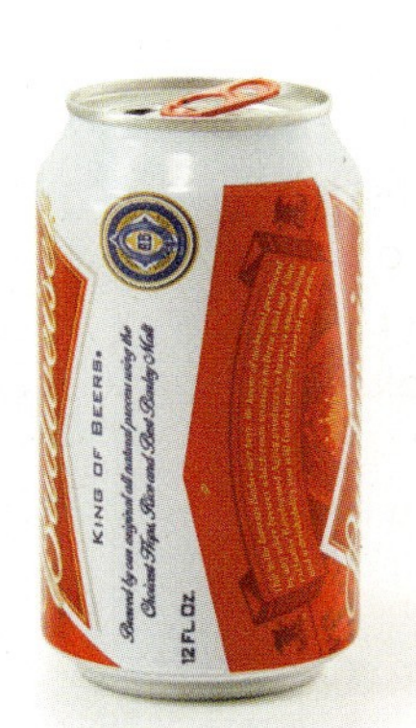

about 5% alcohol

12 fl oz of REGULAR BEER

shown in a 14-oz glass

about 8%–10% alcohol

6–7 fl oz of SOME CRAFT BEERS

about 12% alcohol

5 fl oz of TABLE WINE

"hard liquor" — whiskey, gin, rum, vodka, tequila, etc.

about 40% alcohol

1.5 fl oz shot of 80-PROOF SPIRITS

12 fl oz of REGULAR BEER = 6–7 fl oz of SOME CRAFT BEERS = 5 fl oz of TABLE WINE = 1.5 fl oz shot of 80-PROOF SPIRITS

The percent of "pure" alcohol, expressed here as alcohol by volume (alc/vol), varies by beverage.

? *What type of alcoholic beverage often contains more than one standard unit of alcohol?*

Photo credits: Eli Ensor

which is a highly reactive and toxic compound that can damage cellular components, including DNA. Acetaldehyde is then converted to acetate by the enzyme acetaldehyde dehydrogenase, and acetate then disperses to tissues throughout the body where it is converted to acetyl-coenzyme A, which can be used as a source of energy in the liver and elsewhere in the body. With higher levels of alcohol intake, the excessive amount of acetyl-coenzyme A that is produced in the liver results in high levels of fat synthesis that can cause a fatty liver, and eventually cause liver damage. Most alcohol is metabolized to acetate in the liver, but a small amount can also be metabolized in the stomach by the same two-step process, while even smaller amounts are excreted through our breath, sweat, and urine. **(INFOGRAPHIC 19.6)**

Factors that affect intoxication

Alcohol stays in the blood and body fluids until the liver is able to detoxify all the alcohol that has been consumed. In general, the liver can only metabolize about an ounce of alcohol per hour, regardless of how much has been consumed. Thus it is important to limit intake to prevent excessive alcohol accumulation in the blood. But many factors above and beyond the amount and rate of alcohol consumption can influence how intoxicated a person gets while

INFOGRAPHIC **19.6** **Alcohol Absorption and Metabolism** *Alcohol is metabolized primarily in the liver where the enzyme alcohol dehydrogenase (ADH) converts alcohol to acetaldehyde. Acetaldehyde is then converted to acetate by acetaldehyde dehydrogenase (ALDH). Acetate can either be metabolized as a source of energy—producing carbon dioxide and water, or it can be used to synthesize fat.*

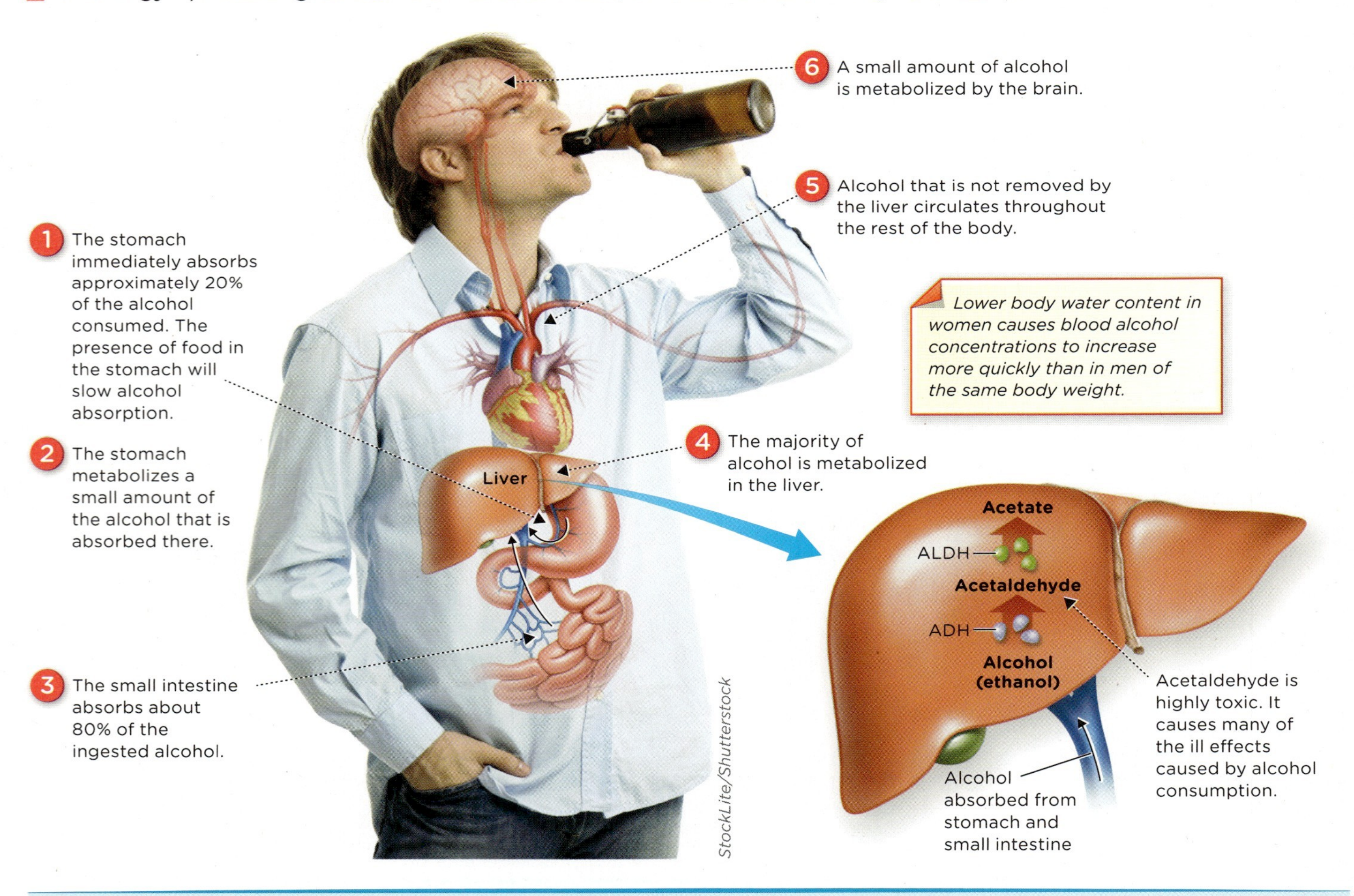

Why are individuals with high rates of ADH activity and low rates of ALDH activity at low risk of alcohol use disorder?

drinking. These factors include the person's sex, body weight, food intake before or during alcohol consumption, his or her use of drugs or prescription medications, and even genes.

Women, for instance, experience a more rapid rise in their blood alcohol levels than do men of the same size with similar alcohol intakes. This heightened effect is primarily a result of differences in body composition. Alcohol disperses in body water, and women—who have proportionally more body fat than men—have lower levels of total body water, causing alcohol to be more concentrated in the smaller volume of water. It is also often stated that the metabolism of alcohol in the stomach—before it ever reaches the blood—is faster in men than in women. However, more recent studies generally find no evidence for this. The presence of food in the stomach can serve to slow alcohol's absorption by the stomach and its passage into the small intestine, thereby reducing the rate at which the alcohol appears in the bloodstream.

The concentration of alcohol in breath and urine mirrors the concentration of alcohol in the blood, so the **blood alcohol concentration** (BAC) of an individual can be determined by measuring the amount of alcohol in the breath. The BAC is the standard means of assessing the extent of a person's alcohol impairment; a BAC of 0.08% is the legal limit for intoxication in the United States for drivers 21 years and older. A woman weighing 130 pounds easily exceeds the legal limit after consuming the equivalent of two alcoholic drinks. **(INFOGRAPHIC 19.7)**

BLOOD ALCOHOL CONCENTRATION
the percentage of alcohol in blood that is used as a measure of the degree of intoxication of an individual

Women and men tolerate alcohol differently partly because of size differences. *Women achieve higher blood alcohol levels than men do after drinking a similar amount of alcohol.*

Asiaselects/Getty Images

INFOGRAPHIC 19.7 Blood Alcohol Content

Use this table to estimate your blood alcohol level following the consumption of alcoholic beverages.

Male / Female	Approximate blood alcohol percentage (by vol.)								
	Body Weight								
Drinks	40 kg	45 kg	55 kg	64 kg	73 kg	82 kg	91 kg	100 kg	109 kg
	90 lb	100 lb	120 lb	140 lb	160 lb	180 lb	200 lb	220 lb	240 lb
1 (Male)	–	0.04	0.03	0.03	0.02	0.02	0.02	0.02	0.02
1 (Female)	0.05	0.05	0.04	0.03	0.03	0.03	0.02	0.02	0.02
2 (Male)	–	0.08	0.06	0.05	0.05	0.04	0.04	0.03	0.03
2 (Female)	0.10	0.09	0.08	0.07	0.06	0.05	0.05	0.04	0.04
3 (Male)	–	0.11	0.09	0.08	0.07	0.06	0.06	0.05	0.05
3 (Female)	0.15	0.14	0.11	0.10	0.09	0.08	0.07	0.06	0.06
4 (Male)	–	0.15	0.12	0.11	0.09	0.08	0.08	0.07	0.06
4 (Female)	0.20	0.18	0.15	0.13	0.11	0.10	0.09	0.08	0.08
5 (Male)	–	0.19	0.16	0.13	0.12	0.11	0.09	0.09	0.08
5 (Female)	0.25	0.23	0.19	0.16	0.14	0.13	0.11	0.10	0.09
6 (Male)	–	0.23	0.19	0.16	0.14	0.13	0.11	0.10	0.09
6 (Female)	0.30	0.27	0.23	0.19	0.17	0.15	0.14	0.12	0.11
7 (Male)	–	0.26	0.22	0.19	0.16	0.15	0.13	0.12	0.11
7 (Female)	0.35	0.32	0.27	0.23	0.20	0.18	0.16	0.14	0.13

Subtract .01% for each 40 minutes of drinking

1 drink =
1.5 oz 80-proof liquor
12 oz 5% beer
5 oz 12% wine

Eli Ensor

Why do blood alcohol levels rise more quickly in women than in men of the same body weight?

Potential benefits of alcohol

Alcohol consumption can provide potential health benefits. However, it is important to recognize that the evidence for the beneficial effects of alcohol is not as strong as the evidence for its harmful effects. Also, the risk-to-benefit ratio of light to moderate drinking is more favorable in those older than 50 years than it is for those who are younger than 50.

What is "moderate" alcohol consumption?

In healthy adults, moderate alcohol consumption—defined by the Dietary Guidelines for Americans as having up to one drink per day for women and up to two for men—is associated with a reduced risk of several chronic diseases and conditions, including heart disease, stroke, diabetes mellitus, abdominal obesity, and dementia, as well as a reduction in overall mortality risk. The most significant health benefits of light to moderate drinking seem to be on the cardiovascular system, with the risk of heart disease typically being reduced by 30% to 35%, with increases in high-density lipoprotein cholesterol contributing significantly to this reduction in risk. It is important to note, however, that excessive alcohol consumption causes hypertension and impairs cardiac function. And binge drinking, even in those who typically drink only lightly, increases the incidence of heart attacks and death. Even just an occasional binge-drinking episode virtually eliminates the protective effects seen with otherwise light to moderate intakes.

And although not depicted in the U.S. MyPlate, many national food guides from around the world include moderate alcohol consumption as a component of a healthful diet in adults. (Note that the definition of moderate consumption refers to the amount consumed on any single day and not the average over several days.) The most beneficial drinking pattern associated with a decrease in the risk of cardiovascular disease seems to be the consumption of one to two glasses of red wine immediately before or during the evening meal, as is practiced in many Mediterranean countries. However, it is not known if the benefits result from the alcohol itself, or from the social bonding or the avoidance of excess consumption that is promoted by this tradition. It is important to consider that the demonstrated benefits may also relate to the lifestyle habits and practices of people who consume moderate amounts of alcohol.

INFOGRAPHIC **19.8 Alcohol Involvement in Accidental Death, Homicide, and Suicide**

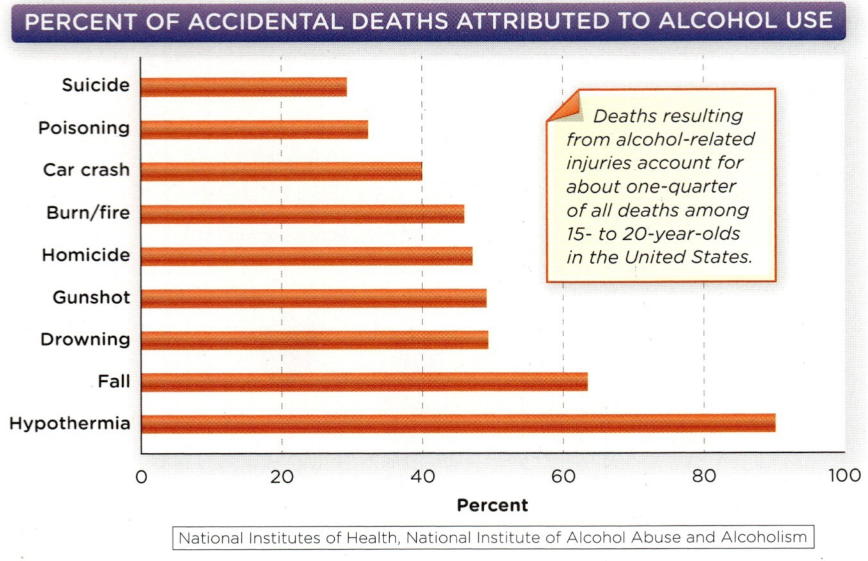

? *What effects of alcohol consumption likely explain the high rates of accidental death that are associated with its use?*

Harmful effects of alcohol

The Dietary Guidelines for Americans emphasize, however, that people should not begin drinking or drink more frequently because of these potential health benefits, because it is not possible to predict in which individuals alcohol abuse will become a problem. Also, alcohol consumption can have harmful effects on our health as well. Alcohol consumption is associated with an increased risk of some types of cancer, and excessive intake is associated with violent crime, drowning, and injuries from falls and motor vehicle crashes. Recent studies actually conclude that there is no safe level of alcohol consumption with regard to one's risk of cancer. Even light drinking is associated with increased risk of cancers of the mouth, throat, esophagus, and breast. And the more alcohol one drinks, the greater the risk. Also, there are many confounding variables and unanswered questions as to the effect of moderate alcohol intake on overall health, chief among them the lifestyle characteristics of moderate drinkers that can contribute to reduced risk. In other words, people who drink moderately may have other healthy lifestyle habits that contribute to the reduction in disease and mortality risk. **(INFOGRAPHIC 19.8)**

Alcohol can have other adverse long-term effects on health. Alcohol affects every single cell in the body, so chronic use and abuse can have widespread detrimental effects on organs and body systems including the brain, cardiovascular system, liver, digestive system, and immune system. Excessive alcohol use is the third leading cause of death for people in the United States each year, and it is the single strongest risk factor for death among males 15 to 59 years of age. Intoxication, too, can lead to impaired brain function because alcohol disrupts the production and function of neurotransmitters. This leads to impaired judgment, reduced reaction time, and loss of balance and motor skills, all of which increase the risk of motor vehicle crashes, violence, and other injuries. Alcohol also increases blood flow to skin and accelerates the loss of heat from the body, greatly increasing the risk of hypothermia if intoxication results in a loss of consciousness while in a cold environment. Finally, alcohol use can lead to alcohol dependence, coma, and death, if the alcohol is consumed too rapidly and in large amounts. **(INFOGRAPHIC 19.9)**

Certain people should never consume any alcohol. These include children and adolescents—yet underage drinking, though illegal, is nevertheless a huge public health issue. People aged 12 to 20 years drink 11% of all alcohol consumed in the United States, and more than 90% of this alcohol is consumed in excessive amounts in relatively short periods. Alcohol consumed during years of critical brain development can cause permanent impairment of cognitive function and increase the risk of fatal and nonfatal injuries. Plus, children who consume alcohol before the age of 15 are five times more likely to become alcohol dependant than are adults who begin drinking at age 21. Kids who drink alcohol are also more likely to engage in risky sexual behaviors, have poorer school performance, and are at higher risk for suicide as well as more likely to commit homicide than other kids.

INFOGRAPHIC 19.9 Health Effects of Heavy Drinking or Recurrent Binge Drinking *There are many ways alcohol can damage the body. For example, alcohol may irritate cells of the mouth and throat, and that could lead to DNA changes and, ultimately, cancer. Alcohol and its byproducts can also damage the liver, leading to inflammation and scarring. As liver cells try to repair the damage, mistakes in DNA replication may lead to cancer.*

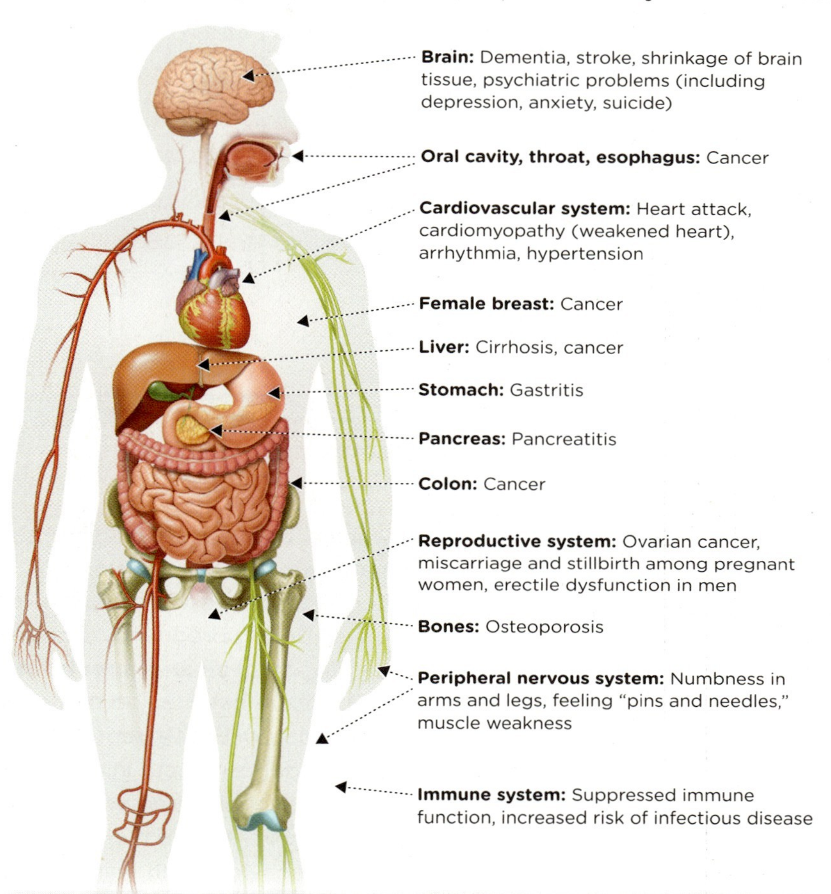

The risk of what two categories of disease seem to be most strongly affected by excess alcohol consumption?

INFOGRAPHIC 19.10 Consequences of Drinking Among College Students Between 18 and 24 Years of Age *Virtually all students experience the effects of drinking whether they drink or not.*

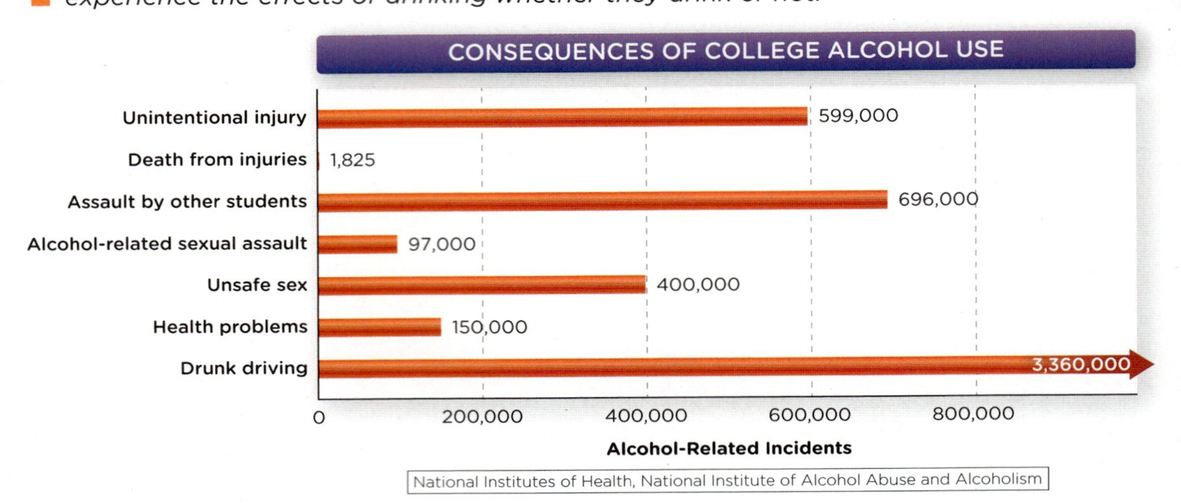

? *How many students do you know who have in some way been negatively affected by someone else's use of alcohol?*

Pregnant women, and women planning to become pregnant, should also abstain from drinking alcohol, as there is no known safe level of alcohol use during pregnancy. Several conditions, including **fetal alcohol spectrum disorders** (FASDs), have been linked to alcohol use during pregnancy. FASDs represent a group of conditions that can occur in a person whose mother drank alcohol during pregnancy. The effects range in severity and can include physical as well as behavioral and learning problems. Even moderate drinking (one to six alcoholic drinks per week) in pregnant women has been shown to decrease the IQ of their child when tested at eight years of age.

Individuals who plan to drive, operate machinery, or take part in other activities that require attention, skill, or coordination, as well as individuals who have medical conditions or take certain medications, should not consume alcohol. It is also wise for people with a personal or family history of alcohol use disorder, alcohol abuse, or who are unable to limit their alcohol intake to moderate levels, to abstain from alcohol use.

FETAL ALCOHOL SPECTRUM DISORDERS
a group of conditions that can occur in a person whose mother drank alcohol during her pregnancy; effects range in severity and can include physical, behavioral, and learning problems

Among college students in particular, binge drinking, a pattern of alcohol consumption that brings the BAC to 0.08% or higher, is a big problem. This pattern of drinking usually corresponds to consuming five or more drinks on a single occasion for men or four or more drinks on a single occasion for women, generally within about two hours. As many as 17% of the U.S. population reports binge drinking, but as many as 40% of college-aged students have engaged in binge drinking in the past month. Even college students who don't drink can still be affected by the drinking of others. **(INFOGRAPHIC 19.10)**

Alcohol abuse, also common among college students, is a pattern of drinking that has negative consequences for one's health, relationships, and academic and work performance. Signs of alcohol abuse include a failure to fulfill responsibilities, drinking despite dangers or consequences, aggressive or violent behavior while intoxicated, legal issues related to alcohol, and increasing dependency on alcohol. Frequent heavy drinking can increase alcohol tolerance by ramping up the activity of an alternate pathway of alcohol metabolism in the liver (not involving alcohol dehydrogenase). As a result, the rate at which alcohol is metabolized is increased, thereby requiring a person to drink more and more alcohol to produce the same effects. In addition, this alternative pathway produces highly reactive chemical compounds (reactive oxygen species) that cause widespread damage to tissues throughout the body. Ongoing alcohol abuse can lead to

INFOGRAPHIC 19.11 What Are the Symptoms of an Alcohol Use Disorder?

See if you recognize any of these symptoms in yourself.

In the past year, have you:

- ☐ had times when you ended up drinking **more, or longer**, than you intended?
- ☐ more than once wanted to **cut down or stop** drinking, or tried to, but couldn't?
- ☐ more than once gotten into situations while or after drinking that **increased your chances of getting hurt** (such as driving, swimming, using machinery, walking in a dangerous area, or having unsafe sex)?
- ☐ had to drink **much more** than you once did to **get the effect** you want? Or found that your **usual number** of drinks had **much less effect** than before?
- ☐ continued to drink even though it was making you feel **depressed or anxious** or adding to **another health problem**? Or after having had a **memory blackout**?
- ☐ spent a **lot of time** drinking? Or being sick or getting over other aftereffects?
- ☐ continued to drink even though it was causing **trouble** with your **family** or **friends**?
- ☐ found that drinking — or being sick from drinking — often **interfered with taking care** of your **home** or **family**? Or caused **job** troubles? Or **school** problems?
- ☐ **given up** or **cut back** on **activities** that were important or interesting to you, or gave you pleasure, in order to drink?
- ☐ more than once gotten **arrested**, been held at a police station, or had other **legal problems** because of your drinking?
- ☐ found that when the effects of alcohol were wearing off, you had **withdrawal symptoms**, such as trouble sleeping, shakiness, restlessness, nausea, sweating, a racing heart, or a seizure? Or sensed things that were not there?

If you ***DON'T*** have symptoms, then **limiting yourself to moderate alcohol consumption and avoiding episodes of binge drinking** will reduce your risk of developing an alcohol use disorder in the future.

According to the National Institute on Alcohol Abuse and Alcoholism, low-risk drinking is:
- Men: no more than 4 drinks on any day and no more than 14 drinks per week.
- Women: no more than 3 drinks on any day and no more than 7 drinks per week.

If you ***DO*** have any symptoms, then alcohol may already be a cause for concern. The more symptoms you have, the more urgent the need for change. A health professional can look at the number, pattern, and severity of symptoms to see whether an alcohol use disorder is present and help you decide the best course of action.

Do you or anyone you know display one or more of these characteristics?

Photo credit: Designsstock/Shutterstock

alcohol use disorder, too. *Alcohol use disorder* is a medical diagnosis given to individuals who have severe problems with alcohol and meet certain diagnostic criteria.

People can determine if they or their friends or loved ones have a problem with alcohol by checking for common signs and symptoms. **(INFOGRAPHIC 19.11)**

Hangover

A common consequence of alcohol consumption is the development of a hangover as the BAC falls following an episode of excessive drinking. Although individuals exhibit large differences in hangover susceptibility, the characteristic symptoms include headache, tiredness, impaired cognitive function, thirst,

INFOGRAPHIC 19.12 Approximate Calorie Content of Various Alcoholic Beverages

Beer: Scotch Ale, Stout
12 ounces
10% alcohol
300 kcal

Beer, Ale
12 ounces
5% alcohol
150 kcal

Light Beer, Lager
12 ounces
5% alcohol
100 kcal

Red and White Wines
5 ounces
12%-14% alcohol
120-130 kcal

Distilled Spirits
1.5 ounces
40% alcohol
100 kcal

Mojito
6 ounces
6% alcohol
140 kcal

Strawberry Daiquiri
12 ounces
10% alcohol
225 kcal

Traditional Martini
2.25 ounces
40% alcohol
125 kcal

Piña Colada
9 ounces
10% alcohol
490 kcal

Margarita
6 ounces
12% alcohol
250 kcal

Photo credits (top—left to right): *Floortje/Getty Images, Brian Macdonald/Getty Images, Foodcollection/Getty Images, C Squared Studios/Getty Images, Thomas Vogel/Getty Images.*
Photo credits (bottom—left to right): *Iatko Kostic/Getty Images, mphillips007/Getty Images, YinYang/Getty Images, Michael Phillips/Getty Images, Jack Andersen/Getty Images.*

dizziness, nausea, and altered mood. The underlying causes of hangover are not well-understood. However, research suggests that alterations in immune system function resulting in the production of a variety of hormone-like molecules (cytokines) and an increase in inflammation are the greatest contributors to a hangover's development. Blood glucose concentrations are also often observed to decrease during a hangover, which may contribute to symptoms of weakness, fatigue, and altered mood. There is also some evidence that elevated concentrations of acetaldehyde in blood may cause tissue damage that contributes to the symptoms of hangover. Current research neither supports the notion that symptoms result from dehydration, nor that rehydration can reduce the severity of symptoms. Furthermore, the consumption of beer and wine does not have a dehydrating effect as the fluid volume of these beverages more than offsets the increased rate of water loss caused by alcohol consumption. Other factors may contribute to hangover severity, including the type of alcohol consumed (darker beverages tend to produce more severe symptoms) and disrupted sleep patterns.

Alcohol and nutritional considerations

There are also nutritional considerations associated with alcohol use. As alcohol contains seven calories per gram, it is a fairly dense source of energy. Yet because it lacks overall nutrient value, alcohol is considered a source of "empty calories." **(INFOGRAPHIC 19.12)** The Dietary Guidelines for Americans recommend that less than 5% of total calorie intake should come from alcohol. Chronic alcohol abuse can compromise nutrition by reducing the secretion of digestive enzymes, impairing nutrient absorption and utilization, contributing to potential nutrient deficiencies. In addition, alcohol often displaces more nutrient-dense foods, resulting in dietary inadequacies.

College drinking can have multiple potential adverse and dangerous consequences, both short term and long term. Even if students are unwilling to consider abstaining, they should strive to drink responsibly by moderating intake, not drinking and driving, and not contributing to excessive intake by others. Awareness is the first step: Individuals should track their alcohol consumption patterns, as doing so can not only help them recognize a problem, but it can also help them set goals and come up with strategies for cutting down. Virtually all college campuses have resources and services that can support students in more responsible alcohol use. One national program that can help is the NIAAA's Rethinking Drinking program.

A SUCCESSFUL TRANSITION TO COLLEGE LIFE

College students are continuously challenged by competing demands, including academic responsibilities and involvement in extracurricular and social activities. Healthy food choices may become low priorities when compared with turning in a paper on time or attending a Friday night party. Furthermore, eating behavior is influenced by environmental characteristics, such as residency (living in a dorm versus an apartment), the particular culture of a school, and the pressures of exams.

The challenges young adults face are substantial. To prioritize healthy eating, young people need nutrition knowledge, self-discipline, and self-control. In addition, students have to make food choices within a university-specific setting (for example, eating in a student dining hall), where they are subject to the availability, accessibility, appeal, and prices of foods. All of these choices are influenced by friends and peers.

How can students make the best transition from college into a healthy adulthood? Campaigns by university administrators can promote responsible alcohol consumption and provide information and advice to enhance healthy food choices and preparation. Students can also benefit from programs that build self-control skills and assist in developing time-management skills. Finally, knowledge of good nutrition can be acquired during the college years and used for a lifetime.

Making food choices. *Eating behavior is influenced by environmental characteristics of a school, including academic pressure.*

David L. Moore-CA/Alamy

CHAPTER 19 **BRING IT HOME**

Determinants of eating behavior

Use the chart below to document your food and beverage intake for at least one 24-hour period. For the purposes of this activity, you do not need to record specific foods or beverages; rather, you will be examining eating patterns and determinants. For each meal or snack, record the time you spent eating; the location (for example, kitchen table, dorm room, cafeteria, restaurant, car); who you were with or if you were alone; your degree of hunger (on a scale of 0 to 3); the reason(s) for food or beverage choice; how you were feeling/emotions; how full you were at the end of the meal or snack; and any additional comments or observations. After completing the chart, answer the questions that follow.

Meal (M) or Snack (S)									
Date and time (Estimated start and end time)									
Location									
With company (list name[s]) or alone?									
Degree of hunger on a scale of 0 = not at all hungry to 3 = very hungry									
Reasons for food choice: C = Convenience/availability T = Taste/texture P = Price (cost) N = Nutrition/health S = Social–what friends/ family are eating O = Other									
Emotions/feelings: N = Neutral H = Happy/excited S = Stressed B = Bored T = Tired L = Lonely/sad O = Other									
Fullness following meal or snack 0 = not at all full 1 = no longer hungry, but not "full" 2 = full and satisfied 3 = physical sensation of fullness									
Comments or observations?									

1. What was the average time you spent eating? Do you feel you eat too quickly? If so, do you think slowing down and paying attention to the taste and texture of food consumed would add to enjoyment and meal satisfaction? Explain.
2. Where and with whom were most of your meals or snacks consumed? Comment on how you think location and who you were with effects food choice, quantity consumed, and enjoyment/meal satisfaction.
3. How did the level of hunger influence food choice and quantity consumed? Do you feel hunger is a primary determinant of when you eat? Or do other factors, like habit or schedule, determine meal or snack timing?
4. What was the primary reason(s) for your food choices? How do you think these reasons affect the healthfulness and nutritional quality of your overall diet?
5. How do you feel emotions affect your overall intake and food choice?
6. Comment on the level of fullness following your meals or snacks. Did you note any patterns related to time you spent eating, beginning level of hunger, or other factors?
7. Look back over your chart and your answers to these questions. Discuss how environmental, physical (hunger), social, and psychological variables guide your food choices and eating patterns.
8. After exploring the factors and situations that influence your food or beverage choices and eating patterns, are there any modifications in timing, location, or environment that you feel would positively influence your overall diet and health?

Take It Further

As discussed, the majority of first-year college students gain weight. Although weight gain on average is significantly less than the clichéd Freshman 15, the changes in eating patterns, choice of and access to food, physical activity, and overall environment present many challenges.

1. Look back at the chart you created and think about the many variables that influence when, where, why, and what we eat. Discuss how these factors may influence a young adult moving from a home environment to a college campus.
2. What five suggestions would you give a new college students to help them avoid weight gain and to choose a balanced, nutrient-dense diet on campus? You can find some ideas by visiting: *Eight Ways to Beat the Freshman 15* at http://www.eatright.org/Public/content.aspx?id=6442471553

KEY IDEAS

The college years are a time of food-related transitions that may present challenges in maintaining a healthy diet and a healthy body weight.

Nutrition studies show that many older teens and young adults consume diets that are energy-dense and nutrient-poor, which may result in inadequate intake of important nutrients such as calcium, iron, and fiber.

College women are at particular risk of iron-deficiency anemia because of suboptimal intake and higher needs than men and younger women.

The Freshman 15 is a term inappropriately used to describe the pattern of weight gain during the first year of college as students adapt to campus life. Studies demonstrate that although weight gain may occur for many students, it averages three to four pounds.

Hunger is a key determinant for eating, but food choice is influenced by many factors including personal preferences, nutrition knowledge, food environment and cost, and social and cultural influences.

Eating disorders are considered mental illnesses characterized by an unhealthy and abnormal relationship with food and body weight that threaten health and interfere with many areas of a person's life.

Anorexia nervosa, which primarily affects adolescent girls and young women, is characterized by distorted body image and excessive dieting that leads to severe weight loss with a pathological fear of becoming fat.

Bulimia nervosa is characterized by frequent episodes of binge eating followed by purging behaviors, such as self-induced vomiting or excessive exercise, to avoid weight gain.

Binge eating disorder is characterized by recurring episodes of eating significantly more food in a short period than most people would eat under similar circumstances, accompanied by feelings of lack of control.

Alcohol is a potentially intoxicating ingredient found in beer, wine, and liquor that is not a nutrient, but provides energy (seven calories per gram) and has druglike effects via the central nervous system.

Alcohol use and abuse in college-aged students is a major public health problem that can result in numerous adverse effects and consequences.

Moderate alcohol consumption in adults (two drinks per day for men and one for women) may have potential health benefits. A "drink" generally equates to the amount of alcohol found in 12 ounces of beer, 5 ounces of wine, or 1.5 ounces of 80-proof distilled liquor.

The liver can only metabolize about one ounce of alcohol an hour, regardless of the amount consumed. Excess intake results in a rise in blood alcohol concentration (BAC) and levels of intoxication.

With similar alcohol intake, BACs rise more quickly in women than men of the same size because women have less total body water content.

The metabolism of ethanol produces acetaldehyde and other highly reactive chemical compounds that can cause tissue damage and impair health.

NEED TO KNOW

Review Questions

1. Teens and young adults who consume energy-dense, nutrient-poor diets have been shown to have suboptimal intake of:
   a. calories.
   b. fiber.
   c. HDL cholesterol.
   d. omega-6 fatty acids.
   e. sodium.

2. Which of the following statements is TRUE regarding iron?
   a. Most women consume the RDA for iron.
   b. Iron requirements for women are the same for men.
   c. Iron requirements are higher for girls aged 14 to 18 years than for women aged 19 to 50 years.
   d. Iron requirements are lower for girls aged 14 to 18 years than for women aged 19 to 50 years.
   e. Iron-deficiency anemia in women is extremely rare.

3. The phrase "Freshman 15" has been used to refer to a pattern of weight gain that may occur during the first year of college. According to studies, the actual average weight gain is:
   a. 3 to 4 pounds.
   b. 6 to 8 pounds.
   c. 10 to 12 pounds.
   d. more than 15 pounds.

4. Which of the following is NOT a characteristic or consequence of anorexia nervosa?
   a. absent or irregular menstrual cycle
   b. bone loss
   c. dehydration
   d. elevated blood pressure
   e. loss of cardiac muscle

5. Binge eating disorder differs from bulimia nervosa in that:
   a. depression often follows a binge.
   b. large quantities of food are consumed in a short period.
   c. there is no purging after a binge.
   d. there is a feeling of lack of control during a binge.

6. All of the following are TRUE of males with eating disorders, EXCEPT:
   a. they represent 5% to 15% of patients with anorexia nervosa.
   b. they represent 5% to 15% of patients with bulimia nervosa.
   c. they represent an estimated 40% of patients with binge eating disorder.
   d. they represent an estimated 40% of patients with any eating disorder.

7. Alcohol contains _____ calories per gram.
   a. 4
   b. 7
   c. 9
   d. 12
   e. None of the above. The body cannot absorb and utilize the calories in alcohol.

8. Moderate drinking is considered to be no more than:
   a. 10% of total calories from alcohol.
   b. one ounce of alcohol for each 20 kg of body weight per day.
   c. one drink for women and two drinks for men per day.
   d. two drinks for women and three drinks for men per day.
   e. The amount that alters blood alcohol concentration to 1% or more.

9. All of the following are TRUE of women as compared with men, EXCEPT:
   a. women experience a more rapid increase in blood alcohol concentration with similar alcohol intakes.
   b. women are more likely to experience the intoxicating effects of alcohol with similar alcohol intakes.
   c. lower levels of body fat in women decrease absorption of alcohol into the bloodstream.
   d. lower levels of total body water in women increase the effects of alcohol.

10. Moderate alcohol consumption is associated with:
   a. a lower incidence of all types of cancer.
   b. a reduced risk of cardiovascular disease.
   c. an increased overall mortality.
   d. greater risk-to-benefit ratios in 20- to 30-year-olds.

11. Most of the alcohol consumed by an individual is:
   a. metabolized in the liver.
   b. metabolized in the stomach.
   c. excreted through respiration and perspiration.
   d. excreted in the urine.
   e. absorbed into adipose tissue.

12. During the first step of alcohol metabolism, the enzyme alcohol dehydrogenase converts alcohol to a toxic compound called:
   a. arachidonic acid.
   b. acetaldehyde.
   c. acetate.
   d. acetyl-coenzyme A.

Dietary Analysis Using SuperTracker

Using SuperTracker to understand nutritional needs during the college years

1. Log onto the United States Department of Agriculture website at www.choosemyplate.gov. At the top of the page, click the "Audience," "Students" and then the "College" link.
2. On the left-hand side of the page, you'll see the "MyPlate On Campus Toolkit" link. Click on this link and scroll down to TIPS TO KEEP YOU GOING on page 5.
3. Identify two tips that you believe are the most useful for the college student to remember. Why did you select these two?
4. Next scroll down to TAKE ACTION ON CAMPUS on page 9. Would you be willing to become a MyPlate On Campus Ambassador? Why or why not?
5. Underage drinking is a problem on college campuses. Identify two health risks of underage drinking. How can colleges discourage underage drinking on campuses?

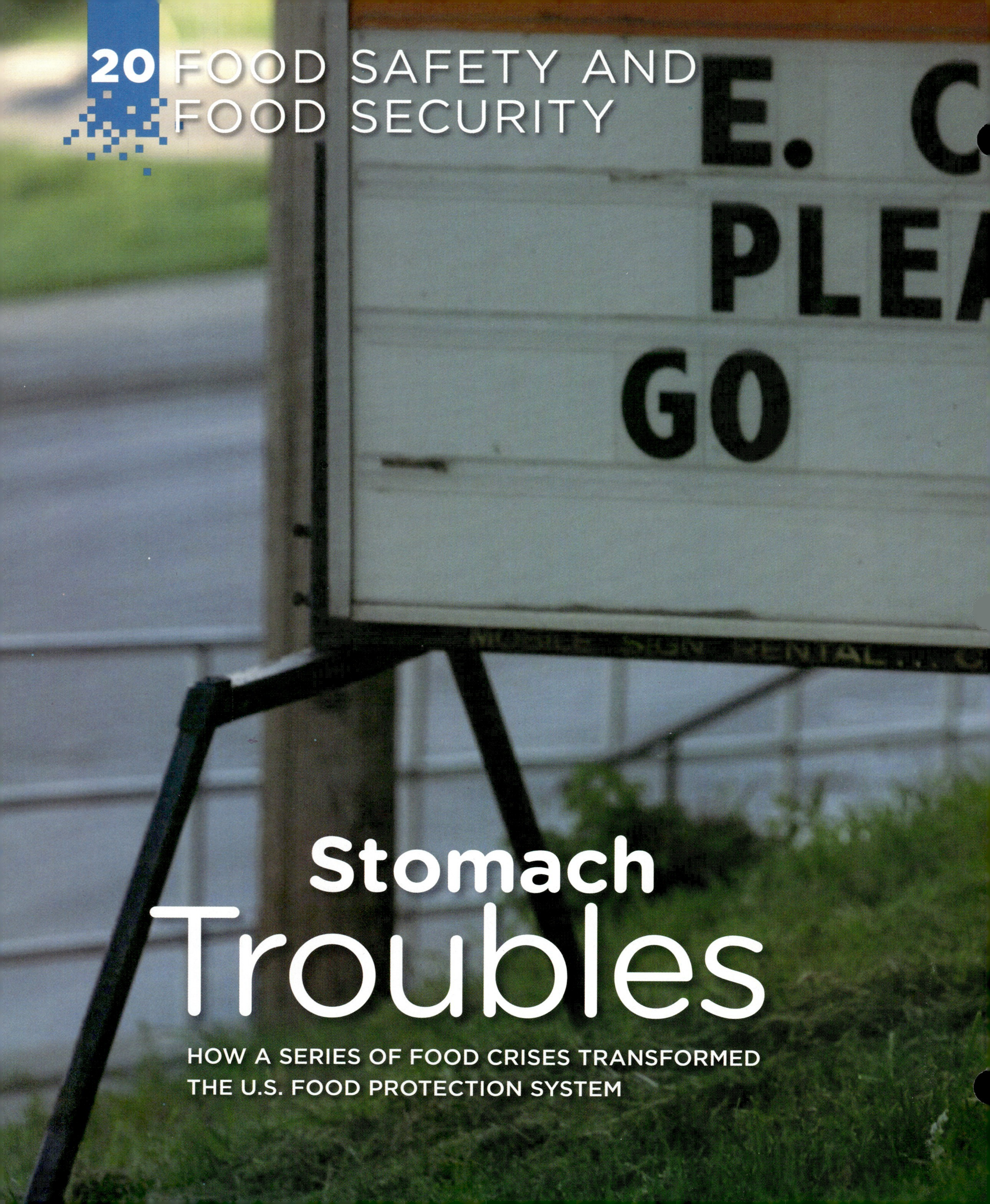

20 FOOD SAFETY AND FOOD SECURITY

Stomach Troubles

HOW A SERIES OF FOOD CRISES TRANSFORMED THE U.S. FOOD PROTECTION SYSTEM

Gabriel B. Tait/KRT/Newscom

LEARNING OBJECTIVES

- Identify the locations of food preparation that are associated with the highest incidence of foodborne illness outbreaks (Infographic 20.1)
- Describe ways in which food can be contaminated at each point from "farm to fork" (production to consumption) (Infographic 20.2)
- Name at least three of the most common pathogens that can cause foodborne illness, and name their common food sources (Infographics 20.3 and 20.4)
- Identify the groups most vulnerable to foodborne illness (Infographic 20.5)
- List the four basic food safety principles with an example of each (Infographics 20.7, 20.8, 20.9, and 20.10)
- Define "organic" and explain how the term is used on food labels (Infographic 20.11)
- Describe how food additives are regulated in the United States and the primary reasons for their use (Infographic 20.12)
- Discuss the status of food security in the United States (Infographic 20.13)
- Discuss the potential benefits and risks of genetically modified foods (Infographic 20.14)
- Identify areas in the world where undernourishment is common (Infographic 20.15)

In late summer of 2006, as Americans flocked to watch Samuel L. Jackson do battle with *Snakes on a Plane*, mourned the demotion of Pluto from planet to giant interstellar rock, and learned to Tweet for the first time, emergency department doctors around the country began to notice an unnerving number of cases of people complaining of severe gastric distress.

This was no ordinary stomachache. People hobbled into the emergency department doubled over in pain, with searing abdominal cramps, rectal

E. coli *bacteria.* *Undercooked beef (especially hamburger), contaminated water, unpasteurized milk and juice, raw fruits and vegetables (e.g., sprouts), can contain* Escherichia coli (E. coli). *Some types of* E. coli *can cause severe illness.*

David M. Phillips/Science Source

bleeding, and bloody diarrhea. These were likely cases of food poisoning, but the severity of the symptoms suggested a more insidious cause than your standard foodborne pathogen.

Laboratory tests eventually revealed that these individuals were suffering from infection with a particular strain of *Escherichia coli* (*E. coli*), known as O157:H7. *E. coli* is a common form of bacteria, found in the intestines of many animals, including humans. Strain O157:H7 has acquired additional genes from other bacteria that make it especially toxic. It is one of several Shiga toxin–producing bacterial strains that can cause severe illness in people and sometimes even death.

When the first of these patients with gastric problems started arriving in Wisconsin emergency departments, it was not yet clear that an outbreak was at hand. But soon, remarkably similar cases began popping up in other states. By autumn, some 200 people in 28 states had been sickened, more than half of whom required hospitalization, and three died. The common factor linking all these individuals? They had all eaten Dole brand spinach from a bag.

FDA scientists eventually traced the *E. coli* outbreak to a spinach packaging plant in San Juan Bautista, California. DNA testing pinpointed the plot of land where the tainted spinach had been grown. "The spinach originated from really just one corner of one field in California," says Caroline Smith DeWaal of the Center for Science in the Public Interest (CSPI), a nonprofit advocacy group. "And the spinach got into a washing facility where it contaminated the water, and then a lot of spinach got contaminated. Then that was shipped all over the country."

As consumers were warned to stop buying bagged spinach, some nagging questions loomed over the entire episode: How could this have happened? Don't we have protections in place to protect our food supply?

The 2006 spinach outbreak brought to light, in the most unfortunate way possible, certain flaws in the nation's food safety control system. What experts in food safety had known for a while became painfully clear to the average American: The nation's food supply was not as safe as many believed. **(INFOGRAPHIC 20.1)**

INFOGRAPHIC 20.1 Distribution of Foodborne Disease Outbreak-Associated Illnesses by the Place Where Food Was Prepared, 2009–2012

A foodborne disease outbreak is defined as the occurrence of two or more cases of a similar illness resulting from eating the same food.

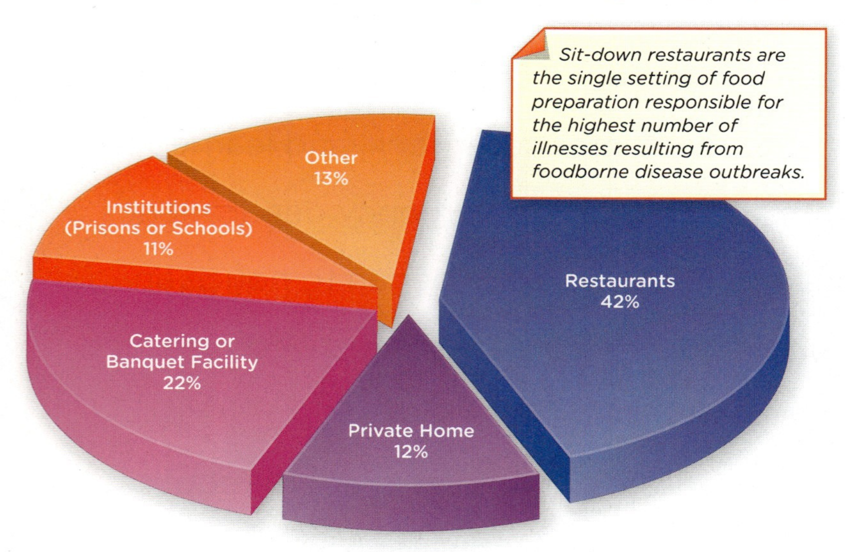

? *Give two likely reasons why restaurants are the location of the majority of foodborne illnesses.*

A BRIEF HISTORY OF FOOD SAFETY IN AMERICA

In the contemporary United States, **food safety** is ensured by an overlapping system of rules, regulations, and practices that preserve the quality of food and prevent contamination with foodborne pathogens, such as bacteria and viruses. The system we have in place today recognizes that the safety of food can be compromised at any point from production to consumption otherwise known as "farm to fork." **(INFOGRAPHIC 20.2)**

Most Americans take it for granted that the food they buy in the supermarket is "safe" to eat, but not many know how that safety is ensured, and even fewer realize just who is responsible for protecting them.

Prior to the early twentieth century, food products (and drugs) in this country were largely unregulated. At a time when more and more Americans were moving away from farms, and buying food in the marketplace, they also had no guarantees about what they were actually eating. There was no monitoring or control over how food was handled from the field to the shelves or bins from which it was sold. It was not uncommon for milk to be "preserved" with formaldehyde, hams to be adulterated with borate, and babies' "soothing syrups" to be laced with morphine.

Food safety advocates had been arguing for years that something needed to be done to protect consumers from hazardous contaminants in food and unsafe food production and handling practices, but it was not until 1905 that the proper incentive for change was reached. That incentive came from a controversial book—*The Jungle* by Upton Sinclair. In this work of "muckraking" literature, Sinclair wrote about the unsanitary conditions of the Chicago meatpacking industry, in often stomach-churning prose.

> There would be meat stored in great piles in rooms; and the water from leaky roofs would drip over it, and thousands of rats would race about on it. . . . These rats were nuisances, and the packers would put poisoned bread out for them; they would die, and then rats, bread, and meat would go into the hoppers together . . . there were things that went into the sausage in comparison with which a poisoned rat was a tidbit. (*The Jungle*, Ch 14)

Citizens were appalled by what they read, as was U.S. President Theodore Roosevelt, who realized he needed to do something to clean up the mess. The revelations led directly to the passage of the nation's first food safety legislations. The Meat Inspection Act and the Pure Food and Drug Act were signed into law on the same day in 1906.

Under the Meat Inspection Act, government inspectors were given the authority to inspect carcasses in slaughterhouses. Using what became known as the "poke-and-sniff method," inspectors literally used sight, touch, and smell to detect rotten or contaminated meat. The Pure Food and Drug Act put into place rules against adulteration (contamination with foreign substances and known poisons) and misbranding (labeling something as one thing when it was another) of foods and prescription drugs.

FDA historian Suzanne Junod, explains the far-reaching consequences of the laws. "The Meat Inspection Act had inspectors in every single meat plant in the country. They

FOOD SAFETY
the policies and practices that apply to the production, handling, preparation, and storage of food in order to prevent contamination and foodborne illness

Early food inspection practices. *This was the scene at a Chicago meat packing house in 1906, when the Meat Inspection Act was enforced. Inspectors used sight, touch, and smell to detect contaminated meat (in this case pork).*

Bettmann/CORBIS

INFOGRAPHIC 20.2 How Does Food Become Contaminated? *Microbes constantly surround us, and there are many opportunities for food to become contaminated during production and preparation.*

Fresh fruit and vegetables may be contaminated if they are washed or irrigated with water that is contaminated with animal manure or human sewage.

Bacteria are present in the intestines of healthy animals, and meat and poultry are often contaminated with these bacteria during slaughter.

Many shellfish concentrate bacteria naturally present in seawater.

PROCESSING AND DISTRIBUTION

Produce can be contaminated during harvest, packing, or the stocking of store shelves when individuals performing these tasks are ill or practice poor hygiene.

Equipment used during processing and distribution may have been previously contaminated by microbes, and then transferred to foods.

GROCERY

Buffet

Buffet

PREPARATION

Food handlers can contaminate foods, or microbes can be transferred from one food to another (cross-contamination) when the same knife, cutting board, or other utensil is used to prepare both foods without the utensils being washed in between.

? *What should you do with your fresh produce before you eat it, even if it has a rind or peel?*

couldn't operate unless there was an inspector there. With the Food and Drug Act, they pretty much put their faith in the ability of science to detect and control adulteration and misbranding issues."

To this day, meat and poultry safety and food and drug safety are handled in different ways by different parts of the government. Meat inspection is handled by the U.S. Department of Agriculture (USDA), while food (produce and packaged foods, for example) and drugs are the domain of the U.S. Food and Drug Administration (FDA).

The poke-and-sniff method of meat inspection was an improvement over having no controls, but it was far from perfect. The main problem with this method is that bacteria are invisible to the naked eye and are not always detectable by touch or smell. Also, meat production has become more industrialized over the years, creating opportunities for contamination. One modern hamburger may contain meat from potentially thousands of cows, for example. In addition, on modern factory farms, livestock are often housed closely together in large pens where they are exposed to fecal matter from other animals, making it more likely that fecal matter on an animal's hide might end up in meat.

These problems came to a head in the winter of 1993, when several hundred adults and children became severely ill and four children died after eating undercooked hamburgers from Jack in the Box fast-food restaurants in several Western states. The hamburgers were contaminated with *E. coli* O157:H7, the deadly strain of Shiga toxin-producing bacteria. Health inspectors traced the outbreak to the sale of the fast-food chain's Monster Burger ("So good it's scary!"), which was being sold at a discounted price as part of a promotion. The high demand for the burgers overwhelmed the cooks, and the burgers were not cooked long enough to kill the bacteria. Unsuspecting customers, many of them children, ate the tainted burgers and got sick.

Foodborne illness (also known as *food poisoning*) is a very common, though largely preventable, condition in the United States.

INFOGRAPHIC 20.3 Top Five Pathogens Causing Foodborne Illnesses in the United States

The percentage of total foodborne illness caused by these pathogens and their contribution to the total number of hospitalizations caused by foodborne illness.

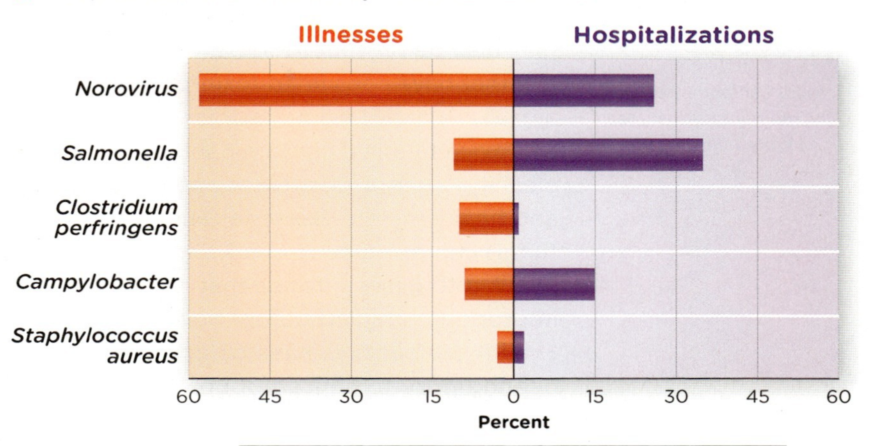

What pathogen causes the most severe illness?

The Centers for Disease Control estimate that each year roughly 1 in 6 Americans (or 48 million people) get sick from eating contaminated food. Of these, 128,000 are hospitalized and 3,000 die. Illness may result from consuming food that contains naturally occurring toxins, or food that is contaminated with toxic chemicals or pathogens (viruses, bacteria, or parasites). Although more than 250 foodborne diseases have been described, the most common are caused by just five pathogens: *norovirus* (a virus) and *Salmonella*, *Clostridium perfringens*, *Staphylococcus*, and *Campylobacter* (all bacteria). Although *E. coli* is not a leading cause of foodborne illness, it does cause the fifth highest number of hospitalizations due to foodborne illnesses. **(INFOGRAPHIC 20.3)**

FOODBORNE INTOXICATION AND INFECTION

Foodborne illnesses fall into two general categories: intoxication and infection. **Foodborne intoxication** is caused by ingestion

FOODBORNE ILLNESS
a largely preventable disease or condition caused by consumption of a contaminated food or beverage that primarily affects the gastrointestinal tract

FOODBORNE INTOXICATION
An illness that is caused by ingesting foods that contain a toxin

CROSS-CONTAMINATION *the transfer of bacteria or other pathogens from one source or surface to another*

FOODBORNE INFECTION *an infection that results in illness caused by consuming foods contaminated with microorganisms that then multiply in the intestines*

of foods that contain a toxin that may be naturally present in the food, introduced by contamination with poisonous chemicals, or produced by bacteria or fungi growing on foods. Toxins occur naturally in some plant foods and some mushrooms. Toxins may also be present in some fish and shellfish that have consumed toxin-producing algae.

Toxic chemical contaminants include cleaning agents, pesticides and herbicides, and heavy metals. Bacteria (and, less often, fungi) can produce toxins when they are allowed to grow in foods that have been improperly handled before being eaten. Even if the bacteria have been killed by heating prior to consumption, the toxin is already present in the food and can cause illness rapidly. The toxin may act as quickly as 30 minutes following the ingestion of the contaminated food.

In the United States, foodborne intoxication is most commonly caused by two strains of bacteria, *Staphylococcus aureus* and *Bacillus cereus.*

The bacterium* Staphylococcus aureus (S. aureus) *lives on the skin. *The likelihood of foods being contaminated with* S. aureus *can be drastically reduced when food handlers wash their hands.*

Taylor Jorjorian/Alamy

S. aureus lives on the skin and in the nasal passages of about 25% of the population, and food handlers are the source of food contamination with this bacterium. When contaminated foods are not promptly refrigerated these bacteria grow on the food and release a heat-stable toxin that can withstand boiling (212°F) for as long as two hours. The likelihood of foods being contaminated with *S. aureus* can be drastically reduced if food handlers practice proper hygiene, particularly adequate hand washing.

B. cereus is a bacterium that is widespread in the environment. When these bacteria grow on contaminated rice products, and other starchy foods such as potatoes and pasta, a heat-stable toxin is produced that causes vomiting. Because *B. cereus* is abundant in soil and on uncooked produce it is often introduced into cooked foods through **cross-contamination** in the kitchen when uncooked or undercooked vegetables or herbs come in contact with cooked foods. When those cooked foods are not promptly refrigerated the transferred bacteria can then grow. Because both *S. aureus* and *B. cereus* produce heat-stable toxins, reheating contaminated foods that were not properly cooled (or kept hot) will often not prevent intoxication.

The majority of foodborne illnesses are **foodborne infections** that result when we consume foods that are contaminated with microorganisms that then multiply in the intestines, causing an infection that results in illness. One of the most common sources of pathogens in foods and water is from contamination with either animal or human feces. When we become ill from consuming foods or water contaminated in this manner it is referred to as the *fecal-oral transmission* of disease. Another way that pathogens are introduced into foods is by cross-contamination, where pathogens from contaminated surfaces, utensils, people, or foods (such as raw meats) are transferred to another object or food. This is of particular concern when contaminated foods are then eaten without being cooked after they are contaminated.

Viruses, parasites, and bacteria

Disease-causing microorganisms (or pathogens) include viruses, parasites, and bacteria. *Viruses* are infectious agents that consist of genetic material surrounded by a protein shell, and are so small that they cannot be seen with a conventional light microscope. Viruses can only multiply inside the living cells of a host and for this reason they are unable to multiply in foods. *Parasites* are organisms that live on or in a host, and obtain the nutrients they need from the host. They vary widely in size, from small single-celled organisms (*protozoans*), to large multicelled organisms such as parasitic worms (helminths) that may exceed 30 feet in length. Like viruses, parasites cannot multiply in foods but they can survive in the environment and enter the body through food.

Bacteria are single-celled microorganisms that can be seen with a conventional microscope. Bacteria increase in number by dividing into two cells, with each cell then growing to full size before they divide again. Unlike either viruses or parasites, bacteria can multiply in food when conditions such as temperature, pH, salt concentrations, moisture content, and oxygen levels are favorable for their growth. The longer these favorable conditions for bacterial growth persist, the more bacteria will be present in the food, and the more severe will be the illness if the contaminated food is ingested.

Symptoms of foodborne illness

Foodborne illnesses can have different symptoms, but since the offending pathogens are ingested and enter the gastrointestinal tract, symptoms commonly begin there. Nausea, vomiting, abdominal cramps, and diarrhea are most common. Although foodborne illnesses may sometimes have a rapid onset and may last only hours or days, they can also lead to more serious and potentially life-threatening complications. Different types of foodborne diseases vary in the time of onset, typical symptoms, duration of symptoms, and potential consequences. (INFOGRAPHIC 20.4)

Those most at risk for foodborne illness

Multiple factors affect the disease-causing potential of foodborne pathogens. Most important, the strength of a person's immune system plays a big role. Healthy college-aged students, for example, may not be affected or have minimal symptoms, whereas a young child may experience more drastic consequences. The people most at risk of contracting a foodborne illness, are those whose immune systems are not functioning at full capacity. They are at risk of lengthier illnesses with increased risk of hospitalization and death. Infants and young children whose immune systems are not yet fully developed, older adults, and individuals with diseases such as HIV/AIDS, cancer, and diabetes have weakened immune systems that allow greater opportunity for foodborne pathogens to multiply. Pregnant women are also more susceptible to foodborne illness because pregnancy suppresses the immune system of the mother. As a result, pathogens may cross the placenta and infect the unborn child whose immune system is underdeveloped.

One particular foodborne pathogen that pregnant women are often warned against is *Listeria monocytogenes*. This pathogen is the third leading cause of death due to foodborne illness in the United States, and one to which pregnant women and their unborn children are particularly susceptible. It is estimated that pregnant women are 10 times more likely to become infected with *Listeria* than the general population (adults 65 years and older are 4 times more likely to be infected). Though the symptoms of illness (called listeriosis) experienced by the expectant mother are typically mild (fever, fatigue, and body aches), this infection can have devastating effects on the pregnancy, potentially causing miscarriage, stillbirth, preterm birth, or life-threatening infection in the newborn. Pregnant women should avoid consuming soft cheeses (feta, Brie, Camembert, and some Mexican-style cheeses), refrigerated smoked seafood, pâtés, or meat spreads; and hot dogs (unless they are steaming hot). These foods are some of the most common food sources of *Listeria*. Individuals in the

INFOGRAPHIC 20.4 Characteristics of Selected Foodborne Illnesses

Pathogen	Onset Time	Duration of Illness	Signs and Symptoms	Common Sources
Bacillus cereus vomiting-type	0.5–6 hrs	~24 hrs	Nausea and vomiting.	Rice products and other starchy foods such as potato and pasta products are typically associated with outbreaks caused by a bacterial toxin that causes vomiting. This bacterium is widespread in the environment and can multiply in foods when temperature has not been properly controlled.
Bacillus cereus diarrheal-type	6–15 hrs	~24 hrs	Diarrhea, abdominal cramps and pain.	A wide variety of foods, including meats, milk, fish, and vegetables, have been associated with outbreaks caused by a bacterial toxin that causes diarrhea.
Clostridium perfringens	6–24 hrs	24 hrs or less	Diarrhea and abdominal cramps.	Meats and gravy. Cooking kills bacteria but not necessarily bacterial spores that are in a dormant state. When cooked food is not properly refrigerated these spores produce cells that grow quickly in the "danger zone."
Campylobacter spp.	2–5 days	2–5 days, occasionally 10 days	Diarrhea, cramps, fever, and vomiting; diarrhea may be bloody, sometimes a low-grade fever.	Raw or unpasteurized milk; raw or undercooked poultry, meat, or shellfish. It is found in approximately half of all raw chickens sold in the United States. One of the most common bacterial causes of diarrheal illness in the United States.
Enterotoxigenic *Escherichia coli* (ETEC)	0.5–2 days	2–4 days	Sudden onset of watery diarrhea and usually vomiting.	Generally caused by foods that are contaminated with water that was contaminated with human sewage (such as produce). Rare in the United States but it is one of the most common causes of "traveler's diarrhea" for Westerners visiting developing countries.
Escherichia coli O157:H7	3–4 days	6–8 days	Severe diarrhea (often bloody diarrhea), abdominal cramps, and vomiting. May result in kidney failure, particularly in children.	Undercooked beef (especially ground beef), unpasteurized milk and juices, contaminated raw fruits and vegetables, or water. Present in feces of infected animals and people.
Listeria monocytogenes	3–70 days	Days to weeks	Fever, muscle aches, sometimes diarrhea; may include headache and confusion. Most common in pregnant women and children.	Hot dogs, deli meats, poultry, refrigerated smoked seafood, soft cheeses made with unpasteurized milk. Listeria grows well even in the refrigerator.
Norovirus	12–48 hrs	1–3 days	Diarrhea, vomiting, nausea, and stomach pain. Diarrhea is more common in adults and vomiting is more common in children.	Produce, shellfish, ready-to-eat foods touched by infected food workers (salads, sandwiches, ice, cookies, fruit), or any other foods contaminated with vomit or feces from an infected person.
Salmonella spp.	12–72 hrs	4–7 days	Diarrhea, fever, abdominal cramps. The most common cause of death resulting from foodborne illness.	Poultry, eggs, meat, unpasteurized milk or juice, raw fruits and vegetables, and nuts.
Staphylococcus aureus	0.5–6 hrs	24–48 hrs	Nausea, vomiting, abdominal cramps, diarrhea.	Salads (such as egg, tuna, chicken, and potato), bakery products (such as cream-filled pastries, cream pies, and chocolate éclairs), and sandwiches. *S. aureus* is commonly found on the skin, and in the nose and throat of healthy people. When a food handler contaminates food and the food is not properly refrigerated, the bacteria grow at room temperature and produce a heat-resistant toxin that causes illness.

When Should You See a Physician About a Diarrheal Illness?

- A fever over 101.5°F
- Blood in stools
- Diarrheal illness that lasts more than three days
- Prolonged vomiting that prevents keeping liquids down (which can lead to dehydration)
- Signs of dehydration (decrease in urination, dry mouth and throat, feeling dizzy when standing up)

What pathogens cause illness by a foodborne intoxication?

Photo credits (top to bottom): Bibiz1/Shutterstock, Evlakhov Valeriy/Shutterstock, Cristian Baitg/Getty Images, Cristian Baitg/Getty Images, Food Collection/Superstock, Carlos Gawronski/Getty Images, Davies and Starr/Getty Images, Stephanie Frey/Shutterstock, Evlakhov Valeriy/Shutterstock, Eli Ensor, Bildagentur Zoonar GmbH/Shutterstock, draganica/Shutterstock

INFOGRAPHIC 20.5 Groups at High Risk for Foodborne Illness and the Foods They Should Avoid *Because those with weakened immune systems are at higher risk of contracting a foodborne illness than the general population they should avoid consuming foods that are the most frequent causes of illness.*

Pregnant women

Pregnancy suppresses the immune system of the mother.

Young children

The immune system of young children is not yet fully developed.

Older adults

As people age the effectiveness of their immune system declines.

People with some diseases or receiving some medical treatments

The immune response can be decreased by some diseases such as HIV/AIDS, diabetes, and some cancers. Also, drugs used to treat patients with transplants, cancer, and some inflammatory diseases decrease immune system function.

FOODS TO AVOID

- Raw or undercooked meat or poultry
- Raw fish, partially cooked seafood (such as shrimp and crab), and refrigerated smoked seafood
- Raw shellfish (including oysters, clams, mussels, and scallops) and their juices
- Unpasteurized (raw) milk and products made with raw milk, like yogurt and cheese
- Soft cheeses made from unpasteurized milk, such as feta, Brie, Camembert, blue-veined, and Mexican-style cheeses (such as Queso Fresco, Panela, Asadero, and Queso Blanco)
- Raw or undercooked eggs or foods containing raw or undercooked eggs, including certain homemade salad dressings, homemade cookie dough and cake batters, and homemade eggnog
- Unwashed fresh vegetables, including lettuce/salads
- Unpasteurized fruit or vegetable juices (these juices will carry a warning label)
- Hot dogs, luncheon meats (cold cuts), fermented and dry sausage, and other deli-style meats, poultry products, and smoked fish — unless they are reheated until steaming hot
- Salads (without added preservatives) prepared on site in a deli-type establishment, such as ham salad, chicken salad, or seafood salad
- Unpasteurized, refrigerated pâtés or meat spreads
- Raw sprouts (alfalfa, bean, or any other sprout)

Source: http://www.fda.gov/Food/FoodborneIllnessContaminants/PeopleAtRisk/ucm352830.htm

other high-risk groups should also avoid these foods, as well as other foods that are common sources of foodborne pathogens. **(INFOGRAPHIC 20.5)**

■ ■ ■

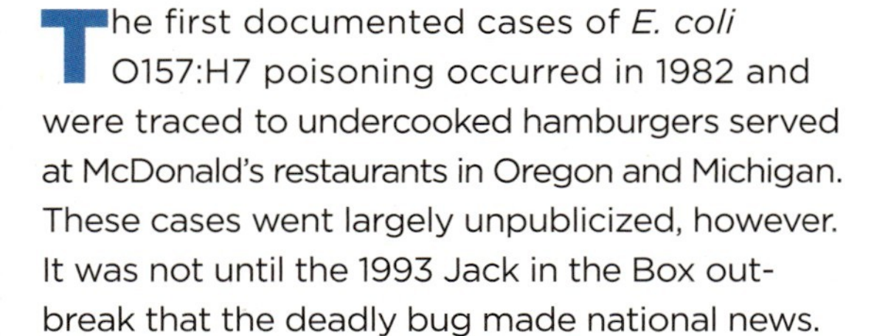

The first documented cases of *E. coli* O157:H7 poisoning occurred in 1982 and were traced to undercooked hamburgers served at McDonald's restaurants in Oregon and Michigan. These cases went largely unpublicized, however. It was not until the 1993 Jack in the Box outbreak that the deadly bug made national news.

The headline-generating incident highlighted the outdated meat inspection system and provided the impetus for Congress to once again take up the issue of food safety. The response was a 1998 law that established a new system of prevention and monitoring for all meatpacking plants in the United States.

Ironically, the principles on which the new system was based were not themselves new. They had been in use since the 1960s to protect one very valuable group of Americans: astronauts.

ENGINEERING FOOD SAFETY: HACCP AND IRRADIATION

In 1963, as NASA geared up for its Gemini and Apollo missions, it realized that protecting astronauts from food poisoning was vital to successful space missions. On Earth a sick stomach might be unpleasant, but in space it could be catastrophic.

"You don't want an astronaut with diarrhea," said Robert Tauxe, a physician with the Centers for Disease Control and Prevention, in an interview with *Frontline* about the 1993 Jack in the Box outbreak.

HACCP (HAZARD ANALYSIS AND CRITICAL CONTROL POINTS) *preventative food safety system that addresses the manufacture, storage, and distribution of food products*

IRRADIATION *technology that uses low doses of radiation to destroy insects and bacteria in foods to improve food safety*

To help prevent what became known as the "two-bucket problem," NASA worked to develop a program of food protection called **HACCP**. This mouthful of an acronym stands for **Hazard Analysis and Critical Control Points**.

In essence, NASA scientists approached food safety as an engineering problem. Instead of focusing on end product inspection, they focused on the prevention of hazards. They analyzed and identified all of the possible points where germs (or other hazards) could enter the food supply, and then monitored these "critical control points" with a science-based program of microbial testing. One analysis, for example, revealed that telephones in food-processing plants were a source of potential contamination and needed to be monitored.

Food irradiation. *Peaches on the left were not irradiated and are spoiling sooner than the irradiated peaches on the right. Food irradiation (the application of ionizing radiation to food) is a technology that reduces or eliminates microorganisms and insects in food. The Food and Drug Administration (FDA) has approved irradiation of meat and poultry, fresh fruits, vegetables, and spices. The FDA determined that the process is safe and effective in decreasing the presence of and/or eliminating harmful bacteria.*

Ted Foxx/Alamy, U.S. Department of Agriculture, Food Safety and Inspection Service

In addition to implementing HACCP, NASA also began, in 1972, to irradiate meat served to astronauts. Food **irradiation** kills bacteria, parasites, and insects by damaging DNA, which improves food safety. However, food irradiation does not kill all bacteria in foods so good food-handling practices must still be used. Any living cells in the food are killed or damaged, as well, and this can prolong shelf life for fruits and vegetables by delaying ripening or inhibiting sprouting (of potatoes, for example). The FDA has approved the irradiation of a variety of foods, including meat and poultry, fresh fruits and vegetables, and spices and seasonings.

Although irradiated foods are generally not widely available, most spices sold wholesale in the United States have been irradiated. The effect of food irradiation has been studied extensively, and it has not been found to noticeably affect the taste or texture of foods, nor does it make foods radioactive or significantly reduce their nutrient content. You can determine if foods have been irradiated by looking for the international symbol for irradiation, as well as the statement "Treated with radiation" or "Treated by irradiation" on the food label. However, individual ingredients (such as spices) in multi-ingredient foods do not need to be labeled.

What started out as a special program for astronauts soon spread to the wider civilian food supply. Beginning in the 1970s, HACCP was adopted by some companies voluntarily, since the companies had an interest in avoiding the bad publicity that can come from an outbreak. It was not until 1998, after the poisonings earlier in the decade, that HACCP became federally required for the meat industry. The system is now employed at the approximately 9,000 slaughterhouses in the United States.

Under the new HACCP system, instead of relying on USDA inspectors to test for contamination using the poke-and-sniff test, meat processing plants are required to test carcasses for invisible pathogens such as *E. coli* and salmonella using scientific methods. This approach helped to make the meat supply safer. But still there was the spinach problem.

Because fruits and vegetables are not covered under USDA jurisdiction, spinach was not subject to HACCP. The FDA had been struggling for years to get the authority to enforce stronger protections on fruits and vegetables. But it was not until a major health crisis occurred, in the form of the spinach *E. coli* epidemic, that the political will was there.

"The thing that those 2006 outbreaks did was put the issue of on-farm food safety on the table," says Smith DeWaal, of the CSPI, which had for years been trying to get Congress to pass a comprehensive overhaul of the country's food safety laws.

INFOGRAPHIC 20.6 The Percent of Reported Foodborne Disease Outbreak-Associated Illnesses by Food Category, 2011–2012

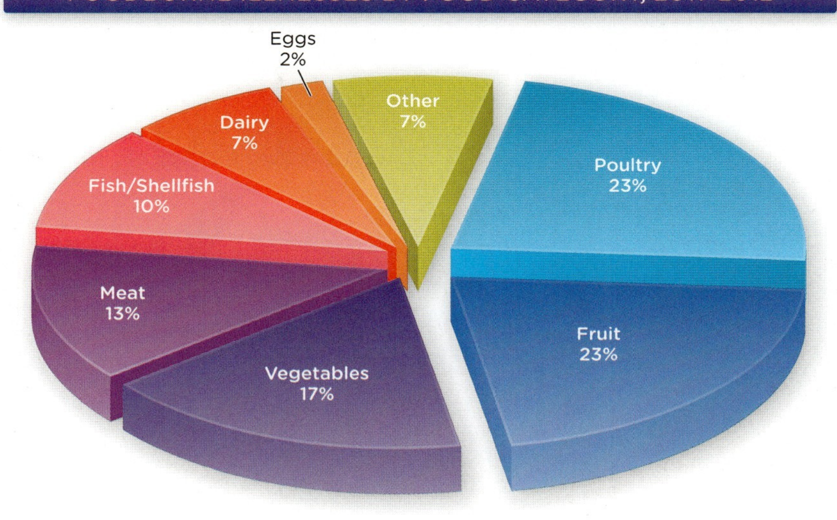

? *What percent of reported foodborne illnesses are caused by contaminated produce?*

FOOD SAFETY AND MODERNIZATION ACT

The outcome of this crisis was the **Food Safety Modernization Act (FSMA)**, designed to apply mandatory preventive controls, or HACCP, throughout the food supply. On-farm food safety is the provision of FSMA that allows the FDA to establish regulations for the production of fresh fruits and vegetables. The timeliness of this provision can be appreciated by recognizing that about 40% of reported foodborne illness currently results from the consumption of contaminated fresh produce. **(INFOGRAPHIC 20.6)**

FSMA was signed into law by President Obama in 2011. It has been hailed as the most sweeping change to food safety legislation in 70 years. FSMA gave the FDA more authority in the regulation of food production facilities and in recalling contaminated foods, and established new standards for safe produce. In addition, FSMA gave the FDA the power to ensure **food defense**. Food defense differs from food safety in that it protects against intentional, rather than unintentional, contamination of food.

Would FSMA have prevented the 2006 spinach outbreak had the laws been in place back then? Possibly, says Smith DeWaal. "FSMA would mean that there were more eyes on the process, more controls at the farm level. . . . you would have a process control plan, a food safety plan in every facility where that spinach would have been washed. That plan might require water testing, so if they had discovered the [contaminated] water earlier, the spinach may have gotten into the facility but may never have gotten released to the public."

These safety mechanisms, while not perfect, she says, would have gone a long way toward avoiding the outbreak.

FOOD SAFETY IN THE HOME

Governments and food companies are not the only ones with responsibility for maintaining food safety; consumers, too, must shoulder a portion of the burden. There are things that we can all do to increase the likelihood that the food we consume is safe. These actions—summarized by the verbs *clean*, *separate*, *cook*, and *chill*—fall under the heading of safe handling

FOOD SAFETY MODERNIZATION ACT (FSMA)
signed into law in 2011, the FSMA aims to ensure that the U.S. food supply is safe from contamination by giving federal regulators more proactive authority in how food is grown, harvested, and processed

FOOD DEFENSE
an effort to protect the food supply against intentional contamination due to sabotage, terrorism, counterfeiting, or other illegal, intentionally harmful means

INFOGRAPHIC 20.7 Four Steps to Food Safety *Following these practices when handling foods will eliminate four of the top five risk factors that contribute to foodborne illnesses, and reduce your risk of getting sick from contaminated foods.*

Disinfect surfaces with a dilute chlorine bleach solution, 1 teaspoon per quart of water. Leave the solution on the surface for about 10 minutes to be effective.

Top 5 Risk Factors Contributing to Foodborne Illnesses

1. Improper hot/cold holding temperatures of potentially hazardous foods.
2. Improper cooking temperatures.
3. Contaminated utensils and equipment.
4. Poor health and hygiene of food preparers and handlers.
5. Food from unsafe sources, such as raw or unpasteurized milk.

? *Identify which foodborne illness risk factors can be eliminated or reduced by each of these four steps to food safety.*

Photo credits (top to bottom): Alexey Stiop/Shutterstock, StepanPopov/Shutterstock, Natalia Hirshfeld/Shutterstock, Radius Images/Alamy, pryzmat/Shutterstock

of foods. The importance of these four steps in improving food safety is demonstrated by the fact that four of the top five causes of foodborne illness are related to inadequate cleaning and improper temperatures. **(INFOGRAPHIC 20.7)**

Pathogens can be found in many places in the kitchen including your hands, so food safety begins by thoroughly washing hands, countertops, cutting boards, and utensils. Proper hand washing will dramatically reduce the incidence of foodborne illnesses. Hands should be scrubbed for 20 seconds using soap and running water, being certain to wash the backs of your hands and under your nails. Hands should be washed frequently during food preparation, as well as after touching pets, using the bathroom, blowing your nose, handling garbage, and handling uncooked meats, poultry, fish, or eggs. It is also important to keep countertops, cutting boards, utensils, and dishes clean by washing with hot, soapy water (or a dilute bleach solution) after preparing each food item: Rinsing them with water is not effective at stopping the spread of foodborne infections.

Fresh fruits and vegetables should also be washed, even if you are planning to peel them before they are eaten. If they are not washed pathogens can be transferred from the outside of the peel to the edible portion as the produce is cut or peeled. Rinse produce under running water but do not use

INFOGRAPHIC 20.8 Seven Tips for Cleaning Fruits and Vegetables

- ✓ Wash your hands for 20 seconds with warm water and soap before and after preparing fresh produce.
- ✓ Cut away any damaged or bruised areas before preparing or eating.
- ✓ Gently rub produce while holding under plain running water. There's no need to use soap or a produce wash.
- ✓ Wash produce BEFORE you peel it, so dirt and pathogens aren't transferred from the knife onto the fruit or vegetable.
- ✓ Use a clean vegetable brush to scrub firm produce, such as melons and cucumbers.
- ✓ Dry produce with a clean cloth or paper towel to further reduce bacteria that may be present.
- ✓ Throw away the outermost leaves of a head of lettuce or cabbage.

Olesya Feketa/Shutterstock

http://www.fda.gov/ForConsumers/ConsumerUpdates/ucm256215.htm

soap or bleach. Firm produce like melons, summer squash, and potatoes should be scrubbed under running water with a clean produce brush. **(INFOGRAPHIC 20.8)**

Avoiding cross-contamination of foods

To reduce our risk of foodborne illness it is also important to avoid cross-contaminating cooked and ready-to-eat foods and fresh produce by keeping them separate from raw meats (including poultry and seafood) and eggs. Avoiding cross-contamination begins even before you arrive home from the store by keeping those foods separate from other foods in your grocery cart. Place raw meats in plastic bags to prevent their juices from dripping onto other foods, and do not put them in the same grocery bags with other foods at the checkout line. When you refrigerate these foods keep them in sealed plastic bags or containers to prevent their juices from contaminating other food in the refrigerator. It only takes one drop of juice from raw chicken to infect a person, and studies have found that approximately 50% of raw chicken products in the United States are contaminated with *Campylobacter* bacteria. During food preparation always clean cutting boards, countertops, knives, and other utensils with hot, soapy water immediately after coming into contact with raw meats and eggs. If possible, use one cutting board for fresh produce and a separate one for raw meats. Excessively worn or nicked cutting boards should be replaced because they can be very hard to clean adequately. Once your food is cooked always serve it on clean dishware, never place cooked food back on dishware that has come in contact with raw meats.

The importance of appropriate food temperatures

To destroy pathogens in meat, poultry, seafood, and egg dishes cook them to a safe minimum internal temperature as measured with a probe thermometer. Ground meats must be cooked to a higher temperature than steaks, chops, or roasts because bacteria in ground meats are distributed throughout the meat rather than being concentrated on the surface of the meat.

Bacteria on the surface of meats are more easily destroyed because the surface of the meat reaches higher temperatures more quickly than internal portions of the meat. Because the bacteria found in raw poultry infiltrates deep into the meat, even whole,

Ground meats must be cooked to a higher temperature than steaks, chops, or roasts because bacteria in ground meats are distributed throughout the meat rather than being concentrated on the surface of the meat. To destroy harmful bacteria, cook ground beef to a safe minimum internal temperature of 160 °F (71.1 °C). Although these burgers look "done" the thermometer indicates they are undercooked.

Charles Stirling/Alamy

unground poultry must be cooked to a higher temperature than what is required for beef, pork, lamb, and veal. The safe minimum internal temperature for casseroles and leftovers is also set high because bacteria can be found throughout these foods, and not just on their surface. When heating foods in the microwave let them sit for about two minutes after heating (or follow the manufacturer's instructions for commercial products) to reduce cold spots in the food and provide more time for bacteria to be destroyed. As long as foods have been handled properly before being cooked, following these cooking recommendations will make them safe to eat. However, if raw meats, ingredients for casseroles, or leftovers have not been kept at an appropriate temperature, bacteria may grow and produce toxins that may not be destroyed by cooking or reheating, causing foodborne illness. (INFOGRAPHIC 20.9)

Promptly chilling perishable foods after purchase or preparation will also reduce bacterial growth in foods, thereby reducing the risk of both foodborne infections and

INFOGRAPHIC **20.9** **USDA Recommended Safe Minimum Internal Temperatures** *Cook all food to these minimum temperatures as measured with a probe thermometer inserted to the center of the food.*

* After you remove meat from a grill, oven, or other heat source, allow it to rest for the specified amount of time. During the rest time, its temperature remains constant or continues to rise, which destroys harmful germs.

http://www.IsItDoneYet.gov

Why is the safe internal temperature of ground meats higher than it is for roasts or steaks of the same meat?

intoxication. Leftovers should be refrigerated in shallow containers to allow for more rapid cooling. When leftovers are refrigerated in deep containers the food at the center can stay within the range of temperatures that promote bacterial growth (the temperature "danger zone") for too long, allowing bacteria to multiply. When properly handled food is stored in the freezer at 0°F it will be safe for an indefinite period, however the flavor and texture of foods will deteriorate over time.

It is imperative to avoid the temperature danger zone. Bacteria grow at temperatures between 41°F and 135°F, and they grow particularly fast at temperatures between 70°F and 120°F. Keeping hot foods hot and cold foods cold is important, as is making sure that perishable foods never spend more than two hours in the temperature danger zone. Because bacteria multiply very quickly at room temperatures, meats and seafood should only be marinated in the refrigerator, and never on the kitchen counter. And never use the uncooked marinade on cooked foods as the bacteria from the raw foods will now be in the marinade as well. Keeping foods out for extended periods, such as at holiday or party buffets, potlucks, and picnics can create significant challenges to keeping food safe. During these types of events it is very important to keep hot foods at 135°F or higher and cold foods at 41°F or cooler. Care should also be exercised when thawing foods to be sure they do not spend a significant amount of time in the temperature danger zone. To avoid this never thaw foods on the kitchen counter, but thaw them in the refrigerator, in cold water, or in the microwave. (INFOGRAPHIC 20.10)

As much as these actions can help prevent food poisoning, they are no substitute for food safety mechanisms at earlier points

INFOGRAPHIC **20.10** **Temperature Danger Zone and Safe Thawing of Foods** *To avoid foodborne illnesses keep hot foods hot and cold foods cold. Bacteria grow at temperatures between 41°F and 135°F, and they grow particularly quickly at temperatures between 70°F and 120°F. Be sure that perishable foods never spend more than two hours in the temperature danger zone.*

Leftovers should be consumed within 3-4 days. Ground meats and uncooked sausages should not be refrigerated for more than 2 days; while chops, roasts, and steaks should be cooked or frozen in 3-5 days.

Chill perishable foods within two hours of preparation or purchase, and within one hour if the air temperature is 90°F or higher.

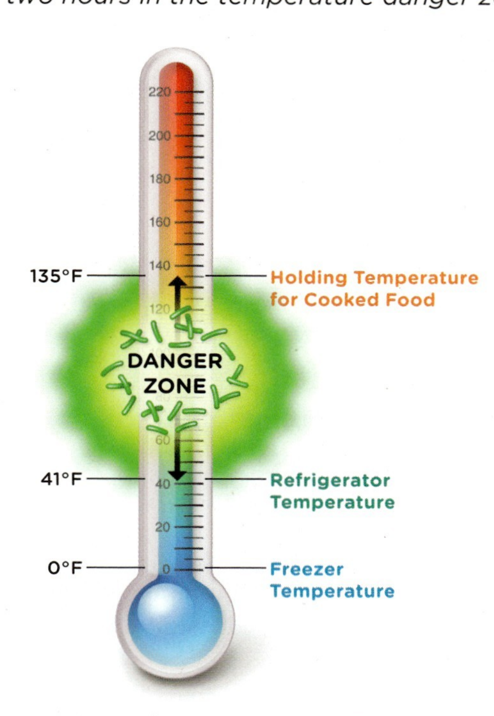

Safe Thawing of Foods

NEVER THAW FOODS ON THE COUNTER

✓ **Refrigerator:** This is the safest way to thaw foods. Make sure the juices from thawing meats and poultry do not drip onto other foods.

✓ **Cold Water:** To thaw foods faster put them in a sealed plastic bag and submerge in cold water. Change the water every 30 minutes. Cook foods immediately after thawing.

✓ **Microwave:** When thawing food in a microwave, plan to cook it immediately after thawing because some areas of the food may become warm and begin to cook during the thawing process (bringing the food into danger zone temperatures).

✓ **Cook without thawing:** Cooking will take about 50% longer than for fully thawed or fresh meat or poultry.

Why isn't it necessary to immediately cook foods that have been thawed in the refrigerator?

ORGANIC
grown and processed using no synthetic fertilizers or pesticides

in the food production chain. For example, the spinach that was contaminated with *E. coli* would not have been made safe by washing, because the bacteria had infiltrated the inside of spinach leaves.

Ironically, the spinach that sickened so many people in 2006 was **organic**. Many people believe that organic produce is safer and perhaps even healthier than regular produce. But when it comes to foodborne

INFOGRAPHIC **20.11** **USDA Organic Foods** *The use of the term "organic" on food labels is regulated by the USDA. Its use on food labels (anywhere other than the ingredients list) indicates that the food has been produced through approved methods.*

100 PERCENT ORGANIC

- All ingredients must be certified organic.
- Any processing aids must be organic (such as cleaning agents for equipment, yeast, and binders).
- Label may include the USDA organic seal and/or 100 percent organic claim.

USDA

ORGANIC

- All agricultural ingredients (not salt, for example) must be certified organic (except as specified).
- Allowed nonorganic ingredients cannot compose more than five percent of the total content by weight or fluid volume (excluding salt and water).
- Label may include the USDA organic seal and/or organic claim.

MADE WITH ORGANIC

- Multi-ingredient foods with this claim must contain at least 70% certified-organic ingredients by weight or fluid volume (excluding salt and water).
- All ingredients—including nonorganic ingredients—must be produced without genetically modified organisms (GMOs), or other prohibited substances.
- If a product meets the above requirements, its label may include a statement like, "Made with organic oats and cranberries." A more generic statement like, "Made with organic ingredients," is not allowed.
- The product may contain only organic forms of the specific ingredients.
- Label may not include the USDA organic seal.
- Organic ingredients must be identified with an asterisk (or other mark) in the listed ingredients.

SPECIFIC ORGANIC INGREDIENTS

- Multi-ingredient products with less than 70% certified-organic content (excluding salt and water) don't need to be certified. Any non-certified product may not include the USDA organic seal or use the word "organic" on the principal display panel.
- May only list certified organic ingredients as organic in the ingredients list, and the percentage of organic ingredients.
- Remaining ingredients are not required to follow the USDA organic regulations.

What is the minimum organic content of a food that is required for the label to display the USDA organic seal?

pathogens that is often not the case. Organic means that no artificial fertilizers or pesticides were used in the growth of the food. For people who want to reduce exposure to synthetic pesticides, eating organic is a good way to accomplish this goal. But organic foods can just as easily become contaminated with harmful bacteria—especially if livestock manure is used as fertilizer. (INFOGRAPHIC 20.11)

Food additives

As part of its role in ensuring a safe food supply, the FDA has primary responsibility for regulating substances that are added to food. **Food additives** are any substances added to food during processing, production, or preparation, including preservatives, flavorings, colorings, leavening agents (to improve consistency), and even vitamins and minerals. They can serve multiple purposes: to maintain or improve safety and freshness; to improve or maintain nutritional value; and to improve taste, texture, and appearance. Food manufacturers are required to list all ingredients in food products on food labels, including additives. (INFOGRAPHIC 20.12)

FOOD ADDITIVES
any substance added to a food product

INFOGRAPHIC **20.12** **Reasons for Using Food Additives** *They provide a number of useful functions that can improve the palatability of foods and even enhance food safety.*

Maintain or Improve Safety and Freshness

- Preservatives reduce product spoilage caused by mold, air, bacteria, fungi, or yeast, and help control contamination that can cause foodborne illness.
- Antioxidants prevent fats and oils (and the foods containing them) from becoming rancid or developing an off-flavor.
- Antioxidants also prevent cut fresh fruits such as apples from turning brown when exposed to air.

Improve or Maintain Nutritional Value

- Vitamins and minerals (and fiber) are added to many foods to make up for those lacking in a person's diet or lost in processing, or to enhance the nutritional quality of a food.

Improve Taste, Texture, and Appearance

- Spices, natural and artificial flavors, and sweeteners are added to enhance the taste of food.
- Food colors maintain or improve appearance and make foods more appealing.
- Emulsifiers, stabilizers, and thickeners give foods pleasing texture and consistency.
- Leavening agents allow baked goods to rise during baking.
- Some additives help maintain the taste and appeal of foods with reduced fat content.

Adapted from **"Overview of Food Ingredients, Additives & Colors," FDA**
http://www.fda.gov/food/ingredientspackaginglabeling/foodadditivesingredients/ucm094211.htm#types

EXAMPLES OF FOOD ADDITIVES AND THEIR USES

Preservatives and Antioxidants	Color Additives	Flavor Enhancers	Fat Replacers	Emulsifiers	Stabilizers and Thickeners
• Ascorbic acid	• FD&C* colors	• MSG (monosodium glutamate)	• Olestra	• Soy lecithin	• Gelatin
• Citric acid	• Annatto extract	• Hydrolyzed soy protein	• Cellulose gel	• Monoglycerides	• Pectin
• Sodium benzoate	• Beta-carotene	• Autolyzed yeast extract	• Carrageenan	• Diglycerides	• Guar gum
• Calcium propionate	• Carmine	• Disodium inosinate	• Modified food starch	• Polysorbates	• Carrageenan
• Sodium nitrate	• Caramel color		• Guar gum		
• BHA					
• BHT					
• EDTA					

*Food, Drug, and Cosmetic act

What type of food additive can increase food availability and variety?

Photo credit: Monkey Business Images/Shutterstock

FOOD SECURITY *always having physical, social, and economic access to sufficient, safe, and nutritious food that meets dietary needs and food preferences for an active and healthy life*

FOOD INSECURITY *lack of secure access to sufficient amounts of safe and nutritious food for normal growth and development and an active and healthy life*

Direct food additives are often added during the processing of food and may be natural additives like spices, herbs, and salt, or manmade substances such as vitamins and chemical preservatives. Indirect additives are not intentionally added to foods, but are introduced into the food during packaging, processing, or storage. Indirect additives are generally present only in minute quantities and include compounds from packaging like paper fibers, adhesives, and compounds found in plastic food containers and metal-based food and beverage cans such as *bisphenol A* (BPA). They can also include sanitizers such as bleach or iodine, and production aids such as corrosion inhibitors and lubricants.

The FDA has the primary legal responsibility for the safety of food additives and oversees the process for approval of food additives. The FDA is also responsible for determining the types of foods in which specific additives can be used, the maximum amount that can be used, and how it will be identified on the food label. Most additives in the food supply are considered "generally recognized as safe" (GRAS) on the basis of historical use in food and published studies on safe consumption. There are hundreds of substances on the GRAS list. Among these are common ingredients like salt, sugar, baking soda, vanilla, yeast, and spices. These substances do not have to go through an approval process. Although most food additives are GRAS, nutritionists generally agree that it is best to emphasize whole foods in one's diet rather than processed foods that often contain multiple food additives.

Although most food additives have no adverse effects, there is now substantial evidence to support the theory that a high intake of artificial food colors (AFC) can cause small but statistically significant increases in hyperactivity in some children. A number of foods frequently consumed by children contain high amounts of AFC, including sodas, fruit and sports drinks, breakfast cereals, baked goods and frostings, and candies. Levels of AFC consumption by children in the United States can easily exceed the amounts that have most consistently been observed to have adverse effects. Some children are more susceptible to these effects than others.

BEYOND FOOD SAFETY: FOOD SECURITY

Making sure that food is safe to consume is a critical component of feeding populations domestically and internationally. Food safety is just one dimension of **food security** which encompasses practices, policies, and provisions to assure that people have access to a safe, secure, nutritious, and sustainable food supply. The two topics are related because many of the food safety issues that confront modern Americans are the result of agricultural practices designed to feed an ever-larger number of people, both at home and abroad.

Both meat production and agriculture in this country are largely industrialized processes, operating on economies of scale. If people were still growing their own food on small farms, for instance, we wouldn't have the potential for mass outbreaks of foodborne illness that threaten the public's health. Some modern agricultural practices actually increase the risk that dangerous bacteria may emerge to threaten human health. The common use of antibiotics to treat livestock fed a diet of grain, for example, is a likely cause of the emergence of antibiotic-resistant bacteria.

Conversely, **food insecurity** exists when people lack secure access to sufficient amounts of safe and nutritious food. According to the USDA, in 2013 14.3% of American households were uncertain of having, or were unable to acquire, enough food to meet the needs of all their members because they had insufficient money or other resources for

INFOGRAPHIC 20.13 Food Insecurity in the United States

The United States enjoys a high level of food security compared with many developing nations. However, a significant portion of U.S. adults and children do not have consistent access to a safe, secure, nutritious, and sustainable food supply.

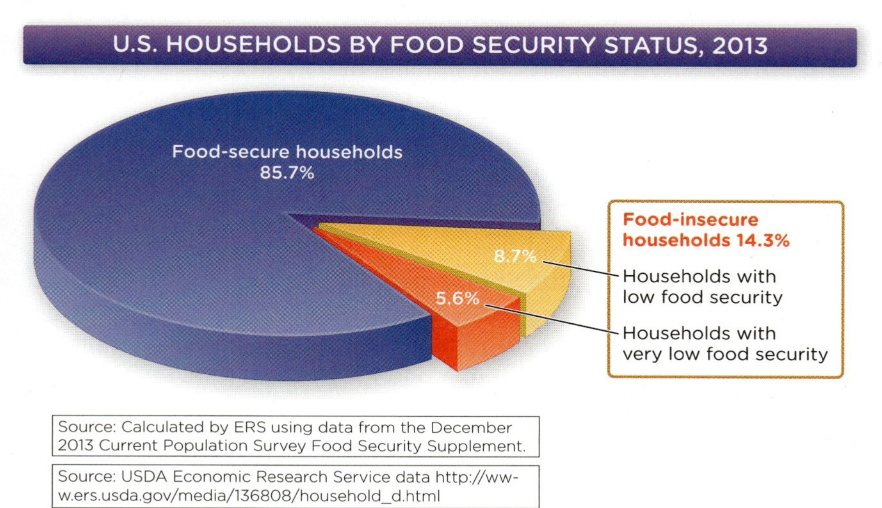

Source: Calculated by ERS using data from the December 2013 Current Population Survey Food Security Supplement.

Source: USDA Economic Research Service data http://www.ers.usda.gov/media/136808/household_d.html

food. (INFOGRAPHIC 20.13) For 49 million or more Americans, poverty causes compromised access to sufficient, varied, and nutritious foods.

In addition to economic status, where people live influences their food security. In Chapter 2 we examined the concept of *food deserts* in the United States where large, mostly poor areas had little or no access to grocery stores that sell a variety of nutrient-dense foods, including vegetables, fruits, whole grains, dairy products, and lean protein

Thanksgiving soup kitchen dinner. *According to the USDA Economic Research Service data, in 2013, rates of food insecurity were substantially higher than the national average for households with children headed by single women or single men, and Black- and Hispanic-headed households. Food insecurity was more common in large cities and rural areas than in suburban areas and areas surrounding large cities.*

Steve Skjold/Alamy

UNDERNUTRITION *inadequate nourishment caused by insufficient dietary intake of one or more essential nutrients or poor absorption and/or use of nutrients in the body*

OVERNUTRITION *excess intake or imbalance of calories and/or essential nutrients relative to need that results in adverse health effects*

GENETICALLY MODIFIED ORGANISM (GMO) *living organisms whose genetic material has been altered through the use of genetic engineering; GMOs are the source of genetically modified foods*

GENETICALLY MODIFIED (GM) FOODS *foods derived from organisms that have had their genetic material (DNA) modified in some way that does not occur naturally; for example, through the introduction of a gene from a different organism*

sources. And somewhat surprisingly, a recent study indicated that more than half of college students at public universities experienced food insecurity at some point during the previous year–potentially due to the cost of nutritious foods, limited income, and less access to and eligibility for food or social support systems than are available to other groups.

Food insecurity is complex and, as food deserts demonstrate, can also contribute to **overnutrition**—or excess intake of energy-dense foods at the expense of nutrient-dense foods—which can lead to obesity and other chronic diseases. Global health authorities warn that chronic diseases associated with overnutrition will overtake **undernutrition** (inadequate intake of essential nutrients and often overall energy) as the leading causes of death in low-income communities. Therefore, healthcare providers, educators, and policy makers must consider both under- and overnutrition in addressing poverty and food insecurity.

In the United States, several nutrition assistance programs provide supplemental food to enhance access, availability, and affordability of nutritious foods. These include the Special Supplemental Nutrition Program for Women, Infants, and Children (WIC); the National School Lunch Program (NSLP); and the Supplemental Nutrition Assistance Program (SNAP), formerly referred to as Food Stamps. SNAP is the largest program in the domestic hunger safety net. And community initiatives are being developed that strive to enhance food security and bring more diverse, healthful foods to underserved populations.

__GMO crops.__ Signs mark different crop varieties in a soybean field, including genetically modified crops. Plants may be genetically engineered (modified) to enhance the growth or nutritional profile of food crops. Food and food ingredients from genetically engineered plants were introduced into the food supply in the 1990s.

Jim West/Alamy

Genetically modified organisms

These measures help ensure access to food, but they don't address the problem of food abundance. To help increase yields, and improve crop resistance to pests, many food companies have turned to genetic engineering—the process of inserting genes into, or modifying the genes of, an organism. Such **genetically modified organisms (GMOs)** are increasingly common on the shelves of U.S. supermarkets (they are more restricted in Europe).

Examples of **genetically modified (GM) foods** include corn, soybeans, rice, and tomatoes. In the United States, although current federal law does not mandate the labeling of GM foods, the FDA is responsible for regulating the safety of GM crops consumed by humans or animals. (INFOGRAPHIC 20.14)

Food supply can vary globally based on many factors. For example, droughts, floods, and political instability can all affect food supply, as can climate change, dependence on fossil fuels, and loss of biodiversity. Climate change is a particularly worrisome factor, potentially affecting not only food supply but also food safety; some evidence suggests that the prevalence

INFOGRAPHIC 20.14 Pros and Cons of Genetically Modified Organisms

POTENTIAL BENEFITS OF GENETICALLY ENGINEERED FOODS:

- Increased vitamin and/or phytochemical content of foods.
- Improved taste or texture of foods.
- Disease- and drought-resistant plants that require fewer environmental resources (water, fertilizer, etc.)
- Decreased use of pesticides.
- Increased supply of food with reduced cost and longer shelf life.
- Faster growing plants and animals.
- Food with more desirable traits, such as potatoes that absorb less fat when fried.
- Medicinal foods that could be used as vaccines or other medications.

POTENTIAL RISKS OF GENETICALLY MODIFIED FOODS:

- Modified plants or animals may have genetic changes that are unexpected and harmful.
- Modified organisms may interbreed with natural organisms and out-compete them, leading to extinction of the original organism or to other unpredictable environmental effects.
- Plants may be less resistant to some pests and more susceptible to others.
- May cause plant foods to produce proteins that could be allergenic in some people.

Elnur/Shutterstock

of several types of foodborne illness increases proportionally with average weekly temperatures.

Sustainability

A 2014 report published by the nonprofit Center for Food Safety argues that food security may be jeopardized by climate change and recommends the following: eating fresh, unprocessed foods; buying local, in-season produce; choosing organic foods; eliminating industrial meat and dairy consumption; and reducing food waste. These efforts, the report argues, may help ensure sustainability of the food supply.

Strictly speaking, **sustainability** means the use of resources at rates that do not exceed the capacity of Earth to replace them. For food, a sustainable system implies safety and security of the food supply, a strong food industry in terms of jobs and growth, and, at the same time, environmental sustainability in terms of biodiversity, water, and soil quality.

According to the 2014 Food and Agricultural Organization of the United *Nations Report on Food Insecurity in the World,* a total of 805 million people, or around one in nine people in the world, were estimated to be suffering from chronic hunger, regularly not getting enough food to conduct an active life. Sub-Saharan Africa remains the region with the highest prevalence of **undernourishment**, with

SUSTAINABILITY *the ability to meet our current needs without compromising the ability of future generations to meet their needs*

UNDER-NOURISHMENT *prolonged inability (for at least one year) to acquire enough food to meet dietary energy requirements*

INFOGRAPHIC 20.15 Percentage of Undernourishment in the Total Population of Developing Countries, 2011–2014

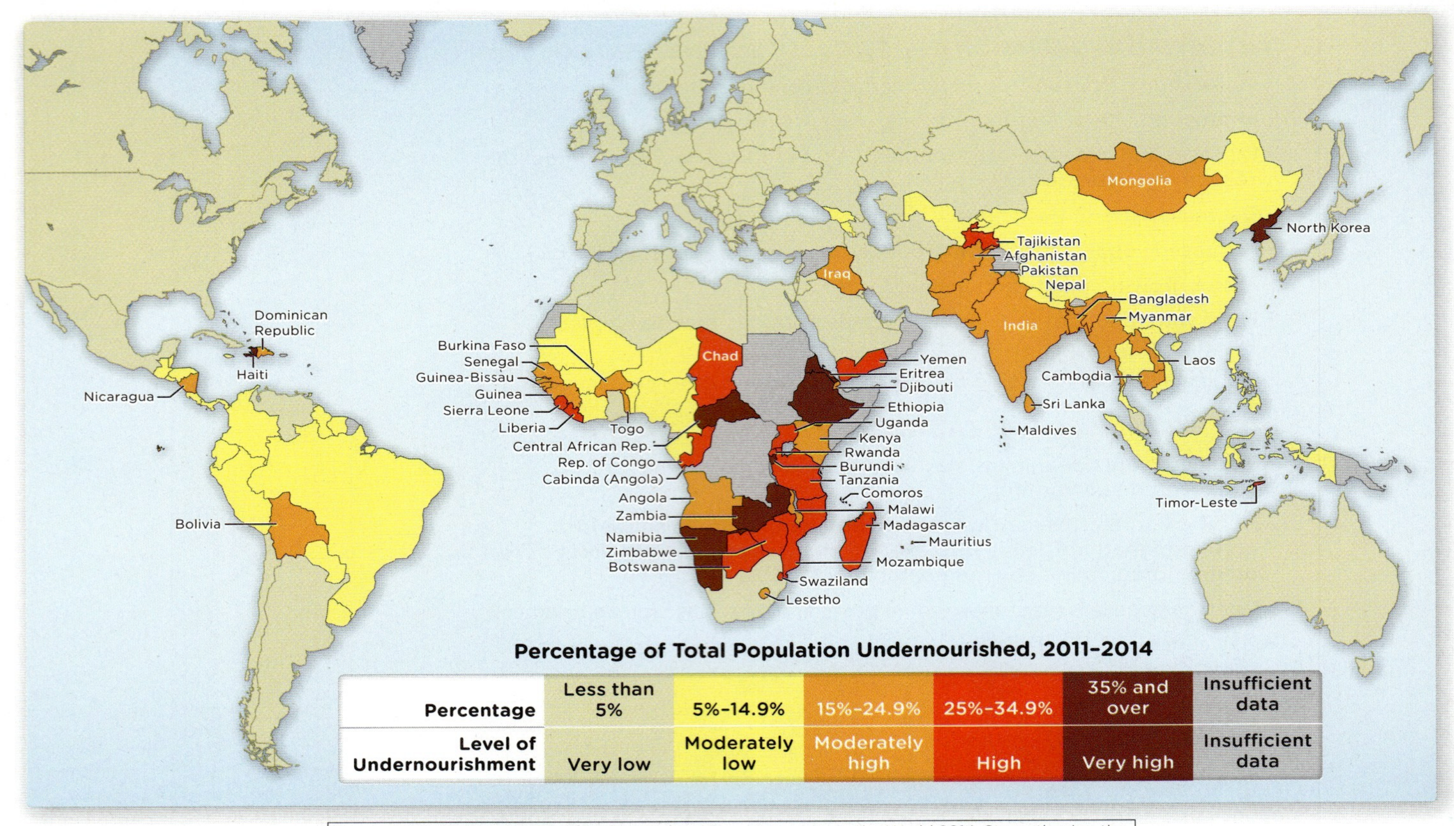

Source: FAO, IFAD and WFP. 2014. The State of Food Insecurity in the World 2014. Strengthening the enabling environment for food security and nutrition. Rome, FAO. Data source: fao.org/economic/ess.

more than one in five people estimated to be undernourished. **(INFOGRAPHIC 20.15)** Undernourishment as a consequence of food insecurity in these areas is of particular concern in children with increased susceptibility to infections and often fatal consequences of illness (like chronic diarrhea), as well as risk of impaired growth and development.

Feeding the world and advancing food security with an expanding population will require commitment, cooperation, and innovation among citizens and scientists. Given population growth and rising incomes that provide opportunity for greater consumption of resources and consumer goods, it is estimated that the demand for food will rise by 70% to 100% by 2050. Many programs and initiatives at the national and international levels are geared toward reducing hunger and boosting food security.

If history is any guide, there will likely be more food crises, involving both safety and security, before a sustainable food culture prevails.

CHAPTER 20 **BRING IT HOME**

Focus on food safety

Keeping food safe to consume and avoiding foodborne illnesses requires proper food handling and hygiene at all points from "farm to fork." From what you've learned in this chapter and by using the resources and tips provided at one or more of the websites below, identify at least two ways the individuals in each of the following scenarios can decrease their risk of foodborne illness. Ideas for the first scenario have been provided as an example.

Partnership for Food Safety Education at http://www.eatright.org/Public/

Foodsafety.gov at http://www.foodsafety.gov/

Center for Disease Control and Prevention (Food Safety) at http://www.cdc.gov/foodsafety/

Academy for Nutrition and Dietetics (Food Safety) at http://www.eatright.org/Public/

1. Supermarket checkout bagger
   - Do not place raw meats or poultry in same bag as fresh produce
   - Wash hands thoroughly and often
2. Shopper at a supermarket
3. Chef at a restaurant
4. Restaurant wait staff
5. Restaurant patron
6. Manager at a cafeteria-style dining facility
7. Server at a cafeteria
8. Parent preparing and packing a school lunch
9. Individual preparing a meal at home
10. College student in their dorm room

Take it Further

Recent studies indicate that the percentage of college students that are food insecure is greater than the percentage of the overall population who are food insecure. List five reasons college students might have challenges obtaining sufficient, nutritious food.

KEY IDEAS

Foodborne illness is a very common, though largely preventable, condition in the United States. The most common foodborne illnesses are caused by five pathogens: a virus called *norovirus* and the bacteria *Salmonella*, *Staphylococcus aureus*, *Clostridium perfringens*, and *Campylobacter*.

Foodborne illnesses can have different symptoms, times of onset, and potential consequences. Most begin in the gastrointestinal tract with nausea, vomiting, abdominal cramps, and/or diarrhea. They can also lead to more serious and potentially life-threatening complications.

Foodborne illnesses fall into two general categories: intoxication, which is caused by the injection of toxins; and infections, which are caused by the ingestion of pathogens that then multiply in the intestine.

Individuals whose immune systems are not functioning at full capacity are at higher risk of foodborne illnesses; this includes infants, young children, pregnant women and their unborn children, older adults, and individuals with chronic diseases such as diabetes, HIV/AIDS, and some cancers.

The Food Safety Modernization Act (FSMA), signed into law in 2011, granted more authority to federal regulators to be proactive in keeping food safe rather than simply responding to issues, outbreaks, or violations.

The USDA Food Safety and Inspection Service in conjunction with the FDA has identified four basic food safety principles that together can reduce the risk of foodborne illness summarized by the verbs *clean*, *separate*, *cook*, and *chill*.

Foods can be cross-contaminated when pathogens are transferred to foods from food handlers, contaminated foods, unclean surfaces and utensils, and pets.

Bacteria grow in the temperature danger zone between 41°F and 135°F. To avoid foodborne illnesses keep hot foods hot and cold foods cold.

Food additives include any substances added to food, including preservatives, flavors, and vitamins and minerals. They serve multiple purposes: to maintain or improve safety and freshness; to improve or maintain nutritional value; and to improve taste, texture, and appearance.

The FDA is responsible for ensuring the safety of food additives. Most additives in the food supply are considered "generally recognized as safe" (GRAS) based on historical use in food and published studies on safe consumption.

Food security refers to having enough safe food for people to consume. Food insecurity exists when people lack secure access to sufficient amounts of safe and nutritious food.

Food insecurity is associated with poverty and unemployment and can increase the risk of both undernutrition and overnutrition because of the lack of access to nutrient-dense foods.

Nutrition assistance programs that provide supplemental food to specific populations include the Special Supplemental Nutrition Program for Women, Infants, and Children; the National School Lunch Program; and the Supplemental Nutrition Assistance Program.

Genetically modified foods (GM foods) are derived from organisms whose genetic material has been modified in a way that does not occur naturally. Genetically modified organisms (GMOs) can have potential benefits as well as risks.

Sustainability means the use of resources at rates that do not exceed the capacity of Earth to replace them. For food, a sustainable system implies safety and security of the food supply, a strong food industry in terms of jobs and growth, and, at the same time, environmental sustainability in terms of biodiversity, water, and soil quality.

NEED TO KNOW

Review Questions

1. All of the following are common causes of foodborne illness, EXCEPT:
   a. *campylobacter*.
   b. *clostridium perfringens*.
   c. *norovirus*.
   d. *pneumococcus*.
   e. *salmonella*.

2. Food intoxication can be caused by all of the following, EXCEPT:
   a. toxins produced by bacteria.
   b. toxins produced by fungi.
   c. natural accumulation of toxins in shellfish.
   d. toxins naturally present in some plants.
   e. toxins produced by viruses.

3. Which of the following pathogens can multiply in foods?
   a. parasites
   b. viruses
   c. bacteria
   d. parasites, viruses, and bacteria

4. Which of the following must be cooked to the highest internal temperature to insure safety?
   a. poultry
   b. pork chop
   c. ground beef
   d. egg dishes

5. The effects of foodborne illness may be most severe and of special concern for:
   a. a 2-year-old child and her mother, who is 5 months pregnant.
   b. a person being treated for chemotherapy.
   c. an 85-year-old grandfather.
   d. All of the above.

6. All of the following are TRUE with regard to certified organic produce, EXCEPT that it is:
   a. grown without the use of synthetic fertilizers and pesticides.
   b. regulated by USDA.
   c. significantly less likely to produce foodborne illness than conventional produce.
   d. not genetically modified or engineered.

7. If an outbreak of foodborne intoxication is caused by the consumption of corned beef that was cooked and sliced yesterday and then reheated today by boiling, which one of the four steps to food safety was most probably not followed?
   a. clean
   b. separate
   c. cook
   d. chill

8. All of the following are TRUE with regard to FSMA, EXCEPT that it:
   a. stands for the Food Safety Modernization Act.
   b. limits the FDA's authority in food production facilities.
   c. is the most sweeping change in food safety legislation in over 70 years.
   d. provides for protection against intentional, as well as unintentional contamination of food.
   e. established regulations for the production of vegetables and fruits.

9. The danger zone for bacterial growth in foods is:
   a. 32°F to 132°F
   b. 35°F to 140°F
   c. 35°F to 165°F
   d. 41°F to 135°F
   e. 41°F to 165°F

10. All of the following are TRUE with regard to food additives, EXCEPT:
    a. they include indirect additives that become part of food during packaging and processing.
    b. they are added only to improve safety and freshness.
    c. they are considered GRAS substances so most don't have to go through an approval process.
    d. the FDA has primary responsibility for ensuring their safety.
    e. They can include salt, spices, and herbs.
11. Federally recognized GRAS substances:
    a. must have gone through a stringent approval process by FDA.
    b. include only about 50 substances to date.
    c. do not have to be included on food labels.
    d. include salt, spices, and herbs.
12. The former U.S. Food Stamps program is now known as the:
    a. Federal Food Supply Program.
    b. Food for Life Program.
    c. Supplemental Nutrition Assistance Program.
    d. Women and Infant Assistance Program.
13. For Americans who experience food insecurity, the *main factor* limiting access to sufficient quantities of safe and nutritious foods is:
    a. poverty.
    b. that fresh fruits and vegetables are not available year round.
    c. knowledge about nutrition.
    d. the prevalence of obesity.

Dietary Analysis Using Choosemyplate

Understanding food safety and food security

Preventing foodborne illness is very important, especially for people who are more susceptible, such as the ill, the young, the old, and pregnant women. In addition, people with limited income need to be aware of strategies that will allow them to purchase nutritious foods on a budget. Using choosemyplate.gov, you will be able to identify the ways to prevent foodborne illness and ways to purchase healthy food on a limited income.

1. Log onto the United States Department of Agriculture website at www.choosemyplate.gov. At the top of the page click on "Popular Topics" and "Food Safety."
2. Identify each of the four important safety practices listed on the Food Safety page. Explain why each of these is important.
3. Do you believe that the average person might also be unaware of some of these recommendations? Explain your answer.
4. What do you think is the single most important factor in preventing foodborne illness? Why do you believe this?
5. Click the "Healthy Eating on a Budget" link in the left column. Click on the "Shop Smart to Fill Your Cart" link, and then click on "Tips for Every Aisle." List one smart shopping tip for each of the food groups. Return to the main website page and click the "Healthy Eating on a Budget" link under Popular Topics at the top of the left column.
6. Finally, click the "Understand the Price Tag" link. What is the difference between the unit price and the retail price?

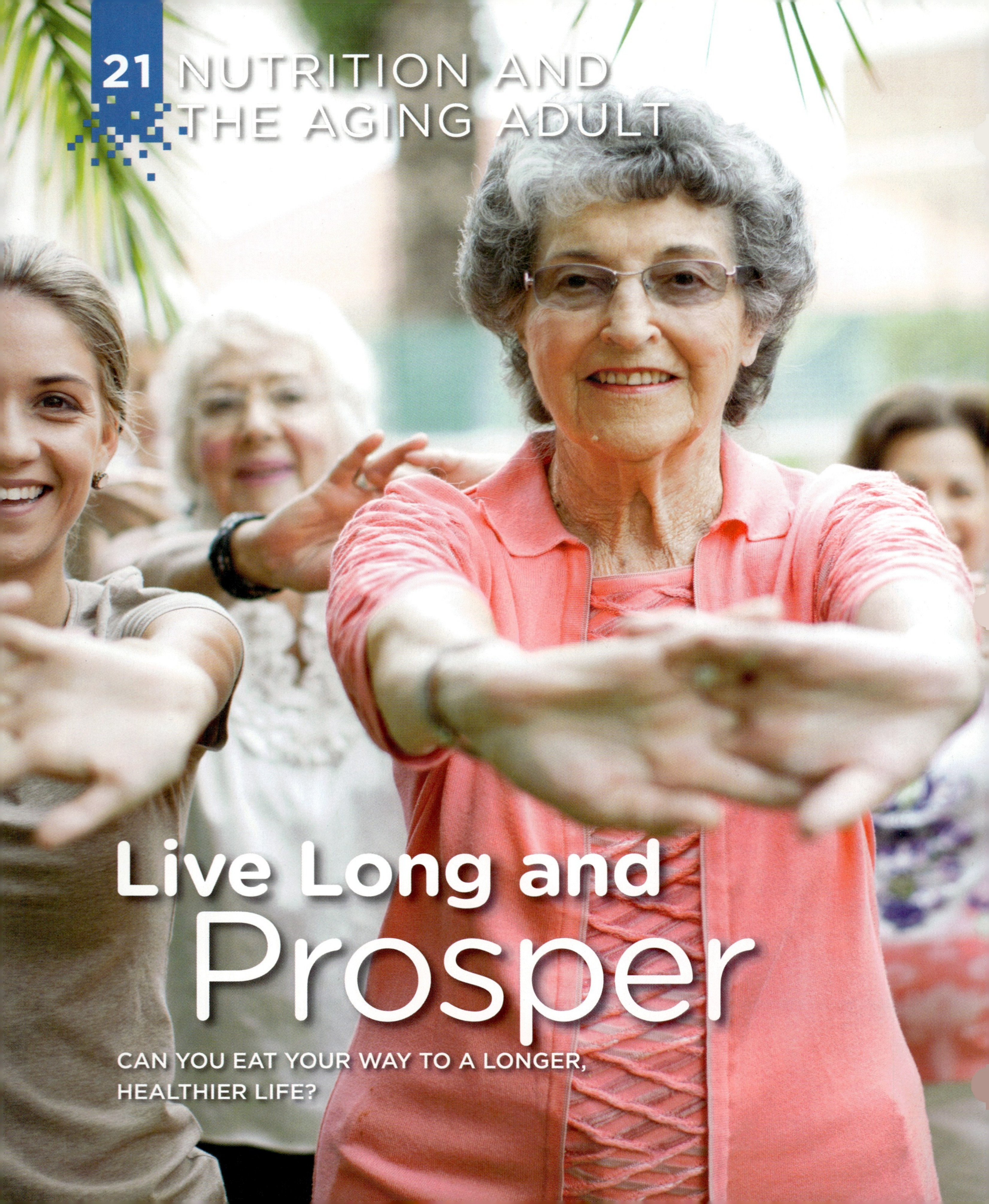

21 NUTRITION AND THE AGING ADULT

Live Long and Prosper

CAN YOU EAT YOUR WAY TO A LONGER, HEALTHIER LIFE?

Zero Creatives/Corbis

LEARNING OBJECTIVES

- Discuss at least four age-related physiological changes (Infographic 21.1)
- Describe changes in the age structure of the population in the United States in the next 25 years (Infographic 21.2)
- Explain how skeletal muscle is needed to heal from illness and injury, the benefits of maintaining adequate muscle mass, and the benefits of exercise (Infographics 21.3 and 21.9)
- Discuss how physical, psychological, economic, and social factors may influence the ability of the aging adult to achieve the recommended nutrient intake (Infographic 21.4)
- Describe three nutrition-related conditions that are common in older adults, their precipitating factors, and potential consequences (Infographic 21.5)
- Identify at least five nutrients of concern in the aging adult and describe ways to obtain adequate intake of these nutrients (Infographics 21.6 and 21.7)
- Identify lifestyle dietary characteristics that may delay the onset of age-related disease (Infographic 21.10)

When Leonard Nimoy died in February 2015, at the age of 83, he was remembered most for his iconic role as the pointy-eared Mr. Spock on the television and movie series *Star Trek.* Part human, part Vulcan, Spock shared with his fellow Vulcans an intensely logical mindset. Vulcans had long since figured out how to prevent poverty, war, and other social problems. They were also masters of their own health, eating a vegetarian diet and living to be 200 years old. Hence the traditional Vulcan salute, "Live long and prosper."

The actor Leonard Nimoy was famous for delivering the Vulcan salute "Live long and prosper," in the Star Trek *science fiction films and television show. He lived to age 83.*

Jeffrey Mayer/Getty Images

Nimoy himself managed to live seven years beyond the average life expectancy for a man in the United States, and was mentally vigorous until the end, writing poetry and Tweeting to his many followers. He might have lived even longer, if he hadn't smoked as a young man, or had quit earlier. Nimoy died of complications from chronic obstructive pulmonary disease, which is often a late, smoldering consequence of smoking.

LONGEVITY *the length or duration of life; often refers to individuals whose lifespan is longer than the population average*

AGING *the accumulation of diverse, harmful changes in cells and tissues with advancing age that is responsible for the increased risk of disease and death*

Star Trek is science fiction, of course, but the search for the secrets to a longer life is a very real focus of contemporary science, involving researchers in fields as diverse as genetics, cell biology, nutrition, psychology, and sociology. Geneticists working in laboratories around the world have already figured out how to double the lifespan of small animals like mice and worms by altering their genes. Venture capitalists are currently investing in companies, like the Bay Area–based Calico, whose explicit aim is to use biotechnologies to extend the human lifespan.

Calorie restriction without nutrient deficiency also increases the longevity of small animals and is one of the most studied interventions to delay aging in experimental animals. When calories are restricted from an early age by 30% (compared with a self-selected diet) longevity is increased by about 40% in a number of organisms—from yeast to laboratory mice and rats. Recent studies have also found that calorie restriction decreases the risk of chronic diseases and increases longevity in rhesus monkeys. Researchers do not expect people to restrict their diets in this manner, but they study calorie restriction because it has such a potent effect on aging. Elders who have a healthy body weight should not contemplate this sort of energy restriction since maintaining a healthy body weight through diet and physical activity is critical for healthy aging. It is hoped, however, that these studies will reveal details about the mechanisms of aging and allow the development of effective measures to slow the aging process.

Short of altering our genetic code or consuming some Vulcan-style elixir, are there ways that humans can increase their **longevity** or lifespan through behavioral measures? Many scientists answer a resounding "Yes." But before we look at ways to increase longevity, let's first describe the concepts of *aging* and *life expectancy*.

WHAT HAPPENS WHEN WE AGE?

The time we spend as children and young adults is characterized by the acquisition of new skills, physical growth and maturity, and psychological development. Although we continue to learn and enjoy new challenges throughout our lives, eventually, as we advance in years, the human body experiences some signs of deterioration or decline that we recognize as **aging**. Aging is the accumulation of diverse harmful changes that occur in cells and tissues that are responsible for decreased physical

INFOGRAPHIC **21.1** **Physiological Changes Associated with Aging** *Although aging does put us at greater risk for health issues, many older adults can be healthy and active well into their advanced years.*

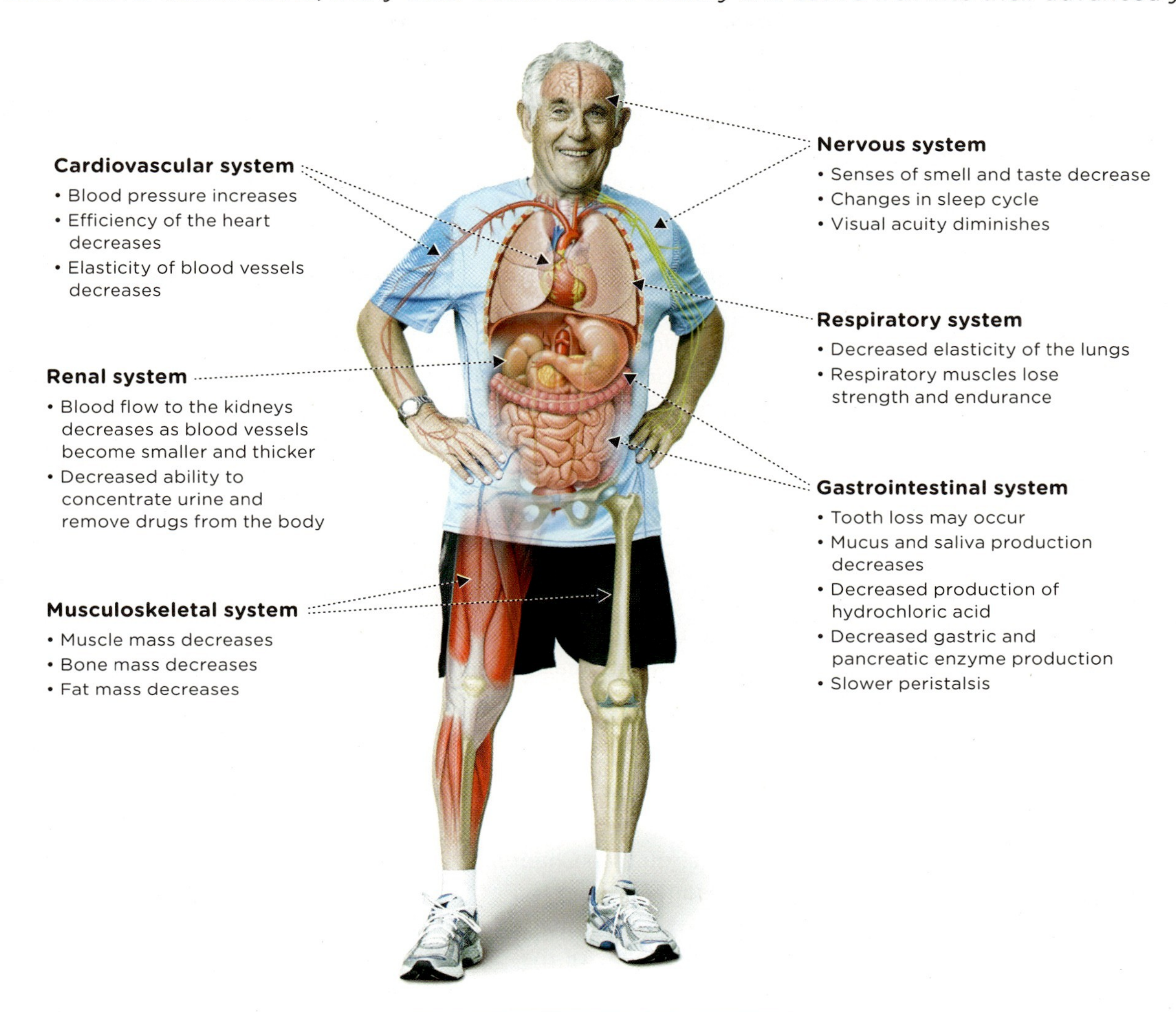

? *How might age-related changes in the body influence the development of nutrition-related health problems? Give at least one example.*

Photo credit: Jakob Helbig/Getty Images

and mental capacity and an increased risk of disease and death. Aging is a complex multifaceted process that varies from person to person; however, over time it affects the cells and all of the major organ systems of the body. **(INFOGRAPHIC 21.1)**

The changes in the body associated with aging are due to changes in individual cells and the organs they make up. As cells age, cellular functioning slows. Also, in some organs, cells die and are not replaced, so the number of cells decreases. The number of cells in the liver and kidneys for example, decreases as the body ages. However, even with this decline in cell number, most functions remain adequate because organs begin with more capacity than the body needs. For example, if half the liver is destroyed, the remaining tissue is more than enough to maintain normal function. Thus, disease and illness, rather than normal aging, usually account for most of the loss of function in old age.

LIFE EXPECTANCY *the average number of years individuals in a specific population are expected to live*

LIFESPAN *span of time between the birth and death of a person*

Over the years a number of theories have been proposed to explain the process of aging. These theories are generally divided into two broad categories: programmed theories that contend that aging follows a biological timetable, and damage or error theories that suggest that aging is caused by an accumulation of molecular and cellular damage.

Major programmed theories of aging include: (1) *programmed longevity*—the switching on and off of specific genes that leads to aging; (2) the *endocrine theory*—where age-related alterations in the regulation of the endocrine system (including stress hormones and insulin) control the rate of aging; and (3) the *immunological theory*—which suggests that the immune system is programmed to decline over time which causes increased vulnerability to disease.

Some of the leading damage or error theories of aging are: (1) *free radicals* cause oxidative damage to proteins, lipids, and DNA; (2) *protein cross-linking* impairs protein function; (3) *DNA damage* results in mutations that cause cells to malfunction; and (4) *epigenetic changes* to DNA alters patterns of gene expression (see Chapter 1). It should be appreciated that aging is a complex process that involves a number of these mechanisms, and many of these mechanisms are interconnected.

One body system that is significantly changed in the aging adult is the cardiovascular system (the heart and blood vessels). Blood flow to the body declines due to many factors, including atrophy of the heart muscle, calcification of the heart valves, loss of elasticity in the artery walls, and fatty deposits in the arteries. The reduced blood flow causes older adults to tire more easily than younger adults since less oxygen is being exchanged, which in turn, reduces kidney and liver function, and provides less nourishment to the cells. This means older people are more likely to experience side effects from drugs, have a slower rate of healing, and an impaired response to stress. Even though organ functions often remain adequate in aging, the decline in optimal function means that older people are less able to handle strenuous physical activity, extreme temperature changes in the environment, and illness.

LIFE EXPECTANCY AND LIFESPAN

An important distinction to be made when discussing aging and longevity is between *life expectancy* and *lifespan*. **Life expectancy** refers to the number of years a person can be expected to live, based on the statistical average. This statistical average is calculated based on an overall population, often beginning from the time of birth, and includes those who die during or shortly after childbirth, those who die during adolescence and adulthood, and those who live well into old age. The average life expectancy for a man in the United States today is 76.4 years; for a woman, it's 81.2 years.

Lifespan, on the other hand, is the number of years that any particular individual lives. Leonard Nimoy had a lifespan of 83 years, for example. The maximum number of years that any known person has lived is 122 years. Therefore, 122 is sometimes cited as the *maximum human lifespan*, although that will change as soon as someone lives longer.

Life expectancy has increased dramatically over the twentieth century. In 1900, life expectancy for a man was 46 years. Today, as you read above, a man's life expectancy is more than 76 years. Statistics like these can be a source of confusion. This does not mean that, in 1900, most men dropped dead at age 46. The greatest difference between then and now is the change in the frequency of infant mortality, with many fewer children dying as a result of infectious and communicable diseases. Because those deaths are included in the calculation of the average life expectancy, they greatly affect the overall measure. Decreases in infant mortality occurred because of improvements in sanitation (principally, cleaner water), which reduced the incidence of communicable diseases like cholera, and the availability of vaccines to prevent diseases such as smallpox, polio, and measles.

The Frenchwoman Jeanne Louise Calment lived to age 122, and is the longest-lived human whose age has been verified by official records. She was famous for her sense of humor and her love of eating chocolate. She died in 1997.

Gamma-Rapho via Getty Images

For those surviving to adulthood, the gains in life expectancy have been more modest over the same period: about nine additional years for women and five additional years for men. Major factors that are responsible for recent increases in the life expectancy of U.S. adults are a decrease in the prevalence of smoking and improved medical care that allows individuals who are affected with chronic diseases to live longer. During the next 25 years, two factors—longer life spans and aging baby boomers—will combine to double the population of Americans 65 years or older to about 72 million. By 2030, older adults will account for roughly 20% of the U.S. population. **(INFOGRAPHIC 21.2)**

INFOGRAPHIC 21.2 The Aging U.S. Population *Combined with the aging of the Baby Boomer generation, the increased longevity of the U.S. population will lead to an expansion of older age groups between 2015 and 2040.*

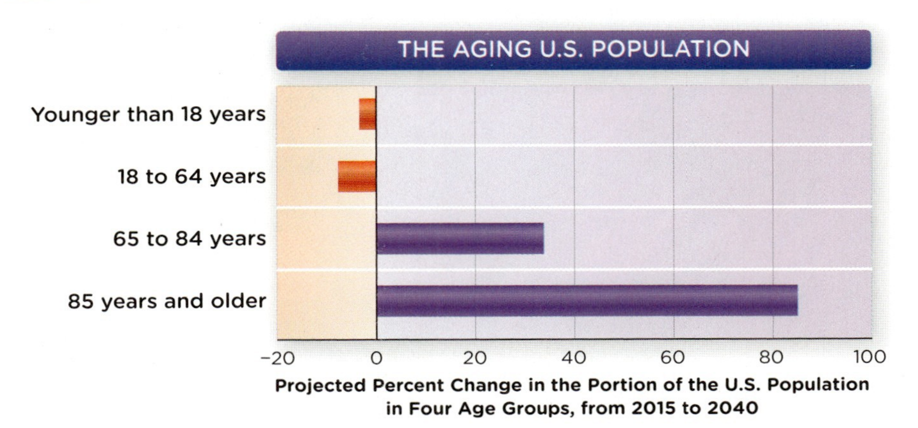

How could nationwide high-quality nutrition reduce the negative effects of an aging population in the United States?

Clearly, a number of factors can influence how long a person lives. The most obvious is disease. The biggest killers today are largely chronic diseases, such as heart disease, cancer, and diabetes. Anything that increases one's odds of getting these diseases, such as smoking or obesity, will have a correspondingly negative impact on one's life expectancy. There are also more subtle influences. People with higher levels of education have a higher life expectancy than those with lower levels. Decreased life expectancy is also associated with being unmarried, low socioeconomic status, and being part of some racial groups.

BLUE ZONES

Let's return to the question of whether there are ways in which humans can increase their longevity through behavioral measures. The answer is yes. Support comes from the study of areas on our planet where humans experience increased longevity. Known as "Blue Zones," these are areas with a higher than average number of individuals who live to be at least 100 years old. Blue Zones have been identified in parts of Japan, Italy, Canada, and Costa Rica.

Among the best studied of the so-called Blue Zones is the Seventh-day Adventist community. The Adventists are a conservative sect of Protestant Christianity that goes back to the 1860s. They believe in living according to Biblical principles and observe a day of rest (the Sabbath) on Saturday. Although not exactly a "zone," since they live all over the world (with a large population in California), this group shares a common way of living, born of their religious beliefs, that links them together. And according to researchers, they are among the longest-living populations on the planet, if not *the* longest.

"We found that Adventists were getting, on average, more than seven years of extra life expectancy, for men, and about four and a half years for women as compared to other Californians," says Gary Fraser, a cardiologist and epidemiologist at Loma Linda Medical School, who has studied the Adventists. "The magnitude of that difference surprised me because it's quite major."

What's behind this allotment of extra years? Fraser's group has identified several key behaviors and lifestyle choices

Zones of greatest longevity. *The map indicates locations, the so-called "Blue Zones," where people tend to live long lives. The Blue Zone concept was popularized by journalist Dan Buettner, in his book* The Blue Zones: 9 Lessons for Living Longer from the People Who've Lived the Longest.

that they believe give Adventists their propensity for longer lifespans. It turns out that Adventists do (and don't do) a whole bunch of things that improve their life expectancy—some obvious, some less so. For instance, they don't smoke. They also rarely drink alcohol. Since both smoking and excessive drinking are associated with a range of diseases, it's not surprising that refraining from these things improves health and increases longevity. Many Adventists are also vegetarians or vegans (refer to Chapter 9 for more on plant-based diets), they emphasize fresh, minimally processed foods, and they tend to regularly consume nuts. These food choices, too, affect health, as we will see.

Through lifestyle choices, this group of people has experienced a big boost in longevity. And if Adventists can do it, researchers say, then so can the rest of us.

The concept of a Blue Zone comes from a 2004 study of especially long-lived individuals in the mountainous region of Sardinia, an island off the coast of Italy. Researchers began circling progressively narrower areas on a map where longevity seemed to be highest, until a (blue) zone of maximum longevity was reached. The Blue Zone concept was popularized by journalist Dan Buettner, in his book *The Blue Zones: 9 Lessons for Living Longer from the People Who've Lived the Longest.* Buettner traveled to far-flung regions of the world with the goal of identifying the common factors that contribute to these astounding feats of longevity.

Many of the so-called Blue Zones are isolated island or mountain communities, such as the archipelago of Okinawa in Japan and the highlands of Sardinia in Italy, where age-old customs remain intact and where migration into and out of the area is rare.

The Adventist community, by contrast, is not found in a geographically restricted area, but that turns out to be a plus for longevity researchers. Unlike the other Blue Zones, where the geographic isolation of gene pools might imply a genetic reason for increased longevity, the Adventists are linked only by behavior. They therefore provide one of the best populations in which to explore the controllable factors, such as nutrition, that may influence longevity.

The first studies of the Adventists were conducted in the 1950s, and even back then it was clear that Adventists tended to live longer than their non-Adventist neighbors. But it has only been in the past decade that

Seventh-day Adventist populations are long-lived. Members of this community do not smoke and rarely drink alcohol.

robuart/Shutterstock

Harvesting food at low tide. *The archipelago of Okinawa in Japan is known for a population with an abundance of centenarians. The woman in this photo is 89 years old and gathering asa seaweed, a food often eaten by Okinawans.*

David McLain/Aurora Photos

SARCOPENIA *age-related reduction in skeletal muscle mass*

scientists have begun to unravel the specific behavioral and lifestyle factors that are behind their longer lives.

ENERGY NEEDS AND PHYSICAL ACTIVITY

Results of epidemiological research like the Adventist studies show that nutrition plays a key role in maintaining the health of the aging adult, and therefore in determining longevity and quality of life. For the most part, the nutritional needs of healthy older adults (51 years and older) are similar to those of healthy adults of younger ages, but there are a few key ways in which their nutritional needs differ. The most important one involves energy (caloric) needs.

As people age, they require fewer calories to maintain their weight and power their activities. That's because total and resting energy expenditure decrease progressively with age. There are several contributing factors to this decline, but the biggest one is a decrease in physical activity—older adults are less physically active than younger folks. In turn, this reduced activity leads to reduced muscle mass, a phenomenon called **sarcopenia**, which further reduces resting energy expenditure (since muscle burns more calories than fat; see Chapter 15). Some of the changes in muscle mass that occur with aging are independent of activity levels, and relate to changes in hormone levels and cellular aging, both of which lead to predictable age-related changes in body composition and body fat distribution. But to a large extent, physical activity can offset losses in muscle mass. **(INFOGRAPHIC 21.3)**

Older adults need fewer calories to meet their typical energy expenditure, but their need for specific macro- and micronutrients does not decrease (and may even increase because of other bodily changes). As a

INFOGRAPHIC **21.3** **Active Aging** *Being physically inactive can be bad for you, no matter your age or health condition. Remaining physically active will promote a higher quality of life throughout your advanced years.*

? *What difficulties could prevent seniors from being physically active and how can these difficulties be overcome?*

Photo credits (top to bottom): Tom Wang/Shutterstock, Monkey Business Images/Shutterstock, Monkey Business Images/Shutterstock

consequence, older adults must make sure to consume a diet that is more nutritionally dense than the diet they may have eaten when they were younger. There is less room for empty calories.

SPECIAL NUTRITIONAL CONCERNS FOR OLDER ADULTS

Despite this need for nutrient-dense foods, many older adults experience bodily changes that impair their ability or desire to meet these requirements. For example, many older people experience a decline in the senses of smell (olfaction) and taste (gustation), which can make food less palatable and therefore affect appetite because food just isn't as appealing as it once was. Aging causes a more dramatic decline in the sense of smell than it does in taste. More than 60% of individuals between the ages of 65 and 80 have a major loss of smell, increasing to more than 75% of those who are older than 80 years. A reduction in mucus production that traps and transfers odorants, and a decrease in olfactory receptors and neurons that detect those odorants, are thought to be major factors leading to this age-related loss of smell. Diminished taste is caused by age-related decreases in the number, size, and sensitivity of taste buds.

With these declines in taste and smell, visual cues play an increasingly important role in stimulating appetite. It has been found that enhancing the dining room and the presentation of food increases food intake in nursing home residents. In addition to changes in taste and smell, loss of teeth and periodontal disease can compromise one's ability to chew, which can also make obtaining adequate nutrition a challenge. (INFOGRAPHIC 21.4)

Food and drug interactions

Nutritional status may be further compromised by the effects of medications on appetite and nutrient absorption. According to the Centers for Disease Control, more than 75% of older Americans (60 years and older) use two or more prescription drugs and 37% use five or more! This prevalence of drug use, both prescription and over-the-counter medications, combined with age-related alterations in physiological function, is a special concern for seniors. Healthcare providers must consider the implications of interactions among medications, dietary supplements, and food. Some drugs can affect the

INFOGRAPHIC **21.4** **Challenges to Healthy Aging** *Many physiological, pathological, socioeconomic, and psychological factors contribute to decreased food intake in the elderly.*

Physiological Factors
- Decreased taste and smell
- Slower gastric emptying and prolonged satiety
- Reduced lean body mass and energy expenditure

Pathological Factors
- Poor dental health
- Difficulty swallowing
- Illness, dietary restrictions
- Dementia
- Medications/GI symptoms

Socioeconomic Factors
- Isolation
- Poverty
- Difficulty purchasing and preparing food

Psychological Factors
- Depression
- Loneliness
- Grief/loss

Juanpablo San Martin/Getty Images

Consuming adequate amounts of vitamin B_{12}, folate, iron, calcium, and vitamin D may be difficult for some older adults. The requirements for calcium and vitamin D increase in older adults in general, and the absorption of vitamin B_{12}, folate, and iron decrease in many older adults.

Pick two of these factors and explain how and why they could result in decreased food intake.

metabolism, absorption, or excretion of certain nutrients (for example, antacids can diminish the absorption of vitamin B_{12}). And some foods can influence the action and effectiveness of medications. Fresh grapefruit and grapefruit juice, for example, contain a compound that decreases the breakdown of some drugs in the small intestine and liver, increasing their concentrations in the blood and enhancing their effects. And for individuals taking anticoagulant medications, atypical intake of vitamin K (an especially large serving of spinach or kale, for example) can decrease the medication's effectiveness.

Changes in the digestive system

Age-related gastrointestinal changes may also occur, especially with regard to the bacterial composition of the gut, which can affect nutrient absorption. Diminished gastric acid secretion and slower motility in the small intestine can result in overgrowth of intestinal bacteria that interferes with absorption of nutrients. Reduced gastric acid production also causes malabsorption of naturally occurring vitamin B_{12}, because gastric acid must release the vitamin from food proteins for it to be absorbed. It is estimated that at least 25% of those older than 60 years are deficient or marginally deficient in vitamin B_{12} and low B_{12} levels are associated with cognitive impairment and dementia. For this reason it is strongly recommended that the elderly receive their vitamin B_{12} from fortified foods or supplements. (INFOGRAPHIC 21.5)

Reduced gastric motility also contributes, along with an age-related weakening of the colon wall and low fiber intake, to the

INFOGRAPHIC **21.5** **Selected Nutrition-Related Conditions Common in Older Adults**

Certain conditions common in older adults may be improved or prevented through appropriate nutrition and physical activity.

Age-related macular degeneration: Leading cause of vision loss with aging. Consuming a high-dose supplement containing vitamins C and E, zinc, copper, and the phytochemicals lutein and zeaxanthin can slow its progression.

Sarcopenia: Involuntary loss of lean body mass in the elderly typically caused by reduced levels of physical activity; less than optimal protein intake may also be a contributing factor.

Bacterial overgrowth: Decreased gastrointestinal tract motility and hydrochloric acid production increase the bacterial population in the small intestine and may increase GI symptoms and reduce the absorption of some nutrients.

Osteoporosis: Lack of adequate weight-bearing exercise and inadequate intake of vitamin D and calcium lead to a loss of bone mass, increasing the risk of bone breakage.

Dementia: High-fat diets and obesity may increase the risk of dementia. Plant-based diets high in a variety of phytochemicals appear to reduce the risk. Adequate intake of vitamins B_6 and B_{12} and folate has been found to reduce age-related cognitive decline in some studies.

Impaired immune function: Aging causes a natural decline in immune function that may be exacerbated by protein, zinc, or vitamin D deficiencies; increased susceptibility to foodborne illnesses.

Atrophic gastritis: Caused by an autoimmune disorder that destroys cells in the stomach that produce gastric acid and intrinsic factor needed for absorption of vitamin B_{12}. Adequate gastric acid is also necessary for the efficient absorption of several minerals.

Diverticulosis: May affect 50% of adults older than 60. Develops when tiny pockets are formed when the lining of the colon protrudes outward through weak spots in the colon wall.

Which of these conditions may be prevented or delayed by proper nutrition earlier in life?

Photo credit: StA-gur Karlsson/Getty Images

Joining a friend for lunch. *Seniors who regularly eat with a companion may avoid the pitfalls of social isolation and more fully enjoy food.*

MBI/Alamy Stock Photo

development of diverticular disease, which affects approximately 50% of people by the age of 65 and higher incidence in older age groups. Changes in the digestive system can also reduce the production of stomach acid, which plays an important defensive role against the aging adult's increased risk of foodborne illness (for more on foodborne illness see Chapter 20).

Because of all these things, the elderly are particularly vulnerable to malnutrition. And that, in turn, can compromise their ability to live long and prosper. It has been reported that about 60% of hospitalized adults 65 years or older and up to 85% of residents of nursing homes are malnourished. Elderly living in the community fare better, but many are still at risk for malnourishment, with slightly fewer than 40% being malnourished or at risk of malnourishment.

Depression and nutritional status

In addition to the varied physiological factors that affect the health and nutritional status of aging adults, psychosocial factors have an impact on their ability and motivation to obtain adequate and appropriate nourishment. Loss, loneliness, and lack of social support may result in increased risk of depression and suboptimal intake when eating alone. Research (and common sense) shows that individuals who are psychologically healthy, resilient, and have a sense of purpose in their lives are more likely to age successfully and experience better quality of life and overall health. This echoes the characteristics found in the Blue Zone populations.

NUTRIENT RECOMMENDATIONS FOR SENIORS

The physiological changes with increasing age can alter how the body absorbs and utilizes nutrients, thus the Daily Reference Intake values (DRIs) for certain vitamins and minerals are different for older adults. In particular, the requirements for calcium and vitamin D—both important for the prevention of osteoporosis (Chapter 13)—are higher. Calcium absorption is reduced in older adults, and vitamin D synthesis can be compromised because of reduced exposure to sunlight. The RDA for calcium is increased from 1,000 mg to 1,200 mg in women 51 and older and in men older than 70. The RDA for vitamin D increases from 600 IUs to 800 IUs in adults (men and women) over the age of 70. There is also a higher requirement for vitamin B_6 with age. A deficiency of this vitamin can lead to cognitive impairment while supplements have been shown to reduce the occurrence of late-life depression.

Protein intake is of particular concern, as one-third of older adults are not meeting the Recommended Dietary Allowance for protein. Furthermore, a number of studies demonstrate that slightly higher protein intakes (approximately 1.0 to 1.5 g/kg/d) in adults older than 65 years can effectively reduce the loss of lean body mass that occurs with age. This may improve functionality, and reduce the risk of disability and death, particularly when combined with resistance exercise. Several studies have also shown that consuming 25 to 30 grams of protein at each meal slows age-associated loss of muscle mass and improves gains in muscle mass in response to resistance training in older individuals.

Fluid recommendations are the same for older adults, but because older adults may experience reduced thirst sensation and increased fluid loss through diminished kidney function and as a side effect from certain medications, they may have a harder time meeting these requirements. With an increased risk of dehydration, food guides for aging adults have added emphasis on sufficient fluid intake. (INFOGRAPHIC 21.6)

Older individuals can benefit from plant-based and animal-based proteins such as these (shown: shrimp, tofu, kidney beans, hummus, chicken, and eggs). Intakes of 1 to 1.5 g/kg of bodyweight per day may reduce the loss of lean body mass in individuals over age 65.

the food passionates/Corbis

Resistance training builds and maintains muscle mass. *This 100-year-old woman lifts weights each morning and lives in an Adventist community in Loma Linda, California.*

David McLain/Getty Images

INFOGRAPHIC 21.6 Healthy Eating for Older Adults *Making wise food choices can help prevent aging-related nutritional problems.*

Protein Foods

Increasing protein intake slightly, to 1.0–1.5 g/kg body weight, may reduce the loss of lean body mass in older adults.

Fluids

As we age, our kidneys have a reduced capacity to conserve water and the sensation of thirst diminishes. For this reason it may be wise for the elderly to consume fluids when not thirsty.

Calcium

Decreased hydrochloric acid production decreases calcium absorption. To promote bone health the RDA for calcium increases from 1,000 mg to 1,200 mg/day at age 51 for women and age 71 for men.

Vitamin D

Because individuals often get outside less often as they age, and the conversion of cholesterol to vitamin D is less efficient, the RDA for vitamin D increases 33% for both men and women at age 71.

Vitamin B_6

The RDA for vitamin B_6 increases at age 51. Low vitamin B_6 status in the elderly is associated with and increased risk of cognitive decline and depression.

Vitamin B_{12}

Reduced production of gastric acid is common in the elderly, which decreases the absorption of naturally occurring (vitamin B_{12}). It is recommended that adults older than 60 years meet their needs for vitamin B_{12} through fortified foods or supplements, which do not depend on gastric acid for absorption.

Iron

The low gastric acid production that is common in the elderly reduces the absorption of non-heme iron, increasing the risk of anemia. However, high iron stores are more common in the elderly than is iron deficiency. For this reason iron supplements are generally not recommended unless an individual has been diagnosed with iron deficiency.

Zinc

Elderly individuals are often found to have moderate zinc deficiency from low intake and decreased absorption due to low gastric hydrochloric acid production. Low zinc status is correlated with impaired immune and cognitive function.

What four nutrients share at least one food that is a good source of all four of them? And what is that food?

Photo credits (top left to bottom right): kpatyhka/Shutterstock, 4kodiak/Getty Images, Evlakhov Valeriy/Shutterstock, koosen/Shutterstock, Maria Toutoudaki/Getty Images, Everything/Shutterstock, Paul Johnson/Getty Images, Datacraft Co Ltd/Getty Images, indigolotos/Shutterstock, Eli Ensor, t_kimura/Getty Images, Eli Ensor, Jacek Chabraszewski/Shutterstock, Foodcollection RF/Getty Images, Dulce Rubia/Shutterstock, ifong/Shutterstock, shutterdandan/Shutterstock, John Scott/Getty Images, Kaan Ates/Getty Images, Kaan Ates/Getty Images, Nicholas Eveleigh/Exactostock/Superstock, AVprophoto/Shutterstock

In addition to those nutrients that are specifically altered in the DRIs, there are several other nutrients that may provide benefits in some cases. These include zinc, which plays a supportive role in immunity, and omega-3 fatty acids, which may reduce symptoms of rheumatoid arthritis, slow the progression of age-related macular degeneration, and reduce the risk of Alzheimer's disease.

As life expectancy increases, age-related declines in cognitive function are projected to quadruple (approximately) the prevalence of dementia worldwide by the year 2050. While increasing age is the strongest predictor of cognitive decline, various lifestyle factors—such as physical activity and diet—also play an important role. Smoking, obesity, and diets high in saturated fat appear to increase the risk of cognitive decline and dementia, while regular physical activity and plant-based diets with ample intakes of vegetables, fruits, nuts, and whole grains are associated with decreased risk. Regular consumption of coffee, black and green teas, cocoa, and wine are also linked to a reduced risk of dementia. Diets that are rich in these foods not only provide abundant amounts of the antioxidant vitamins (C and E), they also provide a rich array of phytochemicals that are believed to be protective from dementia. It is currently thought that most phytochemicals do not exert their protective effects through their often-promoted antioxidant functions, but they likely directly affect a variety of cellular processes that help preserve cognitive function. Lastly, several studies have observed that regular fish consumption is also associated with a reduced risk of dementia, perhaps due to increased intakes of omega-3 fatty acids. It is important to emphasize that these dietary factors are protective when they are part of an overall diet pattern, as very few studies have found dietary supplements to be effective at preserving cognitive function or slowing the progression of dementia.

On the basis of current research, the best diet to delay age-related disease onset is one that provides sufficient, but not excessive calories; is low in saturated fat and high in whole-grain cereals, legumes, fruits and vegetables; and maintains a lean body weight. In consideration of the unique nutritional and physical activity needs associated with advancing years, there is a MyPlate for Older Adults, developed by nutrition scientists at the Jean Mayer USDA Human Nutrition Research Center on Aging, to complement the federal government's MyPlate for adults. (INFOGRAPHIC 21.7)

STUDYING THE ADVENTIST LIFESTYLE FOR CLUES TO LONGEVITY

Though it makes intuitive sense that cutting out known risk factors for disease, like smoking and drinking, would improve health (and it is fairly easy to show this empirically), dissecting the nutritional contributions to health and longevity is a bit more difficult. It requires large, carefully planned studies that are statistically powered to detect subtle effects of dietary variables.

The first study to look at the nutrition of Adventists was the *Adventist Health Study,* which began in 1976. This study analyzed the health-related behaviors of 34,192 Adventist men and women 30 years and older. (The individuals in this study were all non-Hispanic whites, although later studies have included Hispanics and African Americans as well.) Individuals filled out a detailed questionnaire and the researchers made follow-up contact until 1988. The Adventists' dietary choices were assessed by asking about how often they consumed 55 different foods or food groups. Exercise levels were assessed and scored with a similar battery of questions. Study participants were contacted each year to ascertain any hospitalizations and other health information, such as heart attacks or cancers. Height and weight measures provided by participants were used to calculate body mass index (BMI).

From this information, the study investigators then analyzed the relationship between individuals' behaviors and their risk of disease and death. The result? Certain behaviors practiced by a subset of Adventists were associated with clear benefits to health and longevity. Being physically active, eating lots of nuts, being vegetarian, and having a healthy BMI each provided approximately 1.5 to 2.5 years of extra expectancy. The boost to longevity was additive, with longevity increasing to a greater extent as Adventists

INFOGRAPHIC 21.7 MyPlate for Older Adults *The energy needs of older adults generally decline with age because of decreased metabolism and physical activity. However, requirements for vitamins and minerals remain the same, or in some cases even increase. To meet these nutrient needs it is important for older adults to choose nutrient-dense foods that provide high levels of vitamins and minerals per serving.*

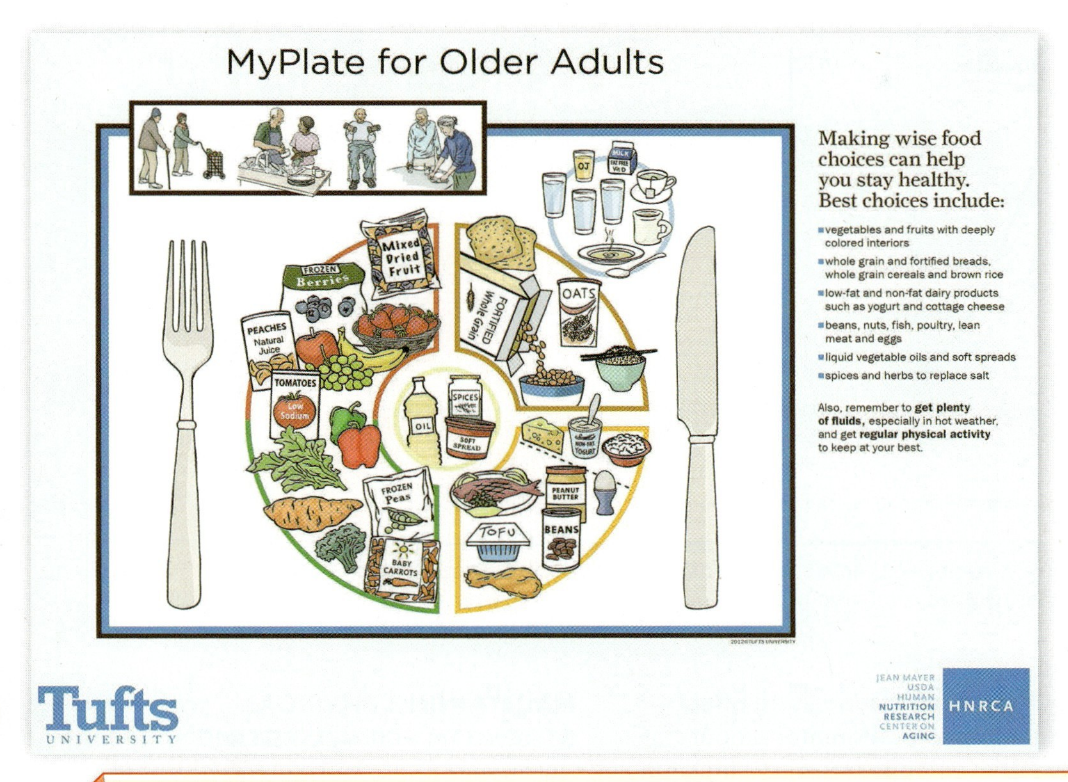

Problems caused by teeth, gums, or dentures can make it hard for some individuals to chew. Softer foods such as cooked vegetables, canned fruits, cooked beans, eggs, and tofu may be easier to consume.

Older adults may lose interest in eating because of changes in their senses of taste and smell. What ways can you think of to overcome these eating challenges?

Photo credit: Copyright 2011 Tufts University. For details about the MyPlate for Older Adults please see http://nutrition.tufts.edu/research/myplate-older-adults.

practiced a greater number of these behaviors. The average life expectancy (at age 30) of Adventist men and women was 81.2 and 83.9 years, respectively. This corresponds to an extra 7.3 years of life for Adventist men and an extra 4.4 years for Adventist women, when compared with other Californians.

When just the Adventist vegetarians were used as the basis of comparison, the difference was even more striking: Adventist vegetarian men and women had a life expectancy that was 9.5 and 6.1 years, respectively, longer than their fellow Californians.

The authors of the study, published in 2001, conclude, "These results strongly suggest that behavioral choices influence the expected age at death by several years, even as much as a decade." (INFOGRAPHIC 21.8)

Why men experience more of a benefit is not entirely clear. There is some evidence that men, in general, tend to make worse dietary choices than women, so their nutritional deficits might be more easily corrected, explains Fraser.

The Adventist studies (there are now several) have unique advantages, not least of which is the fact that the study population is relatively homogeneous with regard to smoking and drinking (they abstain). This allows study investigators to isolate the

INFOGRAPHIC 21.8 Survival of Men and Women Adventists Compared with Californians *Both male and female Seventh-day Adventists have longer life expectancies than the average Californian man and woman, possibly due to lifestyle choices such as abstaining from eating meat, smoking, and drinking.*

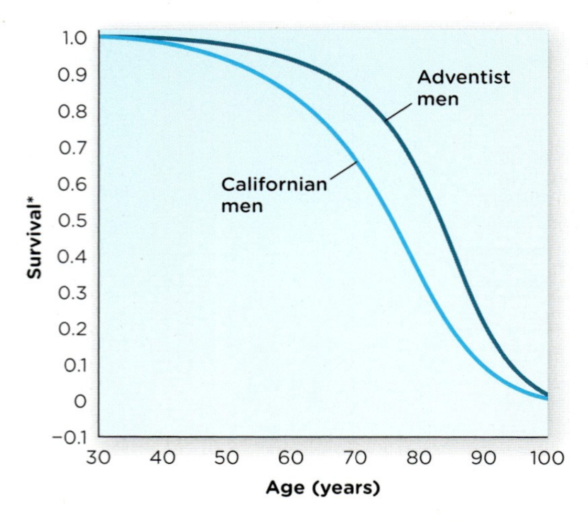

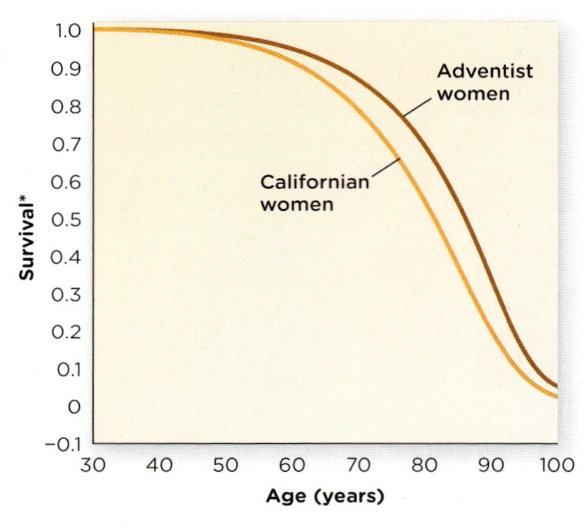

*Survival is the probability of living another year.

How much more likely to survive is a 70-year-old Adventist man than a 70-year-old Californian man? What about a 70-year-old Adventist woman compared with a 70-year-old Californian woman?

nutritional variables that seem to affect health outcomes with a greater amount of certainty. "These people are very similar in other respects, except for the way that they eat," says Fraser.

Exercise benefits all age groups. *There are many ways to enjoy physical activity at any age. These women participate in water aerobics class in a Seventh-day Adventist Community in Loma Linda, California.*

Peter Bohler/Redux

MAINTAINING PHYSICAL STRENGTH FOR A LIFETIME

Nutrition is one important ingredient for a long and healthy life, but it's not the only one. Equally important is staying active. This doesn't have to be through deliberate exercise at a gym. Typical among people in the Blue Zones, including the Adventists, is a consistent pattern of remaining active well into later life through natural means. Whether through gardening or walking or swimming, older residents of Blue Zones maintain an active lifestyle that helps forestall the loss of muscle mass. And this turns out to be a crucial matter of health, for several reasons.

Adequate muscle mass is important for mounting an adequate response to various stresses and reducing the risk of several chronic diseases. Muscle serves as a reservoir of amino acids that can be used during periods of physiological stress such as illness, injury, or surgery to synthesize antibodies as part of an immune response. Adequate muscle mass is also essential for successful recovery from these conditions.

Maintaining adequate skeletal muscle mass (and preventing sarcopenia) also reduces the

INFOGRAPHIC 21.9 The Role of Skeletal Muscle in Health, Illness, and Injury

Adequate skeletal muscle mass is important for recovery from illness and injury, and to reduce the risk of some chronic diseases.

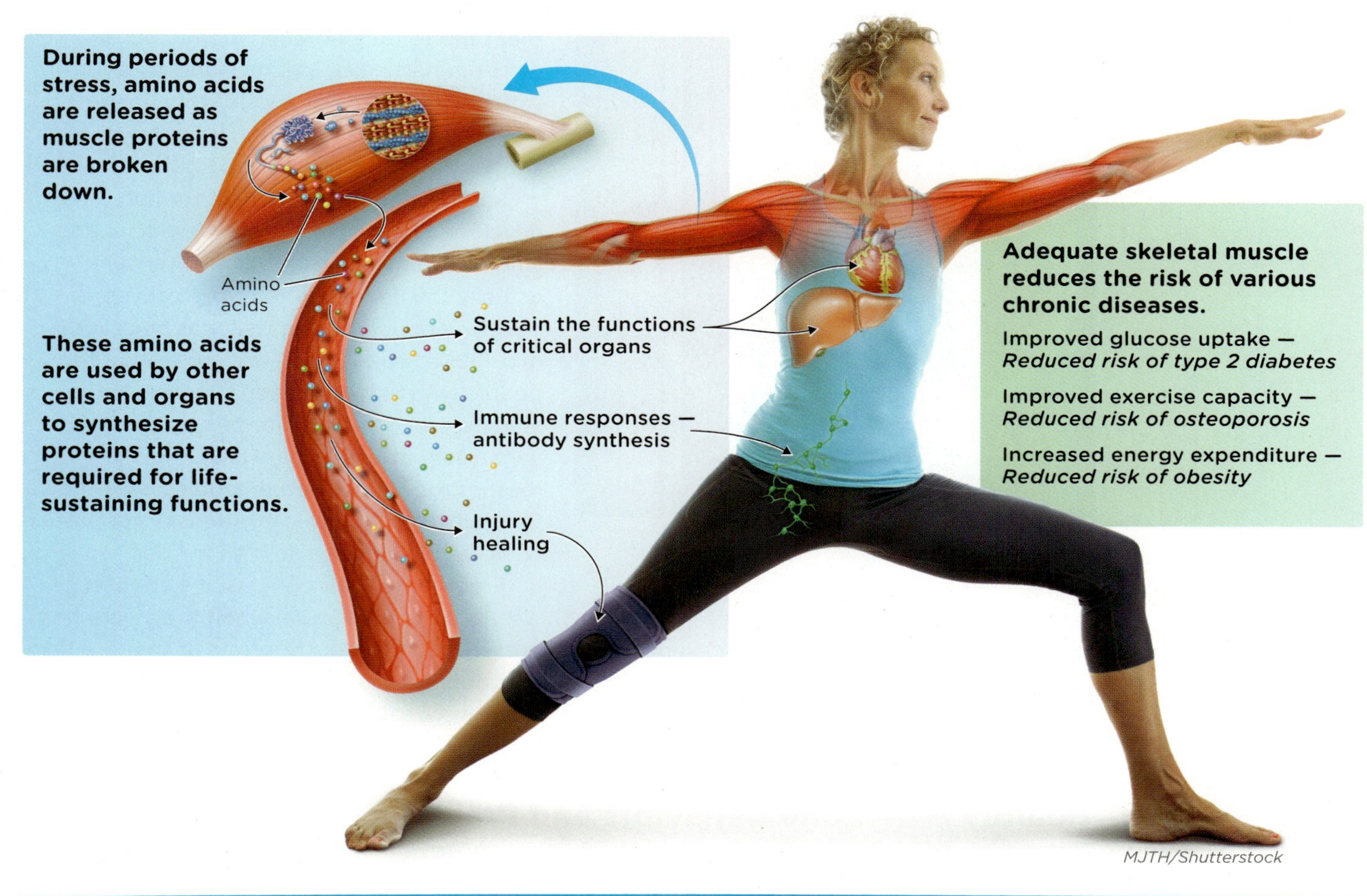

Why does protein malnutrition increase the risk of death following major surgery?

risk of developing type 2 diabetes, since muscle is the primary site of insulin-stimulated glucose uptake from blood and the major means of clearing excess glucose from the blood. Adequate skeletal muscle mass and strength allows for continued physical activity, which is required to maintain bone density and reduce the risk of osteoporosis. Finally, studies continue to confirm that maintenance of muscle mass and strength not only helps prevent falls and injury, but can help keep aging adults independent and promote longer, healthier lives. **(INFOGRAPHIC 21.9)**

There is some evidence that physical activity not only makes a person feel better—boosting mental and physical fitness—but can also forestall some of the physiological effects of aging. Older adults who exercise have bodies that are in some senses physiologically younger than those who don't.

BLUE ZONE SECRETS

Eating nutritious foods and exercising seem obvious, but there are some other features of Blue Zones that seem to contribute to longer life. Most important of these are psychosocial variables like social connectedness and having a "plan" or a "purpose." Residents of Blue Zones are distinctive in the way they value their elders, making them important members of the community. As journalist Buettner said in his Ted Talk, with regard to Sardinia, "The older you get, the more equity you have, the more wisdom you're celebrated for." Older

A meal rich with lentils and nuts. *Seniors can benefit from a diet that contains more protein than the RDA of 0.8 grams per kilogram of bodyweight per day.*

people in Blue Zones also tend to find a clear reason to get up in the morning, even after they stop working. In Okinawa, for example, there is no word for "retirement." Having a plan or purpose is not just good for mental health; it's actually an independent contributor to longevity. "The dietary factors and physical activity and BMI seem to operate independently of the psychosocial factors," Fraser says.

There are also particular classes of foods that tend to be embraced in Blue Zone diets—in particular, legumes and nuts. The Japanese tend to eat a lot of tofu, made from soybeans (a legume), while Adventists eat a lot of both legumes and nuts.

Interestingly, this penchant for nuts was developed long before there were scientific studies supporting their value. "We were actually the first study to publish anything positive about nuts and that was back in the mid-1990s," says Fraser, who notes that the people with higher nut consumption in their study seemed to be protected against coronary artery disease. They don't know all the reasons for that, but other studies have shown that nuts tend to lower low-density lipoprotein cholesterol. They're also high in fiber.

Fraser says their conclusions about nuts were a bit controversial when they were first presented. "Most nutritionists at the time felt that nuts were a kind of snack food and they often were salted and they were fatty and so forth and they were not recommending them

INFOGRAPHIC **21.10** **Healthy Blue Zone Behaviors and Practices** *People who live in Blue Zones around the world share a few common attributes that scientists believe may contribute to their increased longevity. Adopting these habits could extend the life expectancy of the rest of the world.*

? *Which behaviors or practices of residents of Blue Zones are common in your own community? Which are uncommon?*

for a good healthy diet." But Fraser had done a fellowship at the University of Minnesota where some of the seminal work was done on the relationship of different fatty acids to blood cholesterol and heart disease, so he had reason to suspect that nuts would be healthful because they are low in saturated fat and often high in monounsaturated fats. "We were a bit surprised to see the power that nuts had," says Fraser. "Day-to-day, they decrease the risk of heart attacks by about 50 percent." A quarter of Adventists eat nuts five or more times a week. (INFOGRAPHIC 21.10)

Beyond Centenarians

Many Blue Zones have become famous for the numbers of centenarians (people at least 100 years of age) in the population. Although that is one measure of longevity, it's not necessarily the most important one for determining the overall health of a population. These very long-lived individuals represent a very small fraction of the total population, and they could conceivably be individuals with a genetic propensity that conveys longer life. The Adventist study is different because it looked at the benefit of healthful choices on life expectancy of the population as a whole.

"The fact that Adventist men live, on average, seven or so years longer than their Californian counterparts does inevitably mean that we're going to have a higher proportion of centenarians, too," explains Fraser. "But we believe it's more meaningful to not focus on just one small extreme group but to talk about whether the life expectancy of the whole population is being moved." The Adventists get a boost to life expectancy across the lifespan. (INFOGRAPHIC 21.11)

INFOGRAPHIC 21.11 Expected Length of Life (Californian Adventists Compared with International Populations) *Californian Adventists have a life expectancy greater than that of the populations of most nations, with only a select few (such as Iceland and Japan) that approach similar numbers. Californian Adventists who follow a vegetarian diet have even longer life expectancies.*

	Length of Life (Years)			
	Men		Women	
Country (Year)	At Birth	At Age 65	At Birth	At Age 65
Australia (1990)	73.9	15.2	80.0	19.0
Canada (1985–1987)	73.0	14.9	79.7	19.1
Denmark (1989–1990)	72.0	14.1	77.7	17.9
Finland (1989)	70.9	13.8	78.9	17.7
Iceland (1989–1990)	75.7	16.1	80.3	19.3
Japan (1990)	75.9	16.2	81.3	19.9
New Zealand (1987–1989)	71.6	14.1	77.6	17.8
Norway (1990)	73.4	14.6	79.8	18.6
United Kingdom (1985–1987)	71.9	13.4	77.6	17.3
United States (1990)	73.0	14.9	79.7	19.1
Californian Adventists (1980–1988)*	78.5	19.1	82.3	21.6
Vegetarians	80.2	20.3	84.8	22.6

*Hazards for those aged 0 to 29 years are those from Californian State data, as data for these ages are not available for Adventists. Non-Adventist data are taken from international longevity comparisons (1992).

To what age was a Californian Adventist woman who was 65 years old in 1988 expected to live?

What are the main lessons of the Adventist study for individuals wanting to increase their chances of living a long, full life? "I think our results provide pretty strong evidence that a plant-based diet has quite a lot of advantages, whether you're looking at blood pressure or diabetes or life expectancy or coronary heart disease or certain cancers," says Fraser, who is himself a pescatarian (a vegetarian who eats fish).

Fraser doesn't expect the non-Adventist population to become vegetarian any time soon, but he does think it's reasonable to suggest that people cut back on meat consumption. "You can start to talk about things such as meatless days or emphasizing certain vegetable food groups," he notes.

Largely because of their nutritional habits, the Adventists are some of the healthiest and longest-lived people on the planet. Their example shows, pretty conclusively, that making well-informed choices regarding nutrition can have large effects on health. In that sense, their dietary and exercise habits can serve as a model for other communities. If living long and prospering is one of your goals, then you might want to consider changing your lifestyle to more closely mirror that of the Adventists. To quote Mr. Spock, "It's only logical."

CHAPTER 21 **BRING IT HOME**

Assisted living case study

Eleanor is an 84-year-old white female residing in an assisted living facility in North Dakota. She eats two meals each day in the dining hall and has a small kitchen with a refrigerator and microwave in her residence. She is able to place an order once a week for the delivery of foods and beverages from a local supermarket, but has limited finances and limited space for food storage. Eleanor's recent visit to her primary care provider revealed mild dehydration and a weight loss of several pounds since her visit one year ago (Eleanor's BMI is 20 kg/m^2). Eleanor had previously been given a diagnosis of osteoporosis at age 78 with a recent bone scan showing progressive bone loss.

1. Identify at least five nutrition-related concerns from Eleanor's brief profile.
2. What possible challenges may be compromising Eleanor's ability to meet nutritional and fluid requirements?
3. What are some of Eleanor's risk factors for osteoporosis that may be contributing to bone loss?
4. Based on what you've learned about nutrition-related concerns and recommendations in the aging adult, list 3 to 5 specific food or dietary strategies for Eleanor.

Take It Further

The role of diet in healthy aging is of interest to the growing numbers of seniors and is the focus of many ongoing research projects. Visit PubMed at http://www.ncbi.nlm.nih.gov/pubmed. Using the search terms "nutrition" and "aging," browse through recent articles and studies. Choose one article that addresses a nutrient, dietary component, or nutritional consideration with regard to aging. Read the abstract and briefly describe study findings or implications.

KEY IDEAS

Nutrition and physical activity play key roles in maintaining health and preventing chronic disease, as well as in potential longevity and quality of life, in the aging adult.

There are physical, psychological, social, and economic factors that can influence and potentially compromise the dietary intake and nutritional status of the older adult.

Energy needs in older adults generally decline because of decreased physical activity and metabolic rate; however, nutrient needs do not decrease (they may even increase), making it especially important to consume nutrient-dense foods.

Age-related changes in body composition and body fat distribution occur to varying degrees, but physical activity helps offset losses in skeletal muscle mass (sarcopenia).

Age-related gastrointestinal changes may occur, including the bacterial composition of the gut, diminished gastric acid secretion, and slower motility in the small intestine.

Nutrients of concern in older adults due to diminished intake or possible changes in absorption and utilization include protein, calcium, vitamin D, vitamin B_6, vitamin B_{12}, folate, iron, zinc, fiber, and omega-3 fatty acids.

Although fluid requirements are the same for older adults, the risk of dehydration is higher than in younger adults due to diminished thirst sensation, increased fluid loss, and side effects of certain medications.

Poor nutrition and physiological changes with advanced age, along with possible depressed immunity and a decreased ability to fight and recover from illness, surgery, or infection, can lead to suboptimal nutritional status or even malnutrition.

Nutrient inadequacies may contribute to cognitive decline with age. In addition, some studies show that certain dietary nutrients, like omega-3 fatty acids and some phytochemicals, may reduce the risk of dementia.

Based on current research, the best diet to delay age-related disease onset avoids excess calorie intake and is low in saturated fat and high in whole grains, legumes, nuts, fruits, and vegetables.

NEED TO KNOW

Review Questions

1. All of the following are TRUE regarding life expectancy, EXCEPT that it:
   a. is based on a statistical average calculated on a population.
   b. represents the number of years a person is expected to live.
   c. represents the actual number of years an individual lives.
   d. is a calculation that includes those who die during or shortly after birth.

2. The DRI for _____ is higher for older adults than for their younger counterparts.
   a. calories
   b. iron
   c. vitamin A
   d. vitamin B_6
   e. vitamin C

3. The DRI for calcium:
   a. decreases slightly with age with reduced physical activity and reduced lean mass.
   b. increases from 1,000 mg to 1,200 mg for adults 51 years and older.
   c. increases from 1,000 mg to 1,200 mg in females only after age 70.
   d. remains the same after age 30 when peak bone mass is achieved.

4. Why might older adults be prone to vitamin D deficiency?
   a. a suboptimal intake of vitamin D
   b. reduced exposure to sunlight and ultraviolet light
   c. diminished production of vitamin D in the skin
   d. increased vitamin D dietary requirements
   e. all of the above

5. Sarcopenia can be defined as:
   a. small, abnormally shaped red blood cells.
   b. the condition of porous and brittle bones that occurs with age.
   c. the age-related reduction in skeletal muscle mass.
   d. elevated levels of disaccharides in the blood.

6. The elderly at highest risk of malnutrition appear to be those:
   a. living at home.
   b. living with family members.
   c. in acute-care facilities (hospitals).
   d. in long-term care facilities.

7. Gastric acid production and secretion decreases with age, which potentially results in all of the following, EXCEPT:
   a. the digestion and absorption of certain nutrients
   b. diminished growth of intestinal bacteria.
   c. malabsorption of naturally occurring vitamin B_{12} from food.
   d. increased risk of foodborne illness.

8. With regard to protein intake in the older adult, recent studies demonstrate:
   a. the DRI for protein increases by 50% after age 70.
   b. the benefits of protein intake are best with lean animal foods.
   c. there is no evidence of any benefit of protein intakes above the DRI of 0.8 g/kg body weight.
   d. that slightly increased protein intake can reduce loss of lean body mass.

9. All of the following are TRUE with regard to fluid status in the older adult, EXCEPT:
   a. diminished sensation of thirst.
   b. diminished kidney function with increased fluid loss.
   c. increased DRI for water after age 70.
   d. increased risk for dehydration.
   e. the use of certain medications can increase fluid loss.

10. Studies demonstrate that the best diet to delay age-related chronic disease includes which of the following characteristics?
    a. avoids excess calories
    b. high in vegetables
    c. high in legumes
    d. low in saturated fat
    e. all of the above

Appendix 1

Dietary Reference Intakes (DRIs): Estimated Average Requirements Food and Nutrition Board, Institute of Medicine, National Academies										
Life Stage Group	Calcium (mg/d)	CHO (g/d)	Protein (g/kg/d)	Vitamin A (μg/d)[a]	Vitamin C (mg/d)	Vitamin D (μg/d)	Vitamin E (mg/d)[b]	Thiamin (mg/d)	Riboflavin (mg/d)	Niacin (mg/d)
Infants										
0 to 6 mo										
6 to 12 mo			1.0							
Children										
1–3 y	500	100	0.87	210	13	10	5	0.4	0.4	5
4–8 y	800	100	0.76	275	22	10	6	0.5	0.5	6
Males										
9–13 y	1,100	100	0.76	445	39	10	9	0.7	0.8	9
14–18 y	1,100	100	0.73	630	63	10	12	1.0	1.1	12
19–30 y	800	100	0.66	625	75	10	12	1.0	1.1	12
31–50 y	800	100	0.66	625	75	10	12	1.0	1.1	12
51–70 y	800	100	0.66	625	75	10	12	1.0	1.1	12
> 70 y	1,000	100	0.66	625	75	10	12	1.0	1.1	12
Females										
9–13 y	1,100	100	0.76	420	39	10	9	0.7	0.8	9
14–18 y	1,100	100	0.71	485	56	10	12	0.9	0.9	11
19–30 y	800	100	0.66	500	60	10	12	0.9	0.9	11
31–50 y	800	100	0.66	500	60	10	12	0.9	0.9	11
51–70 y	1,000	100	0.66	500	60	10	12	0.9	0.9	11
> 70 y	1,000	100	0.66	500	60	10	12	0.9	0.9	11
Pregnancy										
14–18 y	1,000	135	0.88	530	66	10	12	1.2	1.2	14
19–30 y	800	135	0.88	550	70	10	12	1.2	1.2	14
31–50 y	800	135	0.88	550	70	10	12	1.2	1.2	14
Lactation										
14–18 y	1,000	160	1.05	885	96	10	16	1.2	1.3	13
19–30 y	800	160	1.05	900	100	10	16	1.2	1.3	13
31–50 y	800	160	1.05	900	100	10	16	1.2	1.3	13

NOTE: An Estimated Average Requirement (EAR) is the average daily nutrient intake level estimated to meet the requirements of half of the healthy individuals in a group. EARs have not been established for vitamin K, pantothenic acid, biotin, choline, chromium, fluoride, manganese, or other nutrients not yet evaluated via the DRI process.

[a]As retinol activity equivalents (RAEs). 1 RAE = 1 μg retinol, 12 μg β-carotene, 24 μg α-carotene, or 24 μg β-cryptoxanthin. The RAE for dietary provitamin A carotenoids is two-fold greater than retinol equivalents (RE), whereas the RAE for preformed vitamin A is the same as RE.

[b]As α-tocopherol. α-Tocopherol includes *RRR*-α-tocopherol, the only form of α-tocopherol that occurs naturally in foods, and the *2R*-stereoisomeric forms of α-tocopherol (*RRR*-, *RSR*-, *RRS*-, and *RSS*-α-tocopherol) that occur in fortified foods and supplements. It does not include the *2S*-stereoisomeric forms of α-tocopherol (*SRR*-, *SSR*-, *SRS*-, and *SSS*-α-tocopherol), also found in fortified foods and supplements.

itamin B_6 (mg/d)	Folate (μg/d)[d]	Vitamin B_{12} (μg/d)	Copper (μg/d)	Iodine (μg/d)	Iron (mg/d)	Magnesium (mg/d)	Molybdenum (μg/d)	Phosphorus (mg/d)	Selenium (μg/d)	Zinc (mg/d)
					6.9					2.5
0.4	120	0.7	260	65	3.0	65	13	380	17	2.5
0.5	160	1.0	340	65	4.1	110	17	405	23	4.0
0.8	250	1.5	540	73	5.9	200	26	1,055	35	7.0
1.1	330	2.0	685	95	7.7	340	33	1,055	45	8.5
1.1	320	2.0	700	95	6	330	34	580	45	9.4
1.1	320	2.0	700	95	6	350	34	580	45	9.4
1.4	320	2.0	700	95	6	350	34	580	45	9.4
1.4	320	2.0	700	95	6	350	34	580	45	9.4
0.8	250	1.5	540	73	5.7	200	26	1,055	35	7.0
1.0	330	2.0	685	95	7.9	300	33	1,055	45	7.3
1.1	320	2.0	700	95	8.1	255	34	580	45	6.8
1.1	320	2.0	700	95	8.1	265	34	580	45	6.8
1.3	320	2.0	700	95	5	265	34	580	45	6.8
1.3	320	2.0	700	95	5	265	34	580	45	6.8
1.6	520	2.2	785	160	23	335	40	1,055	49	10.5
1.6	520	2.2	800	160	22	290	40	580	49	9.5
1.6	520	2.2	800	160	22	300	40	580	49	9.5
1.7	450	2.4	985	209	7	300	35	1,055	59	10.9
1.7	450	2.4	1,000	209	6.5	255	36	580	59	10.4
1.7	450	2.4	1,000	209	6.5	265	36	580	59	10.4

[c]As niacin equivalents (NE). 1 mg of niacin = 60 mg of tryptophan.

[d]As dietary folate equivalents (DFE). 1 DFE = 1 μg food folate = 0.6 μg of folic acid from fortified food or as a supplement consumed with food = 0.5 μg of a supplement taken on an empty stomach.

SOURCES: *Dietary Reference Intakes for Calcium, Phosphorous, Magnesium, Vitamin D, and Fluoride* (1997); *Dietary Reference Intakes for Thiamin, Riboflavin, Niacin, Vitamin B_6, Folate, Vitamin B_{12}, Pantothenic Acid, Biotin, and Choline* (1998); *Dietary Reference Intakes for Vitamin C, Vitamin E, Selenium, and Carotenoids* (2000); *Dietary Reference Intakes for Vitamin A, Vitamin K, Arsenic, Boron, Chromium, Copper, Iodine, Iron, Manganese, Molybdenum, Nickel, Silicon, Vanadium, and Zinc* (2001); *Dietary Reference Intakes for Energy, Carbohydrate, Fiber, Fat, Fatty Acids, Cholesterol, Protein, and Amino Acids* (2002/2005); and *Dietary Reference Intakes for Calcium and Vitamin D* (2011). These reports may be accessed via www.nap.edu.

Dietary Reference Intakes (DRIs): Recommended Dietary Allowances and Adequate Intakes, Vitamin
Food and Nutrition Board, Institute of Medicine, National Academies

Life Stage Group	Vitamin A (μg/d)[a]	Vitamin C (mg/d)	Vitamin D (μg/d)[b,c]	Vitamin E (mg/d)[d]	Vitamin K (μg/d)	Thiamin (mg/d)	Riboflavi (mg/d)
Infants							
0 to 6 mo	400*	40*	10	4*	2.0*	0.2*	0.3*
6 to 12 mo	500*	50*	10	5*	2.5*	0.3*	0.4*
Children							
1–3 y	**300**	**15**	**15**	**6**	30*	**0.5**	**0.5**
4–8 y	**400**	**25**	**15**	**7**	55*	**0.6**	**0.6**
Males							
9–13 y	**600**	**45**	**15**	**11**	60*	**0.9**	**0.9**
14–18 y	**900**	**75**	**15**	**15**	75*	**1.2**	**1.3**
19–30 y	**900**	**90**	**15**	**15**	120*	**1.2**	**1.3**
31-50 y	**900**	**90**	**15**	**15**	120*	**1.2**	**1.3**
51–70 y	**900**	**90**	**15**	**15**	120*	**1.2**	**1.3**
> 70 y	**900**	**90**	**20**	**15**	120*	**1.2**	**1.3**
Females							
9–13 y	**600**	**45**	**15**	**11**	60*	**0.9**	**0.9**
14–18 y	**700**	**65**	**15**	**15**	75*	**1.0**	**1.0**
19–30 y	**700**	**75**	**15**	**15**	90*	**1.1**	**1.1**
31–50 y	**700**	**75**	**15**	**15**	90*	**1.1**	**1.1**
51–70 y	**700**	**75**	**15**	**15**	90*	**1.1**	**1.1**
> 70 y	**700**	**75**	**20**	**15**	90*	**1.1**	**1.1**
Pregnancy							
14–18 y	**750**	**80**	**15**	**15**	75*	**1.4**	**1.4**
19–30 y	**770**	**85**	**15**	**15**	90*	**1.4**	**1.4**
31–50 y	**770**	**85**	**15**	**15**	90*	**1.4**	**1.4**
Lactation							
14–18 y	**1,200**	**115**	**15**	**19**	75*	**1.4**	**1.6**
19–30 y	**1,300**	**120**	**15**	**19**	90*	**1.4**	**1.6**
31–50 y	**1,300**	**120**	**15**	**19**	90*	**1.4**	**1.6**

NOTE: This table (taken from the DRI reports, see www.nap.edu) presents Recommended Dietary Allowances (RDAs) in **bold type** and Adeq Intakes (AIs) in ordinary type followed by an asterisk (*). An RDA is the average daily dietary intake level sufficient to meet the nutrient requir ments of nearly all (97–98 percent) healthy individuals in a group. It is calculated from an Estimated Average Requirement (EAR). If sufficient entific evidence is not available to establish an EAR, and thus calculate an RDA, an AI is usually developed. For healthy breastfed infants, an A the mean intake. The AI for other life-stage and gender groups is believed to cover the needs of all healthy individuals in the groups, but lack data or uncertainty in the data prevent being able to specify with confidence the percentage of individuals covered by this intake.

[a]As retinol activity equivalents (RAEs). 1 RAE = 1 μg retinol, 12 μg β-carotene, 24 μg α-carotene, or 24 μg β-cryptoxanthin. The RAE for die provitamin A carotenoids is two-fold greater than retinol equivalents (RE), whereas the RAE for preformed vitamin A is the same as RE.

[b]As cholecalciferol. 1 μg cholecalciferol = 40 IU vitamin D.

[c]Under the assumption of minimal sunlight.

[d]As α-tocopherol. α-Tocopherol includes *RRR*-α-tocopherol, the only form of α-tocopherol that occurs naturally in foods, and the *2R*-stere isomeric forms of α-tocopherol (*RRR*-, *RSR*-, *RRS*-, and *RSS*-α-tocopherol) that occur in fortified foods and supplements. It does not inclu the *2S*-stereoisomeric forms of α-tocopherol (*SRR*-, *SSR*-, *SRS*-, and *SSS*-α-tocopherol), also found in fortified foods and supplements.

[e]As niacin equivalents (NE). 1 mg of niacin = 60 mg of tryptophan; 0–6 months = preformed niacin (not NE).

[f]As dietary folate equivalents (DFE). 1 DFE = 1 μg food folate = 0.6 μg of folic acid from fortified food or as a supplement consumed with food = 0.5 μg of a supplement taken on an empty stomach.

Niacin (mg/d)[e]	Vitamin B_6 (mg/d)	Folate (mg/d)[f]	Vitamin B_{12} (μg/d)	Pantothenic Acid (mg/d)	Biotin (μg/d)	Choline (mg/d)[g]
2*	0.1*	65*	0.4*	1.7*	5*	125*
4*	0.3*	80*	0.5*	1.8*	6*	150*
6	**0.5**	**150**	**0.9**	2*	8*	200*
8	**0.6**	**200**	**1.2**	3*	12*	250*
12	**1.0**	**300**	**1.8**	4*	20*	375*
16	**1.3**	**400**	**2.4**	5*	25*	550*
16	**1.3**	**400**	**2.4**	5*	30*	550*
16	**1.3**	**400**	**2.4**	5*	30*	550*
16	**1.7**	**400**	**2.4**[h]	5*	30*	550*
16	**1.7**	**400**	**2.4**[h]	5*	30*	550*
12	**1.0**	**300**	**1.8**	4*	20*	375*
14	**1.2**	**400**[i]	**2.4**	5*	25*	400*
14	**1.3**	**400**[i]	**2.4**	5*	30*	425*
14	**1.3**	**400**[i]	**2.4**	5*	30*	425*
14	**1.5**	**400**	**2.4**[h]	5*	30*	425*
14	**1.5**	**400**	**2.4**[h]	5*	30*	425*
18	**1.9**	**600**[j]	**2.6**	6*	30*	450*
18	**1.9**	**600**[j]	**2.6**	6*	30*	450*
18	**1.9**	**600**[j]	**2.6**	6*	30*	450*
17	**2.0**	**500**	**2.8**	7*	35*	550*
17	**2.0**	**500**	**2.8**	7*	35*	550*
17	**2.0**	**500**	**2.8**	7*	35*	550*

[g]Although AIs have been set for choline, there are few data to assess whether a dietary supply of choline is needed at all stages of the life cycle, and it may be that the choline requirement can be met by endogenous synthesis at some of these stages.

[h]Because 10 to 30 percent of older people may malabsorb food-bound B_{12}, it is advisable for those older than 50 years to meet their RDA mainly by consuming foods fortified with B_{12} or a supplement containing B_{12}.

[i]In view of evidence linking folate intake with neural tube defects in the fetus, it is recommended that all women capable of becoming pregnant consume 400 μg from supplements or fortified foods in addition to intake of food folate from a varied diet.

[j]It is assumed that women will continue consuming 400 μg from supplements or fortified food until their pregnancy is confirmed and they enter prenatal care, which ordinarily occurs after the end of the periconceptional period—the critical time for formation of the neural tube.

SOURCES: *Dietary Reference Intakes for Calcium, Phosphorous, Magnesium, Vitamin D, and Fluoride* (1997); *Dietary Reference Intakes for Thiamin, Riboflavin, Niacin, Vitamin B_6, Folate, Vitamin B_{12}, Pantothenic Acid, Biotin, and Choline* (1998); *Dietary Reference Intakes for Vitamin C, Vitamin E, Selenium, and Carotenoids* (2000); *Dietary Reference Intakes for Vitamin A, Vitamin K, Arsenic, Boron, Chromium, Copper, Iodine, Iron, Manganese, Molybdenum, Nickel, Silicon, Vanadium, and Zinc* (2001); *Dietary Reference Intakes for Water, Potassium, Sodium, Chloride, and Sulfate* (2005); and *Dietary Reference Intakes for Calcium and Vitamin D* (2011). These reports may be accessed via www.nap.edu.

Dietary Reference Intakes (DRIs): Recommended Dietary Allowances and Adequate Intakes, Elements

Food and Nutrition Board, Institute of Medicine, National Academies

Life Stage Group	Calcium (mg/d)	Chromium (μg/d)	Copper (μg/d)	Fluoride (mg/d)	Iodine (μg/d)	Iron (mg/d)	Magnesium (mg/d)	Manganes (mg/d)
Infants								
0 to 6 mo	200*	0.2*	200*	0.01*	110*	0.27*	30*	0.003*
6 to 12 mo	260*	5.5*	220*	0.5*	130*	**11**	75*	0.6*
Children								
1–3 y	**700**	11*	**340**	0.7*	**90**	**7**	**80**	1.2*
4–8 y	**1,000**	15*	**440**	1*	**90**	**10**	**130**	1.5*
Males								
9–13 y	**1,300**	25*	**700**	2*	**120**	**8**	**240**	1.9*
14–18 y	**1,300**	35*	**890**	3*	**150**	**11**	**410**	2.2*
19–30 y	**1,000**	35*	**900**	4*	**150**	**8**	**400**	2.3*
31–50 y	**1,000**	35*	**900**	4*	**150**	**8**	**420**	2.3*
51–70 y	**1,000**	30*	**900**	4*	**150**	**8**	**420**	2.3*
> 70 y	**1,200**	30*	**900**	4*	**150**	**8**	**420**	2.3*
Females								
9–13 y	**1,300**	21*	**700**	2*	**120**	**8**	**240**	1.6*
14–18 y	**1,300**	24*	**890**	3*	**150**	**15**	**360**	1.6*
19–30 y	**1,000**	25*	**900**	3*	**150**	**18**	**310**	1.8*
31–50 y	**1,000**	25*	**900**	3*	**150**	**18**	**320**	1.8*
51–70 y	**1,200**	20*	**900**	3*	**150**	**8**	**320**	1.8*
> 70 y	**1,200**	20*	**900**	3*	**150**	**8**	**320**	1.8*
Pregnancy								
14–18 y	**1,300**	29*	**1,000**	3*	**220**	**27**	**400**	2.0*
19–30 y	**1,000**	30*	**1,000**	3*	**220**	**27**	**350**	2.0*
31–50 y	**1,000**	30*	**1,000**	3*	**220**	**27**	**360**	2.0*
Lactation								
14–18 y	**1,300**	44*	**1,300**	3*	**290**	**10**	**360**	2.6*
19–30 y	**1,000**	45*	**1,300**	3*	**290**	**9**	**310**	2.6*
31–50 y	**1,000**	45*	**1,300**	3*	**290**	**9**	**320**	2.6*

NOTE: This table (taken from the DRI reports, see www.nap.edu) presents Recommended Dietary Allowances (RDAs) in **bold type** and Adequate Intakes (AIs) in ordinary type followed by an asterisk (*). An RDA is the average daily dietary intake level sufficient to meet the nutrient requirements of nearly all (97–98 percent) healthy individuals in a group. It is calculated from an Estimated Average Requirement (EAR). If sufficient scientific evidence is not available to establish an EAR, and thus calculate an RDA, an AI is usually developed. For healthy breastfed infants, an AI is the mean intake. The AI for other life-stage and gender groups is believed to cover the needs of all healthy individuals in the groups, but lack of data or uncertainty in the data prevent being able to specify with confidence the percentage of individuals covered by this intake.

Molybdenum (μg/d)	Phosphorus (mg/d)	Selenium (μg/d)	Zinc (mg/d)	Potassium (g/d)	Sodium (g/d)	Chloride (g/d)
2*	100*	15*	2*	0.4*	0.12*	0.18*
3*	275*	20*	**3**	0.7*	0.37*	0.57*
17	**460**	**20**	**3**	3.0*	1.0*	1.5*
22	**500**	**30**	**5**	3.8*	1.2*	1.9*
34	**1,250**	**40**	**8**	4.5*	1.5*	2.3*
43	**1,250**	**55**	**11**	4.7*	1.5*	2.3*
45	**700**	**55**	**11**	4.7*	1.5*	2.3*
45	**700**	**55**	**11**	4.7*	1.5*	2.3*
45	**700**	**55**	**11**	4.7*	1.3*	2.0*
45	**700**	**55**	**11**	4.7*	1.2*	1.8*
34	**1,250**	**40**	**8**	4.5*	1.5*	2.3*
43	**1,250**	**55**	**9**	4.7*	1.5*	2.3*
45	**700**	**55**	**8**	4.7*	1.5*	2.3*
45	**700**	**55**	**8**	4.7*	1.5*	2.3*
45	**700**	**55**	**8**	4.7*	1.3*	2.0*
45	**700**	**55**	**8**	4.7*	1.2*	1.8*
50	**1,250**	**60**	**12**	4.7*	1.5*	2.3*
50	**700**	**60**	**11**	4.7*	1.5*	2.3*
50	**700**	**60**	**11**	4.7*	1.5*	2.3*
50	**1,250**	**70**	**13**	5.1*	1.5*	2.3*
50	**700**	**70**	**12**	5.1*	1.5*	2.3*
50	**700**	**70**	**12**	5.1*	1.5*	2.3*

SOURCES: *Dietary Reference Intakes for Calcium, Phosphorous, Magnesium, Vitamin D, and Fluoride* (1997); *Dietary Reference Intakes for Thiamin, Riboflavin, Niacin, Vitamin B_6, Folate, Vitamin B_{12}, Pantothenic Acid, Biotin, and Choline* (1998); *Dietary Reference Intakes for Vitamin C, Vitamin E, Selenium, and Carotenoids* (2000); and *Dietary Reference Intakes for Vitamin A, Vitamin K, Arsenic, Boron, Chromium, Copper, Iodine, Iron, Manganese, Molybdenum, Nickel, Silicon, Vanadium, and Zinc* (2001); *Dietary Reference Intakes for Water, Potassium, Sodium, Chloride, and Sulfate* (2005); and *Dietary Reference Intakes for Calcium and Vitamin D* (2011). These reports may be accessed via www.nap.edu.

Dietary Reference Intakes (DRIs): Recommended Dietary Allowances and Adequate Intakes, Total Water and Macronutrients

Food and Nutrition Board, Institute of Medicine, National Academies

Life Stage Group	*Total* Water[a] (L/d)	Carbohydrate (g/d)	Total Fiber (g/d)	Fat (g/d)	Linoleic Acid (g/d)	α-Linolenic Acid (g/d)	Protein[b] (g/d)
Infants							
0 to 6 mo	0.7*	60*	ND	31*	4.4*	0.5*	9.1*
6 to 12 mo	0.8*	95*	ND	30*	4.6*	0.5*	**11.0**
Children							
1–3 y	1.3*	**130**	19*	ND[c]	7*	0.7*	**13**
4–8 y	1.7*	**130**	25*	ND	10*	0.9*	**19**
Males							
9–13 y	2.4*	**130**	31*	ND	12*	1.2*	**34**
14–18 y	3.3*	**130**	38*	ND	16*	1.6*	**52**
19–30 y	3.7*	**130**	38*	ND	17*	1.6*	**56**
31-50 y	3.7*	**130**	38*	ND	17*	1.6*	**56**
51–70 y	3.7*	**130**	30*	ND	14*	1.6*	**56**
> 70 y	3.7*	**130**	30*	ND	14*	1.6*	**56**
Females							
9–13 y	2.1*	**130**	26*	ND	10*	1.0*	**34**
14–18 y	2.3*	**130**	26*	ND	11*	1.1*	**46**
19–30 y	2.7*	**130**	25*	ND	12*	1.1*	**46**
31–50 y	2.7*	**130**	25*	ND	12*	1.1*	**46**
51–70 y	2.7*	**130**	21*	ND	11*	1.1*	**46**
> 70 y	2.7*	**130**	21*	ND	11*	1.1*	**46**
Pregnancy							
14–18 y	3.0*	**175**	28*	ND	13*	1.4*	**71**
19–30 y	3.0*	**175**	28*	ND	13*	1.4*	**71**
31–50 y	3.0*	**175**	28*	ND	13*	1.4*	**71**
Lactation							
14–18 y	3.8*	**210**	29*	ND	13*	1.3*	**71**
19–30 y	3.8*	**210**	29*	ND	13*	1.3*	**71**
31–50 y	3.8*	**210**	29*	ND	13*	1.3*	**71**

NOTE: This table (taken from the DRI reports, see www.nap.edu) presents Recommended Dietary Allowances (RDA) in **bold type** and Adequate Intakes (AI) in ordinary type followed by an asterisk (*). An RDA is the average daily dietary intake level sufficient to meet the nutrient requirements of nearly all (97–98 percent) healthy individuals in a group. It is calculated from an Estimated Average Requirement (EAR). If sufficient scientific evidence is not available to establish an EAR, and thus calculate an RDA, an AI is usually developed. For healthy breastfed infants, an AI is the mean intake. The AI for other life-stage and gender groups is believed to cover the needs of all healthy individuals in the groups, but lack of data or uncertainty in the data prevent being able to specify with confidence the percentage of individuals covered by this intake.

[a] *Total* water includes all water contained in food, beverages, and drinking water.

[b] Based on g protein per kg of body weight for the reference body weight, e.g., for adults 0.8 g/kg body weight for the reference body weight.

[c] Not determined.

SOURCES: *Dietary Reference Intakes for Energy, Carbohydrate, Fiber, Fat, Fatty Acids, Cholesterol, Protein, and Amino Acids* (2002/2005); and *Dietary Reference Intakes for Water, Potassium, Sodium, Chloride, and Sulfate* (2005). The reports may be accessed via www.nap.edu.

Dietary Reference Intakes (DRIs): Acceptable Macronutrient Distribution Ranges
Food and Nutrition Board, Institute of Medicine, National Academies

Macronutrient	Range (percent of energy)		
	Children, 1–3 y	Children, 4–18 y	Adults
Fat	30–40	25–35	20–35
n-6 polyunsaturated fatty acids[a] (linoleic acid)	5–10	5–10	5–10
n-3 polyunsaturated fatty acids[a] (α-linolenic acid)	0.6–1.2	0.6–1.2	0.6–1.2
Carbohydrate	45–65	45–65	45–65
Protein	5–20	10–30	10–35

[a]Approximately 10 percent of the total can come from longer-chain *n*-3 or *n*-6 fatty acids.

SOURCE: *Dietary Reference Intakes for Energy, Carbohydrate, Fiber, Fat, Fatty Acids, Cholesterol, Protein, and Amino Acids* (2002/2005). The report may be accessed via www.nap.edu.

Dietary Reference Intakes (DRIs): Acceptable Macronutrient Distribution Ranges
Food and Nutrition Board, Institute of Medicine, National Academies

Macronutrient	Recommendation
Dietary cholesterol	As low as possible while consuming a nutritionally adequate diet
Trans fatty Acids	As low as possible while consuming a nutritionally adequate diet
Saturated fatty acids	As low as possible while consuming a nutritionally adequate diet
Added sugars[a]	Limit to no more than 25% of total energy

[a]Not a recommended intake. A daily intake of added sugars that individuals should aim for to achieve a healthful diet was not set.

SOURCE: *Dietary Reference Intakes for Energy, Carbohydrate, Fiber, Fat, Fatty Acids, Cholesterol, Protein, and Amino Acids* (2002/2005). The report may be accessed via www.nap.edu.

Dietary Reference Intakes (DRIs): Tolerable Upper Intake Levels, Vitamins
Food and Nutrition Board, Institute of Medicine, National Academies

Life Stage Group	Vitamin A (μg/d)[a]	Vitamin C (mg/d)	Vitamin D (μg/d)	Vitamin E (mg/d)[b,c]	Vitamin K	Thiamin	Riboflavi
Infants							
0 to 6 mo	600	ND[e]	25	ND	ND	ND	ND
6 to 12 mo	600	ND	38	ND	ND	ND	ND
Children							
1-3 y	600	400	63	200	ND	ND	ND
4-8 y	900	650	75	300	ND	ND	ND
Males							
9–13 y	1,700	1,200	100	600	ND	ND	ND
14–18 y	2,800	1,800	100	800	ND	ND	ND
19-30 y	3,000	2,000	100	1,000	ND	ND	ND
31-50 y	3,000	2,000	100	1,000	ND	ND	ND
51-70 y	3,000	2,000	100	1,000	ND	ND	ND
> 70 y	3,000	2,000	100	1,000	ND	ND	ND
Females							
9–13 y	1,700	1,200	100	600	ND	ND	ND
14–18 y	2,800	1,800	100	800	ND	ND	ND
19–30 y	3,000	2,000	100	1,000	ND	ND	ND
31-50 y	3,000	2,000	100	1,000	ND	ND	ND
51-70 y	3,000	2,000	100	1,000	ND	ND	ND
> 70 y	3,000	2,000	100	1,000	ND	ND	ND
Pregnancy							
14–18 y	2,800	1,800	100	800	ND	ND	ND
19–30 y	3,000	2,000	100	1,000	ND	ND	ND
31-50 y	3,000	2,000	100	1,000	ND	ND	ND
Lactation							
14–18 y	2,800	1,800	100	800	ND	ND	ND
19–30 y	3,000	2,000	100	1,000	ND	ND	ND
31-50 y	3,000	2,000	100	1,000	ND	ND	ND

NOTE: A Tolerable Upper Intake Level (UL) is the highest level of daily nutrient intake that is likely to pose no risk of adverse health effects to almost all individuals in the general population. Unless otherwise specified, the UL represents total intake from food, water, and supplements. Due to a lack of suitable data, ULs could not be established for vitamin K, thiamin, riboflavin, vitamin B_{12}, pantothenic acid, biotin, and carotenoids. In the absence of a UL, extra caution may be warranted in consuming levels above recommended intakes. Members of the general population should be advised not to routinely exceed the UL. The UL is not meant to apply to individuals who are treated with the nutrient under medical supervision or to individuals with predisposing conditions that modify their sensitivity to the nutrient.

[a]As preformed vitamin A only.

[b]As α-tocopherol; applies to any form of supplemental α-tocopherol.

[c]The ULs for vitamin E, niacin, and folate apply to synthetic forms obtained from supplements, fortified foods, or a combination of the two.

Niacin mg/d)[c]	Vitamin B_6 (mg/d)	Folate (μg/d)[c]	Vitamin B_{12}	Pantothenic Acid	Biotin	Choline (g/d)	Carotenoids[d]
ND	ND	ND	ND	ND	ND	ND	ND
ND	ND	ND	ND	ND	ND	ND	ND
10	30	300	ND	ND	ND	1.0	ND
15	40	400	ND	ND	ND	1.0	ND
20	60	600	ND	ND	ND	2.0	ND
30	80	800	ND	ND	ND	3.0	ND
35	100	1,000	ND	ND	ND	3.5	ND
35	100	1,000	ND	ND	ND	3.5	ND
35	100	1,000	ND	ND	ND	3.5	ND
35	100	1,000	ND	ND	ND	3.5	ND
20	60	600	ND	ND	ND	2.0	ND
30	80	800	ND	ND	ND	3.0	ND
35	100	1,000	ND	ND	ND	3.5	ND
35	100	1,000	ND	ND	ND	3.5	ND
35	100	1,000	ND	ND	ND	3.5	ND
35	100	1,000	ND	ND	ND	3.5	ND
30	80	800	ND	ND	ND	3.0	ND
35	100	1,000	ND	ND	ND	3.5	ND
35	100	1,000	ND	ND	ND	3.5	ND
30	80	800	ND	ND	ND	3.0	ND
35	100	1,000	ND	ND	ND	3.5	ND
35	100	1,000	ND	ND	ND	3.5	ND

[d]β-Carotene supplements are advised only to serve as a provitamin A source for individuals at risk of vitamin A deficiency.

[e]ND = Not determinable due to lack of data of adverse effects in this age group and concern with regard to lack of ability to handle excess amounts. Source of intake should be from food only to prevent high levels of intake.

SOURCES: *Dietary Reference Intakes for Calcium, Phosphorous, Magnesium, Vitamin D, and Fluoride* (1997); *Dietary Reference Intakes for Thiamin, Riboflavin, Niacin, Vitamin B_6, Folate, Vitamin B_{12}, Pantothenic Acid, Biotin, and Choline* (1998); *Dietary Reference Intakes for Vitamin C, Vitamine E, Selenium, and Carotenoids* (2000); *Dietary Reference Intakes for Vitamin A, Vitamin K, Arsenic, Boron, Chromium, Copper, Iodine, Iron, Manganese, Molybdenum, Nickel, Silicon, Vanadium, and Zinc* (2001); and *Dietary Reference Intakes for Calcium and Vitamin D* (2011). These reports may be accessed via www.nap.edu.

Dietary Reference Intakes (DRIs): Tolerable Upper Intake Levels, Elements Food and Nutrition Board, Institute of Medicine, National Academies									
Life Stage Group	Arsenic[a]	Boron (mg/d)	Calcium (mg/d)	Chromium	Copper (μg/d)	Fluoride (mg/d)	Iodine (μg/d)	Iron (mg/d)	Magnesium (mg/d)[b]
Infants									
0 to 6 mo	ND[e]	ND	1,000	ND	ND	0.7	ND	40	ND
6 to 12 mo	ND	ND	1,500	ND	ND	0.9	ND	40	ND
Children									
1-3 y	ND	3	2,500	ND	1,000	1.3	200	40	65
4-8 y	ND	6	2,500	ND	3,000	2.2	300	40	110
Males									
9-13 y	ND	11	3,000	ND	5,000	10	600	40	350
14-18 y	ND	17	3,000	ND	8,000	10	900	45	350
19-30 y	ND	20	2,500	ND	10,000	10	1,100	45	350
31-50 y	ND	20	2,500	ND	10,000	10	1,100	45	350
51-70 y	ND	20	2,000	ND	10,000	10	1,100	45	350
> 70 y	ND	20	2,000	ND	10,000	10	1,100	45	350
Females									
9-13 y	ND	11	3,000	ND	5,000	10	600	40	350
14-18 y	ND	17	3,000	ND	8,000	10	900	45	350
19-30 y	ND	20	2,500	ND	10,000	10	1,100	45	350
31-50 y	ND	20	2,500	ND	10,000	10	1,100	45	350
51-70 y	ND	20	2,000	ND	10,000	10	1,100	45	350
> 70 y	ND	20	2,000	ND	10,000	10	1,100	45	350
Pregnancy									
14-18 y	ND	17	3,000	ND	8,000	10	900	45	350
19-30 y	ND	20	2,500	ND	10,000	10	1,100	45	350
61-50 y	ND	20	2,500	ND	10,000	10	1,100	45	350
Lactation									
14-18 y	ND	17	3,000	ND	8,000	10	900	45	350
19-30 y	ND	20	2,500	ND	10,000	10	1,100	45	350
31-50 y	ND	20	2,500	ND	10,000	10	1,100	45	350

NOTE: A Tolerable Upper Intake Level (UL) is the highest level of daily nutrient intake that is likely to pose no risk of adverse health effects to almost all individuals in the general population. Unless otherwise specified, the UL represents total intake from food, water, and supplements. Due to a lack of suitable data, ULs could not be established for vitamin K, thiamin, riboflavin, vitamin B_{12}, pantothenic acid, biotin, and carotenoids. In the absence of a UL, extra caution may be warranted in consuming levels above recommended intakes. Members of the general population should be advised not to routinely exceed the UL. The UL is not meant to apply to individuals who are treated with the nutrient under medical supervision or to individuals with predisposing conditions that modify their sensitivity to the nutrient.

[a]Although the UL was not determined for arsenic, there is no justification for adding arsenic to food or supplements.

[b]The ULs for magnesium represent intake from a pharmacological agent only and do not include intake from food and water.

[c]Although silicon has not been shown to cause adverse effects in humans, there is no justification for adding silicon to supplements.

anganese (mg/d)	Molybdenum (μg/d)	Nickel (mg/d)	Phosphorus (g/d)	Selenium (μg/d)	Silicon[c]	Vanadium (mg/d)[d]	Zinc (mg/d)	Sodium (g/d)	Chloride (g/d)
ND	ND	ND	ND	45	ND	ND	4	ND	ND
ND	ND	ND	ND	60	ND	ND	5	ND	ND
2	300	0.2	3	90	ND	ND	7	1.5	2.3
3	600	0.3	3	150	ND	ND	12	1.9	2.9
6	1,100	0.6	4	280	ND	ND	23	2.2	3.4
9	1,700	1.0	4	400	ND	ND	34	2.3	3.6
11	2,000	1.0	4	400	ND	1.8	40	2.3	3.6
11	2,000	1.0	4	400	ND	1.8	40	2.3	3.6
11	2,000	1.0	4	400	ND	1.8	40	2.3	3.6
11	2,000	1.0	3	400	ND	1.8	40	2.3	3.6
6	1,100	0.6	4	280	ND	ND	23	2.2	3.4
9	1,700	1.0	4	400	ND	ND	34	2.3	3.6
11	2,000	1.0	4	400	ND	1.8	40	2.3	3.6
11	2,000	1.0	4	400	ND	1.8	40	2.3	3.6
11	2,000	1.0	4	400	ND	1.8	40	2.3	3.6
11	2,000	1.0	3	400	ND	1.8	40	2.3	3.6
9	1,700	1.0	3.5	400	ND	ND	34	2.3	3.6
11	2,000	1.0	3.5	400	ND	ND	40	2.3	3.6
11	2,000	1.0	3.5	400	ND	ND	40	2.3	3.6
9	1,700	1.0	4	400	ND	ND	34	2.3	3.6
11	2,000	1.0	4	400	ND	ND	40	2.3	3.6
11	2,000	1.0	4	400	ND	ND	40	2.3	3.6

[d]Although vanadium in food has not been shown to cause adverse effects in humans, there is no justification for adding vanadium to food and vanadium supplements should be used with caution. The UL is based on adverse effects in laboratory animals and this data could be used to set a UL for adults but not children and adolescents.

[e]ND = Not determinable due to lack of data of adverse effects in this age group and concern with regard to lack of ability to handle excess amounts. Source of intake should be from food only to prevent high levels of intake.

SOURCES: *Dietary Reference Intakes for Calcium, Phosphorous, Magnesium, Vitamin D, and Fluoride* (1997); *Dietary Reference Intakes for Thiamin, Riboflavin, Niacin, Vitamin B_6, Folate, Vitamin B_{12}, Pantothenic Acid, Biotin, and Choline* (1998); *Dietary Reference Intakes for Vitamin C, Vitamin E, Selenium, and Carotenoids* (2000); *Dietary Reference Intakes for Vitamin A, Vitamin K, Arsenic, Boron, Chromium, Copper, Iodine, Iron, Manganese, Molybdenum, Nickel, Silicon, Vanadium, and Zinc* (2001); *Dietary Reference Intakes for Water, Potassium, Sodium, Chloride, and Sulfate* (2005); and *Dietary Reference Intakes for Calcium and Vitamin D* (2011). These reports may be accessed via www.nap.edu.

Appendix 2: CONVERSIONS

Measure	Abbreviation	Equivalent
1 gram	g	1,000 milligrams
1 milliliter	ml	1/1,000 of a liter
1 milligram	mg	1,000 micrograms
1 microgram	µg	1/1,000,000 of a gram
1 kilogram	kg	1,000 grams
1 pound	lb	454 grams or 16 ounces
1 kilocalorie	kcal, Cal	1,000 calories
1 kilojoule	kJ	1,000 joules
1 tablespoon	Tbsp	3 teaspoons
1 cup		16 tablespoons

U.S. Liquid Volume Equivalents

8 fluid ounces = 1 cup
1 pint = 2 cups (= 16 fluid ounces)
1 quart = 2 pints (= 4 cups)
1 gallon = 4 quarts (= 16 cups)

U.S. to Metric Conversions

1/5 teaspoon = 1 ml
1 teaspoon = 5 ml
1 tablespoon = 15 ml
1 fluid ounces = 30 ml
1 cup = 240 ml
2 cups (1 pint) = 470 ml
4 cups (1 quart) = .95 liter
4 quarts (1 gallon) = 3.8 liters
1 ounce = 28 grams
1 pound = 454 grams

Metric to U.S. Conversions

1 milliliter = 1/5 teaspoon
5 ml = 1 teaspoon
15 ml = 1 tablespoon
30 ml = 1 fluid ounce
100 ml = 3.4 fluid ounces
240 ml = 1 cup
1 liter = 34 fluid ounces
1 liter = 4.2 cups
1 liter = 2.1 pints
1 liter = 1.06 quarts
1 liter = .26 gallon
1 gram = .035 ounce
100 grams = 3.5 ounces
500 grams = 1.10 pounds
1 kilogram = 2.205 pounds

Appendix 3: KEY FUNCTIONS OF MICRONUTRIENTS

VITAMINS AND MINERALS DIRECTLY INVOLVED IN ENERGY METABOLISM

KEY IDEAS

- Energy is obtained from the metabolism of carbohydrates, fats, and amino acids by several pathways involving numerous enzyme catalyzed reactions; and many of these enzymes require vitamins or minerals for their activity.
- Vitamins provide the essential components of coenzymes that are required for many enzyme catalyzed reactions in energy metabolism. **See Infographic 11.4**.
- Similar to vitamins, minerals often function as critical cofactors that bind to enzymes and are required for enzymes to catalyze chemical reactions.

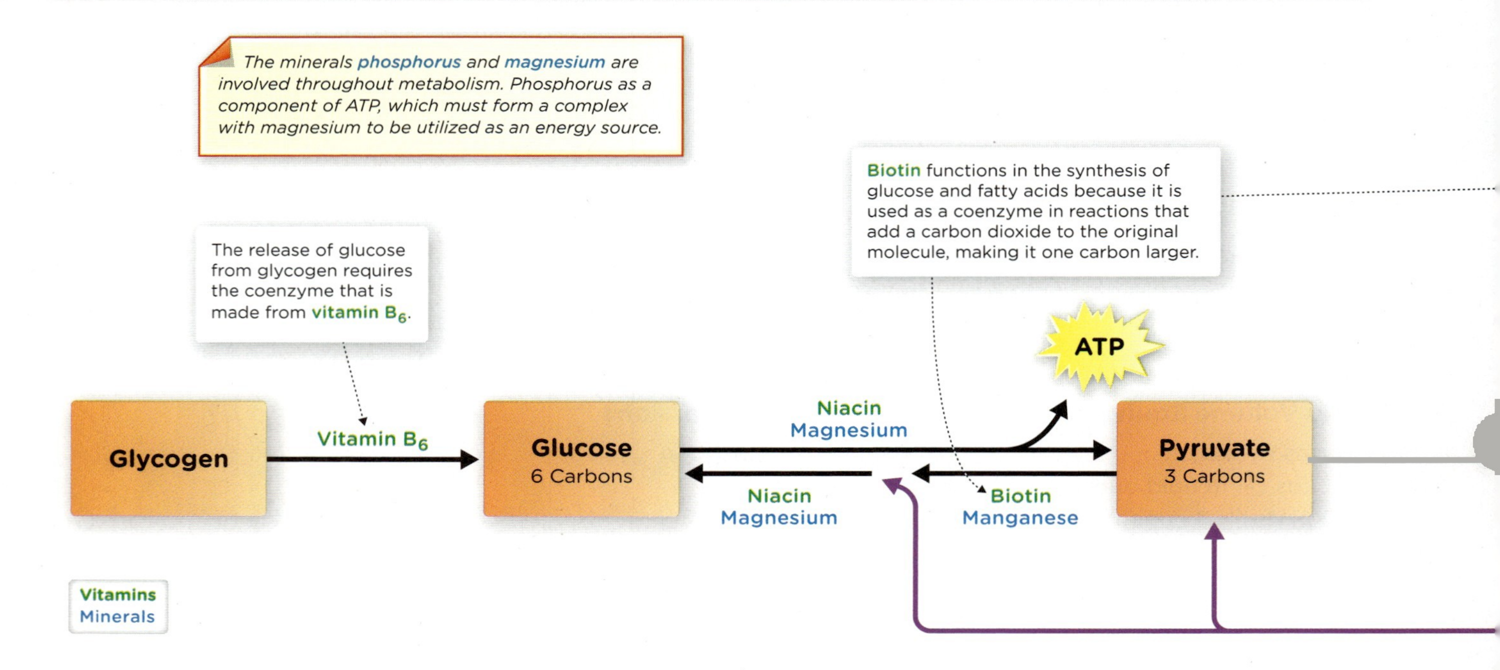

VITAMINS AND MINERALS INVOLVED IN ENERGY METABOLISM

Food source information is provided in the Infographics listed below.

Water-Soluble Vitamins
- Thiamin (p 240), **Infographic 11.5**
- Riboflavin (p 243), **Infographic 11.7**
- Niacin (p 241), **Infographic 11.6**
- Vitamin B_6 (p 243), **Infographic 11.8**
- Folate (p 244), **Infographic 11.11**
- Vitamin B_{12} (p 247), **Infographic 11.12**
- Pantothenic acid (p 244)
- Biotin (p 253)

Major Minerals
- Magnesium (p 291), **Infographic 13.5**
- Phosphorus (p 291)

Trace Minerals
- Manganese (p 328)

1. What two sources of energy would we be unable to use well if we were severely deficient in vitamin B_6?
2. What macronutrients can be used to synthesize fatty acids when they are eaten in excess?
3. Can you anticipate what exhaled waste product is produced when pyruvate is metabolized to acetyl-CoA?
4. Examine the flow through the pathways and answer this question: why can't glucose be synthesized from fatty acids?

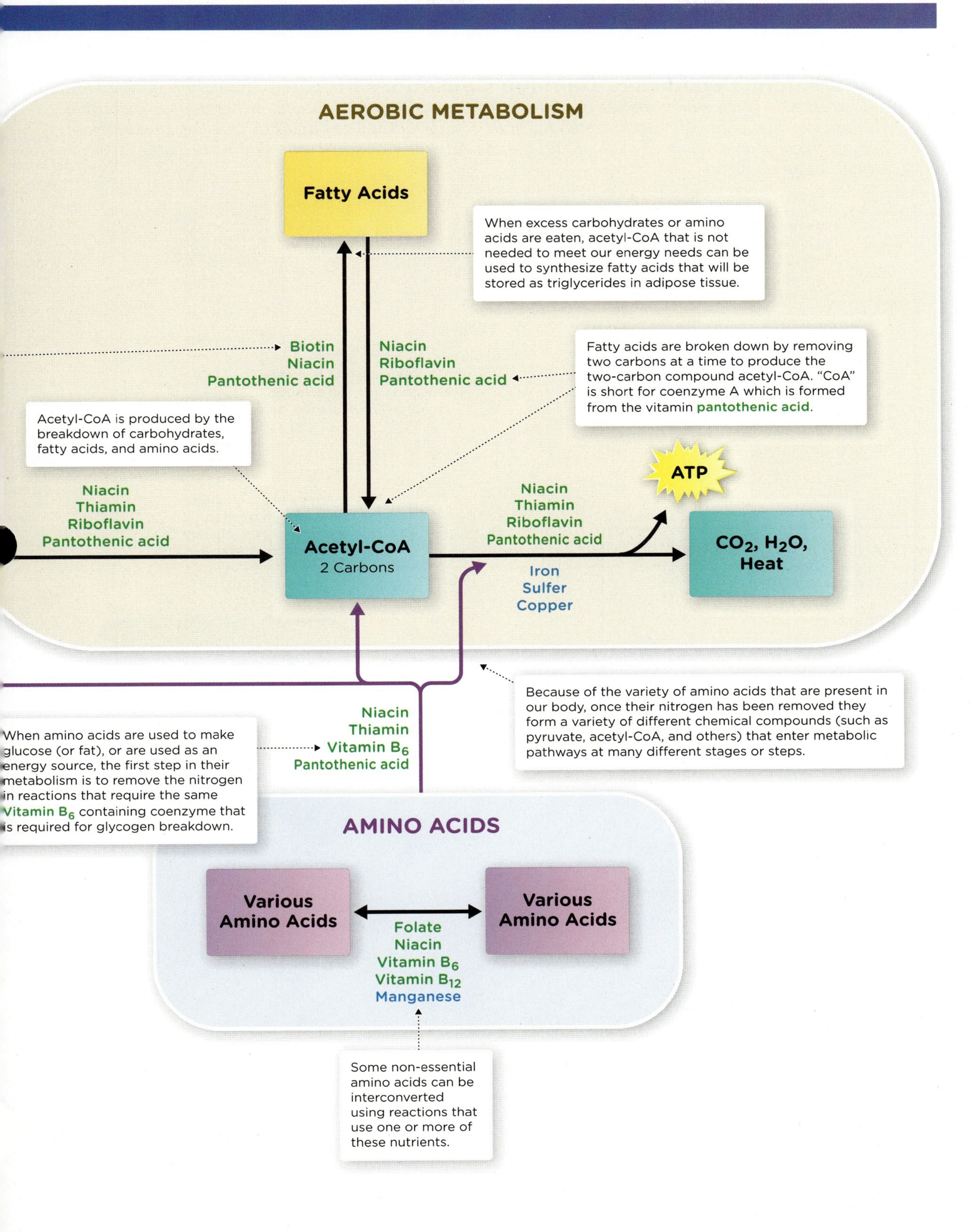

AEROBIC METABOLISM
Fatty Acids
When excess carbohydrates or amino acids are eaten, acetyl-CoA that is not needed to meet our energy needs can be used to synthesize fatty acids that will be stored as triglycerides in adipose tissue.
Biotin
Niacin
Pantothenic acid
Niacin
Riboflavin
Pantothenic acid
Fatty acids are broken down by removing two carbons at a time to produce the two-carbon compound acetyl-CoA. "CoA" is short for coenzyme A which is formed from the vitamin pantothenic acid.
Acetyl-CoA is produced by the breakdown of carbohydrates, fatty acids, and amino acids.
ATP
Niacin
Thiamin
Riboflavin
Pantothenic acid
Acetyl-CoA
2 Carbons
Niacin
Thiamin
Riboflavin
Pantothenic acid
CO_2, H_2O, Heat
Iron
Sulfer
Copper
Because of the variety of amino acids that are present in our body, once their nitrogen has been removed they form a variety of different chemical compounds (such as pyruvate, acetyl-CoA, and others) that enter metabolic pathways at many different stages or steps.
When amino acids are used to make glucose (or fat), or are used as an energy source, the first step in their metabolism is to remove the nitrogen in reactions that require the same Vitamin B_6 containing coenzyme that is required for glycogen breakdown.
Niacin
Thiamin
Vitamin B_6
Pantothenic acid
AMINO ACIDS
Various Amino Acids
Various Amino Acids
Folate
Niacin
Vitamin B_6
Vitamin B_{12}
Manganese
Some non-essential amino acids can be interconverted using reactions that use one or more of these nutrients.

NUTRIENTS WITH ANTIOXIDANT FUNCTION

KEY IDEAS

- Some vitamins and minerals function as antioxidants.
- Antioxidants can donate an electron and neutralize a free radical without becoming unstable and reactive.

Antioxidants Defend Against Oxidative Damage Caused by Free Radicals *Free radicals are molecules containing unpaired electrons, which makes them highly reactive. The free radical either causes oxidative damage by reacting with another molecule and chemically modifying it, or it stabilizes itself by stealing an electron from a nearby molecule, which creates a new free radical and begins a chain reaction. Antioxidants are able to stop the chain reaction by donating an electron.*

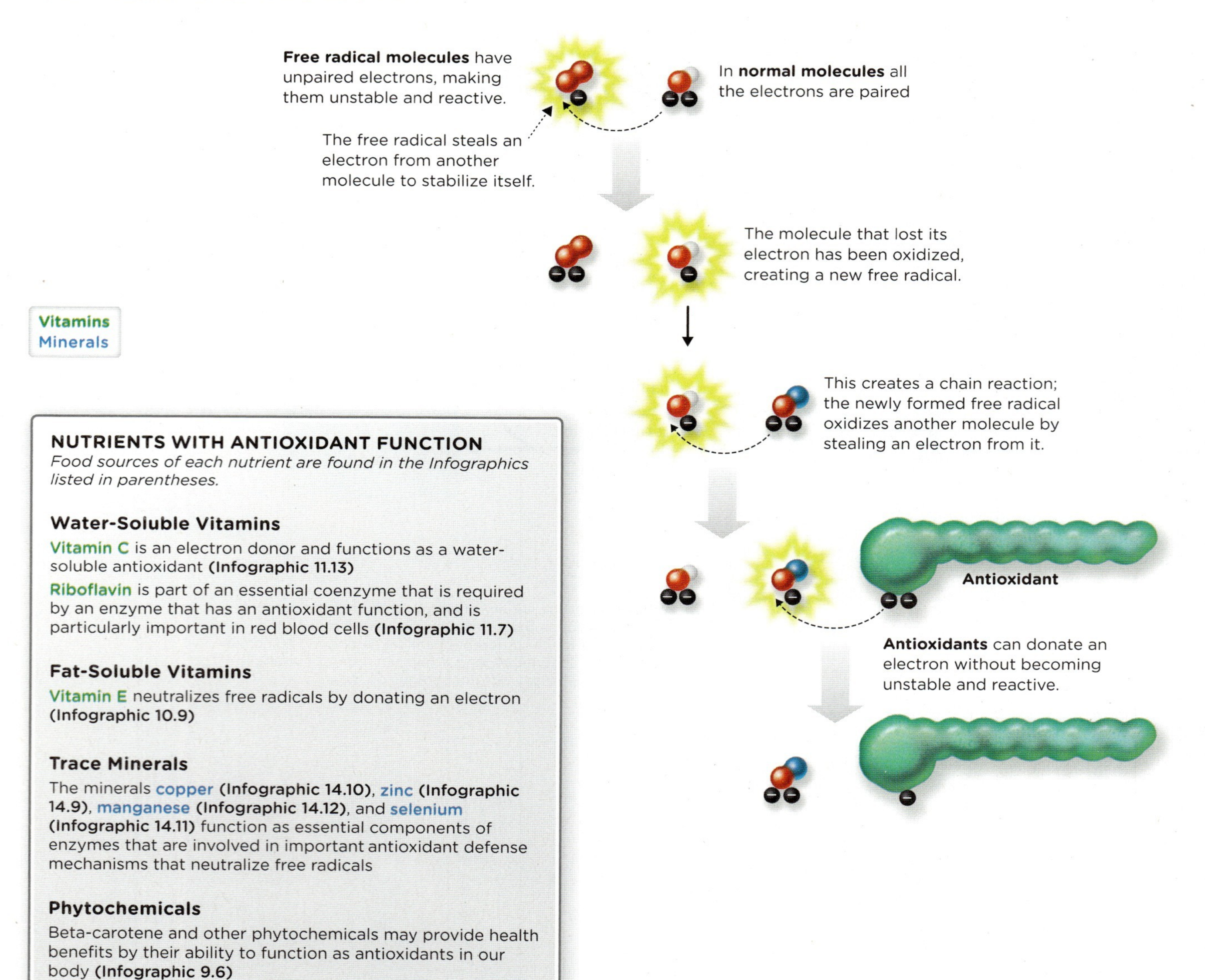

1. Why might those who suntan regularly have increased skin wrinkling as they age?
2. What characteristics do antioxidants possess that allow them to break the chain of oxidative damage?
3. What might be the consequences of free-radical damage to the DNA of a cell?

Sources of Free Radicals and Their Effects *The free radicals we are exposed to can come from environmental sources or are produced by our own bodies. Although free radicals have necessary functions in our body, high levels can cause damage.*

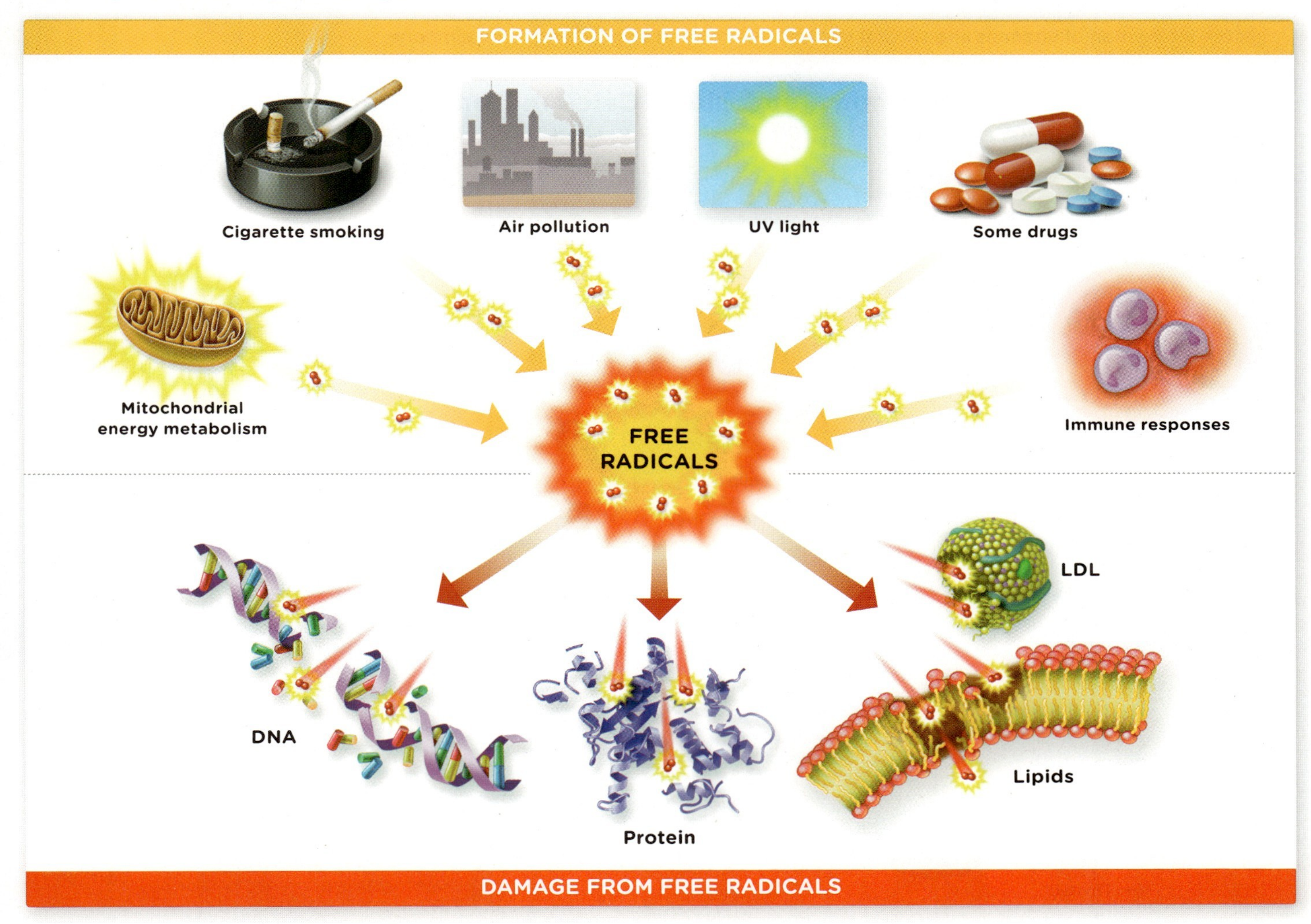

The Antioxidant Functions of Vitamin E *Vitamin E can break the chain of oxidation by donating an electron to free radicals without becoming unstable. Because it is a fat-soluble vitamin it is particularly good at performing this function in cell membranes, and even in lipoproteins like LDL.*

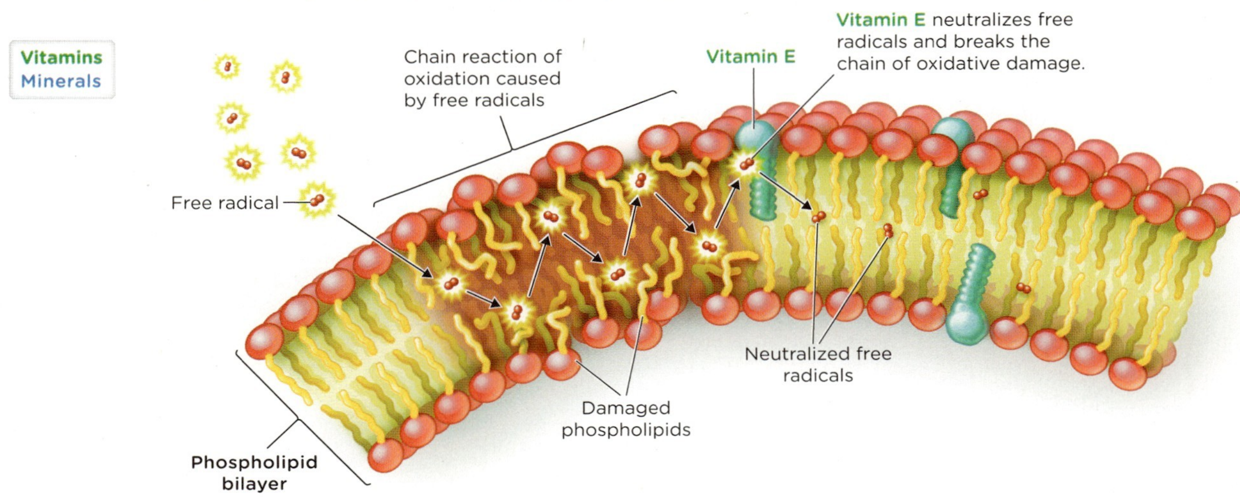

NUTRIENTS INVOLVED IN BONE HEALTH

KEY IDEAS

- Bone is an active organ, it is constantly being remodeled as it is undergoes continual cycles of being broken down and rebuilt.
- All body cells require calcium to function, and its absorption, excretion and release from bone is tightly regulated.
- The interaction of vitamins and several minerals are needed to build and maintain bone.

Vitamin D, Parathyroid Hormone, and Blood Calcium Concentrations *The active hormone form of vitamin D and parathyroid hormone work together to maintain calcium concentrations in blood.*

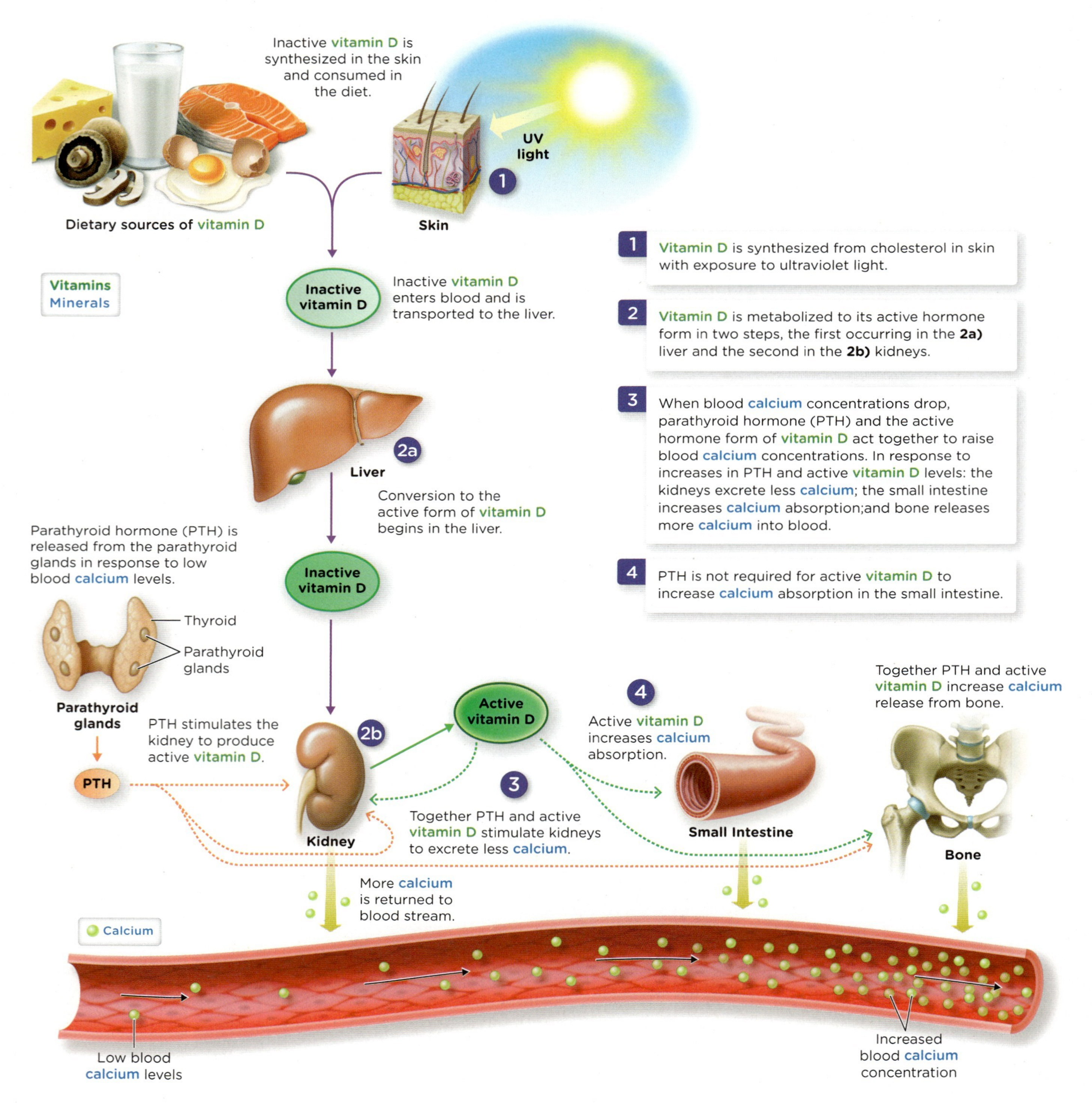

1. Vitamin D is synthesized from cholesterol in skin with exposure to ultraviolet light.
2. Vitamin D is metabolized to its active hormone form in two steps, the first occurring in the **2a)** liver and the second in the **2b)** kidneys.
3. When blood calcium concentrations drop, parathyroid hormone (PTH) and the active hormone form of vitamin D act together to raise blood calcium concentrations. In response to increases in PTH and active vitamin D levels: the kidneys excrete less calcium; the small intestine increases calcium absorption;and bone releases more calcium into blood.
4. PTH is not required for active vitamin D to increase calcium absorption in the small intestine.

Calcium, Phosphorus, and Magnesium are the Primary Mineral Components of Bone

Accounting for 98% of the body's total mineral content by weight, calcium, phosphorus, and magnesium play key roles in the development and maintenance of bone and other calcified tissues.

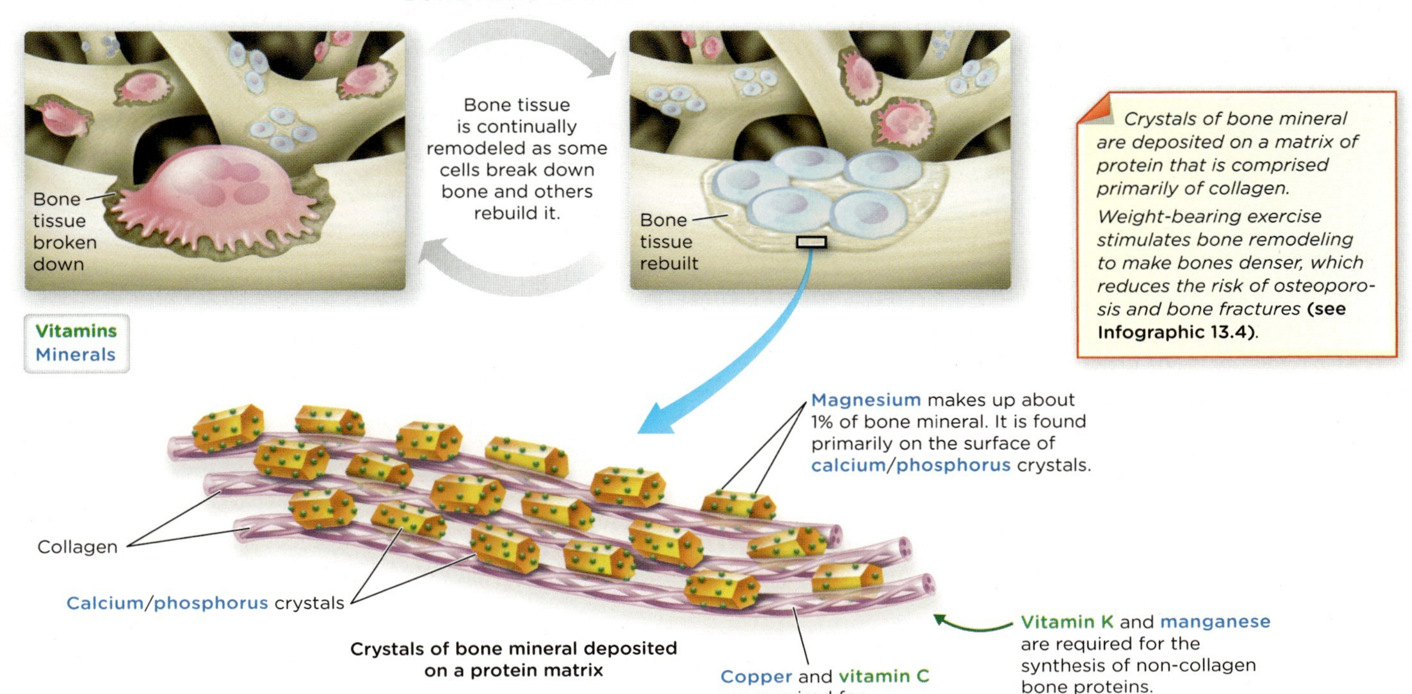

Crystals of bone mineral are deposited on a matrix of protein that is comprised primarily of collagen.

Weight-bearing exercise stimulates bone remodeling to make bones denser, which reduces the risk of osteoporosis and bone fractures **(see Infographic 13.4)**.

VITAMINS AND MINERALS INVOLVED IN BONE HEALTH

Food source information is noted in parentheses.

Fat-Soluble Vitamins

Vitamin D is required for calcium absorption, and proper bone growth and maintenance. **(Infographic 10.3)**
—Severe vitamin D deficiency in children can cause rickets, which results in soft bones and skeletal malformations. See chapter 10.
—In adults vitamin D deficiency can cause osteomalacia, causing bones to become soft and weak as they are depleted of bone mineral.
—Because only a few foods are naturally good sources of vitamin D, fortified foods are an important source of vitamin D for many individuals.

Vitamin K is required for reactions that chemically modify some bone proteins — making them able to bind calcium. **(Infographic 10.11)**

Water-Soluble Vitamins

Vitamin C is required as a cofactor for enzymes that chemically modify the fibrous protein collagen to increase its strength. **(Infographic 11.13)**

Major Minerals

Calcium is the most abundant mineral in the human body and is the main component of bones and teeth. **(Infographic 13.3)**

Phosphorus is the second most abundant mineral in the body and a structural component of bone. Phosphorus is found In most protein-rich foods such as meat and dairy and deficiency is rare.

Magnesium is required for absorption and metabolism of calcium. Magnesium contributes to the strength and firmness of bones and makes teeth harder. Fifty to sixty percent of the total magnesium in the body is found in bone. **(Infographic 13.5)**

Manganese is required for the synthesis of non-collagen bone proteins. **(Page 328)**

Trace Minerals

Copper is required as a cofactor for enzymes that chemically modify collagen to increase its strength. **(Infographic 14.10)**

MICRONUTRIENTS THAT SUPPORT BLOOD HEALTH

KEY IDEAS

All but two of the vitamins and about half of the major and trace minerals play some role in maintaining blood health.

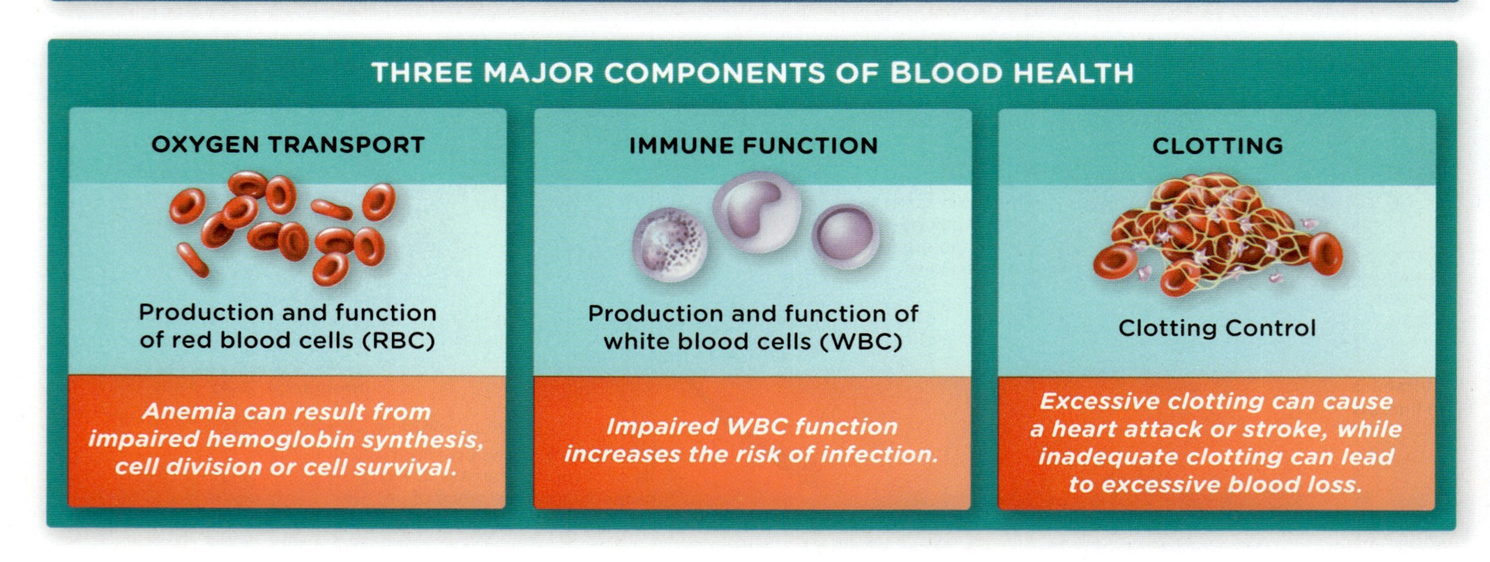

Nutrients Involved in the Production and Function of RBCs

Anemia can be caused by nutrient deficiencies that impair hemoglobin synthesis, the division and maturation of RBC, or RBC survival.

NUTRIENTS REQUIRED FOR HEMOGLOBIN SYNTHESIS *The iron containing heme group has a deep red color, and is the oxygen carrying component of hemoglobin in RBCs.*

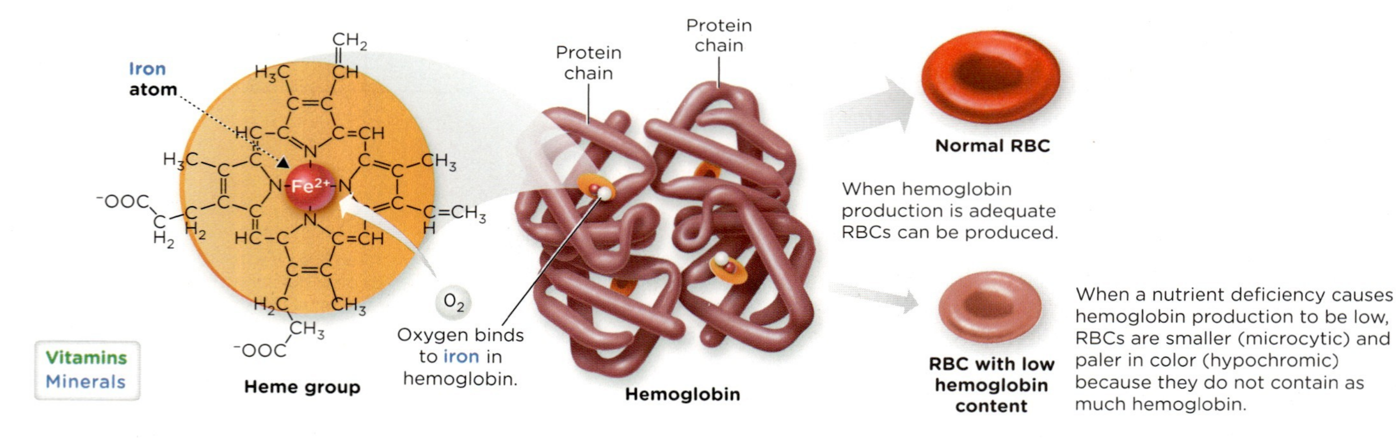

NUTRIENT ROLES IN THE PRODUCTION AND FUNCTION OF RED BLOOD CELLS

- Copper is required to mobilize iron from sites of storage.
- Zinc and sulfur function as cofactors for enzymes involved in heme synthesis.
- Vitamin B_6, niacin, riboflavin, and pantothenic acid are components of coenzymes required by enzymes involved in heme synthesis.

B-Vitamins and trace minerals required for heme synthesis

- Vitamin B_6
- Niacin
- Riboflavin
- Pantothenic acid
- Iron
- Copper
- Zinc
- Sulfur

Why does iron deficiency cause RBCs to be paler in color?

Nutrients Required for Cell Division and Development

Vitamin B_{12}, folate, zinc, and iron are required for DNA synthesis and cell division. A deficiency that disrupts DNA synthesis will have the most detrimental effects on cells in the body that are dividing most rapidly, such as red blood cells.

- Deficiencies of **folate** and **vitamin B_{12}** are the most common causes of anemia resulting from impaired cell division.

Vitamins
Minerals

Folate and vitamin B_{12} adequate

In bone marrow, precursor red blood cells divide and mature normally.

Red blood cell precursor

Folate and vitamin B_{12} deficient

In bone marrow, impaired DNA synthesis will arrest cell division and block the full maturation of red blood cells.

Cells divide normally

In the blood, normal mature red blood cells no longer contain nuclei.

Dr. E. Walker/Science Source

Normal red blood cells

Macrocytic (megaloblastic) cells

Cells are unable to divide adequately

In the blood, these immature red blood cells are larger than normal and they contain large nuclei.

Red Blood Cell Survival

Deficiencies of nutrients that protect red blood cell membranes from oxidative damage can cause anemia because cells are destroyed and removed from blood before their normal lifespan is over.

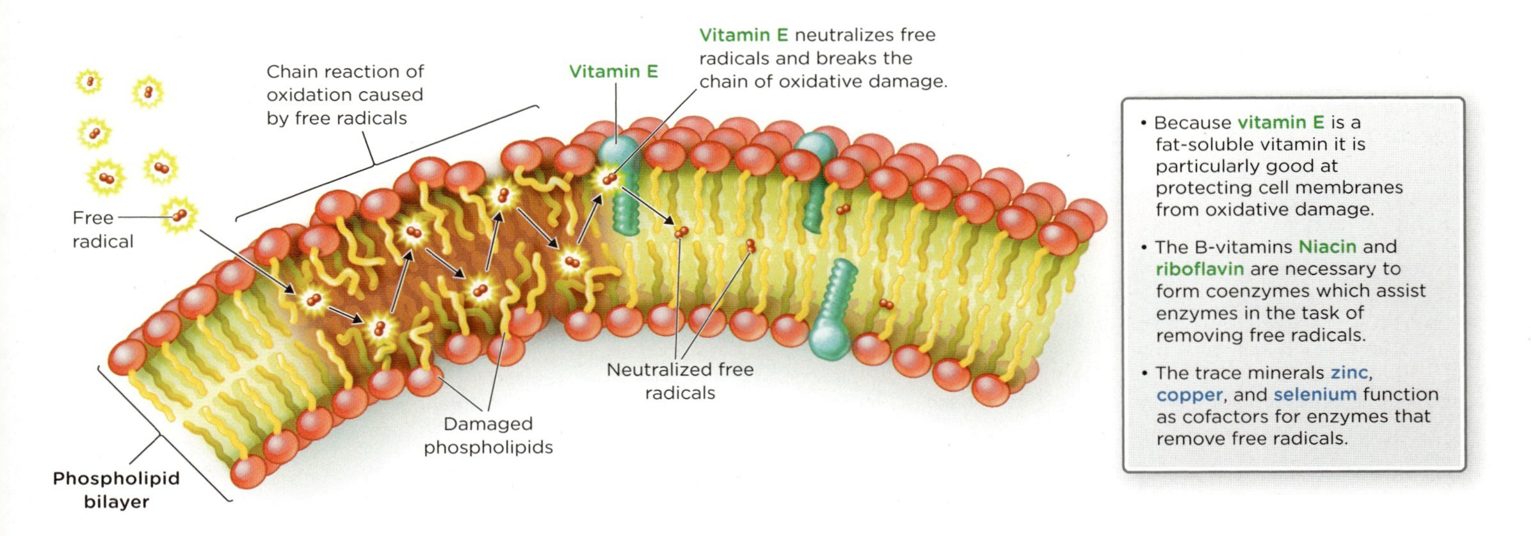

- Because **vitamin E** is a fat-soluble vitamin it is particularly good at protecting cell membranes from oxidative damage.
- The B-vitamins **Niacin** and **riboflavin** are necessary to form coenzymes which assist enzymes in the task of removing free radicals.
- The trace minerals **zinc**, **copper**, and **selenium** function as cofactors for enzymes that remove free radicals.

? An individual is found to have a low number of red blood cells, but the cells all appear to be of normal size and color. What is a likely cause of the anemia?

MICRONUTRIENTS THAT SUPPORT BLOOD HEALTH

Blood Clotting is Dependent on Calcium and Vitamin K

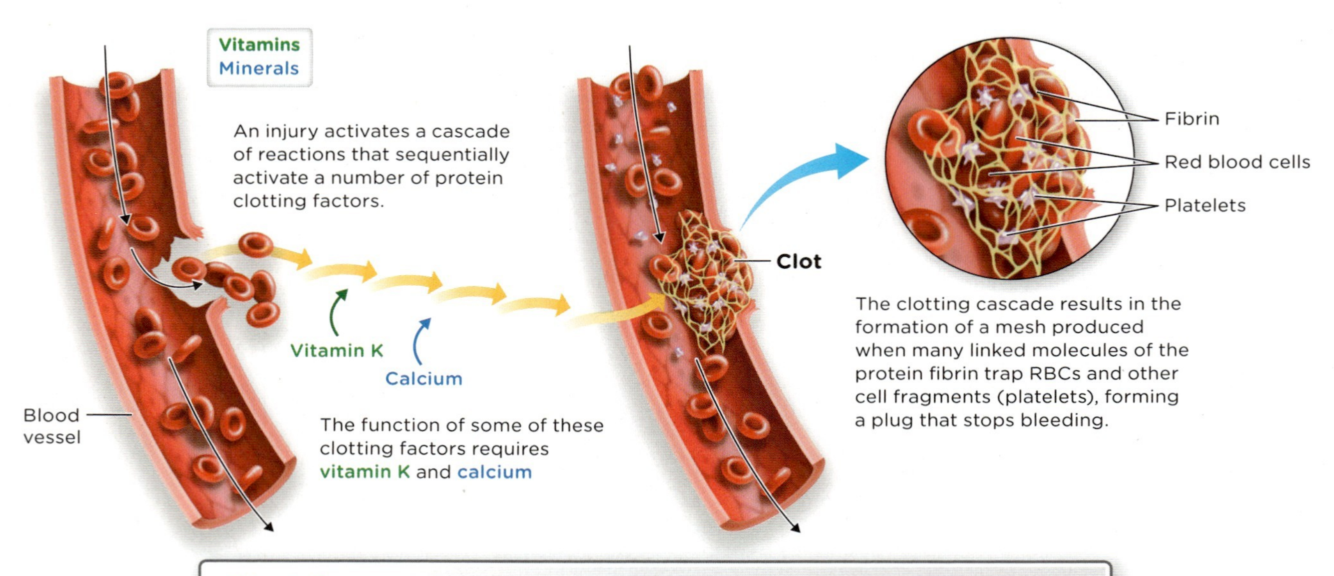

- Vitamin K is required for an enzymatic reaction that modifies several protein clotting factors.
- The vitamin K dependent modification of these clotting factors allows them to bind calcium.
- The binding of calcium to clotting factors is necessary for them to function.
- Because infants have low levels of vitamin K when they are born, they are typically given a vitamin K injection just after birth to prevent uncontrolled bleeding.

Why does a vitamin K deficiency impair clotting?

Micronutrients and Immunity *Together, the variety of cells that make up our immune system require nearly every vitamin, and major and trace mineral to function appropriately. The nutrients are required for: 1) adequate cell division and development, 2) regulation of the immune response, 3) functions designed to kill bacteria and virus infected cells, and 4) protect the immune cells from free radicals that are produced to kill bacteria.*

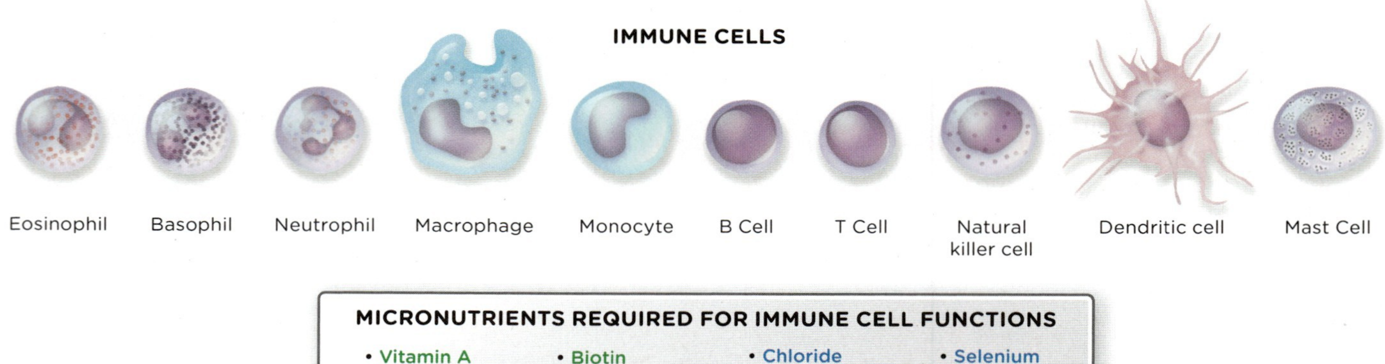

MICRONUTRIENTS REQUIRED FOR IMMUNE CELL FUNCTIONS

- Vitamin A
- Vitamin C
- Vitamin D
- Vitamin B_6
- Vitamin B_{12}
- Biotin
- Niacin
- Riboflavin
- Niacin
- Chloride
- Copper
- Iron
- Manganese
- Magnesium
- Selenium
- Sulfur
- Zinc

NEED TO KNOW REVIEW QUESTION ANSWERS

Chapter 1

1. b
2. a
3. c
4. a
5. e
6. d
7. b
8. a
9. b
10. c

Chapter 2

1. d
2. c
3. a
4. b
5. d
6. a
7. d
8. a
9. b
10. b

Chapter 3

1. d
2. c
3. b
4. c
5. d
6. a
7. b
8. d
9. a
10. b
11. b
12. c

Chapter 4

1. b
2. c
3. c
4. a
5. c
6. d
7. b
8. a
9. c
10. b
11. c
12. a

Chapter 5

1. d
2. a
3. d
4. b
5. c
6. a
7. d
8. b
9. b
10. d
11. c
12. a

Chapter 6

1. a
2. c
3. b
4. d
5. d
6. a
7. b
8. c
9. a
10. d

Chapter 7

1. b
2. e
3. d
4. d
5. d
6. c
7. d
8. a
9. c
10. e

Chapter 8

1. c
2. c
3. d
4. c
5. b
6. d
7. c
8. c
9. e
10. b
11. a

Chapter 9

1. a
2. d
3. b
4. d
5. c
6. b
7. b
8. a
9. e
10. d
11. b
12. a

Chapter 10

1. d
2. d
3. e
4. c
5. d
6. e
7. b
8. a
9. c
10. e
11. a
12. b

Chapter 11

1. b
2. e
3. b
4. a
5. b
6. c
7. d
8. c
9. b
10. d
11. d

Chapter 12

1. b
2. d
3. d
4. a
5. b
6. a
7. c
8. c
9. a
10. c
11. d

Chapter 13

1. c
2. d
3. b
4. d
5. c
6. a
7. b
8. e
9. b
10. d
11. c

Chapter 14

1. d
2. a
3. c
4. b
5. c
6. b
7. d
8. b
9. e
10. b

Chapter 15

1. b
2. c
3. a
4. d
5. a
6. c
7. a
8. d
9. c
10. d
11. b

Chapter 16

1. a
2. d
3. c
4. b
5. d
6. d
7. c
8. c
9. e
10. a
11. b

Chapter 17

1. b
2. a
3. d
4. c
5. a
6. e
7. c
8. c
9. e
10. b

Chapter 18

1. b
2. d
3. d
4. d
5. b
6. a
7. b
8. d
9. d
10. c
11. e

Chapter 19

1. b
2. d
3. a
4. d
5. c
6. d
7. b
8. c
9. c
10. b
11. a
12. b

Chapter 20

1. d
2. e
3. c
4. a
5. d
6. c
7. d
8. b
9. d
10. b
11. d
12. c
13. a

Chapter 21

1. c
2. d
3. b
4. e
5. c
6. d
7. b
8. d
9. c
10. e

Glossary

abdominal obesity excessive fat distributed around the stomach and abdomen; the most often used measure of abdominal obesity is waist size; abdominal obesity in women is a waist size 35 inches or higher, and in men it is a waist size of 40 inches or higher.

absorption the process by which nutrients and other substances are removed from the lumen of the gastrointestinal tract to enter the bloodstream or lymphatic system.

Acceptable Macronutrient Distribution Range (AMDR) the range of energy intakes that should come from each macronutrient to provide a balanced diet.

acesulfame K a non-nutritive sweetener that is long-lasting and heat stable; it is used in a wide variety of products, such as sugar-free beverages and desserts.

acid-base balance the process by which the body maintains homeostasis of body pH; when this does not occur, the body cannot function normally, which can eventually lead to death.

acid reflux the regurgitation of acid content from the stomach into the esophagus; characterized by a burning feeling in the chest called heartburn.

active transport the energy-requiring, carrier-mediated process of transporting a substance across a cell membrane against a concentration gradient.

activity energy expenditure (AEE) the amount of energy expended in physical activity per day.

added sugars sugars that are added to foods during processing, food preparation, or at the table; not those that occur naturally in foods.

adenosine triphosphate (ATP) the primary energy currency of our cells; a molecule composed of adenosine and three phosphate groups; when cells require energy, the bond between the last two phosphates is broken, releasing the stored energy and forming adenosine diphosphate.

adequacy an aspect of healthy eating; consuming foods that provide the calories and essential nutrients necessary to maintain and promote optimal health.

Adequate Intake (AI) nutrients for which the available data are not sufficient to confidently determine an EAR; often the best estimate of the amount that is adequate to meet the needs of the majority of the population based on available data.

aerobic pertaining to, or requiring oxygen (O_2).

aging the accumulation of diverse, harmful changes in cells and tissues with advancing age that is responsible for the increased risk of disease and death.

air displacement a method of determining body composition; assesses a person's body volume and density.

alcohol a potentially intoxicating ingredient, produced through fermentation, found in beer, wine, and liquor; drinking alcohol is in the form of ethyl alcohol, or ethanol; it can be considered a food because it provides energy, but it also has druglike effects, acting as a central nervous system depressant.

alcohol use disorder a medical diagnosis given to individuals who exhibit severe problems with alcohol and meet specific diagnostic criteria.

allergen an antigen that, when exposed to the body, stimulates an abnormal immune response.

alveoli small sacs in the breasts in which milk is produced.

amenorrhea an abnormal absence of menstruation, often due to low energy intake.

amino acid a molecule containing a central carbon atom, an amino group that contains an atom of nitrogen; an acid group, a hydrogen atom, and a side chain. Only the side chain differs for each amino acid, giving each its unique properties.

amylases enzymes that break down starch into smaller polysaccharides and disaccharides.

anabolism a process of building body muscle and tissue in which complex molecules are made from simpler ones, usually requiring energy.

anaerobic occurring in the absence of oxygen.

anaphylaxis a massive immune response to an allergen that causes swelling, changes in blood pressure, and dilated blood vessels in oversensitive individuals and can result in death.

anaphylactic shock a condition caused by decreased oxygen supply to the heart and other body tissues, and by vasodilation as a result of a heightened immune response to an allergen.

antibodies proteins found in blood that are produced in response to foreign substances, such as bacteria, viruses, and allergens, that have entered the body.

anticoagulants medications that prevent blood clotting.

antidiuretic hormone (ADH) a peptide hormone produced by the hypothalamus, with the primary function of decreasing the amount of water excreted by the kidneys; it is stored in and released from the posterior pituitary gland.

antioxidant a substance that prevents damage to cells by inhibiting the oxidation caused by free radicals. An antioxidant gives electrons to other substances and can therefore stabilize or deactivate free radicals.

appetite a desire for food for reasons other than, or in addition to, hunger.

atherosclerosis a type of cardiovascular disease characterized by the narrowing and loss of elasticity of blood vessel walls; it is caused by accumulation of plaque and inflammation of tissue.

ascorbic acid also known as vitamin C; it acts as a coenzyme in biological reactions, aids in hormone production, and is involved in the synthesis of collagen.

aspartame a non-nutritive sweetener that is widely used in sugar-free soft drinks; it can withstand elevated temperatures for only a brief period but is destroyed at baking temperatures.

B vitamins a group of water-soluble vitamins that serve as coenzymes in the conversion of carbohydrates, fat, and protein into energy.

balance an aspect of healthy eating that includes choosing the correct proportions of foods from each of the food groups and the appropriate amounts of calories, macronutrients, vitamins, and minerals.

Bacillus cereus a bacterium; one of the leading causes of foodborne illness in the United States; it produces a heat-stable toxin that, when consumed, causes vomiting.

basal metabolic rate the amount of energy expended in basal metabolism over a fixed period of time, typically expressed as kcal per day.

basal metabolism the energy expenditure required to maintain the essential functions that sustain life.

beriberi a condition caused by thiamin deficiency, which is characterized by extreme weakness, muscle wasting, and eventual heart failure.

beta-carotene a pigment found in plants, and a precursor to vitamin A; also known as provitamin A.

bile a fluid produced in the liver, concentrated and stored in the gallbladder, and secreted into the small intestine in response to food present in stomach; bile promotes the digestion of fat by emulsifying it, which allows lipase easier access.

binders dietary elements, such as phytates and oxalates, that limit bioavailability of other nutrients by chemically binding to minerals.

binge drinking a pattern of alcohol consumption that brings blood alcohol concentration to 0.08% or higher; this typically corresponds to about five or more drinks on a single occasion for men or four or more for women, within a two-hour time period.

binge eating disorder a condition characterized by recurring episodes of eating significantly more food in a short period than most people would eat under similar circumstances, accompanied by feelings of lack of control.

bioavailability the degree to which nutrients can be absorbed and utilized by the body; bioavailability is influenced by many factors, including physiological and dietary conditions.

bioelectrical impedance a method of determining body composition that uses electrical currents to determine body fat content.

blind or blinded study a study in which the researchers or the research subjects (or both) are unaware of whether an intervention or treatment is received until the study is concluded.

blood alcohol concentration (BAC) the percentage of alcohol in blood that is used as a measure of the degree of intoxication of an individual.

body image the way a person perceives their body size, shape, or overall appearance.

body mass index (BMI) an indirect measure of body fat calculated from a person's weight and height.

bolus a masticated, round lump of food, lubricated in the mouth by mixing with saliva.

bone remodeling the process of continuous bone breakdown and rebuilding, which is required for bone maintenance and repair.

bran the section of a whole grain that contains the majority of dietary fiber and a significant amount of B vitamins and minerals.

brush border name for the microvilli-covered surface of the small intestine that functions in the absorption of nutrients.

built environment the aspects of our surroundings that are designed by humans including buildings, transportation systems, and green spaces.

bulimia nervosa a condition characterized by frequent episodes of binge eating followed by purging behaviors to avoid weight gain; purging may occur as vomiting, use of laxatives, or excessive exercise.

caffeine a central nervous system stimulant that has been shown to improve performance in almost all high-intensity activities lasting more than one minute.

calcitriol the active hormonal form of vitamin D that increases the absorption of calcium in the gastrointestinal tract.

calcium the most abundant mineral in the body; 99% is found in the bones and teeth, while the other 1% is located in body cells and fluids, where it is necessary for functions such as blood clotting, nerve transmission, and muscle contraction.

calcium homeostasis the balance between the actions of parathyroid hormone, vitamin D, and the kidneys to tightly control serum calcium levels.

calorie a unit of measure defined as the energy required to raise one gram of water 1° Celsius.

cancer a group of conditions that result from the uncontrolled growth or division of abnormal cells that invade a part or parts of the body.

carbohydrates compounds made up of carbon, hydrogen, and oxygen atoms that are found in foods as either simple sugars or complex carbohydrates; they contain four kilocalories of energy per gram.

carbohydrate loading a technique employed by endurance athletes that involves consuming 10–12 grams per kilogram of body weight each day, for 1 to 2 days prior to competition, while reducing training volume and intensity; the resulting increase in glycogen stores can increase time to exhaustion during intense exercise by 90 minutes or more.

carbon skeleton a chain of carbon atoms that form the backbone of an organic molecule.

carboxyl group the acid group attached to one end of a fatty acid chain.

carbohydrate counting a method to track carbohydrates consumed so that those with diabetes can appropriately balance physical activity and medication to manage blood glucose levels.

cardiorespiratory activities activities that increase the heart rate and promote increased use of oxygen to improve overall body condition and endurance.

cardiovascular disease (CVD) a group of conditions that impair the heart, arteries, veins, and capillaries.

carotenoids compounds—primarily beta-carotene—that have vitamin A activity only after they are converted in the body into one of the active forms of the vitamin (retinol, retinal, or retinoic acid).

catabolism the breakdown of large molecules, such as protein, fat, and glycogen, into smaller ones, causing a loss of skeletal muscle mass.

cells the smallest functional unit of living organisms.

chemical digestion digestion that involves enzymes and other chemical substances released from salivary glands, stomach, pancreas, and gallbladder.

chloride a mineral that is an electrolyte involved in fluid balance; it is often found in table salt (NaCL).

cholecystokinin (CCK) a hormone released from the small intestine in response to dietary fats and proteins in the small intestine; it stimulates the gallbladder to release bile and the pancreas to secrete juice into the lumen of the small intestine.

cholesterol a sterol that is produced by the body; it is a critical component of our cell membranes and is also needed as a precursor for the synthesis of bile acids, vitamin D, and steroid hormones such as estrogen and testosterone.

choline an essential nutrient and water-soluble compound; it forms the neurotransmitter acetylcholine, and is part of two of the most abundant phospholipids in the cell membranes.

chromium a trace mineral; its role in the body is not well understood.

chronic disease diseases that are generally slow in progression and of long duration; some examples include heart disease and diabetes.

chylomicron a very large lipoprotein that transports triglycerides and other dietary lipids away from the small intestine, first in the lymph and then in the blood.

chyme semi-liquefied, partially digested contents that leave the stomach a few teaspoons at a time to enter the small intestine.

circulatory system a system made up of veins, arteries, capillaries, heart and lymphatic vessels; responsible for movement of blood and lymph throughout the body.

cobalamin also known as vitamin B_{12}; it acts as a coenzyme in deriving energy from amino acids, as well as in the reactions involving folate and the conversion of homocysteine to methionine.

coenzyme a compound that binds to a protein (enzyme) and is required for its function or activity.

cofactor a nonprotein chemical compound that activates and assists enzymatic reactions.

complementary proteins two or more incomplete protein sources that together provide adequate amounts and proportions of all the essential amino acids; examples include beans and grains, or beans and nuts.

complete protein a food source that contains all nine essential amino acids in the appropriate proportions.

constipation difficulty, or reduced frequency, of stool passage through intestines.

control group the group in a scientific experiment that does not receive the experimental treatment.

copper a trace mineral that functions as a co-factor in oxygen-dependent enzymes in the body; these enzymes participate in energy metabolism, formation of connective tissue, regulation of iron storage and transport, and antioxidant functions.

creatine phosphate (phosphocreatine) a compound that transfers its phosphate and stored energy to ADP to make more ATP; also known as phosphocreatine.

cretinism a condition characterized by mental retardation, deafness and muteness, stunted growth, delayed sexual maturation, and other abnormalities; it can be caused by severe iodine deprivation during fetal growth.

critical periods developmental events occurring during the first trimester of pregnancy in which cells differentiate and organs and vital systems begin to develop; disruption of growth or development during a critical period may be irreversible.

cross-contamination the transfer of bacteria or other pathogens from one source or surface to another.

curcumin a polyphenol found in turmeric spice; it may have antioxidant and anti-inflammatory effects.

cytosol the intracellular fluid of the cell outside the mitochondria.

DASH diet also known as Dietary Approaches to Stop Hypertension; a food plan that assists with blood pressure control by moderating sodium intake while increasing the intake of potassium, calcium, and magnesium.

denaturation the process by which the shape of a protein is altered; this can be caused by exposure to heat, light, acid, or chemical reaction.

delayed cord clamping the practice of delaying the severance of the umbilical cord by at least a minute after birth; it provides many health benefits, such as increasing the newborn's stores of iron.

dehydration water deficiency resulting from fluid losses that exceed intake.

dental caries also called cavities or tooth decay; the progressive destruction of tooth enamel and ultimately the tooth itself through the action of bacteria on carbohydrates in the mouth.

deoxyribonucleic acid (DNA) nucleic acid that stores the body's genetic information; it is made of a double strand of nucleotide subunits.

diarrhea loose, watery stools on more than three occasions in a 24-hour period.

dietary fat compound found in plant and animal foods that is an energy source, and is necessary for absorption and transport of fat-soluble vitamins.

dietary fiber naturally occurring carbohydrates and lignin from plants that either cannot be digested in the intestinal tract or for which digestion is delayed.

dietary folate equivalent (DFE) a system established to account for the differences in bioavailability between folic acid in dietary supplements and the folate found in food.

Dietary Guidelines for Americans (DGAs) national health guidelines developed by the United States Department of Agriculture and the Department of Health and Human Services that provide information and advice, based on the newest scientific evidence, on how to choose a healthy eating plan.

Dietary Reference Intakes (DRI) quantitative reference values for vitamins, minerals, macronutrients, and energy that are used to assess and plan the diets of healthy people in the United States and Canada.

dietary supplement a food or substance that supplements the diet and contains one or more dietary ingredients or their constituents, including vitamins, minerals, herbs, amino acids, and certain other substances.

digestion the process of breaking food down to its smallest units in order for the nutrients to be absorbed and used by the body.

digestive system the system responsible for digestion, made up of digestive tract (mouth, esophagus, stomach, small intestine and large intestine), and accessory organs (salivary glands, liver, gallbladder, and pancreas).

dipeptide an amino acid chain that contains two amino acids.

disaccharide a carbohydrate that consists of two sugar molecules; examples include maltose, sucrose, and lactose.

diverticular disease condition in which there are small pouches or pockets in the wall or lining of the colon; a single pouch is called a diverticulum.

docosahexaenoic acid (DHA) a 22-carbon omega-3 fatty acid that can be produced in the body from the essential fatty acid alpha-linoleic acid, also found in oily fish.

duodenum the first portion of the small intestine after the stomach.

dual energy x-rays absorptiometry (DEXA) a method of determining body composition; a differential absorption of x-rays of two different energy levels are used to calculate bone mineral mass, lean soft-tissue mass, and fat mass.

eicosapentaenoic acid (EPA) a 20-carbon omega-3 fatty acid that can be produced in the body by the metabolism of the essential fatty acid alpha linolenic acid, or provided in the diet by oily fish.

E. coli a common form of bacteria, found in the intestines of many animals; certain strains are especially toxic and cause severe illness.

eating disorder an unhealthy and abnormal relationship with food and weight that threatens health and interferes with many areas of a person's life.

electrolytes electrically charged minerals (ions) dissolved in body fluids that balance the fluid outside the cells with the fluid inside the cells; electrolyte balance is necessary for the transport of nutrients, muscle contractions, and other processes.

embryo the developing human during the first two to eight weeks of gestation.

emulsification a process that allows lipids—fats—to mix with water; it is an essential part of digestion, as it aids in the absorption of dietary fats.

endosperm the element of a grain that contains the highest amount of starch and protein; this is all that remains when a grain is refined.

energy the capacity to do work; obtained through the breakdown of carbohydrates, proteins, and fats in foods and beverages.

energy density the amount of energy or calories in a given weight of food; generally presented as the number of calories in a gram (kcal/g).

energy metabolism a series of chemical reactions in the body that break down carbohydrates, fats, and proteins to release energy, as well as the chemical reactions that use energy to construct molecules and carry out body processes.

energy substrates macronutrients used by the body to provide energy; typically in the form of glucose or fatty acids.

enriched grains cereal grains that lost nutrients during processing but have vitamins and minerals added back in; refined grains are often enriched.

enrichment a process used to replace some of the B vitamins and iron that are extracted from grains when they are refined.

enzymes protein molecules that catalyze, or speed up, the rate at which a chemical reaction produces new compounds with altered chemical structures; enzyme names end in the suffix "-ase."

epidemiological study a scientific study that compares two groups of people, one of which has been exposed to an environmental factor, and one of which has not.

epigallocatechin gallate (EGCG) a flavonoid found abundantly in white and green tea; it may have anti-cancer, anti-obesity, anti-atherosclerotic, and anti-diabetic effects.

epigenetics a scientific field that examines the cross-generational effects of exposure to nutrients, toxins, and behaviors.

epinephrine (adrenaline) a hormone released from the adrenal glands to help the body prepare for a fight-or-flight response by increasing glucose availability in the blood; it can also be administered as an injection to treat the potentially life-threatening symptoms of anaphylaxis.

epithelial cells cells that form the skin and mucus membranes inside the body, such as those present in our eyes, lungs, and intestines.

ergogenic aids physical, mechanical, nutritional, psychological, or pharmacological substances or treatments believed to improve athletic performance by either directly improving physiological variables associated with exercise performance or removing subjective restraints that may limit physiological capacity.

essential amino acids the nine amino acids that cannot be produced by the human body and therefore must be obtained from food; they include histadine, isoleucine, leucine, lysine, methionine, phenylalanine, threonine, tryptophan, and valine.

essential fatty acids fatty acids required in the diet because they cannot be synthesized by the human body; the two essential fatty acids are linoleic acid and linolenic acid.

essential nutrients nutrients that must be supplied through food because the body itself cannot produce or synthesize sufficient quantities to meet its needs.

Estimated Average Requirement (EAR) the average nutrient intake level estimated to meet the daily requirements of half of the healthy individuals for the different sexes and life-stage groups.

Estimated Energy Requirement (EER) estimated number of calories per day required to maintain energy equilibrium in a healthy adult; this value is dependent on age, sex, height, weight, and level of physical activity.

exercise intentional physical exertion that results in a healthy or healthier level of physical fitness and physical and mental health.

excretion elimination of waste from the body; digestive waste includes any food stuff not absorbed by the intestinal tract.

experiment a scientific procedure conducted to test a hypothesis.

experimental group the group in a scientific experiment that does receive the experimental treatment.

extracellular found or occurring outside the cells of the body.

facilitated diffusion movement of a substance across a cell membrane, down a concentration gradient, with the assistance of a specific transport protein.

fasting hypoglycemia low levels of blood glucose that occurs when people have not eaten, have drunk too much alcohol, or have underlying hormonal conditions or tumors.

fat-soluble vitamins essential micronutrients that are soluble in fat, require the presence of bile for absorption, and are stored in body tissue; these include vitamins A, D, E, and K.

fat substitute an additive that replaces fat in foods that is not absorbed by the body; provides a fat-like texture and produces similar sensation in the mouth.

fats a term for triglycerides, a subclass of lipids, that are the primary form of fat in our bodies and our food.

fatty acids a type of lipid that is the primary component of triglycerides and phospholipids; consists of a hydrocarbon chain, with a carboxyl group on one end and a methyl group on the other.

fecal-oral transmission a pathway for foodborne illness; it occurs when food is contaminated with either animal or human feces.

female athlete triad a condition recognized in female athletes; it is characterized by the interrelated energy restriction, menstrual dysfunction, and bone loss (osteoporosis).

fermentation the metabolic pathway through which microscopic yeast obtain energy from simple sugars in grains or fruit; the end product of this process is often alcohol.

fertility the capability to produce offspring and have a healthy pregnancy.

fetal alcohol spectrum disorders (FASDs) a group of conditions that are caused by use of alcohol during pregnancy; effects range in severity and can include physical, behavioral, and learning problems.

fetal alcohol syndrome mental and physical defects observed in infants born to mothers who consumed a significant amount of alcohol during pregnancy.

fetus the developing human from eight weeks of gestation until birth.

fiber a diverse group of polysaccharides, including cellulose and hemicellulose, found in plants' cell walls; they may be straight chains of glucose or they may be branched chains composed of a variety of sugars. Fiber is abundant in legumes, nuts, whole grains, vegetables, and some berries but humans lack the digestive enzymes to break down plant fiber, so it passes undigested through the digestive tract.

flexibility activities activities that promote the ability to move joints through their whole span of movement.

flexitarian diet a diet that is mostly plant-based, but occasionally includes meat.

fluoride a trace mineral that plays a role in the prevention of dental caries through the hardening and maintenance of tooth enamel; it is also important in stabilizing bone structure.

folate also known as vitamin B_9; it acts as a coenzyme in the synthesis of nucleic acids (DNA and RNA) and metabolism of certain amino acids; the terms folate and folic acid are often used interchangeably.

food additives any substance added to a food product to prevent spoiling, modify flavor or texture, or change nutritional value.

food allergen a substance in food (usually protein) that the body identifies as harmful and that elicits an allergic reaction from the immune system.

food allergy a reproducible adverse reaction to a food that is caused by activation of an immune response.

foodborne illness a largely preventable disease or condition caused by consumption of a contaminated food or beverage that primarily affects the gastrointestinal tract.

foodborne infection an infection caused by consuming foods contaminated with a harmful bacterium, virus, or parasite that then multiplies in the intestines.

foodborne intoxication an illness that is caused by ingesting foods that contain a toxin that may be naturally present in the food, introduced by contamination with poisonous chemicals, or produced by bacteria or fungi growing on foods.

food desert a neighborhood or community with little access to a variety of affordable, healthy food.

Food and Drug Administration (FDA) the government agency responsible for the supply of safe food, regulation of additives, and labeling.

food defense an effort to protect the food supply against intentional contamination due to sabotage, terrorism, counterfeiting, or other illegal, intentionally harmful means.

food environment the social and physical factors that influence the foods we eat.

food insecurity lack of secure access to sufficient amounts of safe and nutritious food for normal growth and development and an active and healthy life.

food intolerance a reproducible adverse reaction to a food that is not a direct result of an immune response; not usually life-threatening.

food jags developmentally "normal" habits or rituals formed by children as they strive for more independence and control.

food labeling the declaration on a food package that describes the nutrient content and serving size of a food.

food safety the policies and practices that apply to the production, handling, preparation, and storage of food in order to prevent contamination and foodborne illness.

Food Safety Modernization Act (FSMA) signed into law in 2011, the FSMA aims to ensure that the U.S. food supply is safe from contamination by giving federal regulators more proactive authority in how food is grown, harvested, and processed.

food security always having physical, social, and economic access to sufficient, safe, and nutritious food that meets dietary needs and food preferences for an active and healthy life.

foremilk breast milk produced at the beginning of feeding; it contains lactose and proteins, but little fat.

fortification the addition of vitamins and/or minerals to a food product.

free radicals substances that are naturally formed in the body or present in the environment that have an unpaired electron; at high levels they damage cells, including DNA, through a process called oxidation.

Freshman 15 a popular term used to describe the pattern of weight gain associated with the first year of college; in reality, weight gain averages three to four pounds during this time.

fructose a monosaccharide found in fruits, vegetables, and honey.

functional fiber nondigestible carbohydrates that are isolated from plants and animals and later added to foods; examples include psyllium and pectin.

functional food a food that provides additional health benefits beyond basic nutrition that may reduce disease risk or promote good health; these are also known as nutraceuticals.

galactose one of the monosaccharides that make up milk sugar (lactose).

gallstones small, pebble-like substances that develop in the gallbladder.

gastric juice secreted by the stomach during digestion; contains hydrochloric acid, proteases, lipase, and mucus.

gastroesophageal reflux disease (GERD) a recurrent and more serious form of acid reflux, accompanied by inflammation and/or erosion of the esophageal lining.

gastrointestinal tract a tubular channel extending from the mouth to the anus where digestion and absorption occur; also called the alimentary tract.

"generally recognized as safe" (GRAS) any substance intentionally added to food that is generally recognized, among qualified experts, as having been adequately shown to be safe under the conditions of its intended use; not subject to premarket review and approval by the FDA.

genetics a scientific field that describes how genes encoded in DNA are passed on between generations.

genetically modified (GM) foods foods derived from organisms that have had their genetic material (DNA) modified in some way that does not occur naturally; this may be achieved by introducing a gene from a different organism.

genetically modified organism (GMO) living organisms whose genetic material has been altered through the use of genetic engineering; GMOs are the source of genetically modified foods.

germ the embryo of the whole grain seed that germinates and grows; contains essential fatty acids and a number of B vitamins and minerals.

gestation the time during which the embryo develops in the uterus—from fertilization to birth.

gestational diabetes a condition of elevated blood glucose levels arising in approximately 18% of all pregnant women, most of whom revert to normal blood glucose levels after delivery.

ghrelin a hormone, produced primarily in the stomach, that stimulates hunger before a meal.

glucagon a hormone produced in the pancreas that increases glucose availability in the blood.

glucose a monosaccharide that circulates in the bloodstream; found in fruits, vegetables, and honey.

glycemic index a number used to rank carbohydrate foods by their ability to raise blood glucose levels compared with a reference standard.

glycemic load the extent of increase in blood glucose levels, calculated by multiplying glycemic index by the carbohydrate content of a food.

glycerol a three-carbon compound that makes up the backbone of a triglyceride molecule.

glycogen a complex carbohydrate that is the storage form of glucose in animals; it is found primarily in skeletal muscle and the liver, and is the most highly branched of all polysaccharides.

glycolysis an anaerobic energy system that breaks down glucose into a three-carbon molecule called pyruvate, producing ATP in the process.

goiter enlargement of the thyroid gland, most often caused by lack of iodine in the diet.

growth charts a series of percentile curves illustrating the distribution of selected body measurements—weight, height, body mass index—in children ages 2 to 20 years.

growth spurts periods of accelerated physical development in children and adolescents with associated changes in height and weight.

HACCP (Hazard Analysis and Critical Control Points) preventative food safety system that addresses the manufacture, storage, and distribution of food products.

health claims a statement on a packaged food or dietary supplement that indicates a link between a food, food component, or dietary supplement and a reduction in the risk of a disease; all health claims must be approved by the FDA.

healthy diet an eating pattern characterized by variety, balance, adequacy, and moderation that promotes health and reduces risk of chronic disease.

heavy drinking typically defined as consuming 15 or more standard alcohol drinks per week for men and 8 or more per week for women; heavy drinking can increase alcohol tolerance by increasing the rate at which alcohol is metabolized.

helminth large, multicellular parasitic worms; those that live in the intestinal tract are called intestinal parasites.

hemochromatosis a hereditary condition that results in increased absorption of iron and high iron stores; it can often result in iron toxicity.

heme iron iron derived from hemoglobin; the most bioavailable form of dietary iron; found in animal foods.

hemoglobin the oxygen-carrying protein in red blood cells.

hemorrhaging uncontrolled bleeding, can be caused by vitamin K deficiency.

hemorrhoids swollen or inflamed veins in anus or lower rectum.

herbal supplement a type of dietary supplement that includes plants (botanicals), single or in combination; typically dried preparations of flowers, leaves, roots, bark, and seeds.

high-density lipoprotein (HDL) a lipoprotein responsible for transporting cholesterol from the bloodstream and tissues back to the liver.

hindmilk breast milk produced towards the end of feeding; it contains more fat than foremilk.

hormones chemical substances that serve as messengers in the control and regulation of body processes.

hunger the biological impulse that drives us to seek out and consume food to meet our energy needs.

hydrochloric acid (HCl) a component of gastric juices that helps unfold proteins during digestion.

hydrodensitometry a method of determining body composition by measuring the difference in body weight in air and body weight under water during brief immersion.

hydrogenation chemical process by which hydrogen molecules are added to unsaturated fatty acids to make them more solid.

hygiene hypothesis the idea that reduced exposure to childhood infections in developed countries may explain drastic increases in the rates of allergies and other immune disorders, such as asthma, type 1 diabetes, celiac disease, and multiple sclerosis.

hypercalcemia a high level of calcium in the blood.

hyperglycemia abnormally high blood glucose levels.

hypertension blood pressure of at least 140/90 mmHg most of the time; a major risk factor for cardiovascular disease and stroke.

hypervitaminosis A vitamin A toxicity typically caused by supplementation; complications include liver damage and birth defects.

hypervitaminosis D a vitamin D toxicity typically as a result of oversupplementation.

hypoglycemia abnormally low blood glucose levels, resulting in symptoms of anxiety, hunger, sweating, and heart palpitations.

hypokalemia decreased blood levels of potassium; it is characterized by fatigue, muscle weakness, abnormal heart rhythms, increased calcium excretion, and reduced insulin production.

hyponatremia a condition characterized by a low serum sodium concentration and clinical signs of confusion, nausea, vomiting, bloating, and swelling around the brain; it may be seen in athletes who consume excess water with no sodium.

hypovitaminosis A a deficiency of vitamin A; in early stages it causes night blindness and scaly skin, but can eventually cause permanent blindness.

hypothesis a proposed explanation for an observation that can be tested through experimentation.

ileum the last and longest portion of the small intestine.

immune response the process by which the body recognizes and protects itself against invading pathogens and foreign substances.

immunoglobin E a class of antibody released in response to an allergen.

incomplete protein a food that lacks or supplies low amounts of one or more essential amino acids.

inflammatory bowel syndrome (IBD) general name for diseases that cause inflammation and irritation of the gastrointestinal tract; examples include Crohn's disease and ulcerative colitis.

ingestion the process of taking food or other substances into the gastrointestinal tract via the mouth.

ingredients list a list of ingredients on a food package presented in descending order of amount, measured by weight, according to the guidelines of the Nutrition Labeling and Education Act.

initiation the first step in the development of cancer, in which a cell's DNA is mutated, making the cell more likely to divide than normal.

insoluble fiber a type of fiber that does not dissolve in water, increases transit time through the GI tract, and contributes "bulk" to stool, fostering regular bowel movements.

insulin a hormone produced in the pancreas that acts to regulate glucose in the bloodstream; insulin lowers blood glucose levels and promotes transport of glucose into the muscle cells and other tissues.

insulin pump a medical device used by individuals with type 1 diabetes that delivers insulin as needed.

insulin resistance a condition in which cells have a decreased sensitivity to insulin, resulting in impaired glucose uptake from blood, increased blood glucose levels, and further insulin release from the pancreas.

interstitial fluid a fluid that surrounds the cells of the body and creates an aqueous environment.

intracellular found within the body's cells.

iodine a trace mineral that is necessary for the creation of thyroid hormones in the body; food sources include table salt, milk, and processed grain products.

ions elements that carry a positive or negative charge; the charge of an ion allows it to participate in chemical reactions and bond with other molecules.

iron the most abundant trace mineral in the human body; it is a crucial component of hundreds of enzymes and other proteins in the body.

iron-deficiency anemia a condition characterized by fatigue, decreased immune function, and impaired development; caused by reduced levels of iron in the blood, which decreases oxygen transport around the body.

irradiation technology that uses low doses of radiation to destroy insects and bacteria in foods to improve food safety.

irritable bowel syndrome (IBS) a group of symptoms that occur together: abdominal pain or discomfort, along with diarrhea and/or constipation.

jejunum the middle portion of the small intestine.

ketoacidosis the formation of ketones from fatty acids that may occur when diabetes is left untreated.

ketone bodies compounds synthesized from fatty acids by the liver when insulin levels are low.

kilocalories (kcal) standard unit to measure energy provided by food.

kwashiorkor a condition caused by an inadequate protein intake with reasonable caloric intake; it is characterized by a swollen belly, caused by edema and a fatty liver.

lactate (lactic acid) a molecule formed by glycolysis when the energy demands of skeletal muscle are high.

lactase an enzyme that breaks down lactose into glucose and galactose.

lactation production and secretion of milk from the mammary glands.

lactose a disaccharide sometimes called "milk sugar," as it is found only in milk, yogurt, and other dairy products; made up of glucose and galactose.

lacto-ovo vegetarian a vegetarian diet consisting of plant foods plus dairy (lacto) and egg (ovo) products.

lactose intolerance a condition characterized by diminished levels of the enzyme lactase and subsequent reduced ability to digest the disaccharide lactose.

large intestine consists of the cecum, colon, and the rectum; extracts electrolytes, some fatty acids, and water from digested food before it is excreted as feces.

lecithin the most abundant phospholipid in the body; it's frequently added to food products like salad dressings as an emulsifier because of its ability to keep water and lipids from separating.

leptin a hormone produced by adipose tissue that plays a role in body fat regulation and long-term energy balance; when fat stores increase, leptin production increases and acts on the brain.

life expectancy the average number of years individuals in a specific population are expected to live.

lifespan span of time between the birth and death of a person.

limiting amino acid the amino acid in the shortest supply relative to its requirement for protein synthesis.

linoleic acid an omega-6 polyunsaturated essential fatty acid.

linolenic acid (alpha-linolenic acid) an omega-3 polyunsaturated essential fatty acid; it is modified in the body to produce EPA and DHA.

lipases enzymes that break down fats (triglycerides) by releasing one or more fatty acids.

lipids structurally diverse group of naturally occurring molecules made up of carbon, hydrogen, and oxygen that are generally insoluble in water, but are soluble in organic solvents; examples include fatty acids, triglycerides, sterols, and phospholipids.

lipoproteins particles formed by the assembly of proteins and phospholipids that transport lipids in lymph and in blood.

Listeria monocytogenes a foodborne pathogen that is the third leading cause of death due to food-borne illness in the United States; pregnant women and their unborn children are particularly susceptible.

listeriosis a serious infection caused by eating food tainted with the bacterium *Listeria monocytogenes,* which can cause premature birth, miscarriage, fetal death, and newborn illness, as the bacteria can cross the placenta.

lobules a group of alveoli (milk-producing sacs) in the breasts.

longevity the length or duration of life; often refers to individuals whose lifespan is longer than the population average.

low-density lipoprotein (LDL) a lipoprotein responsible for transporting primarily cholesterol from the liver through the bloodstream to the tissues.

lumen inner space of the GI tract lined with mucosal cells.

lycopene a carotenoid found in tomatoes, watermelon, and pink grapefruit; it may reduce the risk of developing cataracts, and prostate and ovarian cancers.

lymphatic system a system of vessels in which the products of fat digestion, among other things, are transported from the GI tract to the blood.

macronutrients a term used to describe nutrients that we require relatively large daily amounts; these include carbohydrates, proteins, water, and fats.

magnesium a mineral that is a cofactor in more than 300 chemical reactions in the body; as an ion it plays a role in muscle contraction, nerve impulse conduction, and heart rate maintenance.

major minerals minerals with a daily requirement of 100 milligrams or more; examples include sodium, potassium, chloride, calcium, magnesium, phosphorus, and sulfur.

malnutrition a state of undernutrition or overnutrition caused by inadequate, excessive, or unbalanced intake of calories and/or essential nutrients.

maltase an enzyme that breaks maltose down into two glucose units.

maltose a disaccharide formed in large amounts as a product of starch digestion; however, very little is found in the foods we eat.

manganese a trace mineral that functions as a co-factor for enzymes involved in antioxidant function; it is involved in the metabolism of carbohydrates, cholesterol, and amino acids.

mannitol a sweetener used primarily in making chewable tablets; likely to cause a laxative effect when consumed in large amounts.

marasmus a condition caused by inadequate intake of protein, calories, and other nutrients; it is characterized by emaciation.

maximal oxygen consumption (VO_{2max}) the maximum amount of oxygen that can be used at the peak of intense aerobic exercise.

mechanical digestion physical breakdown of food by mastication (chewing) and mixing with digestive fluids.

Mediterranean diet a dietary pattern rich in fruit, vegetables, nuts, olive oil, and whole grains and low in processed and red meats, dairy products, and sweets; it is followed in Mediterranean countries and has been proven to produce health benefits.

megaloblastic anemia a type of anemia characterized by large, immature, and sometimes irregularly shaped red blood cells; usually caused by folate or vitamin B_{12} deficiency.

messenger ribonucleic acid (mRNA) the type of RNA that carries the genetic code for a specific protein from the nucleus to the cytoplasm where proteins are made.

metabolic syndrome a cluster of risk factors associated with the development of cardiovascular disease and type 2 diabetes.

metabolism is all of the life-sustaining chemical reactions that occur in living organisms that convert one molecule into another molecule.

metastasis during the process of cancer progression, mutated cells acquire additional mutations that allow the cancer cells to migrate to, and invade, other tissues.

methyl group a group of three hydrogen atoms bonded to a carbon atom found at one end (the "omega" end) of the fatty acid chain.

micronutrients term used to describe nutrients essential in our daily diet to maintain good health, but required in only small amounts; these include vitamins and minerals.

microvilli very small projections that protrude from the absorptive mucosal cells of the villi in the small intestine; responsible for increasing the surface area for absorption twenty-fold.

minerals inorganic chemical elements obtained through foods that are essential in human nutrition; there are 16 minerals considered essential in human nutrition, with diverse regulatory and structural functions.

mitochondria small organelles within the cytoplasm of cells where the vast majority of all ATP is produced.

moderation an aspect of healthy eating; not overindulging in any one type of food, food group, or in potentially harmful foods.

modifiable risk factors behaviors that can increase risk of disease; these include smoking, sedentary behavior, obesity, excessive alcohol consumption, high blood pressure, and poor dietary habits.

molybdenum a trace mineral that is a co-factor for four enzymes in the body.

monosaccharide a carbohydrate that consists of only one sugar molecule; examples include glucose, fructose, and galactose.

monounsaturated fatty acids a fatty acid with only one double bond between carbons in the carbon chain; they are abundant in olive oil, canola oil, and nuts.

morning sickness nausea and vomiting often experienced during the first trimester of pregnancy; it is triggered by hormonal changes, and may sometimes cause unique food aversions.

motility contractions of the smooth muscles of the GI tract that mix food with digestive fluids and propel food along the length of the tract.

mucosa tissues that line the gastrointestinal tract, made up of mucosal cells (epithelial cells).

mucus a viscous solution that lubricates and protects the GI tract.

mutation a permanent change in the DNA sequence of a gene.

myocardial infarction damage to heart tissue caused by decreased blood flow to the coronary arteries; also known as a heart attack.

myoglobin a protein that functions to provide oxygen to muscles; contains less iron than hemoglobin.

MyPlate a visual presentation of foods from five food groups "on a dinner plate" to represent the ideal balance that will provide a spectrum of nutrients.

National Health and Nutrition Examination Survey (NHANES) a group of studies designed to assess the health and nutritional status of adults and children in the United States.

National School Lunch Program a federally assisted meal program operating in public and non-profit private schools and residential child care institutions that provides nutritionally balanced, low-cost (or free) lunches to children each school day.

NEAT (nonexercise activity thermogenesis) the energy expended for everything we do that is not sleeping, eating, or sports-like exercise.

negative energy balance the state in which energy intake is less than energy expenditure resulting in weight loss.

neotame a non-nutritive sweetener that is very similar in structure to aspartame, but has much greater stability in solution and can withstand high temperatures encountered during baking.

neural tube defects malformations of the brain, spine or spinal column of a developing embryo; these abnormalities are present at birth.

neuromotor activities activities that incorporate motor skills such as balance, coordination, and agility (also known as functional fitness).

neurotransmitters chemical substances involved in transmitting signals between nerve cells.

niacin also known as vitamin B_3; it plays an important role in energy metabolism, and is commonly found in meats, fish, peanuts, mushrooms, and fortified cereals.

niacin equivalents (NEs) the total amount of niacin that is provided by a food from both the preformed vitamin and that which can be synthesized in the body from tryptophan in the food.

nitrogen balance a reflection of protein (nitrogen) intake versus protein (nitrogen) breakdown; indicates if the body is gaining, losing, or maintaining protein.

nonessential amino acids amino acids that the body can make and therefore need not be obtained through diet.

non-heme iron a form of iron derived from plant foods; it is less bioavailable than heme iron.

nutrients chemical substances obtained from food that are essential for body function; they are needed for metabolism, growth, development, reproduction, and tissue maintenance and repair.

nutrient claims declarations on food packages to indicate a possibly beneficial level of nutrient (e.g., "high fiber," "low fat," etc.), federally regulated to be consistent with labeling laws.

nutrient deficiency a condition resulting from insufficient supply of essential nutrients through dietary inadequacy or impaired absorption or use.

nutrient density the amount of nutrients supplied by a food in relation to the number of calories in that food.

nutrition an interdisciplinary study of factors that affect our food choices, the physiological processes involved in processing and delivering the chemical components of those foods to cells throughout our body, and ultimately how those chemicals affect our health.

Nutrition Facts Panel provides specific information about the calorie content and nutrition for specific components, including serving size, number of servings, and number of calories per serving, as well as information on the amount of dietary fat, cholesterol, dietary fiber, dietary sodium, carbohydrates, dietary proteins, vitamins, and minerals in each serving.

Nutrition Labeling and Education Act (NLEA) an act that allows the FDA to require nutrition labeling of most prepared foods and dietary supplements.

nutritional genomics a field of science that studies the effect of food on genes.

obesity a condition characterized by accumulation of excess body fat, generally associated with adverse health effects such as heart disease, hypertension, type 2 diabetes, arthritis, and gout.

Olestra a fat substitute that has chemical components similar to those of triglycerides, but they are in a different configuration, and cannot be digested or absorbed by the body.

oligosaccharides a short-chain carbohydrate that consists of 3 to 10 monosaccharide units joined together.

omega-3 fatty acid a polyunsaturated fatty acid that has the first double bond at the third carbon molecule from the methyl end of the chain; it is associated with a decreased risk of cardiovascular disease and improved brain function. Dietary sources include walnuts, flax seeds, soy, canola oil, and chia seeds.

omega-6 fatty acid a polyunsaturated fatty acid that has the first double bond at the sixth carbon molecule from the methyl end of the carbon chain; it is needed for normal growth and for synthesis of important hormone-like compounds called eicosanoids. Dietary sources include cooking oils, salad dressings, nuts, and seeds.

organic foods produced through approved methods that integrate cultural, biological, and mechanical practices that foster the recycling of resources, promote ecological balance, and conserve biodiversity.

organic compounds compounds that contain both carbon-carbon and carbon-hydrogen bonds.

oral glucose tolerance test (OGTT) a test used to diagnose prediabetes and diabetes; it measures the body's response to glucose in the bloodstream.

osmosis a process by which water passes between intracellular and extracellular spaces through cellular membranes.

osteomalacia the adult form of rickets, caused by a vitamin D deficiency. It is characterized by softening and weakening of bones accompanied by pain in the pelvis, lower back, and legs.

osteopenia a condition characterized by low bone mineral density.

osteoporosis a bone disease in which the bone density and total mass are decreased, leading to porous bones, increased fragility, and susceptibility to fractures.

overnutrition excess intake or imbalance of calories and/or essential nutrients relative to need that results in adverse health effects.

overweight a moderate amount of excess body fat or an excess amount of body weight that may come from muscles, bone, fat, and water.

oxidation a loss of electrons; in the body, this results in damage to cells, and can eventually lead to diseases like cancer and heart disease.

pancreatic juice released by the pancreas during digestion; contains bicarbonate and enzymes.

pantothenic acid also known as vitamin B_5; it has critical functions in energy metabolism and is also required for the synthesis of fatty acids, cholesterol, steroid hormones, and two neurotransmitters.

parasites organisms that live on or in a host, and obtain the nutrients they need from the host; they vary widely in size.

parathyroid hormone (PTH) a hormone released from the parathyroid gland in response to low serum calcium levels.

pathogens disease-causing microorganisms, such as viruses, parasites, and bacteria.

pellagra a disease caused by niacin deficiency; its characteristic symptoms are scaly skin, mouth sores, dermatitis, diarrhea, and dementia.

pepsin an enzyme produced by the cells lining the stomach that is involved in protein digestion.

peptide short chain of amino acids attached together.

peptide bond the bond that forms between two amino acids in the formation of a protein molecule.

percent Daily Value (%DV) an estimation of the amount of a specific nutrient contained in one serving, expressed as a percentage of the Daily Value, based on a daily intake of 2,000 kcal; DVs were developed specifically for nutrition labels.

peristalsis rhythmic, wavelike contractions of the smooth muscle of the GI tract.

pernicious anemia a condition caused by a failure to produce intrinsic factor, resulting in vitamin B_{12} malabsorption and megaloblastic anemia.

pescatarian diet a semi-vegetarian diet that excludes meats and poultry, but includes plant foods, dairy foods, eggs, fish, and shellfish.

phospholipids a molecule that is both hydrophobic and hydrophilic, and is required to form cell membranes; an example is lecithin, which can be found in egg yolks, liver, and some plant products.

phosphorus the second most abundant mineral in the body; it plays a critical role in bone health and energy metabolism, and is an essential component of bone and cartilage, phospholipids, DNA, and RNA.

photosynthesis the process by which plants convert the energy of the sun into chemical energy in the form of carbohydrates.

physical activity any bodily movement produced by skeletal muscles that causes energy expenditure.

physical fitness the ability to perform moderate to vigorous activity without undue fatigue.

phytochemicals compounds found in plant foods that are physiologically active and beneficial to human health; they are not considered to be essential nutrients.

pica an eating disorder characterized by the desire to eat nonfood substances such as dirt, clay, or paint chips; pica may be a sign of anemia.

placebo a medication or treatment used in scientific experiments that contains no active properties.

placebo effect apparent effect experienced by a patient in response to a "fake" treatment due to the patient's expectation of an effect.

placenta organ within the uterus that allows for exchange between maternal and fetal circulations, via the umbilical cord.

plant-based diet a diet that emphasizes whole plant foods, limits processed foods, and may or may not include foods of animal origin.

plaque a waxy accumulation of cholesterol and triglycerides on the lining of the blood vessels; this can block the passage of blood flow, eventually leading to tissue damage or tissue death.

polypeptide an amino acid chain that contains many amino acids; several polypeptides can combine to form a protein.

polyphenols the most abundant and diverse photochemical in our diet; they are particularly rich in berries, coffee, tea, red wine, cocoa powder, nuts, and spices.

polysaccharide a long-chain carbohydrate that consists of more than 10 monosaccharides joined together; they tend not to have a sweet taste and can be found in foods such as whole grain breads, dried beans, and starchy vegetables.

polyunsaturated fatty acids a fatty acid with two or more double bonds between carbons in the carbon chain; they are found in corn, safflower, sunflower, sesame, and soybean oils.

positive energy balance the state when energy intake exceeds energy expenditure that leads to weight gain.

potassium a mineral that is the primary electrolyte within cells; it works together with sodium and chloride to maintain fluid balance.

prebiotics nondigestible carbohydrates broken down by colon bacteria that foster the growth of good bacteria.

prediabetes a condition of higher-than-normal blood glucose levels, but not high enough to be diagnosed as diabetes; characterized by a fasting plasma glucose level of 100 mg to 125 mg per dl of blood.

preformed vitamins vitamins already present in their active form.

pregnancy the condition of being pregnant, encompassing the time from fertilization through birth.

probiotics live, beneficial bacteria found in fermented foods that can restore or maintain a healthy balance of "friendly" bacteria in the GI tract.

processed food any food that is altered from its raw form through processing such as canning, cooking, freezing, or milling; processing often involves adding ingredients such as sodium-containing additives and preservatives.

prostaglandins compounds derived from long-chain fatty acids that are released during injury or stress.

proteases enzymes released from the pancreas that break down proteins.

proteins large molecules consisting of carbon, hydrogen, oxygen, and nitrogen assembled in one or more chains of amino acids.

protein quality a measure of how well a protein meets our needs for protein synthesis; based on the proportion of essential amino acids present.

protein synthesis the process of building peptide chains and proteins from amino acids using information provided by genes; synthesis is a two-step process of transcription and translation.

protein turnover the continuous breakdown and re-assembly of proteins in the body.

provitamins the inactive form (or precursor) of a vitamin that requires conversion to the active form in order to fulfill biological functions in the body.

puberty dynamic period of development with periodic growth spurts that result in changes in body size, shape, composition, and sex-specific maturation.

pyridoxine also known as vitamin B_6; it functions as a coenzyme in the release of glucose from stored glycogen and in amino acid metabolism.

pyruvate a compound produced from the breakdown of glucose; lactate is produced from pyruvate and is produced faster than it can be metabolized in mitochondria.

qualified health claim a claim that describes the relationship between an ingredient and a reduced risk of a disease or condition, when the evidence linking a food, food component, or supplement to a reduced risk of a disease is emerging but does not meet the standard for a health claim.

quasi-vegetarian diet a diet that excludes red meat, but includes other animal products.

quercetin a flavonoid found in red and yellow onions, kale, capers, and cranberries; it has been shown to have anti-inflammatory effects, and may reduce the risk of heart disease and cancer.

randomized controlled trials a scientific experiment that rigorously compares experimental interventions with controls, and randomly assigns people to each category to offset any potential bias.

reactive hypoglycemia low levels of blood glucose that occur after eating large amounts of carbohydrates, causing a huge release of insulin and rapid drop in blood glucose.

Recommended Dietary Allowance (RDA) the recommended nutrient intake levels that meet the daily needs and decrease risks of chronic disease in almost all healthy people for different sexes and life-stage groups.

refined grain cereal grains that have been milled, a process that removes the bran and germ, leaving only the endosperm; white flour is an example of this.

registered dietitian a food and nutrition expert who has met the minimum academic and professional requirements to qualify for the credential.

resistance activities activities performed against an opposing force that increase muscle strength, improve body composition, and promote healthy bone density.

resistant starch a starch that remains intact after cooking, is not broken down by human digestive enzymes, and is not absorbed from the intestines. Diets high in resistant starch may improve insulin sensitivity and prolong the sensation of fullness.

respiration the process of transporting oxygen from the air to the cells within tissues (inhalation) and transporting carbon dioxide from cells to the air (exhalation).

retinal a form of vitamin A that is critical for vision and is derived from the conversion of retinol in the body.

retinoic acid a form of vitamin A derived from retinal; it is essential for growth and development.

retinol the active form of vitamin A in animal and fortified foods; it's also the form of vitamin A that is stored in the body and transported between tissues.

retinol activity equivalents (RAE) a measure of vitamin A activity; it accounts for differences in bioavailability between preformed vitamin A (retinol) and provitamin A carotenoids.

rhodopsin a pigment in the retina that absorbs light and triggers nerve impulses to the brain for vision.

riboflavin also known as vitamin B_2; it plays an important role in energy metabolism, and is found in dairy products.

riboflavinosis a condition caused by a riboflavin deficiency; it is characterized by cracks and redness of the lips and corners of the mouth, swelling of tissues in the mouth, and a sore throat.

rickets a condition caused by vitamin D deficiency, characterized by "bowed" legs due to impaired bone mineralization, softening of bones, skeletal malformations, and muscular weakness.

risk factor any characteristic, condition, or behavior that increases the likelihood of developing a particular disease.

saccharin a non-nutritive sweetener widely used in sugar-free soft drinks and as a tabletop sweetener; it can be used in baking without losing its sweetness.

saliva fluid secreted from salivary glands in the mouth to moisten food and provide lubrication.

satiation the sense of fullness while eating that leads to the termination of a meal.

satiety the effect that a meal has on our interest in food after and between meals and when we feel hungry again.

saturated fatty acid a fatty acid that contains no double bonds between carbons in the carbon chain and carries the maximum number of hydrogen atoms; they are relatively solid at room temperature. Examples include animal fats, coconut oil, and palm kernel oil.

scientific method a specific series of steps that involves a hypothesis, measurements and data gathering, and interpretation of results.

scurvy a disease caused by a deficiency of vitamin C resulting in bleeding gums, bruising, joint pain, and impaired wound healing.

segmentation process in which circular muscles in the small intestine contract to mix intestinal contents with digestive fluids and bring nutrients in the intestinal fluid in contact with the intestine's absorptive surface.

select nutrients of concern nutrients that a particular demographic may not consume in adequate quantities.

selenium a trace mineral that functions as a co-factor for several antioxidant enzymes and with other selenium-dependent enzymes in the activation of thyroid hormone; it is essential for normal growth, development, and metabolism.

semi-vegetarian diet a mostly plant-based diet that restricts the consumption of some meats.

side chain a component of an amino acid that distinguishes one amino acid from another; they vary in length and composition.

signs objective evidence of disease that are observed by health care professionals, such as a rash or abnormal blood tests.

simple carbohydrates short carbohydrates made up of one or two sugar units (monosaccharides and disaccharides); also known as sugars.

simple diffusion movement of a substance across a cell membrane, down a concentration gradient.

skinfold thickness a method of determining body composition; calipers are used to measure the skinfold thickness at multiple sites, and these measurements are used to calculate an estimate of body fat.

small for gestational age (SGA) a birth weight that is below the 10th percentile of gestational age; SGA infants are at higher risk of health complications during and after childbirth.

small intestine the primary site for the digestion of food and the absorption of nutrients; it's split into three sections, the duodenum, the jejunum, and the ileum.

sodium a mineral that is commonly consumed in table salt, also known as sodium chloride (NaCl); sodium is the primary electrolyte responsible for maintaining fluid balance between cells and throughout the body.

solid fats foods or food ingredients that contain high amounts of saturated fats, which make them solid at room temperature. Reducing dietary solid fats is an important way to reduce saturated fat intake and excess calories.

solubility the ability of a substance (solute) to dissolve in a solution (solvent).

soluble fiber a type of fiber that dissolves in water and often forms a viscous gel that acts to slow digestion and lower blood cholesterol and risk of heart disease; it is also often readily fermented by bacteria in the colon.

solute a dissolved substance.

solvent a liquid substance that is capable of dissolving another substance such as water.

sorbitol a nutritive sweetener used in sugarless gums, chocolate candies, and ice cream; it is not metabolized by bacteria in the mouth and therefore does not promote tooth decay. Excess amounts can cause a laxative effect.

sphincter a ringlike muscle that relaxes or contracts to open or close a bodily passageway.

Staphylococcus aureus one of the leading causes of foodborne illness in the United States; it is a bacterium that, under certain conditions, releases a heat-stable toxin.

starches complex carbohydrates that are abundant in grains, legumes, and starchy vegetables; they consist of long chains of glucose joined together by digestible bonds. Examples include amylose and mylopectin.

sterols complex lipids with four interconnected carbon rings with a hydrocarbon side chain; a precursor for synthesis of steroid hormones.

stevia a non-nutritive sweetener that is isolated from the leaves of the South American plant; it is heat stable and used primarily in beverages, as a tabletop sweetener, and in yogurt.

stroke a cerebral event that occurs when blood vessels supplying the brain are damaged or blocked.

structure/function claims a statement on the label of a packaged food or dietary supplement about how that product might affect the human body's structure or function.

sucrase an enzyme that breaks sucrose into fructose and glucose.

sucralose a non-nutritive sweetener made from sucrose by replacing 3 —OH groups with chlorine. It is used as a tabletop sweetener and it is widely used in beverages where it is remarkably stable over long periods.

sucrose a disaccharide made up of fructose and glucose; also known as table sugar.

sulfur a mineral that is present in protein-rich foods; it is present in the vitamins thiamin and biotin, as well as the amino acids cysteine and methionine.

Supplements Facts Panel package label that must indicate that the product is a supplement, not a conventional food, and must include serving size and amount of product per serving size, the percent of Daily Value that a particular ingredient or nutrient provides per serving, and a list of the product's dietary ingredients.

sustainability the use of resources at rates that do not exceed the capacity of Earth to replace them; for food, a sustainable system implies safety and security of the food supply, a strong food industry in terms of jobs and growth, and, environmental sustainability in terms of biodiversity, water, and soil quality.

symptoms subjective evidence of disease that are experienced by the individual that only they can perceive, such as a stomachache or fatigue.

tagatose a monosaccharide that is almost identical to fructose, but provides fewer calories because it is poorly absorbed; used to provide both bulk and sweetness in foods such as ice cream, cakes, and candies. Also known as Naturlose.

taste buds taste receptor cells found on the tongue within the papillae that are involved in sensing foods on the basis of specific flavors, such as sweet, sour, salty, bitter, and umami.

thermic effect of food (TEF) the energy needed to digest, absorb, and metabolize nutrients in food.

thiamin also known as vitamin B_1; it's needed to provide energy from the breakdown of glucose, fatty acids, and some amino acids, as well as the production of sugars needed for the synthesis of RNA and DNA.

thyroid gland a gland located in the neck that releases the iodine-containing thyroid hormones, which are involved in metabolic processes.

thyroid stimulating hormone (TSH) a hormone released from the anterior pituitary that stimulates the thyroid gland to produce and secrete thyroid hormones.

tissue a collection of cells that have a similar origin.

tocopherols a group of fat-soluble vitamin E molecules, primarily found in vegetable oils.

Tolerable Upper Intake Level (UL) the highest amount of a specific nutrient that most people can consume daily without risk of side effects.

total carbohydrates total amount of starch, sugars, and dietary fiber in one serving.

total energy expenditure (TEE) the total amount of energy expended through basal metabolism, thermic effect of food, and activity energy expenditure.

trace minerals minerals with a daily requirement of less than 100 milligrams; examples include iron, zinc, copper, iodine, selenium, molybdenum, fluoride, manganese, and chromium.

trans fatty acids fatty acids created by adding hydrogen to liquid vegetable oils (partial hydrogenation) to make them more solid; intake of trans fatty acids is associated with an increased risk of chronic diseases, such as heart disease and cancer.

transcription the first step in protein synthesis; the process by which information encoded in genes (DNA) is used to make messenger RNA.

translation the second step in protein synthesis; process by which the information in messenger RNA is translated into a protein.

triglycerides the storage form of fat, made up of three fatty acid chains attached to the three carbons on a glycerol molecule.

trimester one-third of the normal gestation period of a pregnancy; weeks 1 to 13 are considered the first trimester, weeks 14 to 26 are the second semester, and weeks 27 to 40 make up the third trimester.

tripeptide an amino acid chain that contains three amino acids.

type 1 diabetes mellitus an autoimmune disease characterized by elevated blood glucose levels, caused by destruction of the cells in the pancreas that normally produce insulin.

type 2 diabetes mellitus a condition characterized by elevated blood sugar levels due to reduced insulin sensitivity (or resistance) and some impairment of insulin secretion from the pancreas.

ulcer irritation or perforation of stomach (gastric) or small intestinal (duodenal) mucosal wall, caused by *Helicobacter pylori* infection, decreased mucus production, or impaired removal of stomach acid.

ultratrace minerals minerals that are found in the body, but not considered "essential" at this time; they include arsenic, boron, nickel, silicon, and vanadium.

umbilical cord a ropelike structure that supplies the fetus with nutrients and oxygen and removes waste.

undernourishment the inability, for a year or more, to acquire enough food to meet dietary energy requirements; undernourishment impairs growth and development and increases susceptibility to infections.

undernutrition inadequate nourishment caused by insufficient dietary intake of one or more essential nutrients or poor absorption and/or use of nutrients in the body.

unsaturated fatty acids a fatty acid that has at least one double bond between carbons in the carbon chain and has fewer than the maximum possible number of hydrogen atoms; they are generally liquid and room temperature and are most often found in plant foods such as seeds, nuts, grains, and vegetable oils.

uterus a muscular organ that holds the developing fetus.

variety an aspect of healthy eating involving choosing many different foods from all food groups, ensuring a broad range of nutrients consumed.

vegan diet a diet that eliminates all foods of animal origin.

vegetarian diet a diet consisting of plant-based foods, which excludes all meats, fish, and shellfish, but may include dairy products and eggs.

very low-density lipoprotein (VLDL) a lipoprotein that originates in the liver and is responsible for transporting primarily triglycerides to adipose tissue, cardiac muscle, and skeletal muscles.

villi fingerlike projections that protrude from the absorptive mucosal cells of the small intestine into the lumen of the GI tract; responsible for increasing the available surface area for absorption.

virus an infectious agent that consists of genetic material surrounded by a protein shell; viruses can only multiply inside the living cells of a host, and are thus unable to multiply in foods.

visceral fat fat in the abdominal area that surrounds the body's internal organs, which has been shown to be an independent health risk.

vitamins organic compounds that are required in small quantities for specific functions in the body.

vitamin A a fat-soluble vitamin that is integral to vision and cell differentiation; in some situations, vitamin A functions as a hormone.

vitamin D a fat-soluble vitamin that is involved in bone growth and maintenance, cell development, and immunity; vitamin D also functions as a hormone.

vitamin E a fat-soluble vitamin that functions as an antioxidant.

vitamin K a fat-soluble vitamin that is involved in blood clotting and bone formation.

vomiting forcible ejection of contents of stomach through the mouth; may be self-induced or due to sickness or food-borne illness.

water an essential nutrient that has critical functions in the body.

water-soluble vitamins vitamins that disperse easily in water-based solutions; these include the B vitamins, vitamin C, and the vitamin-like nutrient choline.

Wernicke–Korsakoff syndrome a condition caused by thiamin deficiency; it generally affects alcoholics, and can result in severe neurological disturbances.

whole grain cereal grains, or foods made from cereal grains, that contain all the essential parts, including the energy-rich endosperm, the oil-rich germ, and the fiber-rich bran coating of the entire grain seed in its original proportions.

xylitol a sweetener used in mouthwash, sugarless gums, and candies; it does not promote tooth decay, and causes a cooling sensation in the mouth when used in chewing gums and hard candies.

zinc a trace mineral required for the function of thousands of proteins and enzymes in the body.

REFERENCES

Chapter 1

Academic Medical Centre Department Clinical Epidemiology and Biostatistics. (2012). *Dutch Famine Study*. Retrieved from http://www.hongerwinter.nl. Accessed October 3, 2015.

Austin, G. L., Ogden, L. G., Hill, J. O. (2011, April). Trends in carbohydrate, fat, and protein intakes and association with energy intake in normal-weight, overweight and obese individuals: 1971–2006. *American Journal of Clinical Nutrition, 93*(4): 836–843. doi: 10.3945/ajcn.110.000141.

Academy of Nutrition and Dietetics. (n.d.) Becoming an RD/DTR: The Basics for Students from the Academy of Nutrition and Dietetics. Retrieved from http://www.eatright.org/BecomeanRDorDTR/content.aspx?id=8142. Accessed October 3, 2015.

Dietary Reference Intakes for Calcium and Vitamin D. (2011). *Institute of Medicine of the National Academies.* Retrieved from http://iom.nationalacademies.org/~/media/Files/Report%20Files/2010/Dietary-Reference-Intakes-for-Calcium-and-Vitamin-D/Vitamin%20D%20and%20Calcium%202010%20Report%20Brief.pdf. Accessed October 7, 2015.

Dwyer, J. (2006, February). Starting down the right path: nutrition connections with chronic diseases of later life. *American Journal of Clinical Nutrition, 88*(2): 4155–4205.

Epigenetics. (2007, July 24). *PBS NOVA*. Retrieved from http://www.pbs.org/wgbh/nova/body/epigenetics.html. Accessed October 7, 2015.

Heron, M. (2013, December 20). Deaths: Leading Causes for 2010. *National Vital Statistics Report, 62*(6). Retrieved from http://www.cdc.gov/nchs/data/nvsr/nvsr62/nvsr62_06.pdf. Accessed October 7, 2015.

Interpreting Science in a Social Media World (2010, May). *Food Insight*. Retrieved from http://www.foodinsight.org/Newsletter/Detail.aspx?topic=Interpreting_Science_in_a_Social_Media_World. Accessed October 7, 2015.

Kyle, U. G., & Pichard, C. (2006, July). The Dutch Famine of 1944–1945: a pathophysiological model of long-term consequences of wasting disease. *Current Opinion in Clinical Nutrition and Metabolic Care, 9*(4): 388–394.

Lumney, L. H., Stein, A. D., & Susser, E. (2011, April). Prenatal famine and adult health. *Annual Review of Public Health, 32*(10), 237–262.

Nutrigenetic Testing: Tests Purchased from Four Websites Mislead Consumers: Hearing before the Special Committee on Aging, *U.S. Senate* (2006, July 27) (Testimony of Gregory Kutz). Retrieved from http://www.gao.gov/assets/120/114612.pdf. Accessed October 7, 2015.

Ravelli, G., Stein, A. S., & Susser, M. W. (1976, August 12). Obesity in young men after famine exposure in utero and early infancy. *New England Journal of Medicine, 295*(7): 349–353.

Roseboom, T. J., van der Meulen, J. H., Osmond, C., Barker, D. J., Ravelli, A. C., & Bleker, O. P. (2011, July). Adult survival after prenatal exposure to the Dutch famine 1944–45. *Paediatric and Perinatal Epidemiology, 15*(3): 220–225.

United States Department of Agriculture. (2014). Diet and Disease. Retrieved from http://fnic.nal.usda.gov/diet-and-disease. Accessed October 7, 2015.

United States Department of Health and Human Services. (2014, May 9). Healthy People 2020. Retrieved from http://www.healthypeople.gov/2020/default.aspx. Accessed October 7, 2015.

Waterland, R. A., & Randy, L. J. (2003), August). Transposable elements: targets for early nutritional effects on epigenetic gene regulation. *Molecular and Cellular Biology, 23*(15): 5293–5300.

Wolff, G. L., Kodell, R. L., Moore, S. R., & Cooney, C. A. (1998, August). Maternal epigenetics and methyl supplements affect *agouti* gene expression in A^{vy}/a mice. *Federation of American Societies for Experimental Biology Journal, 12*(11): 949–957.

World Health Organization. (2003). Diet Nutrition and the Prevention of Chronic Diseases. *WHO Technical Report Series, (916)*.

Chapter 2

Aaron, K. J., & Saunders, P. W. (2013). The role of dietary salt and potassium intake in cardiovascular health and disease: a review of the evidence. *Mayo Clinic Proceedings, 88*: 987–995.

Davis, C., & Saltos, E. (199, May). Dietary recommendations and how they have changed over time. *Agriculture Information Bulletin, 750*: 33–50. Washington, DC: U.S. Department of Agriculture.

Food and Drug Administration Safety and Innovation Act (FDASIA): Request for Comments on the Development of a Risk-Based Regulatory Framework and Strategy for Health Information Technology. (2013, May 30). *Federal Register, 78*(104): 32390–32391.

Fung, B. (2012, September 14). People who read nutrition labels are skinnier. *The Atlantic*. Retrieved from http://www.theatlantic.com/health/archive/2012/09/people-who-read-nutrition-labels-are-skinnier/262393/. Accessed October 7, 2015.

Japanese Food Guide Spinning Top. (2014). *Japanese Ministry of Health, Labour and Welfare & the Ministry of Agriculture, Forestry and Fisheries.* Retrieved from http://www.mhlw.go.jp/english/. Accessed October 5, 2015.

Lee, H. (2012, April). The role of local food availability in explaining obesity risk among young school-aged children. *Social Science and Medicine, 74*(8): 1193–1203.

Loureiro, M. L., Yen, S. T., & Nayga, R. M. (2012, March 27). The effects of nutritional labels on obesity. *Agricultural Economics, 43*(3): 333–342.

Manger Bouger Programme National Santé (2014). Retrieved from http://www.mangerbouger.fr/. Accessed October 7, 2015.

National Health and Nutrition Examination Survey (NHANES) (April 29, 2014). Centers for Disease Control and Prevention. Retrieved from http://www.cdc.gov/nchs/nhanes.htm/. Accessed October 7, 2015.

NuVal Homepage (2014). Retrieved from http://www.nuval.com. Accessed October 7, 2015.

Oldways Preservation Trust (2014). Retrieved from http://oldwayspt.org/. Accessed October 7, 2015.

Otten, J. J., Hellwig, J. P., & Meyers, L. D. (eds). (2006). *Dietary Reference Intakes: The Essential Guide to Nutrient Requirements*. Washington, DC: The National Academies Press.

Rogers, K. (2013, March 27). The Legacy of Lifestyle. *Scientific American*. Retrieved from http://blogs.scientificamerican.com/guest-blog/2013/03/27/the-legacy-of-lifestyle/. Accessed October 7, 2015.

U.S. Department of Agriculture (2014). SuperTracker. Retrieved from https://www.supertracker.usda.gov. Accessed October 7, 2015.

U.S. Department of Agriculture (2014). Retrieved from http://www.usda.gov/wps/portal/usda/usdahome/. Accessed October 7, 2015.

U.S. Food and Drug Administration (1998, June 11). Guidance and Regulation. Retrieved from http://www.fda.gov/Food/GuidanceRegulation/default.htm. Accessed October 7, 2015.

U.S. Food and Drug Administration (2005, September 28). Labeling and Nutrition. Retrieved from http://www.fda.gov/Food/IngredientsPackagingLabeling/LabelingNutrition/default.htm. Accessed October 7, 2015.

U.S. Department of Agriculture (2014). *ChooseMyPlate.gov*. Retrieved from http://www.choosemyplate.gov. Accessed October 7, 2015.

Walker, R. (2010, November 12). Walgreens tackles "food deserts." *The New York Times*. Retrieved from http://www.nytimes.com/2010/11/14/magazine/14fob-consumed-t.html?_r=0&adxnnl=1&adxnnlx=1400136651-yAHUP4vXhHO1ekYU0CojuA. Accessed October 7, 2015.

Watters, E. (2006, November 22). DNA is not destiny: The new science of epigenetics. *Discover Magazine*. Retrieved from http://healthypeople.gov/2020/default.aspx/

Wojcicki, J. M., & Heyman, M. B. (2012, May 28). Adolescent nutritional awareness and use of food labels: Results from the national nutrition health and examination survey. *BMC Pediatrics, 12*(55). doi:10.1186/1471-2431-12-55.

World Health Organization (2014). Retrieved from http://www.who.int/en/. Accessed October 7, 2015.

Wrigley, N., Warm, D., & Margetts, B. (2003). Deprivation, diet, and food-retail access: findings from the Leeds 'food deserts' study. *Environment and Planning Association, 35*(1): 151–188.

Chapter 3

Bieskieierski, J. R., Peters, S. L., Newnham, E. D., Rosella, O., Muir, J. R., & Gibson, R. (2013, August). No effects of gluten in patients with self-reported non-celiac gluten sensitivity after dietary reduction of fermentable, poorly absorbed, short-chain carbohydrates. *Gastroenterology, 145*(2): 320–328. doi:10.1053/j.gastro.2013.04.051.

Catassi, C., et al. (2013, September 26). Non-Celiac Gluten sensitivity: the new frontier of gluten related disorders. *Nutrients, 5*(10): 3893–3853. doi: 10.3390/nu5103839.

Di Sabatino, A., & Corazza, G. R. (2012, February 21). Nonceliac gluten sensitivity: sense or sensibility? *Annals of Internal Medicine, 156*(4): 309–311. doi: 10.7326/0003-4819-156-4-201202210-00010.

Grundmann, O., & Yoon, S. L. (2010, January 13). Irritable bowel syndrome: Epidemiology, diagnosis, and treatment: An update for health-care practitioners. *Journal of Gastroenterology and Hepatology, 25*(4): 691–699. doi: 10.1111/j.1440-1746.2009.06120.x

Kellow, N. J., Coughlan, M. T., & Reid, C. M. (2014, April 14). Metabolic benefits of dietary prebiotics in human subjects: a systematic review of randomized controlled trials. *British Journal of Nutrition, 111*(7): 1147–1161. doi: 10.1017/S0007114513003607.

Losowsky, M. S. (2008, April). A History of Coeliac Disease. *Digestive Diseases, 26*(2): 112–120. doi:10.1159/000116768

Ludvigsson, J. F., Rubio-Tapia, A., van Dyke, C. T., Melton, L. J., Zinsmeister, A. R., Lahr, B. D., & Murray, J. A. (2013, May). Increasing incidence of celiac disease in a North American population. *American Journal of Gastroenterology, 108*(5): 818–824. doi: 10.1038/ajg.2013.60.

Probiotics (2015, January) National Center for Complementary and Alternative Medicine (NCCAM). Retrieved from http://nccam.nih.gov/health/probiotics. Accessed October 7, 2015.

Puschnik, A., Lau, L., Cromwell, E. A., Balmaseda, A., Zompi, S., & Harris, E. (2013, June 13). Correlation between dengue-specific neutralizing antibodies and serum avidity in primary and secondary dengue virus 3 natural infections in humans. *PLoS Neglected Tropical Diseases, 7*(6): e2274. doi: 10.1371/journal.pntd.0002274

Riddle, M. S., Murray, J. A., & Porter C. K. (2012, August). The incidence and risk of celiac disease in a healthy US adult population. *American Journal of Gastroenterology, 107*(8): 1248–1255. doi: 10.1038/ajg.2012.130.

Roberfroid, M. (2007). Prebiotics: The Concept Revisited. *The Journal of Nutrition, 137*(3): 830S–837S.

Rubio-Tapia, A., et al. (2009, February 4). Increased Prevalence and Mortality in Undiagnosed Celiac Disease. *Gastroenterology, 137*(1): 88–93. doi: http://dx.doi.org/10.1053/j.gastro.2009.03.059

Rubio-Tapia, A., Ludvigsson, J. F., Brantner, T. L., Murray, J. A., & Everhart, J. E. (2012, October). The prevalence of celiac disease in the United States. *American Journal of Gastroenterology, 107*: 1538–1544. doi:10.1038/ajg.2012.219.

Sellitto, M., et al. (2012, March 14). Proof of concept of microbiome-metabolome analysis and delayed gluten exposure on celiac disease autoimmunity in genetically at-risk infants. *PLOS One, 7*(3). doi: 10.1371/journal.pone.0033387.

U.S. Department of Health and Human Services and U.S. Department of Agriculture (2016). *2015–2020 Dietary Guidelines for Americans.* 8th Edition. Retrieved from http://health.gov/dietaryguidelines/2015/guidelines/. Accessed January 15, 2016.

U.S. Department of Health and Human Services (2013, September 10). Digestive Diseases Statistics for the United States. Retrieved from http://digestive.niddk.nih.gov/statistics. Accessed October 7, 2015.

Volta, U., Calo, G., Tovoli, F., & De Giorgio, R. (2013, September). Non-celiac gluten sensitivity: questions still need to be answered despite increasing awareness. *Cellular and Molecular Immunology, 10*(5): 383–392. doi: 10.1038/cmi.2013.28.

Chapter 4

The Academy of Nutrition and Dietetics (2015). Position of the Academy of Nutrition and Dietetics: Health Implications of Dietary Fiber. *Journal of the Academy of Nutrition and Dietetics, 115*: 1861-1870. doi: 10.1016/j.jand.2015.09.003.

The Academy of Nutrition and Dietetics (2012). Position of the Academy of Nutrition and Dietetics: Use of Nutritive and Nonnutritive Sweeteners. *Journal of the Academy of Nutrition and Dietetics, 112*: 739–748. doi: 10.1016/j.jand.2012.03.009.

American Heart Association (2014, May 14). Sugars, Added Sugars, and Sweeteners. Retrieved from http://www.heart.org/HEARTORG/GettingHealthy/NutritionCenter/HealthyDietGoals/Sugars-and-Carbohydrates_UCM_303296_Article.jsp. Accessed October 7, 2015.

Cho, S. S., Qi, L., Fahey, G. C., & Klurfeld, D. M. (2013, June 26). Consumption of cereal fiber, mixtures of whole grains and bran, and whole grains and risk reduction in type 2 diabetes, obesity, and cardiovascular disease. *The American Journal of Clinical Nutrition, 98*: 594–619. doi: 10.3945/ajcn.113.067629

Ferretti, R. J., & Levander, O. A. (1974). Effects of milling and processing on the selenium content of grains and cereal products. *Journal of Agriculture and Food Chemistry, 22*(6): 1049–1051. doi: 10.1021/jf60196a057.

Food and Drug Administration (2006, February 15). FDA Provides Guidance on "Whole Grain" for Manufacturers. [Press Release]. Retrieved from http://www.fda.gov/newsevents/newsroom/pressannouncements/2006/ucm108598.htm. Accessed October 7, 2015.

Fornal, Ł., Soral-Śmietana, M., Śmietana, Z., & Szpendowski, J. (1987), Chemical Characteristics and Physicochemical Properties of the Extruded Mixtures of Cereal Starches. *Starch/Stärke, 39*: 75–78. doi: 10.1002/star.19870390303.

Mobley, A. R., Slavin, J. L., & Hornick, B. A. (2013, September). The future of recommendations on grain foods in dietary guidance. *Journal of Nutrition, 143*(9): 1527S-1532S. doi: 10.3945/jn.113.175737.

Mozaffarian, R. S., Lee, R. M., Kennedy, M. A., Luwig, D. S., Mozaffarian, D., & Gortmaker, S. L. (2013, December). Identifying whole grain foods: a comparison of different approaches for selecting more healthful whole grain products. *Public Health Nutrition, 16*(12): 2255–2264. doi: 10.1017/S1368980012005447.

Otten, J. J., Hellwig, J. P, & Meyers, L. D. (eds). (2006). Dietary Carbohydrates: Sugars and Starches. *The Essential Guide to Nutrient Requirements* (265–338). Washington, DC: The National Academies Press.

Otten, J. J., Hellwig, J. P., & Meyers, L. D. (eds). (2006). Dietary, Functional, and Total Fiber. *The Essential Guide to Nutrient Requirements* (339–421). Washington, DC: The National Academies Press.

U.S. Department of Agriculture (2014). Nutritive and Nonnutritive Sweetener Resources. *Food and Nutrition Information Center*. Retrieved from http://fnic.nal.usda.gov/food-composition/nutritive-and-nonnutritive-sweetener-resources. Accessed October 7, 2015.

U.S. Department of Health and Human Services and U.S. Department of Agriculture (2016). *2015–2020 Dietary Guidelines for Americans.* 8th Edition. Retrieved from http://health.gov/dietaryguidelines/2015/guidelines/. Accessed January 15, 2016.

Yao, B., et al. (2014, February). Dietary fiber intake and risk of type 2 diabetes: a dose-response analysis of prospective studies. *European Journal of Epidemiology, 29*(2): 79–88. doi: 10.1007/s10654-013-9876-x.

Chapter 5

The Academy of Nutrition and Dietetics (2013). Position of the Academy of Nutrition and Dietetics: Oral Health and Nutrition. *Journal of the Academy of Nutrition and Dietetics, 113*(5): 693–701.

Alberti, K. G., et al. (2009, October 20). Harmonizing the metabolic syndrome: a joint interim statement of the

International Diabetes Federation Task Force on Epidemiology and Prevention; National Heart, Lung, and Blood Institute; American Heart Association; World Heart Federation; International Atherosclerosis Society; and International Association for the Study of Obesity. *Circulation, 120*(16): 1640–1645. doi: 10.1161/CIRCULATIONAHA.109.192644.

American Diabetes Association (2008, January). Nutrition Recommendations and Interventions for Diabetes: A position statement of the American Diabetes Association. *Diabetes Care, 31*(Supplement 1): S61–S78. doi: 10.2337/dc08-S061.

American Diabetes Association (2013). Carbohydrate Counting. Retrieved from http://www.diabetes.org/food-and-fitness/food/what-can-i-eat/understanding-carbohydrates/carbohydrate-counting.html. Accessed October 7, 2015.

American Diabetes Association (2013). Common Diabetes Terms. Retrieved from http://www.diabetes.org/diabetes-basics/common-terms. Accessed October 7, 2015.

American Diabetes Association (2013). Diagnosing Diabetes and Learning About Prediabetes. Retrieved from http://www.diabetes.org/diabetes-basics/diagnosis. Accessed October 7, 2015.

American Diabetes Association (2015). Fast Facts: Data and Statistics and Diabetes. Retrieved from http://www.professional.diabetes.org/facts. Accessed October 7, 2015.

American Dietetic Association (2009). Position of the American Dietetic Association and American Society for Nutrition: Obesity, Reproduction, and Pregnancy Outcomes. *Journal of the American Dietetics Association, 109*(5): 918–927. doi: 10.1016/j.jada.2009.03.020

American Heart Association (2013). Diabetes: Statistical Fact Sheet 2013 Update. Retrieved from http://www.heart.org/idc/groups/heart-public/@wcm/@sop/@smd/documents/downloadable/ucm_319585.pdf. Accessed October 7, 2015.

Atkinson, F. S., Foster-Powell, K., & Brand-Miller, J. C. (2008, December). International Tables of Glycemic Index and Glycemic Load Values: 2008. *Diabetes Care, 31*(12): 2281–2283. doi: 10.2337/dc08-1239.

Brand-Miller, J., McMillian-Price, J., Steinbeck, K., & Caterson, I. (2009, August). Dietary glycemic index: health implications. *Journal of the American College of Nutrition, 28*(supplement): 446S–449S.

Callejas, D., et al. (2013, February 1). Treatment of diabetes and long-term survival following insulin and glucokinase gene therapy. *Diabetes, 62*(5): 1718–1729. doi: 10.2337/db12-1113

CDC Division of Diabetes Translation (2011, November). Maps of Diagnosed Diabetes and Obesity in 1994, 2000, and 2010. Retrieved from http://www.cdc.gov/diabetes/statistics/slides/maps_diabetesobesity94.pdf. Accessed October 7, 2015.

Centers for Disease Control and Prevention (2014). 2014 National Diabetes Statistics Report. Retrieved from http://www.cdc.gov/diabetes/data/statistics/2014StatisticsReport.html. Accessed October 7, 2015.

Centers for Disease Control and Prevention (2014, April 17). Diabetes Public Health Resource. Retrieved from http://www.cdc.gov/diabetes. Accessed October 7, 2015.

Ceska, R. (2007, September). Clinical implications of the metabolic syndrome. *Diabetes and Vascular Disease Research, 4*(Supplement 3): S2–S4.

Colberg, S. R., et al. (2010, December). Exercise and type 2 diabetes, the American College of Sports Medicine and the American Diabetes Association: joint position statement. *Diabetes Care, 33*(12): e147-e167. doi: 10.2337/dc10-9990.

Davis, N., Forges, B., Wylie-Rosett, J. (2009, June). Role of obesity and lifestyle interventions in the prevention and management of type 2 diabetes. *Minerva Medica, 100*(3): 221–228.

Diabetes: MedlinePlus (2014). MedlinePlus. Retrieved from http://www.nlm.nih.gov/medlineplus/diabetes.html. Accessed October 7, 2015.

The diabetes pandemic (2011, July 9). *The Lancet, 879*(9786): 99. doi: 10.1016/S0140-6736(11)61068-4.

The discovery of insulin (2013). *Nobelprize.org*. Retrieved from http://www.nobelprize.org/educational/medicine/insulin/discovery-insulin.html. Accessed October 7, 2015.

Esfahani, A., et al. (2009, August). The glycemic index: physiological significance. *Journal of the American College of Nutrition, 28*(supplement): 439S–445S.

Esfahani, A., Wong, J. M., Mirrahimi, A., Villa, C. R., Kendall, C. W. (2011, January). The application of the glycemic index and glycemic load in weight loss: A review of the clinical evidence. *International Union of Biochemistry and Molecular Biology Life, 63*(1): 7–13. doi: 10.1002/iub.418.

Fitzgerald, K. (2013, April 3). Direct association between Type 2 Diabetes and obesity found. *Medical News Today*. Retrieved from http://www.medicalnewstoday.com/articles/258532.php. Accessed October 7, 2015.

Foster-Powell, K., Hold, S., & Brand-Miller, J. (2002, January). International table of glycemic index and glycemic load values: 2002. *American Journal of Clinical Nutrition, 76*(1): 5–56.

Glucose test—blood (2012, June 2). Retrieved from http://www.nlm.nih.gov/medlineplus/ency/article/003482.htm. Accessed October 7, 2015.

Han, S., Crowther, C. A., Middleton, P., Heatley, E. (2013, March 28). Different types of dietary advice for women with gestational diabetes mellitus. *The Cochrane Database of Systematic Reviews, 3*: CD009275. doi: 10.1002/14651858.CD009275.pub2.

Higgins, J. A. (2004, May–June). Resistant starch: metabolic effects and potential health benefits. *Journal of the AOAC International, 87*(3): 761–768.

Higgins, J. A., Higbee, D. R., Donahoo, W. T., Brown, I. L., Bell, M. L., Bessesen, D. H. (2004). Resistant starch consumption promotes lipid oxidation. *Nutrition & Metabolism, 1*(1): 8. doi: 10.1186/1743-7075-1-8

International Diabetes Federation (2005). IDF Worldwide Definition of the Metabolic Syndrome. Retrieved from http://www.idf.org/metabolic-syndrome. Accessed October 7, 2015.

Joslin Diabetes Center (n.d.). The History of Insulin. Retrieved from http://www.joslin.org/info/History_of_Insulin.html. Accessed October 7, 2015.

Joslin Diabetes Center (n.d.). Oral Diabetes Medications Summary Chart. Retrieved from http://www.joslin.org/info/oral_diabetes_medications_summary_chart.html. Accessed October 7, 2015.

Kahn, C. R., & Rafferty, J. F. (2000, February). The Joslin Diabetes Center. *Molecular Medicine, 6*(2): 65–68.

Kahn, S. E., Hull, R. L., & Utzschneider, K. M. (2006, December 14). Mechanisms linking obesity to insulin resistance and type 2 diabetes. *Nature, 444*: 840–846. doi:10.1038/nature05482.

Kilpatrick, E. S., Rigby, A. S., & Atkin, S. L. (2009, October). The diabetes control and complications trial: the gift that keeps on giving. *Nature Reviews Endocrinology, 5*(10): 537–545. doi: 10.1038/nrendo.2009.179.

Mattar, R., de Campos Mazo, D. F., & Carrilho, F. J. (2012). Lactose intolerance: diagnosis, genetic, and clinical factors. *Journal of Clinical and Experimental Gastroenterology, 5*:113–121. doi: 10.2147/CEG.S32368.

National Center for Chronic Disease Prevention and Health Promotion (2012). *Diabetes Report Card 2012.* Atlanta, GA: Centers for Disease Control and Prevention, US Department of Health and Human Services.

National Dairy Council (2009, October 20). Lactose intolerance rates may be significantly lower than previously believed. Retrieved from http://www.nationaldairycouncil.org/PressandMedia/NewsAlertArchives/Pages/LACTOSEINTOLERANCERATESMAYBESIGNIFICANTLYLOWERTHANPREVIOUSLYBELIEVED.aspx. Accessed October 7, 2015.

Nicklas, T. A., et al. (2009, September/October). Prevalence of self-reported lactose intolerance in a multiethnic sample of adults. *Nutrition Today, 44*(5): 222–227. doi: 10.1097/NT.0b013e3181b9caa6

Rickels, M. R., et al. (2013, August). Improvement in B-cell secretory capacity after human islet transplantation according to the CIT07 Protocol. *Diabetes, 62*(8): 2890–2897. doi: 10.2337/db12-1802

Sattley, M. (2008, December 17). The History of Diabetes. *Diabetes Health*. Retrieved from http://www.diabeteshealth.com/the-history-of-diabetes. Accessed October 7, 2015.

Savaiano, D. A, Boushey, C. J., & McCabe, G. P. (2006, April). Lactose intolerance symptoms assessed by meta-analysis: a grain of truth that leads to exaggeration. *Journal of Nutrition, 136*(4): 1107–1113.

Schmidt, S., Schelde, B., & Norgaard, K. (2013, March 21). Effects of advanced carbohydrate counting in patients with type 1 diabetes: a systematic review. *Diabetic Medicine.* doi: 10.1111/dme.12446.

Schwartz, S., Fabricatore, A. N., & Diamond, A. (2012). Weight reduction in diabetes. *Advances in Experimental Medicine and Biology, 771*: 438–458.

Schwingshackl, L., & Hoffmann, G. (2013, August). Long-term effects of low glycemic index/load vs. high glycemic index/load diets on parameters of obesity and obesity-associated risks: a systematic review and meta-analysis. *Nutrition, Metabolism, and Cardiovascular Diseases, 23*(8): 699–706. doi: 10.1016/j.numecd.2013.04.008.

Snell-Bergeon, J. K., Wadwa, R. P. (2012, June). Hypoglycemia, diabetes, and cardiovascular disease. *Diabetes Technology and Therapeutics, 14*(Supplement 1): S51–S58. doi: 10.1089/dia.2012.0031.

Tabak, A. G., Herder, C., Rathmann, W., Brunner, E. J., & Kivimaki, M. (2012, June 16). Prediabetes: a high-risk state for diabetes development. *The Lancet, 379*(9833): 2279–2290. doi: 10.1016/S0140-6736(12)60283-9.

Wilson, J. (2013, November 6). Cells offer hope for Type 1 diabetes. CNN Health. Retrieved from http://www.cnn.com/2013/11/06/health/islet-cell-transplantation-diabetes/index.html. Accessed October 7, 2015.

Wilt, T. J., et al. (2010, February). Lactose intolerance and health. *Evidence Report and Technology Assessment,* (192): 1–410.

Yandell, Kate (2013, February 15). Dogs cured of Type 1 diabetes. *The Scientist Magazine.* Retrieved from http://www.the-scientist.com/?articles.view/articleNo/34394/title/Dogs-Cured-of-Type-1-Diabetes. Accessed October 7, 2015.

Chapter 6

Akoh, C. C. (1998, March). Fat replacers. *Food Technology, 52*(3): 47–53.

American Heart Association (2013, May). About Cholesterol. Retrieved from http://www.heart.org/HEARTORG/Conditions/Cholesterol/AboutCholesterol/About-Cholesterol_UCM_001220_Article.jsp. Accessed October 7, 2015.

American Heart Association (2014, May). Know your Fats. Retrieved from http://www.heart.org/HEARTORG/Conditions/Cholesterol/PreventionTreatmentofHighCholesterol/Know-Your-Fats_UCM_305628_Article.jsp. Accessed November 6, 2015.

American Heart Association (2014, January). Meet the Fats. Retrieved from http://www.heart.org/HEARTORG/GettingHealthy/FatsAndOils/MeettheFats/Meet-the-Fats_UCM_304495_Article.jsp. Accessed October 7, 2015.

Astrup, A. (2005, February). The role of dietary fat in obesity. *Seminars in Vascular Medicine, 5*(1): 40–47.

Calder, P. C. (2012, July). The role of marine omega-3 (n-3) fatty acids in inflammatory processes, atherosclerosis and plaque stability. *Molecular Nutrition and Food Research, 56*(7): 1073–1080. doi: 10.1002/mnfr.201100710.

Centers for Disease Control and Prevention (2014). Diet/Nutrition FastStats. Retrieved from http://www.cdc.gov/nchs/fastats/diet.htm. Accessed October 7, 2015.

Deckelbaum, R. J., & Torrejon, C. (2012, March). The omega-3 fatty acid nutritional landscape: health benefits and sources. *Journal of Nutrition, 142*(3): 587S–591S. doi: 10.3945/jn.111.148080.

Eckel, R. H., Kris-Etherton, P., Lichtenstein, A. H., Wylie-Rosett, J., Groom, A., Stitzel, K. F., Yin-Piazza, S. (2009, February). Americans' awareness, knowledge, and behaviors regarding fats: 2006–2007. *Journal of the American Dietetic Association, 109*(2): 288–296. doi: 10.1016/j.jada.2008.10.048.

Institute of Medicine of the National Academies (2005). *Dietary Reference Intakes for Energy, Carbohydrate, Fat, Fatty Acids, Cholesterol, Protein, and Amino Acids.* Washington, DC: The National Academies Press.

International Food Information Council Foundation (2013, May). 2013 Food and Health Survey: Consumer Attitudes Towards Food Safety, Nutrition and Health. Retrieved from http://www.foodinsight.org/LinkClick.aspx?fileticket=spavtJtVkzM%3d&tabid=1482. Accessed October 7, 2015.

Lin, C. T., & Yen, S. T. (2010, April). Knowledge of dietary fats among US consumers. *Journal of the American Dietetic Association, 110*(4): 613–618. doi: 10.1016/j.jada.2009.12.020.

National Center for Complementary and Alternative Medicine (2009). Omega-3 Supplements: an Introduction. Retrieved from http://nccam.nih.gov/health/omega3/introduction.htm. Accessed October 7, 2015.

National Center for Health Statistics (2013). *Health, United States, 2012: With a Special Feature on Emergency Care.* Hyattsville, MD.

Phang, M., Sinclair, A. J., Lincz, L. F., & Garg, M. L. (2012, February). Gender—specific inhibition of platelet aggregation following omega-3 fatty acid supplementation. *Nutrition, Metabolism, and Cardiovascular Diseases, 22*(2): 109–114. doi: 10.1016/j.numecd.2010.04.012.

Ryan, K. K., Woods, S. C., & Seeley, R. J. (2012, January 11). Central nervous system mechanisms linking the consumption palatable high-fat diets to the defense of greater adiposity. *Cell Metabolism, 15*(2): 137–149. doi: 10.1016/j.cmet.2011.12.013/

Siriwardhana, N., Kalupahana, N. S., & Moustaid-Moussa, N. (2012). Health benefits of n-3 polyunsaturated fatty acids: eicosapentaenoic acid and docosahexaenoic acid. *Advances in Food and Nutrition Research, 65*: 2011–2022. doi: 10.1016/B978-0-12-416003-3.00013-5.

U.S. Department of Agriculture (2014). DRI Tables. Retrieved from http://fnic.nal.usda.gov/dietary-guidance/dietary-reference-intakes/dri-tables-and-application-reports. Accessed October 7, 2015.

U.S. Department of Agriculture and U.S. Department of Health and Human Services (2010, December). Food and Food Components to Reduce. *Dietary Guidelines for Americans, 2010* (7th ed.). Washington, DC: U.S. Government Printing Office.

U.S. Department of Health and Human Services and U.S. Department of Agriculture (2016). *2015–2020 Dietary Guidelines for Americans*. 8th Edition. Retrieved from http://health.gov/dietaryguidelines/2015/guidelines/. Accessed January 15, 2016.

Chapter 7

39 Years and Counting: Bogalusa Heart Study Charts Risk Factors in Child Health (2012, January 10). https://tulanesphtm.wordpress.com/2012/01/10/39-years-and-counting-bogalusa-heart-study-charts-risk-factors-in-child-health. Accessed October 7, 2015.

Academy of Nutrition and Dietetics (2007, September). Dietary Fatty Acids for Healthy Adults. *Journal of the Academy of Nutrition and Dietetics 107*(9): 1599–1611.

Academy of Nutrition and Dietetics (n.d.). Disorders of Lipid Metabolism: Evidence-based Nutrition Practice Guideline. Retrieved from http://andevidencelibrary.com/topic.cfm?cat=4527&auth=1. Accessed October 7, 2015.

American Heart Association (2013, November 12). AC/AHA Publish New Guideline for Management of Blood Cholesterol: Update focuses on lifestyle, statin therapy for patients who most benefit. Retrieved from http://newsroom.heart.org/news/acc-aha-publish-new-guideline-for-management-of-blood-cholesterol. Accessed October 7, 2015.

American Heart Association (2013). American Heart Association 2020 Impact Goal. Retrieved from http://www.heart.org/idc/groups/heart-public/@wcm/@sop/@smd/documents/downloadable/ucm_319831.pdf. Accessed October 7, 2015.

American Heart Association (2014). Drug Therapy for Cholesterol. Retrieved from http://www.heart.org/HEARTORG/Conditions/Cholesterol/PreventionTreatmentofHighCholesterol/Drug-Therapy-for-Cholesterol_UCM_305632_Article.jsp. Accessed October 7, 2015.

American Heart Association (2014). Fish and Omega-3 Fatty Acids. Retrieved from http://www.heart.org/HEARTORG/GettingHealthy/NutritionCenter/HealthyDietGoals/Fish-and-Omega-3-Fatty-Acids_UCM_303248_Article.jsp. Accessed October 7, 2015.

American Heart Association (2014). Understanding the new Prevention Guidelines. Retrieved from http://www.heart.org/HEARTORG/Conditions/Understanding-the-New-Prevention-Guidelines_UCM_458155_Article.jsp. Accessed October 7, 2015.

American Heart Association (2014). What Your Cholesterol Levels Mean. Retrieved from http://www.heart.org/HEARTORG/Conditions/Cholesterol/AboutCholesterol/What-Your-Cholesterol-Levels-Mean_UCM_305562_Article.jsp. October 7, 2015.

Baum, S. J., et al. (2012, May–June). Fatty Acids in cardiovascular health and disease: a comprehensive update. *Journal of Clinical Lipidology, 6*(3): 216–234. doi: 10.1016/j.jacl.2012.04.077.

Berenson, G. S. (2001, November). Bogalusa Heart Study: a long term community study of a rural biracial population. *American Journal of Medical Science, 322*(5): 293–300.

Centers for Disease Control and Prevention (2015, March). Facts About Cholesterol. Retrieved from http://www.cdc.gov/cholesterol/facts.htm. Accessed October 30, 2015.

Cheskin, L. J., Miday, R., Zorich, N., & Filoon, T. (1998, January 14). Gastrointestinal symptoms following consumption of olestra or regular triglyceride potato chips: a controlled comparison. *Journal of the American Medical Association, 279*(2): 150–152.

Coronary Heart Disease Risk Factors (2015, October). MedlinePlus. Retrieved from http://www.nhlbi.nih.gov/health/health-topics/topics/hd/atrisk. Accessed October 30, 2015.

de Lorgeril, M., & Salen, P. (2012, May 21). New insights into the health effects of dietary saturated and omega-6 and omega-3 polyunsaturated fatty acids. *BMC Medicine, 10*:50. doi: 10.1186/1741-7015-10-50.

Deckelbaum, R. J., & Torrejon, C. (2012, March). The omega-3 fatty acid nutritional landscape: health benefits and sources. *Journal of Nutrition, 142*(3): 587S–591S. doi: 10.3945/jn.111.148080.

Doell, D., Folmer, D., Lee, H., Honigfort, M., & Carberry, S. (2012). Updated estimate of trans fat intake by the US population. *Food Additives and Contaminants: Part A, Chemistry, Analysis, Control, Exposure & Risk Assessment, 9*(6): 861–872. doi: 10.1080/19440049.2012.664570

Dugan, A. (2013, August 6). Fast Food Still Major Part of U.S. Diet. Retrieved from http://www.gallup.com/poll/163868/fast-food-major-part-diet.aspx. Accessed October 30, 2015.

Go, A. S. (2013). Heart Disease and Stroke Statistics- 2013 Update. *Circulation, 127*: e6–e245. doi: 10.1161/CIR.0b013e31828124ad.

Grundy, S. M., et al. (2005, October 25). Diagnosis and management of the metabolic syndrome: an American Heart Association/National Heart, Lung and Blood Institute Scientific Statement. *Circulation, 112*(17): 2735–2752.

Hansson, G. K., & Hermansson, A. (2011, March). The immune system in atherosclerosis. *Nature Immunology, 12*(3): 204–212. doi: 10.1038/ni.2001.

Harris, W. S., et al. (2009). Omega-6 fatty acids and risk for cardiovascular disease: a science advisory from American Heart Association Nutrition Subcommittee of the Council on Nutrition, Physical Activity, and Metabolism; Council on Cardiovascular Nursing; and Council on Epidemiology and Prevention. *Circulation, 119*:902–907. doi: 10.1161/CIRCULATIONAHA.108.191627

Hendrich, S. (2010). (n-3) Fatty Acids: Clinical Trials in People with Type 2 Diabetes. *Advances in Nutrition, 1*: 3–7. doi: 10.3945/an.110.1003

Higdon, J. (2005, December). Essential Fatty Acids. *Linus Pauling Institute*. Retrieved from http://lpi.oregonstate.edu/infocenter/othernuts/omega3fa/. Accessed October 30, 2015.

Keany, J. F., Curfman, G. D., & Jarcho, J. A. (2014, January 16). A pragmatic view of the new cholesterol treatment guidelines. *New England Journal of Medicine, 370*: 275–278. doi: 10.1056/NEJMms1314569

Kotwal, S., Jun, M., Sullivan, D., Perkovic, V., & Neal, B. (2012, November). Omega 3 Fatty acids and cardiovascular outcomes: systematic review and meta-analysis. *Circulation: Cardiovascular Quality and Outcomes, 5*(6): 808–818. doi: 10.1161/CIRCOUTCOMES.112.966168.

Krychtiuk, K. A., Kastl, S. P., Speidl, W. S., & Wojta J. (2013). Inflammation and coagulation in atherosclerosis. *Hamostaseologie 33*(4):269–282. doi: 10.5482/HAMO-13-07-0039.

Lawrence, G. D. (2013, May). Dietary fats and health: dietary recommendations in the context of scientific evidence. *Advances in Nutrition, 4*: 294–302. doi: 10.3945/an.113.003657.

Massiera, F., et al. (2010, April 20). A Western-like fat diet is sufficient to induce a gradual enhancement in fat mass over generations. *Journal of Lipid Research, 51*: 2352–2361. doi: 10.1194/jlr.M006866.

Mereddu, G. F., Brandimarte, F., & DeLuca, L. (2012, September). High-density lipoprotein levels and risk of cardiovascular events: a review. *Journal of Cardiovascular Medicine, 13*(9): 575–586. doi: 10.2459/JCM.0b013e32834bb3c8.

Miller, M., et al. (2011, April 18). Triglycerides and cardiovascular disease: a scientific statement from the American Heart Association. *Circulation, 123*: 2292–2333. doi: 10.1161/CIR.0b013e3182160726.

Mosby, T. T., Cosgrove, M., Sarkardei, S., Platt, K. L., & Kaina, B. (2012, October). Nutrition in adult and childhood cancer: role of carcinogens and anti-carcinogens. *Anticancer Research, 32*(10): 4171–4192.

Mozaffarian, D., Appel, L. J., & van Horn, L. (2011). Recent advances in preventive cardiology and lifestyle medicine. *Circulation, 123*: 2870–2891.

National Center for Chronic Disease Prevention and Health Promotion (n.d.). Can Lifestyle Modifications Using Therapeutic Lifestyle Changes (TLC) Reduce Weight and the Risk for Chronic Disease? Research to Practice Series, No. 7. Retrieved from http://www.cdc.gov/nutrition/downloads/r2p_life_change.pdf. Acccessed October 30, 2015.

National Center for Complementary and Alternative Medicine (2009). Omega-3 Supplements: An Introduction. Retrieved from http://nccam.nih.gov/health/omega3/introduction.htm. Accessed October 30, 2015.

National Cholesterol Education Program (2001). ATP III At-A-Glance: Quick Desk Reference. Retrieved from http://www.nhlbi.nih.gov/guidelines/cholesterol/atglance.htm. Accessed October 30, 2015.

National Heart, Lung, and Blood Institute (2002). *Third Report of the National Cholesterol Education Program (NCEP) Expert Panel on Detection, Evaluation, and Treatment of High Blood Cholesterol in Adults (Adult Treatment Panel III)* Retrieved from http://circ.ahajournals.org/. Accessed October 30, 2015.

National Heart, Lung, and Blood Institute (2011). What are coronary heart disease risk factors? Retrieved from http://www.nhlbi.nih.gov/health/health-topics/topics/hd/. Accessed October 30, 2015.

National Heart, Lung, and Blood Institute (2011). What is Metabolic Syndrome? Retrieved from http://www.nhlbi.nih.gov/health/health-topics/topics/ms/. Accessed October 30, 2015.

National Heart, Lung, and Blood Institute (2013). About the Heart Truth. Retrieved from http://www.nhlbi.nih.gov/educational/hearttruth/about/index.htm. Accessed October 30, 2015.

National Heart, Lung, and Blood Institute (2014). Risk Assessment Tool for Estimating Your 10-year Risk of Having a Heart Attack. Retrieved from http://cvdrisk.nhlbi.nih.gov/calculator.asp. Accessed October 30, 2015.

National Heart, Lung and Blood Institute & Boston University (2014). *Framingham Heart Study.* Retrieved from http://www.framinghamheartstudy.org. Accessed October 30, 2015.

Nurses' Health Study 2 (2013). Retrieved from http://www.nhs3.org/index.php/our-story/20-nurses-health-study-2. Accessed October 30, 2015.

The Nurse's Health Study (n.d.). Retrieved from http://www.channing.harvard.edu/nhs/. Accessed October, 30 2015.

Oh, K., Hu, F. B., Manson, J. E., Stampfer, M. J., & Willett, W. C. (2005). Dietary fat intake and risk of coronary heart disease in women: 20 years of follow-up of the Nurses' Health Study. *American Journal of Epidemiology, 161*(7): 672–679. doi: 10.1093/aje/kwi085

Phillips, C. M., Kesse-Guyot, E., McManus, R., Hercberg, S., Lairon, D., Planells, R., & Roche, H. M. (2012, May). High dietary saturated fat intake accentuates obesity risk associated with the fat mass and obesity-associated gene in adults. *Journal of Nutrition, 142*(5): 824–831. doi: 10.3945/jn.111.153460.

Rees, K., Dyakova, M., Ward, K., Thorogood, M., & Brunner, E. (2013, March 28). Dietary advice for reducing cardiovascular risk. *Cochrane Database Systematic Review, 3*: CD002128. doi: 10.1002/14651858.CD002128.pub4.

Remig, V., Franklin, B., Margolis, S., Kostas, G., Nece, T., & Street, J. C. (2010, April). Trans fats in America: a review of their use, consumption, health implications, and regulation. *Journal of American Dietetic Association, 110*(4): 585–592. doi: 10.1016/j.jada.2009.12.024.

Schwingshackl, L., & Hoffman, G. (2012, December). Monounsaturated fatty acids and risk of cardiovascular disease: synopsis of the evidence available from systematic reviews and meta-analyses. *Nutrients 4*(12): 1969–2007. doi: 10.3390/nu4121989.

Simopoulos, A. P. (2002, October). The importance of the ration of omega-6/omega-3 essential fatty acids. *Biomedical Pharmacotherapy, 56*(8): 365–379.

Siri-Tarino, P. W., Sun, Q., Hu, F. B., & Krauss, R. M. (2010, March). Meta-analysis of prospective cohort studies evaluating the association of saturated fat with cardiovascular disease. *American Journal of Clinical Nutrition, 91*(3): 535–546. doi: 10.3945/ajcn.2009.27725.

Swidey, Neil. (2013, July 28). Walter Willett's food fight. *The Boston Globe Magazine*. Retrieved from http://www.bostonglobe.com/magazine/2013/07/27/what-eat-harvard-walter-willett-thinks-has-answers/5WL3MIVdzHCN2ypfpFB6WP/story.html. Accessed October 30, 2015.

Taata, Y., et al. (2013). Fish intake and risks of total and cause-specific mortality in 2 population-based cohort studies of 134,296 men and women. *American Journal of Epidemiology, 178*(1): 46–57.

Understanding cholesterol testing and results (2014, November 26). Medline Plus. Retrieved from http://www.nlm.nih.gov/medlineplus/ency/patientinstructions/000386.htm. Accessed November 6, 2015.

U.S. Department of Health and Human Services and U.S. Department of Agriculture (2016). *2015–2020 Dietary Guidelines for Americans.* 8th Edition. Retrieved from http://health.gov/dietaryguidelines/2015/guidelines/. Accessed January 15, 2016.

U.S. Department of Health and Human Services (2005, December). Your Guide to Lowering Your Cholesterol with TLC. Retrieved from http://www.nhlbi.nih.gov/health/public/heart/chol/chol_tlc.pdf. Accessed October 30, 2015.

World Heart Federation (n.d.). Cardiovascular disease risk factors. Retrieved from http://www.world-heart-federation.org/heart-facts/fact-sheets/cardiovascular-disease-risk-factors. Accessed October 30, 2015.

Yang, Q., et al. (2012, March 28). Trends in cardiovascular health metrics and associations with all-cause and CVD mortality among US adults. *Journal of the American Medical Association, 307*(12): 1273–1283. doi: 10.1001/jama.2012.339.

Chapter 8

Austin, G. L., Ogden, L. G., Hill, J. O. (2011, April). Trends in carbohydrate, fat, and protein intakes and association with energy intake in normal-weight, overweight and obese individuals: 1971–2006. *American Journal of Clinical Nutrition, 93*(4): 836–843. doi: 10.3945/ajcn.110.000141.

Bosse, J. D., & Dixon, B. M. (2012, September 8). Dietary protein to maximize resistance training: a review and examination of protein spread and change theories. *Journal of the International Society of Sports Nutrition, 9*(1): 42. doi: 10.1186/1550-2783-9-42.

Calculate the amount of protein that should be consumed daily based on the latest guidelines (n.d.). Retrieved from http://www.globalrph.com/protein-calculator.htm. Accessed October 30, 2015.

Calvez, J., Poupin, N., Chesneau, C., Lassale, C., & Tomé, D. (2012, March). Protein intake, calcium balance, and health consequence. *European Journal of Clinical Nutrition, 66*(3): 281–295. doi: 10.1038/ejcn.2011.196.

Cao, J. J., & Nielsen, F. H. (2010, November). Acid diet (high-meat protein) effects on calcium metabolism and bone health. *Current Opinions in Clinical Nutrition and Metabolic Care, 13*(6): 698–702. doi: 10.1097/MCO.0b013e32833df691.

Daily Protein Intake Per Capita (2011). Retrieved from http://chartsbin.com/view/1155. Accessed October 30, 2015.

Darling, A. L., Millward, D. J., Torgerson, D. J., Hewitt, C. E., & Lanham-New, S. A. (2009, December). Dietary protein and bone health: a systematic review and meta-analysis. *American Journal of Clinical Nutrition, 90*(6): 1674–1692. doi: 10.3945/ajcn.2009.27799.

Dideriksen, K., Reitelseder, S., & Holm, L. (2013, March 13). Influence of amino acids, dietary protein, and physical activity on muscle mass development in humans. *Nutrients 5*(3): 852–876. doi: 10.3390/nu5030852.

Feskens, E. J., Sluik, D., & van Woudenbergh, G. J. (2013, April). Meat consumption, diabetes, and its complications. *Current Diabetes Reports, 13*(2): 298–306. doi: 10.1007/s11892-013-0365-0.

Folgoni, V. L. (2008). Current protein intake in America: analysis of the National Health and Nutrition Examination Survey, 2003–2004. *American Journal of Clinical Nutrition, 87*(5): 1554S–1557S.

Fox, E. A., McDaniel, J. L., Breitbach, A. P., & Weiss, E. P. (2011, June 21). Perceived protein needs and measured protein intake in collegiate male athletes: an observational study. *Journal of the International Society of Sports Nutrition, 8*: 9. doi: 10.1186/1550-2783-8-9

Heaney, R. P., & Layman, D. K. (2008, May). Amount and type of protein influences bone health. *American Journal of Clinical Nutrition, 87*(5): 1567S–1570S.

Josse, A. R., Atkinson, S. A., Tarnopolsky, M. A., & Phillips, S. M. (2012, January). *Journal of Clinical Endocrinology and Metabolism, 97*(1): 251–260. doi: 10.1210/jc.2011-2165.

Kerstetter, J. E., Kenny, A. M., & Insogna, K. L. (2011, February). Dietary protein and skeletal health: a review of recent human research. *Current Opinions in Lipidology, 22*(1): 16–20. doi: 10.1097/MOL.0b013e3283419441.

Lagiou, P., Sadin, S., Lof, M., Trichopoulos, D., Adami, H., Weiderpass, E. (2012, June 26). Low carbohydrate- high protein diet and the incidence of cardiovascular diseases in Swedish women: prospective cohort study. *BMJ 344*: 4206. doi: 10.1136/bmj.e4026.

Micha, R., Michas, G., & Mozaffarian, D. (2012, December). Unprocessed red and processed meats and risk of coronary artery disease and type 2 diabetes—an updated review of the evidence. *Current Atherosclerosis Reports, 14*(6): 515–524. doi: 10.1007/s11883-012-0282-8.

Millward, D. J., Layman, D. K., Tomé, D., & Schaafsma, G. (2008, May). Protein quality assessment: impact of expanding understanding of protein and amino acid needs for optimal health. *American Journal of Clinical Nutrition, 87*(5): 1576S–1581S.

Paddo-Jones, D., Westman, E., Mattes, R. D., Wolfe, R. R., Astrup, A., & Westerterp-Plantenga, M. (2008, May). Protein, weight management, and satiety. *American Journal of Nutrition, 87*(5): 1558S–1561S.

Pan, A., et al. (2012, April 9). Red meat consumption and mortality: results from 2 prospective cohort studies. *Archives of Internal Medicine, 172*(7): 555–563. doi: 10.1001/archinternmed.2011.2287.

Phillips, S. M. (2012, August). Dietary protein requirements and adaptive advantages in athletes. *British Journal of Nutrition, 108*(Supplement 2): S158–S167. doi: 10.1017/S0007114512002516.

Phillips, S. M., & Van Loon, L. J. (2011). Dietary protein for athletes: from requirements to optimum adaption. *Journal of Sports Science, 29*(Supplement 1): S29–S38. doi: 10.1080/02640414.2011.619204.

Rohrmann, S., et al. (2013, March 7). Meat consumption and mortality—results from the European Prospective Investigation into Cancer and Nutrition. *BMC Medicine, 11*:63. doi: 10.1186/1741-7015-11-63.

Salehi, M., Moradi-Lakeh, M., Salehi, M. H., Nohomi, M., & Kolahdooz, F. (2013, May). Meat, fish, and esophageal cancer risk: a systematic review and dose-response meta-analysis. *Nutrition Review, 71*(5): 257–267. doi: 10.1111/nure.12028.

Santesso, N., et al. (2012, July). Effects of higher- versus lower-protein diets on health outcomes: a systematic review and meta-analysis. *European Journal of Clinical Nutrition, 66*(7): 780–788. doi: 10.1038/ejcn.2012.37

U.S. Department of Agriculture Dietary Advisory Committee (2010). Part D. Section 4: Protein. *Report of the Dietary Guidelines Advisory Committee on the Dietary Guidelines for Americans, 2010* (D4-1–D4-43).

U.S. Department of Health and Human Services and U.S. Department of Agriculture (2016). *2015–2020 Dietary Guidelines for Americans*. 8th Edition. Retrieved from http://health.gov/dietaryguidelines/2015/guidelines/. Accessed January 15, 2016.

Westerterp-Plantenga, M. S., Lemmens, S. G., & Westerterp, K. R. (2012, August). Dietary protein—its role in satiety, energetics, weight loss, and health. *British Journal of Nutrition, 108*(Supplement 2): S105–S112. doi: 10.1017/S0007114512002589.

Chapter 9

54 Interesting Facts about Vegetarianism and Veganism (2013, March). Retrieved from http://facts.randomhistory.com/facts-about-vegetarianism-and-veganism.html. Accessed October 30, 2015.

Barnard, N. D., Katcher, H. I., Jenkins, D. J., Cohen, J., & Turner-McGrievy, G. (2009, May). Vegetarian and vegan diets in type 2 diabetes management. *Nutrition Review, 67*(5): 255–263. doi: 10.1111/j.1753-4887.2009.00198.x.

Boeing, H., et al. (2012, September). Critical review: vegetables and fruit in the prevention of chronic diseases. *European Journal of Nutrition, 51*(6): 637–663. doi: 10.1007/s00394-012-0380-y.

Burkert, N. T., Muckenhuber, J., Großschädl F., Rasky, E., & Freidl, W. (2014, February 7). Nutrition and health—the association between eating behavior and various health parameters: a matched sample study. *PLoS One, 9*(2): e88278. doi: 10.1371/journal.pone.0088278.

Chan, J. M., Gong, Z., Holly, E. A., & Bracci, P. M. (2013). Dietary patterns and risk of pancreatic cancer in a large population-based case-control study in the San Francisco Bay Area. *Nutrition and Cancer, 65*(1): 157–164. doi: 10.1080/01635581.2012.725502.

Craig, W. J. (2009, May). Health effects of vegan diets. *American Journal of Clinical Nutrition, 89*(5): 1627S–1633S. doi: 10.3945/ajcn.2009.26736N.

Craig, W. J. (2010, December). Nutrition concerns and health effects of vegetarian diets. *Nutrition in Clinical Practice, 25*(6): 613–620. doi: 10.1177/0884533610385707.

Craig, W. J., Mangels, A. R., & American Dietetic Association (2009, July). Position of the American Dietetic Association: vegetarian diets. *Journal of the American Dietetic Association, 109*(7): 1266–1282.

Dangour, A. D., Lock, K., Hayter, A., Aikenhead, A., Allen, E., & Uauy, R. (2010, July). Nutrition-related health effects of organic foods: a systematic review. *American Journal of Clinical Nutrition, 92*(1): 203–210. doi: 10.3945/ajcn.2010.29269.

Estruch, R., et al. (2013, April 4). Primary prevention of cardiovascular disease with a Mediterranean diet. *New England Journal of Medicine, 368*(14): 1279–1290. doi: 10.1056/NEJMoa1200303.

Ferdowsian, H. R., & Barnard, N. D. (2009, October 1). Effects of plant-based diets on plasma lipids. *American Journal of Cardiology, 104*(7): 947–956. doi: 10.1016/j.amjcard.2009.05.032.

Forman, J., Silverstein, J., Committee on Nutrition, Council on Environmental Health & American Academy of Pediatrics (2012, November). Organic foods: health and environmental advantages and disadvantages. *Pediatrics, 130*(5): e1406–e1415. doi: 10.1542/peds.2012-2579.

Giacosa, A., et al. (2013, January). Cancer prevention in Europe: The Mediterranean diet as a protective choice. *European Journal of Cancer Prevention, 22*(1): 90–95. doi: 10.1097/CEJ.0b013e328354d2d7.

Huang, T., Yang, B., Zheng, J., Li, G., Wahlqvist, M. L., & Li, D. (2012). Cardiovascular disease mortality and cancer incidence in vegetarians: a meta-analysis and systematic review. *Annals of Nutrition and Metabolism, 60*(4): 233–240. doi: 10.1159/000337301.

Liu, R. H. (2003, September). Health benefits of fruit and vegetables are from additive and synergistic combinations of phytochemicals. *American Journal of Clinical Nutrition, 78*(Supplement 3): 517S–520S.

Lonnerdal, B. (2009, April 8). Soybean ferritin: implications for iron status of vegetarians. *The American Journal of Clinical Nutrition, 89*(5): 1680S–1685S. doi: 10.3945/ajcn.2009.26736W.

McEvoy, C. T., Temple, N., & Woodside, J. V. (2012, December). Vegetarian diets, low-meat diets, and health: a review. *Public Health Nutrition, 15*(12): 2287–2294. doi: 10.1017/S1368980012000936.

Messina, M., & Messina, V. (2010, August). The role of soy in vegetarian diets. *Nutrients, 2*(8): 855–888. doi: 10.3390/nu2080855.

Mosby, T. T., Cosgrove, M., Sarkardei, S., Platt, K. L., & Kaina, B. (2012, October). Nutrition in adult and childhood cancer: role of carcinogens and anti-carcinogens. *Anticancer Research, 32*(10): 4171–4192.

National Institutes of Health, Office of Dietary Supplements (2011). Vitamin B12: Dietary Supplement Fact Sheet. Retrieved from http://ods.od.nih.gov/factsheets/VitaminB12-HealthProfessional/. Accessed October 30, 2015.

Newport, Frank. (2012, July 26). In U.S., 5% Consider Themselves Vegetarians. Retrieved from http://www.gallup.com/poll/156215/consider-themselves-vegetarians.aspx. Accessed October 30, 2015.

Organic Nutrition Labels: What does the "organic" label really mean? (2012). Retrieved from http://usda-fda.com/organic-nutrition.htm. Accessed October 30, 2015.

Orlich, M. J., Singh, P. N., Sabaté, J., Jaceldo-Siegl, K., Fan, J., Knutsen, S., Beeson, W. L., & Fraser, G. E. (2013, July 8). Vegetarian dietary patterns and mortality in Adventist Health Study 2. *Journal of the American Medical Association Internal Medicine, 173*(13): 1230–1238. doi: 10.1001/jamainternmed.2013.6473.

Pettersen, B. J., Anousheh, R., Fan, J., Jaceldo-Siegl, K., & Fraser, G. E. (2012, October). Vegetarian diets and blood pressure among white subjects: results from the Adventist Health Study-2 (AHS 2). *Public Health Nutrition, 15*(10): 1909–1916. doi: 10.1017/S1368980011003454.

Robinson, R. (2013). *Eating on the Wild Side*. New York: Little Brown and Company.

Smith-Spangler, C., et al. (2012, September 4). Are organic foods safer or healthier than conventional alternatives? a systematic review. *Annals of Internal Medicine, 157*(5): 348–366. doi: 10.7326/0003-4819-157-5-201209040-00007.

Tantamango-Bartley, Y., Jaceldo-Siegl, K., Fan, J., & Fraser, G. (2013, February). Vegetarian diets and the incidence of cancer in a low-risk population. *Cancer Epidemiology Biomarkers & Prevention, 22*(2): 286–294. doi: 10.1158/1055-9965.EPI-12-1060.

Tonstad, S., Butler, T., Yan, R., & Fraser, G. E. (2009, May). Type of vegetarian diet, body weight, and prevalence of type 2 diabetes. *Diabetes Care, 32*(5): 791–796. doi: 10.2337/dc08-1886.

U.S. Department of Agriculture (n.d). *Vegetarian Nutrition* http://fnic.nal.usda.gov/lifecycle-nutrition/vegetarian-nutrition. Accessed October 30, 2015.

U.S. Department of Agriculture Agricultural Marketing Service (2014). National Organic Program. Retrieved from http://www.ams.usda.gov/AMSv1.0/nop. October 30, 2015.

U.S. Department of Agriculture Center for Nutrition Policy and Promotion (2011, June). Healthy Eating for Vegetarians. *DG Tip Sheet No. 8*. Retrieved from http://www.choosemyplate.gov/sites/default/files/tentips/DGTipsheet8HealthyEatingForVegetarians.pdf. Accessed October 30, 2015.

U.S. Department of Health and Human Services and U.S. Department of Agriculture (2016). *2015–2020 Dietary Guidelines for Americans*. 8th Edition. Retrieved from http://health.gov/dietaryguidelines/2015/guidelines/. Accessed January 15, 2016.

Vegetarian & Vegan Diet Pyramid (2015). Retrieved from http://oldwayspt.org/resources/heritage-pyramids/vegetarian-diet-pyramid/overview. Accessed October 30, 2015.

The Vegetarian Research Group (2014). Retrieved from http://www.vrg.org. Accessed October 30, 2015.

Weaver, C. M. (2009, May). Should dairy be recommended as part of a healthy vegetarian diet? *American Journal of Clinical Nutrition, 89*(5): 1634S–1637S. doi: 10.3945/ajcn.2009.26736O.

Zazpe, I., Sanchez-Tainta, A., Toledo, E., Sanchez-Villegas, A., & Martinez-Gonzalez, M. A. (2014, January). Dietary patterns and total mortality in a Mediterranean cohort: the SUN project. *Journal of the Academy of Nutrition and Dietetics, 114*(1): 37–47. doi: 10.1016/j.jand.2013.07.024.

Chapter 10

Bjelakovic, G., Nikolova, D., & Giuud, C. (2013, September 6). Meta-regression analyses, meta-analyses, and trial sequential analyses of the effects of supplementation with beta-carotene, vitamin A, and vitamin E singly or in different combinations on all-cause mortality: do we have evidence for lack of harm? *PLoS One, 8*(9): e74558. doi: 10.1371/journal.pone.0074558.

Cavalier, E., Delanaye, P., Shapelle, J. P., & Souberbielle, J. C. (2009). Vitamin D: current status and perspectives. *Clinical Chemistry and Laboratory Medicine, 47*(2) 120–127. doi 10.1515/CCLM.2009.036.

Diffey, B. L. (2010, August). Is casual exposure to summer sunlight effective at maintaining adequate vitamin D status? *Photodermatology, Photoimmunology, Photomedicine, 26*(4): 172–176. doi: 10.1111/j.1600-0781.2010.00518.x.

Diffey, B. L. (2010, June). Modelling the seasonal variation of vitamin D due to sun exposure. *British Journal of Dermatology, 162*(6): 1342–1348. doi: 10.1111/j.13652133.2010.09697

Ferland, G. (2012). The discovery of vitamin K and its clinical applications. *Annals of Nutrition and Metabolism, 61*(3): 213–218. doi: 10.1159/000343108.

Forrest, K. Y., & Stuhldreher, W. L. (2011, January). Prevalence and correlates of vitamin D deficiency in US adults. *Nutrition Research, 31*(1): 48–54. doi: 10.1016/j.nutres.2010.12.001.

Guiterrez-Mazariegos, J., Theodosiou, M., Campo-Paysaa, F., & Schubert, M. (2011, August). Vitamin A: a multifunctional tool for development. *Seminars in Cell and Developmental Biology, 22*(6): 603–610. doi: 10.1016/j.semcdb.2011.06.001.

Gunta, S. S., Thadhani, R. I., & Mak, R. H. (2013, June). The effect of vitamin D status on risk factors for cardiovascular disease. *Nature Reviews Nephrology, 9*(6): 337–347. doi: 10.1038/nrneph.2013.74.

Hariri, M., et al. (2013, May). B Vitamins and antioxidants intake is negatively correlated with risk of stroke in Iran. *International Journal of Preventative Medicine, 4*(Supplement 2): S284–S289.

Hojo, K., Watanabe, R., Mori, T., & Takemoto, N. (2007, September). Quantitative measurement of tetrahydromenaquinone-9 in cheese fermented by propionibacteria. *Journal of Dairy Science, 90*(9): 4078–4083.

Ku, Y. C., Liu, M. E., Ku, C. S., Liu, T. Y., & Lin, S. L. (2013, September 26). Relationship between vitamin D deficiency and cardiovascular disease. *World Journal of Cardiology, 5*(9): 337–346.

Lanska, D. J. (2010). Chapter 29: historical aspects of the major neurological vitamin deficiency disorders: overview and fat-soluble vitamin A. *Handbook of Clinical Neurology, 95*: 435–444. doi: 10.1016/S0072-9752(08)02129-5.

Manoury, E., Jourdon, K., Boyaval, P., & Fourcassie, P. (2013, March). Quantitative measurement of vitamin K2 (menaquinones) in various fermented dairy products using a reliable high-performance liquid chromatography method. *Journal of Dairy Science, 96*(3): 1335–1346. doi: 10.3168/jds.2012-5494.

Martineau, A., & Joliffe, D. (2014, July). Vitamin D and human health: from the gamete to the grave: report on a meeting held at Queen Mary University of London, 23rd–25th April 2014. *Nutrients, 6*(7) 2759–2919. doi: 10.3390/nu6072759.

Meydani, M. (2001, February). Vitamin E and atherosclerosis: beyond prevention of LDL oxidation. *Journal of Nutrition, 131*(2): 3665–3685.

Meydani, M., Han, S. N., & Wu, D. (2005). Vitamin E and immune response in the aged: molecular mechanisms and clinical implications. *Immunology Review, 205*:269–284.

Mitri, J. Muraru, M. D., & Pittas, A. G. (2011, September). Vitamin D and type 2 diabetes: a systematic review. *European Journal of Clinical Nutrition, 65*(9): 1005–1015. doi: 10.1038/ejcn.2011.118.

Muscogiuri, G., et al. (2012, February). Can vitamin D deficiency cause diabetes and cardiovascular diseases? Present evidence and future perspectives. *Nutrition, Metabolism, and Cardiovascular Diseases, 22*(2): 81–87. doi: 10.1016/j.numecd.2011.11.001.

National Cancer Institute (2014). Antioxidants and Cancer Prevention. Retrieved from http://www.cancer.gov/cancertopics/factsheet/prevention/antioxidants. Accessed October 30, 2015.

National Institutes of Health Office of Dietary Supplements (2013). Vitamin A—Health Professional Fact Sheet. Retrieved from http://ods.od.nih.gov/factsheets/VitaminA-HealthProfessional/. Accessed October 30, 2015.

National Institutes of Health Office of Dietary Supplements (2011). Vitamin D—Health Professional Fact Sheet. Retrieved from http://ods.od.nih.gov/factsheets/VitaminD-HealthProfessional/. Accessed October 30, 2015.

National Institutes of Health Office of Dietary Supplements (2013). Vitamin D—Health Professional Fact Sheet. Retrieved from http://ods.od.nih.gov/factsheets/VITAMINE-HealthProfessional/. Accessed November 4, 2015.

Piro, A., Tagarelli, G., Lagonia, P., Tagarelli, A., & Quattrone, A. (2010). Casimir Funk: his discovery of the vitamins and their deficiency disorders. *Annals of Nutrition and Metabolism, 57*(2): 85–88. doi: 10.1159/000319165.

Semba, R. D. (2012, October). The discovery of vitamins. *International Journal for Vitamin and Nutrition Research, 82*(5): 310–315. doi: 10.1024/0300-9831/a000124.

Semba, R. D. (2012). On the 'discovery' of vitamin A. *Annals of Nutrition and Metabolism, 61*(3): 192–198. doi: 10.1159/000343124.

Sommer, A., Vyas, K. S. (2012, November). A global clinical view on vitamin A and carotenoids. *American Journal of Clinical Nutrition, 96*(5): 1204S–6S. doi: 10.3945/ajcn.112.034868.

Rajakumar, K. (2003, August). Vitamin D, cod-liver oil, sunlight, and rickets: a historical perspective. *Pediatrics, 112*(2): e132–e135.

Ricciarelli, R., Argellati, F., Pronzato, M. A., & Domenicotti, C. (2007, October–December). Vitamin E and neurodegenerative diseases. *Molecular Aspects in Medicine, 28*(5–6): 591–606.

Rosenfeld, L. (1997, April). Vitamine—vitamin. The early years of discovery. *Clinical Chemistry, 43*(4): 680–685.

U.S. Department of Health and Human Services and U.S. Department of Agriculture (2016). *2015–2020 Dietary Guidelines for Americans.* 8th Edition. Retrieved from http://health.gov/dietaryguidelines/2015/guidelines/. Accessed January 15, 2016.

Vasanthi, H. R., Parameswari, R. P., & Das, D. K. (2011). Tocotrienols and its role in cardiovascular health—a lead for drug design. *Current Pharmaceutical Design 17*(21): 2170–2175.

Vitamin A. (2013). Retrieved from http://www.nlm.nih.gov/medlineplus/ency/article/002400.htm. Accessed November 4, 2015.

Wada, S. (2012, January). Cancer preventive effects of vitamin E. *Current Pharmaceutical Biotechnology, 13*(1): 156–164.

Weber, D., & Grune, T. (2012, February). The contribution of β-carotene to vitamin A supply of humans. *Molecular Nutrition and Food Research, 56*(2): 251–258. doi: 10.1002/mnfr.201100230.

Wolf, G. (2005, March 1). The discovery of the antioxidant function of vitamin E: the contribution of Henry A. Mattill. *The Journal of Nutrition, 135*(3): 363–366.

Wolf, G. (2004, June 1). The discovery of vitamin D: the contribution of Adolf Windaus. *The Journal of Nutrition: 134*(6): 1299–1302.

World Health Organization (2009). Global prevalence of vitamin A deficiency in population at risk 1995–2005. *WHO Global Database on Vitamin A Deficiency*. Geneva: World Health Organization.

Yang, C. S., Suh, N., & Kong, A. T. (2012, May 1). Does Vitamin E prevent or promote cancer? *Cancer Prevention Research 5*: 701. doi: 10.1158/1940-6207.CAPR-12-0045.

Chapter 11

Berry, R. J., Bailey, L., Mulinare, J., Bower, C. & Folic Acid Working Group (2010, March). Fortification of flour with folic acid. *Food and Nutrition Bulletin, 31*(1 Supplement): S22–S35.

Clarke, R., et al. (2007, November). Low vitamin B-12 status and risk of cognitive decline in older adults. *The American Journal of Clinical Nutrition, 86*(5):1384–1391.

Combs, G. F. (2012). *The Vitamins: Fundamental Aspects in Nutrition and Health* (4th ed). Waltham, MA: Academic Press.

Comerford, K. B. (2013, November 6). Recent developments in multivitamin/mineral research. *Advances in Nutrition, 4*(6): 644–656. doi: 10.3945/an.113.004523.

Crider, K. S., Bailey, L. B., & Berry, R. J. (2011, March). Folic acid food fortification—its history, effect, concerns, and future directions. *Nutrients, 2*(2):370–384. doi: 10.3390/nu3030370.

Cziraky, M. J., Watson, K. E., & Talbert, R. L. (2008, October). Targeting low HDL-cholesterol to decrease residual cardiovascular risk in the managed care setting. *Journal of Managed Care Pharmacy, 14*(8 Supplement): S3–S28.

De-Regli, L. M., Fernandez-Gaxiola, A. C., Dowswell, T., & Peña-Rosa, J. P. (2010, October 6). Effects of safety of periconceptional folate supplementation for preventing birth defects. *Cochrane Database of Systematic Reviews, 10*: CD007950. doi: 10.1002/14651858.CD007950.pub2.

Fusco, D., Colloca, G., Lo Monaco, M. R., & Cesari, M. (2007). Effects of antioxidant supplementation on the aging process. *Journal of Clinical Interventions in Aging, 2*(3): 377–387.

Halliwell, B. (2012, May). Free radicals and antioxidants: updating a personal view. *Nutritional Review, 70*(5): 257–265. doi: 10.1111/j.1753-4887.2012.00476.x.

Hemila, H., & Chalker, E. (2013, January 31). Vitamin C for preventing and treating the common cold. *Cochrane Database of Systematic Reviews, 1*: CD000980. doi: 10.1002/14651858.CD000980.pub4.

Higdon, J. (2002). Folic Acid. *Linus Pauling Institute.* Retrieved from http://lpi.oregonstate.edu/infocenter/vitamins/fa/#drug_interact. Accessed November 4, 2015.

International Food Information Council Foundation (2011, July). *Functional Foods.* Retrieved from http://www.foodinsight.org/Content/3842/Final%20Functional%20Foods%20Backgrounder.pdf. Accessed November 4, 2015.

Institute of Medicine (1998). *Dietary Reference Intakes for Thiamin, Riboflavin, Niacin, Vitamin B6, Folate, Vitamin B12, Pantothenic Acid, Biotin, and Choline.* Washington, DC: The National Academies Press.

Institute of Medicine of the National Academies (2005). Dietary Reference Intakes: Vitamins. *Dietary Reference Intakes for Energy, Carbohydrate, Fiber, Fat, Fatty Acids, Cholesterol, Protein, and Amino Acids*. Washington, DC: The National Academies Press.

Malouf, R., & Grimley Evans, J. (2008, October 8). Folic acid with or without vitamin B12 for the prevention and treatment of healthy elderly and demented people. *Cochrane Database of Systematic Reviews, 4*: CD004514. doi: 10.1002/14651858.CD004514.pub2.

Marti-Carvajal, A. J., Sola, I., Lathyris, D., Karakitsiou, D. E., & Simanccas-Racines, D. (2013, January 31). Homocysteine-lowering interventions for preventing cardiovascular events. *Cochrane Database of Systematic Reviews, 1*: CD006612. doi: 10.1002/14651858.CD006612.pub3.

McCaddon, A. (2013, May). Vitamin B12 in neurology and ageing: clinical and genetic aspects. *Biochimie, 95*(5): 1066–1076. doi: 10.1016/j.biochi.2012.11.017.

Morris, M. S. (2012, November 1). The role of B vitamins in preventing and treating cognitive impairment and decline. *Advances in Nutrition, 3*(6): 801–812. doi: 10.3945/an.112.002535.

National Institutes of Health Office of Dietary Supplements (2012). Folate—Health Professional Fact Sheet. Retrieved from http://ods.od.nih.gov/factsheets/Folate-HealthProfessional/. Accessed November 4, 2015.

National Institutes of Health Office of Dietary Supplements (2011). Vitamin B6—Health Professional Fact Sheet. Retrieved from http://ods.od.nih.gov/factsheets/VitaminB6-HealthProfessional/. Accessed November 4, 2015.

National Institutes of Health (2011). Vitamin B12—Health Professional Fact Sheet. Retrieved from http://ods.od.nih.gov/factsheets/VitaminB12-HealthProfessional/. Accessed November 4, 2015.

National Institutes of Health Office of Dietary Supplements (2013). Vitamin C—Health Professional Fact Sheet. Retrieved from http://ods.od.nih.gov/factsheets/VitaminC-HealthProfessional/. Accessed November 4, 2015.

O'Leary, F., Allman-Farinelli, M., & Samman, S. (2012, December 14). Vitamin B12 status, cognitive decline and dementia, a systematic review of prospective cohort studies. *British Journal of Nutrition, 108*(11): 1948–1961. doi: 10.1017/S0007114512004175.

Padayatty, S. J., et al. (2003, February). Vitamin C as an antioxidant: evaluation of its role in disease prevention. *Journal of the American College of Nutrition, 22*(1): 18–35.

Pernicious Anemia (2012). Retrieved from http://www.nlm.nih.gov/medlineplus/ency/article/000569.htm. Accessed November 4, 2015.

Reynolds, E. (2006, November). Vitamin B12, folic aid, and the nervous system. *The Lancet Neurology, 5*(11): 949–960. doi: 10.1016/S1474-4422(06)70598-1.

Ross, A. C. (2013). *Modern Nutrition in Health and Disease* (11th ed). Philadelphia: Lippincott Williams & Wilkins.

Valko, M., Leibfritz, D., Moncol, J., Cronin, M., Mazur, M., & Telser, J. (2006). Free radicals and antioxidants in normal physiological functions and human diseases. *The International Journal of Biochemistry and Cell Biology, 39*(1): 44–84.

Wang, Z. M., et al. (2012, October). Folate and risk of coronary heart disease: a meta-analysis of prospective studies. *Nutrition, Metabolism, and Cardiovascular Diseases, 22*(10): 890–899. doi: 10.1016/j.numecd.2011.04.011.

Zeisel, S. H., & da Costa, K. A. (2009, November). Choline: an essential nutrient for public health. *Nutrition Review, 67*(11): 615–623. doi: 10.1111/j.1753-4887.2009.00246.x.

Chapter 12

Academy of Nutrition and Dietetics (2013). Position of the Academy of Nutrition and Dietetics: functional foods. *Journal of the Academy of Nutrition and Dietetics, 113*(8): 1096–1103.

Albanes, D. (2009, January 7). Vitamin supplements and cancer prevention: where do randomized controlled trials stand? *Journal of the National Cancer Institute, 101*(1): 2–4.

Aleccia, J. (2012, October 17). Daily multivitamin cuts men's cancer risk by 8 percent, large study finds. *NBC News*. Retrieved from http://www.nbcnews.com/health/health-news/daily-multivitamin-cuts-mens-cancer-risk-8-percent-large-study-f1C6519472. Accessed November 4, 2015.

American Dietetic Association (2009). Position of the American Dietetic Association: Nutrient Supplementation. *Journal of the American Dietetic Association, 109*(12): 2073–2085. doi: 10.1016/j.jada.2009.10.020.

Bailey, R. L., Gahche, J. J., Miller, P. E., Thomas, P. R., & Dwyer, J. T. (2013, March 11). Why US adults use dietary supplements. *Journal of the American Medical Association Internal Medicine, 173*(5): 355–361. doi: 10.1001/jamainternmed.2013.2299.

Baily, R. L., Gahche, J. J., Thomas, P. R., & Dwyer, J. T. (2013, December). Why US children use dietary supplements. *Pediatric Research, 74*(6): 737–741. doi: 10.1038/pr.2013.160.

Baron, J. A., et al. (2003, March 6). A randomized trial of aspirin to prevent colorectal adenomas. *New England Journal of Medicine, 348*(10): 891–899.

Baume, N., Mahler, N., Kamber, M., Mangin, P, & Saugy, M. (2005, January 6). Research of stimulants and anabolic steroids in dietary supplements. *Scandinavian Journal of Medicine and Science in Sports, 16*(1): 41–48. doi: 10.1111/j.1600-0838.2005.00442.x.

Bjelakovic, G., Nikolova, D., Gluud, L. L., Simonetti, R. G., & Gluud, C. (2007, February 28). Mortality in Randomized Trials of Antioxidant Supplements for Primary and Secondary Prevention: Systematic Review and Meta-analysis. *The Journal of the American Medical Association, 297*(8): 842–857. doi: 10.1001/jama.297.8.842.

Briones, T., & Bruya, M. A. (1990, February). The professional imperative: research utilization in the search for scientifically based nursing practice. *Focus on Critical Care, 17*(1): 78–81.

Cohen, P. A., Benner, C., & McCormick, D. (2012, January). Use of pharmaceutically adulterated dietary supplement, Pai You Guo, among Brazilian-born women in the United States. *Journal of General Internal Medicine, 27*(1): 51–56. doi: 10.1007/s11606-011-1828-0.

Cohen, P. A., Travis, J. C., & Venhuis, B. J. (2013, October 14). A methamphetamine analog (N, α-diethyl-phenylethylamine) identified in a mainstream dietary supplement. *Drug Testing and Analysis*. Retrieved from 10.1002/dta.1578.

Dwyer, J., Nahin, R. L, Rogers, G. T., Barnes, P. M., Jacques, P. M., Sempos, C. T., & Bailey, R. (2013, June). Prevalence and predictors of children's dietary supplement use: the 2007 National Health Interview Survey. *American Journal of Nutrition, 87*(6): 1331–7. doi: 10.3945/ajcn.112.052373.

Gazanio, J. M., et al. (2012, November 14). Multivitamins in the prevention of cancer in men: the Physician's Health Study II randomized controlled trial. *The Journal of the American Medical Association, 308*:18. doi 10.1001/jama.2012.14641.

Geyer, H., Parr, M. K., Koehler, K., Mareck, U., Schanzer, W., & Thevis, M. (2008, July). Nutritional supplements cross-contaminated and faked with doping substances. *Journal of Mass Spectrometry, 43*(7): 892–902. doi: 10.1002/jms.1452.

Goodman, G., et al. (2004, October). The beta-carotene and retinol efficacy trial: incidence of lung cancer and cardiovascular disease mortality during 6-year follow-up after stopping β-carotene and retinol supplements. *Journal of the National Cancer Institute, 96*(23): 1743–1750. doi: 10.1093/jnci/djh320.

Hamburg, M. A. (2010, December 15). Letter to Dietary Supplement Manufacturers. Retrieved from http://www.fda.gov/downloads/Drugs/ResourcesForYou/Consumers/BuyingUsingMedicineSafely/MedicationHealthFraud/UCM236985.pdf. Accessed November 4, 2015.

Herbs and Supplements (2014). Retrieved from http://www.nlm.nih.gov/medlineplus/druginfo/herb_All.html. Accessed November 4, 2015.

Kabat, G. (n.d.). Natural does not mean safe. Retrieved from http://www.slate.com/articles/health_and_science/medical_examiner/2012/11/herbal_supplement_dangers_fda_does_not_regulate_supplements_and_they_can.htm. Accessed November 4, 2015.

Kamber, M., Baume, N., Saugy, M., & Rivier, L. (2001, June). Nutritional supplements as a source for positive doping cases? *International Journal of Sports Nutrition and Exercise Metabolism, 11*(2): 258–263.

Klein, E. A., et al. (2011, October 12). Vitamin E and the risk of prostate cancer: The Selenium and Vitamin E Cancer Prevention Trial. *The Journal of the American Medical Association, 306*(14): 1549–1556. doi: 10.1001/jama.2011.1437.

Kumbhani, D. J., & Bhatt, K. L. (2006). Physicians' Health Study II (PHS II). *Medscape Cardiology*. Retrieved from http://www.medscape.com/viewarticle/583792. Accessed November 4, 2015.

Marcus, D. M., & Grollman, A. P. (2012, July 9). The consequences of ineffective regulation of dietary supplements. *Archives of Internal Medicine, 172*(13): 1035–1036. doi: 10.1001/archinternmed.2012.2687.

Mursu, J., Robien, K., Harnack, L., Park, K., & Jacobs, D. (2011). Dietary supplements and mortality rate in older women: the Iowa Women's Health Study. *Archives of Internal Medicines, 171*(18): 1625–1633. doi: 10.1001/archinternmed.2011.445.

Myung, S. K., et al. (2013, January 18). Efficacy of vitamin and antioxidant supplements in prevention of cardiovascular disease: systematic review and meta-analysis of randomized controlled trials. *BMJ, 346*(f10). doi: 10.1136/bmj.f10.

National Cancer Institute (n.d.). Alpha-Tocopherol, Beta-Carotene Cancer Prevention (ATBC) Study. Retrieved from http://atbcstudy.cancer.gov. Accessed November 15, 2015.

National Center for Complementary and Alternative Medicine (NCCAM) (2014). Herbs at a Glance. Retrieved from http://nccam.nih.gov/health/herbsataglance.htm. Accessed November 4, 2015.

National Center for Complementary and Alternative Medicine (NCCAM) (2009). Omega-3 Supplements: An Introduction. Retrieved from http://nccam.nih.gov/health/omega3/introduction.htm. Accessed November 4, 2015.

National Center for Complementary and Alternative Medicine (NCCAM) (2013). Probiotics. Retrieved from http://nccam.nih.gov/health/probiotics. Accessed November 4, 2015.

National Institutes of Health Office of Dietary Supplements (2011). Botanical Dietary Supplements—Health Professional Fact Sheet. Retrieved from http://ods.od.nih.gov/factsheets/BotanicalBackground-HealthProfessional/. Accessed November 4, 2015.

Parker-Pope, Tara (2011, October 11). More Evidence Against Vitamin Use. *The New York Times*. Retrieved from http://well.blogs.nytimes.com/2011/10/11/more-evidence-against-vitamin-use/. Accessed November 4, 2015.

Physicians' Health Study (2012). Retrieved from http://phs.bwh.harvard.edu. Accessed November 4, 2015.

Pipe, A., & Ayotte, C. (2002). Nutritional supplements and doping. *Clinical Journal of Sport Medicine, 12*: 245–249.

Rabin, R. C. (2012, October 22). Curbing Enthusiasm on Daily Multivitamins. *The New York Times*. Retrieved from http://well.blogs.nytimes.com/2012/10/22/curbing-the-enthusiasm-on-daily-multivitamins/. Accessed November 4, 2015.

Roberfroid, M. (2007, March). Prebiotics: The concept revisited. *The Journal of Nutrition, 137*(3): 8305–8375.

Sesso, H. D., et al. (2012, November 7). Multivitamins in the prevention of cardiovascular disease in men: The Physicians' Health Study II Randomized Controlled Trial. *Journal of the American Medical Association, 308*(17): 1751–1760. doi: 10.1001/jama.2012.14805.

U.S. Food and Drug Administration (2013, January). Guidance for industry: a food labeling guide (9. Appendix A: definitions of nutrient content claims). Retrieved from http://www.fda.gov/Food/GuidanceRegulation/GuidanceDocumentsRegulatoryInformation/LabelingNutrition/ucm064911.htm. Accessed November 4, 2015.

U.S. Food and Drug Administration (2013). Health claims meeting significant scientific agreement (SSA). Retrieved from http://www.fda.gov/food/ingredientspackaging labeling/labelingnutrition/ucm2006876.htm. Accessed November 4, 2015.

U.S. Food and Drug Administration (2013, October 11). Inspections, Compliance, Enforcement, and Criminal Investigations. Retrieved from http://www.fda.gov/ICECI/EnforcementActions/WarningLetters/2013/ucm371203.htm. Accessed November 4, 2015.

U.S. Food and Drug Administration (2013, December). Label claims for conventional foods and dietary supplements. Retrieved from http://www.fda.gov/Food/Ingredients PackagingLabeling/LabelingNutrition/ucm111447.htm. Accessed November 4, 2015.

U.S. Food and Drug Administration (2013). Summary of Qualified Health Claims Subject to Enforcement Discretion. Retrieved from http://www.fda.gov/Food/IngredientsPackagingLabeling/LabelingNutrition/ucm073992.htm. Accessed November 4, 2015.

U.S. Food and Drug Administration (2015). Dietary Supplements. Retrieved from http://www.fda.gov/Food/DietarySupplements/default.htm. Accessed November 4, 2015.

Use of CoZ10, Digestive Enzymes, Probiotics, and B-Vitamins on the Rise According to ConsumerLab.com Survey (2013). Retrieved from http://www.consumerlab.com/news/highlights_vitamin_supplements_survey/1_31_2013/. Accessed November 4, 2015.

Wood, S. (2012, November 5). Big bucks, no bang: PHS II shows no benefits of vitamins for preventing CVD. *Medscape Cardiology.* Retrieved from http://www.medscape.com/viewarticle/773970?t=1. Access November 4, 2015.

Wu, C., Wang, C., & Kennedy, J. (2011, November). Changes in herb and dietary supplement use in the US adult population: a comparison of the 2002 and 2007 National Health Interview surveys. *Clinical Therapeutics, 33*,11: 1749–1758.

Ye, Y., Li, J., & Yuan, Z. (2013). Effect of antioxidant vitamin supplementation on cardiovascular outcomes: a meta-analysis of randomized controlled trials. *PLoS One, 8*(2): e56803. doi: 10.1371/journal.pone.0056803.

Young, A. (2013, October 25). Firm in outbreak probe has history of run-ins with FDA. *USA Today*. Retrieved from http://www.usatoday.com/story/news/nation/2013/10/24/usplabs-has-history-of-fda-run-ins-ceo-with-criminal-history/3179113/. Accessed November 4, 2015.

Chapter 13

Aaron, K. J., & Sanders, P. W. (2013, September). Role of dietary salt and potassium intake in cardiovascular health and disease: a review of the evidence. *Mayo Clinic Proceedings, 88*(9): 987–995. doi: 10.1016/j.mayocp.2013.06.005.

Academy of Nutrition and Dietetics (2012, September). Position of the Academy of Nutrition and Dietetics: The Impact of Fluoride and Health. *Journal of the Academy of Nutrition and Dietetics, 112*(9): 1443–1453.

Academy of Nutrition and Dietetics (2013, April). Position of the Academy of Nutrition and Dietetics: Nutrition Security in Developing Nations: Sustainable Food, Water, and Health. *Journal of the Academy of Nutrition and Dietetics, 113*(4): 581–595.

American Dietetic Association (2009). Position of the American Dietetic Association: Food and Water Safety. *Journal of the American Dietetic Association, 109*(8): 1449–1460.

American Heart Association (2015). Sodium and Salt. Retrieved from http://www.heart.org/HEARTORG/GettingHealthy/NutritionCenter/HealthyDietGoals/Sodium-Salt-or-Sodium-Chloride_UCM_303290_Article.jsp. Accessed November 4, 2015.

American Heart Association (2015). A Primer on Potassium. Retrieved from http://sodiumbreakup.heart.org/sodium-411/what-about-potassium/. Accessed November 4, 2015.

Armstrong, L. E., Maresh, C. M., Castellani, J. W., Bergeron, M. F., Keneflick, R. W., & Riebe, D. (1994, September). Urinary indices of hydrations status. *International Journal of Sport Nutrition, 4*(3): 265–279.

Azouly, A, Garzon, P., & Eisenberg, M. J. (2011, March). Comparison of the mineral content of tap water and bottled waters. *Journal of General Internal Medicine, 16*(3): 168–175. doi: 10.1111/j.1525-1497.2001.04189.x.

Bashyam, H. (2007, July 9). Lewis Dahl and the genetics of salt-induced hypertension. *The Journal of Experimental Medicine, 204*(7): 1507.

Benini, O., D'Alessandro, C., Gianfaldoni. D., & Cupisti, A. (2011). Extra-phosphate load from food additives in commonly eaten foods: a real and insidious danger for renal patients. *Journal of Renal Nutrition*, 21(4): 303–308.

Boesler, M. (2013, July 12). You are paying 300 times more for bottled water than tap water. *Business Insider*. Retrieved from http://www.businessinsider.com/bottled-water-costs-2000x-more-than-tap-2013-7. Accessed November 4, 2015.

Booth, A., & Camacho, P. (2013, November). A closer look at calcium absorption and the benefits and risks of dietary versus supplemental calcium. *Postgraduate Medicine, 125*(6): 73–81. doi: 10.3810/pgm.2013.11.2714.

Calcium (2015). Retrieved from http://www.nlm.nih.gov/medlineplus/calcium.html. Accessed November 4, 2015.

Centers for Disease Control and Prevention (2011, July 11). Press Release: High Sodium, Low Potassium Diet Linked to Increased Risk of Death. Retrieved from http://www.cdc.gov/media/releases/2011/p0711_sodiumpotassiumdiet.html. Accessed November 4, 2015.

Centers for Disease Control and Prevention (2015). High Blood Pressure (Hypertension) Information. Retrieved from http://www.cdc.gov/bloodpressure/. Accessed November 4, 2015.

Centers for Disease Control and Prevention (2014). Intake of Calories and Selected Nutrients for the United States Population, 1999–2000. *National Health and Nutrition Examination Survey.* Retrieved from http://www.cdc.gov/nchs/data/nhanes/databriefs/calories.pdf. Accessed November 4, 2015.

Chang AR, Lazo M, Appel LJ, Gutiérrez OM, Grams ME. (2014, February). High dietary phosphorus intake is associated with all-cause mortality: results from NHANES III. *The American Journal of Clinical Nutrition,* 99(2):320-7. doi: 10.3945/ajcn.113.073148

Chloride in diet (2015, April 2). Retrieved from http://www.nlm.nih.gov/medlineplus/ency/article/002417. Accessed November 4, 2015.

Couzy, F., Keen, C., Gershwin, M. E., & Mareschi, J. P. (1993, January–March). Nutritional implications of the interactions between minerals. *Progress in Food and Nutrition Science, 17*(1): 65–87.

Dietary Sodium (2013). Retrieved from http://www.nlm.nih.gov/medlineplus/dietarysodium.html. Accessed November 4, 2015.

Franco, V., & Oparil, S. (2006, June). Salt sensitivity, a determinant of blood pressure, a cardiovascular disease and survival. *Journal of the American College of Nutrition, 25*(3 Supplement): 247S–255S.

Food and Nutrition Board, Institute of Medicine, National Academies (n.d.). Dietary reference intakes (DRIs): Tolerable Intake Levels, Vitamins. Retrieved from http://iom.edu/Activities/Nutrition/SummaryDRIs/-/media/Files/Activity%20Files/Nutrition/DRIs/ULs%20for%20Vitamins%20and%20Elements.pdf. Accessed November 4, 2015.

Food and Nutrition Board, Institute of Medicine, National Academies (2004). Dietary reference intakes (DRIs): Electrolytes and Water. Retrieved from http://www.iom.edu/Global/News%20Announcements/-/media/442A08B899F44DF9AAD083D86164C75B.ashx. Accessed November 4, 2015.

Food and Nutrition Board, Institute of Medicine, National Academies (n.d.). Dietary reference intakes (DRIs): Elements. Retrieved from http://www.iom.edu/-/media/Files/Activity%20Files/Nutrition/DRIs/DRI_Elements.pdf. Accessed November 4, 2015.

Food and Nutrition Board, Institute of Medicine, National Academies (n.d.). Dietary reference intakes (DRIs): Vitamins. Retrieved from http://iom.edu/Activities/Nutrition/SummaryDRIs/-/media/Files/Activity%20Files/Nutrition/DRIs/RDA%20and%20AIs_Vitamin%20and%20Elements.pdf. Accessed November 4, 2015.

Fulgoni, V. L., Keast, D. R., Bailey, R. L., & Dwyer, J. (2011, August 24). Foods, fortificants, and supplements: Where do Americans get their nutrients? *The Journal of Nutrition, 141*(10): 1847–1854. doi: 10.3945/jn.111.142257.

Gelski, J. (2013, November 19). Potassium's multiple positives. *Food Business News*. Retrieved from http://www.foodbusinessnews.net/articles/news_home/Supplier-Innovations/2013/11/Potassiums_multiple_positives.aspx?ID={A19A8347-8C66-4683-8562-27E5E038D0CC}&cck=1

Gibson, B. R., Gibbs, M., & Ferguson, E. (2010), June 31). A review of phytate, iron, zinc, and calcium concentrations in plant-based complementary foods used in low-income countries and implications for bioavailability. *Food and Nutrition Bulletin*, (2 Suppl): S134–46.

High blood pressure (2012). Retrieved from http://www.nlm.nih.gov/medlineplus/ency/article/000468.htm. Accessed November 4, 2015.

Hotz, C, & Gibson, R. S. (2007, April). Traditional food-processing and preparation practices to enhance the bioavailability of micronutrients in plant-based diets. *The Journal of Nutrition, 137*(4): 1097–1100.

Ions (2012, August 30). Retrieved from http://www.nlm.nih.gov/medlineplus/ency/article/002385.htm. Accessed November 4, 2015.

Jacques, E. (2010, October 9–15). Promoting healthy drinking habits in children. *Nursing Times, 108*(41): 20–1.

Jequier, E., & Constant, F. (2010, February). Water as an essential nutrient: the physiological basis of hydration. *European Journal of Clinical Nutrition, 64*(2): 115–123. doi: 10.1038/ejcn.2009.111.

Karaguzel, G., & Holick, M. F. (2010, December). Diagnosis and treatment of osteopenia. *Reviews in Endocrine and Metabolic Disorders, 11*(4): 237–51. doi: 10.1007/s11154-010-9154-0.

Karppanen, H., & Mervaala, E. (2006, September–October). Sodium intake and hypertension. *Progress in Cardiovascular Diseases, 49*(2): 59–75.

Leon, J. B., Sullivan, C. M., & Sehgal, A. R. (2013, July). The prevalence of phosphorus-containing food additives in top-selling foods in grocery stores. *The Journal of Renal Nutrition, 23*(4): 265–270. doi: 10.1053/j.jrn.2012.12.003.

Magnesium in diet (2013, February 18). Retrieved from http://www.nlm.nih.gov/medlineplus/ency/article/002423.htm. Accessed November 4, 2015.

Mente, A., et al. (2014, August 14). Association of urinary sodium and potassium excretion with blood pressure. *New England Journal of Medicine, 371*: 601–611. doi: 10.1056/NEJMoa1311989.

Minerals (2014). Retrieved from http://www.nlm.nih.gov/medlineplus/minerals.html. Accessed November 4, 2015.

Muckelbauer, R., Sarganas, G., Gruneis, A., & Muller-Nordhorn, J. (2013, August). Association between water consumption and body weight outcomes: a systematic review. *American Journal of Clinical Nutrition, 98*(2): 282–299. doi: 10.3945/ajcn.112.055061.

National Cancer Institute (2009, May). Calcium and Cancer Prevention: Strength and Limits of the Evidence. Retrieved from http://www.cancer.gov/cancertopics/factsheet/prevention/calcium. Accessed November 4, 2015.

National Cancer Institute (2014). Sources of Sodium among the US Population, 2005–6. Retrieved from http://appliedresearch.cancer.gov/diet/foodsources/sodium/. Accessed November 4, 2015.

National Center for Chronic Disease Prevention and Health Promotion (2014, February). Sodium: Q&A. Retrieved from http://www.cdc.gov/salt/pdfs/sodium_qanda.pdf. Accessed November 4, 2015.

National Heart, Lung, and Blood Institute (2015, September 16). What is the DASH Eating Plan? Retrieved from http://www.nhlbi.nih.gov/health/health-topics/topics/dash/. Accessed November 4, 2015.

National Institutes of Health Office of Dietary Supplements (2013, November). Calcium—Health Professional Fact Sheet. Retrieved from http://ods.od.nih.gov/factsheets/Calcium-HealthProfessional/. Accessed November 4, 2015.

National Institutes of Health Office of Dietary Supplements (2013, November 4). Magnesium—Health Professional Fact Sheet. Retrieved from http://ods.od.nih.gov/factsheets/Magnesium-HealthProfessional/. Accessed November 4, 2015.

National Research Council (2005). *Dietary Reference Intakes for Water, Potassium, Sodium, Chloride, and Sulfate*. (Washington, DC: The National Academies Press).

Nimni, M. E., Han, B., & Cordoba, F. (2007, November 6). Are we getting enough sulfur in our diet? *Nutrition and Metabolism, 4*:24.

O'Donnell, M., et al. (2014, August 14). Urinary sodium and potassium excretion, mortality, and cardiovascular events. *New England Journal of Medicine, 371*: 612–623. doi: 10.1056/NEJMoa1311889.

Oparil, S. (2014). Low sodium intake—cardiovascular health benefit or risk? *New England Journal of Medicine, 371*: 677–679. doi: 10.1056/NEJMe1407695.

Otten, J. J., Hellwig, J. P., & Meyers, L. D., (eds) (2006). Phosphorus. *Dietary Reference Intakes: The Essential Guide to Nutrient Requirements* (146–189) (Washington, DC: The National Academies Press).

Otten, J. J., Hellwig, J. P., & Meyers, L. D., (eds) (2006). Magnesium. *Dietary Reference Intakes: The Essential Guide to Nutrient Requirements* (190–249) (Washington, DC: The National Academies Press).

Otten, J. J., Hellwig, J. P., & Meyers, L. D., (eds) (2006). Potassium. *Dietary Reference Intakes: The Essential Guide to Nutrient Requirements* (187–268) (Washington, DC: The National Academies Press).

Otten, J. J., Hellwig, J. P., & Meyers, L. D., (eds) (2006). Sodium and Chloride. *Dietary Reference Intakes: The Essential Guide to Nutrient Requirements* (Washington, DC: The National Academies Press).

Parcell, S. (2002, February). Sulfur in human nutrition and applications in medicine. *Alternative Medical Review, 7*(1): 22–44.

Park, S., Sherry, B., O'Toole, T., & Huang, Y. (2011, August). Factors associated with low drinking water intake among adolescents: The Florida Youth Physical Activity and Nutrition Survey, 2007. *Journal of the American Dietetic Association, 111*(8): 1211–1217. doi: 10.1016/j.jada.2011.05.006.

Phosphorus in diet (2015, February 2). Retrieved from http://www.nlm.nih.gov/medlineplus/ency/article/002424.htm. Accessed November 5, 2015.

Popkin, B. M., D'anci, K. E., & Rosenberg, I. H. (2010, August). Water, Hydration, and Health. *Nutritional Review, 68*(8): 439–458.

Potassium in diet (2014, May 13). Retrieved from http://www.nlm.nih.gov/medlineplus/ency/article/002413.htm. Accessed November 4, 2015.

Reddy, M. B., & Love, M. (1999). The impact of food processing on the nutritional quality of vitamins and minerals. *Advances in Experimental Medicine and Biology, 459*: 55–106.

Sengupta, P. (2013, August). Potential health impacts of hard water. *International Journal of Preventative Medicine, 4*(8): 866–875.

Shifting the Balance of Sodium and Potassium in Your Diet (n.d.). Retrieved from http://www.hsph.harvard.edu/nutritionsource/sodium-potassium-balance/. Accessed November 4, 2015.

Stolarz-Skrzypek, K., Bednarski, A., Czarnecka, D., Kawecka-Jaszcz, K., & Staessen, J. A. (2013, April). Sodium and potassium and the pathogenesis of hypertension. *Current Hypertension Reports, 15*(2): 122–130. doi: 10.1007/s11906-013-0331-x.

U.S. Department of Health and Human Services (1998). *Lowering Your Blood Pressure with DASH* (Washington, DC).

U.S. Food and Drug Administration (2013). Bottled Water Everywhere: Keeping it Safe. Retrieved from http://www.fda.gov/forconsumers/consumerupdates/ucm203620.htm. Accessed November 4, 2015.

U.S. Food and Drug Administration (2013). Lowering Salt in Your Diet. Retrieved from http://www.fda.gov/forconsumers/consumerupdates/ucm181577.htm. Accessed November 4, 2015.

U.S. Geological Survey (2013, August). Water Hardness and Alkalinity Retrieved from http://water.usgs.gov/owq/hardness-alkalinity.html. Accessed November 4, 2015.

U.S. Department of Health and Human Services and U.S. Department of Agriculture (2016). *2015–2020 Dietary Guidelines for Americans.* 8th Edition. Retrieved from http://health.gov/dietaryguidelines/2015/guidelines/. Accessed January 15, 2016.

Valtin, Heinz (2002, November 1). "Drink at least eight glasses of water a day." Really? Is there scientific evidence? *American Journal of Physiology, 283*: R993–R1004. doi: 10.1152/ajpregu.00365.2002.

World Health Organization (2013, January 31). WHO issues new guidance on dietary salt and potassium. Retrieved from http://www.who.int/mediacentre/news/notes/2013/salt_potassium_20130131/en/. Accessed November 4, 2015.

Yang, Q., et al. (2011, July 11). Sodium and potassium intake and mortality among US adults: prospective data from the Third National Health and Nutrition Examination Survey. *Archives of Internal Medicine, 171*(13): 1183–1991. doi: 10.1001/archinternmed.2011.257.

Chapter 14

Abdollah, M., Farshchi, A., Nikfar, S., & Seyedifar, M. (2013). Effect of chromium on glucose and lipid profiles in patients with type 2 diabetes; a meta analysis of randomized trials. *Journal of Pharmacy and Pharmaceutical Sciences, 16*(1): 99–114.

Alexander, J. (2007). Selenium. *Novartis Found Symposium, 282*: 143–149.

Althuis, M. C., Jordan, N. E., Ludington, E. A., & Wittes, J. T. (2002, July). Glucose and insulin responses to dietary chromium supplements: a meta analysis. *The American Journal of Clinical Nutrition, 76*(1): 148–155.

Bailey, C. H. (2014, January). Improved meta-analytic methods show no effect of chromium supplements on fasting glucose. *Biological Trace Element Research, 157*(1): 1–8. doi: 10.1007/s12011-013-9863-9.

Blanton, C. (2014, January). Improvements in iron status and cognitive function in young women consuming beef or non-beef lunches. *Nutrients, 6*(1): 90–110. doi: 10.3390/nu6010090.

Bohac, Lucie (2011, August). The food industry can play an important role in correcting iodine deficiency. *IDD Newsletter, 39*(3): 12.

Brewer, G. J. (2009, June). The risks of copper toxicity contributing to cognitive decline in the aging population and to Alzheimer's disease. *The Journal of the American College of Nutrition, 28*(3): 238–242.

Brower, G. J. (2012, March–April). Copper excess, zinc deficiency, and cognition loss in Alzheimer's disease. *Biofactors, 38*(1): 107–113. doi: 10.1002/biof.

Brown, K. H., Hambidge, K. M., Ranum, P., & Zinc Fortification Working Group (2010, March). Zinc fortification of cereal flours: current recommendations and research needs. *Food Nutrition Bulletin, 31*(Supplement 1): S62–74.

Caldwell, K. L., Pan, Y., Mortensen, M. E., Makhmudov, A., Merrill, L., & Moye, J. (2013, August). Iodine status in pregnant women in the National Children's Study and in U.S. women (15–44 years). National Health and Nutrition Examination Survey 2005–2010. *Thyroid, 23*(8): 927–937.

Centers for Disease Control and Prevention (2001). Recommendations for using fluoride to prevent and control dental caries in the United States. *MMWR, 50*(No. RR-14).

Chan, S., Gerson, B., & Subramaniam, S. (1998, December). The role of copper, molybdenum, selenium, and zinc in nutrition and health. *The Journal of Laboratory and Clinical Medicine, 18*(4): 673–685.

Charlton, K., & Skeaff, S. (2011, November). Iodine fortification: why, when, what, how, and who? *Current Opinions in Clinical Nutrition and Metabolic Care, 14*(6): 618–24. doi: 10.1097/MCO.0b013e32834b2b30.

Chromium in diet (2015, February 2). Retrieved from http://www.nlm.nih.gov/medlineplus/ency/article/002418.htm. Accessed November 4, 2015.

Cogswell, M. E., Looker, A. C., Pfeiffer, C. M., Cook, J. D., Lacher, D. A., Beard, J. L., Lynch, S. R., & Grummer-Strawn, L. M. (2009, May). Assessment of iron deficiency in US preschool children and nonpregnant females of child bearing age: National Health and Nutrition Examination Survey, 2003–2006. *American Journal of Clinical Nutrition, 89*(5): 1334–42. doi: 10.3945/ajcn.2008.27151.

Collins, J. F., & Klevay, L. M. (2011, November). Copper. *Advances in Nutrition, 2*(6): 520–522. doi: 10.3945/an.111.001222.

Copper in diet (2015, February 2). Retrieved from http://www.nlm.nih.gov/medlineplus/ency/article/002419.htm. Accessed November 4, 2015.

Cuajungco, M. P., & Faget, K. Y. (2003, January). Zinc takes the center stage: its paradoxical role in Alzheimer's disease. *Brain Research Reviews, 41*(1): 44–56.

Dasgupta, P. K., Liu, Y., & Dyke, J. V. (2008, February). Iodine nutrition: iodine content of iodized salt in the United States. *Environmental Science and Technology, 42*(4): 1315–1323.

Emord, J. W. (2005, August 25). Qualified Health Claims: Letter of Enforcement Discretion—Chromium Picolinate and Insulin Resistance (Docket No. 2004Q-0144). *Food and Drug Administration*. Retrieved from http://www.fda.gov/Food/IngredientsPackagingLabeling/LabelingNutrition/ucm073017.htm#ftnref1. Accessed November 4, 2015.

Erdman, J. W., Macdonald, I. A., & Zeisel, S. H. (2012, September). *Present Knowledge in Nutrition* (10th ed). (Hoboken, NJ: Wiley-Blackwell).

Finley, J. W., & Davis, C. D. (1999). Manganese deficiency and toxicity: are high or low dietary amounts of manganese causes for concern? *Biofactors, 10*(1): 15–24.

Franklin, T. J., & Twose, P. A. (1997, July 1). Reduction in beta-adrenergic response of culture glioma cells following depletion of intracellular GTP. *European Journal of Biochemistry, 77*(1): 113–117.

Gibson, R. S., Heath, A. L., & Szymlek-Gay, E. A. (2014, May 28). Is iron and zinc nutrition a concern for vegetarian infants and young children in industrialized countries? *American Journal of Clinical Nutrition, 100*(Supplement 1): 459S–468S.

Grabrucker, A. M., Rowan, M., & Garner, C. C. (2011, September). Brain-delivery of zinc-ions as potential treatment for neurological diseases: mini review. *Drug Delivery Letters, 1*(1): 13–23.

Higdon, J. (2003, April). Copper. *Micronutrient Information Center.* Retrieved from http://lpi.oregonstate.edu/infocenter/minerals/copper/. Accessed November 5, 2015.

Higdon, J. (2015, April). Fluoride. *Micronutrient Information Center*. Retrieved from http://lpi.oregonstate.edu/mic/minerals/fluoride. Accessed November 5, 2015.

Higdon, J. (2010). Manganese. *Micronutrient Information Center*. Retrieved from http://lpi.oregonstate.edu/infocenter/minerals/manganese/. Accessed November 5, 2015.

Higdon, J. (2014). Molybdenum. *Micronutrients Information Center.* Retrieved from http://lpi.oregonstate.edu/infocenter/minerals/molybdenum/. Accessed November 5, 2015.

Hotz, C., & Brown, K. H. International Zinc Nutrition Consultative Group (IZiNCG) Technical Document #1 Assessment of the Risk of Zinc Deficiency in Populations and Options for Its Control. *Food and Nutrition Bulletin, 25*, 1 (supplement 2).

Hunt, J. R. (2003, September). Bioavailability of iron, zinc, and other trace minerals from vegetarian diets. *The American Journal of Clinical Nutrition, 78*(3 Supplement): 633S–639S.

Hunt, J. R. (2002, May). Moving toward a plant-based diet: are iron and zinc at risk? *Nutritional Review, 60*(5 Pt. 1): 127–134.

Hurrell, R., & Egli, I. (2010, May). Iron bioavailability and dietary reference values. *American Journal of Clinical Nutrition, 91*(5): 1461S–1467S. doi: 10.3945/ajcn.2010.28674F.

Iodine in diet (2015, February 2). Retrieved from http://www.nlm.nih.gov/medlineplus/ency/article/002421.htm. Accessed November 5, 2015.

Iron in diet (2015, February 2). Retrieved from http://www.nlm.nih.gov/medlineplus/ency/article/002422.htm. Accessed November 5, 2015.

Iron overdose (2014, January 19). Retrieved from http://www.nlm.nih.gov/medlineplus/ency/article/002659.htm. Accessed November 5, 2015.

Kejizer, M., Pickering, J. W., & van Germert, M. J. (1991). Laser beam diameter for port wine stain treatment. *Lasers in Surgery and Medicine, 11*(6): 601–605.

Kieliszek, M., & Blazejak, S. (2013, May). Selenium: significance, and outlook for supplementation. *Nutrition, 29*(5): 713–718. doi: 10.1016/j.nut.2012.11.012

Laboda, G. (1990, November). Life-threatening hemorrhage after placement of an endosseous implant: report of case. *Journal of the American Dental Association, 121*(5): 599–600.

Landman, G., Bilo, H., Houweling, S. T., & Kleefstra, N. (2014, April 14). Chromium does not belong in the diabetes treatment arsenal: Current evidence and future perspectives. *World Journal of Diabetes, 5*(2): 160–164. doi: 10.4239/wjd.v5.i2.160.

Lee, S., Ananthakrishnan, S., & Pearce, E. N. (2014, April 3). Iodine Deficiency. Retrieved from http://emedicine.medscape.com/article/122714-overview. Accessed November 5, 2015.

Leung, A. M., Braverman, L. E., & Pearce, E. N. (2012, November). History of U.S. Iodine Fortification and Supplementation. *Nutrients, 4*(11): 1740–1746.

Lim, K. H., Riddell, L. J., Nowson, C. A., Booth, A. O., & Szymlek, Gay, E. A. (2013, August 13). Iron and zinc nutrition in the economically-developed world: a review. *Nutrients, 5*(8): 3184–3211. doi: 10.3390/nu5083184.

Maret, W. (2013, January 1). Zinc biochemistry: from a single zinc enzyme to a key element of life. *Advances in Nutrition, 4*(1): 82–91. doi: 10.3945/an.112.003038.

Masharani, U., Gjerde, C., McCoy, S., Maddux, B. A., Hessler, D., Goldfine, I. D., & Youngren, J. F. (2012, November 30). Chromium supplementation in non-obese non-diabetic subjects is associated with a decline in insulin sensitivity. *BMC Endocrine Disorders, 30*(12): 31. doi: 10.1186/1472-6823-12-31.

Minerals (2013, April). Retrieved from http://www.nlm.nih.gov/medlineplus/minerals.html. Accessed November 5, 2015.

Moshfegh, A., Goldman, J., & Cleveland, L. (2005). *What we eat in America*, NHANES 2001–2002: Usual Nutrient Intakes from Food Compared to Dietary Reference Intakes. US Department of Agriculture, Agricultural Research Service.

National Institute of Dental and Craniofacial Research (2014). The story of fluoridation. Retrieved from http://www.nidcr.nih.gov/oralhealth/topics/fluoride/thestoryoffluoridation.htm. Accessed November 5, 2015.

National Institutes of Health Office of Dietary Supplements (2013, November 4). Chromium—Health Professional Fact Sheet. Retrieved from http://ods.od.nih.gov/factsheets/Chromium-HealthProfessional/. Accessed November 5, 2015.

National Institutes of Health Office of Dietary Supplements (2011, June 24). Iodine Fact Sheet for Consumers https://ods.od.nih.gov/factsheets/Iodine-Consumer. Accessed November 5, 2015.

National Institutes of Health Office of Dietary Supplements (2014, April 8). Iron—Health Professional Fact Sheet. Retrieved from http://ods.od.nih.gov/factsheets/Iron-HealthProfessional/. Accessed November 5, 2015.

National Institutes of Health Office of Dietary Supplements (2013, November 22). Selenium—Health Professional Fact Sheet. Retrieved from http://ods.od.nih.gov/factsheets/Selenium-HealthProfessional/. Accessed November 5, 2015.

National Institutes of Health Office of Dietary Supplements (2013, June 5). Zinc—Health Professional Fact Sheet. Retrieved from http://ods.od.nih.gov/factsheets/Zinc-HealthProfessional/. Accessed November 5, 2015.

Nisen, Max (2013, November 19). How iodized salt made Americans smarter and boosted the US economy. *Business Insider*. Retrieved from http://www.businessinsider.com/effects-of-iodization-of-salt-2013-11#ixzz2ngdSFT7x. Accessed November 6, 2015.

Olivares, M., Pizarro, F., & Ruz, M. (2006). Inhibition of iron absorption by zinc: effect of physiological and pharmacological doses: TL015. *Pediatric Research, 60*: 636. doi: 10.1203/00006450-200611000-00027.

Praveena, S., Pasula, S., & Sameera, K. (2013, September). Trace elements in diabetes mellitus. *Journal of Clinical Diagnostic Research, 7*(9): 1863–1865. doi: 10.7860/JCDR/2013/5464.3335.

Preedy, V. R., Burrow, G. N., & Watson, R. R. (2009, March 17). *Comprehensive Handbook of Iodine: Nutrition, Biochemical, Pathological, and Therapeutic Aspects* (Waltham, MA: Academic Press).

Prohaska, J. R. (2011, March). Impact of copper limitation on expression and function of multicopper oxidases(ferroxidases). *Advances in Nutrition, 2*(2): 89–95. doi: 10.3945/an.110.000208.

Rayman, M. P. (2012, March). Selenium and human health. *The Lancet, 379*(9822): 1256–1268. doi: 10.1016/S0140-6736(11)61452-9.

Rios-Castillo, I., Olivares, M., Brito, A., Romana, D., & Pizarro, F. (2012, December 1). One-month of calcium supplementation does not affect iron bioavailability: A randomized controlled trial. *Nutrition, 30*(1): 44–48.

Ross, A. C., Caballero, B., Cousins, R. J., Tucker, K. L., & Ziegler, T. R. (2012, December). *Modern Nutrition in Health and Disease* (11th ed). (Phladelphia: Lippincott Williams and Wilkins).

Saltzman, E., & Karl, J. P. (2013). Nutrient deficiencies after gastric bypass surgery. *Annual Review of Nutrition, 33*: 183–203. doi: 10.1146/annurev-nutr-071812-161225.

Selenium in diet (2013). Retrieved from http://www.nlm.nih.gov/medlineplus/ency/article/002414.htm. Accessed November 5, 2015.

Swanson, C. A., et al. (2012, June). Summary of an NIH workshop to identify research needs to improve the monitoring of iodine status in the United States and to inform the DRI. *Journal of Nutrition, 142*(6): 1175S–1185S. doi: 10.3945/jn.111.156448.

Teucher, B., Olivares, M., & Cori, H. (2004, November). Enhancers of iron absorption: ascorbic acid and other organic acids. *International Journal of Vitamin and Nutrition Research, 74*(6): 403–419.

Trumbo, P. R. (2013, November 6). Evidence needed to inform the next dietary reference intakes for iodine. *Advances in Nutrition, 4*(6): 716–722. doi: 10.3945/an.113.004804.

Turek, M. J., & Fazel, N. (2009, March). Zinc deficiency. *Current Opinions in Gastroenterology, 25*(2): 136–143. doi: 10.1097/MOG.0b013e328321b395.

U.S. Department of Agriculture (n.d.). Dietary Reference Intakes: Elements. Retrieved from http://www.iom.edu/Global/News%20Announcements/~/media/48FAAA2FD9E74D95BBDA2236E7387B49.ashx. Accessed November 5, 2015.

U.S. Environmental Protection Agency (2013). Basic information about fluoride in drinking water. Retrieved from http://water.epa.gov/drink/contaminants/basicinformation/fluoride.cfm. Accessed November 5, 2015.

Vincent, J. B. (2013). Chromium: Is it essential, pharmacologically relevant, or toxic? *Metal Ions in Life Science, 13*: 171–198. doi: 10.1007/978-94-007-7500-8_6.

Zimmerman, M. B. (2009, June). Iodine deficiency. *Endocrine reviews, 30*(4): 376–408. doi: 10.1210/er.2009-0011.

Zimmerman, M. B. (2008). Iodine requirements and the risks and benefits of correcting iodine deficiency in populations. *Journal of Trace Elements in Medical Biology, 22*(2): 81–92. doi: 10.1016/j.jtemb.2008.03.001.

Zimmerman, M. B., Jooste, P. L., & Pandav, C. S. (2008, October 4). Iodine-deficiency disorders. *The Lancet, 372*(9645): 1251–1262. doi: 10.1016/S0140-6736(08)61005-3.

Chapter 15

Academy of Nutrition and Dietetics (2009, February). Position of the American Dietetics Association: weight management. *Journal of the American Dietetic Association, 109*(2): 330–346.

Academy of Nutrition and Dietetics (2015, January 30). Healthy Weight Gain. Retrieved from http://www.eatright.org/resource/health/weight-loss/your-health-and-your-weight/healthy-weight-gain. Accessed November 5, 2015.

Ainsworth, B. E., et al. (2011). Compendium of Physical Activities: a second update of codes and MET values. *Medicine and Science of Sports and Exercise, 43*: 1575–1581.

Almiron-Roig, E., Solis-Trapala, I., Dodd, J., & Jebb, S. A. (2013, December). Estimating food portions. Influence of unit number, meal type, and energy density. *Appetite, 71*: 95–103: doi: 10.1016/j.appet.2013.07.012.

Barazzoni, R., Silva, V., & Singer, P. (2014, April). Clinical biomarkers in metabolic syndrome. *Nutrition in Clinical Practice, 29*(2): 215–221. doi: 10.1177/0884533613516168.

Bray, G. A. (2013). Energy and fructose from beverages sweetened with sugar or high-fructose corn syrup pose a health risk for some people. *Advances in Nutrition, 3*: 220–225. doi: 10.3945/an.112.002816.

Camps, S. G., Verhoef, S. P., & Westerterp, K. R. (2013, May). Weight loss, weight maintenance, and adaptive thermogenesis. *American Journal of Clinical Nutrition, 97*(5): 990–994. doi: 10.3945/ajcn.112.050310.

Castaneda, T. R., Tong, J., Datta, R., Culler, M., & Tschop, M. H. (2010, January). Ghrelin in the regulation of body weight and metabolism. *Frontiers in Neuroendocrinology, 31*(1): 44–60. doi: 10.1016/j.yfrne.2009.10.008.

Centers for Disease Control and Prevention (2011). Assessing your weight. Retrieved from http://www.cdc.gov/healthyweight/assessing/Index.html. Accessed November 5, 2015.

Centers for Disease Control and Prevention (2015, June 19). Basics about childhood obesity. Retrieved from http://www.cdc.gov/obesity/childhood/basics.html. Accessed November 5, 2015.

Centers for Disease Control and Prevention (2013). Overweight and Obesity. Retrieved from http://www.cdc.gov/obesity/. Accessed November 5, 2015.

Centers for Disease Control and Prevention (2013, October). Prevalence of Obesity Among Adults: United States, 2011–2012. *NCHS Data Brief, 131*. Retrieved from http://www.cdc.gov/nchs/data/databriefs/db131.htm. Accessed November 5, 2015.

Centers for Disease Control and Prevention (2015, June 16). Obesity Causes and Consequences. Retrieved from http://www.cdc.gov/obesity/adult/causes/index.html. Accessed November 5, 2015.

Choquet, H., & Meyre, D. (2011, May). Genetics of Obesity: What have we learned? *Current Genomics, 12*(3): 169–179. doi: 10.2174/138920211795677895

Cummings, D. E., & Overduin, J. (2007, January 2). Gastointestinal regulation of food intake. *Journal of Clinical Investigation, 117*(1): 13–23. doi: 10.1172/JCI30227.

De Pergola, G., Silvestris, F. (2013). Obesity as a major risk factor for cancer. *Journal of Obesity, 2013*: 291546. doi: 10.1155/2013/291546.

Dekker, M. J., Su, Q., Baker, C., Rutledge, A. C., & Adell, K. (2010, November). Fructose: a highly lipogenic nutrient implicated in insulin resistance, hepatic steatosis, and the metabolic syndrome. *American Journal of Physiology, Endocrinology, and Metabolism, 299*(5): E685-E694. doi: 10.1152/ajpendo.00283.2010.

Ello-Martin, J. A., Ledikwe, J. H., & Rolls, B. J. (2005, July). The influence of food portion size and energy density on energy intake: implications for weight management. *American Journal of Clinical Nutrition, 82*(1 Supplement): 236S–241S.

Finelli, C., Gioia, S., & La Sala, N. (2012). Physical activity: an important adaptive mechanism for body-weight control. *ISRN Obesity, 2012*: 675285. doi: 10.5402/2012/675285.

Federal Trade Commission (2012, July). Weighing the Claims in Diet Ads. Retrieved from http://www.consumer.ftc.gov/articles/0061-weighing-claims-diet-ads. Accessed November 5, 2015.

Flegal, K. M., Kit, B. K., Orpana, H., & Graubard, B. I. (2013, January 2). Association of all-cause mortality with overweight and obesity using standard body mass index categories: a systematic review and meta-analysis. *Journal of the American Medical Association, 309*(1): 71–82. doi: 10.1001/jama.2012.113905.

Greenberg, A. S., & Obin, M. S. (2006). Obesity and the role of adipose tissue in inflammation and metabolism. *The American Journal of Clinical Nutrition, 83*(2): 461S–465S.

Guyenet, S. J., & Schwartz, M. W. (2012, March). Clinical review: Regulation of food intake, energy balance, and body fat mass: implications for the pathogenesis and treatment of obesity. *Journal of Clinical Endocrinology and Metabolism, 97*(3): 745–755. doi: 10.1210/jc.2011-2525.

Hall, K. D., Heymsfield, S. B., Kemnitz, J. W., Klein, S., Schoeller, D. A., & Speakman, J. R. (2012, April). Energy balance and its components: implications for body weight regulation. *American Journal of Clinical Nutrition, 95*(4): 989–994. doi: 10.3945/ajcn.112.036350.

Hamdy, O. (2014) Obesity. Retrieved from http://emedicine.medscape.com/article/123702-overview. Accessed November 5, 2015.

Hamdy, O., Porramatikul, S., Al-Ozairi, E. (2006, November). Metabolic obesity: the paradox between visceral and subcutaneous fat. *Current Diabetes Reviews, 2*(4): 367–373.

Harrington, M., Gibson, S., & Cottrell, R. C. (2009, June). A review and meta-analysis of the effect of weight loss on all-cause mortality risk. *Nutrition Research Reviews, 22*(1): 93–108. doi: 10.1017/S0954422409990035.

Harris, J. L., et al. (2009). Priming effects of food advertising on eating behavior. *Health Psychology, 28*(4): 404–413. doi: 10.1037/a0014399.

Henry, C. J. K. (2000, June). Mechanisms of changes in basal metabolism during aging. *European Journal of Clinical Nutrition, 54*(3): S77–S91.

Hill, J. O., Wyatt, H. R., Reed, G. W., & Peters, J. C. (2003, February 7). Obesity and the environment: Where do we go from here? *Science, 299*(5608): 853–855. doi: 10.1126/science.1079857.

Ho-Pham, L. T., et al. (2011, June). More on Body Fat Cutoff Points. *Mayo Clinic Proceedings, 86*(6): 584–587.

Janssen, I., Katzmarzyk, P. T., & Ross, R. (2004, March). Waist circumference and not body mass index explains obesity-related health risk. *American Journal of Clinical Nutrition, 79*(3): 379–384.

Jensen, M. D., et al. (2013, November 12). 2013 AHA/ACC/TOS Guideline for the Management of Overweight and Obesity in Adults: A Report of the American College of Cardiology/American Heart Association Task Force on Practice Guidelines and the Obesity Society. *Circulation*. Retrieved from http://circ.ahajournals.org/content/early/2013/11/11/01.cir.0000437739.71477. ee.full.pdf. Accessed November 5, 2015.

Johannsen, D. L., Knuth, N. D., Huizenga, R., Rood, J. D., Ravussin, E., & Hall, K. D. (2012, July). Metabolic slowing with massive weight loss despite preservation of fat-free mass. *Journal of Clinical Endocrinology and Metabolism, 97*(&): 2489–2496. doi: 10.1210/jc.2012-1444.

Johannsen, D. L., & Ravussin, E. (2008, October). Spontaneous physical activity: relationship between fidgeting and body weight control. *Current Opinions on Diabetes and Obesity, 15*(5): 409–415. doi: 10.1097/MED.0b013e32830b10bb.

Johnston, B. C., et al. (2014, September 3). Comparison of Weight Loss Among Named Diet Programs in Overweight and Obese Adults: A Meta Analysis. *Journal of the American Medical Association, 312*(9): 923–933. doi: 10.1001/jama.2014.10397.

Johnstone, A. M., Murison, S. D., Duncan, J. S., Rance, K. A., & Speakman, J. R. (2005, November). Factors influencing variation in basal metabolic rate include fat-free mass, fat mass, age, and circulating thyroxine but not sex, circulating leptin, or triiodothyronine. *American Journal of Clinical Nutrition, 82*(5): 841–848.

King, J. A., Wasse, L. K., Stensel, D. J., & Nimmo, M. A. (2013, September). Exercise and ghrelin. A narrative overview of research. *Appetite, 68*: 83–91. doi: 10.1016/j.appet.2013.04.018.

Kotz, C. M., & Levine, J. A. (2005, September). Role of nonexercise activity thermogenesis (NEAT) in obesity. *Minnesota Medicine, 88*(9): 54–57.

Lavie, C. J., et al. (2010, November). Use of Body Fatness Cutoff Points. *Mayo Clinic Proceedings, 85*(11): 1057–1061.

Ledikwe, J. H., Ello-Martin, J. A., & Rolls, B. J. (2005, April). Portion sizes and the obesity epidemic. *Journal of Nutrition, 135*(4): 905–909.

Lee, S. Y., & Gallagher, D. (2008, September). Assessment methods in human body composition. *Current opinion in clinical nutrition and metabolism care, 11*(5): 566–572.

Levine, J. A. (2007). Nonexercise activity thermogenesis—liberating the life-force. *Journal of Internal Medicine, 262*:273-287. doi: 10.1111/j.1365-2796.2007.01842.x

Levine, J. A., Baukol, P., & Pavlidis, I. (1999). The energy expended in chewing gum. *New England Journal of Medicine, 341*:2100.

Levine, J. A., Eberhardt, N. L., & Jensen, M. D. (1999, January 8). Role of nonexercise activity thermogenesis in resistance to fat gain in humans. *Science, 283*(5399): 212–214. doi: 10.1126/science.283.5399.212.

Levine, J. A., Schleusner, S. J., & Jensen, M. D. (2000). Energy expenditure of nonexercise activity. *American Journal of Clinical Nutrition, 72*: 1451–4.

Levine, J. A., Vander Weg, M. W., Hill, J. O., & Klesges, R. C. (2006). Non-exercise activity thermogenesis. *Arteriosclerosis, thrombosis, and vascular biology, 26*: 729–736. doi: 10.1161/ 01.ATV.0000205848.83210.73.

Ling, H., Lenz, T. L., Burns, T. L., & Hilleman, D. E. (2013, December). Reducing the risk of obesity: defining the role of weight loss drugs. *Pharmacotherapy, 33*(12): 1308–1321. doi: 10.1002/phar.1277.

Manini, T. M. (2010, January). Energy expenditure and aging. *Thermodynamics and Ageing, 9*(1): 1–11.

Mantzoros, C. S., et al. (2011, October). Leptin in human physiology and pathophysiology. *American Journal of Physiology, Endocrinology, and Metabolism, 301*(4): E567–E584. doi: 10.1152/ajpendo.00315.2011.

Martinez, J. A. (2000, August). Body weight regulation: causes of obesity. *Proceedings of the Nutrition Society, 59*(3): 337–345.

Matsuzawa, Y., Shimomura, I., Nakamura, T., Keno, Y., Kotani, K., & Tokunaga, K. (1995, September 6). Pathophysiology and pathogenesis of visceral fat obesity. *Obesity Research, 3*(S2): 187s–194s. doi: 10.1002/j.1550-8528.1995.tb00462.x.

McManus, A. M. (2007, September). Physical activity—a neat solution to an impending crisis. *Journal of Sports Science and Medicine, 6*(3): 368–373.

Myint, P. D., Kwok, C. S., Luben, R. N., Wareham, N. J., & Khaw, K. (2014, June 24). Body fat percentage, body mass index and waist-to-hip ratio as predictors of mortality and cardiovascular disease. *Heart*, doi: 10.1136/heartjnl-2014-305816. Retrieved from http://heart.bmj.com/content/early/2014/06/24/heartjnl-2014-305816.short. Accessed November 5, 2015.

National Heart, Lung, and Blood Institute (2012). How can overweight and obesity be prevented. Retrieved from http://www.nhlbi.nih.gov/health/health-topics/topics/obe/prevention.html. Accessed November 5, 2015.

National Heart, Lung, and Blood Institute (2012). What causes overweight and obesity? Retrieved from http://www.nhlbi.nih.gov/health/health-topics/topics/obe/causes.html. Accessed November 5, 2015.

National Heart, Lung, and Blood Institute (2014). Calculate Your Body Mass Index. Retrieved from http://www.nhlbi.nih.gov/guidelines/obesity/BMI/bmicalc.htm. Accessed November 5, 2015.

National Weight Control Registry (n.d.). Retrieved from http://www.nwcr.ws. Accessed November 5, 2015.

Nelson, J. K., & Zeratsky, K. (2013, May 1). Is NEAT part of your weight-control plan? Retrieved from http://www.mayoclinic.org/healthy-living/nutrition-and-healthy-eating/expert-blog/neat-non-exercise-activity-thermogenesis/bgp-20056175. Accessed November 5, 2015.

Nestle, M., & Nesheim, M. (2012). *Why Calories Count: From Science to Politics* (University of California Press).

Ogden, C. L., Carroll, M. D., Kit, B. K., & Flegal, K. M. (2013, October). Prevalence among adults: United States, 2011–2012. *NCHS Data Brief, 131*: 1–8.

Otten, J. J., Hellwig, J. P., & Meyers, L. D. (eds) (2006). Energy. *Dietary Reference Intakes: The Essential Guide to Nutrient Requirements* (107–264). (Washington, DC: The National Academies Press).

Papas, M. A., Alberg, A. J., Ewing, R., Helzlsouer, K. J., Gary, T. L., & Klassen, A. C. (2007). The built environment and obesity. *Epidemiological Review, 29*: 129–143.

Picot, J., Jones, J., Colquitt, J. L., Gospodarevskaya, E., Loveman, E., Baster, L., & Clegg, A. J. (2009, September). The clinical effectiveness and cost-effectiveness of bariatric (weight loss) surgery for obesity: a systematic review and economic evaluation. *Health Technology Assessment, 13*(41):215–357. doi: 10.3310/hta13410.

Qi, L., & Cho, Y. A. (2013, June). Gene-environment interaction and obesity. *Nutritional Review, 66*(12): 684–694. doi: 10.1111/j.1753-4887.2008.00128.x.

Raavussin, E. (2005, January 28). A NEAT way to control weight? *Science, 307*(5709): 530–531. doi: 10.1126/science.1108597.

Reese, M. (2008, January). Underweight: A Heavy Concern. *Today's Dietitian, 10*(1): 56.

Reuda-Clausen, C. F., Padwal, R. S., & Sharma, A. M. (2013, August). New pharmacological approaches for obesity management. *National Review of Endocrinology, 9*(8): 467–478. doi: 10.1038/nrendo.2013.113.

Rippe, J. M., & Saltzman, E. (2013, September). Sweetened beverages and health: current state of scientific understandings. *Advances in Nutrition, 4*: S27–S29. doi: 10.3945/an.113.004143.

Roberts, S. B., & Rosenberg, I. (2006). Nutrition and aging: Changes in the regulation of energy metabolism with aging. *Physiological Review, 86*: 651–667. doi: 0.1152/physrev.00019.2005

Salis, J. F., Floyd, M. F., Rodriguez, D. A., & Saelens, B. E. (2012). Role of built environments in physical activity, obesity, and cardiovascular diseases. *Circulation, 125:* 729–737. doi: 10.1161/CIRCULATIONAHA.110.969022.

Schoeller, D. A. (2009, April 21). The energy balance equation: looking back and looking forward are two very different views. *Nutrition Reviews, 67*(5): 249–254. doi: 10.1111/j.1753-4887.2009.00197.x.

Speakman, J. R., et al. (2011, November). Set points, settling points, and some alternative models: theoretical options to understand how genes and environments combine to regulate body adiposity. *Disease Models and Mechanisms, 4*(6): 733–745. doi: 10.1242/dmm.008698.

Talley, N. J., et al. (eds). (2014). *GI Epidemiology: Diseases and Clinical Methodology,* Second Edition. Hoboken, NJ: Wiley-Blackwell.

Teh, K. C., & Aziz, A. R. (2002). Heart rate, oxygen uptake, and energy cost of ascending and descending the stairs. *Medicine and Science in Sports and Exercise, 34*: 695–699.

Thomas, J. G., Bond, D. S., Phelan, S., Hill, J. O., & Wing, R. R. (2013, January). Weight-loss maintenance for 1-year in the National Weight Control Registry. *American Journal of Preventive Medicine, 46*(1): 17–23.

Thorgood, M., Appleby, P. N., Key, T. J., & Mann, J. (2003). Relation between body mass index and mortality in an unusually slim cohort. *Journal of Epidemiology and Community Health, 57*: 130–133. doi: 10.1136/jech.57.2.130.

Vasselli, J. R. (2012, September). The role of dietary components in leptin resistance. *Advances in Nutrition, 3*: 736–738, doi: 10.3945/an.112.002659.

Vucenik, I., & Stains, J. P. (2012, October). Obesity and cancer risk: evidence, mechanisms, and recommendations. *Annals of the New York Academy of Sciences, 1271*: 37–43. doi: 10.1111/j.1749-6632.2012.06750.x.

Walley, A. J., Asher, J. E., & Froguel, P. (2009, July). The genetic contribution to non syndromic human obesity. *National Review of Genetics, 10*(7): 431–442. doi: 10.1038/nrg2594.

Wandell, P. E., Carisson, A. C., & Theobald, H. (2009, May). The association between BMI value and long-term mortality. *International Journal of Obesity, 33*(5): 577–582. doi: 10.1038/ijo.2009.36.

Webb, D. (n.d.). How many calories equal one pound? *Today's Dietitian.* Retrieved from http://www.todaysdietitian.com/news/exclusive0612.shtml. Accessed November 5, 2015.

Weight Loss Surgery (2014). Retrieved from http://www.nlm.nih.gov/medlineplus/weightlosssurgery.html. Accessed November 5, 2015.

Weinsier, R. L., Hunter, G. R., Heini, A. F., Goran, M. I., & Sell, S. M. (1998, August). The etiology of obesity: relative contribution of metabolic factors, diet, and physical activity. *American Journal of Medicine, 105*(2): 145–150.

Wing, R. R., & Phelan, S. (2005, July). Long-term weight loss maintenance. *The American Journal of Clinical Nutrition, 82*(1): 222S–225S.

What you should know about popular diets (2014). Retrieved from http://www.nutrition.gov/weight-management/what-you-should-know-about-popular-diets. Accessed November 5, 2015.

World Health Organization (2015, January). Obesity and overweight. Retrieved from http://www.who.int/mediacentre/factsheets/fs311/en/. Accessed November 5, 2015.

Young, J. L., & Nestle, M. (2012). Reducing portion sizes to prevent obesity: a call to action. *American Journal of Preventive Medicine, 43*(5): 565–568. doi: 10.1016/j.amepre.2012.07.024.

Chapter 16

American College of Sports Medicine (2007, February). Exercise and fluid replacement. *Medicine and Science in Sports and Exercise, 39*(2): 337–390. doi: 10.1249/mss.0b013e31802ca597.

American College of Sports Medicine (2011). The female athlete triad. Retrieved from http://www.acsm.org/docs/brochures/the-female-athlete-triad.pdf. Accessed November 5, 2015.

American College of Sports Medicine (2011). Selecting and effectively using hydration for fitness. Retrieved from http://www.acsm.org/docs/brochures/selecting-and-effectively-using-hydration-for-fitness.pdf. Accessed November 5, 2015

American College of Sports Medicine (2011). Selecting and effectively using sports drinks, carbohydrate gels, and energy bars. Retrieved from http://www.acsm.org/docs/brochures/selecting-and-effectively-using-sports-drinks-carbohydrate-gels-and-energy-bars.pdf. Accessed November 5, 2015.

American Dietetic Association, Dietitians of Canada, & American College of Sports Medicine (2009, March). Nutrition and athletic performance. *Medicine and Science in Sports and Exercise, 41*(2): 709–731. doi: 10.1249/MSS.0b013e31890eb86.

Bauman, A., et al. (2011, August). The descriptive epidemiology of sitting. A 20-country comparison using the International Physical Activity Questionnaire (IPAQ). *American Journal of Preventive Medicine, 41*(2): 228035. doi: 10.1016/j.amepre.2011.05.003.

The Benefits of Physical Activity (n.d.). *The Nutrition Source*. Retrieved from http://www.hsph.harvard.edu/nutritionsource/staying-active. Accessed November 5, 2015.

Centers for Disease Control and Prevention (2008). 2008 Physical Activity Guidelines for Americans: Fact Sheet for Health Professionals on Physical Activity Guidelines for Older Adults. Retrieved from http://www.cdc.gov/nccdphp/dnpa/physical/pdf/PA_Fact_Sheet_OlderAdults.pdf. Accessed November 5, 2015.

Centers for Disease Control and Prevention (2014, June 4). How much physical activity do adults need? Retrieved from http://www.cdc.gov/physicalactivity/everyone/guidelines/adults.html. Accessed November 5, 2015.

Centers for Disease Control and Prevention (2015, June 4). How much physical activity do children need? Retrieved from http://www.cdc.gov/physicalactivity/everyone/guidelines/children.html. Accessed November 5, 2015.

Centers for Disease Control and Prevention (2015, June 4). Physical activity and health: the benefits of physical activity. Retrieved from http://www.cdc.gov/physicalactivity/everyone/health/. Accessed November 5, 2015.

Colombani, P. C., Mannhart, C., & Mettler, S. (2013, January 28). Carbohydrates and exercise performance in non-fasted athletes: a systematic review of studies mimicking real-life. *Nutrition Journal,* 12–16. doi: 10.1186/1475-2891-12-16.

DeMarco, H. (n.d.). Pre-Event Meals. *ACSM Current Comment*. Retrieved from http://www.acsm.org/docs/current-comments/preeventmeals.pdf. Accessed November 5, 2015.

Donnelly, J. E., Blair, S. N., Jakicic, J. M., Manore, M. M., Rankin, J. W., & Smith, B. K. (2009, February). Appropriate physical activity intervention strategies for weight loss and prevention of weight regain for adults. *Medicine and Science in Sports and Exercise, 41*(2): 459–471. doi: 10.1249/MSS.0b013e3181949333.

Garber, C. E., et al. (2011, July). Quantity and quality of exercise for developing and maintaining cardiorespiratory, musculoskeletal, and neuromotor fitness in apparently healthy adults: guidance for prescribing exercise. *Medicine and Science in Sports and Exercise, 43*(7): 1334–1359. doi: 10.1249/MSS.0b013e318213fefb

Katzmarzyk, P. T., & Lee, I. (2012). Sedentary behavior and life expectancy in the USA: a cause-deleted life table analysis. *BMJ Open, 2*(4). doi: 10.1136/bmjopen-2012-000828.

Knight, J. A. (2012, Summer). Physical inactivity: associated diseases and disorders. *Annals of Clinical Laboratory Science, 42*(3): 320–337.

Levin, J. A., Vander Weg, M. W., Hill, J. O., & Klesges, R. C. (2006, April). Non-exercise activity thermogenesis: the crouching tiger hidden dragon of societal weight gain. *Arteriosclerosis, Thrombosis, and Vascular Biology, 26*(4): 729–736.

Loprinzi, P. D., Cardinal, B. J., Loprinzi, K. L., & Lee, H. (2012). Benefits and environmental determinants of physical activity in children and adolescents. *Obesity Facts, 5*(4): 597–610. doi: 10.1159/000342684.

Lukaski, H. C., Haymes, E., & Kanter, M. (n.d.). Vitamin and mineral supplements and exercise. *ACSM Current Comment*. Retrieved from http://www.acsm.org/docs/current-comments/vitaminandmineralsupplementsandexercise.pdf. Accessed November 5, 2015.

Maughan, R. J. (1999, December). Nutritional ergogenic aids and exercise performance. *Nutrition Research Reviews, 12*(2): 255–280. doi: 10.1079/095442299108728956.

Moore, L. J., Midgeley, A. W., Thurlow, S., Thomas, G., & McNaughton, L. R. (2010, January). Effect of the glycaemic index of a pre-exercise meal on a metabolism and cycling time trial performance. *Journal of Science and Medicine in Sport, 13*(3): 182–188. doi: 10.1016/j.jsams.2008.11.006.

Nattiv, A., Loucks, A. B., Manore, M. M., Sanborn, C. F., Sundgot-Borgen, J., & Warren, M. P. (2007, October). The female athlete triad. *Medicine and Science in Sports and Exercise, 39*(10): 1867–1882. doi: 10.1249/mss.0b013e318149f111.

Poirier, P., & Després, J. P. (2001, August). Exercise in weight management of obesity. *Cardiology Clinics, 19*(3): 459–470.

Spriet, L. L., & Graham, T. E. (n.d.). Caffeine and exercise performance. *ACSM Current Comment*. Retrieved from http://www.acsm.org/docs/current-comments/caffeineandexercise.pdf. Accessed November 5, 2015.

Teramoto, M., & Bungum, T. J. (2010, July). Mortality and longevity of elite athletes. *Journal of Science and Medicine in Sport, 13*(4): 410–416. doi: 10.1016/j.jsams.2009.04.010.

U.S. Department of Agriculture, Agriculture Research Service. (2008). Nutrition Intakes from Food: Mean Amounts and Percentages of Calories from Protein, Carbohydrate, and Alcohol, One Day, 2005–2006. Retrieved from www.ars.usda.gov/ba/bhnrc/fsrg. Accessed November 5, 2015.

U.S. Department of Agriculture (2010). Part D. Section 5: Carbohydrates. *Report of the DGAC on the Dietary Guidelines for Americans, 2010.* Retrieved from http://www.cnpp.usda.gov/publications/dietaryguidelines/2010/dgac/report/d-5-carbohydrates.pdf. Accessed November 5, 2015.

U.S. Department of Agriculture (n.d.). Why is physical activity important? Retrieved from http://www.choosemyplate.gov/physical-activity-why.html. Accessed November 5, 2015.

U.S. Department of Health and Human Services (2014). 2008 Physical Activity Guidelines for Americans: Summary. Retrieved from http://www.health.gov/paguidelines/guidelines/summary.aspx. Accessed November 5, 2015.

Warburton, D. E., Nicol., C. W., & Bredin, S. S. (2006, March 14). Health benefits of physical activity: the evidence. *Canadian Medical Association Journal, 174*(6): 801–809. doi: 10.1503/cmaj.051351.

Woolf, K., & Manore, M. M. (2006, October). B-vitamins and exercise: does exercise alter requirements? *International Journal of Sports Nutrition and Exercise Metabolism, 16*(5): 453–484.

Chapter 17

Academy of Nutrition and Dietetics (2014, July). Position of the Academy of Nutrition and Dietetics: Nutrition and Lifestyle for a Healthy Pregnancy Outcome. *Journal of the Academy of Nutrition and Dietetics, 114*(7): 1099–1103.

American Academy of Pediatrics (2012, February 27). Breastfeeding and the Use of Human Milk. *Pediatrics, 129*(3): e827–e841. doi: 10.1542/peds.2011-3552.

American Academy of Pediatrics (2011). Promoting Healthy Nutrition. *Bright Futures Guidelines for Health Supervision of Infants, Children, and Adolescents.* Retrieved from http://brightfutures.aap.org/pdfs/Guidelines_PDF/6-Promoting_Healthy_Nutrition.pdf. Accessed November 6, 2015.

American College of Obstetricians and Gynecologists (2011, July). Vitamin D: screening and supplementation during pregnancy. *Committee Opinion, 295*(118): 197–198.

Bakker, R., Steegers, E. A., Obradov, A., Raat, H., Hofman, A., & Jaddoe, V. W. (2010, June). Maternal caffeine intake from coffee and tea, fetal growth, and the risks of adverse birth outcomes: the Generation R study. *American Journal of Clinical Nutrition, 91*(6): 1691–1698. doi: 10.3945/ajcn.2009.28792.

Brown, H. L., & Graves, C. R. (2013, March). Smoking and marijuana use in pregnancy. *Clinical Obstetrics and Gynecology, 56*(1): 107–113. doi: 10.1097/GRF.0b013e318282377d.

Centers for Disease Control and Prevention (2015, April 28). Folic Acid Recommendations. Retrieved from http://www.cdc.gov/ncbddd/folicacid/recommendations.html. Accessed November 6, 2015.

Cheng, J., Su., H., Zhu, R., Wang, X., Peng, M., Song, J., & Fan, D. (2014, February). Maternal coffee consumption during pregnancy and risk of childhood acute leukemia: a metaanalysis. *American Journal of Obstetrics and Gynecology, 210*(2): 151. doi: 10.1016/j.ajog.2013.09.026.

Committee on Obstetric Practice & American College of Obstetricians and Gynecologists (2012, December). Committee Opinion No. 543: Timing of umbilical cord clamping after birth. *Obstetrics & Gynecology, 120*(6): 1522–1526. doi: 10.1097/01.AOG.0000423817.47165.48.

Greenberg, J., Bell, S. J., & Ausdal, W. V. (2008, Fall). Omega-3 Fatty Acid Supplementation During Pregnancy. *Reviews in Obstetrics & Gynecology, 1*(4): 162–169.

Hackshaw, A., Rodeck, C., & Boniface, S. (2011, September-October). Maternal smoking in pregnancy and birth defects: a systematic review based on 173,687 malformed cases and 11.7 million controls. *Human Reproduction Update, 17*(5): 589–604. doi: 10.1093/humupd/dmr022.

Homko, C. J., Sivan, E., Reece, E. A., & Boden, G. (1999). Fuel metabolism during pregnancy. *Seminars in Reproductive Endocrinology, 17*(2): 119–125.

Hoyt, A. T., et al. (2013, August). Maternal caffeine consumption and small for gestational age births: results from a population-based case-control study. *Maternal and Child Health Journal, 18*(6): 1540–1551. doi: 10.1007/s10995-013-1397-4.

Institute of Medicine. (2009, May). Weight gain during pregnancy: Reexamining the guidelines. Report Brief. Retrieved from http://iom.edu/~/media/Files/Report%20Files/2009/Weight-Gain-During-Pregnancy-Reexamining-the-Guidelines/Report%20Brief%20-%20Weight%20Gain%20During%20Pregnancy.pdf. Accessed November 6, 2015.

King, J. C. (2006). Maternal obesity, metabolism, and pregnancy outcomes. *Annual Review of Nutrition, 26*: 271–291.

Kramer, M. S. (2013). The epidemiology of low birthweight. *Nestle Nutrition Workshop Series, 74*:1–10, doi: 10.1159/000348382.

Lippi, G., & Franchini, M. (2011, January). Vitamin K in neonates: facts and myths. *Blood Transfusion, 9*(1): 4–9.

Low birth weight. (2014). March of Dimes. Retrieved from http://www.marchofdimes.com/baby/low-birthweight.aspx. Accessed November 6, 2015.

Pena-Rosas, J. P., De-Regli, L. M., Dowswell, T., & Viteri, F. E. (2012, December 12). Daily oral iron supplementation during pregnancy. *Cochrane Database Systematic Review, 12*:CD004736. doi: 10.1002/14651858.CD004736.pub4.

Pineless, B. L., et al. (2014, April). Systematic review and meta-analysis of miscarriage and maternal exposure to tobacco smoke during pregnancy. *American Journal of Epidemiology, 179*(7): 807–823.

Rau, B. J., Castillo, P. R., Gianopoulos, J., & Boyd, L. (2011, December). World's tiniest preemies are growing up and doing fine. Retrieved from http://www.eurekalert.org/pub_releases/2011-12/luhs-wtp120811.php. Accessed November 6, 2015.

Safi, J., Joyeux, L., & Chalouhi, G. E. (2012). Periconceptional folate deficiency and implications in neural tube defects. *Journal of Pregnancy, 2012*: 295083. doi: 10.1155/2012/295083.

U.S. Department of Agriculture (2015, July). Health and Nutrition Information for Pregnant and Breastfeeding Women. Retrieved from http://www.choosemyplate.gov/moms-pregnancy-breastfeeding. Accessed November 6, 2015.

U.S. Food and Drug Administration (2014, June). Fish: What Pregnant Women and Parents Should Know. Retrieved from http://www.fda.gov/food/foodborneillnesscontaminants/metals/ucm393070.htm. Accessed November 6, 2015.

Vahratian, A. (2009, March). Prevalence of overweight and obesity among women of childbearing age. *Maternal and Child Health Journal, 13*(2): 268–273.

Chapter 18

Abrams, S. A., & Tiosano, D. (2014, February). Update on vitamin D during childhood. *Current Opinions in Endocrinology, Diabetes, and Obesity, 21*(1): 51–55. doi: 10.1097/01.med.0000436252.53459.ef.

Academy of Nutrition and Dietetics (2008, June). Nutrition Guidance for Healthy Children Aged 2 to 11 Years. *Journal of the American Dietetics Association, 108*(6): 1038–1047.

Alliance for a Healthier Generation. Retrieved from https://www.healthiergeneration.org. Accessed November 6, 2015.

American Academy of Pediatrics (2013, November 1). Children, Adolescents, and the Media. *Pediatrics, 132*(5): 958–961. doi: 10.1542/peds.2013-2656.

Baker, R. D., & Greer, F. R. (2010, November 1). Diagnosis and prevention of iron deficiency and iron-deficiency anemia in infants and young children (0-3 years of age.) *Pediatrics, 126*(5): 1040–1050. doi: 10.1542/peds.2010-2576

Centers for Disease Control and Prevention (2010, September 9). Growth Charts. Retrieved from http://www.cdc.gov/growthcharts/. Accessed November 6, 2015.

Centers for Disease Control and Prevention (2015, May 15). About BMI for Children and Teens. Retrieved from http://www.cdc.gov/healthyweight/assessing/bmi/childrens_bmi/about_childrens_bmi.html. Accessed November 6, 2015.

Centers for Disease Control and Prevention (2015, June 19). Basics about childhood obesity. Retrieved from http://www.cdc.gov/obesity/childhood/basics.html. Accessed November 6, 2015.

Centers for Disease Control and Prevention (2015, June 19). Childhood Obesity Causes and Consequences. Retrieved from http://www.cdc.gov/obesity/childhood/problem.html. Accessed November 6, 2015.

Centers for Disease Control and Prevention (2015, October 27). Strategies to Prevent Obesity. Retrieved from http://www.cdc.gov/obesity/childhood/solutions.html. Accessed November 6, 2015.

Centers for Disease Control and Prevention (2014). Childhood obesity facts: Prevalence of childhood obesity in the United States, 2011–2012. Retrieved from http://www.cdc.gov/obesity/data/childhood.html. Accessed November 6, 2015.

Cogswell, M. E., et al. (2009, May). Assessment of iron deficiency in US preschool children and non pregnant females of childbearing age: National Health and Nutrition Examination Survey 2003–2006. *American Journal of Clinical Nutrition, 89*(5): 1334–1342. doi: 0.3945/ajcn.2008.27151.

Collins, C., Duncanson, K., & Burrows, T. (2014, January 6). A systematic review investigating associations between parenting style and child feeding behaviors. *Journal of Human Nutrition and Dietetics, 27*(6): 557–568. doi: 10.1111/jhn.12192.

de Silva, D., et al. (2014, May). Primary prevention of food allergy in children and adults: systematic review. *European Journal of Allergy and Clinical Immunology, 69*(5): 581–589. doi: 10.1111/all.12334.

FARE-Food Allergy Research and Education (2014). Retrieved from http://www.foodallergy.org. Accessed November 6, 2015.

Food Research and Action Center (2015). National School Lunch Program. Retrieved from http://frac.org/federal-foodnutrition-programs/national-school-lunch-program/. Accessed November 6, 2015.

Food Research and Action Center (2015). School meal nutrition standards. Retrieved from http://frac.org/federal-foodnutrition-programs/national-school-lunch-program/school-meal-nutrition-standards/. Accessed November 6, 2015.

Greer, F. R. & Krebs, N. F. (2006, February 1). Optimizing Bone Health and Calcium Intakes of Infants, Children, and Adolescents. *Pediatrics, 117*(2): 578–585. doi: 10.1542/peds.2005-2822

Hoelscher, D. M., Kirk, S., Ritchie, L., Cunningham-Sabo, L., & Academy Positions Committee (2013, October). Position of the Academy of Nutrition and Dietetics: Interventions for the prevention and treatment of pediatric overweight and obesity. *Journal of the Academy of Nutrition and Dietetics, 113*(1): 1375–1394. doi: 10.1016/j.jand.2013.08.004.

Institute of Medicine of the National Academies (2005). Dietary Reference Intakes: Macronutrients. *Dietary Reference Intakes for Energy, Carbohydrate, Fiber, Fat, Fatty Acids, Cholesterol, Protein, and Amino Acids* (Washington, DC: The National Academies Press).

Kranz, S., Brauchia, M., Slavin, J. L., & Miller, K. B. (2012). What do we know about dietary fiber intake in children and health? The effects of fiber intake on constipation, obesity, and diabetes in children. *Advances in Nutrition, 3*: 47–53. doi: 10.3945/an.111.001362.

Larson, N., & Story, M. (2013, April). A review of snacking patterns among children and adolescents: what are the implications of snacking for weight status? *Childhood Obesity, 9*(2): 10–15. doi: 10.1089/chi.2012.0108.

Lipsky, L. M., & Iannotti, R. J. (2012). Associations of television viewing with eating behaviors in the 2009 Health Behavior in School-aged Children Study. *Archives of Pediatrics and Adolescent Medicine, 166*(5): 465–472. doi: 10.1001/archpediatrics.2011.1407.

National Institute of Allergy and Infectious Diseases (2012). Understanding Food Allergy. Retrieved from http://www.niaid.nih.gov/topics/foodallergy/understanding/Pages/default.aspx. Accessed November 6, 2015.

National Institute of Allergy and Infectious Diseases (2010). Guidelines for the Diagnosis and Management of Food Allergy in the United States. Retrieved from http://www.niaid.nih.gov/topics/foodallergy/clinical/Pages/default.aspx. Accessed November 6, 2015.

National Institute of Dental and Craniofacial Research (2014). Dental caries (tooth decay) in children (age 2 to 11). Retrieved from http://www.nidcr.nih.gov/DataStatistics/FindDataByTopic/DentalCaries/DentalCariesChildren2to11.htm. Accessed November 6, 2015.

Ogden, C. L., Kit, B. K., Caroll, M. D. & Park, S. (2011). Consumption of sugar drinks in the United States, 2005–2008. *NCHS data brief, 71*. (Hyattsville, MD: National Center for Health Statistics).

Otten, J. J., Hellwig, J. P., & Meyers, L. D. (eds) (2006). Macronutrients and Healthful Diets. *Dietary Reference Intakes: The Essential Guide to Nutrient Requirements*. Food and Nutrition Board, Institute of Medicine.

Patrick, H., & Nicklas, T. A. (2005, April). A review of family and social determinants of children's eating patterns and diet quality. *Journal of the American College of Nutrition, 24*(2): 83–92.

Pearson, N., & Biddle, S. J. (2001, August). Sedentary behavior and dietary intake in children, adolescents, and adults. A systematic review. *American Journal of Preventive Medicine, 41*(2): 178–188. doi: 10.1016/j.amepre.2011.05.002

Ramsay, S. A., Branen, L. J., & Johnson, S. L. (2012, February). How much is enough? Tablespoon per year of age approach meets nutrient needs for children. *Appetite, 58*(1): 163–167. doi: 10.1016/j.appet.2011.09.028.

Reedy, J., & Krebs-Smith, S. M. (2010, October). Dietary sources of energy, solid fats, and added sugars among children and adolescents in the United States. *Journal of the American Dietetics Association, 110*(1): 1477–1484.

Rideout, V. J., Foehr, U. G., & Roberts, D. F. (2010, January). Generation M2 : Media in the Lives of 8–18 Year Olds. A Kaiser Family Foundation Study. Retrieved from http://files.eric.ed.gov/fulltext/ED527859.pdf. Accessed November 6, 2015.

Robert Wood Johnson Foundation. Retrieved from http://www.rwjf.org/en.html. November 6, 2015.

Rogol, A. D., Clark, P. A., & Roemmich, J. N. (2000, August). Growth and pubertal development in children and adolescents: effects of diet and physical activity. *American Journal of Clinical Nutrition, 72*(2): 521S–528S.

Satter, E. (n.d.). Ellyn Satter's Division of Responsibility in Feeding. Retrieved from http://ellynsatterinstitute.org/dor/divisionofresponsibilityinfeeding.php. Accessed November 6, 2015.

Sicherer, S. H., & Sampson, H. A. (2014, February). Food allergy: epidemiology, pathogenesis, diagnosis, and treatment. *The Journal of Allergy and Clinical Immunology, 133*(2): 291–307.

U.S. Department of Agriculture & United States Department of Health and Human Services (2010, December). *Dietary Guidelines for Americans, 2010* (7th ed). (Washington, DC: US Government Printing Office).

U.S. Department of Agriculture, Food and Nutrition Service (2014). Child Nutrition Programs. Retrieved from http://www.fns.usda.gov/school-meals/child-nutrition-programs. Accessed November 6, 2015.

U.S. Department of Agriculture, Food and Nutrition Service (2015). Child Nutrition Tables. Retrieved from http://www.

fns.usda.gov/pd/child-nutrition-tables. Accessed November 6, 2015.

U.S. Department of Agriculture, Food and Nutrition Service (2015, September 9). National School Lunch Program. Retrieved from http://www.fns.usda.gov/nslp/national-school-lunch-program-nslp. Accessed November 6, 2015.

U.S. Department of Agriculture, Food and Nutrition Service (2015, August 11). School Meals: Nutrition Standards. Retrieved from http://www.fns.usda.gov/school-meals/nutrition-standards-school-meals. Accessed November 6, 2015.

Wagner, C. L., & Greer, F. R. (2008, November 1). Prevention of rickets and vitamin D deficiency in infants, children, and adolescents. *Pediatrics, 122*(5): 1142–1152. doi: 10.1542/peds.2008-1862.

Chapter 19

Academy of Nutrition and Dietetics (2012, August). Food and Nutrition for Older Adults: Promoting Health and Wellness. *Journal of the American Dietetic Association, 112*(8): 1255–1277.

Academy of Nutrition and Dietetics (2013, April). Eight Ways to Beat the Freshman 15. Retrieved from http://www.eatright.org/Public/content.aspx?id=6442471553. Accessed November 6, 2015.

American Dietetic Association (2011). Position of the American Dietetic Association: Nutrition Intervention in the Treatment of Eating Disorders. *Journal of the American Dietetic Association, 111*(8): 1236–1241.

American Psychiatric Association (2013). Feeding and Eating Disorders. Retrieved from http://www.dsm5.org/Documents/Eating%20Disorders%20Fact%20Sheet.pdf. Accessed November 6, 2015.

American Psychological Association (2013). Treatment for anorexia and bulimia. Retrieved from http://www.apa.org/topics/eating/treatment.aspx. Accessed November 6, 2015.

Anorexia nervosa (2014, March 10). Retrieved from http://www.nlm.nih.gov/medlineplus/ency/article/000362.htm. Accessed November 6, 2015.

Arcules, J., Mitchell, A. J., Wales, J., & Nielsen, S. (2011, July). Mortality rates in patients with anorexia nervosa and other eating disorders. A meta-analysis of 36 studies. *Archives of General Psychiatry, 68*(7): 724–731. doi: 10.1001/archgenpsychiatry.2011.74.

Bernstein, M., & Munoz, N. (2012, August). Position of the Academy of Nutrition and Dietetics: Food and Nutrition for Older Adults: Promoting Health and Wellness. *Journal of the Academy of Nutrition and Dietetics, 112*(8): 1255–1277.

Boek, S., Bianco-Simeral, S., Chan, K., & Goto, K. (2012, July–August). Gender and race are significant determinants of students' food choices on a college campus. *Journal of Nutrition Education and Behavior, 44*(4): 372–378. doi: 10.1016/j.jneb.2011.12.007.

Bulimia (2015, March 4). Retrieved from http://www.nlm.nih.gov/medlineplus/ency/article/000341.htm. Accessed November 6, 2015.

Centers for Disease Control and Prevention (2014). Fact Sheets—Binge Drinking. Retrieved from http://www.cdc.gov/alcohol/fact-sheets/binge-drinking.htm. Accessed November 6, 2015.

Centers for Disease Control and Prevention (2014). Fact Sheets—Underage Drinking. Retrieved from http://www.cdc.gov/alcohol/fact-sheets/underage-drinking.htm. Accessed November 6, 2015.

Centers for Disease Control and Prevention (2015, April 16). Facts about Fetal Alcohol Spectrum Disorders (FASDs). Retrieved from http://www.cdc.gov/NCBDDD/fasd/facts.html. Accessed November 6, 2015.

Centers for Disease Control and Prevention (2015, May 25). Healthy Aging. Retrieved from http://www.cdc.gov/aging/index.htm. Accessed November 6, 2015.

Centers for Disease Control and Prevention (2013). *The State of Aging and Health in America 2013*. (Atlanta, GA: CDC, USDA).

Chedraui, P., & Perez-Lopez, F. R. (2013, August). Nutrition and health during mid-life: searching for solutions and meeting challenges for the aging population. *Climacteric, 16*(Supplement 1): 85-95. doi: 10.3109/13697137.2013.802884.

Courtney, K., & Polich, J. (2009, January). Binge drinking in young adults: data, definition, and determinants. *Psychology Bulletin, 135*(1): 142–156. doi: 10.1037/a0014414.

Crombie, A. P., Illich, J. Z., Dutton, G. R., Panton, L. B., & Abood, D. A. (2009, February). The freshman weight gain phenomenon revisited. *Nutrition Review, 67*(2): 83–94. doi: 10.1111/j.1753-4887.2008.00143.x.

Delinsky, S. S., & Wilson, G. T. (2008, January). Weight gain, dietary restraint, and disordered eating in the freshman year of college. *Eating Behavior, 9*(1): 82–90. doi: 10.1016/j.jneb.2007.01.001.

Everitt, A. V., et al. (2006). Dietary approaches that delay age-related diseases. *Journal of Clinical Interventions in Aging, 1*(1): 11–31.

Ferrara, C. M. (2009, February). The college experience: physical activity, nutrition, and implications for intervention and future research. *Journal of Exercise Physiology, 12*(1): 24–35.

Food and Nutrition Board, Institute of Medicine, National Academies (n.d.). Dietary reference intakes (DRIS): Estimated average requirements. Retrieved from http://iom.edu/Activities/Nutrition/SummaryDRIs/~/media/Files/Activity%20Files/Nutrition/DRIs/New%20Material/5DRI%20Values%20SummaryTables%2014.pdf. Accessed November 6, 2015.

Foundation for Advancing Alcohol Responsibility (2013). Binge Drinking Statistics. Retrieved from http://responsibility.org/college-students-and-drinking/college-binge-drinking-statistics/. Accessed November 6, 2015.

Gunzerath, L., Faden, V., Zakhari, S., & Warren, K. (June, 2004). National Institute on Alcohol Abuse and Alcoholism report on moderate drinking. *Alcoholism: Clinical and Experimental Research, 28*(6): 829–847.

Halterman, J. S., Kaczorowski, J. M., Aligne, C. A., Auinger, P., & Szilagyi, P. G. (2001, June). Iron deficiency and cognitive achievement among school-aged children and adolescents in the United States. *Pediatrics, 107*(6): 1381–1386.

Iacovino, J. M., Gredysa, D. M., & Wilfley, D. E. (2012, August). Psychological treatments for binge eating disorder. *Current Psychiatry Reports, 14*(4): 432–446.

Jelski, W., Kozlowski, M., Laudanski, J., Nikilinski, J., & Szmitkowski, M. (2009, April). The activity of class I, II, III, and IV alcohol dehydrogenase (ADH) isoenzymes and aldehyde dehydrogenase (ALDH) in esophageal cancer. *Digestive Diseases and Sciences, 54*(4): 725–730.

Jones, A. W. (2010). Evidence-based survey of the elimination rates of ethanol from blood with applications in forensic casework. *Forensic Science International, 200*:1–20.

Kechagias, A., Jonsson, K. A., Borch, K., & Jones, A. W. (2001, April). Influence of age, sex, and *Helicobacter pylori* infection before and after eradication on gastric alcohol dehydrogenase activity. *Alcoholism: Clinical Experiments and Research, 25*(4): 508–512.

Killip, S., Bennet, J. M., & Chamber, M. D. (2007, March). Iron deficiency anemia. *American Family Physician, 75*(5): 671–678.

Lacaille, L. J., Dauner, K. N., Krambeer, R. J., & Pedersen, J. (2011). Psychosocial and environmental determinants of eating behaviors, physical activity, and weight change among college students: a qualitative analysis. *Journal of the American College of Health, 59*(6): 531–538. doi: 10.1080/07448481.2010.523855.

Lai, C. L., Chao, Y. C., Chen, Y. C., Liao, C. S., Chen, M. C., Liu, Y. C., & Yin, S. J. (2000, November). No sex and age influence on the expression pattern and activities of human gastric alcohol and aldehyde dehydrogenases. *Alcoholism: Clinical Experiments and Research, 24*(11): 1625–1632.

Li, K. K., Concepcion, R. Y., Lee, H., Cardinal, B. J., Ebbeck, V., Woekel, E., & Readdy, R. T. (2012, May–June). An examination of sex differences in relation to the eating habits of and nutrient intakes of university students. *Journal of Nutrition Education and Behavior, 44*(3): 246–250. doi: 10.1016/j.jneb.2010.10.002.

Ling, J., Stephens, R., & Hefferrnan, T. M. (2010, June). Cognitive and psychomotor performance during alcohol hangover. *Current Drug Abuse Review, 3*(2): 80–87.

Lloyd-Richardson, E. E., Bailey, S., & Wing, Rena (2009, March). A prospective study of weight gain during the college freshman and sophomore years. *Preventive Medicine, 48*(3): 256–261.

Lucas, D. R., Brown, R. A., Wassef, M., & Giles, T. D. (2005). Alcohol and the cardiovascular system. *Journal of the American College of Cardiology, 45*: 1916–1924.

Matsumoto, M., Yokoyama, H., Shiraishi, H., Suzuki, H., Kato, S., Miura, S., & Ishii, H. (2001, June). Alcohol dehydrogenase activities in the human gastric mucosa: effects of Helicobacter pylori infection, sex, age, and the part of the stomach. *Alcoholism: Clinical Experiments and Research, 25*(6 suppl): 29S–34S.

Millonig, G., Wang, Y., Homann, N., et al. (2011, February 1). Ethanol-mediated carcinogenesis in the human esophagus implicates CYP2E1 induction and the generation of carcinogenic DNA-lesions. *International Journal of Cancer, 128*(3): 533–540.

My Plate for Older Adults (2011). Retrieved from http://www.nutrition.tufts.edu/research/myplate-older-adults. Accessed November 6, 2015.

National Council on Alcoholism and Drug Dependence (n.d.). Alcohol Abuse Self-Test. Retrieved from http://www.ncadd.org/index.php/learn-about-alcohol/alcohol-abuse-self-test. Accessed November 6, 2015.

National Eating Disorders Association (n.d.). Anorexia nervosa. Retrieved from http://www.nationaleatingdisorders.org/anorexia-nervosa. Accessed November 6, 2015.

National Eating Disorders Association (n.d.). Binge eating disorder. Retrieved from http://www.nationaleatingdisorders.org/binge-eating-disorder. Accessed November 6, 2015.

National Eating Disorders Association (2013). Binge Eating Disorder is added to the DSM-V. Retrieved from http://www.nationaleatingdisorders.org/sites/default/files/ResourceHandouts/InfographicRGB.pdf. Accessed November 6, 2015.

National Eating Disorders Association (n.d.). Bulimia nervosa. Retrieved from http://www.nationaleatingdisorders.org/bulimia-nervosa. Accessed November 6, 2015.

National Eating Disorders Association (n.d.). Other specified feeding or eating disorder. Retrieved from http://www.nationaleatingdisorders.org/other-specified-feeding-or-eating-disorder. Accessed November 6, 2015.

National Institute on Alcohol Abuse and Alcoholism (n.d.). Alcohol and Health. Retrieved from http://www.niaaa.nih.gov/alcohol-health. Accessed November 6, 2015.

National Institute on Alcohol Abuse and Alcoholism (n.d.). College Drinking. Retrieved from http://www.niaaa.nih.gov/alcohol-health/special-populations-co-occurring-disorders/college-drinking. Accessed November 6, 2015.

National Institute on Alcohol Abuse and Alcoholism (2010, September). Beyond Hangovers: Understanding Alcohol's Impact on Your Health. Retrieved from http://pubs.niaaa.nih.gov/publications/Hangovers/beyondHangovers.htm. Accessed November 6, 2015.

National Institute of Mental Health (2011). What are eating disorders? Retrieved from http://www.nimh.nih.gov/health/publications/eating-disorders/index.shtml. Accessed November 6, 2015.

Nova, E., Baccan, G. C., Veses, A., Zapatera, B., & Marcos, A. (2012). Potential health benefits of moderate alcohol consumption : current perspectives in research. *Proceedings of the Nutrition Society, 71*: 307–315. doi:10.1017/S0029665112000171.

O'Keefe, J. H., Bhatti, S. K., Bajwa, A., DiNicolantonio, J. J., & Lavie, C. J. (2014, March). Alcohol and cardiovascular health: the dose makes the poison...or the remedy. *Mayo Clinic Proceedings, 89*(3): 382–393.

O'Keefe, J. H., Bybee, K. A., & Lavie, C. J. (2007, September 11). Alcohol and cardiovascular health: the razor-sharp double-edged sword. *Journal of the American College of Cardiology, 50*(11): 1009–1014.

Penning, R., van Nuland, M., Filervoet, L. A., Olivier, B., & Vester, J. C. (2010, June). The pathology of alcohol hangover. *Current Drug Abuse Review, 3*(2): 68–75.

Pica (2014, February 24). Retrieved from http://www.nlm.nih.gov/medlineplus/ency/article/001538.htm. Accessed November 6, 2015.

Poli, A., et al. (2013, June). Moderate alcohol use and health: a consensus document. *Nutrition, Metabolism, and Cardiovascular Diseases, 23*(6): 487–504.

Racette, S. B., Seusinger, S. S., Strube, M. J., Highstein, G. R., & Deusinger, R. H. (2008, January-February). *Journal of Nutrition Education and Behavior, 40*(1): 39–41. doi: 10.1016/j.jneb.2007.01.001.

Rehm, J., & Shield, K. (2014). Alcohol consumption. *World Cancer Report.* Lyon, France: International Agency for Research on Cancer.

Robinson, E., Thomas, J., Averyard, P., & Higgs, S. (2014, March). What everyone else is eating: a systematic review and meta-analysis of the effect of informational eating norms on eating behavior. *Journal of the Academy of Nutrition and Dietetics, 114*(3): 414–429.

Segura-Garcia, C., et al. (2012, December). Orthorexia nervosa: a frequent eating disordered behavior in athletes. *Eating and Weight Disorders, 17*(4): 226–233. doi: 10.3275/8272.

Smink, F., Hoeken, D., & Hoek, H. W. (2012, August). Epidemiology of eating disorders: incidence, prevalence, and mortality rates. *Current Psychiatry Reports, 14*(4): 406–414.

Smith-Jackson, T., & Reel, J. J. (2012). Freshmen women and the "freshman 15": perspectives on prevalence and causes of college weight gain. *Journal of the American College of Health, 60*(1): 14–20. doi: 10.1080/07448481.2011.555931.

Stookey, J. D. (1999, February). The diuretic effects of alcohol and caffeine and total water intake misclassification. *European Journal of Epidemiology, 15*(2): 181–188.

Swanson, S. A., Crow, S. J., Le Grange, D., Swendsen, J., & Merikangas, K. R. (2011, July). Prevalence and correlates of eating disorders in adolescents: Results from the national comorbidity survey replication adolescent supplement. *Archives of General Psychiatry, 68*(7): 714–723. doi: 10.1001/archgenpsychiatry.2011.22.

Swift, R., & Davidson, D. (1998). Alcohol hangover: mechanisms and mediators. *Alcohol Health Research World, 22*(1): 54–60.

U.S. Department of Agriculture (2014). Aging. Retrieved from https://fnic.nal.usda.gov/lifecycle-nutrition/aging. Accessed November 6, 2015.

U.S. Department of Agriculture (2014). Elderly nutrition program. Retrieved from http://www.nutrition.gov/food-assistance-programs/elderly-nutrition-program. Accessed November 6, 2015.

U.S. Department of Agriculture (n.d.). Estimated Calorie Needs per Day by Age, Gender, and Physical Activity Level. Retrieved from http://www.cnpp.usda.gov/sites/default/files/usda_food_patterns/EstimatedCalorieNeedsPerDayTable.pdf. Accessed November 6, 2015.

U.S. Department of Agriculture (2014). Seniors. Retrieved from http://www.nutrition.gov/life-stages/seniors. Accessed November 6, 2015.

Varga, M., Dukay-Szabo, S., Tury, F., & van Furth, E. F. (2013, June). Evidence and gaps in the literature on orthorexia nervosa. *Eating and Weight Disorders, 18*(2): 103–111. doi: 10.1007/s40519-013-0026-y.

Vella-Zarb, R. A., & Elgar, F. J. (2009, September–October). The 'freshman 5': a meta-analysis of weight gain in the freshman year of college. *Journal of American College of Health, 58*(2): 161–166. doi: 10.1080/07448480903221392.

Weitzen, T. E., Corenlia-Carlson, T., Fitzpatrick, M. E., Kennington, B., Bean, P., & Jefferies, C. (2012, September 17). Treatment issues and outcomes for males with eating disorders. *Eating Disorders: The Journal of Treatment and Prevention, 20*(5): 444–459. doi: 10.1080/10640266.2012.715527.

World Health Organization (2014). Retrieved from http://www.who.int/topics/alcohol_drinking/en/. Accessed November 6, 2015.

Chapter 20

Academy of Nutrition and Dietetics (April, 2013). Nutrition Security in Developing Nations: Sustainable Food, Water, and Health. *Journal of the Academy of Nutrition and Dietetics, 113*(4): 581–595.

Barfoot, P., & Brookes, G. (2014, March). Key global environmental impacts of genetically modified (GM) crop use 1996–2012. *GM Crops and Food, 5*(2): 149–160.

Benbrook, C. (2013, February 19). Are organic foods safer or healthier? *Annals of Internal Medicine, 158*(4): 296–297. doi: 10.7326/0003-4819-158-4-201302190-00017.

Berghaus, R. D., Thayer, S. G., Law, B. F., Milk, R. M., Hofacre, C. L., & Singer, R. S. (2013, July). Enumeration of Salmonella and Campylobacter spp in environmental farm samples and processing plant carcass rinses from commercial broiler chicken flocks. *Applied Environmental Microbiology, 79*(13): 4106–4114. doi: 10.1128/AEM.00836-13.

Center for Food Safety (2014, February 27). New Report Connects Climate Change and Food Insecurity. [Press release]. Retrieved from http://www.centerforfoodsafety.org/press-releases/2948/new-report-connects-climate-change-and-food-insecurity#. Accessed November 6, 2015.

Centers for Disease Control and Prevention (2011). CDC 2011 Estimates of Foodborne Illness in the United States. Retrieved from http://www.cdc.gov/foodborneburden/2011-foodborne-estimates.html. Accessed November 6, 2015.

Centers for Disease Control and Prevention (2015). *E. coli (Escherichia coli)*. Retrieved from http://www.cdc.gov/ecoli/. Accessed November 6, 2015.

Centers for Disease Control and Prevention (2014). *Campylobacter*. Retrieved from http://www.cdc.gov/nczved/divisions/dfbmd/diseases/campylobacter/. Accessed November 6, 2015.

Centers for Disease Control and Prevention (2015, February 3). Global Water, Sanitation, and Hygiene (WASH). Retrieved from http://www.cdc.gov/healthywater/global/. Accessed November 6, 2015.

Centers for Disease Control and Prevention (2013). Norovirus. Retrieved from http://www.cdc.gov/norovirus/index.html. Accessed November 6, 2015.

Centers for Disease Control and Prevention (2013). Trends in Foodborne Illness in the United States, 2012. Retrieved from http://www.cdc.gov/features/dsfoodnet2012/. Accessed November 6, 2015.

Centers for Disease Control and Prevention (2015). A-Z Index for Foodborne Illness. Retrieved from http://www.cdc.gov/foodsafety/diseases/. Accessed November 6, 2015.

Centers for Disease Control and Prevention (2015, October 8). *Clostridium perfringens*. Retrieved from http://www.cdc.gov/foodsafety/clostridium-perfingens.html. Accessed November 6, 2015.

Centers for Disease Control and Prevention (2014). Estimates of Foodborne Illness in the United States. Retrieved from http://www.cdc.gov/foodborneburden/index.html. Accessed November 6, 2015.

Centers for Disease Control and Prevention (2015, October 15). Listeria (Listeriosis). Retrieved from http://www.cdc.gov/listeria/index.html. Accessed November 6, 2015.

Centers for Disease Control and Prevention (2015, September 4). Salmonella. Retrieved from http://www.cdc.gov/salmonella/index.html. Accessed November 6, 2015.

Cody, M. M., & Stretch, T. (2014, November). Position of the Academy of Nutrition and Dietetics: Food and Water Safety. *Journal of the Academy of Nutrition and Dietetics, 114*(11): 1819–1829. doi: http://dx.doi.org/10.1016/j.jand.2014.08.023.

Coleman-Jensen, A., Nord, M., & Singh, A. (2013, September). Household Food Security in the United States in 2012. *Economic Research Report No. 41*.

Dangour, A. D., Lock, K., Hyter, A., Aikenhead, A., Allen, E., & Uauy, R. (2010, July). Nutrition-related health effects of organic foods: a systematic review. *American Journal of Clinical Nutrition, 92*(1): 203–210. doi: 10.3945/ajcn.2010.29269.

DiLeo, M. V., Bakker, M., Chu, E. Y., Hoekenga, O. (2014, March 28). An assessment of the relative influences of genetic background, functional diversity at major regulatory genes, and transgenic constructs on the tomato fruit metabolome. *Plant Genome, 7*(1). doi: 10.3835/plantgenome2013.06.0021.

Dona, A., & Arvanitoyannis, I. S. (2008, November 6). Health risks of genetically modified foods. *Critical Reviews in Food Science and Nutrition, 48*(2). doi: 10.1080/10408390701855993.

European Commission (2015). Sustainable Food. Retrieved from http://ec.europa.eu/environment/eussd/food.htm. Accessed November 16, 2015.

Fisher, M. (2014, April 3). A new approach to detecting unintended changes in GM Foods. *Crop Science Society of America*. Retrieved from https://www.crops.org/science-news/new-approach-detecting-changes-gm-foods. Accessed November 6, 2015.

FAO, IFAD, & WFP (2014). The State of Food Insecurity in the World 2014: Strengthening the enabling environment for food security and nutrition. Retrieved from http://www.fao.org/3/a-i4030e.pdf. Accessed November 6, 2015.

Food additives (2014, May 6). Retrieved from http://www.nlm.nih.gov/medlineplus/ency/article/002435.htm. Accessed November 6, 2015.

Food and Agriculture Organization of the United Nations (2011). *FAO in the 21st Century: Ensuring Food Security in a Changing World* (Rome, Italy).

Food and Agriculture Organization of the United Nations (2015). Hunger Portal. Retrieved from http://www.fao.org/hunger/en/. Accessed November 6, 2015.

Food and Agriculture Organization of the United Nations (2013). *The State of Food Insecurity in the World.* Retrieved from http://www.fao.org/docrep/018/i3458e/i3458e.pdf. Accessed November 6, 2015.

Food Research and Action Center (2014). Hunger and Poverty in the United States. Retrieved from http://frac.org/reports-and-resources/hunger-and-poverty/. Accessed November 6, 2015.

Forman, J., Silverstein, J., Committee on Nutrition, Council on Environmental Health & American Academy of Pediatrics (2012, November). Organic foods: health and environmental advantages and disadvantages. *Pediatrics, 130*(5): e1406–e1415. doi: 10.1542/peds.2012-2579.

Global Food Security Index: Overview (2015, May). Retrieved from http://foodsecurityindex.eiu.com/Index. Accessed November 6, 2015.

Hanning, I. B., O'Bryan, C. A., Crandall, P. G. & Ricke, S. C. (2012). Food Safety and Food Security. *Nature Education Knowledge, 3*(10):9.

Havelaar, A. H., et al. (2013, June 17). WHO Initiative to Estimate the Global Burden of Foodborne Diseases. *The Lancet, 381*: 599. doi: 10.1016/S0140-6736(13)61313-6.'

Hoefkens, C., et al. (2010, November). Consuming organic versus conventional vegetables: the effect on nutrient and contaminant intakes. *Food Chemistry and Toxicology, 48*(11): 3058–3066. doi: 10.1016/j.fct.2010.07.044.

Holben, D. H., & American Dietetic Association (2010, September). Position of the American Dietetic Association: food insecurity in the United States. *Journal of the American Dietetic Association, 110*(9): 1368–1377.

Hug, K. (2008). Genetically modified organisms: do the benefits outweigh the risks? *Medicina (Kaunas), 44*(2): 87–99.

Killan, L. (2012, September). Food safety: an integral part of food security. *International Atomic Energy Agency Bulletin, 53*(3).

McCann, D., Barrett, A., Cooper, A., et al. (2008, February). ADHD and Food Additives Revisited. *AAP Grand Rounds, 19*(2): 17. doi: 10.1542/gr.19-2-17.

McComas, K. A., Besley, J. C., & Steinhardt, J. (2014, July). Factors influencing U.S. consumer support for genetic modification to prevent crop disease. *Appetite, 78*(8): 8–14. doi: 10.1016/j.appet.2014.02.006.

National Institute of Environmental Health Sciences (2014). Bisphenol A (BPA). Retrieved from https://www.niehs.nih.gov/health/topics/agents/sya-bpa/. Accessed November 6, 2015.

Organic Farming Crucial to Food Security, Addressing Climate Change (2013, January 18). Retrieved from https://www.sustainablebusiness.com/index.cfm/go/news.display/id/24475. Accessed November 6, 2015.

Patton-López, M. M., López-Cevallos, D. F., Cancel-Tirado, D. I., & Vazquez, L. (2014). Prevalence and correlates of food insecurity among students attending a midsize rural university in Oregon. *Journal of Nutrition Education and Behavior.* 46(3), 209–214. doi:10.1016/j.jneb.2013.10.007

Scheinfeld, N. S., Mokashi, A., & Lin, A. (2014). Protein-Energy Malnutrition. Retrieved from http://emedicine.medscape.com/article/1104623-overview. Accessed November 6, 2015.

Smith-Spangler, C., et al. (2012, September). Are organic foods safer or healthier than conventional alternatives?: a systematic review. *Annals of Internal Medicine, 157*(5): 348–366. doi: 10.7326/0003-4819-157-5-201209040-00007.

Stevens, L. J., Burgess, J. R., Stochelski, M. A., & Kuczek, T. (April, 2015). Amounts of artificial food dyes and added sugars in foods and sweets commonly consumed by children. *Clinical Pediatrics*. April 2015 vol. 54 no. 4309–321.

Stevens, L. J., Burgess, J. R., Stochelski, M. A., & Kuczek, T. (2014, February). Amounts of artificial food colors in commonly consumed beverages and potential behavioral implications for consumption in children. *Clinical Pediatrics, 53*(2): 133–40.

Stevens, L. J., Kuczek, T., Burgess, J. R., Stochelski, M. A., Arnold, L. E., & Galland, L. (2013, May). Mechanisms of behavioral, atopic, and other reactions to artificial food colors in children. *Nutritional Review, 71*(5): 269–281.

Tanumilhardjo, S. A., et al. (2007, November). Poverty, obesity, and malnutrition: an international perspective recognizing the paradox. *Journal of the American Dietetic Association, 107*(11): 1966–1972.

Todd, J, & Morrison, R. M. (2014, March 4). Less eating out, improved diets, and more family meals in the wake of the Great Recession. *Food Choices and Health*. Retrieved from http://www.ers.usda.gov/amber-waves/2014-march/less-eating-out,-improved-diets,-and-more-family-meals-in-the-wake-of-the-great-recession.aspx#.U4w1lxawbP1. Accessed November 6, 2015.

U.S. Department of Agriculture Economic Research Service (2014, September 3). Key Statistics and Graphics. Retrieved from http://www.ers.usda.gov/topics/food-nutrition-assistance/food-security-in-the-us/key-statistics-graphics.aspx. Accessed November 6, 2015.

U.S. Environmental Protection Agency (2015). Pesticide and Food: Health Problems Pesticides May Pose. Retrieved from http://www.epa.gov/pesticides/food/risks.htm. November 6, 2015.

U.S. Food and Drug Administration (2014, November). Bisphenol A (BPA): Use in Food Contact Application. Retrieved from http://www.fda.gov/newsevents/publichealthfocus/ucm064437.htm. Accessed November 6, 2015.

U.S. Food and Drug Administration (2001, January). DRAFT Guidance for Industry: Voluntary Labeling Indicating Whether Foods Have or Have Not Been Developed Using Bioengineering; Draft Guidance. Retrieved from http://www.fda.gov/Food/GuidanceRegulation/GuidanceDocumentsRegulatoryInformation/LabelingNutrition/ucm059098.htm. Accessed November 6, 2015.

U.S. Food and Drug Administration (2015, September 3). Food Facts—Food Irradiation. Retrieved from http://www.fda.gov/Food/ResourcesForYou/Consumers/ucm261680.htm. Accessed November 6, 2015.

U.S. Food and Drug Administration (2014). FDA Food Safety Modernization Act (FSMA). Retrieve from http://www.fda.gov/Food/GuidanceRegulation/FSMA/. Accessed November 6, 2015.

U.S. Food and Drug Administration (2015, October 17). FDA's Role in Regulating Safety of GE Foods. Retrieved from http://www.fda.gov/forconsumers/consumerupdates/ucm352067.htm. Accessed November 6, 2015.

U.S. Food and Drug Administration (2015, September 2). Foodborne Illness-Causing Organisms in the U.S. What You Need to Know. http://www.fda.gov/Food/FoodborneIllness-Contaminants/FoodborneIllnessesNeedToKnow/default.htm. Accessed November 6, 2015.

U.S. Food and Drug Administration (2015, September 18). Food Defense. Retrieved from http://www.fda.gov/food/fooddefense/. Accessed November 6, 2015.

U.S. Food and Drug Administration (2015). Hazard Analysis and Critical Control Points (HACCP). Retrieved from http://www.fda.gov/food/guidanceregulation/haccp/default.htm. Accessed November 6, 2015.

U.S. Food and Drug Administration (2014). How Safe Are Color Additives? Retrieved from http://www.fda.gov/forconsumers/consumerupdates/ucm048951.htm. Accessed November 6, 2015.

U.S. Food and Drug Administration (2014). Overview of Food Ingredients, Additives, and Colors. Retrieved from http://www.fda.gov/Food/IngredientsPackagingLabeling/FoodAdditivesIngredients/ucm094211.htm. Accessed November 6, 2015.

U.S. Department of Agriculture (2015, September 8). Definitions of Food Security. Retrieved from http://www.ers.usda.gov/topics/food-nutrition-assistance/food-security-in-the-us/definitions-of-food-security.aspx#.U4w1KhawbP2. Accessed November 6, 2015.

U.S. Department of Agriculture (2014). Food Security in the U.S.: Key Statistics & Graphics. Retrieved from http://www.ers.usda.gov/topics/food-nutrition-assistance/food-security-in-the-us/key-statistics-graphics.aspx#.U4w1lBawbP3. Accessed November 6, 2015.

U.S. Department of Agriculture (2015) Organic Regulations. Retrieved from http://www.ams.usda.gov/rules-regulations/organic. Accessed November 6, 2015.

U.S. Department of Agriculture (2015). Nutrition Assistance Programs. Retrieved from http://fnic.nal.usda.gov/nutrition-assistance-programs. Accessed November 6, 2015.

U.S. Department of Agriculture Food and Nutrition Service (2015). Supplemental Nutrition Assistance Program (SNAP). Retrieved from http://www.fns.usda.gov/snap/supplemental-nutrition-assistance-program-snap. Accessed November 6, 2015.

U.S. Department of Agriculture Food and Nutrition Service (2015). Women, Infants, and Children. Retrieved from http://www.fns.usda.gov/wic/women-infants-and-children-wic. Accessed November 6, 2015.

Unnevehr, L. J. (ed) (2003). Food Safety in Food Security and Food Trade. *Focus, 10*(1).

Williams, A., & Oyzarzabal, O. A. (2012, August 24). Prevalence of campylobacter spp. in skinless, boneless retail broiler meat from 2005 through 2011 in Alabama, USA. *BMC Microbiology, 12*: 184. doi: 10.1186/1471-2180-12-184.

World Health Organization (2015). Food, genetically modified. Retrieved from http://who.int/topics/food_genetically_modified/en/. Accessed November 6, 2015.

World Water Council (2015). Water Supply and Sanitation. Retrieved from http://www.worldwatercouncil.org/library/archives/water-supply-sanitation/. Accessed November 6, 2015.

Chapter 21

Alzheimer's Disease International. (2013). Policy brief for heads of government: the global impact of dementia 2013–2050. London: Alzheimer's Disease International.

Arab L, et al. (2013). Epidemiologic Evidence of a Relationship between Tea, Coffee, or Caffeine Consumption and Cognitive Decline. *Advances in Nutrition.* 4: 115–122.

Barnard ND, et al. (2014). Dietary and lifestyle guidelines for the prevention of Alzheimer's Disease. *Neurobiology of Aging* 35: S74–S78.

Buettner, D. (2009). How to Live to Be 100+. Retrieved from http://www.ted.com/talks/dan_buettner_how_to_live_to_be_100. Accessed November 6, 2015.

Buettner, D. (2005). The Secrets of Long Life. *National Geographic.* Retrieved from http://ngm.nationalgeographic.com/2005/11/longevity-secrets/buettner-text. Accessed November 6, 2015.

Colman, R. J., et al. (2014, April 1). Caloric restriction reduces age-related and all-cause mortality in rhesus monkeys. *Nature Communications, 5*: 3557. doi: 10.1038/ncomms4557.

Fraser, G. (2003). *Diet, Life Expectancy, and chronic Disease: Studies of Seventh-Day Adventists and Other Vegetarians.* New York: Oxford University Press.

Fraser, G. E., Orlich, M. J., & Jaceldo-Siegl, K. (2015, April). Studies of chronic diseases in Seventh-Day Adventists. *International Journal of Cardiology, 184*: 573.

Fraser, G. E., & Shavlik, D. J. (2001). Ten years of life: Is it a matter of choice? *Archives of Internal Medicine, 161*(13): 1645–1652.

Gu, Q., Dillon, C. F., & Burt, V. L. (2010, September). Prescription Drug Use Continues to Increase: U.S. Prescription Drug Data for 2007-2008. *National Center for Health Statistics Data Breif, 42*. Retrieved from http://www.cdc.gov/nchs/data/databriefs/db42.htm. Accessed November 6, 2015.

Hirani, V., et al. (2015, March 26). Sarcopenia is associated with incident disability, institutionalization in community-dwelling older men: the Concord Health and Ageing in Men Project. *Journal of the American Medical Directors Association, pii*: S1525-8610(15)00151-6. doi: 10.1016/j.jamda.2015.02.006.

Hu N, et al. (2013) Nutrition and the Risk of Alzheimer's Disease. *BioMed Research International.* http://www.ncbi.nlm.nih.gov/pubmed/23865055. Accessed November 6, 2015.

Orlich, M. J., & Fraser, G. E. (2014, July). Vegetarian diets in the Adventist Health Study 2: A Review of Initial Published Findings. *American Journal of Clinical Nutrition, 100*(Supplement 1): 353S–358S.

Otaegui-Arrazola A, et al. (2014, February) Diet, cognition, and Alzheimer's disease: food for thought. *European Journal of Nutrition,* 53:1–23.

Singh, P. N., Haddad, E., Tonstad, S., & Fraser, G. E. (2011, June). Does excess body fat maintained after the seventh decade decrease life expectancy? *Journal of the American Geriatric Society, 59*(6): 1003–11. doi: 10.1111/j.1532-5415.2011.03419.x.

Shipton, M. J., & Thachil, J. (2015, April). Vitamin B12 Deficiency-a 21st Century Perspective. *Clinical Medicine, 15*(2): 145–50. doi: 10.7861/clinmedicine.15-2-145.

van de Rest, O., Berendsen, A.A., Haveman-Niles, A., deGroot, L.C. (2015) Dietary patterns, cognitive decline, and dementia: A systematic review. *Advances in Nutrition,* 6:154–168.

Vauzour D. (2014) Effect of flavonoids on learning, memory and neurocognitive performance: relevance and potential implications for Alzheimer's disease pathophysiology. *Journal of the Science of Food and Agriculture,* 94: 1042–1056.

INDEX

Note: Page numbers followed by f indicate infographics.

Dietary Reference Intakes (DRIs): Recommended Dietary Allowances and Adequate Intakes, Elemen

Food and Nutrition Board, Institute of Medicine, National Academies

Life Stage Group	Calcium (mg/d)	Chromium (µg/d)	Copper (µg/d)	Fluoride (mg/d)	Iodine (µg/d)	Iron (mg/d)	Magnesiu (mg/d)
Infants							
0 to 6 mo	200*	0.2*	200*	0.01*	110*	0.27*	30*
6 to 12 mo	260*	5.5*	220*	0.5*	130*	**11**	75*
Children							
1–3 y	**700**	11*	**340**	0.7*	**90**	**7**	**80**
4–8 y	**1,000**	15*	**440**	1*	**90**	**10**	**130**
Males							
9–13 y	**1,300**	25*	**700**	2*	**120**	**8**	**240**
14–18 y	**1,300**	35*	**890**	3*	**150**	**11**	**410**
19–30 y	**1,000**	35*	**900**	4*	**150**	**8**	**400**
31–50 y	**1,000**	35*	**900**	4*	**150**	**8**	**420**
51–70 y	**1,000**	30*	**900**	4*	**150**	**8**	**420**
> 70 y	**1,200**	30*	**900**	4*	**150**	**8**	**420**
Females							
9–13 y	**1,300**	21*	**700**	2*	**120**	**8**	**240**
14–18 y	**1,300**	24*	**890**	3*	**150**	**15**	**360**
19–30 y	**1,000**	25*	**900**	3*	**150**	**18**	**310**
31–50 y	**1,000**	25*	**900**	3*	**150**	**18**	**320**
51–70 y	**1,200**	20*	**900**	3*	**150**	**8**	**320**
> 70 y	**1,200**	20*	**900**	3*	**150**	**8**	**320**
Pregnancy							
14–18 y	**1,300**	29*	**1,000**	3*	**220**	**27**	**400**
19–30 y	**1,000**	30*	**1,000**	3*	**220**	**27**	**350**
31–50 y	**1,000**	30*	**1,000**	3*	**220**	**27**	**360**
Lactation							
14–18 y	**1,300**	44*	**1,300**	3*	**290**	**10**	**360**
19–30 y	**1,000**	45*	**1,300**	3*	**290**	**9**	**310**
31–50 y	**1,000**	45*	**1,300**	3*	**290**	**9**	**320**

NOTE: This table (taken from the DRI reports, see www.nap.edu) presents Recommended Dietary Allowances (RDAs) in **bold type** and Adequate Intakes (AIs) in ordinary type followed by an asterisk (*). An RDA is the average daily dietary intake level sufficient to meet the nutrient requirements of nearly all (97–98 percent) healthy individuals in a group. It is calculated from an Estimated Average Requirement (EAR). If sufficient scientific evidence is not available to establish an EAR, and thus calculate an RDA, an AI is usually developed. For healthy breastfed infants, an AI is the mean intake. The AI for other life stage and gender groups is believed to cover the needs of all healthy individuals in the groups, but lack of data or uncertainty in the data prevent being able to specify with confidence the percentage of individuals covered by this intake.